THE SECOND CATARACT FORTRESS OF DORGINARTI

Team members of the 1964 season at Dorginarti

Back row: John C. Lorence, Mielburn D. Thurman, Otto J. Schaden
Front row: Louise Storts, Bruce G. Trigger, Richard H. Pierce, James E. Knudstad, Wenche Pierce
Not present: Rudolph H. Dornemann, Sylvia Ericson, Alfred J. Hoerth

CAMPAGNE INTERNATIONALE POUR LA SAUVEGARDE
DES MONUMENTS DE LA NUBIE

EXCAVATIONS AT SERRA EAST AND DORGINARTI
JAMES E. KNUDSTAD, DIRECTOR

THE SECOND CATARACT FORTRESS OF DORGINARTI

by

LISA A. HEIDORN

with contributions by

CAROL MEYER AND JOANNA THEN-OBŁUSKA

INSTITUTE FOR THE STUDY OF ANCIENT CULTURES
OF THE UNIVERSITY OF CHICAGO

NUBIAN EXPEDITION • VOLUME XII

Library of Congress Control Number: 2022950943
ISBN-13 (hardback): 978-1-61491-083-1
ISBN-13 (Adobe PDF): 978-1-61491-087-9
ISBN-13 (Adobe Digital Editions): 978-1-61491-084-8

Printed in the United States of America.

Series Editors
Thomas G. Urban, Charissa Johnson,
Steven Townshend, and Andrew Baumann

with the assistance of
Connie Gundry Tappy and Alexandra Witsell

Editors' Note
While this publication was at press, the Oriental Institute changed its name to Institute for the Study of Ancient Cultures. References to the Oriental Institute and Oriental Institute Museum registration numbers still appear throughout the book.

Cover Illustrations
Panorama of Dorginarti, from the southwest
(courtesy of Rosa Frey and James E. Knudstad)
and objects found at Dorginarti
(from top to bottom, left to right: Chian amphora OIM E24372,
stone with fortress name OIM E24326, faience figurine OIM E24332,
fish plaque OIM E24398, arrowhead OIM E24421,
arrowhead Khartoum 14277, and arrowhead OIM E24349B)

Cover Design
James M. Slate

Layout
PerfecType, Nashville, TN

The paper used in this publication meets the minimum requirements of American National Standard for Information Services—Permanence of Paper for Printed Library Materials, ANSI/NISO Z39.48-1992. ♾

Table of Contents

List of Figures

List of Maps

List of Plans

List of Plates

List of Abbreviations

BM	British Museum
ca.	*circa*, about, approximately
cat.	catalog
cm	centimeter
D	diameter, Dorginarti
diss.	dissertation
E	east
e.g.	*exempli gratia*, for example
et al.	*et alia*, and others
exc.	excavated
fasc.	fascicle
fig(s).	figure(s)
g	gram(s)
H	height
HD	hole diameter
i.e.	*id est*, that is
IMG	image
JE	Journal d'Entrée (Cairo Museum)
kg	kilogram(s)
Khartoum	Khartoum field registration number
km	kilometer(s)
L	length
LAH	Lisa A. Heidorn
m	meter(s)
mm	millimeter(s)
max.	maximum
MB	marl bowl
MFA	Museum of Fine Arts (Boston)
MJ	marl jar
N	north
n.d.	no date
n(n).	note(s)
n.s.	new series
no(s).	number(s)
OIM E	Oriental Institute Museum, Egyptian (registration number)
p(p).	page(s)
pl(s).	plate(s)
PhD	*philosophiae doctor*, doctor of philosophy (degree)
pXRF	portable X-ray fluorescence
RR	red rim
RS	red slip
S	south
SB	silt bowl
SJ	silt jar
SNM	Sudan National Museum
sp.	species
Th	thickness

unexc.	unexcavated
vol(s).	volume(s)
vs.	versus
yr	year
W	width, west

Bibliography

Abu al-ʿIzz, Muhammad Safi al-Din

1971 *Landforms of Egypt.* Translated by Yusuf A. Fayid. Cairo: American University in Cairo Press.

Adams, William Y.

1964 "An Introductory Classification of Meroitic Pottery." *Kush* 12: 126–73.

1977 *Nubia: Corridor to Africa.* Princeton: Princeton University Press.

1986 *Ceramic Industries of Medieval Nubia.* Part II. Lexington: University Press of Kentucky.

2004 "Site 6-G-9 and the Problem of Early Kushite Settlement in Lower Nubia." *Sudan & Nubia* 8: 64–69.

2005 *The West Bank Survey from Faras to Gemai.* Vol. 2, *Sites of Meroitic and Ballaña Age.* Sudan Archaeological Research Society Publication 13. BAR International Series 1335. Oxford: Archaeopress.

n.d. "Pottery Wares of the Ptolemaic and Roman Periods at Qasr Ibrim: Preliminary Ware Descriptions." Unpublished manuscript.

Addison, Frank

1949 *Jebel Moya.* The Wellcome Excavations in the Sudan. 2 vols. London: Oxford University Press.

Aharoni, Miriam, and Yohanan Aharoni

1976 "The Stratification of Judahite Sites in the 8th and 7th Centuries B.C.E." *Bulletin of the American Schools of Oriental Research* 224: 73–90.

Aharoni, Yohanan

1976 "Forerunners of the Limes: Iron Age Fortresses in the Negev." *Israel Exploration Journal* 17, no. 1: 1–17.

Ahmed, Khidir Abdelkarim

1985 *Meroitic Settlement in the Central Sudan: An Analysis of Sites in the Nile Valley and the Western Butana.* Cambridge Monographs in African Archaeology 8. BAR International Series 197. Oxford: BAR International.

Alekseeva, Ekaterina M.

1978 *Ancient Beads of the Northern Black Sea Littoral.* Vol. 2. Moscow: Izdatel'stvo Nauka.

Amiran, Ruth B. K.

1969 *Ancient Pottery of the Holy Land from Its Beginnings in the Neolithic Period to the End of the Iron Age.* Jerusalem: Massada Press.

Amiran, Ruth B. K., and Immanuel Dunayevsky

1958 "The Assyrian Open-Court Building and Its Palestinian Derivatives." *Bulletin of the American Schools of Oriental Research* 149: 25–32.

Anderson, Julie, and Salah eldin Mohamed Ahmed

2009 "What Are These Doing Here above the Fifth Cataract?!! Napatan Royal Statues at Dangeil." *Sudan & Nubia* 13: 78–86.

Andrews, Carol

1990 *Ancient Egyptian Jewelry.* London: British Museum Publications.

Anthes, Rudolf

1965 *Mit Rahineh 1956.* Philadelphia: University Museum.

Arveiller-Dulong, Véronique, and Marie-Dominique Nenna

2011 *Les verres antiques du Musée du Louvre.* Vol. 3, *Parures, instruments et éléments d'incrustation.* Paris: Musée du Louvre.

Aston, David A.

1996a *Egyptian Pottery of the Late New Kingdom and Third Intermediate Period (Twelfth–Seventh Centuries BC): Tentative Footsteps in a Forbidding Terrain.* Studien zur Archäologie und Geschichte Altägyptens 13. Heidelberg: Heidelberger Orientverlag.

1996b "A Group of Twenty-Fifth Dynasty Pots from Abydos." *Mitteilungen des Deutschen Archäologischen Instituts Kairo* 52: 1–10.

1996c "Sherds from a Fortified Townsite near Abu ʿId." *Cahiers de la céramique égyptienne* 4: 19–46.

1998 *Die Keramik des Grabungsplatzes Q I.* Part 1, *Corpus of Fabrics, Wares and Shapes.* Forschungen in der Ramses-Stadt. Grabungen des Pelizaeus-Museums Hildesheim in Qantir–Pi-Ramesse 1. Mainz am Rhein: Philipp von Zabern.

1999 *Elephantine XIX: Pottery from the Late New Kingdom to the Early Ptolemaic Period.* Archäologische Veröffentlichungen, Deutsches Archäologisches Institut, Abteilung Kairo 95. Mainz am Rhein: Verlag Philipp von Zabern.

2004 "Amphorae in New Kingdom Egypt." *Ägypten und Levante* 14: 175–213.

2009 *Burial Assemblages of Dynasty 21–25: Chronology, Typology, Developments.* Österreichische Akademie der Wissenschaften,

Denkschriften der Gesamtakademie 54. Contributions to the Chronology of the Eastern Mediterranean 21. Vienna: Österreichischen Akademie der Wissenschaften.

2010 "The Pottery from Sector L 49." In *Heracleopolis Magna (Ehnasya el Medina, Egypto)*, vol. 1, *La necropolis "real" del Tercer Periodo Intermedio y su reutilizacion*, edited by Maria del Carmen Pérez Die, pp. 28–83. Madrid: Ministerio de Cultura.

2014 *Pottery Recovered Near the Tombs of Seti I (KV 17) and Siptah (KV 47) in the Valley of the Kings*. Aegyptiaca Helvetica 24. Basel: Schwabe Verlag.

Aston, David A., and Bettina Bader

1998 "Einige Bemerkungen zum späten Neuen Reich in Matmar." *Mitteilungen des Deutschen Archäologischen Instituts Kairo* 54: 19–48.

Aston, David A., and David Jeffreys

2007 *Excavations at Kom Rabia (Site RAT): Post-Ramesside Levels and Pottery*. Survey of Memphis 3. Egypt Exploration Society Excavation Memoir 81. London: Egypt Exploration Society.

Azim, Michel

1975 "Quatre campagnes de fouilles sur la forteresse de Saï, 1970–1973. 1ère partie: L'installation pharaonique." *Cahiers de Recherches de l'Institut de Papyrologie et d'Égyptologie de Lille* 3: 93–126.

Bailey, Donald M.

1991 *Excavations at el-Ashmunein*. Vol. 4, *Hermopolis Magna: Buildings of the Roman Period*. British Museum Expedition to Middle Egypt. London: British Museum Press.

Barbour, Kenneth Michael

1961 *The Republic of the Sudan: A Regional Geography*. London: University of London Press.

Bates, Oric

1917 *Ancient Egyptian Fishing*. Harvard African Studies 1. Cambridge, MA: Harvard University Press.

Baud, Michel

2011 "Premières données sur le palais royal de Mouweis." In *La pioche et la plume: Autour du Soudan, du Liban et de la Jordanie. Hommages archéologiques à Patrice Lenoble*, edited by Vincent Rondot, F. Alpi, and F. Villeneuve, pp. 339–57. Paris: Presses de l'Université Paris-Sorbonne.

Baud, Michel, Aminata Sackho-Autissier, and Sophie Labbé-Toutée

2010 *Meroe: Un empire sur le Nil*. Milan: Officina Libraria and Paris: Musée du Louvre Éditions.

Bavay, Laurent

1998 "La céramique dans le secteur du parvis de la porte monumentale." In *Tanis: Travaux récents sur le Tell Sân el-Hagar*, edited by Philippe Brissaud and Christiane Zivie-Coche, pp. 316–32. Mission française des fouilles de Tanis 1987–1997. Paris: Éditions Noêsis.

Beit-Arieh, Itzhaq, Bruce C. Cresson, and Pirhiya Beck

2007 *Horvat ʿUza and Horvat Radum: Two Fortresses in the Biblical Negev*. Monograph Series of the Institute of Archaeology of Tel Aviv University 25. Tel Aviv: Emery and Claire Yass Publications in Archaeology.

Ben-Dor Evian, Shirly

2011 "Egypt and the Levant in the Iron Age I–IIA: The Ceramic Evidence." *Tel Aviv* 38: 94–119.

Ben-Shlomo, David

2014 "Ceramic Objects: Marked Pottery, Mud Objects, and Various Ceramic Artifacts." In *The Smithsonian Institution Excavation at Tell Jemmeh, Israel, 1970–1990*, edited by David Ben-Shlomo and Gus W. Van Beek, pp. 838–56. Smithsonian Contributions to Anthropology 50. Washington, DC: Smithsonian Institute Scholarly Press.

Ben-Shlomo, David, Itzhaq Shai, and Aren M. Maeir

2004 "Late Philistine Decorated Ware ('Ashdod Ware'): Typology, Chronology, and Production Centers." *Bulletin of the American Schools of Oriental Research* 335: 1–34.

Ben-Shlomo, David, and Gus W. Van Beek, eds.

2014 *The Smithsonian Institution Excavation at Tell Jemmeh, Israel, 1970–1990*. Smithsonian Contributions to Anthropology 50. Washington, DC: Smithsonian Institute Scholarly Press.

Berlejung, Angelika

2012 "The Assyrians in the West: Assyrianization, Colonialism, Indifference, or Development Policy." In *Congress Volume Helsinki 2010*, edited by Martti Nissinen, pp. 21–60. Vetus Testamentum Supplements 148. Leiden: Brill.

Bettles, Elizabeth A.

2003 *Phoenician Amphora: Production and Distribution in the Southern Levant*. BAR International Series 1183. Oxford: Archaeopress.

Bietak, Manfred

2005 "Neue Paläste aus der 18. Dynastie." In *Structure and Significance: Thoughts on Ancient Egyptian Architecture*, edited by Peter Jánosi, pp. 131–68. Österreichische Akademie der Wissenschaften, Denkschriften der Gesamtakademie 33. Untersuchungen der Zweigstelle Kairo des Österreichischen Archäologischen

Instituts 25. Vienna: Verlag der Österreichischen Akademie der Wissenschaften.

Bikai, Patricia M.
1987 *The Phoenician Pottery of Cyprus.* Nicosia: A. G. Levantis Foundation.

Binder, Michaela
2011 "The 10th–9th Century BC: New Evidence from Cemetery C of Amara West." *Sudan & Nubia* 15: 39–53.

Binder, Michaela, Neal Spencer, and Marie Millet
2011 "Cemetery D at Amara West: The Ramesside Period and Its Aftermath." *British Museum Studies in Ancient Egypt and Sudan* 16: 47–99.

Blanquet, Claire-Hélène
1992 "Typologie de la bouteille du nouvel an." In *Amosiadès: Mélange offerts au Professeur Claude Vandersleyen par ses anciens etudiants,* edited by Claude Obsomer and Ann-Laure Oosthoek, pp. 49–54. Louvain-la-Neuve: Université catholique de Louvain.

Bonnet, Charles
1999 "The Funerary Traditions of Middle Nubia." In *Recent Research in Kushite History and Archaeology: Proceedings of the VIIIth International Conference for Meroitic Studies,* edited by Derek A. Welsby, pp. 1–17. British Museum Occasional Paper 31. London: British Museum Press.
2009 "Un ensemble religieux nubien devant une forteresse égyptienne du début de la XVIIIe dynastie: Mission archéologique suisse à Doukki Gel–Kerma (Soudan)." *Genava* n.s. 57: 95–108.
2011 "Les destructions perpétrées durant la campagne de Psammétique II en Nubie et les depots consécutifs." In *Statues égyptiennes et kouchites démembrées et reconstituées: Hommage à Charles Bonnet,* edited by Dominique Valbelle and Jean-Michel Yoyotte, pp. 21–32. Paris: Presses de l'Université Paris–Sorbonne.

Bonnet, Charles, and Saleh Eddin Mohamed Ahmed
1984 "Un bâtiment résidentiel d'époque napatéenne." *Genava* n.s. 32: 35–42.

Bonnet, Charles, and Dominique Valbelle
2003 "Un dépôt de statues royales du début du VIe siècle av. J.-C. à Kerma." *Comptes rendus de l'Académie des Inscriptions et Belle-Lettres* 147: 749–71.

Boulet, Stéphanie
2015 "Étude céramologique préliminaire des campagnes de fouille de la chapelle d'Osiris Ounnefer Neb-Djefaou 2013–2014." *Cahier de Karnak* 15: 63–79.
2017 "Nouvelles discussions sur l'apparition et l'usage du tour rapide en Égypte." *Bulletin de la céramique égyptienne* 27: 289–98.

Boulet, Stéphanie, and Catherine Defernez
2014 "Ceramic Production in the Theban Area from the Late Period: New Discoveries in Karnak." In *Thebes in the First Millennium BC,* edited by Elena Pischikova, Julia Budka, and Kenneth Griffin, pp. 603–24. Newcastle upon Tyne: Cambridge Scholars Publishing.

Bourriau, Janine
1981 *Umm el-Gaʿab: Pottery from the Nile Valley before the Arab Conquest.* Cambridge: Cambridge University Press.
1985 "Technology and Typology of Egyptian Ceramics." In *Ancient Technology to Modern Science,* vol. 1, *Ceramics and Civilization,* edited by W. D. Kingery, pp. 227–42. Columbus, OH: American Ceramics Society.
2003 "Imported Amphorae in the Third Intermediate and Late Dynastic Periods from the Excavations 2000–2001." In Ulrich Hartung et al., "Tell el-Faraʿin-Buto. 8. Vorbericht." *Mitteilungen des Deutschen Archäologischen Instituts Kairo* 59: 224–33.
2010 *Kom Rabia: The New Kingdom Pottery.* Survey of Memphis 4. Egypt Exploration Society Excavation Memoirs 93. London: Egypt Exploration Society.

Bourriau, Janine, and Peter French
2007 "Imported Amphorae from Buto Dating from c. 750 BC to the Early 6th Century AD." *Cahiers de la céramique égyptienne* 8: 115–33.

Bourriau, Janine, and Hans-Åke Nordström
1990 "The Vienna System." In *An Introduction to Ancient Egyptian Pottery,* fasc. 2, *Ceramic Technology: Clays and Fabrics,* edited by Dorothea Arnold and Janine Bourriau, pp. 168–82. Sonderschrift des Deutschen Archäologischen Instituts Kairo 17. Mainz am Rhein: Philipp von Zabern.

Bradley, Rebecca
1984 "Meroitic Chronology." In *Meroitische Forschungen 1980,* edited by Fritz Hintze, pp. 195–211. Meroitica 7. Berlin: Akademie-Verlag.

Briend, Jacques, and Jean-Baptiste Humbert
1980 *Tell Keisan (1971–1976): Une cité phénicienne en Galilée.* Orbis Biblicus et Orientalis, Series Archaeologica 1. Fribourg: Éditions Universitaires and Göttingen: Vandenhoeck & Ruprecht.

Brissaud, Philippe, and Laurence Cotelle
1987 "À propos de tessons retaillés en forme de jetons." In *Cahiers de Tanis,* vol. 1, *Mission française des fouilles de Tanis,* edited by Philippe

Brissaud, pp. 101–6. Éditions Recherche sur les Civilisations Mémoire 75. Paris: Éditions Recherche sur les Civilisations.

Brunton, Guy

1930 *Qau and Badari.* Vol. 3. British School of Archaeology in Egypt 50. London: Bernard Quaritch.

1948 *Matmar.* British Museum Expedition to Middle Egypt Occasional Paper 2. London: Bernard Quaritch.

Budge, E. A. Wallis

1909 *A Guide to the Egyptian Galleries (Sculpture).* London: British Museum. Accessed September 3, 2015. https://archive.org/details/guidetoegyptiang00brit.

Budka, Julia

2007 "Documentation and Excavation of Dome and Crevice Graves in Kirbekan." In *Proceedings of the Second International Conference on the Archaeology of the Fourth Nile Cataract, Berlin, August 4th–6th, 2005,* edited by Claudia Näser and Mathias Lange, pp. 73–90. Meroitica 23. Wiesbaden: Harrassowitz Verlag.

2009 "Benchmarks, Team Marks and Pot Marks from the Asasif (Western Thebes)." In *Pictograms or Pseudo Script? Non-textual Identity Marks in Practical Use in Ancient Egypt and Elsewhere,* edited by Ben J. J. Haring and Olaf E. Kaper, pp. 67–91. Egyptologische Uitgaven 25. Leuven: Peeters.

2010a "Kushite Tomb Groups in Late Period Thebes." In *Between the Cataracts: Proceedings of the 11th Conference for Nubian Studies, Warsaw University, 27 August–2 September 2006,* edited by Włodzimierz Godlewski and Adam Łajtar, pp. 503–18. Polish Archaeology in the Mediterranean Supplement Series 2.2/2: Session Papers. Warsaw: Warsaw University Press.

2010b "The Use of Pottery in Funerary Contexts During the Libyan and Late Period: A View from Thebes and Abydos." In *Egypt in Transition: Social and Religious Development of Egypt in the First Millennium BCE,* edited by Ladislav Bareš, Filip Coppens, and Květa Smolárikova, pp. 22–72. Prague: Charles University.

2014 "The New Kingdom in Nubia: New Results from Current Excavations on Sai Island." *Egitto e Vicino Oriente* 37: 55–87.

Bunnens, Guy

1985 "Le luxe phénicien d'après les inscriptions royales assyriennes." In *Phoenicia and Its Neighbours,* edited by Eric Gubel and Edward Lipiński, pp. 121–33. Studia Phoenicia 3. Leuven: Peeters.

Burckhardt, John Lewis

1978 *Travels in Nubia.* 2nd ed. reprint. New York: AMS Press. First published London: John Murray, 1822.

Burstein, Stanley

1984 "Psamtek I and the End of Nubian Domination in Egypt." *Journal of the Society for the Study of Egyptian Antiquities* 14: 31–34.

Butzer, Karl W., and Carl L. Hansen

1968 *Desert and River in Nubia: Geomorphology and Prehistoric Environments at the Aswan Reservoir.* Madison: University of Wisconsin Press.

Buzon, Michelle R.

2014 "Tombos during the Napatan Period (~750–660 BC): Exploring the Consequences of Sociopolitical Transitions in Ancient Nubia." *International Journal of Paleopathology* 7: 1–7.

Caminos, Ricardo A.

1974 *The New Kingdom Temples of Buhen.* 2 vols. Egypt Exploration Society Archaeological Survey of Egypt Memoir 33–34. London: Egypt Exploration Society.

1994 "Queen Katimala's Inscribed Panel in the Temple of Semna." In *Hommages à Jean Leclant,* vol. 2, *Nubie, Soudan, Éthiopie,* edited by Catherine Berger, Gisèle Clerc, and Nicholas Grimal, pp. 73–80. Bibliothèque d'étude 106. Cairo: Institut français d'archéologie orientale.

Carrano, Julia, Jeffrey R. Ferguson, Gary H. Girty, Stuart T. Smith, and Carl J. Carrano

2008 "A Chemical and Mineralogical Comparison of Nubian and Egyptian Style Ceramics and the Implications for Culture Contact: Preliminary Report." *Sudan & Nubia* 12: 90–98.

Carrano, Julia L., Gary H. Girty, and Carl J. Carrano

2009 "Re-examining the Egyptian Colonial Encounter in Nubia through a Compositional, Mineralogical, and Textural Comparison of Ceramics." *Journal of Archaeological Science* 36: 785–97.

Chaix, Louis

2010 "Animal Exploitation during Napatan and Meroitic Times in the Sudan." In *Between the Cataracts: Proceedings of the 11th Conference for Nubian Studies, Warsaw University, 27 August–2 September 2006,* edited by Włodzimierz Godlewski and Adam Łajtar, pp. 519–25. Polish Archaeology in the Mediterranean Supplement Series 2.2/1: Session Papers. Warsaw: Warsaw University Press.

Charrondiere, U. Ruth, David Haytowitz, and Barbara Stadlmayr

2012 FAO/INFOODS Density Database Version 2.0 (2012). Accessed January 10, 2016. http://www.fao.org/docrep/017/ap815e/ap815e.pdf.

Clairmont, Christoph W.

1963 *The Excavations at Dura-Europos: Final Report 4.* Part 5, *The Glass Vessels.* New Haven: Dura-Europos Publications.

Clapham, Alan, and Peter Rowley-Conwy

2006 "Qasr Ibrim's Archaeobotany." *Egyptian Archaeology* 29: 6–7.

Clark, J. Desmond, James L. Phillips, and Preston S. Staley

1974 "Interpretations of Prehistoric Technology from Ancient Egyptian and Other Sources, Part 1: Ancient Egyptian Bows and Arrows and Their Relevance for African Prehistory." *Paléorient* 2: 323–88.

Clarke, Somers

1916 "Ancient Egyptian Frontier Fortresses." *Journal of Egyptian Archaeology* 3: 155–79.

Cohen, Rudolph

1979 "The Iron Age Fortresses in the Central Negev." *Bulletin of the American Schools of Oriental Research* 236: 61–79.

Cohen, Rudolph, and Yigal Yisrael

1995 *On the Road to Edom: Discoveries from ʿEn Hazeva.* Jerusalem: Israel Museum.

1996 "The Excavations at ʿEin Hazeva/Israelite and Roman Tamar." *Qadmoniot* 29, no. 2: 78–92.

Collombert, Philippe

2003–8 "Par-delà bien et mal: L'inscription de la reine Katimala à Semna." *Kush* 19: 185–218.

Copley, Mark S., Susan Jim, Vicky Jones, Pam Rose, Alan Clapham, David N. Edwards, Mark Horton, Peter Rowley-Conwy, and Richard P. Evershed

2004 "Short- and Long-Term Foraging and Foddering Strategies of Domesticated Animals from Qasr Ibrim, Egypt." *Journal of Archaeological Science* 31: 1273–86.

Crawford, Osbert G. S., and Frank Addison

1951 *Abu Geili, and Saqadi & Dar el Mek.* Wellcome Excavations in the Sudan 3. London: Oxford University Press.

Crivelli, Eduardo, Claudia Kohen, and Silvia Lupo

2011 "Two Foundation Deposits in a Domestic Context at Tell el-Ghaba, a First Millennium Settlement in the Eastern Delta, Egypt." *Journal of the American Research Center in Egypt* 48: 205–20.

Darnell, John Coleman

2006 *The Inscription of Queen Katimala at Semna: Textual Evidence for the Origins of the Napatan State.* Yale Egyptological Studies 7. New Haven: Yale Egyptological Seminar.

Defernez, Catherine

2001 *La céramique perse à Tell el-Herr: Étude chrono-typologique et comparative.* 2 vols. Cahiers de Recherches de l'Institut de Papyrologie et d'Égyptologie de Lille Supplement 5. Lille: Université Charles-de-Gaulle.

2015 "Investigations récentes et mobilier importé des époques saïte et perse dans le temple d'Amon-Rê de Karnak." In *La présence grecque dans la vallée de Thèbes,* edited by Gilles Gorre and Antigone Marangou, pp. 45–65. Rennes: Presses universitaires de Rennes.

Defernez, Catherine, and Stéphanie Boulet

2013 "Étude céramologique préliminaire de la campagne de fouilles de la chapelle d'Osiris Ounnefer Neb-Djefaou 2012." Accessed March 15, 2015. http://www.cfeetk.cnrs.fr/index.php?page=rapport-2012-neb-djefaou.

Defernez, Catherine, and Fabien Isnard

1998 "La céramique provenant de la structure elliptique." In *Tanis: Travaux récents sur le Tell Sân el-Hagar,* edited by Philippe Brissaud and Christiane Zivie-Coche, pp. 155–218. Mission française des fouilles de Tanis 1987–1997. Paris: Éditions Noêsis.

Der Manuelian, Peter

1994 *Living in the Past: Studies in the Archaism of the Egyptian Twenty-Sixth Dynasty.* London and New York: Kegan Paul International.

Ditze, Barbara

2007 "Gedrückt–Geritzt–Gekratzt: Die Gefäße mit Topfmarken." In *Die Keramik des Grabungsplatzes Q I,* part 2, *Schaber–Marken–Scherben,* edited by Edgar B. Pusch and Manfred Bietak, pp. 270–507. Grabung des Pelizaeus-Museums Hildesheim in Qantir–Pi-Ramesse 5. Hildesheim: Roemer- und Pelizaeus-Museum.

Dodson, Aidan, and Dyan Hilton

2004 *The Complete Royal Families of Ancient Egypt.* London: Thames & Hudson.

D'Onofrio, Anna Maria, ed.

2007 *Tallies, Tokens and Counters from the Mediterranean to India: Proceedings of the Meeting Held at the Universitá degli Studi Napoli "L'Orientale," Naples 31st May 2004.* Naples: Università degli Studi Napoli L'Orientale.

Drzewiecki, Mariusz

2011 "Fortress Commandants of the Kingdom of Kush." *Egyptian Archaeology* 43: 25–27.

2012 *Fortresses of Sudan, 2012: Preliminary Report.* Institute of Prehistory, Adam Mickiewicz University of Poznań, Poland. Khartoum:

National Corporation for Antiquities and Museums, Sudan.

Dunham, Dows
1950 *El Kurru*. Royal Cemeteries of Kush 1. Cambridge, MA: Harvard University Press.
1955 *Nuri*. Royal Cemeteries of Kush 2. Boston: Museum of Fine Arts.
1963 *The West and South Cemeteries at Meroe*. Royal Cemeteries of Kush 5. Boston: Museum of Fine Arts.
1967 *Uronarti, Shalfak, Mirgissa*. Second Cataract Forts 2. Boston: Museum of Fine Arts.

Dunham, Dows, and Jozef M. A. Janssen
1960 *Semna, Kumma*. Second Cataract Forts 1. Boston: Museum of Fine Arts.

Dupont, Pierre
1998 "East Greek Trade Amphoras." In *East Greek Pottery*, edited by Robert M. Cook and Pierre Dupont, pp. 142–91. London: Routledge.

Dupont, Pierre, and Jean-Claude Goyon
1992 "Amphores grecques archaïques de Gurna: À propos d'une publication récente." In *Sesto Congresso Internazionale di Egittologia: Atti*, vol. 1, edited by Jean Leclant, pp. 153–66. Turin: International Association of Egyptologists.

Edmonds, James M.
1940 "A Quartz Arrowhead from Zankor." *Sudan Notes and Records* 23: 193.

Edwards, David N.
1996 "Sorghum, Beer, and Kushite Society." *Norwegian Archaeological Review* 29: 65–77.

Edwards, David N., and Anthony J. Mills
2020 *The Archaeological Survey of Sudanese Nubia 1963–69: The Pharaonic Sites*. Oxford: Archaeopress.

Eger, Jana, Jana Helmbold-Doyé, and Tim Karberg
2010 "Osttor und Vorwerk der Festung Gala Abu Ahmed: Bericht über die archäologischen Arbeiten der Kampagnen 2008/9 und 2009." *Der Antike Sudan: Mitteilungen der Sudanarchäologischen Gesellschaft zu Berlin* 21: 71–85.

Eichhorn, Barbara
2013 "Phytoliths on Grinding Stones and Wood Analysis." *Sudan & Nubia* 17: 28–33.

Eide, Tormod, Tomas Hägg, Richard Holton Pierce, and László Török, eds.
1994 *Fontes Historiae Nubiorum: Textual Sources for the History of the Middle Nile Region between the Eighth Century* BC *and the Sixth Century* AD. Vol. 1, *From the Eighth to the Mid-Fifth Century* BC. Bergen: University of Bergen, Department of Classics.

Eigner, Dieter
2013 "A Stone Building in the Desert Sands—Some Remarks on the Architecture of the Gala Abu Ahmed Fortress." In *The Power of Walls: Fortifications in Ancient Northeastern Africa. Proceedings of the International Workshop, University of Cologne 4th–7th August 2011*, edited by Friederike Jesse and Carola Vogel, pp. 309–19. Colloquium Africanum 5. Cologne: Heinrich-Barth Institut.

Eigner, Dieter, and Friederike Jesse
2009 "Im Westen viel Neues: Die Grabungen 2008/09 in der Festung Gala Abu Ahmed." *Der Antike Sudan: Mitteilungen der Sudanarchäologischen Gesellschaft zu Berlin* 20: 141–58.

Eladany, Abeer Helmy
2012 "A Study of a Selected Group of Third Intermediate Period Mummies in the British Museum." PhD diss., University of Manchester. Accessed February 2, 2016. https://www.escholar.manchester.ac.uk/jrul/search/.

Elat, Moshe
1978 "The Economic Relations of the Neo-Assyrian Empire with Egypt." *Journal of the American Oriental Society* 98: 20–34.

Emery, Walter B., Harry S. Smith, and Anne Millard
1979 *The Fortress of Buhen: The Archaeological Report*. Egypt Exploration Society Excavation Memoir 49. London: Egypt Exploration Society.

Erichsen, Wolja Christian
1942 "Erwähnung eines Zuges nach Nubien unter Amasis in einem demotischen Text." *Klio* 34: 56–61.

Fairman, Herbert Walter
1948 "Preliminary Report on the Excavations at Amarah West, Anglo-Egyptian Sudan, 1947–8." *Journal of Egyptian Archaeology* 34: 3–11.

Fantalkin, Alexander
2001 "Mezad Hashavyahu: Its Material Culture and Historical Background." *Tel Aviv* 28, no. 1: 3–165.

Fantalkin, Alexander, and Oren Tal
2009 "Re-discovering the Iron Age Fortress at Tell Qudadi in the Context of Neo-Assyrian Imperialistic Policies." *Palestine Exploration Quarterly* 141, no. 3: 188–206.

Faust, Avraham, and Ehud Weiss
2005 "Judah, Philistia, and the Mediterranean World: Reconstructing the Economic System of the Seventh Century B.C.E." *Bulletin of the American Schools of Oriental Research* 338: 71–92.

Fiedler, Manuel, and Friederike Jesse
2011 "Griechische Keramik aus der Festung Gala Abu Ahmed im Nordsudan." *Archäologischer Anzeiger: Jahrbuch des Deutschen Archäologischen Instituts* 2011, no. 2: 67–80.

Finkelstein, Israel
1992 "Horvat Qitmīt and the Southern Trade in the Late Iron Age II." *Zeitschrift des Deutschen Palästina-Vereins* 108: 156–70.
2005 "A Low Chronology Update: Archaeology, History and the Bible." In *The Bible and Radiocarbon Dating: Archaeology, Text and Science*, edited by Thomas E. Levy and T. Higham, pp. 31–42. London: Equinox.
2010 "Kadesh Barnea: A Reevaluation of Its Archaeology and History." *Tel Aviv* 37: 111–25.

Finkelstein, Israel, and Lily Singer-Avitz
2001 "Ashdod Revisited." *Tel Aviv* 28, no. 2: 231–59.

Flache, Michael
2012 "The Round Structures of Gala Abu Ahmed Fortress in Lower Wadi Howar, Sudan." *Sudan & Nubia* 16: 44–51.

Food and Agriculture Organization of the United Nations
2004 *Human Energy Requirements: Report of a Joint FAO/WHO/UNU Expert Consultation.* FAO Food and Nutrition Technical Report Series 1. Rome: FAO.

Forbes, Robert J.
1965 *Studies in Ancient Technology.* Vol. 3. 2nd ed. Leiden: Brill.

French, Peter
1986 "Late Dynastic Pottery from the Vicinity of the South Tombs." In *Amarna Reports*, vol. 4, edited by Barry Kemp, pp. 147–88. Egypt Exploration Society Occasional Publications 4. London: Egypt Exploration Society.
1992 "A Preliminary Study of Pottery in Lower Egypt in the Late Dynastic and Ptolemaic Periods." *Cahiers de la céramique égyptienne* 3: 83–93.
2004 "Distinctive Pottery from the Second Half of the 6th Century B.C." *Cahiers de la céramique égyptienne* 7: 91–97.
2007 "A Pottery Assemblage of the First Half of the 6th Century BC." In Ulrich Hartung et al., "Tell el-Faraʿin–Buto 9. Vorbericht." *Mitteilungen des Deutschen Archäologischen Instituts Kairo* 63: 101–19.

French, Peter, and Holeil Ghaly
1991 "Pottery Chiefly of the Late Dynastic Period, from Excavations by the Egyptian Antiquities Organization at Saqqara, 1987." *Cahiers de la céramique égyptienne* 2: 93–124.

Fuller, Dorian
2004 "Early Kushite Agriculture: The Archaeobotanical Evidence from Kawa." *Sudan & Nubia* 8: 70–74.

Fuscaldo, Perla
2005 *Tell el-Ghaba: A Saite Settlement in North Sinai, Egypt (Argentine Archaeological Mission, 1995–2004).* Vol. 1, *Introduction, Catalogue.* Colleción Estudios 5. Buenos Aires: Conseja Nacional de Investigaciones Cientificas y Técnicas and Instituto Multidisciplinario de Historia y Ciencias Humanas.
2006 "The Egyptian Painted Wares from Tell el-Ghaba, North Sinai (Early 26th Dynasty)." In *Timelines: Studies in Honour of Manfred Bietak*, vol. 2, edited by Ernst Czerny, Irmgard Hein, Hermann Hunger, Dagmar Malman, and Angela Schwab, pp. 111–17. Orientalia Lovaniensia Analecta 149. Leuven: Peeters.

Gal, Zvi, Yardena Alexandre, and Uri Baruch
2000 *Horbat Rosh Zayit: An Iron Age Storage Fort and Village.* Israel Antiquities Authority Reports 8. Jerusalem: Israel Antiquities Authority.

Gallorini, Carla
1998 "Incised Marks on Pottery and Other Objects from Kahun: Systems of Communication in Egypt during the Late Middle Kingdom." PhD thesis, Institute of Archaeology, University College London. Accessed September 11, 2015. http://discovery.ucl.ac.uk/1317898/.

Gardberg, Carl Jacob
1970 *Late Nubian Sites: Churches and Settlements.* Scandinavian Joint Expedition to Sudanese Nubia 7. Helsinki: Oy Tilgmann Ab.

Gassner, Verena
2011a "Fabrics from Chios." In *Fabrics of the Central Mediterranean* (version 06/06/2011). http://www.facem.at/project-papers.php.
2011b "Fabrics of East Aegean Amphorae." In *Fabrics of the Central Mediterranean* (version 06/06/2011). http://www.facem.at/project-papers.php.

Geiser, Peter
1986 *The Egyptian Nubian: A Study in Social Symbiosis.* Cairo: American University in Cairo Press.

Geus, Francis
1975 "Le cimetière de la XXVe dynastie (MF-NE)." In *Mirgissa*, vol. 2, *Les nécropoles*, by Jean Vercoutter, pp. 479–501. Paris: Direction générale des relations culturelles, scientifiques et techniques avec la collaboration du Centre national de la récherche scientifique.
1979 *Rapport annuel d'activité 1977–1978.* Service des antiquités du Soudan, Section française

de recherche archéologique. Khartoum: el-Gurashi.

Gibbon, Victoria E., and Michelle R. Buzon

2016 "Morphometric Assessment of the Appendicular Skeleton in the New Kingdom and Napatan Components from Tombos in Upper Nubia." *International Journal of Osteoarchaeology* 26: 324–36.

Gilboa, Ayelet

1999 "The Dynamics of Phoenician Bichrome Pottery: A View from Tel Dor." *Bulletin of the American Schools of Oriental Research* 316: 1–22.

2015 "Dor and Egypt in the Early Iron Age: An Archaeological Perspective of (Part of) the Wenamun Report." *Ägypten und Levante* 25: 247–74.

Gilboa, Ayelet, and Yuval Goren

2015 "Early Iron Age Phoenician Networks: An Optical Mineralogy Study of Phoenician Bichrome and Related Wares in Cyprus." *Ancient West & East* 14: 73–110.

Gilboa, Ayelet, Ilan Sharon, and Jeffrey Zorn

2004 "Dor and Iron Age Chronology: Scarabs, Ceramic Sequence and C14." *Tel Aviv* 31: 32–59.

Gilboa, Ayelet, Paula Waiman-Barak, and Richard Jones

2015 "On the Origin of Iron Age Phoenician Ceramics at Kommos, Crete: Regional and Diachronic Perspectives across the Bronze Age to Iron Age Transition." *Bulletin of the American Schools of Oriental Research* 374: 80–89.

Gorre, Gilles, and Antigone Marangout

2015 *La présence grecque dans la vallée de Thèbes.* Rennes: Presses universitaires de Rennes.

Grant, James

1885 *Cassell's History of the War in the Sudan.* Vol. 3. London, Paris, New York, and Melbourne: Cassell & Company. Google-digitized at HathiTrust. Accessed March 5, 2011. http://hdl.handle.net/2027/mdp.39015050321978.

Gratien, Brigitte

2014 "Kerma North of the Fourth Cataract." In *The Fourth Cataract and Beyond: Proceedings of the 12th International Conference for Nubian Studies*, edited by Julie R. Anderson and Derek A. Welsby, pp. 95–101. British Museum Publications on Egypt and Sudan 1. Leuven: Peeters.

Graves-Brown, Carolyn

2010 "The Ideological Significance of Flint in Dynastic Egypt." PhD diss., Institute of Archaeology, University College London. Accessed November 15, 2015. https://discovery.ucl.ac.uk/id/eprint/1306709/1/1306709.pdf.

2015 "Flints and Forts: The Role of Flint in Late Middle–New Kingdom Egyptian Weaponry." In *Walls of the Prince: Egyptian Interactions with Southwest Asia in Antiquity. Essays in Honour of John S. Holladay Jr.*, edited by Timothy P. Harrison, Edward B. Banning, and Stanley Klassen, pp. 37–59. Culture and History of the Ancient Near East 77. Leiden: Brill.

Griffith, Francis Llewellyn

1922 "Oxford Excavations in Nubia VIII–XVII: Napata, Sanam Temple, Treasury, and Town." *Annals of Archaeology and Anthropology (Liverpool)* 9: 67–124.

1923 "Oxford Excavations in Nubia XVIII–XXV: The Cemetery of Sanam." *Annals of Archaeology and Anthropology (Liverpool)* 10: 73–171.

Grimal, Nicholas-Christophe

1981 *Études sur la propaganda royale égyptienne.* Vol. 1, *La stèle triomphale de Pi(ʿankh)y au Musée du Caire: JE 48862 and 47086-47089.* Mémoires de l'Institut français d'archéologie orientale du Caire 104. Cairo: Institut français d'archéologie orientale.

Grose, David Frederick

1989 *The Toledo Museum of Art: Early Ancient Glass.* New York: Hudson Hills Press.

Grossmann, Peter

2003 "Zu dem angeblichen befestigten Kloster über der Südmauer von al-Kâb." *Bulletin de la Société archéologie d'Alexandrie* 47: 113–25.

Haaland, Randi

2007 "Porridge and Pot, Bread and Oven: Food Ways and Symbolism in Africa and the Near East from the Neolithic to the Present." *Cambridge Archaeological Journal* 17, no. 2: 165–82.

Haynes, Joyce

2011 "Assemblages de fragments de sculptures du Gebel Barkal et de Giza retrouvés durant les travaux de rénovation des caves du Museum of Fine Arts de Boston." In *Statues égyptiennes et kouchites démembrées et reconstitutées: Hommage à Charles Bonnet*, edited by Dominique Valbelle and Jean-Michel Yoyotte, pp. 33–52. Paris: Presses de l'Université Paris–Sorbonne.

Heidorn, Lisa A.

1991 "The Saite and Persian Period Forts at Dorginarti." In *Egypt and Africa: Nubia from Prehistory to Islam*, edited by W. Vivian Davies, pp. 205–19. London: British Museum Press.

1992a "The Fortress of Dorginarti and Lower Nubia during the Seventh to Fifth Centuries B.C." PhD diss., Department of Near Eastern Languages and Civilizations, University of Chicago. Accessed February 15, 2016. https://chicago.academia.edu/LisaHeidorn.

1992b "The Persian Claim to Kush in Light of Evidence from Lower Nubia." In *Life in a Multi-cultural Society: Egypt from Cambyses to Constantine and Beyond*, edited by Janet H. Johnson, pp. 147–48. Studies in Ancient Oriental Civilization 51. Chicago: Oriental Institute of the University of Chicago.

1994 "Historical Implications of the Pottery from the Earliest Tombs at El Kurru." *Journal of the American Research Center in Egypt* 31: 115–31.

1997 "The Horses of Kush." *Journal of Near Eastern Studies* 56, no. 2: 105–14.

2013 "Dorginarti: Fortress at the Mouth of the Rapids." In *The Power of Walls: Fortifications in Ancient Northeastern Africa*, edited by Friederike Jesse and Carola Vogel, pp. 293–307. Colloquium Africanum 5. Cologne: Heinrich-Barth Institut.

2018a "The Boston Museum of Fine Arts Pottery from the Twenty-Fifth Dynasty Tombs at el-Kurru and Nuri." In *Nubian Archaeology in the XXIst Century: Proceedings of the Thirteenth International Conference for Nubian Studies, Neuchâtel, 1st–6th September 2014*, edited by Matthieu Honegger, pp. 317–32. Orientalia Lovaniensia Analecta 273. Leuven: Peeters.

2018b "The Sixth Century BC Imported Amphorae from Dorginarti." *Cahiers de la céramique égyptienne* 11: 189–207.

Heinzelin, Jean de

1968 "Geological History of the Nile Valley." In *The Prehistory of Nubia*, edited by Fred Wendorf, pp. 19–55. Dallas: Fort Burgwin Research Center.

Herzog, Zeʾev

2002 "The Fortress Mound at Tel Arad: An Interim Report." *Tel Aviv* 29, no. 1: 3–109.

Herzog, Zeʾev, Miriam Aharoni, Anson F. Rainey, and Shmuel Moshkovitz

1984 "The Israelite Fortress at Arad." *Bulletin of the American Schools of Oriental Research* 254: 1–34.

Hewes, Gordon W.

1964 "Gezira Dabarosa: Report of the University of Colorado Nubian Expedition, 1962–63 Season." *Kush* 12: 174–87.

Hikade, Thomas

2001 "Silex-Pfeilspitzen in Ägypten." *Mitteilungen des Deutschen Archäologischen Instituts Kairo* 57: 109–25.

Höckmann, Ursula, and Sabine Weber, eds.

2012 *Griechische Keramik des 7. und 6. Jhs. v. Chr. aus Naukratis und anderen Orten in Ägypten*. Archäologische Studien zu Naukratis 3. Worms: Wernersche Verlagsgesellschaft.

Hoerth, Alfred

1964a Dorginarti Field and Pottery Notebook, Nos. 1 and 2. Oriental Institute Museum Archives, University of Chicago.

1964b "Salvage Operations in Sudanese Nubia." In *Oriental Institute Annual Report 1963–1964*, pp. 15–16. Chicago: Oriental Institute.

Holladay, John S.

1982 *Cities of the Delta*. Vol. 3, *Tell el-Maskhuta*. American Research Center in Egypt Reports 6. Malibu: Undena Publications.

2004 "Judaeans (and Phoenicians) in Egypt in the Late Seventh to Sixth Centuries B.C." In *Egypt, Israel, and the Ancient Mediterranean World: Studies in Honor of Donald B. Redford*, edited by Gary N. Knoppers and Antoine Hirsch, pp. 405–37. Leiden: Brill.

Hölscher, Uvo

1910 *Das hohe Tor von Medinet Habu: Eine baugeschichtliche Untersuchung*. Wissenschaftliche Veröffentlichung der Deutschen Orient-Gesellschaft 12. Leipzig: J. C. Hinrichs.

1934 *General Plans and Views*. The Excavation of Medinet Habu 1. Oriental Institute Publications 21. Chicago: University of Chicago Press.

1954 *Post-Ramessid Remains*. The Excavation of Medinet Habu 5. Oriental Institute Publications 66. Chicago: University of Chicago Press.

Holthoer, Rotislav

1977 *New Kingdom Pharaonic Sites: The Pottery*. Scandinavian Joint Expedition to Sudanese Nubia 5/1. Lund: Berlings.

Hope, Colin A.

1989 "Amphorae of the New Kingdom." In *Pottery of the Egyptian New Kingdom: Three Studies*, pp. 87–110. Victoria College Archaeology Research Unit Occasional Paper 2. Burwood: Victoria College Press.

Hope, Colin A., Mark A. J. Eccleston, Olaf E. Kaper, Sylvie Marchand, and Deborah Darnell

2000 "Kegs and Flasks from the Dakhleh Oasis." *Cahiers de la céramique égyptienne* 6: 189–234.

Horton, Mark

1991 "Africa in Egypt: New Evidence from Qasr Ibrim." In *Egypt and Africa: Nubia from Prehistory to Islam*, edited by W. Vivian Davies, pp. 264–77. London: British Museum Press.

Ireland, A. W.

1948 "The Climate of the Sudan." In *Agriculture in the Sudan*, edited by John Douglas Tothill, pp. 62–83. Oxford: Oxford University Press.

Jacquet-Gordon, Helen

2012 *Le trésor de Thoutmosis Ier: La céramique.* Karnak-Nord 10. Fouilles de l'Institut français d'archéologie orientale du Caire 65/1–2. Cairo: Institut français d'archéologie orientale.

Jaeger, Bertrand

1982 *Essai de classification et datation des scarabées Menkhéperrê.* Orbis Biblicus et Orientalis, Series Archaeologica 2. Fribourg: Éditions Universitaires and Göttingen: Vandenhoeck & Ruprecht.

1993 *Les scarabées à noms royaux du Museo Civico Archeologico de Bologna.* Bologna: Comune di Bologna.

Jánosi, Peter

1996 "Die Fundamentplattform eines Palastes (?) der späten Hyksoszeit in ʿEzbet Helmi (Tell el-Dabʿa)." In *Haus und Palast im alten Ägypten: International Symposium in Cairo, April 8 to 11, 1992*, edited by Manfred Bietak, pp. 93–98. Untersuchungen der Zweigstelle Kairo des Österreichischen Archäologischen Institutes 14. Vienna: Verlag der Österreichischen Akademie der Wissenschaften.

Jansen-Winkeln, Karl

1989 "Zur Schiffsliste aus Elephantine." *Göttinger Miszellen* 109: 31.

2006 "The Chronology of the Third Intermediate Period: Dyns. 22–24." In *Ancient Egyptian Chronology*, Handbook of Oriental Studies, Section 1: The Near and Middle East, edited by Erik Hornung, Rolf Krauss, and David A. Warburton, pp. 234–64. Leiden: Brill.

Jaritz, Horst

1980 "Naga Abu ʿId." *Bulletin de la céramique égyptienne* 5: 12.

1986 "On Three Townsites in the Upper Thebaid." *Cahiers de Recherches de l'Institut de Papyrologie et d'Égyptologie de Lille* 8: 37–42.

Jesse, Friederike

2007 "Pastoral Groups in the Southern Libyan Desert: The Handessi Horizon (c. 2400–1100 BC)." In *Archaeology of Early Northeastern Africa: In Memory of Lech Krzyżaniak*, edited by Karla Kroeper, Marek Chłodnicki, and Michał Kobusiewicz, pp. 987–1004. Studies in African Archaeology 6. Poznán: Poznán Archaeological Museum.

2013 "Far from the Nile—the Gala Abu Ahmed Fortress in Lower Wadi Howar (Northern Sudan)." In *The Power of Walls: Fortifications in Ancient Northeastern Africa*, edited by Friederike Jesse and Carola Vogel, pp. 321–52. Colloquium Africanum 5. Cologne: Heinrich-Barth Institut.

2016 "Between the New Kingdom and Napata—the Pottery Assemblage of the Gala Abu Ahmed Fortress in Wadi Howar, Northern Sudan." In *Vienna 2—Ancient Egyptian Ceramics in the 21st Century: Proceedings of the International Conference Held at the University of Vienna, 14th–18th of May, 2012*, edited by Bettina Bader, Christian M. Knoblauch, and E. Christiana Köhler, pp. 299–318. Orientalia Lovaniensia Analecta 245. Leuven: Peeters.

Jesse, Friederike, and Rudolph Kuper

2006 "Napata in the West? The Gala Abu Ahmed Fortress in Lower Wadi Howar (NW-Sudan)." *Archéologie du Nil Moyen* 10: 135–59.

Johnston, Alan

1993 "Pottery from Archaic Building Q at Kommos." *Hesperia* 62: 339–82.

Junge, Friedrich

1987 *Elephantine XI: Funde und Bauteile.* Archäologische Veröffentlichungen, Deutschen Archäologisches Institut Kairo 49. Mainz am Rhein: Philipp von Zabern.

Kahlheber, Stefanie

2013 "The Fruit and Seed Remains." *Sudan & Nubia* 17: 33–41.

Kahn, Dan'el

2001 "The Inscription of Sargon II at Tang-i Var and the Chronology of Dynasty 25." *Orientalia* 70: 1–18.

2007 "Judean Auxiliaries in Egypt's Wars against Kush." *Journal of the American Oriental Society* 127, no. 4: 507–16.

2008 "Some Remarks on the Foreign Policy of Psammetichus II in the Levant (595–589 BC)." *Journal of Egyptian History* 1, no. 1: 139–57.

Kaiser, Werner, Frank Andraschko, David A. Aston, Horst Jaritz, Achim Krekeler, Ingrid Nebe, Walter Niederberger, Maja Pigur, and Cornelius von Pilgrim

1990 "Stadt und Tempel von Elephantine: 17./18. Grabungsbericht." *Mitteilungen des Deutschen Archäologischen Instituts Kairo* 46: 185–249.

Kaiser, Werner, Günter Dreyer, Horst Jaritz, Achim Krekeler, Jan Lindemann, Cornelius von Pilgrim, Stephan Seidlmayer, and Martin Ziermann

1988 "Stadt und Tempel von Elephantine: 15./16. Grabungsbericht." *Mitteilungen des Deutschen Archäologischen Instituts Kairo* 44: 135–82.

Kardulias, P. Nick

1992 "Estimating Population at Ancient Military Sites: The Use of Historical and Contemporary

Analogy." *American Antiquity* 57, no. 2: 276–87.

Keeley, Lawrence H., Marisa Fontana, and Russell Quick

2007 "Baffles and Bastions: The Universal Features of Fortifications." *Journal of Archaeological Research* 15, no. 1: 55–95.

Kemp, Barry

1977a "A Building of Amenophis III at Kôm el-ʿAbd." *Journal of Egyptian Archaeology* 63: 71–82.

1977b "The Palace of Apries at Memphis." *Mitteilungen des Deutschen Archäologischen Instituts Kairo* 33: 101–8.

1986 "Large Middle Kingdom Granary Buildings (and the Archaeology of Administration)." *Zeitschrift für Ägyptische Sprache und Altertumskunde* 113: 120–36.

2006 *Ancient Egypt: Anatomy of a Civilization.* 2nd ed. London: Routledge.

Killick, David

2004 "Review Essay: What Do We Know about African Iron Working?" *Journal of African Archaeology* 2: 97–112.

Kirwan, Lawrence P.

1955 "Account of the Excavations, 1935–6." In *The Temples of Kawa*, vol. 2, *History and Archaeology of the Site*, edited by M. F. Laming Macadam, pp. 207–37. London: Oxford University Press.

Kitchen, Kenneth A.

1986 *The Third Intermediate Period in Egypt (1100–650 B.C.).* 2nd ed. with supplement. Warminster: Aris & Phillips.

Klemm, Dieter, and Rosemarie Klemm

2013 *Gold and Gold Mining in Ancient Egypt and Nubia: Geoarchaeology of the Ancient Gold Mining Sites in the Egyptian and Sudanese Eastern Deserts.* Berlin: Springer Verlag.

Klemm, Dieter, Rosemarie Klemm, and Andreas Murr

2001 "Gold of the Pharaohs—6000 Years of Gold Mining in Egypt and Nubia." *Journal of African Earth Sciences* 33: 643–59.

Klimaszewska-Drabot, Edyta

2003 "Proto-Kushite (?) Ceramics, an Archaeological Enigma." In *Survey and Excavations between Old Dongola and Ez-Zuma,* edited by Bogdan Żurawski, pp. 438–41. Nubia 2. Southern Dongola Reach Survey 1. Warsaw: Editions Wydawnictwo Neriton.

Klotz, David

2015 "Darius I and the Sabaeans: Ancient Partners in Red Sea Navigation." *Journal of Near Eastern Studies* 74, no. 2: 267–80.

Knudstad, James

1963–64 Logbook of the Oriental Institute Sudan Expedition. Oriental Institute Museum Archives, University of Chicago.

1964a Architectural drawings. Oriental Institute Museum Archives, University of Chicago.

1964b Letters to Robert McCormick Adams, Director of the Oriental Institute of the University of Chicago. Oriental Institute Museum Archives, University of Chicago.

1964c Letters to Thabit Hassan Thabit, Commissioner of Archaeology, Sudan Antiquities Organization. Oriental Institute Museum Archives, University of Chicago.

1964d Survey Notes for Serra East and Dorginarti. Oriental Institute Museum Archives, University of Chicago.

1966 "Serra East and Dorginarti: Report on the 1963–64 Excavations of the Chicago Oriental Institute Sudan Expedition." *Kush* 14: 165–86.

Knudstad, James, and Richard H. Pierce

1964 Dorginarti Object Register, Division List. Oriental Institute Museum Archives, University of Chicago.

Kormysheva, Eleonora, and Hassan Hussein Idriss

2006 *Gods and Divine Symbols of the Ancient Sudanese Civilisation: Catalogue of the Sudan National Museum in Khartoum.* Moscow: Russian Academy of Sciences, Institute of Oriental Studies.

Krekeler, Achim

1996 "Stadtgrabung am Westkom von Elephantine: Wohnbauten des 1. Jahrtausends v. Chr." In *Haus und Palast im alten Ägypten: International Symposium in Cairo, April 8 to 11, 1992*, edited by Manfred Bietak, pp. 107–15. Österreichische Akademie der Wissenschaften, Denkschriften der Gesamtakademie 14. Vienna: Verlag der Österreichischen Akademie der Wissenschaften.

Kucharczyk, Renata

2011 "Glass from Area F on Kom El-Dikka (Alexandria). Excavations 2008." *Polish Archaeology in the Mediterranean* 20: 56–69.

Laemmel, Sabine

2008 "Preliminary Report on the Pottery from Area Q IV at Qantir/Pi-Ramesse." *Ägypten und Levante* 18: 173–202.

2011 "Le matériel céramique du temple d'Horus de Mesen et des sondages au centre du Tell Sân el-Hagar de la Troisième Période Intermédiaire à la Basse Époque." *Cahier de la Société française des fouilles de Tanis* 4: 1–77.

Lahitte, Miriam

2013 "Gala Abu Ahmed, Perlen und Fragmente aus Straußeneischale." *Der Antike Sudan: Mitteilungen der Sudanarchäologischen Gesellschaft zu Berlin* 24: 75–103.

Lambrino, Marcelle F.

1938 *Les vases archaïques d'Histria.* Bucharest: Fundatia Regele Carol I.

Lawall, Mark L., Nina A. Lejpunskaja, Pavel D. Diatroptov, and Tat'jana L. Samojlova

2010 "L Transport Amphoras." In *The Lower City of Olbia (Sector NG) in the 6th Century BC to the 4th Century AD*, edited by Valentina V. Krapivina and Sergey D. Kryzicki, pp. 355–405. Centre for Black Sea Studies 13. Aarhus: Aarhus University Press.

Lawrence, Arnold W.

1965 "Ancient Egyptian Fortifications." *Journal of Egyptian Archaeology* 51: 69–94.

Leclère, François

2014 "Tell Dafana: Identity, Exploration, and Monuments." In *Tell Dafana Reconsidered: The Archaeology of an Egyptian Frontier Town*, edited by François Leclère and Jeffrey Spencer, pp. 1–40. British Museum Research Publication 199. London: British Museum Press.

Leclère, François, and Jeffrey Spencer, eds.

2014 *Tell Dafana Reconsidered: The Archaeology of an Egyptian Frontier Town.* British Museum Research Publication 199. London: British Museum Press.

Lehmann, Gunnar

1996 *Untersuchung zur späten Eisenzeit in Syrien und Libanon: Stratigraphie und Keramikformen zwischen ca. 720 bis 300 v. Chr.* Altertumskunde des Vorderen Orients 5. Münster: Ugarit-Verlag.

1998 "Trends in the Local Pottery Development of the Late Iron Age and Persian Period in Syria and Lebanon, ca. 700–300 B.C." *Bulletin of the American Schools of Oriental Research* 311: 7–37.

2002 "Iron Age." In *Tel Kabri: The 1986–1993 Excavation Seasons*, edited by Aharon Kempinski, Na'ama Scheftelowitz, and Ronit Oren, pp. 178–222. Tel Aviv: Emery and Claire Yass Publications in Archaeology.

2006 Review of *Phoenician Amphora: Production and Distribution in the Southern Levant*, by Elizabeth A. Bettles. *Bulletin of the American Schools of Oriental Research* 344: 94–97.

Lehmann, Gunnar, and Ann E. Killebrew

2010 "Palace 6000 at Megiddo in Context: Iron Age Central Hall Tetra-Partite Residencies and the 'Bīt Ḫilāni' Building Tradition in the Levant." *Bulletin of the American Schools of Oriental Research* 359: 13–33.

Lehmann, Manuela

2014 "Tower Houses in Tell el-Dabʿa: The Late and Ptolemaic Period." In *Les maisons-tours en Égypte durant la Basse Époque, les périodes ptolémaïque et romaine: Actes de la table-ronde de Paris, Université Paris–Sorbonne (Paris IV), 29–30 novembre 2012*, edited by Séverine Marchi, pp. 57–68. Paris: Université de Paris–Sorbonne.

Leitz, Christian

2002 *Lexikon der ägyptischen Götter und Götterbezeichnungen.* Leuven: Peeters.

Lemaire, André

1987 "Les phéniciens et le commerce entre la mer Rouge et la mer Méditerranée." In *Phoenicia and the East Mediterranean in the First Millennium B.C.*, edited by Eric Gubel and Edward Lipiński, pp. 49–60. Orientalia Lovaniensia Analecta 22. Studia Phoenicia 3. Leuven: Peeters.

Lemorini, Cristina, and Enzo Cocca

2013 "Functional Perspectives on the Lithic Projectile. Points from Gobero: Experimental and Use-Wear Approaches." In *Gobero: The No-Return Frontier: Archaeology and Landscape at the Saharo-Sahelian Borderland*, edited by Elena A. Garcea, pp. 195–208. Journal of African Archaeology Monograph Series 9. Frankfurt-am-Main: Africa Magna Verlag.

Levy, Yosi, Martin Peilstöcker, and Anat Ginzburg

2004 "An Iron Age Fortress in the Sand Dunes of Rishon Letzion." *Qadmoniot* 128: 92–94.

Linseele, Veerle, and Nadja Pöllath

2015 "Local Foods and Traded Goods—the Faunal Remains from the Napatan Outpost at Gala Abu Ahmed (Sudan, First Millennium BC)." *African Archaeological Review* 32: 537–90.

Lister, Florence

1967 *Ceramic Studies of the Historic Periods in Ancient Nubia.* University of Utah Department of Anthropology Anthropological Papers 86. Salt Lake City: University of Utah Press.

Lodomez, Guy

2005 "Le vizier Ânkh-Osorkon de la Troisième Période Intermédiaire (Paris, Musée Rodin, Co. 3386 et Le Caire, Musée Égyptien, JE 91300)." *Chronique d'Égypte* 80: 76–86.

Lohwasser, Angelika

2004 "Die Kleinfunde aus Gala Abu Ahmed im unteren Wadi Howar." *Der Antike Sudan: Mitteilungen der Sudanarchäologischen Gesellschaft zu Berlin* 15: 143–67.

2012 *Aspekte der napatanischen Gesellschaft: Archäologisches Inventar und funeräre Praxis im Friedhof von Sanam—Perspektiven einer kulturhistorischen Interpretation.* Contributions to the Archaeology of Egypt, Nubia, and the Levant 1. Österreichische Akademie der Wissenschaften, Denkschriften der Gesamtakademie 67. Vienna: Verlag der Österreichischen Akademie der Wissenschaften.

2014 "Neue Skarabäen mit *mn-hpr-rꜥ* aus der 25. Dynastie." In *Skarabäen des 1. Jahrtausends: Ein Workshop in Münster am 27. Oktober 2012*, edited by Angelika Lohwasser, pp. 175–97. Orbis Biblicus et Orientalis 269. Fribourg: Academic Press and Göttingen: Vandenhoeck & Ruprecht.

2018 "Das Paneel der Katimala in Semna." *Der Antike Sudan: Mitteilungen der Sudanarchäologischen Gesellschaft zu Berlin* 29: 89–96.

López Grande, Maria José, and Fernando Quesada

1993 "Two Third Intermediate/Late Period Pottery Deposits at Herakleopolis Magna." In *Sesto Congresso Internazionale di Egittologia: Atti*, vol. 1, edited by Jean Leclant, pp. 415–25. Turin: Società Italiana per il Gas.

Lupo, Silvia, and Maria Beatriz Cremonte

2011 "Upper Egyptian Vessels at Tell el-Ghaba, North Sinai: Luxury Objects of the Local Elite." *Bulletin of the Australian Centre for Egyptology* 22: 115–28.

Lyons, Diane, and A. Catherine D'Andrea

2003 "Griddles, Ovens, and Agricultural Origins: An Ethnoarchaeological Study of Bread Baking in Highland Ethiopia." *American Anthropologist* n.s. 105: 515–30.

Macadam, Miles Frederick Laming

1949 *The Temples of Kawa*. Vol. 1, *The Inscriptions*. London: Oxford University Press.

1955 *The Temples of Kawa*. Vol. 2, *History and Archaeology of the Site*. London: Oxford University Press.

Mackensen, Michael, and Mohamed el-Bialy

2013 "The Late Roman Fort at Nag el-Hagar near Kom Ombo: Report on the Fourth Season of the Egyptian-Swiss Mission." In *The First Cataract of the Nile: One Region—Diverse Perspectives*, edited by Dietrich Raue, Stephan J. Seidlmayer, and Philipp Speiser, pp. 111–21. Sonderschrift des Deutschen Archäologischen Instituts Kairo 36. Berlin: De Gruyter.

Magee, Peter

2015 "When Was the Dromedary Domesticated in the Ancient Near East?" *Zeitschrift für Orient-Archäologie* 8: 252–77.

Maillot, Marc

2014 "The Palace of Muweis in the Shendi Reach: A Case Study." In *Les maisons-tours en Égypte durant la Basse Époque, les périodes ptolémaïque et romaine: Actes de la table-ronde de Paris, Université Paris–Sorbonne (Paris IV), 29–30 novembre 2012*, edited by Séverine Marchi, pp. 69–84. Paris: Université de Paris–Sorbonne.

2015 "The Meroitic Palace and Royal City." *Sudan & Nubia* 19: 80–87.

Maley, Jean

1975 "Introduction à la géologie des environs de la deuxième cataracte du Nil, au Soudan." In *Mirgissa*, vol. 2, *Les nécropoles*, edited by Jean Vercoutter, pp. 123–54. Paris: Direction générale des relations culturelles, scientifiques, et techniques avec la collaboration du Centre national de la recherche scientifique.

Manzo, Andrea

2007 "Tokens, Pottery Discs, and Other Administrative Devices: Two Case Studies between Nubia and Ethiopia." In *Tallies, Tokens and Counters from the Mediterranean to India: Proceedings of the Meeting Held at the Università degli Studi di Napoli "L'Orientale," Naples 31st May 2004*, edited by Anna Maria D'Onofrio, pp. 51–57. Naples: Università degli Studi di Napoli L'Orientale.

Marangout, Antigone

2015 "Des Saïtes aux Perses: Sur les traces des amphores grecques à Thèbes." In *La présence grecque dans la vallée de Thèbes*, edited by Gilles Gorre and Antigone Marangou, pp. 67–91. Rennes: Presse universitaires de Rennes.

Marchi, Séverine, ed.

2014 *Les maisons-tours en Égypte durant la Basse Époque, les périodes ptolémaïque et romaine: Actes de la table-ronde de Paris, Université Paris–Sorbonne (Paris IV), 29–30 novembre 2012*. Paris: Université de Paris–Sorbonne.

Markowitz, Yvonne J.

2012 "Nubian Adornment." In *Ancient Nubia: African Kingdoms on the Nile*, edited by Marjorie M. Fisher, Peter Lacovara, Salima Ikram, and Sue D'Auria, pp. 186–99. Cairo: American University in Cairo Press.

Marouard, Grégory

2014 "Maisons-tours et organisation des quartiers domestiques dans les agglomérations du Delta: L'exemple de Bouto de la Basse Époque aux premiers lagides." In *Les maisons-tours en Égypte durant la Basse Époque, les périodes ptolémaïque et romaine: Actes de la table-ronde de Paris, Université Paris–Sorbonne*

(Paris IV), 29–30 novembre 2012, edited by Séverine Marchi, pp. 105–33. Paris: Université de Paris–Sorbonne.

Masson, Aurélia
2007 "Le quartier des prêtres du temple de Karnak: Rapport préliminaire de la fouille de la Maison VII, 2001–2003." *Karnak* 12, no. 2: 593–656.
2011 "Persian and Ptolemaic Ceramics from Karnak: Change and Continuity." *Cahiers de la céramique égyptienne* 9: 269–310.
2013–15 "New Year's Flasks." In *Naukratis: Greeks in Egypt*. British Museum Online Research Catalogue. Accessed November 15, 2015. http://www.britishmuseum.org/research/online_research_catalogues/ng/naukratis_greeks_in_egypt/material_culture_of_naukratis/new_years_flasks.aspx.

Masson, Olivier
1978 *Carian Inscriptions from North Saqqara and Buhen*. Texts from Excavations Memoirs 5. London: Egypt Exploration Society.

Master, Daniel M.
2003 "Trade and Politics: Ashkelon's Balancing Act in the Seventh Century B.C.E." *Bulletin of the American Schools of Oriental Research* 330: 47–64.

McKenzie, Bruce A., and Larry Van Fossen
1995 "Managing Dry Grain in Storage." *Agricultural Engineer's Digest* 20. Purdue University Cooperative Extension Service. Accessed January 10, 2016. http://extension.purdue.edu/extmedia/AED/AED-20.html.

Mełecka-Drozd, Natalia
2012 "The Emergence and Development of Architecture on the Casemate Foundation Platforms in the Nile Delta." *Recherches Archéologiques* n.s. 4: 69–96.

Meshel, Zeev, Israel Carmi, and Dror Segal
1995 "^{14}C Dating of an Israelite Biblical Site at Kuntillet Ajrud (Horvat Teman)." *Radiocarbon* 37, no. 2: 205–12.

Miellé, Lauriane
2014 "Nubian Traditions on the Ceramics Found in the Pharaonic Town on Sai Island. " In *The Fourth Cataract and Beyond: Proceedings of the 12th International Conference for Nubian Studies*, edited by Julie R. Anderson and Derek A. Welsby, pp. 387–92. British Museum Publications on Egypt and Sudan 1. Leuven: Peeters.

Miller, R. L.
1991 "Counting Calories in Egyptian Ration Texts." *Journal of the Economic and Social History of the Orient* 34: 257–69.

Minault-Gout, Anne, and Florence Thill
2012 *Saï*. Vol. 2, *Le cimetière des tombes hypogées du Nouvel Empire (SAC 5)*. 2 parts. Fouilles de l'Institut français d'archéologie orientale 69. Cairo: Institut français d'archéologie orientale.

Mohamed Ahmed, Saleh El-Din
1991 *L'agglomération napatéenne de Kerma: Enquête archéologique et ethnographique en milieu urbain*. Paris: Éditions Récherche sur les civilisations.

Molinero Polo, Miguel Ángel
1995 "Los recipientes de piedra." In *Excavaciones en Ehnasya el-Medina (Heracleópolis Magna)*, vol. 2, edited by Maria Joseé López Grande, Miguel Ángel Molinero Polo, and Fernando Qesada Sanz, pp. 205–77. Madrid: Instituto de conservación y restauración de bienes culturales.

Mommsen, Hans, Mike R. Cowell, Phillip Fletcher, Duncan Hook, Uvo Schlotzhauer, Alexandra Villing, Sabine Weber, and Dyfri Williams
2006 "Neutron Activation Analysis of Pottery from Naukratis and Other Related Vessels." In *Naukratis: Greek Diversity in Egypt*, edited by Alexandra Villing and Udo Schlotzhauer, pp. 69–76. British Museum Research Publication 162. London: British Museum Press.

Monnier, Franck
2010 *Les forteresses égyptiennes du Prédynastique au Nouvel Empire*. Connaissance de l'Égypte ancienne 11. Brussels: Éditions Safran.
2014 "Une iconographie égyptienne de l'architecture défensive." *Égypte Nilotique et Méditerranéenne* 7: 173–219.

Morkot, Robert G.
2016 "North-East Africa and Trade at the Crossroads of the Nile Valley, the Mediterranean and the Red Sea." In *Dynamics of Production in the Ancient Near East 1300–500 BC*, edited by Juan Carlos Moreno García, pp. 257–74. Oxford: Oxbow Books.

Morris, Ellen F.
2005 *The Architecture of Imperialism: Military Bases and the Evolution of Foreign Policy in Egypt's New Kingdom*. Probleme der Ägyptologie 22. Leiden: Brill.

Müller-Winkler, Claudia
1987 *Die ägyptischen Objekt-Amulette: Mit Publikation der Sammlung des Biblischen Instituts der Universität Freiburg, Schweiz*. Orbis Biblicus et Orientalis, Series Archaeologica 5. Freiburg: Universitätsverlag and Göttingen: Vandenhoeck & Ruprecht.

Munsell Color
1988 *Munsell Soil-Color Charts.* Baltimore: Munsell Color.

Myśliwiec, Karol
1987 *Keramik und Kleinfunde aus der Tempel Sethos I.* Archäologische Veröffentlichungen, Deutsches Archäologisches Institut Kairo 57. Mainz am Rhein: Philipp von Zabern.

Na'aman, Nadav
1979 "The Brook of Egypt and Assyrian Policy on the Border of Egypt." *Tel Aviv* 6: 68–90.
2001 "An Assyrian Residence at Ramat Rahel?" *Tel Aviv* 28: 260–80.

Na'aman, Nadav, and Yifat Thareani-Sussely
2006 "Dating the Appearance of Imitations of Assyrian Ware in Southern Palestine." *Tel Aviv* 33, no. 1: 61–82.

Nordström, Hans-Åke
2014 *The West Bank Survey from Faras to Gemai.* Vol. 1, *Sites of Early Nubian, Middle Nubian and Pharaonic Age.* Sudan Archaeological Research Society Publication 21. British Archaeological Reports International Series 2650. Oxford: Archaeopress.

Nunn, John F.
1996 *Ancient Egyptian Medicine.* Norman: University of Oklahoma Press.

Ogden, Jack
1995 "The Gold Jewellery." In *Excavations at Tawilan in Southern Jordan*, edited by Crystal-M. Bennett and Piotr Bienkowski, pp. 69–78. Oxford: Oxford University Press.

Oren, Eliezer D.
1984 "Migdol: A New Fortress on the Edge of the Eastern Nile Delta." *Bulletin of the American Schools of Oriental Research* 256: 7–44.

Orzechowska, Mariola
2003 "Preliminary Report on the Pottery from the Soniyat Temple." In *Survey and Excavations between Old Dongola and Ez-Zuma*, edited by Bogdan Żurawski, pp. 442–57. Nubia 2. Southern Dongola Reach Survey 1. Warsaw: Editions Wydawnictwo Neriton.

Osman, Ali, and David N. Edwards
2011 *The Archaeology of a Nubian Frontier: Survey on the Nile Third Cataract, Sudan.* Bristol: Mahaus Publishing.

Osypińska, Marta
2003 "Faunal Remains from the Southern Dongola Reach." In *Survey and Excavations between Old Dongola and Ez-Zuma*, edited by Bogdan Żurawski, pp. 490–511. Nubia 2. Southern Dongola Reach Survey 1. Warsaw: Editions Wydawnictwo Neriton.

Paice, Patricia
1986–87 "A Preliminary Analysis of Saite and Persian Pottery at Tell el-Maskhuta." *Bulletin of the Egyptological Seminar* 8: 95–107.

Paulette, Tate S.
2015 "Grain Storage and the Moral Economy in Mesopotamia (3000–2000 BC)." PhD diss., University of Chicago. Accessed January 23, 2016. https://oi.uchicago.edu/research/research-archives-library/dissertations/grain-storage-and-moral-economy-mesopotamia-3000.

Petrie, William Matthew Flinders
1888 *Nebesheh (Am) and Defenneh (Tahpanhes).* Tanis 2. Egypt Exploration Society Fund Memoir 4. London: Trübner & Co.
1896 *Naukratis.* Vol. 1. Egypt Exploration Society Memoir 3. London: Trübner & Co.
1906 *Hyksos and Israelite Cities.* British School of Archaeology in Egypt and Egyptian Research Account 12. London: Office of the School of Archaeology, University College / Bernard Quaritch.
1909 *Qurneh.* British School of Archaeology in Egypt and Egyptian Research Account 16. London: School of Archaeology in Egypt.
1928 *Gerar.* British School of Archaeology in Egypt and Egyptian Research Account 43. London: School of Archaeology in Egypt.

Phillips, Jacke
2003 "An Overview of the Ceramics." In *Survey and Excavations between Old Dongola and Ez-Zuma*, edited by Bogdan Żurawski, pp. 387–437. Nubia 2. Southern Dongola Reach Survey 1. Warsaw: Editions Wydawnictwo Neriton.
2010 "Preliminary Analysis of 'Mat'—'Basket'—Impressed Ceramics from the Southern Dongola Reach Survey." In *Between the Cataracts: Proceedings of the 11th Conference for Nubian Studies, Warsaw University, 27 August–2 September 2006*, edited by Włodzimierz Godlewski and Adam Łajtar, pp. 227–36. Polish Archaeology in the Mediterranean Supplement Series 2.2/1: Session Papers. Warsaw: Warsaw University Press.

Pierce, Richard H.
1964a Dorginarti Daybook 1–3. Oriental Institute Museum Archives, University of Chicago.
1964b Field Notes. Oriental Institute Museum Archives, University of Chicago.

Pierce, Wenche M.
1964 Field Notes. Oriental Institute Museum Archives, University of Chicago.

Pliny the Elder
1855 *Natural History, Book VI.* Edited by John Bostock. London: Taylor and Francis. http://

data.perseus.org/citations/urn:cts:latinLit:phi0978.phi001.perseus-eng1:6.35.

Pope, Jeremy W.

2013 "Epigraphic Evidence for a 'Porridge-and-Pot' Tradition on the Ancient Middle Nile." *Azania: Archaeological Research in Africa* 48, no. 4: 473–97.

2014 *The Double Kingdom under Taharqo.* Culture and History of the Ancient Near East 69. Leiden: Brill.

Porter, Bertha, and Rosalind L. B. Moss

1962 *Topographical Bibliography of Ancient Egyptian Hieroglyphic Texts, Reliefs and Paintings.* Vol. 7, *Nubia, the Deserts and Outside Egypt.* Oxford: Griffith Institute / Ashmolean Museum.

Pratico, Gary

1985 "Tell el-Kheleifeh: A Reappraisal." *Bulletin of the American Schools of Oriental Research* 259: 1–32.

1993 *Nelson Glueck's 1938–1949 Excavations at Tell El-Kheleifeh: A Reappraisal.* Annual of the American Schools of Oriental Research Archaeological Reports 3. Atlanta: Scholars Press.

Raedler, Christine

2007 "Keramikschaber aus den Werkstätten der Ramses-Stadt." In *Die Keramik des Grabungsplatzes Q I,* part 2, *Schaber—Marken—Scherben,* edited by Edgar B. Pusch and Manfred Bietak, pp. 11–266. Grabung des Pelizaeus-Museums Hildesheim in Qantir–Pi-Ramesse 5. Hildesheim: Roemer- und Pelizaeus-Museum.

Randall-MacIver, D., and C. Leonard Woolley

1911 *Buhen.* University of Pennsylvania, Egyptian Department of the University Museum, Eckley B. Coxe Jr. Expedition to Nubia 7. Philadelphia: University Museum.

Raxter, Michelle H.

2011 "Egyptian Body Size: A Regional and Worldwide Comparison." PhD diss., University of South Florida. Accessed March 15, 2016. http://scholarcommons.usf.edu/etd/3305.

Ray, John D.

1982 "The Carian Inscriptions from Egypt." *Journal of Egyptian Archaeology* 68: 181–98.

Regev, Dalit

2004 "The Phoenician Transport Amphora." In *Transport Amphorae and Trade in the Eastern Mediterranean: Acts of the International Colloquium at the Danish Institute at Athens, September 26–29, 2002,* edited by Jonas Eiring and John Lund, pp. 337–52. Monographs of the Danish Institute at Athens 5. Athens: Danish Institute at Athens.

Reisner, George A.

1929 "Excavations at Semna and Uronarti by the Harvard-Boston Expedition." *Sudan Notes and Records* 12: 143–61.

Rice, Prudence

1987 *Pottery Analysis: A Sourcebook.* Chicago: University of Chicago Press.

Ritner, Robert

2009 *The Libyan Anarchy: Inscriptions from Egypt's Third Intermediate Period.* Atlanta: Society of Biblical Literature.

Rodziewicz, Mieczysław D.

2005 *Elephantine XXVII: Early Roman Industries on Elephantine.* Archäologische Veröffentlichungen, Deutsches Archäologisches Institut Kairo 107. Mainz am Rhein: Philipp von Zabern.

Rose, Pamela

2008 "Early Settlement at Qasr Ibrim." In *Between the Cataracts: Proceedings of the 11th Conference for Nubian Studies, Warsaw University, 27 August–2 September 2006,* edited by Włodzimierz Godlewski and Adam Łajtar, pp. 195–209. Polish Archaeology in the Mediterranean Supplement Series 2.1: Main Papers. Warsaw: Warsaw University Press.

2019 "Early Kushite Ceramics of the Earlier 1st Millennium BC in Lower and Upper Nubia." In *Handbook of Ancient Nubia,* vol. 2, edited by Dietrich Raue, pp. 675–96. Berlin: De Gruyter.

Roth, Jonathan P.

1999 *The Logistics of the Roman Army at War (264 B.C.–A.D. 235).* Columbia Studies in the Classical Tradition 23. Leiden: Brill.

Rowley-Conwy, Peter A.

1988 "The Camel in the Nile Valley: New Radiocarbon Accelerator (AMS) Dates from Qasr Ibrim." *Journal of Egyptian Archaeology* 74: 245–48.

1989 "Nubia AD 0–550 and the 'Islamic' Agricultural Revolution: Preliminary Botanical Evidence from Qasr Ibrim, Egyptian Nubia." *L'Archéologie du Nil Moyen* 3: 131–37.

Ruby, Jay W.

1964 "Preliminary Report of the University of California Expedition to Dabenarti, 1963." *Kush* 12: 54–56.

Ruffieux, Philippe

2007 "Ensembles céramiques napatéens découverts durant les campagnes 2005–2006 et

2006–2007 à Doukki Gel (Kerma)." *Genava* n.s. 55: 223–39.

2009 "Poteries découvertes dans un temple égyptien de la XVIIIe dynastie à Doukki Gel (Kerma)." *Genava* n.s. 57: 121–34.

Rzepka, Sławomir, Jozef Hudec, Anna Wodzinska, Łukasz Jarmużek, L. Hulková, Veronika Dubcová, Malwina Piorun, and Alena Sefcakova

2014 "Tell el-Retaba from the Second Intermediate Period till the Late Period: Results of the Polish-Slovak Archaeological Mission, Seasons 2011–2012." *Ägypten und Levante* 24: 39–120.

Rzepka, Sławomir, Claire Malleson, Jozef Hudec, Misiewicz Krzysztof, Miron Bogacki, Łukasz Jarmużek, Anna Wodzinska, and Wisław Małkowski

2011 "New Kingdom and the Third Intermediate Period in Tell el-Retaba: Results of the Polish-Slovak Archaeological Mission, Seasons 2009–2010." *Ägypten und Levante* 21: 129–84.

Sagona, Antonio G.

1982 "Levantine Storage Jars of the 13th to the 4th Century BC." *Opuscula Atheniensia* 7: 73–110.

Sakamoto, Tsubasa

2016 "Stratigraphy and Absolute Chronology of Jebel Moya: A Note on Michael Brass' Recent Interpretation." *Der Antike Sudan: Mitteilungen der Sudanarchäologischen Gesellschaft zu Berlin* 27: 133–38.

Sandford, Kenneth S., and William J. Arkell

1933 *Paleolithic Man and the Nile Valley in Nubia and Upper Egypt: A Study of the Region during Pliocene and Pleistocene Times.* Oriental Institute Publications 17. Prehistoric Survey of Egypt and Western Asia 2. Chicago: University of Chicago Press.

Sanhueza-Pino, Laura

2014 "Löwen, Sphingen und Greife auf ägyptischen Skarabäen der Spätzeit." In *Skarabäen des 1. Jahrtausends: Ein Workshop in Münster am 27. Oktober 2012*, edited by Angelika Lohwasser, pp. 105–24. Orbis Biblicus et Orientalis 269. Fribourg: Academic Press and Göttingen: Vandenhoeck & Ruprecht.

Sauneron, Serge, and Jean Yoyotte

1952 "Campagne nubienne de Psammétique II." *Bulletin de l'Institut français d'archéologie orientale* 50: 157–207.

Säve-Söderbergh, Torgny

1989 *Middle Nubian Sites.* Scandinavian Joint Expedition to Sudanese Nubia 4. Uddevalla: Bohusläningens Boktryckeri.

Säve-Söderbergh, Torgny, and Lana Troy

1991 *New Kingdom Pharaonic Sites.* Parts 2–3, *The Finds and the Sites.* Scandinavian Joint Expedition to Sudanese Nubia 5. Uppsala: Almqvist & Wiksell Tryckeri.

Sayce, Archibald Henry

1910 "Karian, Egyptian and Nubian-Greek Inscriptions from the Sudan." *Proceedings of the Society of Biblical Archaeology* 32: 261–67.

Schiff Giorgini, Michela

1971 *Soleb.* Vol. 2, *Les nécropoles.* Florence: Sansoni.

Schipper, Bernd U.

2011 "Egyptian Imperialism after the New Kingdom: The 26th Dynasty and the Southern Levant." In *Egypt, Canaan and Israel: History, Imperialism, Ideology and Literature. Proceedings of a Conference at the University of Haifa, 3–7 May 2009*, edited by Shay Bar, Dan'el Kahn, and J. J. Shirley, pp. 268–89. Culture and History of the Ancient Near East 52. Leiden: Brill.

Schmitz, Philip

2010 "The Phoenician Contingent in the Campaign of Psammetichus II against Kush." *Journal of Egyptian History* 3, no. 2: 321–37.

Schneider, Thomas

2010 "Contributions to the Chronology of the New Kingdom and the Third Intermediate Period." *Ägypten und Levante* 20: 373–403.

Schrader, Sarah, Michele T. Buzon, and Joel Irish

2014 "Illuminating the Nubian 'Dark Age': A Bioarchaeological Analysis of Dental Nonmetric Traits during the Napatan Period." *HOMO: Journal of Comparative Human Biology* 65: 267–80.

Schreiber, Gábor

2002 "Pottery of 'Lotus-Flower and Crosslined-Band' Style: A Marl-Based Ware Group." In *A Tribute to Excellence: Studies Offered in Honor of Ernö Gaál, Ulrich Luft, and László Török*, edited by Tamás A. Bács, pp. 405–20. Studia Aegyptiaca 17. Budapest: Université Eötvös Lorand de Budapest.

2003 *Late Dynastic and Ptolemaic Painted Pottery from Thebes (4th–2nd c. BC).* Dissertationes Pannonicae 3/6. Budapest: Institute of Archeological Sciences, Eötvös Lorand University.

2008 *The Mortuary Monument of Djehutymes II: Finds from the New Kingdom to the Twenty-Sixth Dynasty.* Studia Aegyptiaca Series Maior 2. Budapest: Archaeolingua Alapítvány.

Seiler, Anne

2003 "Die spätzeitliche Keramik." In Daniel Polz, Elke Mählitz, Ute Rummel, and Anne Seiler,

"Bericht über die 9. bis 12. Grabungskampagne in der Nekropole von Draʿ Abu el-Naga/Theben-West." *Mitteilungen des Deutschen Archäologischen Instituts Kairo* 59: 363–69.

Sharon, Ilan, and Anabel Zarzecki-Peleg

2006 "Podium Structures with Lateral Access: Authority Ploys in Royal Architecture in the Iron Age Levant." In *Confronting the Past: Archaeological and Historical Essays on Ancient Israel in Honor of William G. Dever*, edited by Seymour Gitin, J. Edward Wright, and J. P. Dessel, pp. 145–67. Winona Lake, IN: Eisenbrauns.

Shinnie, Peter L., and Rebecca Bradley

1951 "Preliminary Report on the Excavations at Amarah West, 1948–49 and 1949–50." *Journal of Egyptian Archaeology* 37: 5–11.

1980 *The Capital of Kush*. Vol. 1, *Meroë Excavations, 1965–1972*. Meroitica 4. Berlin: Akademie-Verlag.

Singer-Avitz, Lily

1999 "Beersheba—a Gateway Community in Southern Arabian Long-Distance Trade in the Eighth Century BCE." *Tel Aviv* 26: 3–74.

2002 "Arad: The Iron Age Pottery Assemblages." *Tel Aviv* 29: 110–214.

2006 "The Date of Kuntillet ʿAjrud." *Tel Aviv* 33: 196–228.

2007 "On Pottery in Assyrian Style: A Rejoinder." *Tel Aviv* 34: 182–203.

Sjöström, Isabella Welsby, and Ross Thomas

2011 "The Kushite Pottery Sequence at Kawa: Parallels at Home and Abroad." *Sudan & Nubia* 15: 64–71.

Smith, Harry S.

1966 "Kor: Report on the Excavations of the Egypt Exploration Society at Kor, 1965." *Kush* 14: 187–243.

Smith, Stuart T.

2008 "Tombos and the Transition from the New Kingdom to the Napatan Period in Upper Nubia." In *Between the Cataracts: Proceedings of the 11th Conference for Nubian Studies, Warsaw University, 27 August–2 September 2006*, edited by Włodzimierz Godlewski and Adam Łajtar, pp. 95–115. Polish Archaeology in the Mediterranean Supplement Series 2:1. Warsaw: Warsaw University Press.

Smith, Stuart T., and Michele R. Buzon

2014 "Identity, Commemoration, and Remembrance in Colonial Encounters: Burials at Tombos during the Egyptian New Kingdom Nubian Empire and Its Aftermath." In *Remembering the Dead in the Ancient Near East*, edited by Benjamin W. Porter and Alexis T. Boutin, pp. 185–215. Boulder: University Press of Colorado.

Smoláriková, Květa

2008a "Ceramics from the Ramesside Temple." In *Kom Firin*, vol. 1, *The Ramesside Temple and the Site Survey*, edited by Neal Spencer, pp. 58–63. British Museum Research Publications 170. London: British Museum.

2008b *Saite Forts in Egypt: Political-Military History of the Saite Dynasty*. Prague: Czech Institute of Egyptology.

Soper, Robert

1985 "Roulette Decoration on African Pottery: Technical Considerations, Dating and Distributions." *African Archaeological Review* 3: 29–51.

Spaer, Maud

2001 *Ancient Glass in the Israel Museum: Beads and Other Small Objects*. Jerusalem: Israel Museum.

Spalinger, Anthony

1977 "Egypt and Babylonia: A Survey (c. 620 B.C.–550 B.C.)" *Studien zur altägyptischen Kultur* 5: 221–24.

1978 "Psammetichus, King of Egypt: II." *Journal of the American Research Center in Egypt* 15: 49–57.

Spataro, Michela, and Alexandra Villing

2009 "Scientific Investigation of Pottery Grinding Bowls from the Archaic and Classical Eastern Mediterranean." *British Museum Technical Research Bulletin* 3: 89–100.

Spencer, Alan Jeffrey

1993 *Excavations at El-Ashmunein*. Vol. 3, *The Town*. British Museum Expedition to Middle Egypt. London: British Museum Press.

1999 "Casemate Foundations Once Again." In *Studies on Ancient Egypt in Honour of H. S. Smith*, edited by Anthony Leahy and John Tait, pp. 295–300. London: Egypt Exploration Society.

2011 "The Egyptian Temple and Settlement at Naukratis." *British Museum Studies in Ancient Egypt and Sudan* 17: 31–49.

Spencer, Neal

2014 "Amara West: Considerations on Urban Life in Colonial Kush." In *The Fourth Cataract and Beyond: Proceedings of the 12th International Conference for Nubian Studies*, edited by Julie R. Anderson and Derek A. Welsby, pp. 457–85. British Museum Publications on Egypt and Sudan 1. Leuven: Peeters.

Spencer, Neal, Anna Stevens, and Michaela Binder

2014 *Amara West: Living in Egyptian Nubia*. London: Trustees of the British Museum.

Spencer, Patricia

1997 *Amara West*. Vol. 1, *The Architectural Report*. Excavation Memoir 63. London: Egypt Exploration Society.

2002 *Amara West*. Vol. 2, *The Cemetery and the Pottery Corpus*. Excavation Memoir 69. London: Egypt Exploration Society.

Stager, Lawrence A., Daniel M. Master, and J. David Schloen

2011 *Ashkelon*. Vol. 3, *The Seventh Century* B.C. Winona Lake, IN: Eisenbrauns.

Steiner, Herwig

2008 "Ägyptische Festungen und Stadtgründungen in Nubien von der prädynastischen Zeit bis zum Ende des Neuen Reiches." PhD diss., University of Vienna. Accessed January 17, 2016. https://othes.univie.ac.at/2193/.

Stern, Ephraim, and Yitzhak Magen

1984 "A Pottery Group of the Persian Period from Qadum in Samaria." *Bulletin of the American Schools of Oriental Research* 253: 9–27.

Stewart, Hilary

2008 *Indian Fishing: Early Methods on the Northwest Coast*. New York: Douglas McIntyre.

Sullivan, Elaine A.

2013 *A Glimpse into Ancient Thebes: Excavations at South Karnak (2004–2006)*. BAR International Series 2538. Oxford: Archaeopress.

Thareani, Yifat

2002 "The 'Archaeology of the Days of Manasseh' Reconsidered in the Light of Evidence from the Beersheba Valley." *Palestine Exploration Quarterly* 139, no. 2: 69–77.

2016 "The Empire and the 'Upper Sea': Assyrian Control Strategies along the Southern Levantine Coast." *Bulletin of the American Schools of Oriental Research* 375: 77–102.

Then-Obłuska, Joanna

2013 "A Medieval Transcultural Medium: Beads and Pendants from Makurian and Post-Makurian Dongola in Nubia." *Polish Archaeology in the Mediterranean* 22: 679–720.

2014 "The Code of the Hidden Beads—from the Kerma to the Islamic Period according to the Fourth Cataract Material from the Gdańsk Archaeological Museum Collection." In *The Fourth Cataract and Beyond: Proceedings of the 12th International Conference for Nubian Studies*, edited by Julie R. Anderson and Derek Welsby, pp. 1069–90. British Museum Publications on Egypt and Sudan 1. Leuven: Peeters.

2015a "Beads and Pendants from Sedeinga, Nubia." *BEADS: Journal of the Society of Bead Researchers* 27: 29–45.

2015b "Cross-Cultural Bead Encounters at the Red Sea Port Site of Berenike, Egypt: Preliminary Assessment (Seasons 2009–2012)." *Polish Archaeology in the Mediterranean* 24, no. 1: 735–77.

2016 "Meroitic Beadwork: An Overview Based on Finds from Saï in Ancient Nubia." *Polish Archaeology in the Mediterranean* 25: 691–724.

Thill, Florence

2006–7 "Les réoccupations '(pré)napatéennes' dans le cimetière égyptien 8B5/SAC5 de Saï." *Cahiers de Recherches de l'Institut de Papyrologie et d'Égyptologie de Lille* 26: 353–69.

Tietze, Christian

1986 "Amarna: Analyse der Wohnhäuser und soziale Struktur der Stadtbewohner (Teil II)." *Zeitschrift für Ägyptische Sprache und Altertumskunde* 113: 55–78.

Török, László

2009 *Between Two Worlds: The Frontier Region between Ancient Nubia and Egypt 3700* BC–AD *500*. Probleme der Ägyptologie 29. Leiden: Brill.

Trautman, Milton A., and Eric H. Willis

1966 "Isotopes, Inc. Radiocarbon Measurements V." *Radiocarbon* 8: 161–203.

Trigger, Bruce G.

1965 *History and Settlement in Lower Nubia*. Yale University Publications in Anthropology 69. New Haven: Yale University Department of Anthropology.

Tzochev, Chavdar

2011 "Archaic Amphora Imports from Thracian Sites around the Bay of Bourgas." In *Production and Trade of Amphorae in the Black Sea (PATABS II): Acts of the International Round Table Held in Kiten, Nessebar and Sredetz, September 26–30, 2007*, edited by Chavdar Tzochev, Totko Stoyanov, and Anelia Bozkova, pp. 73–86. Sofia: National Archaeological Institute / Bulgarian Academy of Sciences.

United Nations World Food Programme

2016 "Logistics Capacity Development: Post-Harvest Food Loss Reduction in Uganda through Improved Storage and Handling at the Start of the Supply Chain." Standard Project Report 2015. Uganda, Special Operations SO 200836 (2016). Accessed February 1, 2017. http://www1.wfp.org/operations?country=2105.

Ussishkin, David

1995 "The Rectangular Fortress at Kadesh-Barnea." *Israel Exploration Journal* 45, nos. 2–3: 118–27.

2010 "'En Hazeva: On the Gate of the Iron Age II Fortress." *Tel Aviv* 37: 246–53.

Van Beek, Gus W.

1987 "Arches and Vaults in the Ancient Near East." *Scientific American* 257, no. 1: 78–85.

van Gerven Oei, Vincent, and Alexandros Tsakos

2017 "The Etymology of the Toponym 'Dorginarti' (Notes on Medieval Nubian Toponymy 6)." *Dotawo: A Journal of Nubian Studies* 4: 275–78.

Vercoutter, Jean

1966 "Semna South Fort and the Records of Nile Levels at Kumma." *Kush* 14: 125–64.

1970 *Mirgissa*. Vol. 1. Paris: Direction générale des relations culturelles, scientifiques, et techniques avec la collaboration du Centre national de la recherche scientifique.

1975 *Mirgissa*. Vol. 2, *Les nécropoles*. Paris: Direction générale des relations culturelles, scientifiques, et techniques avec la collaboration du Centre national de la recherche scientifique.

Vila, André

1977 *Le district d'Amara Est*. Fasc. 8 of *Prospection archéologique de la vallée du Nil au sud de la cataracte de Dal (Nubie soudanaise)*. Paris: Centre national de la recherche scientifique.

1978 *L'île d'Arnyatta, le district d'Abri (Est et Ouest), le district de Tabaj (Est et Ouest)*. Fasc. 9 of *Prospection archéologique de la vallée du Nil au sud de la cataracte de Dal (Nubie soudanaise)*. Paris: Centre national de la recherche scientifique.

1980 *La nécropole de Missiminia*. Part 1, *Les sépultures napatéennes*. Fasc. 12 of *Prospection archéologique de la vallée du Nil au sud de la cataracte de Dal (Nubie soudanaise)*. Paris: Centre national de la recherche scientifique.

Villing, Alexandra

2006 "'Drab Bowls' for Apollo: The Mortaria of Naukratis and Exchange in the Archaic Eastern Mediterranean." In *Naukratis: Greek Diversity in Egypt*, edited by Alexandra Villing and Udo Schlotzhauer, pp. 31–46. British Museum Research Publication 162. London: British Museum Press.

Villing, Alexandra, and Udo Schlotzhauer, eds.

2006 *Naukratis: Greek Diversity in Egypt: Studies on East Greek Pottery and Exchange in the Eastern Mediterranean*. British Museum Research Publication 162. London: British Museum Press.

Vincentelli, Irene

2006 *Hillat el-Arab: The Joint Sudanese-Italian Expedition in the Napatan Region, Sudan*. Sudan Archaeological Research Society Publication 15. BAR International Series 1570. Oxford: Archaeopress.

2011 "The Treasury and Other Buildings at Sanam (Sudan)." In *La pioche et la plume: Autour du Soudan, du Liban et de la Jordanie. Hommages archéologiques à Patrice Lenoble*, edited by Vincent Rondot, Frédéric Alpi, and François Villeneuve, pp. 269–82. Paris: Presses de l'Université Paris-Sorbonne.

Vittmann, Günter

2001 "Ägyptisch-Karisches." *Kadmos* 40: 39–59.

2003 *Ägypten und die Fremden im erstern vorchristlichen Jahrtausend*. Kulturgeschichte der Antiken Welt 97. Mainz am Rhein: Philipp von Zabern.

Vogel, Carola

1998 "Archäologische und epigraphische Belege zur Nilschiffahrt im Bereich des Zweiten Katarakts." *Studien zur altägyptischen Kultur* 26: 261–72.

2004 *Ägyptische Festungen und Garnisonen bis zum Ende des Mittleren Reiches*. Hildesheimer Ägyptologische Beiträge 46. Hildesheim: Gerstenberg.

2010 "Storming the Gates? Entrance Protection in the Military Architecture of Middle Kingdom Nubia." In *Cities and Urbanism in Ancient Egypt: Papers from a Workshop in November 2006 at the Austrian Academy of Sciences*, edited by Manfred Bietak, Ernst Czerny, and Irene Forstner-Müller, pp. 299–320. Österreichische Akademie der Wissenschaften, Denkschriften der Gesamtakademie 60. Vienna: Verlag der Österreichischen Akademie der Wissenschaften.

von Pilgrim, Cornelius

2009 "Elephantine—(Festungs-)Stadt am ersten Katarakt." In *Cities and Urbanism in Ancient Egypt: Papers from a Workshop in November 2006 at the Austrian Academy of Sciences*, edited by Manfred Bietak, Ernst Czerny, and Irene Forstner-Müller, pp. 257–70. Österreichische Akademie der Wissenschaften, Denkschriften der Gesamtakademie 60. Vienna: Verlag der Österreichischen Akademie der Wissenschaften.

Waiman-Barak, Paula

2020 "Petrographic Analysis of Iron Age IIA Decorated Phoenician Containers and a Cypriot Flask." In *Tel Reḥov: A Bronze and Iron Age City in the Beth-Shean Valley*, vol. 4, *Pottery Studies, Inscriptions and Figurative Art*, edited by Amihai Mazar and Nava Panitz-Cohen, pp. 324–32. Qedem Monographs 62. Jerusalem: Institute of Archaeology.

Waiman-Barak, Paula, Ayelet Gilboa, and Yuval Goren
2014 "A Stratified Sequence of Early Iron Age Egyptian Ceramics at Tel Dor, Israel." *Ägypten und Levante* 24: 315–42.

Waldbaum, Jane C.
2011 "Greek Pottery." In *Ashkelon*, vol. 3, *The Seventh Century B.C.*, edited by Lawrence E. Stager, Daniel M. Master, and J. David Schloen, pp. 127–338. Winona Lake, IN: Eisenbrauns.

Waldbaum, Jane C., and Jodi Magness
1997 "The Chronology of Early Greek Pottery: New Evidence from Israel." *American Journal of Archaeology* 101, no. 1: 23–40.

Webb, Virginia
1978 *Archaic Greek Faience.* Warminster: Aris & Phillips.

Weber, Sabine
2012 "Untersuchung zur archaischen griechischen Keramik aus anderen ägyptischen Fundorten." In *Griechische Keramik des 7. und 6. Jhs. v. Chr. aus Naukratis und anderen Orten in Ägypten*, edited by Ursula Höckmann and Sabine Weber, pp. 196–432. Archäologische Studien zu Naukratis 3. Worms: Wernersche Verlagsgesellschaft.

Welsby, Derek A.
1998 "Survey and Excavations at Kawa, the 1997/8 Season." *Sudan & Nubia* 2: 15–20.
1998–2002 "Survey and Excavations at Kawa." *Kush* 18: 183–90.
2002a "Kushite Buildings at Kawa." *British Museum Studies in Ancient Egypt* 1: 32–45.
2002b "The Kushite Town and Cemetery at Kawa, the 2001–2002 Season: Survey and Excavations." *Sudan & Nubia* 6: 32–37.
2005 "The Kingdom of Kush: Urban Defences and Military Installations." In *Image, Craft and the Classical World: Essays in Honour of Donald Bailey and Catherine Johns*, edited by Nina Crummy, pp. 20–54. Monographies Instrumentum 29. Montagnac: M. Mergoil.
2011 "Excavations in the Kushite Town and Cemetery at Kawa, 2010–11." *Sudan & Nubia* 15: 54–63.
2014 "Excavations within the Kushite Town and Cemetery at Kawa 2013–14." *Sudan & Nubia* 18: 48–58.
2015 "The Qatar-Sudan Archaeologial Project—Excavations and Other Activities at Kawa in the 2014–15 Season." *Sudan & Nubia* 19: 71–79.

Welsby, Derek A., and Julie R. Anderson
2004 *Sudan: Ancient Treasures. An Exhibition of Recent Discoveries from the Sudan National Museum.* London: British Museum Press.

Welsby, Derek A., and Isabella Welsby Sjöström
2006–7 "The Dongola Reach and the Fourth Cataract: Continuity and Change during the 2nd and 1st Millennium BC." *Cahiers de Recherches de l'Institut de Papyrologie et d'Égyptologie de Lille* 26: 379–98.

Wenig, Steffen
1981 "Nubien." In *Lexikon der Ägyptologie*, vol. 4, pp. 526–32. Wiesbaden: Harrassowitz.

Wenke, Robert
1984 *Archaeological Investigations at El-Hibeh 1980: Preliminary Report.* American Research Center in Egypt Reports 9. Malibu: Undena Publications.

Wheeler, Noel F.
1961 "Diary of the Excavation of Mirgissa Fort." *Kush* 9: 87–170.

Wildung, Dieter
1996 *Sudan: Antike Königreiche am Nil.* Tübingen: E. Wasmuth.

Williams, Bruce Beyer
1986 *Excavations between Abu Simbel and the Sudan Frontier.* Part 1, *The A-Group Royal Cemetery at Qustul: Cemetery L.* Oriental Institute Nubian Expedition 3. Chicago: Oriental Institute of the University of Chicago.
1990 *Twenty-Fifth Dynasty and Napatan Remains at Qustul.* Part 7, *Cemeteries W and V.* Oriental Institute Nubian Expedition 7. Chicago: Oriental Institute of the University of Chicago.
1991 *Meroitic Remains from Qustul Cemetery Q, Ballana Cemetery B, and a Ballana Settlement.* Oriental Institute Nubian Expedition 8. Chicago: Oriental Institute of the University of Chicago.

Wilson, Karen
1982 *Mendes: Preliminary Report on the 1979 and 1980 Seasons.* Cities of the Delta 2. American Research Center in Egypt Reports 5. Malibu: Undena Publications.

Wodzińska, Anna
2010 *A Manual of Egyptian Pottery.* Vol. 3, *Second Intermediate Period–Late Period.* Ancient Egypt Research Associates Field Manual Series 1. Boston: Ancient Egypt Research Associates Inc. Accessed September 1, 2012. http://www.aeraweb.org/.
2011a "New Kingdom and Third Intermediate Period in Tell el-Retaba." *Ägypten und Levante* 21: 129–83.

2011b "Pottery and Chronology: Preliminary Remarks on Ceramic Material from Tell el-Retaba." In *Under the Potter's Tree: Studies on Ancient Egypt Presented to Janine Bourriau on the Occasion of Her 70th Birthday,* edited by David A. Aston, Bettina Bader, Carla Gallorini, Paul Nicholson, and Sarah Buckingham, pp. 1015–36. Orientalia Lovaniensia Analecta 204. Leuven: Peeters.

2015 "Part II: The Pottery." In Sławomir Rzepka et al., "Tell el-Retaba from the Second Intermediate Period till the Late Period." *Ägypten und Levante* 24: 94–120.

Wolff, Samuel R.

1996 "Archaeology in Israel." *American Journal of Archaeology* 100: 725–68.

Woolley, C. Leonard, and D. Randall-MacIver

1910 *Karanog: The Romano-Nubian Cemetery.* Philadelphia: University Museum.

World Food Programme

2016 Food Aid Information System. Accessed January 18, 2016. http://www.wfp.org/fais/nutritional-reporting/food-composition-table.

Yahalom-Mack, Naama, Adi Eliyahu-Behar, Mario A. S. Martin, Assaf Kleiman, Ruth Shahack-Gross, Robert S. Homsher, Yuval Gadot, and Israel Finkelstein

2017 "Metalworking at Megiddo during the Late Bronze and Iron Ages." *Journal of Near Eastern Studies* 76, no. 1: 53–74.

Yellin, Janice

1995 "Egyptian Religion and Its Ongoing Impact on the Formation of the Napatan State." *Cahiers de Recherches de l'Institut de Papyrologie et d'Égyptologie de Lille* 17: 243–63.

Yoyotte, Jean

1951 "Le martelage des noms royaux éthiopiens par Psammétique II." *Revue d'Égyptologie* 8: 215–39.

Žabkar, Louis V., and Joan J. Žabkar

1982 "Semna South: A Preliminary Report on the 1966–68 Excavations of the University of Chicago Oriental Institute Expedition to Sudanese Nubia." *Journal of the American Research Center in Egypt* 19: 7–50.

Zauzich, Karl-Theodor

1992 "Ein Zug nach Nubien unter Amasis." In *Life in a Multi-cultural Society: Egypt from Cambyses to Constantine and Beyond,* edited by Janet H. Johnson, pp. 361–64. Studies in Ancient Oriental Civilization 51. Chicago: Oriental Institute of the University of Chicago.

Zibelius, Karola

1972 *Afrikanische Orts- und Völkernamen in hieroglyphischen und hieratischen Texten.* Beihefte zum Tübinger Atlas des Vorderen Orients, Reihe B (Geisteswissenschaften) 1. Wiesbaden: Ludwig Reichert.

Zibelius-Chen, Karola

1988 *Die ägyptische Expansion nach Nubien: Eine Darlegung der Grundfaktoren.* Beihefte zum Tübinger Atlas des Vorderen Orients, Reihe B (Geisteswissenschaften) 78. Wiesbaden: Ludwig Reichert.

1989 "Überlegungen zur ägyptischen Nubienpolitik in der Dritten Zwischenzeit." *Studien zur altägyptischen Kultur* 16: 329–45.

2007 Review of *The Inscription of Queen Katimala at Semna: Textual Evidence for the Origins of the Napatan State,* by John Coleman Darnell. *Bibliotheca Orientalis* 64, nos. 3–4: 377–86.

Zorn, Jeffrey R.

1994 "Estimating the Population Size of Ancient Settlements: Methods, Problems, Solutions, and a Case Study." *Bulletin of the American Schools of Oriental Research* 295: 31–48.

Zukerman, Alexander

2014 "Bread Ovens and Related Installations." In *The Smithsonian Institution Excavation at Tell Jemmeh, Israel, 1970–1990,* edited by David Ben-Shlomo and Gus W. Van Beek, pp. 642–50. Washington, DC: Smithsonian Institution Scholarly Press.

Zukerman, Alexander, and David Ben-Shlomo

2011 "Mortaria as a Foreign Element in the Material Culture of the Southern Levant during the 8th–7th Centuries BCE." *Palestine Exploration Quarterly* 143, no. 2: 87–105.

Żurawski, Bogdan, ed.

2003 *Survey and Excavations between Old Dongola and Ez-Zuma.* Nubia 2. Southern Dongola Reach Survey 1. Warsaw: Editions Wydawnictwo Neriton.

Zych, Iwona

2011 "Finds." In *Berenike 2008–2009: Report on the Excavations at Berenike, including a Survey in the Eastern Desert,* edited by Steven E. Sidebotham and Iwona Zych, pp. 117–62. Warsaw: Polish Centre of Mediterranean Archaeology.

Acknowledgments

The staff and volunteers at the site of Dorginarti, who are listed on the following page, are to be thanked for their perseverance in the final days of their mission in Nubia. Without their diaries, excavation records, and photographs, as well as the plans drawn by James Knudstad, this publication would not have been possible. I apologize in advance for any mistakes introduced into the interpretations of the material.

Thanks are due to Gil Stein, former director of the Oriental Institute (now the Institute for the Study of Ancient Cultures), for realizing the importance of publishing the findings of the Nubian Expedition in their entirety and making both resources and office space available for the project. The staff of the Oriental Institute Museum, including Helen McDonald, Susan Allison, Laura D'Alessandro, and former museum curator Jack Green, are also to be thanked for their help. I am grateful to the volunteers who assisted in sorting out the materials in our museum stores and who aided in their registration, and I also thank Natasha Ayers for initially sorting through some of the materials. John Larson provided access to the archival records, and Austin Kramer photographed most of the registered objects held in Chicago. Laurence (Larry) Lissak undertook a massive project to photograph the potsherds, stone objects, and metallurgical equipment, and although not all of his fine photographs could be included here, they will be included in the Nubian project's future database. Finally, I wish to thank the various editors involved in the publication process, including Thomas Urban, Charissa Johnson, Steve Townshend, Connie Tappy, Alexandra Witsell, and especially Andrew Baumann, who finally brought the project to completion.

I drew and inked the bulk of the pottery and objects in the 1980s, with additional drawings in the 2010s. Pegg Romm volunteered at the right moment to take the numerous scans that Bruce Williams and I had made and to sort through them. I wish to thank Carol Abraczinskas (who worked with us in the 1980s), Carol Meyer, and Nadejda Reshetnikova for drawing various sherds and objects.

Larry Lissak helpfully volunteered his trigonometric expertise to convert the angled distances between the coordinates for our architect, Nadejda Reshetnikova. Nadejda produced the architectural drawings from the information supplied by Knudstad, Larry, and me. She also created 3D reconstructions of the fortress that will be published online. I wish to thank her for the many hours of Skyping and emailing between our respective offices, as well as for her fine work and cheerful disposition.

I thank Thilo Rehren, Edgar Pusch, and Martina Renzi of the University College London's Institute for Archaeo-Metallurgical Studies in Qatar for their willingness to undertake an analysis of the metal objects, crucible fragments, and tuyères from Dorginarti. Martina's trip to Chicago enabled this project to be completed more comprehensively. It was a pleasure working with her. The analysis could not be presented here since it remains unfinished.

The contributions of both Carol Meyer, who analyzed the glass fragments from the fortress and identified the types of stone used for many of the objects, and Joanna Then-Obłuska, who undertook an analysis of the beads, are gratefully acknowledged. The information gleaned from these objects has been greatly increased because of their expertise. Many thanks also to Robert Ritner, Bruce Williams, W. Vivian Davies, and Karola Zibelius-Chen for their help in trying to interpret the name of the fortress, found in a probable foundation deposit below the earliest official's residence, and to Brian Muhs for his suggestions on the script of the ostraca and dipinto.

My work on the site report was initiated at the insistence of Bruce Williams, whose commitment to the publication of the Nubian Expedition materials is well known. The efforts he made to ease my study of the site's photographs and excavation diaries while I was working full-time outside academe led to my transition back to full-time research on the Dorginarti material.

Work on the Nubian Expedition volumes has been made possible by grants from the American Research Center in Egypt's Antiquities Endowment Fund and the Schiff Giorgini Foundation. The research and development of the manuscript for this final publication were made possible through a generous grant from The Shelby White and Leon Levy Program for Archaeological Publications.

Finally, I would especially like to thank the anonymous peer reviewer for greatly improving the accuracy and fullness of the data. The reviewer's fresh perspective on the arrangement of the book, and the editing, was very helpful.

ORIENTAL INSTITUTE STAFF AND VOLUNTEERS

Director and architect: James E. Knudstad
Egyptologist, epigrapher, and codirector: Dr. Richard H. Pierce
Field assistant and pottery processor: Wenche Pierce
Staff archaeologist: Dr. Bruce G. Trigger
Archaeological field assistant and epigraphic assistant: Otto Johann Schaden
Staff photographer and field assistant: Louise Storts
Archaeological field assistant: Mielburn D. Thurman
Archaeological field assistant: Rudolph H. Dornemann
Staff archaeologist: Alfred J. Hoerth
Artist and field assistant (volunteer): John C. Lorence
Field assistant (volunteer): Sylvia Ericson

SUDANESE OFFICIALS

Sudanese commissioner of archaeology: Sayed Thabit Hassan Thabit
Senior inspector of antiquities in Wadi Halfa: Sayed Nigm ed-Din Mohammed Sherif
Chief clerk of the Wadi Halfa Museum and the communities of Abd el-Gadir: Hagg Gemal

EXCAVATION DATES

January 4–June 8, 1964

EXCAVATION FUNDING

The 1964 work at Dorginarti was funded by a contract with the United States Department of State, under Public Law 480, and was requisitioned through the Oriental Institute of the University of Chicago.

Preface

The final report for the excavations at the fortress of Dorginarti has been many years in the making, starting with my initial study and 1992 dissertation and continuing now with the complete excavation report. The Oriental Institute of the University of Chicago took part in the UNESCO[1] Nubian salvage project necessitated by the building of the Aswan High Dam between 1960 and 1970, a multiyear campaign of survey and excavation activities throughout Lower Nubia. The published results from the extensive campaign have contributed greatly to the history of this region of southern Egypt and northern Sudan, and now the excavation report for Dorginarti can be added to the list of publications stemming from it.

During this time, and more important to consider, the residents of the region were relocated to places in Egypt and elsewhere in the Sudan or to newly created settlements at higher elevations, such as that at Wadi Halfa. It was not just a loss of the archaeological remains and Nubian history along the river valley that was of concern, but also the loss of property and homeland for many Nubians.

When the Oriental Institute expedition finished its work at Serra East (1961–64), it concentrated on recovering the remains at Dorginarti. The fortress was a tumbled, waterlogged mess as a result of millennia of flood damage (pl. 1a), and it was assumed that it represented another Middle or New Kingdom establishment similar to the other Second Cataract forts, of which so much was already known. The site was not completely excavated in the limited time available, but the amount of work accomplished was admirable.

These excavations took place from the beginning of January 1964 under the overall direction of archaeologist and architect James E. Knudstad. Richard H. Pierce, the Egyptologist and epigrapher of the Oriental Institute expedition, oversaw the work at the site from January until the beginning of April with the assistance of Wenche Pierce. When Dr. Pierce's contract ended at the beginning of April and the work at Serra East was finishing, James Knudstad moved operations over to Dorginarti (pl. 1b). The excavation continued until June 8, 1964.

Between 52 and 131 workers were busy at Dorginarti on any given day, including Egyptian, Nubian, Shilluk, and Nuba laborers, all under the direction of four to seven Egyptian Quftis. This number of workers, and the mix of cultures, was a handful to manage on a daily basis. The job of moving the debris and silt from atop the outlines of the fortification and excavating the remains was immense.

The pottery encountered was unlike any known types, but the intrusive Meroitic burials atop the Central Sector and the scatter of pottery resulting from disturbance by thieves led to the initial belief that Level II was perhaps Meroitic. The date of Levels III and IV was believed to be late New Kingdom, and some of the pottery does indeed have comparisons with materials now dated to the early Third Intermediate Period. But late New Kingdom and first-millennium pottery was not well known at the time. William Y. Adams knew the Nubian pottery sequence better than most salvage archaeologists, and he ventured that it could date to the late New Kingdom or Napatan period.[2] But there was no time to process the pottery in the manner of a normal excavation, for this was fast-paced salvage work. It is fortunate that the remains have waited some sixty years for their final publication, because knowledge of pottery development during this period is now better as a result of more recent excavations in both Egypt and Nubia during the 1990s and 2000s.

The results of the excavation have been presented in a number of publications, but only the initial site report by Knudstad in *Kush* 14 (1966), an unpublished but widely distributed dissertation, and a few short articles have considered the architecture and associated remains in any detail.[3] While the present volume

1 UNESCO is the United Nations Educational, Scientific and Cultural Organization.

2 Pierce 1964a, February 11; Knudstad 1963–64, p. 234. The Napatan period encompasses the time when the cemeteries and capital of the Kushites were based at Napata—that is, from at least the ninth century BC until sometime during the fourth century. In this later period, they transferred their capital to Meroe. Since the pottery before, during, or just after the Twenty-Fifth Dynasty—and before the Meroitic period—is poorly understood, it is referred to herein as *Napatan* unless a firmer dating is known.

3 Knudstad 1966, pp. 178–86; Heidorn 1992a and 2013. See also Hoerth 1964b. For the Level II architecture and pottery, see Heidorn 2018b.

considers the earlier remains at the site, the Level I Christian building and ten Meroitic graves hollowed out of the earlier walls will be presented in a future Nubian Expedition publication. The phasing designations of the original excavators—Levels I, II, III, and IV—are retained throughout the book.

The work on the Dorginarti materials suffered a gap of thirty years between my dissertation and this final site report, but, as mentioned above, this time allowed the publication of other research and excavation reports that permit a more precise dating for the site. Further refinement in the dating of early first-millennium BC pottery, both the wheel-made and handmade assemblages, is still necessary. This publication will, it is hoped, contribute to a more comprehensive understanding of the architecture and material culture of Lower Nubia and Egypt during the first half of the first millennium BC and, specifically, during the tenth through eighth centuries BC.

The materials presented in this book include the selection of artifacts brought back from the Sudan and stored in the University of Chicago's Oriental Institute Museum as of 2021. Thus, this set of data is incomplete but still remarkable considering the large amount of materials salvaged and transported back to Chicago. Not all the material had received museum registration numbers as of 2017. A number of objects are also stored in the National Museum of Sudan in Khartoum, where most of them were relocated and where they were drawn or photographed by Nadejda Reshetnikova, Bruce Williams, or me. Materials were also left on the site after work was completed and, like the site, have disappeared in the reservoir created by the building of the Aswan High Dam.

Introduction

The earliest ceramics found at the Second Cataract fortress of Dorginarti date from approximately the tenth to the early ninth centuries BC, with the bulk of the pottery dating to the early to mid-eighth century. There are two pottery horizons in the early phases at the site. Only a handful of New Kingdom sherds show that the island was visited at earlier periods. In the 1980s, using pottery parallels from Egypt, I dated the Level III/IV ceramic assemblage to the earlier Twenty-Sixth Dynasty, with reservations that it might stretch back even earlier in time.[1] However, due to the wider availability of well-stratified published pottery from first-millennium BC contexts in Egypt, Sudan, and the Levant over the past two decades, and due to further work on the pottery from the royal cemeteries of Kush at the Museum of Fine Arts, Boston, this pottery dating has been revised.

As discussed in chapter 5, the pottery mirrors the assemblages from the late dynastic deposits at Amarna, the fort south of Edfu near Abu ʿId, and some of the pottery from the Phase IIb and especially the Phase III contexts at Elephantine (see map 1).[2] The published Third Intermediate Period pottery from the chapel of Osiris Wennefer Neb-djefau at Karnak, and that from the temple of Mut at Karnak, is also comparable, as is the Third Intermediate Period pottery from Tanis.[3] These sites have parallels for the earliest forms, as well as for those from the early to mid-eighth century phase at Dorginarti. The establishment of Dorginarti thus predates the height of the Twenty-Fifth Dynasty in the later eighth and early seventh centuries BC. Also pointing to a more precise date for the eighth century assemblage is the relative lack of parallels to the marl jars found in the royal tombs at Kurru and Nuri that date to about 700 BC;[4] to the wheel- and handmade beakers found in Nubian tombs dating to both the Twenty-Fifth Dynasty and before;[5] or to the vessels from the Ptah temple at Karnak, also attributed to the Twenty-Fifth Dynasty.[6]

The reader should note that the pottery from Levels III and IV was, until recently, considered one assemblage but is now considered to belong to an earlier and a later phase at the site.[7] Within the Level III deposits there are vessels of two stylistic phases, those reflecting New Kingdom traditions and those reflecting the initial stage of the marl pottery industry thought to have begun in the mid-eighth century BC in Upper Egypt. Extensive flooding, both during the active life of the fortress and later, brought in forceful streams of water and quantities of sand and rocks that damaged architectural remains and undoubtedly displaced sherds and objects. Disturbances of the earlier architectural remains also occurred during Level II, in addition to the Meroitic burial and Level I Christian activity at the site. Building and other activities disturbed earlier deposits that may have then become mixed with later materials. In any case, most of the pottery in the Oriental Institute Museum is not from complete vessels, and sherds can migrate between levels for the reasons just described.

The sizable amount of handmade pottery found at the site, though not precisely datable, is comparable to vessels found in both Lower and Upper Nubia—for example, from surveys and excavations at sites along the Debba Bend of the river and in the Fourth Cataract region—where the parallels are attributed to the

1 Heidorn 1991, 1992a, and 1992b.

2 French 1986; Aston 1996c and 1999.

3 Boulet and Defernez 2014, pp. 608–10, fig. 31-1A–D. They, however, date this assemblage to the end of the Twenty-First or beginning of the Twenty-Second Dynasty, and although there is no independent evidence, this dating has been considered. There is some overlap with their Twenty-Fifth Dynasty forms, however (e.g., Boulet and Defernez 2014, p. 610, fig. 31-1E–G, K–L). See also Sullivan 2013. For Tanis, see especially Laemmel 2011, pp. 7–45, pls. 1–34.

4 Dunham 1950; Heidorn 2018a.

5 For example, a marl wheel-made beaker was found in situ in the burial chamber of Kurru 72, attributed to the reign of Shebitqo, while silt wheel- or handmade examples were found before the door blocking for Kurru 3, a queen of Taharqo; see Dunham 1950, p. 28, fig. 10b, and p. 105, fig. 35c. See also, for example, the burnished red-slipped beakers from the warrior's tomb at Tombos, dated by a scarab with the name of Shabaqo: Smith 2008, pp. 108–11, fig. 12.

6 Boulet and Defernez 2014, pp. 611–15, fig. 31-2A–E.

7 The area by the north breach in the West Sector was heavily swept by floodwaters, and it appears that earlier residual sherds were found here; another clear example is seen in some of the deposits along the south wall toward the southeast corner of the West Sector, where the Level IV architecture was also exposed by water erosion. Stratified pottery from various contexts is shown in appendix I.

proto-Kushite or Napatan periods (see map 2). There are close analogies from the fortress of Gala Abu Ahmed in the Wadi Howar and some similarities to the handmade pottery from tombs at Sanam.[8]

The possible reasons for the establishment and abandonment of the fortress in its early phases are discussed in chapter 8, while the architecture and layout of the interior buildings of Levels III and IV are discussed in chapter 4. The pottery and objects from Levels III and IV are discussed together in chapters 5 and 7, while the stratigraphic association of vessel types is also presented in appendix I.

The establishment of the Level III and Level IV fortress at the north end of the Second Cataract was only one of a series of instances when the area was reinforced with military and civilian settlements along the river. During these periods, the Nile was the main route connecting trade and political centers to the south and north of Lower Nubia. But whether the establishment was that of a Lower Nubian power, a proto-Kushite stronghold of the early ancestors of the Twenty-Fifth Dynasty family, or Egyptian is difficult to assess. It certainly appears Egyptian in its architectural detail and pottery.

The primary role of a fortress is to control and protect a strategic area, but it is also a symbol of political and economic power, which is a key element in any strategy to retain diplomatic leverage over other powers.[9] The continuing struggle to dominate a region where state power is tenuous, and to control its resources and transport routes, usually gives rise to the building of monumental military installations and other territorial markers. But the utility of a fortress wanes or disappears when political power shifts, when new borders or routes come into being, or when new types of weaponry render their structural design obsolete. It must be stressed that the building and maintenance of the fortifications on the island of Dorginarti required enormous expenditures that were obviously justified by its tactical importance.

The pottery from Level II is clearly differentiated from that of the preceding phases but is once again dominated by northern forms and separated from the earlier phases by more than a century. This is discussed in chapter 6, where architectural and pottery parallels also suggest a connection to Egypt, the East Greek colonies, Phoenicia, and the southern Levant during the sixth century BC. Many objects from this later reoccupation are found in chapter 7, where they are identified as belonging to Level II activity. Very little handmade pottery was found in association with the later sherd deposits, and none was found with the Level II sherd scatters around the North Gate and on the River Stairs. But these fragments did not differ significantly in form or decorative traits from those found in the earlier levels, and they are here placed with the handmade sherds from Levels III and IV.

The picture that emerges from the archaeological record presented in this book and cited in its footnotes shows that Lower Nubia remained important after the withdrawal of Egyptian control in the late second millennium BC and before the advent of the Twenty-Fifth Dynasty.

---

8 For references to parallels from Egyptian sites, as well as Nubian comparisons from the Debba Bend Survey, Gala Abu Ahmed, and Sanam cemetery, see chapter 5.

9 Keeley, Fontana, and Quick 2007, p. 81.

1

Landscape and Resources

Before it was submerged, multiple rock outcrops in the river in the Second Cataract region made sailing through it dangerous, even when flooded. The island of Dorginarti was at the northernmost end of the Kabuka Rapids, a part of the Batn el-Hagar that stretched 8 km south to the fortress and settlement at Mirgissa (fig. 1.1). The riverbanks through this stretch were active throughout time. Desert routes converged on the river here from the oases to the west and from the mineral-rich Red Sea Mountains to the east, making it an ideal place for recovery, resupply, and the exchange of goods.

In the Middle Kingdom, a slipway for boats was constructed north of the fortress at Mirgissa in order to bypass this difficult stretch of cataract.[1] Similarly, during their expedition up the Nile in 1884–85, British, Egyptian, and Canadian military forces constructed a portage for ships and their loads in the vicinity of Wadi Halfa to bypass the river during periods when the water was low. The method of getting people, ships, and goods around the cataract was the same in antiquity as that described by James Grant, a member of the relief expedition sent to aid General Gordon in Khartoum:

> By the 30th of October no less than 140 boats filled with troops were taken up past the Second Cataract at Wady Halfa, and it was calculated that the remainder of the flotilla would go up at the rate of fifty a day. This work was greatly aided by the efforts of that gallant and energetic officer Lord Charles Beresford, who constructed a portage of 2,480 yards in length, by which the most serious difficulties of the cataracts were avoided. The boats were passed along by parties of men thirty-five strong, with the aid of rollers and levers to get them across the portage. He had seven of these parties at work, and the average time they occupied in getting a boat across was an hour and a half. The work was heavy, and the heat was great; but the will and energy of our men were found to be equal to the occasion.[2]

Grant goes on to note that

> the Engineers brought their boats up the terrible Bahr-el-Hajar cataracts without taking them out of the water which was deemed a most creditable feat; but the Canadians suggested that to do this would be impracticable for the more heavily-laden boats, and in order to enable the bulk of the expedition to pass quickly up these cataracts, it would be necessary to station parties of men with camels to portage the cargoes overland, while the empty boats were dragged up by the troops.[3]

Throughout history, one could sail down the river in larger boats when the water level in the river was high; one could "run the rapids" in smaller boats or drag them over the rocks when the water level was lower; or one could haul cargo and boats overland when the level was too low to allow any riverine passage at all. Landings where boats could be repaired and reloaded after passing through the cataract must have existed here, given the rough navigating conditions.

TOPOGRAPHY OF THE ISLAND

The fortress of Dorginarti[4] was situated close to the steep north bank of the island with no more than 400 m separating it from the west shore, according to the Nile

1 Vercoutter 1970, pp. 204–14; Vogel 1998, pp. 266–72.

2 Grant 1885, p. 30.

3 Grant 1885, p. 34.

4 The name of the island of Dorginarti is most likely a rendering of *dork-in-arti*, or "island of the gazelle"; see van Gerven Oei and Tsakos 2017. The consensus is that the name incorporates a Greek loanword into Old Nubian (*dorkas*, "gazelle").

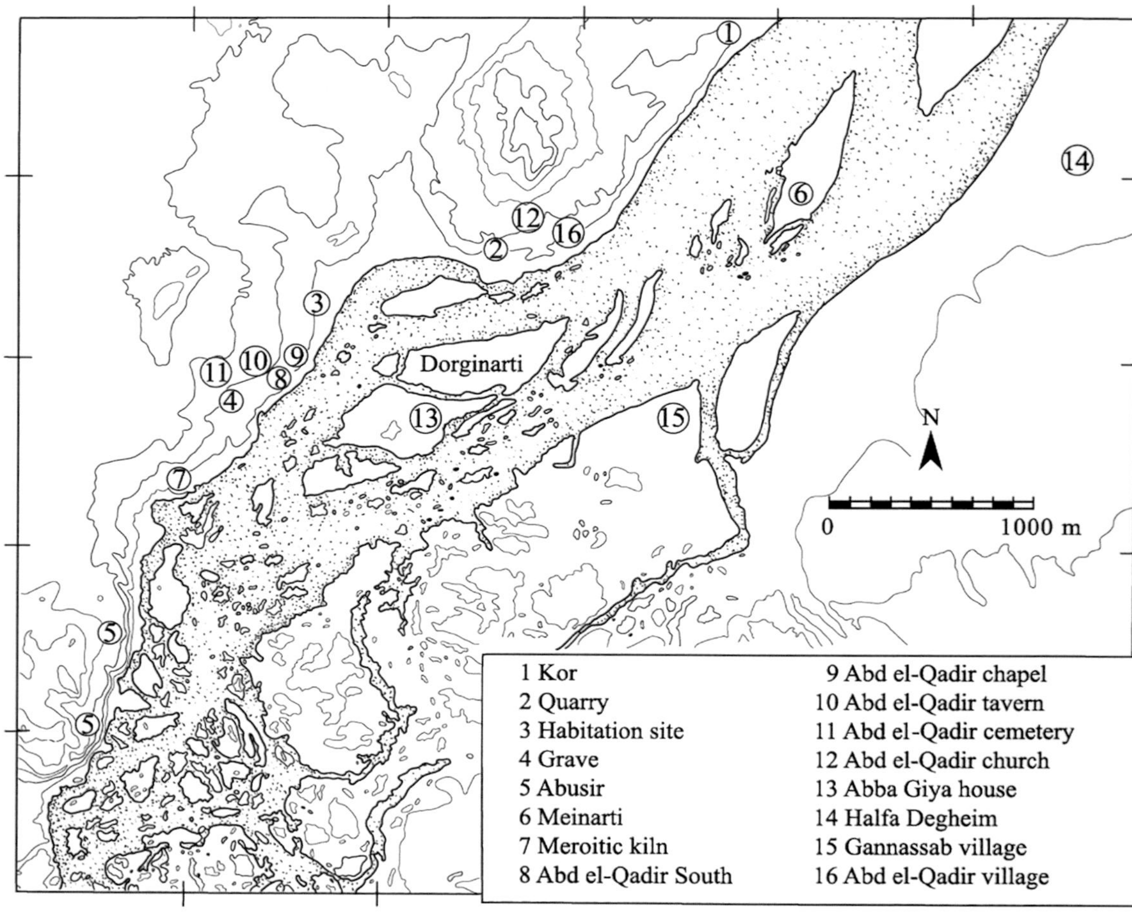

Figure 1.1. Second Cataract map (after James Knudstad)

configuration of the 1960s (pl. 2). The island of Abba Giya, to the south of Dorginarti and connected to it during periods of low water, possessed a higher stone outcrop at its center, but only the remains of a minor Christian building were found there.[5] Both the island's unique topography and the need for an outlook downstream and over the west channel of the Nile determined where the defenses were eventually placed. The location also provided a view of all activities on the west bank, where boats would have been landed or launched after sailing through the cataract or after having been hauled overland around it. Facilitating and guarding riverine traffic was the fortress's primary function.[6] Other locations in the vicinity obviously did not afford the same advantages.

The main gate on the west was positioned to take advantage of a harbor, and the excavators found remains of a built roadway leading westward and downward to a scatter of sherds above the riverbank.[7] The River Stairs and North Gate afforded protected access to the Nile for water, and from here, too, it was a quick trip by boat to the west bank.

The triangular outline of the fortress followed the shape of the high ridge of cataract rock at the northwest of the island, with the highest knoll at the northeast supporting the Central and East Sectors. This outcrop served for surveillance of the immediate area and was the location of both the multistoried Level III and Level IV official's residence and the Level II citadel.

LOWER NUBIAN LANDSCAPE

A reconstruction of the patterns of use and political history of the region entails the consideration of its

5 Knudstad 1963–64, p. 316.

6 Adams 1977, p. 184.

7 Knudstad 1966, p. 179. It was noted that the only other remains on the island were a scattering of sherds on rocky ground at the island's west end, undoubtedly one of the landing places in the river channel that ran west of the island. See also Knudstad 1963–64, p. 335, where he mentions a rock layer to the west of the southwest corner bastion, which appears to have been the foundation for a mud-paved road leading west from the West Gate.

geography, agricultural potential, and possible contemporary settlement patterns before the flooding of the region by the Aswan High Dam. The following paragraphs describe the region before the creation of Lake Nubia (Sudan), which is called *Lake Nasser* in Egypt.

Lower Nubia consisted of the Nile Valley and the adjacent desert region located between Gebel el-Silsila on the north, about 65 km north of the First Cataract of Egypt, and the town of Wadi Halfa on the south, located at the Second Cataract just across the Egyptian border in modern Sudan.[8] The southern boundary of this region was the inhospitable Batn el-Hagar, an arid region that separated the Second Cataract from the fertile areas around the Dal and Third Cataracts in Upper Nubia. The southern border of Upper Nubia, a region that includes the area south of the Second Cataract, has been fluid. It has shifted southward and northward over the course of history depending on the political situation.[9]

Igneous and metamorphic rocks created the rocky land and river outcrops at the First and Second Cataracts, but between these two areas the Nile ran through deposits of Nubian sandstone.[10] Narrow strips of alluvial land adjoined the river on both sides,[11] with the Nubian and Libyan deserts lying to the east and west of the river valley, respectively. The sandy or gravelly plains and low hills to the west were almost completely arid, while on the east beyond the sometimes steep cliffs, the rocky hills and their drainage courses provided collection areas for the infrequent rains.[12] Finally, the Nile's tributary wadis provided some water for additional areas of cultivation at the valley's edge.

South of the Second Cataract the topmost sandstone deposit receded, exposing barren granite formations known as the Basement Complex. Between Kerma and Saras, the Nile followed an indirect course, having cut into the least-resistant lines of the hard deposits. Barbour described the stretch of land between Kerma and the area immediately north of Wadi Halfa in the following terms:

> The total width of the habitable strip is rarely more than two kilometers. . . . The subregion is clearly divisible into reaches of greater and less productivity and wealth, which are largely determined by the nature of the rocks across which the river flows. Where these are strongly resistant to erosion, as for example the igneous rocks of the Basement Complex, the river bed is narrow, often broken by islands and cataracts, and there are few terraces of alluvial soil.[13]

Patches of alluvial land also lay along the river between Wadi Halfa and the First Cataract.[14] The banks of the river were higher in this area, and the water of the Nile did not always inundate the land above the First Cataract.[15] But the topography was less hilly than it was south of the Second Cataract and provided some flat areas for cultivation.[16] John Burckhardt noted that the east bank of the Nile from Korosko to Aswan was more fertile than the west bank. As he described it:

> The eastern bank is, throughout, better adapted for cultivation than the western; and wherever the former is of any breadth, it is covered with the rich alluvial earth deposited by the Nile. On the western side, on the contrary, the sands of the desert are impetuously carried to the very brink of the river, by the north-west winds which prevail during

8 The description presented here refers to the region as it was before 1964, when Lake Nasser flooded Lower Nubia. The Aswan Dam, which created Lake Nasser, was originally built in 1898–1902. Three subsequent enlargements of the dam (in 1908–10, 1929–34, and 1959–69) necessitated that salvage surveys be undertaken to recover the archaeological sites that were threatened by rising water levels.

9 See, for example, Wenig 1981, p. 526; Adams 1977, pp. 20–21.

10 For good descriptions of the region's geology, see Barbour 1961, pp. 32–36; Butzer and Hansen 1968; and de Heinzelin 1968. The geology of the Second Cataract area is described in detail in Maley 1975.

11 The Nile has cut more sharply into the bedrock in Nubia, and the fertile area along the banks is narrow and discontinuous. Because deposition and erosion are in close balance, the floodplain does not accumulate as in the north; see Trigger 1965, p. 10.

12 On these features, see Abu al-ʿIzz 1971, p. 93. The average height of the west ridge is 170 m above sea level, and that of the east ridge 210 m. Both heights are dissected by wadis. In the region between Wadi Halfa and Aswan, the western wadis obtain water from a limestone plateau situated 500 m above sea level; see Abu al-ʿIzz 1971, pp. 97–98, and Barbour 1961, pp. 131–32. For the climate of the northern Sudan, see Barbour 1961, pp. 38–51, and Ireland 1948.

13 Barbour 1961, p. 133. Burckhardt's description of this area is similar; see Burckhardt 1978, pp. 38–79.

14 Trigger (1965, p. 14) divides Lower Nubia into a number of ecological areas in which more fertile areas alternate with poorer districts. The size of the population of these various regions depends on the extent of fertile land along that section of the river valley. However, the river has changed course over time in some places, and the present areas of fertility may not agree with past ones; see, for example, Williams 1986, pp. 6–7.

15 Burckhardt (1978, p. 127) noted that, above the First Cataract, the flood waters of the Nile were never high enough to overflow the banks and thus the fields were irrigated by waterwheels.

16 Barbour 1961, p. 133. These patches of alluvial land are discontinuous because of the lateral projections of the cliffs along the narrow river valley. The river is also quite narrow; see Abu al-ʿIzz 1971, p. 92 (Kalabsha and Silsila) and Burckhardt 1978, p. 20.

> the winter and spring seasons; and it is . . . only in those places where the course of the sandy torrent is arrested by the mountain, that the narrow plain admits of cultivation.[17]

At Faras, the country was quite open on both sides of the river, but it was covered with sand.[18] Although the area had been very fertile and prosperous in ancient times, it was obviously not being cultivated when Burckhardt visited the region in the early nineteenth century.[19] But he observed that the eastern road between Debeira and Ashkeit ran through a fertile area covered by date palms.[20] Immediately to the south, and extending to Wadi Halfa, was a sandy plain where the cultivable area was broader than usual for the Lower Nubian river valley—as was true also at Koshtemna and Girsha.[21] As far south as Mirgissa the river was "studded with islands covered with lush green vegetation, albeit rocky, but beyond this the scattered fields of Nubia disappear as one enters the wasteland known as the Batn el-Hagar."[22]

BOTANICAL AND FAUNAL RESOURCES FROM ARCHAEOLOGICAL SITES

While no flotation or other samples were taken from Dorginarti, the botanical evidence from Napatan levels (700–400 BC) at Qasr Ibrim consisted mostly of samples of six-row hulled barley, but also of lesser quantities of tetraploid wheat emmer, and less still of millet and wild sorghum.[23] Evidence of lentils, date palms, figs, Egyptian balsam, doum palms, the ben-oil tree, linseed, and castor beans was also found in these contexts at Ibrim.[24] Vegetables and fruits included watermelon seeds, a cucurbit (probably cucumber), perhaps both onion and garlic, and spices such as coriander. The existence of this evidence is not surprising since these botanical specimens are all known from pharaonic contexts farther north.

The early Kushite evidence from Kawa (eighth to fifth centuries BC) indicates the existence of dates, watermelon or colocynth seeds, and also broad beans and lupin, the latter of which only appears at Qasr Ibrim in the post-Meroitic period.[25] Emmer wheat, barley, lentils, and grass peas were also found. Domesticated sorghum and millets were perhaps present at Kawa in the periods before circa 400 BC, but this date is earlier than elsewhere in Africa or at Qasr Ibrim.[26] Analysis on plant materials from the site of Gala Abu Ahmed, in the Wadi Howar and west of the Nile, identified domesticated emmer wheat and various wild grasses, vegetable or herb plants (amaranth, sedge), and fruit remains; the radiocarbon dates clustered between approximately 1200 and 900 BC.[27] Botanical remains from grinding stones at the site were also analyzed, revealing evidence for their use in the processing of wild grasses, palm, and woody plants.[28]

The analysis of bones from Qasr Ibrim shows that goats, sheep, and cattle were more or less common during the Napatan period and later, although in varying proportions.[29] Faunal analysis from the fortress of Gala Abu Ahmed, along the Wadi Howar in the western desert, shows that mollusks were brought to the site from the Nile and the Red Sea for eating or making into beads. The excavators uncovered the bones of one Nile perch, a few bird bones, and the shells of ostrich eggs, although the eggs were most likely used for making beads and not eating.[30] Wild mammal bones were present as well. Gazelles were the most common game animals at the site, while young goats and sheep, as well as cattle, were the most common domesticated food animals.

There is also evidence from the Napatan buildings at Kerma that shows the presence of young caprines and cattle in particular, but also sheep and goats. Finally, the evidence from the temple of Soniyat, along the Debba Bend, shows that cattle, caprines, pigs, birds, and fish were eaten there.[31]

17 Burckhardt 1978, p. 20.

18 Burckhardt 1978, p. 37.

19 Trigger (1965, p. 14) notes that Faras was a rich locality in ancient times. Note the rich archaeological remains from the Middle Kingdom fort, the New Kingdom temple and settlement, and the Meroitic and later remains.

20 Burckhardt 1978, pp. 37–38.

21 Burckhardt 1978, p. 129.

22 Trigger 1965, p. 14.

23 Rowley-Conwy 1989, p. 135. See also Clapham and Rowley-Conwy 2006.

24 Rowley-Conwy 1989, pp. 131–34.

25 Fuller 2004.

26 Fuller 2004, p. 73.

27 Kahlheber 2013.

28 Eichhorn 2013.

29 Copley et al. 2004, p. 1274. The Napatan samples are dated between 700 and 300 BC. The bone samples were taken from sheep, goats, and cattle from levels dating to the Napatan through the Christian and Islamic periods and were sampled to detect the types of grasses and grains they ate.

30 Linseele and Pöllath 2015.

31 References for both of these analyses are found in Linseele and Pöllath 2015, p. 540. See also Chaix 2010. For the analysis of remains from the temple at Soniyat, see Osypińska 2003, p. 492.

MINERAL RESOURCES

There is no certain evidence that gold was actively sought in the eastern desert during the early first millennium, which is surprising.[32] However, the secondary use of querns and handstones at a few New Kingdom gold mines in Nubia has led to the suggestion that the reuse of gold-mining equipment occurred during the Kushite period, particularly at sites close to the Nile Valley.[33] The pottery at these putative Napatan sites consisted only of handmade Nubian vessels, which could not be dated.[34] Panning for gold or its extraction from wadi rock would undoubtedly have continued, but to what extent is unknown.

Evidence for the extraction of other minerals that could be obtained from the deserts along the Nile River in Lower Nubia is thus far lacking for the early first millennium BC.

AGRICULTURAL POTENTIAL OF THE SECOND CATARACT REGION

The agricultural potential of the Second Cataract region was never great, but it could have yielded enough vegetables, dates, and grain to support a small riverine population and periodic nomadic encampments.[35] Cultivation along the banks of the river in the vicinity of the Second Cataract, and in the area immediately to the north, could have supplied settlements with some produce and animal products.[36] The arid climate may not have allowed the cultivation of much barley or emmer,[37] but grains and vegetables could have been grown in artificial basins in the Second Cataract alluvium.[38] Luxuries such as wine, as well as staples, were brought either from elsewhere in Nubia or from Egypt.

SUMMARY

Because of the limited amount of arable land, any increase in settlement and prosperity in Lower Nubia at any period in the region's history would have depended on its role as a node along the north–south and east–west transportation routes.[39] Both the settled and nomadic populations of this area were likely involved in raiding, trading, and supply operations and in providing both services and mercenaries to the garrison communities along the Nile or in Egypt itself. The frontier between Egypt and Kush would thus have required a strong line of defense in the area of the Second Cataract and the Batn el-Hagar to facilitate the flow of trade and the movement of people.

32 For an overview of all the resources obtained from Nubia, see Zibelius-Chen 1988, pp. 69–135.

33 Klemm and Klemm 2013, pp. 15, 611. For the distribution of possible Kushite, or perhaps Meroitic, gold-mining sites, see the map in Klemm and Klemm 2013, p. 612, fig. 7.5.

34 Klemm and Klemm 2013, p. 611; Klemm, Klemm, and Murr 2001, p. 656.

35 The productivity of the various regions of Lower Nubia is reflected in the 1960 Egyptian census figures for the area cited in Geiser 1986, pp. 231–32 (appendix K). Estimates vary, but one source noted that the entire Nubian population in 1970 was about 1 million, with 150,000–200,000 people living in Egyptian Nubia alone; see Geiser 1986, p. 86, n. 8.

36 Vercoutter (1970, pp. 165–66, 170–71) believes that the cultivable areas south of Aswan—notably from Ballana and Qustul to the Second Cataract, and from Dal Cataract to Dongola—were fertile enough to support a considerable population.

37 See the discussion in Emery, Smith, and Millard 1979, p. 101.

38 Proposed for earlier periods by Vercoutter (1966, p. 153, n. 99), who discusses evidence that points to the continued use of the same technique, also observed by Frédéric Cailliaud in the early 1800s.

39 This role for Dorginarti is discussed in chapter 8.

2

Early Observations and Excavation of the Site

Early travelers' accounts do not mention the ruined fortress on the island of Dorginarti, perhaps because it presented a less dramatic sight than the massive defenses preserved at places like Mirgissa, Buhen, and Meinarti or because it was less accessible and unseen, having been covered with vegetation.[1] Tamarisk, acacia, and other vegetation concealed the sand- and stone-covered remains of Dorginarti until the early 1900s, with the earliest reference to the site being found in the Buhen publication of D. Randall-MacIver and Leonard C. Woolley, published in 1911.[2] The island is not named in this account, but the description of it—near Meinarti, with a well-preserved and massive brick wall—indicates that the island to which Randall-MacIver and Woolley refer is Dorginarti. The remains were thought to be those of a fortress built by the Egyptians of the Eighteenth Dynasty to guard the northern outlet from the cataract.

It was not until 1916 that Somers Clarke, the author of a study on Nubian fortresses, published a more precise description of the island's remains:

> [Dorginarti] is about a mile south of [Meinarti]. On it is a vast solid wall of crude brick. The fortress is elongated in plan, the length being from north to south, and the east and west sides being approximately parallel. The bricks vary much in size; some are more akin to tiles, 0.35 × 0.35 × 0.10 m., in other places the more usual dimensions, 0.32 × 0.15 × 0.10 m., are found.[3]

Clarke stated that the walls of Dorginarti were 8 m thick and that courses of timber were bonded into them. He also writes that the fortress walls had been eroded by the waters of the Nile and that a stone revetment had been constructed to protect them from the swift current of the flooded river. Lastly, he gives the rough measurements of the fortress as 194 m from north to south and 80 m from east to west.[4] Some of Clarke's statements (for example, the measurements of the fortress, its axis, and the assertion that the walls were built with timber bonding like that found in other Lower Nubian forts of the Middle and New Kingdoms) do not accord with the description of the island found in the diary of Noel F. Wheeler, who participated in the Harvard University and Boston Museum of Fine Arts expedition to Mirgissa in 1931 and 1932.[5] Wheeler gave a detailed description of the fortress, as follows:

> This island is situated at the exit of the Second Cataract near the west bank of the river and flanking the main channel. . . . The island includes several parts which would be separated at high water. In January the island is about 1000 meters long from north to south and about 600 meters wide; at high water it would be about 200 meters long by 80 wide. The fort itself is about 80 meters long by 50 wide. The walls have a thickness comparable with the Middle Kingdom forts, but the general impression is that it is of New Kingdom date, and probably late New Kingdom. The bricklaying is poor and the construction shows a crudity not found in anything of the Middle Kingdom from Semna to Mirgissa. The batter of the walls is much coarser, the lines are less true, and the wall in one place makes

1 Burckhardt (1978, p. 39) mentions remains on an island in the Wadi Halfa area; although he gives no specifics, the reference is probably to the more impressive ruins on Meinarti since he was traveling along the east bank.

2 Randall-MacIver and Woolley 1911, p. 7.

3 Clarke 1916, pp. 164–65. Most of the forts he discusses in his article were thought to date to the Eleventh and Twelfth Dynasties.

4 Clarke 1916, p. 165.

5 Wheeler 1961, pp. 151–52. The diary entry is published in more detail in Dunham 1967, pp. 177–78.

a sweeping curve where a Middle Kingdom builder would almost certainly have made an angle.[6]

Wheeler speculated that the water-dispersed cataract rock on the north and south sides of the fortress may have been the remains of a rubble foundation or breakwater.[7] The following passage is also from his diary:

> The plan of the fort is roughly rectangular, but the east wall is curved at its southern end and the south wall forms an angle greater than 90° with it. There are traces of subsidiary buildings both inside and outside the fort, and there appear to have been some sort of buttresses on the outer faces of the walls, although they are not regular. In the curved section the bricklaying follows the curve throughout the thickness of the wall. Potsherds found here are not very enlightening: many are Greco-Roman, but some might well be of New Kingdom date.[8]

Although Wheeler correctly noted that the timber bonding typically found in the brick walls of such fortresses was lacking, his measurements of the fortress were incorrect.[9] It is apparent from the sketch of the fortress he made in his diary that he included only the west section and omitted the central and east portions, perhaps because they were not visible on the surface. However, the fortress interior alone measures about 136 m from east to west and 50 m from north to south in the West Sector (narrowing to 25 m in the East Sector). The length from the exterior walls of the West Gate to the exterior of the east end is 175 m.[10]

It is apparent that none of the informants above undertook a serious study of the remains at the fortress site. However, the Greco-Roman dating that Wheeler assigned to some of the potsherds he examined at Dorginarti is an important indication of his recognition that the sherds on the surface were later in date than the materials found previously at Middle and New Kingdom fortresses in Lower Nubia.

THE EXCAVATION

It was not until the Oriental Institute of the University of Chicago participated in the Nubian salvage project led by UNESCO from 1959 to 1969 that excavation at Dorginarti was undertaken. The excavation lasted from January 4 to June 8, 1964.

When the team arrived at the site, the water-scoured structure was so ruinous that they doubted intelligible strata could be uncovered. According to the director of the Oriental Institute project, James Knudstad, the site was "stratigraphically complex and fussy . . . a veritable monstrosity of mudbrick laminations and repair work," appearing as a mass of impassive brick before excavation.[11] Wind-blown sand and Nile flooding over the millennia had eroded the multiple rebuilding layers of its walls and scoured the middle of the West Sector free of its remains. The force of the floodwaters from the southwest channel of the river caused breaches in all the enclosure walls except at the east end.

Before work at the site began, Knudstad described its state as follows:

> Heavy and high-standing mudbrick walls define a roughly triangular fort resting on the higher north side of the island. The long sides of the fort run approximately east-west from a short side on the west to an opposite and acute angle on the east. . . . The west half of the interior is low lying and largest; breaks in the north, south, and west outer walls suggested flood damage in this half. Small walls abutting the exterior walls were exposed on the surface within this half also. This area is closed on the east by a heavy cross-wall. . . . The surface of this gave evidence of two periods of construction superimposed upon the bulk of fort brickwork. To the east of the central high area is again a lower and triangular one within outer walls forming the east angle of the fort. The outer fortification walls are clearly buttressed without, and massive repair seems evident in the brickwork and variety of brick sizes. Exterior to the fort on all sides is a spread of tumbled cataract rock, possibly a ruinous embankment placed there against flooding.[12]

The excavation of the fortress began with a team of about 52 men (increasing at one point to 131) and four Quftis (increasing at times to seven). The initial excavation director was Richard Pierce, who was assisted by Wenche Pierce and Mel Thurman.

At the beginning of February, Knudstad sent a letter to the director of the Oriental Institute requesting more money to continue excavations at Dorginarti after the project at Serra East reached its scheduled

6 Dunham 1967, p. 177. Wheeler noted brick sizes measuring 34 × 17 × 9 cm and 35 × 35 × 10 cm.

7 Dunham 1967, p. 178.

8 Dunham 1967, p. 178.

9 Three wood planks were used in reconstructions at the east end of the fortress wall, while wood timbers were seen only in the massive later repair of the northwest corner; see chapter 3, p. 15.

10 Rough measurements for the three sectors of the fortress add up to an area of about 3,754 m^2.

11 Knudstad 1964b, letter to Robert McCormick Adams, February 5.

12 Knudstad 1964c, letter to Sayed Thabit Hassan Thabit, January 12.

completion date on March 12.[13] He noted in his letter that Dorginarti was an archaeologically complex site meriting more attention and that he was willing to stay on until June.

The Pierces left at the beginning of April, and Knudstad continued the excavation of Dorginarti with Al Hoerth and Rudy Dornemann assisting. The excavation proceeded at a rapid pace with about seventy-five workmen until June 8. In his final letter to the Sudanese Antiquities Organization commissioner, T. H. Thabit, Knudstad reported:

> Almost all of the preserved portions of the initial plan in the west interior have been recovered. Preserved features are thought to continue into only two or three small, and as yet unexcavated, spots. This can be said for the smaller east interior as well, where preservation has been found to be fairly complete in plan. Tests beneath the earlier level of the renovated central building in the fort have revealed nothing more substantial than random living debris mixed with windblown sand.[14]

The earlier levels noticed below the River Stairs, and the walls running under the inner wall relining, were left unexcavated.[15] Although additional excavation would perhaps have clarified the earliest levels below the building walls shown on a rendition of the final plan (plan 1), the main periods of occupation at the fortress had been exposed by June, and 60–70 percent of the fortress was mapped. The areas that were not excavated completely were too eroded to map correctly or were minor features overlain with heavy, later construction.[16]

In the earliest levels, such as under the main wall relinings on the north, the pottery—although consisting largely of small, undistinguished, and worn sherds—did not seem to differ substantially in date from that of the main occupational levels.[17]

The fortress interior was divided into three sectors: Central, West, and East. A north–south dividing wall separated the West Sector from the Central Sector, while the official residence was built atop the bedrock heights and separated by function from the storage facilities in the East Sector. These sectors were divided into archaeological units (also referred to as *loci*) after surface clearance exposed walls and structures, and they were further divided by strata (sometimes referred to as *levels*) according to the different archaeological deposits and floors uncovered.[18] Strata numbers are indicated by parentheses—for example, (1) is level 1 in an area—and were not necessarily equivalent from one locus to another; for example, contemporary floor levels did not always have the same number.

The mapping of the site was established by creating a benchmark at the northwest corner of the fortification with a steel pipe set in concrete, and points I–VI were set out on an axis from west to east through the central part of the fortress. Points VII–XIV were established at various places and distances along the walls. The elevation was not fixed above sea level but was site specific and set at 3000 m.[19]

---

13 Knudstad 1964b, letter to Robert McCormick Adams, February 5.

14 Knudstad 1964c, letter to Sayed Thabit Hassan Thabit, June 9.

15 Knudstad 1966, pp. 183–84. Note also his comments in a letter to Sayed Thabit Hassan Thabit, dated May 25, 1964: "The upper end of the stairs lies well above brick paving in the gateway behind it; it may consequently be the renovation of an earlier stairs beneath" (Knudstad 1964b).

16 Knudstad 1964b, letter to Robert McCormick Adams, May 28. He was describing the final work at the fortress.

17 However, in his logbook for June 4, Knudstad (1963–64) said, "digging below L IV in western interior and the pottery . . . begins to look different."

18 Preface to Knudstad and Pierce 1964.

19 Knudstad 1964d.

3

The Fortifications

DETAILS OF CONSTRUCTION

As mentioned in the previous chapter, the length of the fortress, as measured from the exteriors of the east and west enclosure walls, was approximately 175 m, while the enclosed space was 136 m from east to west (plan 1). The interior dimension of the West Sector was about 50 m from north to south, narrowing to 25 m in the East Sector. The interior space measured approximately 3,754 m^2.

The development of the outer walls could not be satisfactorily studied without taking them apart, and this could not be done.[1] Knudstad noted that the repairs and modifications of the outer walls and the inner buildings seem to have been frequent but irregular and that the stratigraphic connections between them were often missing. Finally, the relinings of the outer fortification walls did not seem to fall into definite phases across the fortress. The River Stairs, he noted, belonged "in part if not wholly to later modifications of the north gate, which were removed [before being drawn] and are not shown on the plan."[2] The stairs rose to a height 60–70 cm above the brick paving of the gateway as a result of several modifications in the area, but this later material was not excavated to find an earlier stairway.

The archaeological phases of the fortress are more complicated than the designations Level I, II, III, and IV suggest, since a few areas within the more clearly defined architectural elements of Level III show phases of rebuilding, including modifications to the earlier Level IV architecture, indicating reuse of the area. The Level IV remains were fragmentary, and no phasing could be distinguished because this level was not completely excavated. These stages could not always be tied together from one area to another. The original phase designations of the excavators are kept throughout this book.

Unlike at the Middle Kingdom fortress of Serra East, where the bricks were gray, the Dorginarti bricks were yellowish or tan and of less uniform size, being larger (34–40 cm long).[3] They were laid in the first main wall with thick slabs of mortar, not neat, flat-pressed pads as in the Serra East brickwork. The mortar was not apparent in any walls except the initial curtain walls, where layers of reeds or reed matting were also incorporated, but these were not easily traceable and no pattern was noticed.[4] Wood timbers were rarely found, but when they were, they were noted as being larger in diameter than those at Serra East. They were seen only in the massive later repair of the northwest corner as ties between the wall and bastions, and one beam was also found arranged horizontally in the outer enclosure wall at test trench D 17.[5] Three wooden planks were also found on two walls at the east end, where they were laid flush against the walls.[6] A small amount of sandstone was used for stairs and paving of the West Gate and undoubtedly came from the pharaonic quarry opposite the site near Abd el-Gadir, as noted in the excavation diaries.[7]

The type of brick bonding used in the walls was not noted for all areas, but the photos show a mix of styles for the later reconstructions, with tilted rowlocks to

1 Knudstad 1966, p. 181.

2 Knudstad 1966, p. 181.

3 Knudstad 1963–64, pp. 56–57, 233–34.

4 Knudstad 1963–64, p. 57; Knudstad 1964a, p. 57.

5 The wooden beam was located 1 m above the wall base. See chapter 4, part 1, and the discussion of D 17 and the north wall breach.

6 Knudstad 1964a, p. 28.

7 Knudstad 1966, p. 184, citing the survey of site 5-O-6 published in Nordström 2014, pp. 121–23.

level the courses of brick (pl. 3a). Knudstad noted that the bricks were generally laid perpendicular to the length of the main walls, with alternating headers and stretchers at the face of the wall.[8]

The most ubiquitous brick size evident at Dorginarti measured 35–37 × 17–18 cm. Bricks of this size were used in the main enclosure walls and the early interior buildings, and all these constructions should be more or less contemporary.[9] A smaller size was 33–34 cm long. A combination of the standard and shorter brick sizes was used, for example, for construction along the north wall toward the North Gate and throughout portions of the south wall. The larger 38–40 × 18–20 cm bricks used in the thinner fortification relinings and other interior constructions were later and contemporary with one another. The plans and discussion throughout this volume note the brick sizes where they have been recorded.

The renovations incorporating the larger brick size included the mudbrick construction mass found atop the northwest corner battlements, which covered over the wall walk and the top of the bastion. This phase may have coincided with the reconstruction observed at the top of the crenelated bastion. A number of pottery sherds dating to the sixth century BC were found atop this bastion, although how the enclosure wall was used during Level II remains uncertain. Scatters of later sherds were found in very few locations, being concentrated around the Central Sector, East Sector, North Gate, and River Stairs.

Details of the larger plan are included in the discussion below, but plan 1 shows the entirety of the fortification walls and gates, and figure 3.1 shows the bastion identifications used in this discussion and the location of the three glacis exploration trenches. The locus numbers and descriptive units are listed in table A in appendix III.

FOUNDATIONS AND GLACIS CONSTRUCTION

Trenches excavated outside the walls on the north, southeast, and southwest showed that the stone embankment underneath and abutting these walls stood about 5 m above ground level in places where bedrock was not present.[10] The glacis and embankment sloped away steeply from the walls and bastions down to the sand or bedrock surface of the island (see fig. 3.6 below for the glacis trench profiles). Such stone foundations were usually used as a dry or damp course protection against groundwater and saturation, and this, rather than a defense against military attack, was their main function at Dorginarti.[11] Knudstad stated that "the form of the stone glacis beyond, and possibly surrounding, the fort is beyond recovery and must be surmised from what we know of it [from the] trenches cut down its face to the water-level."[12]

Cataract stones surrounded all sides of the fortress, but because the bedrock rises up in the Central Sector a deep layer of stones was unnecessary. The bedrock continued north of the fortification before it sloped down to the Nile, running alongside the west edge of the River Stairs and forming a protective wall. A smaller outcrop of bedrock was also present in the middle of the large, fragmentary building (bedrock building) on the interior and to the south of the West Gate.[13]

The prepared embankment for the fortification was also noticeable on the west exterior, where the stones thinned out toward the riverbank. This was the only location where the stone preparation for the walls ran under both of the main enclosure walls. Cataract stones were also used for constructing a mud-paved roadway beyond the West Gate and down to the riverbank, which is discussed below. More details concerning a test below the fortification's two main walls in the area at the north breach in the wall are also discussed below.

FOUNDATION ELEVATIONS FOR THE INTERIOR RELININGS AND EXTERIOR WALLS

A comparison of wall-base elevations for the interior fortress relinings show that the north wall relining sloped down gently from west to east, with the relining at D 1 being about 20 cm higher than its base level in room D 68. It then rose up near the north–south dividing wall between the West and Central Sectors to a level equivalent to that at D 1. This slight difference was due to the collapse of the wall through water permeation toward the north breach. The original topography of the ground in these areas, however, is

8 Knudstad 1964a, p. 57.

9 Knudstad 1963–64, p. 338.

10 Knudstad 1964c, letter to Sayed Thabit Hassan Thabit, May 25.

11 Glacis were usually built to prevent wall destruction from projectiles, to make it harder for siege machines to be set up near the fort, or to prevent enemies from attempting to build tunnels beneath the walls, but military reasons for glacis construction may have been secondary on a flood-prone island such as Dorginarti.

12 Knudstad 1964c, letter to Sayed Thabit Hassan Thabit, June 9.

13 Knudstad 1964a, p. 48.

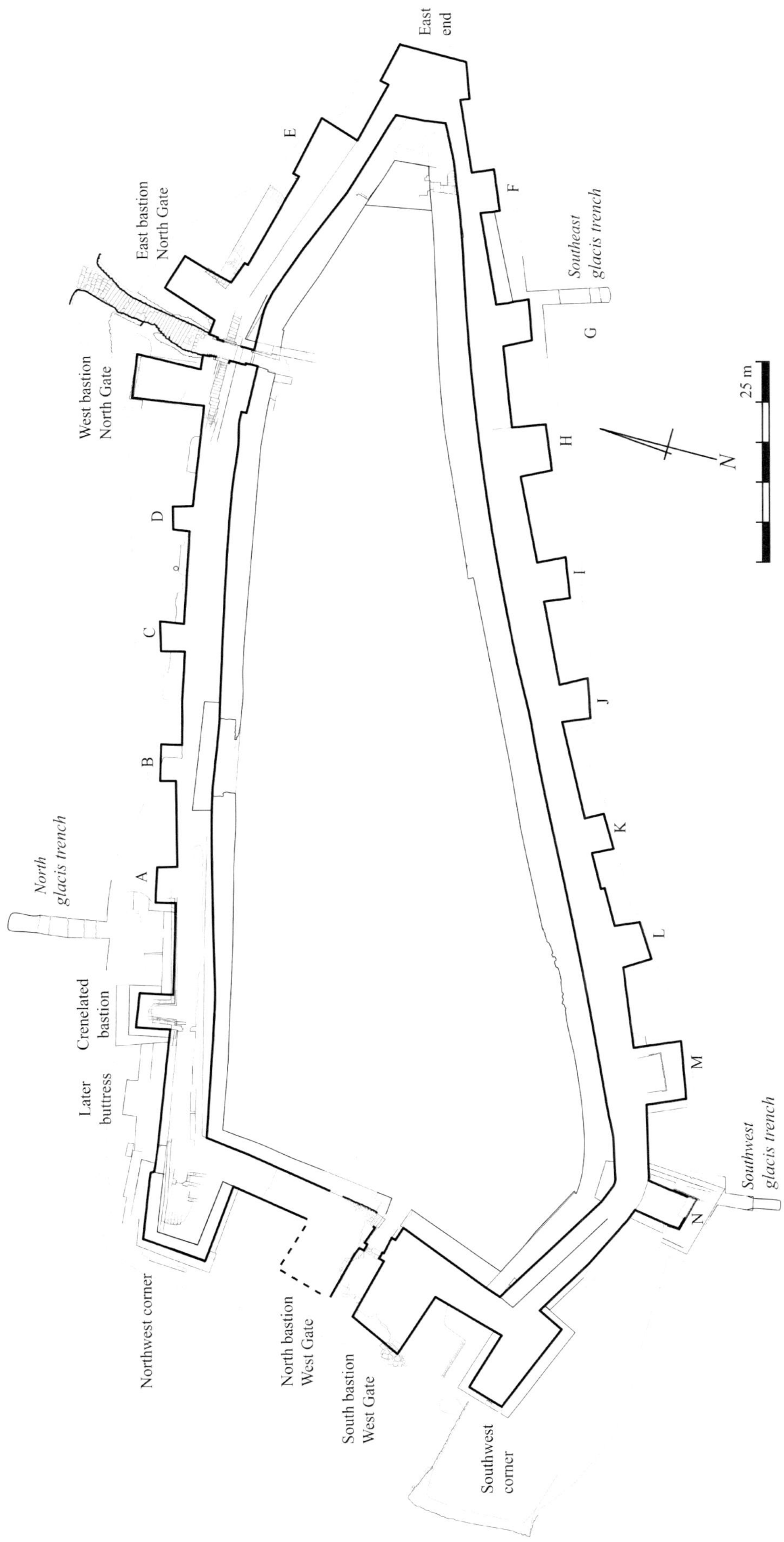

Figure 3.1. Enclosure walls with bastion identifications

unknown. As noted in chapter 2, the elevation was not fixed above sea level but was site specific, set at 3000.[14]

D 1 2523

D 6 2519

D 68 2503

D 45 2504 (east side); 2495 (west side)

Dividing wall 2525

The inner wall foundations along the south of the West Sector were not well preserved. One recorded level, within locus D 101, had an elevation of 2512.

The innermost main wall had a foundation elevation that was 13–22 cm lower than the third relining on the interior (2482 vs. 2495 and 2504).[15] The levels were recorded north of D 45 in the test excavation D 17, lying just west of the north wall breach, which explored the glacis and wall foundations. The exterior of the main fortification wall between bastions B and C had an elevation that was 51 cm above the original ground level of alluvium under it, with 0.5 m of glacis stones between the base and the ground level.

Innermost wall of two core walls 2482

Top of the alluvium under innermost wall 2468

Exterior wall of two core walls 2463

Alluvium below glacis under exterior of wall 2412

THE FORTIFICATION WALLS, BASTIONS, AND GATES

The wall reached up to 8 m thick in areas, including the two core walls, the thick interior wall or relining built after them, and the thinner relinings added later. They were preserved up to 5–6 m high in some areas, although generally the preservation was poor to nonexistent. The shifting of the wall bases and their foundations due to flooding and water saturation weakened them soon after the initial construction of the outer enclosure wall, and therefore it became necessary to repeatedly increase their tensile strength by thickening them, first by building the inner core wall and then by adding the third inner wall.[16] Thinner relinings were later added to the exterior and interior walls. Some walls still preserved their batter, with angles ranging 71–86 degrees for interior and exterior relining walls, which were noted by Knudstad as following the batter of the main walls.[17]

Knudstad reconstructed portions of the fortifications according to the outlines of glacis and stone packing, noting that the lines were accurate to about 10 cm.[18]

Repeated flooding after the fortress fell into disuse caused breaches in the main north, south, and west walls (pl. 3b). The architectural remains at the West Gate were almost at ground level due to this erosion (pl. 4a); cataract stones were piled in front of the gate and had been swept all the way onto the Central Sector platform (pl. 4b). The entire perimeter of the defenses was, however, traceable before excavation because the remains of the glacis and packing of cataract stones against the wall and bastion bases were still visible.

Relinings were added to both the interior and exterior sides of the earlier walls, but sometimes only in places where maintenance was necessary. Maintenance activity is particularly noticeable on the fortress walls lining the West Sector of the fortress, where floodwaters had eroded the central section clear of structures, and in the Central Sector.

The initial wall was built atop a foundation of cataract stones, and the solid rectangular bastions were then added to this wall.[19] A second wall was then built against the interior of the first wall, resting at the same level but with no stone underpinning, and since neither wall exhibited a plastered face it was supposed that they were built in close succession. The glacis and embankment stones were piled against the outer base of this double wall and its bastions.[20]

The two initial sections of the core wall, despite this protection, subsequently slumped after flooding had dampened and shifted the lower courses of brickwork soon after construction. A thick lamination was then added on the interior, which also subsequently collapsed.[21] The latest exterior relinings, and the later Level II fortress walls built atop the north and south

14 Knudstad 1964d.

15 Knudstad 1964a, p. 44 obverse.

16 Adding wall laminations (i.e., linings) was undoubtedly a common method for strengthening enclosure walls at other Egyptian fortifications. See, for example, Rzepka et al. 2011, pp. 143–46, and Rzepka et al. 2014, pp. 73–75.

17 Knudstad 1963–64, p. 57.

18 Knudstad 1964a, p. 53 reverse.

19 Knudstad 1966, p. 185.

20 See the "North Wall to the East" section below for a description of the excavation areas where these construction details were observed.

21 Knudstad 1963–64, p. 331. He notes that he was "fairly convinced that inner . . . lamination [i.e., relining] came not immediately after the other walls."

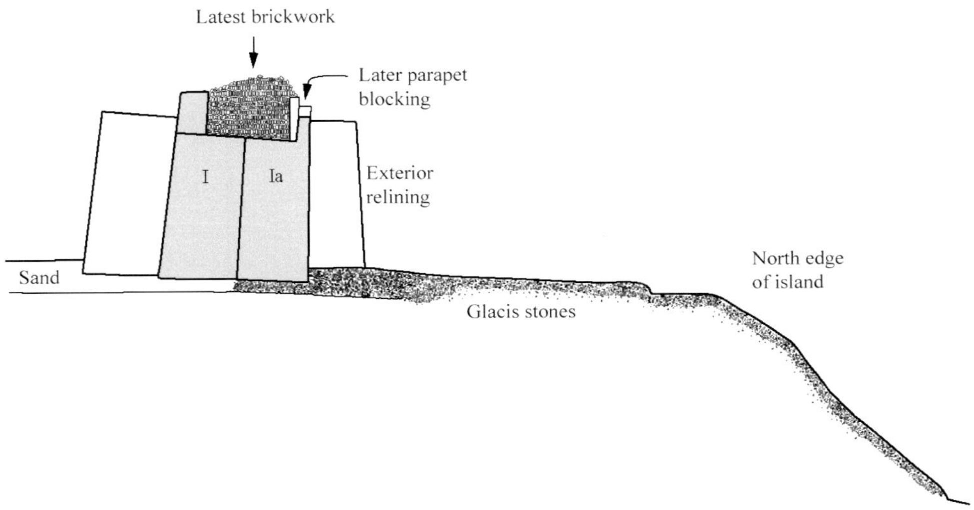

Figure 3.2. Typical section of the north wall, looking west (after James Knudstad), with the latest wall addition shown in white

fortress walls, did not exhibit the angle of the earlier walls (fig. 3.2).

Vertical joints, or seams, between the walls and relinings were sometimes traced longitudinally and were especially noticeable along sections of the south wall.[22] The bricks in the earlier wall constructions were the standard size, measuring 35–37 × 17–18 × 8–9 cm, while the thinner exterior and interior relinings, including rebuilt sections of the wall rampart at the northwest corner, featured bricks measuring 38–40 × 18–20 × 8–10 cm, or even a mix of two or three sizes.

The north core walls had at least two relinings on the exterior and one on the interior, but there were three relinings at the northwest bastion, while two additional relinings rested higher up along the west end of that bastion, representing repairs to the top of the wall.[23] The west wall had at least one relining on its interior. The south wall exhibited one lamination on the exterior and two or more on the interior, but the relining on the exterior was not always clearly traced. Three vertical sections of the main wall came together at the southwest corner, although only the outermost core wall was bonded to the west wall.[24]

The two walls comprising the core walls were for the most part unplastered on both faces, but the third lamination on the interior had a plastered face that was preserved in a number of areas.[25]

Ten horizontal beam holes for timber bonding were found on the north interior wall of the northwest corner where heavy repairs had been undertaken late in the life of the fortress, but the wood no longer survived.[26] The beams were estimated to have been 15 cm in diameter and ran at angles northeast and northwest into the enclosure wall and its relining. The excavators also uncovered evidence of the use of timber to reinforce the interior of the fortress wall in the East Sector, where three vertical planks were laid flat and flush with the face of the walls before the walls were relined.[27]

Fifteen rectangular bastions of varying sizes were added at some point to the outermost fortification wall, two longer bastions framing the North Gate and two sets of double bastions at the southwest and northwest corners flanking the West Gate.[28] It is unclear whether the bastions were bonded consistently into the outer main wall. Each of the flanked gates was fairly simple, with a direct entrance into the fortress through a passageway between the side bastions. The passageway widened toward the exterior, and each bastion had a side stairway affording access to the top of the wall. A trapezoidal bastion capped the east end. A shallower support buttress was eventually added between the crenelated bastion and the northwest corner bastion during a later reinforcement of the enclosure walls at

22 Knudstad 1964a, p. 55.

23 Knudstad 1963–64, pp. 232–33.

24 Knudstad 1963–64, p. 233; Knudstad 1964a, p. 50.

25 This information was taken from Knudstad 1964a. Only the inner main wall at the east end had a plastered face before it was covered over by the addition of the third, inner wall.

26 Knudstad 1966, p. 185; Knudstad 1964a, p. 57. The beam holes are visible on pl. 23b.

27 Knudstad 1964a, p. 28.

28 There were battlements atop the wall and bastion; however, Knudstad called them *buttresses*, which is a term used in earlier publications on Dorginarti. Note Keeley, Fontana, and Quick 2007, especially pp. 67–68.

the northwest corner. The bastions have been given a letter or phrase designation to simplify the discussion (fig. 3.1).

Northwest Corner Details

The complexity of the renovations is apparent in the agglomeration of walls exposed at the northwest corner of the fort, where floodwaters had the greatest impact and caused the most damage. Some of these relinings may have been built in a single construction program or at least were not separated by much time, but the time and resources spent in shoring up the fortification along this stretch of wall was considerable.

For example, the relinings around the crenelated bastion were contiguous and abutted each other on the west of the bastion (pl. 5a). A thick relining was built at a later stage, with a shallow buttress inserted halfway between the crenelated bastion and the northwest corner to bolster the new wall addition and to protect the larger bastion nearby. The east face of the bastion was given a thick lamination that also abutted an east–west relining of the fortification wall in this area (pl. 5b). A view of the area at the bastion also shows the wall layers from the northwest (pl. 6a).[29] The two thicker wall layers at the head of the bastion may be seen at the front, and the two walls lining the west face are clearly visible (the inner lining had split but was one wall). This bastion and its architectural details will be discussed below.

The exterior of the northwest corner bastion showed at least two to three relinings around its south, west, and north faces (pl. 6b).[30] The latest relinings utilized the larger brick size, while the main walls and the bastion itself were constructed of bricks that were 36–37 cm in length. It was at this location that evidence for ten wooden beams was preserved, laid into the wall horizontally to bind the relinings to the main walls. It was also at this spot that evidence of slabs of mortar and matting or reed layers was found, but neither could be traced further. The north side was a jumble of linings, with brick faces that did not always line up with each other, and here the rebuilding used bricks that were 33–35 cm long. The shallow buttress built later between the corner and the crenelated bastion remained standing 67 cm above its foundation course (elevation 2490).

The excavation of the northwest wall area took time because the excavation team confronted a later brick construction mound that filled in the top of the wall (see figs. 3.2–3.3). Its purpose is unknown. The earlier construction details along this northwest stretch of wall, however, were preserved because of this brick encasement.[31]

The preserved wall walk atop this area extended from the northwest corner bastion eastward beyond bastion A. It had been paved with bricks. Evidence of plastering was occasionally preserved. Fragments of waist-high parapets were found along the exterior side of the wall, atop the northwest corner bastion and along part of the walkway to the crenelated bastion, atop this bastion, and in a small section at the west side of bastion A. Evidence of either a wall walk or parapet walls was lacking beyond this stretch, or anywhere else atop the fortifications.

The original pavement of the northwest corner bastion was at elevation 2902 (fig. 3.3). The floor that was found east of here was at 2893, the paving just east of the crenelated bastion was at 2901, and the easternmost preserved paving by bastion A was at 2899. However, the flooring found within the confines of the crenelated bastion was at elevation 2869, or quite a bit lower than the other floor levels of the original(?) wall walk. The higher levels may indicate later renovations to the wall pavements.

At the extension out to the northwest corner bastion, and at the juncture of the north and west wall lines, two sets of brick doorjambs were set at right angles to each other.[32] One set of brick jambs framed a staircase leading down to the plastered brick paving of the northwest corner bastion. This was the earlier construction of the two stairways. The topmost step (elevation 2959) was 57 cm above the brick paving atop the northwest corner bastion (2902). The later brick construction to the south filled in this stairway. The other stairway or ramp led up, reusing the earlier construction for its west side. A floor surface to the east (2964) was associated with this later stairway, the first step being 27 cm above that floor. The top of the unexcavated brick mass to the west was only 5 cm lower than the floor area to the west of the stairway. Reconstruction of the area was perhaps undertaken to heighten the bastion and to provide access to its top via a stairway.

Neighboring this area, along the south face of the fortification wall, were the remains of a waist-high plastered gap set into the wall of the parapet near the

29 The fortified aspect of this particular bastion is discussed below.

30 Knudstad 1964a, pp. 56–57.

31 Knudstad 1966, p. 185.

32 Knudstad 1964a, p. 61.

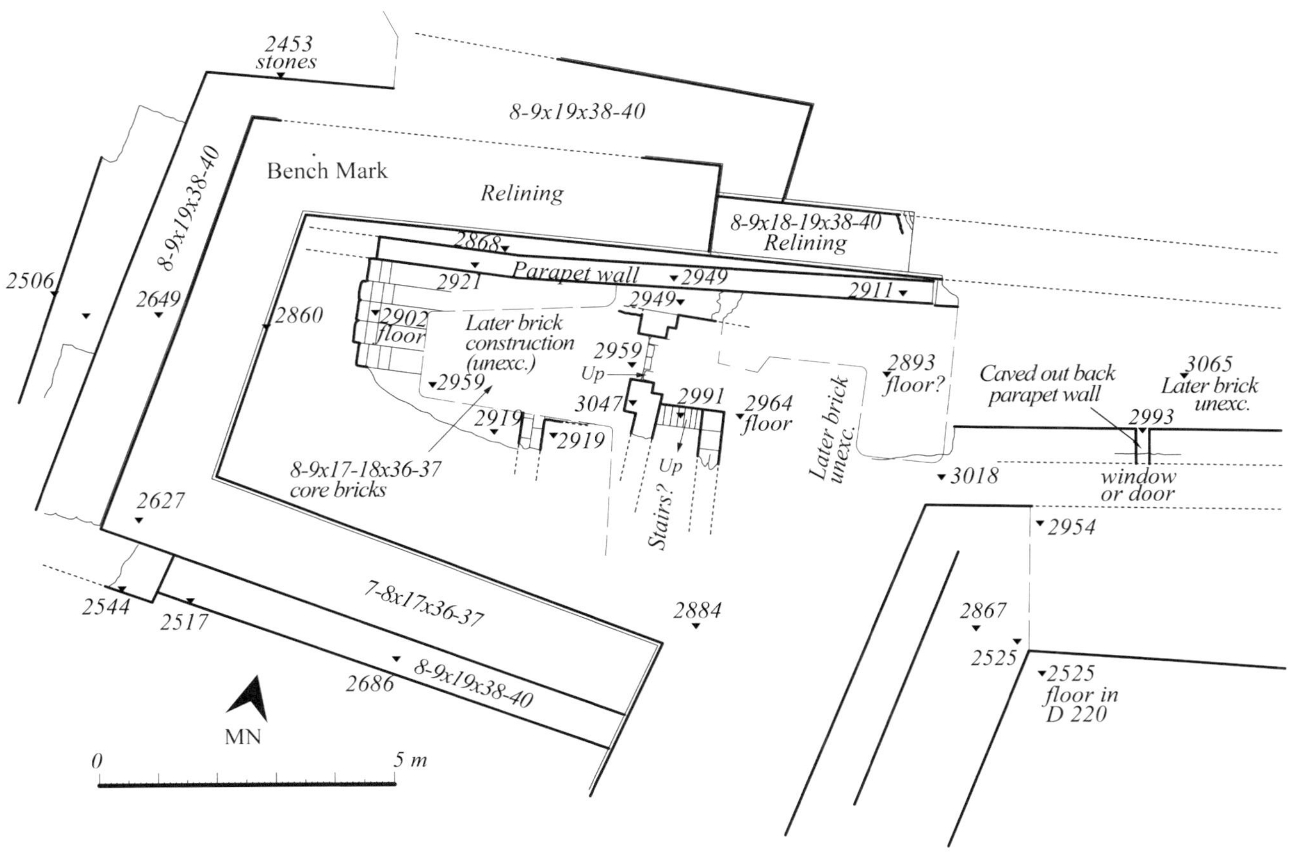

Figure 3.3. Northwest corner bastion and wall walk

corner of the west and north enclosure walls.[33] The construction and the parapet were built at the south edge of the inner wall of the fortification, with the third relining wall to its south. The area was also covered over by later brick construction, some of which was not excavated. The parapet was caved in at this point and the remains were not well preserved, but an opening 63 cm wide between the parapet walls was found. An elevation of 2993 was recorded atop the uppermost brick within the opening. The nearby pavement of the walkway was 92 cm below this sill. Its precise function remains uncertain.

Crenelated Bastion

The heavy repairs and overlying brick had preserved even more stretches of battlement atop the earliest two walls. The crenelated bastion had four semicircular mudbrick crenelations preserved, with a brick laid flat between each raised section (pl. 7a). The individual merlons were preserved on both the east and west sides of the bastion, one complete and one fragmentary. A possible fifth, fragmentary merlon was found atop the north side parapet outside and to the east of the bastion (pl. 7b). The merlons measured 60–67 cm long and 37 cm wide, with 25–35 cm spaces between them.[34] They were preserved in part, with the tops standing 1.04–1.11 m above the bastion floor elevation, so they would not have offered much protection for a person standing upright behind them.[35] The specimen at the southwest juncture with the main wall was not entirely semicircular but rather was flattened at its south edge.[36]

33 Knudstad 1964a, p. 60. Knudstad suggested that the opening may have been a window or narrow door.

34 Knudstad 1964a, p. 63.

35 This is assuming that an archer, for example, was shooting from an upright stance and not from a kneeling or half-kneeling position.

36 Rounded merlons have a long tradition in Egypt; see the tower and battlement parallels discussed by Hölscher (1910, pp. 56–64), who describes Egyptian and Levantine parallels to the rounded merlons of the fortified temple and palace at Medinet Habu.

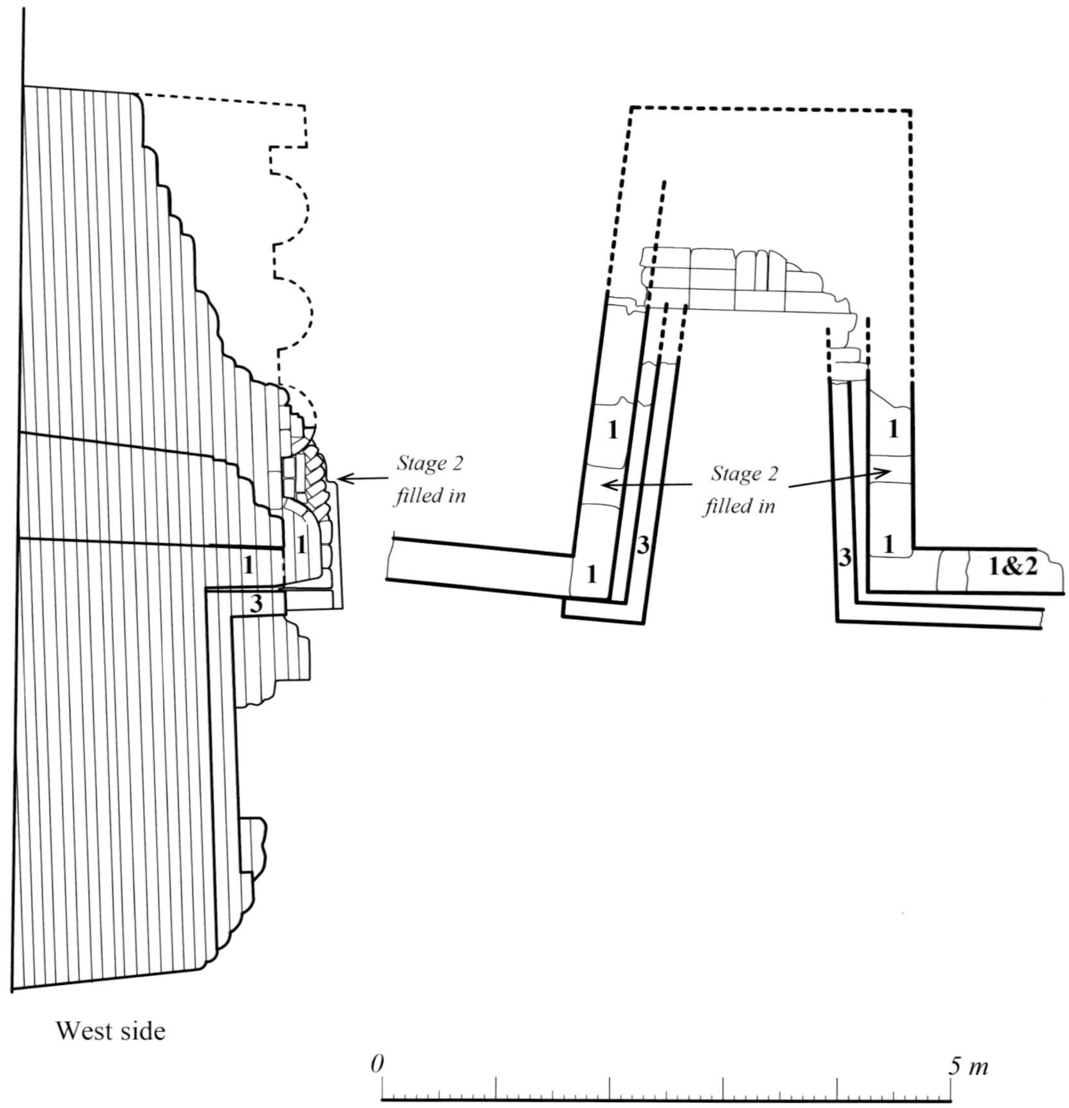

Figure 3.4. Renovations to the crenelated bastion: *1*, original construction with plastered surfaces; *2*, addition to wall atop crenelations, perhaps two plastered finish layers; *3*, later relining inside parapet wall, battered and plastered on exterior only

There was evidence of the area's having undergone three renovations (fig. 3.4).[37] Additions to the wall, the filling in of the crenels, and the covering of the merlons occurred in the second stage of reconstruction, with possibly two plastering phases evident on the interior wall surfaces. A third phase consisted of parapet wall relinings, which extended over the height of the preserved crenelations (pl. 8a). These relinings were battered slightly and had been plastered on their outer surfaces. Some of these reconfigurations are also discussed above under the details for the northwest corner.

The original construction and the later renovations incorporated bricks measuring 36–37 × 17–18 × 8–9 cm, while the exterior wall relinings in the area used the larger, 40 cm long bricks.

A doorway on the wall walk to the west of the crenelated bastion was added at a later date, although it was built before the parapet relinings were done (fig. 3.5).[38] The brick sizes were not noted. All constructions here, including the doorway, had been covered over by the later brick that filled in the top of the wall.

North Wall to the East

A 1.5 m relining ran east along the main wall from the crenelated bastion and over to the next bastion, abutting the west relining of bastion A. The east section of

37 Knudstad 1966, p. 185; Knudstad 1964a, p. 57.

38 Knudstad 1964a, p. 63.

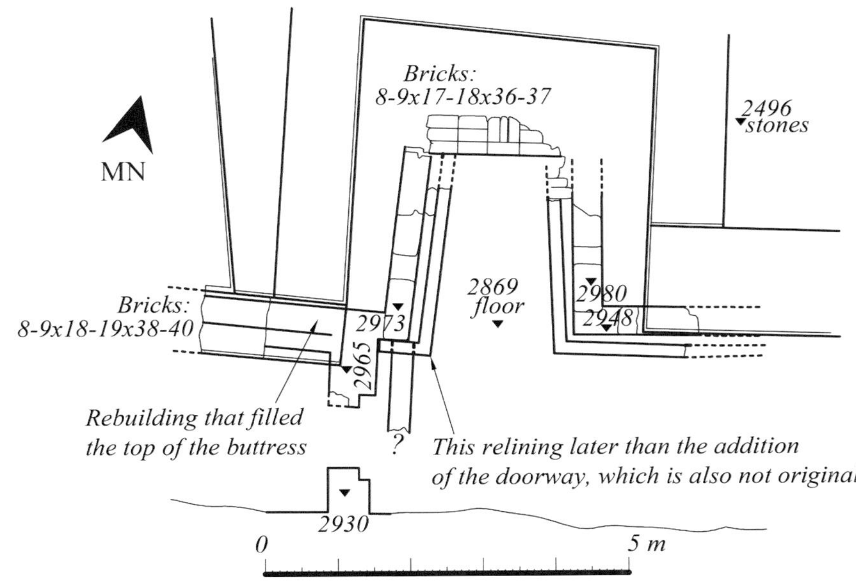

Figure 3.5. Crenelated bastion and later doorway addition

the bastion had slumped to the north.[39] The relinings consisted of larger bricks measuring 38–40 × 19 cm. Atop the wall, the north parapet wall extended over to this bastion before terminating, with the wall-walk paving ending just to the east of the same bastion.[40] The bastion construction incorporated bricks that were 36–38 × 18–20 cm long, some of them on the larger end of brick sizes at the fortress and perhaps part of a later rebuilding. The height at the collapsed north edge of the bastion was preserved 2.61 m below the wall-walk paving to its south (2899), but it was still 1.32 m above the glacis surface on its east.

The excavators undertook an examination of the glacis layer underpinning the outer fortification wall between the crenelated bastion and bastion A. The area along the entire north wall had been mostly excavated 1.5–5.0 m away from the walls and bastions, except at the River Stairs and the north glacis trench (pl. 8b). The glacis trench extended out about 17 m from the nearby bastions (fig. 3.6b). The location of this trench in relation to the fortress walls is shown in figure 3.1 and on plan 1. More than 5 m of alluvial deposits lay atop the glacis stones on the north. The top edge of the slope was 5.66 m above the base of the trench, which never reached the bottom of the glacis emplacement.

The top part of the north glacis trench extended over to the base of the wall and its exterior relining, uncovering a glacis stone layer 65 cm in depth and resting on earth at elevation 2432. The base of the outer fortification wall rested on the stone foundation at elevation 2462, with stones piled up against it. The relining wall, constructed of bricks of the larger size, lay atop the cataract stones.

To the west of the trench was a wall fragment 1.8 m long, preserved at only 5 cm in height. The remains of a packed-earth or brick floor(?) rested 15 cm above the glacis stones.

Bastion B was located at the north breach and had lost its east side to flood erosion, although the lines of glacis stones defining the edges remained in place. Details of the construction methods of the enclosure walls were discovered when two small areas were explored near the north breach to the east of this bastion. The excavators sank test pits on both the interior and the exterior of the outer wall still remaining at the north breach.[41] Their examination revealed that the inner edge of the stone glacis foundation of the outer wall test pit did not continue under the second main wall to the south, which rested on sand. The layer of glacis stones for the outer wall foundation rested on earth 51 cm below the base of the wall—the wall being founded at elevation 2463—but the preserved top of the glacis indicated a total glacis depth of 81 cm on the exterior side. The glacis had been laid down as a foundation for the outermost wall and then was piled up against the exterior of the wall after it was built.

The base of the same wall in the inner wall test pit was founded at elevation 2496, or 33 cm higher than its exterior side.[42] It was at this limit that the glacis stones

39 Knudstad 1964a, p. 58.

40 Knudstad 1964a, p. 60.

41 Knudstad 1964a, pp. 44, 59; Knudstad 1963–64, p. 305.

42 The inner relining of the main enclosure was founded at 2504 on the east side and at 2495 on the west. The top of the

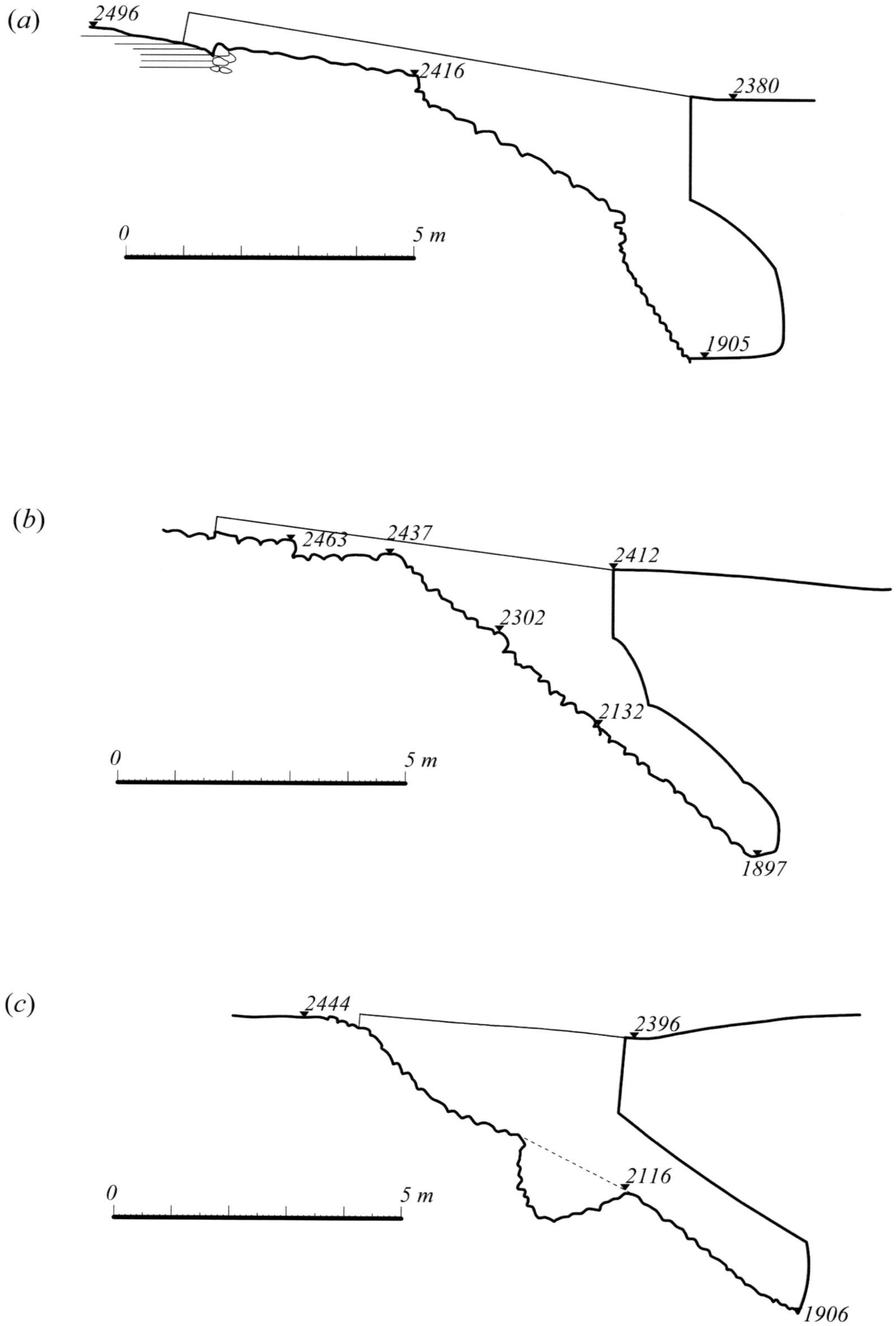

Figure 3.6. Glacis trench profiles

(*a*) Southeast glacis trench at bastion G

(*b*) North glacis trench between crenelated bastion and bastion A

(*c*) Southwest glacis trench at bastion N

appeared to stop.[43] The innermost enclosure wall was founded on sand at 2497. The adjoining wall faces were unplastered, although plaster was found on the interior side of the inner wall at the north breach. The bricks employed for the inner curtain wall construction were 35–36 cm long, while those for the outer wall were mostly 36–37 cm in length but were a mix of the two sizes to the east of the breach.

The outer wall test pit revealed a thin layer of ash running under the fortress wall (4 cm), soil and ashy layers (40 cm, including a 23 cm gray-black deposit at the bottom), and then burned alluvium (6 cm) sitting atop an alluvial deposit.[44] In the topmost layers of the test pit a large wooden beam, approximately 15 cm thick, was found to have been built into the outer core wall.[45]

Bastion B was preserved at a height of 58 cm above its foundation elevation of 2457. Part of the relining between bastions B and C had collapsed, but Knudstad speculated that the remains might also have represented slumped wall bricks.[46] The top elevation of bastion C was 21 cm above the glacis surface to its north. Once again, evidence was uncovered of a later relining between it and bastion D, although this relining was only about 1 m thick on the east. The lining had been cut through and the area reconfigured at some point. An oven was preserved in the middle of the bay.[47] Bastion D remained standing 43 cm above the cataract glacis below it.

Moving eastward along the north wall, one would have encountered the west bastion of the North Gate, which extended farther north to provide protection for the River Stairs (pl. 9). The bastion stood 62 cm high at its north edge and was not well preserved on the north, but the lines of stone packing around its walls have allowed its reconstruction.[48] Two levels of cataract stone were found piled against the bastion in this location, with a level of intentional(?) mud fill between them.[49]

Western Wall Stairway at the North Gate

The outermost main wall segment at this spot contained a stairway with eighteen steps leading up into it at a slight angle (fig. 3.7; pl. 10a).[50] The topmost part was not well preserved. The bottom bricks were flush with the brick-paved passageway of the North Gate and are included in the step total. The entrance to the staircase was about 1.6 m wide before it narrowed to 1.0 m or less. The risers were uniformly one brick width high, with six or seven bricks laid lengthwise on end to form each step. The steps were 32–53 cm deep over the entire length of the staircase. Twelve steps from the bottom, the stairway took a jog to the left and then to the right before tapering to 62 cm in width. The topmost preserved brick was at elevation 2838, although the area looks like the underpinnings of more stairs rather than a step. The stairs rose 3.72 m from the entrance at the bottom to the topmost bricks that had remained in place. There were traces of wall plastering at the top of the staircase.

Along both the bottom and top parts of the stairway were narrow strips along one side or the other, either to the right (bottom part) or to the left (top part). These strips were 20 cm wide, formed of headers laid flat, their surfaces being more or less level with the stairs.

The wall walk to the west of this area at bastion A was at elevation 2899, or some 61 cm higher than the level of the topmost preserved brick. If each stair tread was approximately 20 cm above the previous one, which was the case between the eleventh and fourteenth treads, one can assume that the top of the wall would have been reached after another three steps.

Eastern Wall Stairway at the North Gate

The eastern stairway entered the main wall at an angle and penetrated both the inner and outer walls of the fortification (fig. 3.8). This stairway was opposite the western staircase and may have been part of the same construction phase. It was preserved to a height of eight stairs, varying from 35 to 65 cm in depth and laid in the same fashion as those in the western stairwell.[51] The risers were much shallower than those of

alluvium below this section was recorded at elevation 2468 toward the interior and 2440 at the exterior side of the inner relining. See the discussion of test trench D 17 in chapter 4, part 1.

43 Knudstad 1964a, p. 59.

44 The ceramics stemming from deposits under the wall included a thick-walled bowl with a rounded rim top and a silt jar with a slightly everted rim and red rim band. They are not closely datable, although the silt jar is not a regular type in the early Level IV pottery assemblage.

45 Knudstad 1964a, p. 59.

46 Knudstad 1964a, p. 59, where he questions whether the bricks belonged to a relining or perhaps were part of the wall.

47 Knudstad 1964a, p. 35.

48 Knudstad 1964a, p. 35 reverse.

49 Knudstad 1963–64, p. 312.

50 Knudstad 1963–64, p. 34.

51 The reading of 2489 at the staircase base was not indicated as a floor and may thus be an excavated level; see Knudstad

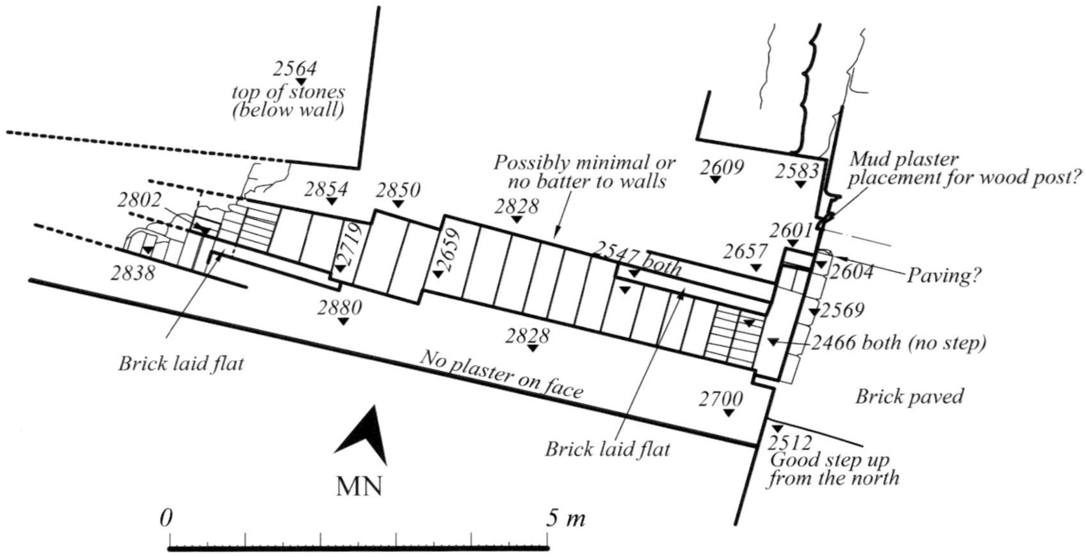

Figure 3.7. Western wall stairs at the North Gate

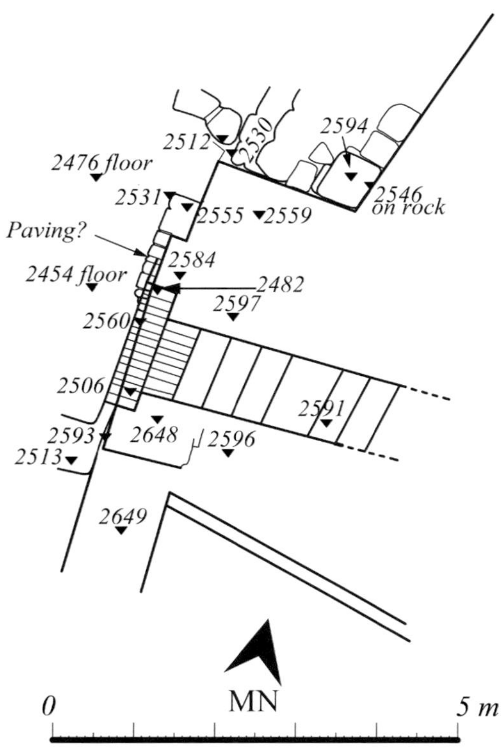

Figure 3.8. Eastern wall stairs at the North Gate

the western stairs. The initial tread was 1.6 m wide and was preserved unevenly, with the south elevation being at 2506 and the north some 24 cm lower, resting atop a level of brick underpinning the step. Plate 10b shows that the original bricks of the lower steps were very eroded and uneven, perhaps due to the later renovations of the North Gate. The staircase was 1.6 m wide at its entrance before narrowing to 1.0 m.

North Gate and Passageway

The passageway through the North Gate was slightly narrower at its south end, at least in its original plan, with the south entrance being less than 1.5 m wide but broadening to about 2.0 m on the north end (fig. 3.9).

Brick paving, set atop cataract stones, overlay the passageway and was part of the original layout in the area. The 1 m wide walls extending into the fortress on the east and west sides were founded 2–17 cm below the brick paving and were contemporary with it. The floor started at an elevation of 2601 at the entrance from the East Sector and steadily sloped down to the original level of the River Stairs, reaching 2489 by the opposing wall stairways and 2476 just to their north. The height atop the first stone step, from a later reconstruction of the River Stairs, was 2539, or 63 cm above the earlier floor of the passage (pl. 11a).

There was a cut 25 cm deep in the paving south of the side stairways for a sill set atop the stone foundations, although the stone or wood sill itself was absent.[52] A door may have stood directly north of the side stairways, since a mud-plastered emplacement, perhaps intended to hold a wooden post, was found directly north of the western stairs.[53] Two low steps—or perhaps only sharp slopes worn in the passageway—were situated between the sill and the wall staircases,

1964a, p. 33.

52 Knudstad 1964a, p. 31.

53 Knudstad 1964a, p. 34.

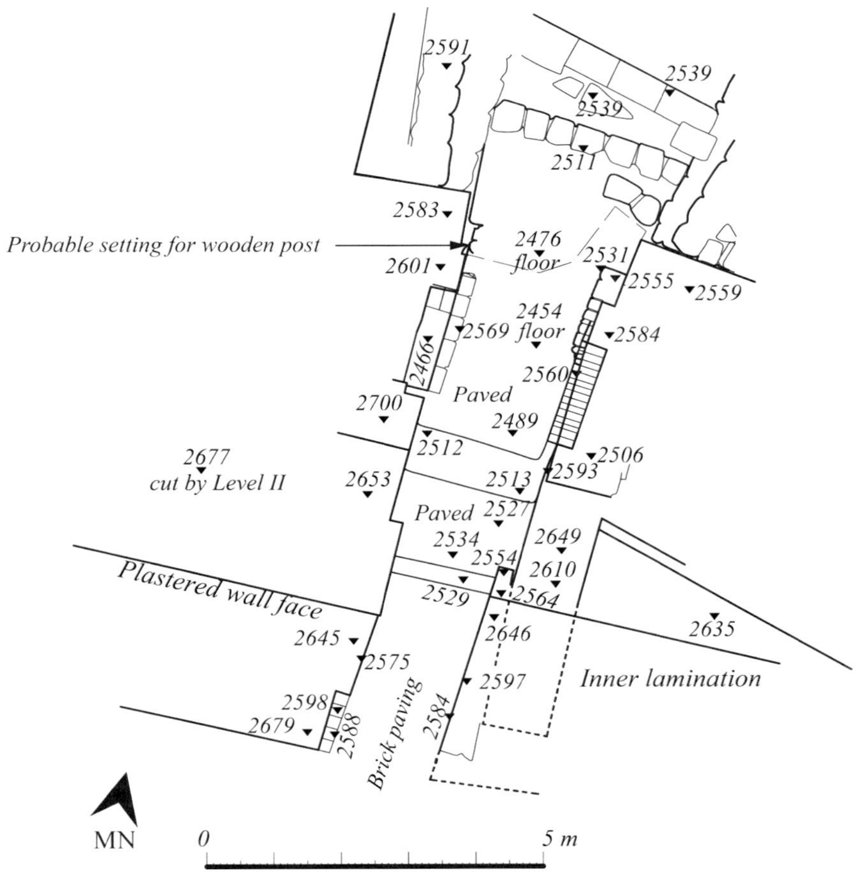

Figure 3.9. North Gate passageway

with the southernmost step 14 cm higher than the middle level of 2513, and the last step 24 cm above the lower level by the wall stairwells.

All original bricks used to construct this passageway and the staircases were 35–37 cm long, with bricks 39 cm long occasionally mixed in as a reminder of the later reconstruction of the area (see below). A later packed-mud floor, preserved only by the entrance from the interior into the North Gate passage, was 5 cm above the brick paving of the passageway.

The rebuilding of the area within and next to the North Gate took place in at least three phases, as seen on plate 11a looking at the east side of the gate passageway. However, not all the relevant details were recorded before their removal. An interior relining wall was perhaps added to the double enclosure wall to the east of the gate.[54] It had been built with a slight inward angle to meet the gate area, and the change of angle left a triangular gap between the two wall faces, which were neither plastered nor finished. Another wall end next to the gate rested atop the lower wall relining at a level 13 cm above the base of the earlier wall (2596 vs. 2584). This was perhaps a renovation of the earlier wall face along the gate passageway.

The east screen wall projecting out into the East Sector was built on an elevation of 2584, which matches those of the inner wall relining. It is uncertain whether the screen wall cut through the higher east–west wall or was cut through by that wall, but in any case the screen wall continues on to meet the inner enclosure wall next to the triangular gap and may have been bonded with it. The wall of a later silo in the East Sector lay atop this screen wall, and the south extension of the wall was dismantled at this later date.

The west entrance to the North Gate also had a shorter screen wall extending about 2 m into the interior. It abutted or was cut by the inner relining, and its south end was finished. The thinner relining wall on the north side of the Central Sector, built of later 40 × 19 cm bricks, was not preserved on the west screen wall. An oven and a hearth were located here after the stub wall was built, being founded 11–17 cm higher, and may have been covered over by the later

54 Knudstad 1964a, p. 31, with a question mark next to where the relining may have been either cut through by the side wall or built over.

relining of the Central Sector.[55] The comparable third inner relining wall at the west side of the North Gate was founded at 2575 at a point along the gate passageway, with the screen wall based higher up at 2599 and 2601.

Later Modifications of the North Gate

The North Gate and River Stairs were renovated late in the life of the fortress, or perhaps were renovated more than once. The later modification of the area included new walls along the passageway and the heightening of the River Stairs, which covered an earlier stairway or its foundations (pl. 11b). The height at the top of this later set of steps was 60–70 cm above the excavated passageway surface. The remodeling may have occurred during Level III.[56] The Level II use of the area seems to be indicated by the existence of a path atop the debris in the passageway. The later stairway was certainly in use at that time, since many East Greek and Levantine amphorae were found smashed atop the stone steps. There was no time to complete the excavation of the area to determine the plan of the earlier River Stairs that were contemporary with the passageway surface.[57]

Both side stairways were filled in with bricks during a later reconstruction of the North Gate, perhaps during the renovation just described. The bricks that remain as evidence for the latest reconfigurations of the passageway were found on both sides of the passageway near the blocked stairways.[58]

The renovations of the entire North Gate area were fairly extensive, since the passageway and stairs served as access to the river throughout the history of the fortress. However, no firm conclusions can be drawn about its precise appearance through time or how many times it was refurbished.

River Stairs

The River Stairs excavated in 1964—which, as mentioned above, represented a later phase of the North Gate—rose to a height of 60–70 cm above the brick paving of the earlier passageway (fig. 3.10).[59] Thirty-four stone treads led down to the river, reaching more than 15 m in length and descending some 6.5 m in elevation. The stairs were constructed of sandstone blocks or slabs, undoubtedly from the quarry identified on the opposite bank just west of Abd el-Qadir (pl. 12).[60]

The River Stairs were protected for most of their length by two flanking bastions and by walls formed of glacis stones and bedrock (pl. 13a). The top of the glacis on the east was 2.11 m above the bottom step slightly to its north, and the bedrock on the west stood 1.64 m above the same step.[61] Below the twenty-third step from the top, the west side had been cut into the bedrock layer that extended down to the sand along the riverbank. The bottommost step was below the water level on June 20, 1964, but no other stones were found below it.

Walls at the North Gate

The outer core wall east of the North Gate measured 2.4–2.7 m in thickness, while the inner wall was about 2.1 m thick. There was no evidence of plaster on their adjoining sides, and there was probably none on the inner wall face.[62] The interior wall of the double enclosure, or the middle wall behind the inner relining, became the only wall at the east end of the fort. There, a plastered wall face was found.[63]

The east bastion of the North Gate stretched northward almost 7 m, with its north face determined by the remaining piles of glacis rock. It was unclear to the excavators whether the brick remains to the east of this bastion, standing only three to four courses high, represented the bottom courses of a relining or a pavement over the glacis rock. It had been either eroded away or cut through at its east edge.

The walls and relinings on the east and west sides of the North Gate do not precisely line up because the walls curved around to follow the planned outline of the fortress and the topography.

East End

Due to water saturation, the fortress wall at the east end had been largely reduced to a powdery consistency described as "featureless clay" (pl. 13b).[64] But the line of the wall and bastion E were apparent from the remaining rock piled up against the outer faces. The inner segment of the core wall was the only enclosure wall

55 Knudstad 1964a, p. 31.

56 Knudstad 1964a, p. 32.

57 Knudstad 1963–64, pp. 334–35.

58 Knudstad 1964a, p. 32, where he questions whether this later renovation was Level III.

59 Knudstad 1966, p. 181.

60 Knudstad 1966, p. 184, referring to Site 5-O-6 in Nordström 2014, p. 12, table I.

61 Knudstad 1964a, p. 35.

62 Knudstad 1964a, pp. 31, 33.

63 Knudstad 1964a, p. 27.

64 Knudstad 1964a, p. 27.

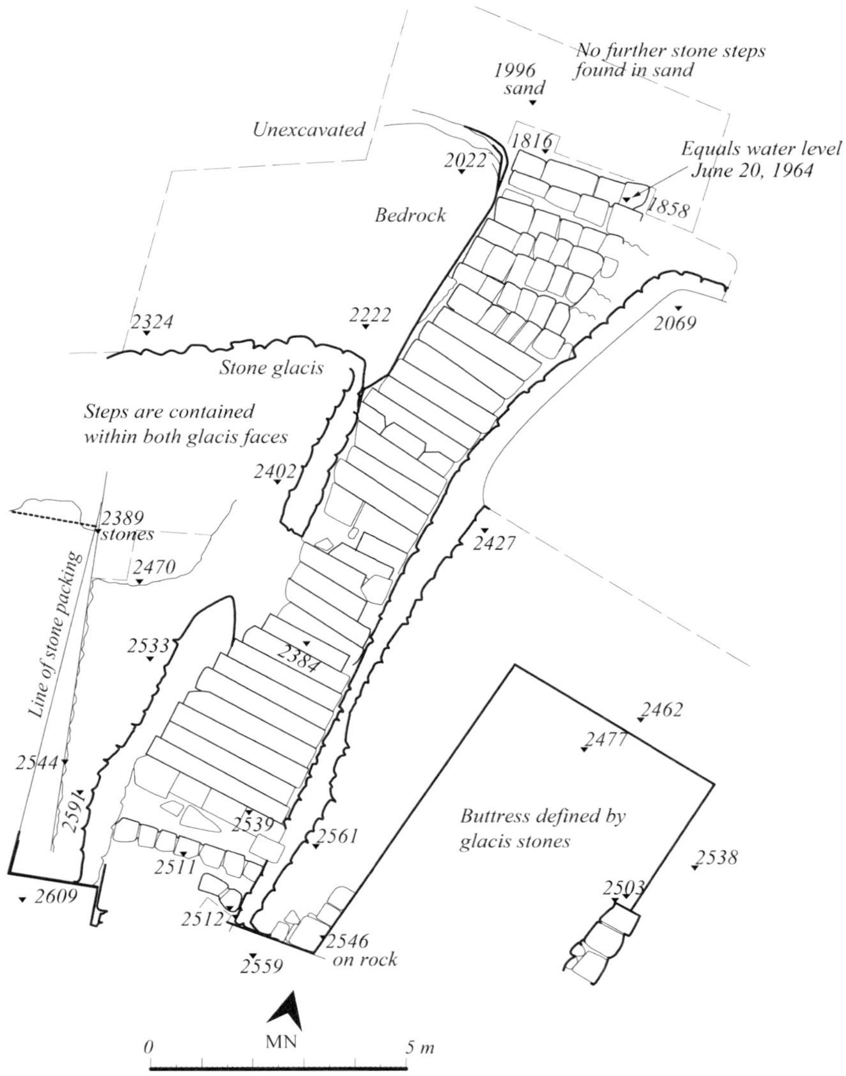

Figure 3.10. River Stairs

for the interior at the easternmost end. It was plastered on its inner face and, in this area, rested on stone foundations. The east end of the fortress was capped by a flattened diagonal bastion and observational platform, the top of which extended 8 m to the west.

The empty space at the east end of the fortress was originally, or at some later time, filled in with brickbat and a brick platform.[65] A plastered finish on the interior of the north and east walls suggests that this space had been functional before it was blocked up. Whether part of the original design or a later renovation, a viewing platform in this location would have allowed an advantageous view looking toward the west bank and downstream.

More evidence for the use of wood in the construction of the fortification walls was found at the interior of the east end in the area just discussed.[66] In this case, the wood in question was not timbers placed as ties into the wall but rather planks that lay vertically against, and flush with, the wall face. There were two of these planks on the interior of a wall on the east, and another plank was found along the north wall edge at a spot where a relining abutted it on the west.

Southern Wall Stairs

Additional information regarding access to and use of the wall ramparts came to light at the east end of the fortress, to the north of bastion F. A stairway was discovered here that led up into the wall from the east end rooms, D 33 and D 34, which are both discussed

65 The drawings note “brick on brickbat fill.”

66 Knudstad 1964a, p. 28.

in chapter 4, part 2 (pl. 14a).[67] About seven steps were traced going up in a southerly direction into the original wall. The stairway then turned eastward, where another five steps ran up along the plastered face of the outermost main wall. The top three steps had collapsed, and no additional steps were found. The stairway had been overbuilt or filled in, and its west edge was untraceable.

The depth of each step was between 35 and 37 cm, and the recorded heights of the risers were between 23 and 32 cm. As in the gateway stairways, it appeared that the stairs had been constructed of bricks laid on their long ends upon rowlocks to level them. The course of the stairway was marked at its entrance by brick jambs.

As mentioned above, the mudbrick in this section of the fortress was decomposed and therefore difficult to excavate, so no identifiable architectural features were found atop the walls in this area. But the stairway was filled in during a later rebuilding phase, which probably coincided with construction of the wall that closed off D 33 and D 34 at the easternmost end of the fortress.

South Wall

The double-walled enclosure was also present on the south side and was similarly relined with a thick interior addition, the inner surface of which was covered with plaster. The wall heights in this area were recorded as 2937 (interior edge 2966) by bastion N, 2992 by bastion M, 2843 by bastion L, 2697 by bastion K, and then rising up to a height of 2883 by bastions I and J (fig. 3.1; plan 1). The top of the wall dropped to the east and then rose 1.86 m again beyond the breaches and toward the Central Sector.[68]

A relining of the wall between bastions F and G rested on the glacis stones packed against the main wall, but it is uncertain what brick sizes were used to construct the relining. Nothing unusual was found at the bastions to the south of the Central Sector, both poorly preserved but with glacis stone outlines (pl. 14b).

The southeast glacis trench (fig. 3.6a) was excavated at the east corner of bastion G, with the elevation atop the glacis at 2496 and reaching bottom at an elevation of 1905. The trench went below the water level, although the stone embankment continued farther down and was not reached.[69] The location of this trench in relationship to the fortress is shown in figure 3.1.

Bastion H was 77 cm lower than the height of the enclosure wall and stood only 14 cm above the glacis.

The outer relining of the section of the wall between bastions J and L was built of bricks of both standard and larger sizes (i.e., of brick lengths of 35 and 40 cm). Between these bastions, the wall was very eroded and almost completely gone (pl. 15a). The wall base was displaced, having slumped or tilted up on sand layers brought in by floodwaters.[70] The brick coursing of the walls on both sides of this section sloped down to the east and to the west.

Bastion M displayed one exterior relining constructed with 36–37 cm bricks, and a thinner second relining on its west side composed of larger bricks. The wall stub on its east side was also a fragment of an outermost relining, where bricks 35–37 cm long were used. A living space identified by a scatter of handmade sherds and a possible in situ column base were located to the west of this bastion, in locus D 25.[71]

Bastion N also rested on a foundation of cataract rocks, and the rocks were then piled up against the outer faces (pl. 15b). Against this particular bastion, three construction phases were observed (fig. 3.11).[72] A short, thick reinforcement around the base of the bastion enclosed the layer of foundation stones below it, and it stood approximately 1.40 m high.[73] A second, higher relining present on the east and west sides rested on another layer of rocks placed atop the reinforcement wall. It was preserved along the north part of the bastion and abutted the main wall. A thinner relining was then built against the sides of the reinforcement walls. This thinner relining was perhaps constructed before the higher relining was built, since the rock foundation for the latter may have been laid after the thinner lining was already in place. There is no evidence of any enclosure wall relinings at this spot. Floods and water saturation obviously played a role in the repeated attempts to reinforce this bastion.

The buildings found between this bastion and the bastion at the southwest corner will be described in more detail in chapter 4, part 1. However, it should be

67 Knudstad 1964a, p. 28.

68 Knudstad 1964a, p. 55; Knudstad 1963–64, p. 307.

69 Knudstad 1963–64, p. 328; Knudstad 1964a, p. 27 reverse.

70 Knudstad 1963–64, p. 307.

71 The extramural evidence will be discussed in chapter 4, part 1. The column may be evidence of a building earlier than the fortress; see Pierce 1964a, February 6, and Knudstad 1964a, p. 2.

72 Knudstad 1964a, p. 52, which includes the southwest glacis trench. The drawing of the rebuilding phases is shown in Knudstad 1964a, p. 52 reverse.

73 Knudstad 1963–64, p. 313, with a sketch of the construction phases.

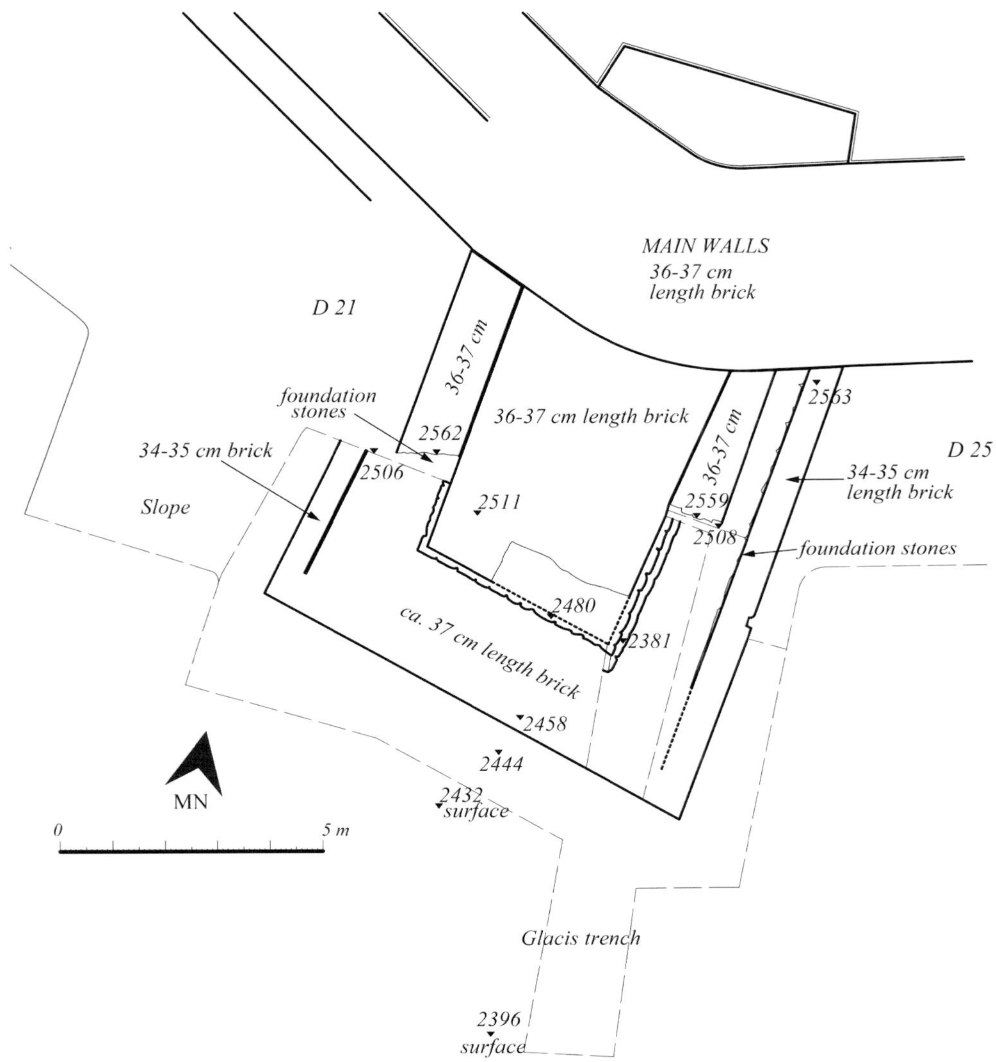

Figure 3.11. Reconstruction phases at bastion N

noted that a wall relining was preserved at the west side of the bay, which was later partly covered by another layer of stones piled atop the glacis to reinforce it. The walls of D 18, D 19, and D 21 were situated atop the later layer of rocks. The relining was built of bricks that were 34–35 cm long and was preserved for only about 6 m. It may have been partly dismantled to accommodate the structures and for reuse of the bricks.

The southwest glacis trench was cut near the east edge of bastion N. The north part of the glacis trench excavation cut through the wall built to reinforce the glacis around the bastion. The trench location and profile are shown in figures 3.1 and 3.6c.

Southwest Corner

Wall lines were traced atop the wall segments running from between the breaches and up to the southwest corner bastion of the fortress (fig. 3.12; plan 1). The outer main wall bonded with the west wall, while the inner main wall and the inner relining abutted it.[74] The later relining along the interior of the west wall abutted the interior relining to the south. Once again the main walls and bastions, and their relinings, were constructed of a mix of bricks 35–37 cm or 36–37 cm long, while the interior relinings were composed of a mix of bricks 37–40 cm long. The main wall of the fortification in this area preserved a plastered face.

The excavators found no indication of any constructions atop the southern enclosure, though a gap between the enclosure walls and the interior relining (intentionally built between the inner main wall and the interior relining) was located north of bastion N.[75] Another space left by the southwest corner bastion, D 125, was also located between the innermost core

74 Knudstad 1963–64, p. 233.

75 Knudstad 1964a, p. 50 reverse.

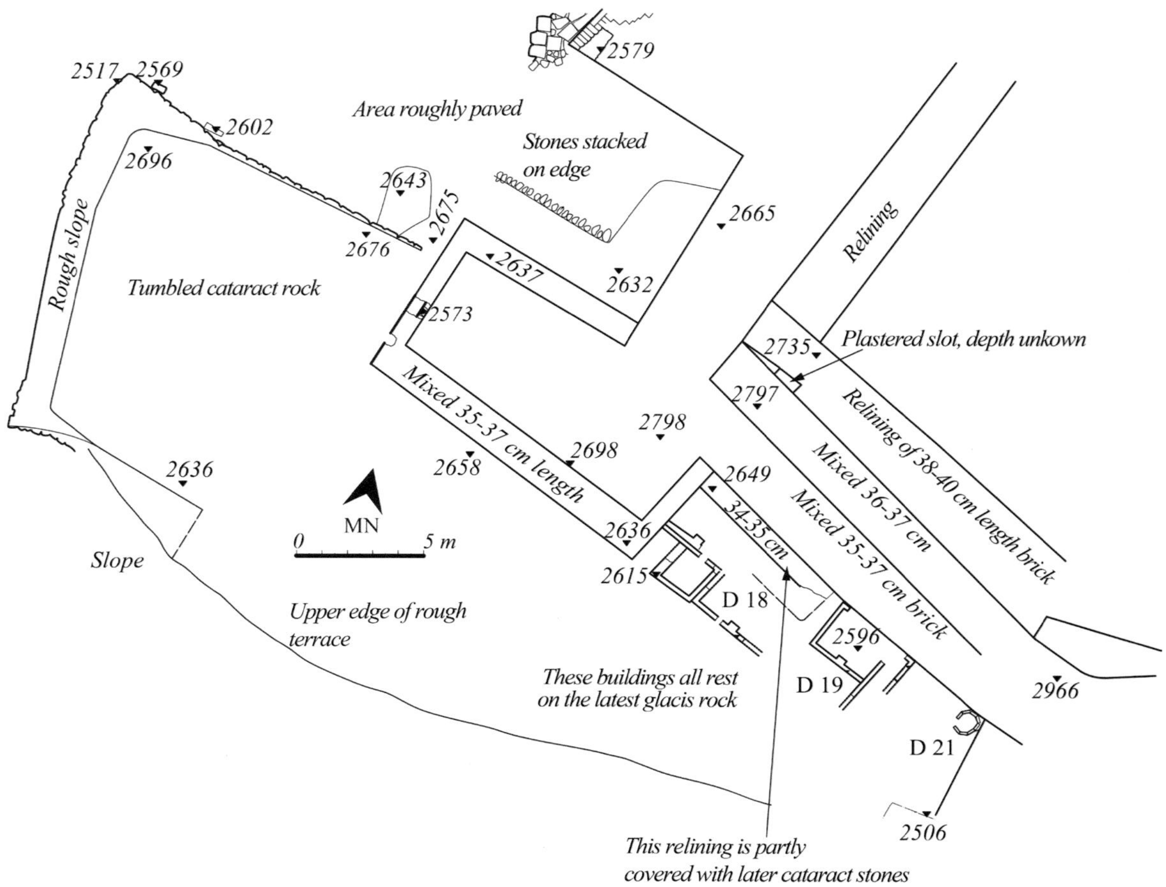

Figure 3.12. Southwest corner bastion and relinings

wall and the inner relining.[76] It was excavated down 2.18 m below the top of the inner enclosure wall and may have continued to the base of the walls, although no complete excavation was undertaken. There is no suggestion as to its function. The sherds of a finely burnished handmade bowl (OIM E36380), the sherds of at least one other handmade bowl (OIM E50451–E50454), a red-rimmed wheel-made bowl, a stone pounder, and a rolled-up piece of leather with straps (a bag or quiver?) and wood fragments were found in D 125 (OIM E24423). This appears to have been an intentional deposit.[77]

A relining surrounded the southwest corner bastion and was built with bricks of the same size, 35–37 cm, as found in the bastion. The bastion remained standing 1.4 m high at its easternmost edge (2798), while its south side was preserved 1.0 m lower and only 40 cm above the surface below it. The south and north walls of the bastion were perhaps finished off with plaster.

The area between the south bastion of the West Gate and the southwest corner bastion exhibited a mud-paved area at the back of the bay that had been edged by a line of stones running east to west. A raised stone pad or terrace was found nearby, extending 14 m to the west of the corner bastion before turning south for 13 m. The terrace had corners at its northwest and southwest edges. It was bordered on the north by a fragment of mud paving, and two stones block pavers were uncovered at the northwest corner. At the northwest corner it sloped westward at an angle of 30.96 degrees, while a rough slope was found along its west side and part of its south side.

West Gate and West Wall

Two bastions once flanked the West Sector gate (fig. 3.13). The southernmost example was mostly preserved, while the north bastion was gone. Floodwaters would have surged through this area, so the state of

---

76 Knudstad 1964a, p. 50.

77 The cooking pot, OIM E36380, is discussed in chapter 5. The leather, OIM E24423, was described as "fine leather, folded, with several fragments of leather straps, some with hair remaining on one side." Wood fragments were found with it. See also a partial list of the items in Hoerth 1964a.

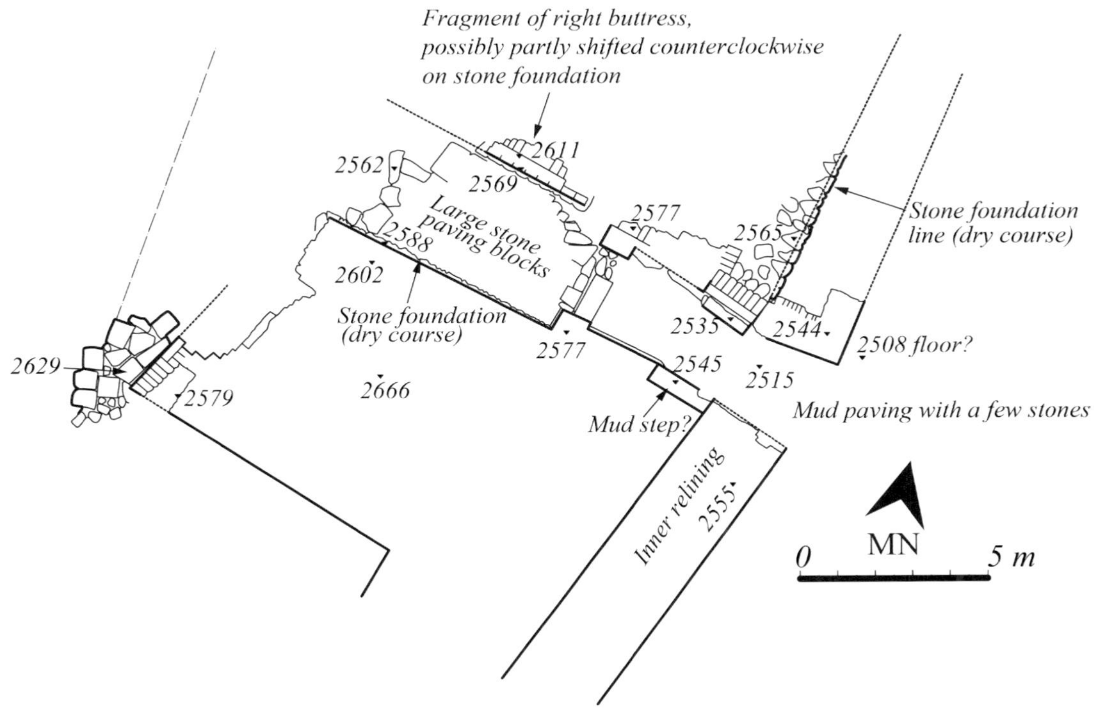

Figure 3.13. West Gate

preservation here was abysmal (pl. 16). The innermost wall relining survived at a low height to the south, with a fragment of relining and some stone foundation lines remaining to the north. The line of stones defining the inner edge of the main wall and the outer edge of the interior relining is one of two places where stone foundations were found for both of the two main walls (see below).[78] The wall relining and the main wall pick up toward the northwest corner. The inner relining here was composed solely of bricks 35–36 cm long rather than of the mix of larger and smaller sizes found south of the West Gate.

A passageway ran between the two bastions and was paved with brick and a few stones on the east and with large, uncut sandstone blocks on the west, all set on a stone footing (see pl. 4a).[79] It was about 13.0 m long and varied in width between 2.0 m (on its interior) and 3.5 m. A stone sill had been placed between two brick jambs about halfway toward the exterior, although no door socket was found. The passage then widened toward the west, but architectural details at the west end were absent. The elevations of the passage actually sloped down to the entrance into the West Sector, but whether this was original or due to floodwater displacement is uncertain. The elevation at the east end, atop a mud paving with a few stones in it, was 2515; the stone sill was founded at 2510, with its top at 2527; and the surface of the sandstone paving's west end was 2562. The passageway was preserved sloping down 47 cm into the fortress at one point along its length, but went 54 cm downward compared to the possible floor patch found at the north side of the fortress interior.

Steps once led up onto the wall to the north and south and were constructed within the inner main wall and adjacent to the inner relining (i.e., the third wall addition). Only one mudbrick stair tread remained on the south side, but two stone steps remained on the north, with the maximum surviving height at 40 cm above the surface of the passageway. The cataract stones, laid for the foundation of both main walls and the passageway, were exposed by the loss of the structural remains on the north side. At this location, therefore, the stone foundations for the fortress ran under both of the main walls and not just the outermost wall as found along the north side.[80]

Outside the gate was a roadway covered with tumbled stones sloping down to the west for about 10 m before being lost under the sand. There was a natural harbor to the west, where sherd evidence showed

78 Knudstad 1964a, p. 49.

79 Knudstad 1964a, p. 49.

80 This was also the case for the inner third wall along the south side at the East Sector; see chapter 4, part 2.

contemporary use of the shoreline, although the excavators found no trace of structures there. A number of thin-walled buildings were located up against the south relining of the northwest bastion, which was composed of the larger bricks found in many later restorations and structures, and a raised terrace area lay to the west of the fortification's southwest corner. These are described in chapter 4, part 1.

SUMMARY OF THE FORTIFICATIONS AND COMPARISONS

The fortress required considerable resources for its construction and maintenance. Its prime location, where its occupants could control river traffic and defend an important node in what is presumed to have been an administrative, trade, and military network, justified its upkeep.[81] The strategic importance of the Second Cataract area is reflected in the earlier settlement and fortification at Buhen, the earlier fortress at Mirgissa, and the line of fortifications along the river through the Second Cataract. The critical location of Dorginarti was apparent again in the sixth century BC, when the Level II fortress was constructed within, and atop, the enclosure walls of the Central Sector. How many generations the Levels III–IV fortress lasted, and whether it was continuously occupied, remains uncertain, but its pottery may suggest two phases of occupation that did not overlap in time.

Crenelations, wall walks, and parapets are most often found in visual depictions of military fortifications, since enclosure walls and other fortification features are not usually well preserved archaeologically.[82] The presence of these features at Dorginarti is therefore remarkable.[83]

The steeply sloped glacis may originally have been intended to thwart enemies approaching, or tunneling under, the wall. But the stone embankment and glacis also protected the mudbrick wall foundations from flood erosion. In any case, this was not entirely successful.

The gates were built in a simple style, with direct entrances and no side chambers.[84] Evidence of one or two possible wooden doors was found within the gate passageways. The narrowing of the passageways in the middle of both the North Gate and the West Gate served a defensive function, but in both cases the passages widened out toward the exterior. The West Gate passageway sloped down progressively into the fort, but perhaps erosion was responsible. The number of gates was restricted to two to limit the vulnerability that exterior access exposed. The West Gate accommodated most of the military or civilian traffic into or out of the West Sector, while the North Gate served for official visits and the delivery of products and equipment to the East and Central Sectors. The gates are similar to those at Gala Abu Ahmed and Gebel Sahaba, both at least partly contemporary with Dorginarti and discussed below.

The resemblance of the Dorginarti fortress to a few earlier Egyptian fortifications in Nubia is mostly to be seen in the irregular ground plan, which followed topographical features, but this resemblance in plan holds little value in determining the date or deciding who built, maintained, and manned the fort.[85] The wall construction, the gates, the interior buildings, and the overall arrangement of the interior are dissimilar to these same features at the Second Cataract fortresses built during the Middle and New Kingdoms. There are no clear Nubian architectural features or any material assemblage that would suggest that this was a stronghold built by the early Twenty-Fifth Dynasty kings or their precursors.[86] The fortified enclosure walls of the religious areas at Kerma from the New Kingdom through the Napatan period, for example, are unlike the fortifications at Dorginarti.[87] The walls at Kerma

81 The relinings of, and additions to, the enclosure walls, for instance, indicate the maintenance required for a fortress situated on a flood-prone island.

82 The Middle and New Kingdom fortress at Buhen preserved evidence of arrow loopholes, for which there is no evidence in Dorginarti's Levels III–IV fortification wall. But there was no preserved archaeological evidence of crenelations at Buhen; see the excavation report in Emery, Smith, and Millard 1979, pp. 6, 22–36.

83 Keeley, Fontana, and Quick 2007. Dry ditches and moats were typical for early fortifications, although an island fortress did not require one, being already surrounded by water. See also Vogel 2010, p. 300, n. 8. The standard architectural features of Egyptian fortifications, as known through images of forts from the earliest times through the New Kingdom, are presented in Monnier 2014.

84 Keeley, Fontana, and Quick 2007, pp. 62–67; Vogel 2010. See also Vogel 2004, p. 125.

85 Irregular ground plans are evident in the second wave of Middle Kingdom fortresses built by Sesostris III, on islands and heights in or near the Nile; see Vogel 2004, pp. 70–77, and Kemp 2006, pp. 231–39. The similar battlements reconstructed for the fortress at Buhen are not based on archaeological evidence. Although there was no trace of crenelations, there is evidence for arrow loopholes on the outer walls. However, similar loopholes were found only in the Level II fortification at Dorginarti.

86 For Kerma, see below. One feature that may be restricted to the Nubian region is the use of supporting piers on the interiors of thin house walls at Dorginarti; see chapter 4, part 1, for discussion and parallels.

87 For example, see the fortified walls built of mud and wood, with rounded or rectilinear bastions—in this case surrounding

have closely spaced semicircular bastions or buttresses, built of mud and wood, and reflect an entirely different architectural tradition. They often surround buildings built on a circular plan and incorporating evidence of posthole constructions.

Although the history of Lower Nubia during the early first millennium BC remains opaque, it is reasonable to assume, given both the significant expenditures needed to construct and maintain it and the nature of its archaeological remains, that Egyptians built the fortress and kept it supplied from Egypt. Otherwise it must be understood that the supply lines with Upper and Lower Egypt were already in place under an earlier Nubian or Kushite power and continued under Alara and Kashta, and that these southerners were in firm control of the region and had the military manpower and resources to build and maintain the fortification for a period of some two hundred years.[88]

The fortification elements discovered at Qasr Ibrim are dissimilar in type and construction to those at Dorginarti. The simple gateway at the southwest side of the site, which was later enclosed by a solid tower with mudbrick exterior, was part of a fragmentary wall running along the top of the gebel to the west and south of the entrance.[89] The probable enclosure wall consisted of mudbrick skin walls around stones set in mortar, then filled in with stone slabs set on a diagonal. Later modifications to the fortifications included the addition of two buttresses or bastions on the exterior, and a later gateway presented a baffled entrance running along the enclosure wall above it, thus exposing whoever attempted to enter. Although marl jar and silt bowl sherds found in loci below the Taharqo temple and contemporary with the earlier gateways and the enclosure wall were similar to some Level III types from Dorginarti,[90] the plan and construction techniques of the fortification were dissimilar. The radiocarbon dates obtained from vegetal matter in the wall bricks imply a date between 1300 and 1010 BC (95.4% probability).[91]

The fortification at Dorginarti and its features can be compared to a number of Nubian fortresses of comparable date.[92] Typical characteristics of most of them are the presence of two gates and the staircases leading up through the gates to the tops of the walls. But some of these fortifications also exhibit vertical stone masonry and crossed walls at the corners of their enclosures, that is, extensions beyond the join of the two walls and at right angles to each other. These features are absent at Dorginarti, where the wall construction was entirely of mudbrick. Also characteristic of most of them is the presence of multiple bastions, including at Dorginarti and at the fortresses of Gala Abu Ahmed, Dabenarti, and, to a lesser extent, Gebel Sahaba.

The fortress of Gala Abu Ahmed was built atop a sandstone spur where the topography determined its trapezoidal outline.[93] The construction techniques of the fortification walls differ from those at Dorginarti. At Gala Abu Ahmed, they were built of drystone masonry with casing walls and a core of stones or stone slabs. The enclosure wall masonry consisted mostly of vertically placed slabs, and although no brickwork was preserved, bricks may have been used in some of the fortress features. Crossed walls strengthened the corners of the enclosure, and bastions were preserved in the middle of the walls and at the gates. Two gateways led into the fort, being built between bastions, and each had a stairway leading up to a wall walk. Parapets ran along the outer face of the enclosure walls and the bastions, mirroring the construction at Dorginarti. A stone-built porch or terrace was found on both sides of the east gateway and, like the stone terrace to the south of Dorginarti's West Gate, may have served as a mustering place for visitors (e.g., merchants) or soldiers. While the interior details of Gala Abu Ahmed will not be discussed in detail, it should be noted that a large building and silos were found within the enclosure. Many of the objects and pottery from the site are similar to those from Dorginarti.[94] Gala Abu Ahmed also yielded later Greek imports, which are discussed in chapter 6.

Gebel Sahaba, a fortress discovered during the Scandinavian Joint Expedition's surveys on the east bank, is also reminiscent of the fortress at Dorginarti.[95]

---

a sacred area—in Bonnet 2009. This wall lasted through the Twenty-Fifth Dynasty and was perhaps destroyed around 590 BC during a Saite campaign in the region; see Bonnet 2011, p. 23.

88 These points are discussed further in chapter 8.

89 Horton 1991, pp. 264–68; Rose 2008, pp. 198–99.

90 Personal observation. I would like to thank Mark Horton and Pamela Rose for allowing me to study the pottery from the 1984 excavations by James Knudstad around the Taharqo temple, and from the 1998 excavations under the same temple and the bastion gateway.

91 Rose 2008, p. 200, table 1, nos. 2–6.

92 The features of Nubian fortresses of this general period are presented in Eigner 2013, p. 313, table 1. See also Jesse and Kuper 2006, pp. 143–45; Welsby 2005.

93 Jesse and Kuper 2006, pp. 143–45. See also Eigner 2013, with fig. 3; Jesse 2013, pp. 321–23, and fig. 2.

94 Eger, Helmbold-Doyé, and Karberg 2010, pp. 77–84. See also Jesse 2013, pp. 332–35. Pottery and object parallels with Gala Abu Ahmed are discussed in chapters 5 and 7.

95 Säve-Söderbergh and Troy 1991, pp. 319–23, pls. 216–19.

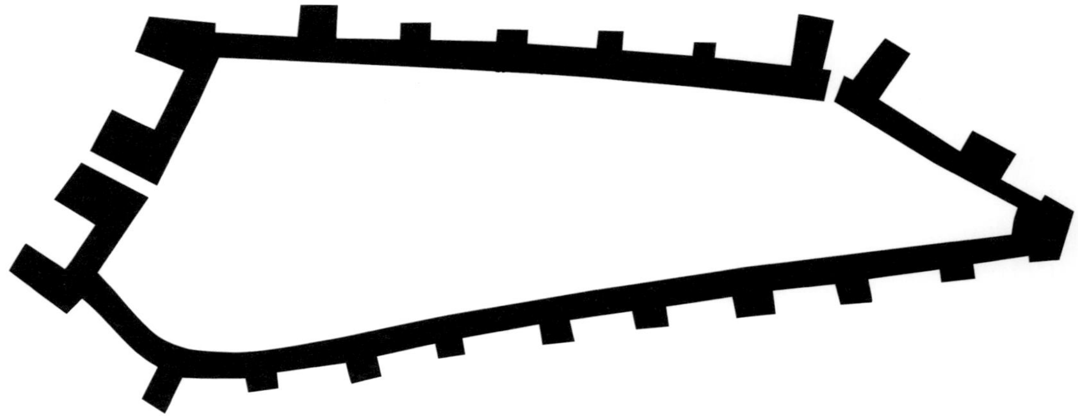

Figure 3.14. Reconstruction of the original enclosure walls

It featured large bastions at the corners, east end, and gates, while smaller buttresses or bastions were built along the middle of the three longer walls. Stairways led to the top of the walls at the gateways. The fortress was built atop a bedrock height above the Nile and was trapezoidal in shape. Its construction techniques were more like those at Qasr Ibrim, with mudbrick skin walls built around a core of stone and mud mortar; the mudbrick casing walls of its buttresses were also filled with stone rubble. The few ceramics found in the Gebel Sahaba fortress—besides those of a much later date—were similar to wheel- and handmade potsherds from Dorginarti. In addition, the sherd disks and a hollow-based stone arrowhead found there are similar to those from Dorginarti and Gala Abu Ahmed and to the hollow-based stone arrowheads from the early ancestral tombs at el-Kurru.[96]

Based on the evidence presented above, it is assumed that at least some phases of the fortresses at Gala Abu Ahmed, Qasr Ibrim, and Gebel Sahaba were contemporary with the fortress at Dorginarti.

After the original fortification wall of Dorginarti was reconstructed, using the initial two curtain walls and the bastions (fig. 3.14), the fortress bore only a slight resemblance to both the trapezoidal fort at Gebel Sahaba and the ruinous fort on the island of Dabenarti. The description of the archaeological remains and construction of the fortress at Dabenarti is meager.[97] The fortress is therefore not discussed here, since there is little evidence for its dating and its plan is not very similar to the fortress on the island of Dorginarti, although its walls do resemble other first-millennium BC fortifications in Nubia.[98]

A possible Kushite fortress has recently been found on the east side of Karni Island, in the Fifth Cataract of the Nile.[99] The irregular fortress has bastions along its walls and two gates, the east one lined by bastions with stairways leading up into the interior, but no such evidence is preserved at the west gate. A stone revetment lined the east enclosure walls as protection against high Nile floods. Trenches had been cut into the bedrock for both the enclosure wall and the buttresses, which were all constructed of stone blocks. No pottery from this fortress has been published, although the surveyors noted handmade sherds with incised decoration and wheel-made storage jars made from gray clay, which may indicate imported Upper Egyptian marl vessels.[100]

The evidence for Third Intermediate Period and Twenty-Fifth Dynasty fortifications in Egypt is less well documented. The fortification at Elephantine during the Third Intermediate Period, contemporary with the fortress at Dorginarti, is fragmentary.[101] The

96 Säve-Söderbergh and Troy 1991, p. 220, nos. 113, 143; note also the grave vessel in Gardberg 1970, pp. 45–47, pl. 21. The survey records were most kindly copied and sent to me by Dr. Säve-Söderbergh in 1988, and I would like to thank him, Lana Troy, and Rotislav Holthoer for sharing the materials with me on my visit to Uppsala in 1988. The pottery and object parallels are presented in chapters 5 and 7.

97 Ruby 1964.

98 The right-angled bastions at the corners are similar to other fortifications; see Jesse and Kuper 2006, pp. 143–45.

99 Drzewiecki 2012, pp. 12–15, pls. 18–22. See also Drzewiecki 2011.

100 I would like to thank Mariusz Drzewiecki for sending me photographs and drawings of the survey sherds collected by the Polish mission (August 12, 2015). The sherds contain wheel- and handmade silt types, of Egyptian or Nubian manufacture, and a wheel-made marl neck sherd; but they are not dated with any degree of certainty due to the sparseness of clearly identifiable rim forms.

101 von Pilgrim 2009, pp. 260–61, 267, fig. 3.

wall trace follows the topography of the island to enclose the settlement and temples within, on the same line as the late Middle Kingdom wall. At the southwest edge a later wall preserved a series of buttresses, but no date was assigned to this segment.

The enclosure of a fortified town at Abu ʿId, south of Edfu, which was surveyed during the 1980s, has pottery that is similar to that from Dorginarti,[102] but the plan and features of this large fortified site are dissimilar. It encloses an extensive area with evidence of four gateways, one with an archway, and only one buttress at the northeast corner of the enclosure wall.[103] The south gate had the remains of one stairway leading up. The acropolis of the site featured a large residence. The site has only been surveyed, never excavated, so even the contemporaneity of the different elements is uncertain. The pottery published from the surface survey of the site is similar to many sherds from Dorginarti and is extensively referenced in chapter 5.

The fortified wall at el-Hibeh, constructed atop a limestone scarp on the west bank of the Nile in Middle Egypt, was only partly excavated in one season, and therefore the enclosure and battlements of the fortification are not well defined.[104] Its fragmentary enclosure wall follows the edge of a limestone scarp above the Nile and is preserved only on the east and southeast sides. No gates or buttresses can be seen in the ground plan. The fortress borders the area between Hermopolis (modern Ashmunein) and Heracleopolis, and all three sites have yielded pottery from the Third Intermediate Period and Saite era. Fortified enclosure walls have not been found at the latter two sites.[105] The fortifications in the Delta expose no architectural parallels at this point in time.

102 See chapter 5 for the comparative ceramics from Abu ʿId published in Aston 1996c. The fortress was surveyed by Horst Jaritz; see Jaritz 1980 and 1986, pp. 37–39.

103 Jaritz 1986, p. 37.

104 Wenke 1984. See also the description of the wall in Lawrence 1965, p. 91.

105 However, see chapter 6 for the fortified and platformed citadel from Ashmunein.

4

The Level III and Level IV Architecture

Part 1. Buildings and Structures within the West Sector and outside the Fortress

GENERAL NOTES

A great deal of the West Sector was cleared during the five and a half months of work at Dorginarti. A coherent plan of the lower level of structures in the sector, which also ran under at least the interior curtain wall and represents part of the earlier Level IV settlement, was not obtained.[1] The earliest remains were never reached because of the time and resource constraints the excavators confronted.

Clearance of the top levels in the Central Sector and the tracing of the lines of the Level II platform and main fortification walls had taken precedence at the beginning of the excavation. But attention soon shifted to the debris-covered surface of the West Sector. An upper level consisting of wall fragments, a silo, and ovens was exposed near the surface, representing either late Level III or Level II buildings. A later interior relining was constructed along the northwest wall, but its remains were fragmentary and did not continue past the corner. No floor surfaces were found in association with the majority of the upper features, having been eroded away.

The top and intermediate levels to the north, along with these levels in the southwest and southeast corners of the West Sector, do not necessarily belong to the same phase and formed no coherent settlement arrangement. Thus we will discuss these upper and intermediate levels at the beginning of each area's section, then move on to the Level III remains, and finally to any noticeable Level IV architectural vestiges. The locus numbers and descriptive loci are also presented in table A in appendix III. The top level in the West Sector is shown on plan 2, while the intermediate level is shown on plan 3. The Level III and Level IV walls in the sector are shown on plan 4, while details are shown in illustrations in the text below. The extramural buildings are shown on the plan of the entire fortress in plan 1.[2] Lastly, a highly tentative reconstruction of the West Sector is presented in figure A.II.3 in appendix II, with the possible buildings and roadways that may have existed but for which there is little evidence.

First, however, we will begin with a general analysis of the types of construction.

DETAILS OF CONSTRUCTION

After the surface was cleared, the excavation exposed a maze of thin-walled rooms, residences, and work areas, which had been poorly built.[3] The walls were usually one brick wide, or 18–22 cm thick, the bricks being added loosely along or atop each other without much evidence for the use of mortar.[4] The room sizes varied considerably, and no set plans for the construction of house units can be detected. The larger, more regular rooms along the north wall ranged from 7.84 m^2 to 16 m^2 in area, while the smaller side rooms averaged about 4 m^2. The room sizes were smaller along the north wall by the breach, ranging between 1.8 m^2 and 3.78 m^2 (D 41–D 43).[5]

1 Knudstad 1963–64, pp. 313, 330.

2 Knudstad 1963–64, pp. 298, 301. The constructions along the north and south fortification walls were "randomly added thin walls and rooms loosely filling the whole of the areas excavated."

3 Knudstad 1963–64, p. 307.

4 Knudstad 1964c, letter to Sayed Thabit Hassan Thabit, April 30: "The rooms, rather than showing any pre-conceived arrangement of regularity in plan, are loosely built against each other from the outer walls inward, and are of small dimensions, less than 4 m^2."

5 For the estimated population of the West Sector, discussed in appendix II, nine of the houses along the north wall in the West Sector were measured, with total areas that ranged from

The houses often shared walls. The use of brick support pilasters, or small buttresses, was common, with the supports being built on the interior of walls to strengthen them and to support roofing materials.[6] Wall segments were added later to block off areas and were often strengthened at the place of abutment by L-shaped or cornered ends.

When there were walls that were more deeply founded than most Level III examples, they were usually at least 19–28 cm below the floor levels of the Level III structures (e.g., in D 1, D 6, D 60, and D 67). Thin, quickly built walls do not need deep foundations, suggesting that the deeper walls were from an earlier phase of the settlement than many of the walls shown on the plan. These were earlier Level IV structural remains that had often been reused in Level III.

There are no indications that the area had been leveled prior to construction. The floor elevations between rooms, passageways, and other spaces varied widely. The base level of the inner fortress relining, however, was fairly level across the stretch of the north wall. These elevations are discussed both in the text below and in chapter 3 on the fortification walls.

There is also no evidence for stairways within the buildings. The two staircases preserved along the north wall were situated outside the houses in what seem to have been communal areas, while three of the four staircases along the south wall were situated in indeterminate locations since there were few associated walls.

Most of the interior buildings shown on the final plan were placed against the plastered surface of the major, third wall addition, and therefore they had no separate wall on their north sides. The walls may have been constructed abutting the wall, or they may have been cut through when the third relining wall was added. The thick wall addition itself rested on layers of ashy flooring that ran under it and then abutted the base of the earlier core walls, at least in the area of the north breach and just to its west.[7] The inner lining was cleared at the north breach to expose more structural remains, which may have continued under the innermost of the two core walls.[8] There was no evidence to prove this, however, and the curtain walls were not removed.

11.8 m^2 (D 70) to 28.6 m^2 (house D 59), with an average size of 21 m^2.

6 There was no mention of roofing materials being found in the room debris, except in the fill atop one floor in the Level III residence in the Central Sector.

7 Pierce 1964a, January 18.

8 Knudstad 1963–64, p. 330.

OVENS

Ovens, hearths, and bins were thinly interspersed throughout the West Sector. They were not always found within each building unit, but in common areas between the structures. Grinding stones for the preparation of grain were found inside the individual house units where food preparation occurred, rather than in the areas around these ovens. Ovens were also seemingly absent from areas near the silos.[9] Remnants of one or two ovens were found near the Level III grain storage area in the southwest corner of the West Sector, and two or three were found near the Level IV silo in D 77, but there was no large concentration of ovens to suggest that bread was baked before distribution. It may be that certain people took responsibility for the grinding of grain and the preparation and baking of bread in neighborhood ovens, built either by them or by groups of neighbors. The state of preservation of the site, however, does not allow a more comprehensive analysis of the daily activities associated with food preparation. We know nothing of the architectural composition of the central part of the West Sector. However, it may be that some food preparation was done on the west bank of the Nile, across from Dorginarti. A series of nine ovens (5-O-13) was found next to the 5-O-6 quarry at Abd el-Gadir, where sandstone slabs were removed for the construction of the fortress. The dating of the pottery at these sites, however, remains uncertain.[10]

The excavators left no detailed description of most ovens they encountered (see below), although many of them had preserved ventilation holes. Based both on ovens described elsewhere and on photographs taken at Dorginarti, the oven walls should have been composed of clay mixed with straw and other temper, perhaps also sherds, and erected atop stone foundations or debris.[11] The walls of Middle Bronze through Persian-period ovens at Tell Jemmeh, Israel, for example, showed evidence of a coiled construction and a domed top, and the ovens often had earthen floors.[12] They typically measured 60–80 cm in diameter. These

9 For a discussion of the silos, see appendix II.

10 Nordström 2014, pp. 121–23. The sherds and faience beads were not illustrated.

11 For a good description of oven construction and other cooking installations, see Zukerman 2014, p. 645. The ovens at Tell Jemmeh were typically 60–80 cm in diameter and probably reached about 60 cm high, with a central opening in their roofs of 40 cm in diameter. A few examples preserved evidence of being domed. All had a small 10–20 cm opening just above the level of the floor.

12 Zukerman 2014, p. 645.

ovens may have been used for baking bread or roasting meat, but such ovens are not always distinguishable from metal-smelting kilns.[13]

The construction of inner wall and outer insulation can be seen on the oven from D 48, belonging to an intermediate level at the northwest corner of the West Sector (a later Level III phase), which was described as having vertical brick packing around its walls (pl. 17a).[14] Oven D 115, in an intermediate level at the southwest corner of the West Sector, was superimposed atop an earlier oven, and the construction of its inner mud wall and brick packing can be clearly seen (pl. 17b).

The superimposition of ovens in the West Sector was due to their short lifespan. Without frequent repair, they may have lasted only about three years depending on their usage.[15] Two superimposed ovens were found in D 76, while the ovens in D 208 were all founded next to one another at different levels (pl. 18a–b). Their foundation levels differed by as much as 37 cm, but they appear to have been rebuilt for concurrent use. The three ovens in D 208 were preserved at the same height, their tops having been intentionally leveled for later rebuilding in the area.

REUSE OF WALLS AND RECONFIGURATION OF AREAS

The ruins of Dorginarti reflect a constant state of flux, perhaps to accommodate sporadic occupations or intermittent arrivals of troop contingents. Constructional adjustments included the repair or building up of earlier walls, the walling off of areas with short partitions, the dismantling of walls, the blockage of doorways—in short, the reconfiguration of any space available to suit the needs of the soldiers, support staff, and family members. The deterioration of the fortress's enclosure walls also brought about changes, such as the addition of a thick relining to the interior side of the double-walled fortification to shore up slumping sections after serious flooding had occurred. This would have entailed moving occupants from houses built before the relining wall was added and then reconfiguring structures and areas to house them anew.

FUNCTIONAL DIFFERENCES

The north half of the Level III settlement encompassed residential buildings and domestic space, with small ovens, bins, and hearths interspersed unevenly throughout. Activities in the residential areas included the grinding or pulverization of grain and/or minerals, as shown by the presence of grinding stones and pounders, a few crucible and tuyère fragments for the repair of metal objects, bread plates, and large stands for utilitarian vessels to hold water or to store food. Stone net weights attest to fishing activity, and stone arrowheads indicate the presence of archers. There was no indication of debitage in or around the fortress to show that arrowhead manufacture occurred on site.

The south part of the West Sector accommodated small workrooms (or grain magazines), large silos, bins, ovens, a few possible houses, and two larger buildings. The two larger structures—at the southwest and southeast ends of the zone—may have served either religious or administrative purposes, but their precise function remains uncertain. The objects found in them were not helpful in this regard, and their interiors were never fully excavated.

The Walls of the Earlier Settlement

Various walls were found under the buildings along the north wall, representing structures that the excavators attributed to the initial occupation of the fortress—that is, Level IV. The elevations within and around the structures under the inner brick wall addition at the north breach were not the lowest elevations recorded along the north wall, but they were lower than those of the Level III buildings and floors up against the wall nearby. Possible structural components were cut through by the wall relining and were sometimes left standing and reused, and a few features were also found under the third, inner relining. These are mentioned below.

It is certainly possible that a settlement existed on the island prior to the construction of the fortress enclosure, but despite the existence of a few earlier sherds, there is no structural evidence for a major occupation of the island before the enclosure was built. However, much of the architecture was left standing at the end of the season and was not fully excavated. A tentative reconstruction of the original fortification walls, before the thick inner wall lining was added, was briefly presented in chapter 3.

CLEARANCE ALONG THE NORTH WALL IN THE WEST SECTOR

While work proceeded in the Central Sector, surface debris along the north and south walls in the West

13 Zukerman 2014, p. 647.

14 Knudstad 1964a, p. 37.

15 Zukerman 2014, p. 650.

Sector was also cleared. The center of the west half of the fortress was found to have been swept clean of all buildings and most occupational debris (pl. 19):

> The entire center of the West Sector revealed the same sandy deposition. Along the fortification walls, deposition from living debris filled the rooms to almost their preserved height. These room walls tapered down in height as they went into the center of the fort and those rooms whose walls were preserved to a height of only ten centimeters were filled with sand. Walls disappeared in the center of the fort completely and test trench D 2 also has sandy deposition.[16]

Three or four building levels were, however, well preserved in the northwest, southwest, and southeast corners of the West Sector, where water and wind erosion had been less severe. The occupational remains and their overlying sand deposits had accumulated to a depth of 2 m in some places.[17]

The initial surface clearance along the wall began with the highest standing remains at the northwest corner and along the north wall relinings. This operation immediately exposed occupational debris and walls, so the excavation proceeded according to architectural loci.[18]

Three strata were excavated after a surface clearance in D 1, above the area of house D 1, with the topmost two yielding seven querns or fragments of querns and a grindstone of river cobble with a naturally formed handle.[19]

D 7 was the surface clearance east of D 6, but only the top layer was removed before walls were encountered. The rooms discovered below it will be discussed under D 66–D 67.

The beginning stages of the West Sector clearance also included the excavation of a 2 × 2 m test pit, D 2, in the center of the west interior to determine whether any structures or archaeological finds remained.[20] The excavation went through fairly sterile layers of surface sand; then dry, hard river silt; and then more hard or clayey sand layers (pl. 20, right). The excavators intended to continue the test pit down to bedrock, but they abandoned the idea after 1.8 m because it was clear that the central part of the interior had been scoured clean of useful archaeological materials. Sherds from three nested, handmade bowls were found in the northeast corner of D 2 at about 1 m in depth, while some weathered handmade sherds were found in the southeast corner of the pit at about 65 cm in depth.[21]

A long north–south trench was also excavated just east of the West Gate, but it was found to be completely devoid of archaeological remains except for a few sherds, so the operation was halted at 30 cm below the surface (pl. 20, left).[22]

Throughout the West Sector was a dark-colored layer of alluvium, an "ash" layer that was initially thought to represent a living surface. It was later realized that this darker band of alluvium was in fact a deposit left by repeated water fluctuations at the site and the saturation of brickwork and fill. But the initial interpretation, based on a possible ash level and building reconstructions below this layer, influenced the numbering of floors in some of the locus descriptions. This must be kept in mind when reading the following discussion.

UPPER LEVEL ALONG THE NORTH WALL

The walls of a circular brick silo were exposed during the surface clearance of D 4 in the northwest corner (plan 2), and the locus number was then used to identify the silo.[23] The initial clearance also exposed a higher inner relining along the north fortification wall, which abutted both this silo and a short wall fragment about 4 m to the east. These latest features and the relining were built using the larger 40 × 20 cm bricks. The thin wall relining had been eroded or cut through, and it was preserved only on the east side of the short wall.[24] The relining was founded at elevations 2711 and 2682 to the west and at 2677 to the east, and it remained standing at a height of 1.23–1.54 m. The wall of D 4 was based at 2675.

These thin relinings were battered, the segment to the east by the short wall fragment set back at 9.1 degrees. It was noted that the incline followed the batter

16 Hoerth 1964a, no. 2, p. 3.

17 Knudstad 1964c, letter to Sayed Thabit Hassan Thabit, May 12.

18 Knudstad 1963–64, p. 174; Knudstad and Pierce 1964, introduction.

19 Pierce 1964b, pp. 7–11.

20 Pierce 1964b, pp. 12–13; Knudstad 1963–64, p. 177.

21 Pierce 1964a, January 15. William Y. Adams noted that the sherds looked like A-Group or early C-Group types, but the handmade sherds from Dorginarti, particularly weathered and broken sherds, resemble their earlier precursors.

22 Knudstad 1963–64, p. 325. The sherds found in the trench all belonged to a handmade bowl with a dark-red slip on its exterior, incised decoration, and a repair hole—OIM E49525, discussed in chapter 5.

23 Knudstad 1963–64, p. 176; Knudstad 1964a, p. 36. The most distinctive sherds from the upper level in D 4, the intermediate levels below it, and the bottom levels are presented in appendix I.

24 Knudstad 1964a, p. 36.

of the main enclosure walls. The surface had been finished with plaster.

Silo D 4

D 4 was preserved for about two-thirds of its 3.26 m diameter, and its brick walls remained for a maximum height of 61 cm. The silo was placed immediately next to the north relining wall, standing only 7 cm from it. Remains of a plastered finish were observed on its exterior and possibly also on its interior.

Each of the three stratigraphic levels below silo D 4 had its own unique architectural components. The bottom of the wall of D 4 was 1.68 m above the floor of D 220, the lowest structural remains below the silo.[25]

All the constructions in the topmost level at the northwest corner, and those in the intermediate level below it, were constructed of bricks measuring approximately 40 × 20 × 8 cm. No contemporary floors were identified for these walls, which sat on ashy debris.

INTERMEDIATE LEVEL ALONG THE NORTH WALL

D 1 Surface Clearance

Surface clearance farther to the east and along the wall at D 1 yielded a blue faience scarab fragment (OIM E24310), a hieratic ostracon (OIM E24308), and a goat burial. The goat's body, laid on its side, was well preserved and was assumed to have been a modern deposit. The goat was 86 cm below the surface on loose sandy fill with scattered sherds and 10 cm below the base of the nearby wall stub. The ostracon was in the fill above it. The artificial surface created by the burial was identified as floor 1.

The intent of the initial excavation of the fill in D 1 was to find the floor associated with the intermediate level wall to its west (plan 3). But no floor level was found throughout the 1.15 m of fill, although a quern, pounder, and numerous sherds were encountered.[26] Some of the sherds were from the Christian use of the site. At 50 cm below the surface, however, the walls of a building were exposed. This structure then became identified as D 1.

The excavators identified the dark fill layer caused by water permeation as strata 2 and 3—two illusory "floors" that had been created within the fill between the walls of D 1. These two strata together were found to be 45 cm in depth, going down 1.6 m below the surface, where the Level III room floor was discovered. The excavation then expanded to include all the area within the room (discussed below).

D 10 Surface Clearance

Ten to 12 m farther east from D 1 was a long north–south wall made of large bricks and running toward the center of the sector for about 9 m.[27] The bricks used for its construction measured 38 × 20 × 7–8 cm, and the wall was 40 cm wide. It was built over the earlier walls of Level III buildings, and the elevations of the wall foundations varied depending on whether or not there were earlier walls just below its base. This long wall ran in a line above the Level III wall between rooms D 10 and D 11 (below D 10 and D 11 surface clearance), with its foundation level being at elevation 2544 on the north, 2591 atop the walls of the earlier D 60, 2554 atop an earlier oven in D 60, and 2549 atop debris over the area of room D 71 below. The wall can be seen in the photographs of the Level III buildings in the area (pl. 21a). No floor surface was associated with it.

An oven was located on the west side of the wall and along the north enclosure wall. It was 78 cm in diameter, with a stoking or ventilation hole on the south side. The oven was founded 39 cm higher than the base of the long wall.[28]

Seven meters to the east of the long wall described above was a 1.20 m wall that ran south from the fortress wall and was preserved to a height of 31 cm above its base.[29] It was 40 cm wide, like the long wall to its west, and was founded at elevation 2568. The bricks used in its construction measured 40 × 19 × 8 cm.

Unfortunately, throughout the West Sector the topmost strata were poorly preserved, and the isolated structural fragments described above do not form any recognizable architectural complexes. The fragmentary walls to the east were considerably wider than those of house D 47, described next, and more substantial than the earlier Level III house walls. All these architectural fragments were on an intermediate level and built atop Level III.

The later reorganization of the settlement evidenced by the topmost or intermediate levels along

25 However, the floors in D 220 are possibly earlier Level IV floors, being some 12–25 cm below the nearby Level III floors in D 221, D 200, and D 216.

26 Pierce 1964a, January 6–7, and 1964b, pp. 6–11.

27 Knudstad 1964a, p. 43 reverse.

28 The base of this topmost oven was left unexcavated in room D 10, as shown in photographs of the area and on the plan for the Level III and Level IV buildings.

29 Knudstad 1964a, p. 44 reverse.

the north wall constituted a wider use of the area either later in Level III or during Level II. A few sherds that can be attributed to the Level II use of the fortress were scattered across this area, although not in any concentration.

House D 47

Below silo D 4 was a large building that included loci D 47, D 48, D 50, and D 200–D 202 (pl. 21b).[30] The building was trapezoidal in shape and measured about 9.5 × 10.5 m along the exterior walls, as shown on plan 3. All its walls were built on earlier debris and were one brick in width—that is, 20–22 cm wide. The structure was built with the larger 40 × 20 cm bricks. The wall bases were level with the floor elevations, although only one good fragment of floor was found in D 47. The floor elevation in D 47 was between 65 and 83 cm below the wall bases of D 4 and between 73 and 85 cm above the Level III floors below it.

The structure included a number of rooms and perhaps also D 200, a locus at the south that continued down into Level III. Room D 50 at the north was oriented east–west and had a small oven, D 48, and a bin at its west end. The north wall of the room was built up against the fortification wall relining, which was atypical for the earlier houses in Level III. The south bounding wall of D 50 may have run farther east. No evidence of a doorway was preserved. The east–west wall was rather crookedly built, with a support pilaster preserved on its north side. This space may have been a courtyard.

The oven, D 48, was 72 cm in diameter and had a vent or stoke hole 9 cm above its base on its east side. Vertical brick packing surrounded the mud wall. A short wall fragment abutting the south side of the oven represented the remains of a bin. There was only one level of finds from this structural phase, except for D 48 (2), which was the deposit within the oven and bin, in which charcoal and a pounder were among the objects recovered.[31]

A good doorway led into D 47, where the clay floor was preserved. Between this room and D 201 there was also a door, situated at the west end of a double wall dividing D 47 from D 201 and D 202. No evident doorway was found for D 202, but its north and east walls were fragmentary. The double wall between D 47, D 201, and D 202 was perhaps the result of an effort to strengthen it or to close off D 47 from D 202. A double wall also stood to the east of D 47. Some of the walls in D 47, D 50, and D 201 showed evidence of a plaster finish. The east edge of the complex of rooms was quite fragmentary, but the wall fragments were aligned. The space between the east wall of the complex and the double walls at the east side of D 47 was less than 3 m wide.

The entrance in or out of this house was located somewhere along the fragmentary remains of the east or south wall. The broken walls of a later oven, resting on ash debris, were also found outside the structure on the east side of D 47. The oven was about 86 cm in diameter and sat on a level 68 cm above the nearby wall base and 51 cm above the wall stubs at D 1 to the east.

The space south of the house was labeled D 200, but no contemporary walls were uncovered in that area. The fragmentary oven described above, and a cornered set of wall fragments exposed during the surface clearance of D 1, were located east of house D 47 and sat on higher debris.

Silo D 213

Room 47 and the oven and bin of D 48 lay atop another circular brick silo, D 213, in the third level of architectural remains at the northwest corner.[32] The construction was built with the larger 38–40 cm bricks. Plaster was found on the floor of the silo and on both the interior and exterior of its 20 cm thick walls. The silo was situated 40–50 cm from the wall relinings in the corner.

The plastered floor of D 213 was situated 20 cm below the clay floor of D 47 above it and 53–65 cm above the uneven floors of D 220 and D 221 in the level below.

Pottery sherds and a stone arrowhead (Khartoum field number 24393) were excavated in the fill below the silo. This fill was excavated to the point where the cross wall between rooms D 220 and D 221 was encountered. The fill in question constituted the fourth level at the northwest corner.

---

30 Knudstad 1964a, p. 37. These walls were initially considered levels 2 and 3 of the four strata at the northwest corner, but the southernmost walls were subsequently considered contemporary and the levels were combined.

31 Finds from D 48 (2), the fill in the oven and bin, were charcoal (OIM E45997), a pounder and rubbing stone or polisher (OIM E50777, E50790), and a faience bead (OIM E50830). The other loci of the house included the following: the fill of D 201 (1) contained sherds and a stone arrowhead (OIM E24395), a faience plaque with an incised fish (OIM E24398), a few pottery disks (OIM E47471, E47472, E47473), and lumps of charcoal (OIM E49002). The other finds throughout the area were sherds (of course) and a pottery disk incised with crossed lines (OIM E47460).

32 Knudstad 1964a, p. 39.

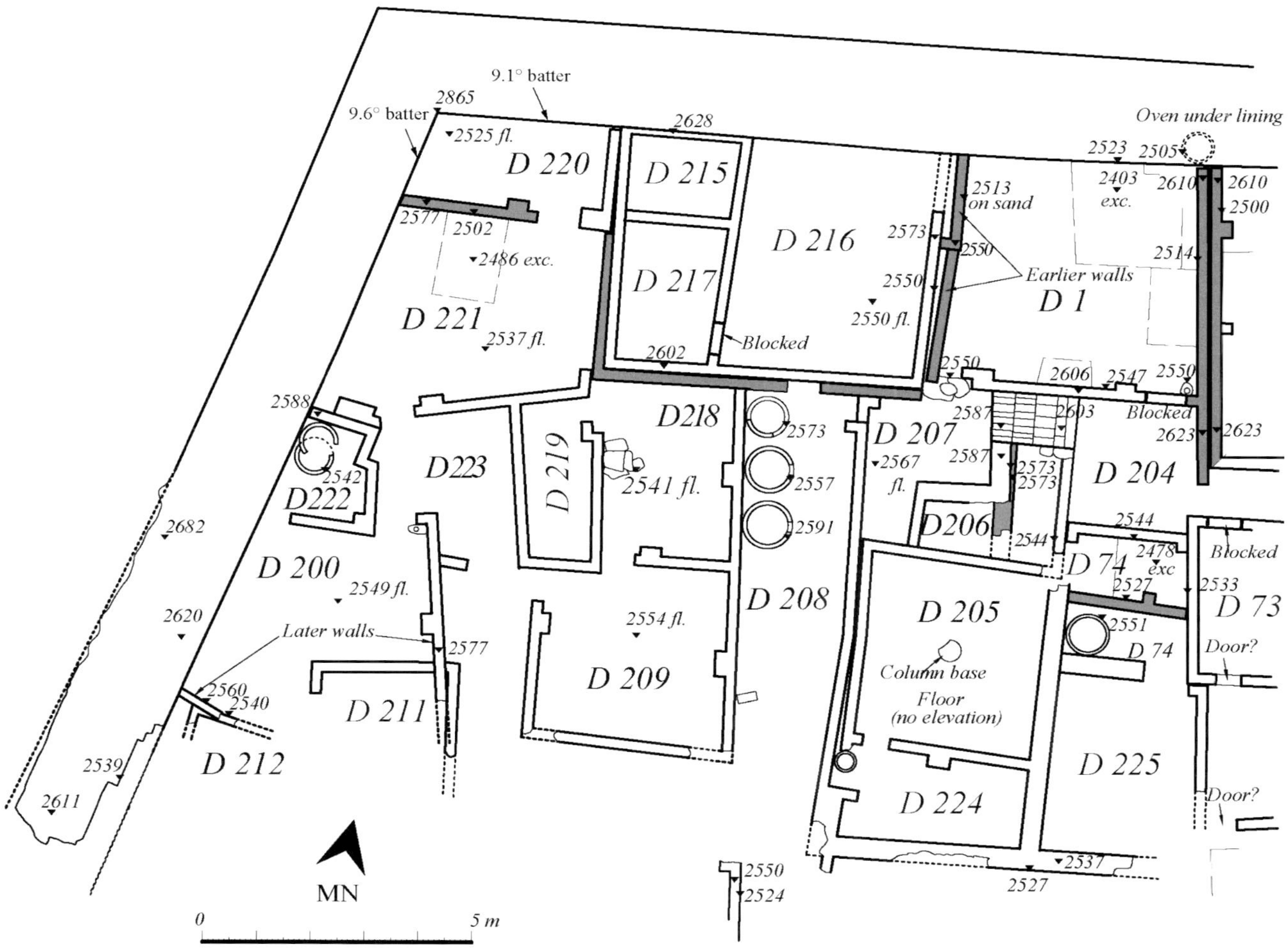

Figure 4.1. Houses along the north wall at the northwest corner

NORTH RESIDENCES OF LEVEL III (AND LEVEL IV REMAINS)

The plan for the West Sector also shows the walls attributed to the earlier Level IV buildings that ran under the innermost wall addition. Objects from the structures along the north wall included numerous querns and pounders, which were much more numerous on the north side of the sector than in the preserved buildings standing along the south wall. The figures in this section show earlier walls, either left in place or reused, in dark gray.

House D 221

As already mentioned, the fill below silo D 213 was excavated down to the point where the dividing wall between rooms D 220 and D 221 appeared some 13 cm below the floor of silo D 213 (fig. 4.1; pl. 22a).[33] The upper walls around the oven in D 222 had also started to appear during the excavation of D 201, since they were only 3 cm below the west wall foundation in that room. The floor level at the doorsill in D 201, however, was 21 cm above the top of the wall, where the ash fill from the lower remains was already evident.

A door at the east end of the wall between D 220 and D 221 provided access between the rooms. The floor level at the northwest corner of D 220 (elevation 2525) was 12 cm lower than the floor level in the middle of room D 221 (2537). A brick support pilaster was found on the wall near the doorway. Like most houses of Level III along the main fortification wall, the enclosure wall relining was used as the north wall of the room.

The front wall of the house was built after other walls in the area were already in place, since it abutted another wall at its southeast corner, using a cornered

33 All objects from this fill layer are identified as D 214 (1). They included typical pottery mostly associated with the main occupations at the fortress (Levels III–IV). Among the finds were the sherds of a beautifully burnished black handmade bowl with a finely executed rim decoration (pl. 72e).

end for support (pl. 22b). The cornered end abutted what was probably an earlier wall that had been left standing. Neither foundation elevations nor elevations of preserved heights were recorded for some of the walls in the area, which makes the photographs of these buildings important for determining which walls were reused or merely left in place during later reconstruction when the enclosure wall addition was built.

This front wall was skewed diagonally running west. A support pilaster was built along it near the doorway into D 223, and an additional support block was built up against the north wall of the kitchen in D 222 on the other side of the doorway. The main fortification wall functioned as the west edge of the house.

The base of the dividing wall between D 220 and D 221 was 35 cm below the floor level of D 221, which means that it probably belonged to a previous building located beneath the addition to the inner enclosure but had been rebuilt and reused. However, an excavation pit carried down 16 cm below the wall base in D 221 uncovered no earlier floor level.[34]

The east wall of D 221 was built separately from the east wall of D 220, and both of these walls run alongside the west wall of the adjacent house to the east. The walls on the east were probably also reused Level IV walls. No elevations were recorded from them.[35]

Access to house D 221 was gained through a doorway between it and D 223, which was perhaps a courtyard or foyer in front of the house. D 223 opened out onto a passageway meant to provide access to these corner buildings from a main street in the West Sector. A short wall stub closed off the space just before the entrance into D 223.

The items recovered from D 221 (1) included a quern fragment (left in the field), a pottery disk incised with crossed lines (OIM E49781), sherds of marl jars with potmarks (OIM E49917, E49926), the rim of a handmade cooking bowl, a clay bottle plug (OIM E24413), and a marl flask sherd.

Areas D 222, D 200, and D 211–D 212

The enclosed space, D 222, contained two superimposed clay ovens and was discovered to be completely filled with a black ash deposit that was also found in the level above it, as mentioned previously (pl. 23a). The space was obviously built in a haphazard manner, perhaps to strengthen the wall by adding corners but also to accommodate the doorway into room D 221.

A doorway, with a stone socket on its right-hand side, opened out at the north end of the north–south wall enclosing D 200, leading into D 223. This wall, along the east side of D 200, featured a small support pilaster along its interior. A floor level was found at elevation 2549, some 12 cm above the floor in D 221 to the north.

The two sets of earlier walls defining D 211 and D 212 on the south were only minimally preserved. The base of the angled wall of D 212 was situated 20 cm below the base of the wall fragment running atop it; the earlier wall base was level with that of the inner fortification wall to its west.[36] An earlier architectural fragment ran under the north–south wall associated with the east edge of the room or courtyard in D 200, but part of this earlier wall was probably reused as the south wall for D 200 (pl. 22a–b).

Not much more can be said about the plan of the poorly preserved area nearest the West Gate, except that a possible floor surface was found just to the north of the gate. The area was obviously reconfigured at some point in time, with the earlier walls probably representing Level IV structures.

House D 209

The structure defined by rooms D 209, D 218, and D 219 was much better organized, with an entry into the front room from the passageway leading back into the complex of rooms by house D 221 (pl. 23b). Where they were preserved, it exhibited bonded walls that were fairly parallel. There were support pilasters in the middle of the walls except for the south wall and those of the smaller room, D 219. In this area, as in all areas of the West Sector, the south walls disappeared as they extended toward the center of the sector.

There was a doorway between D 209 and D 218, where the floor sloped down 13 cm. Four stone supports arranged in a square rested against the west wall of the room. The arrangement measured 75 × 50 cm. No evidence of burning is mentioned in the field notes, but the objects found here suggest it was a hearth.

The objects recovered from D 218 (1) included three pottery tuyère fragments (representing three separate tuyères); a small stone ball (OIM E50836); two pottery disks (OIM E49778, E49780), one with an incised star; and a pottery spool(?) (OIM E50833). A large pot stand

34 The test pit, D 221 (2), was meant to provide a sample of the pottery from below the lowest floor levels in the room, from which came a reused sherd of a typical red-rimmed bowl, one edge ground down for reuse as a tool, and a circular silt disk with incised crossed lines.

35 Knudstad 1964a, p. 46.

36 The dividing wall between D 220 and D 221, mentioned earlier, was 38 cm below these Level III foundations and may be assigned to an original Level IV structure.

and crucible fragments were found near the floor in D 209 (OIM E42960), while the fill above the floor yielded a faience Menkheperre scarab (OIM E24396). The excavation below the floor, in D 209 (2), yielded a pottery disk with incised crossed lines (OIM E49723). Some of the objects indicate that the area may have functioned as a workspace for repairing metal items or melting them down.

Beyond the hearth was a doorway, 47 cm wide, leading into the small room D 219. This doorway was narrower than the doorways in the buildings along the north wall, the average width of which was about 62 cm. A one-handled jug exhibiting a red-slipped exterior with vertical burnish strokes came from this room (OIM E49524), along with various bowl sherds.

Room D 219 shared its west wall with rooms D 221 and D 223, while the north wall of the adjoining room, D 218, was apparently bonded(?) with the earlier wall left standing between houses D 216 and D 221. A north–south wall divided the house from passageway D 208.

To the south of the house were slight wall remains that were not assigned a locus number. At least one doorway in this poorly preserved area opened out onto a continuation of the passageway of D 208 to the east. A stone door socket was found at the north, on a level even with the wall base in the area, at an elevation of 2524, some 30 cm below the floor in D 209. Comparable elevations in the northwest corner were the floor in D 220 (2525), the south wall base of the house D 224 (2527), and the earlier wall in D 74, to the east (2527).

The north–south trench excavated across the center of the West Sector began at the southernmost edge of this area.

Passageway and Ovens in D 208

Three ovens appear on the final plan at the north end of a lane leading back toward an earlier doorway of house D 216, although in the later phase there does not appear to have been a doorway into the house (see pl. 18b).[37] The ovens were built in succession on different levels. The middle oven rested at elevation 2557; the oven wall to its north was 16 cm higher, while the latest oven on the south side rested on a level that was 34 cm higher. They appear to have been rebuilt, however, after which they may all have been active at the same time. The tops of the ovens were leveled during a later renovation of the area.

The ovens ranged between 78 and 93 cm in diameter. The preserved stoke holes on the two southernmost ovens pointed eastward toward the passageway. A quern was found lying upside down to the east of the three ovens, at a level associated with the earliest oven (found in D 208 [2]; left in the field).

As mentioned above, the west wall of D 208 functioned as the dividing wall between the passageway and house D 209. The wall on the east side of D 208 ran parallel with and alongside the adjacent wall of house D 224. A narrow doorway(?) or opening was situated at the north end of the latter wall, measuring less than 50 cm in width and opening into D 207. A lower wall stub found within the doorway may have served to support a threshold leading into the area of D 207 when a later staircase was built over the walls here, or it may have been a remnant of a Level IV feature.

Room D 216 and Level IV Remains

Room D 216 eventually became part of house D 1 to its east (see below), although at one time it was part of a house composed of D 216, D 215, and D 217. There had been an earlier doorway in the south wall, although another wall built within the room had no doorway. The latter stood higher than the wall with the doorway (pl. 24). Although the elevations of the foundations of these walls and their preserved heights are unknown, the photographs of the area confirm this observation. The two walls ran alongside each other. The higher wall, however, appears to have been bonded with all other walls in the room. A doorway possibly led into the room from D 1 to the east, where the wall was incompletely preserved at its north end. If so, the complex of rooms was very large.

Rooms D 215 and D 217 to the west are visible in the photographs because of the preserved height of their walls. The building of the third relining wall cut through the north wall of D 215, leaving it only 10 cm wide. It is unusual for these buildings on the north to have a separate north wall, apart from the enclosure wall, but this slice of wall was left standing after the inner enclosure wall addition. It remained standing 51 cm above the nearby D 220/D 221 cross wall and was 55 cm above the east wall of D 216.

There was no apparent access to D 215. In addition, the doorway in D 217 was blocked off at some point in time. It is possible that the west wall of D 215 was cut away (as seen in the photograph on pl. 24) so that the room could be accessed at a later time from D 220, unless of course this happened during the excavation of the area.

37 The photograph shows the ovens at the beginning of their excavation; a later wall remnant lies to the left of them.

D 216 and the two associated rooms perhaps constituted a three-roomed house in an earlier phase of Level III or in Level IV. There was no apparent access into D 215. In addition, the doorway in D 217 was blocked off at some point in time, and these rooms were not interconnected. It is possible that the west wall of D 215 was cut away (as seen in the photograph on pl. 24) and the space became a storage area with an elevated access for the occupants of house D 221 to the west—unless, of course, the wall was cut into during the excavation of the area rather than during a Level III reconfiguration. After the blocking of the main entry into D 216, the latter, which was likely a Level IV room, may then have been handed over to, or taken by, the occupants of D 1.

In addition to potsherds, the fill of D 215 (1) yielded a quern, which was left in the field. A crucible fragment was also discovered in the fill (OIM E42961). Two querns, one of them sandstone (OIM E50782), and a handstone were found in D 216 (1).

Staircase D 207

Area D 207 underwent a number of rebuilding phases, but the eventual reconfiguration supported a staircase rising from west to east. This staircase may have provided access to the tops of the houses and over to the north fortification wall in a later Level III use of the area.[38] The Z-shaped foundation for the construction was added atop an earlier wall in D 206/D 207. Both the earlier wall top and the foundation of the staircase were at elevation 2573, while a floor level in D 207 and the bottom step rested at 2567.

The staircase was 97 cm wide, although little of its original extent survived. The preserved top of the stairs, which was at about the same preserved height as that of the south wall of house D 1 to the north, had four steps ascending 36 cm from the floor in D 207. There is no recorded elevation for the blocking stub or possible raised threshold between D 207 and D 208 shown on the plan, but it is clearly visible in the photograph on plate 24. The floor level before the stairs was 17 cm higher than the floor in D 216. It is unclear how the different levels of ovens in D 208 are related to the staircase or to the floor next to it, but the base of the earliest of the ovens was 10 cm below this floor level, that of the north oven was 6 cm higher, and that of the latest oven was 24 cm higher.

Besides sherds, the most distinctive find in D 207 was a spherical object made of stone, perhaps a weight (OIM E50835). Numerous sherds were found in D 206 (2), which was a deposit of fill excavated to uncover the earlier wall below the stair construction.

House D 224

At the south end of passageway D 208 stood house D 224, with the larger room D 205 to its north (pl. 25a). In the room identified as D 205, a stone column base was uncovered in the center of the main hall, which is unique among the architectural features that came to light in the West Sector. The east and south walls of the house were also wider than those of a typical residence at Dorginarti, with widths of 37–40 cm, but a thinner partition wall between the rooms had the typical support pilaster built into its middle stretch, and the inner wall on the west was also thinner. The latter wall was adjacent to and ran alongside another wall extending over the length of D 208. Both were preserved at the same height. The south wall of the building extended eastward and also formed the south wall of D 225.

It is uncertain whether or not the westernmost wall on the west side of house D 224 was bonded with the thicker wall to the south, since the bricks were very eroded and powdery. The oven shown near the doorway between D 205 and D 224, which was not excavated, was part of an earlier level in the area. The possible doorway into the house, which is not indicated in Knudstad's final plans, was possibly located in the west wall by the earlier oven.[39] A larger, thick-walled structure, the plan of which was not preserved, may once have stood at this spot before being modified during the Level III occupation.[40]

Pottery was found in the fill of these two rooms, but the only object recorded was a red bead from D 205 (1) (OIM E50820).

Area D 225 and Kitchen D 74

The wider walls of house D 224 continued here (fig. 4.1). The south wall extended eastward and was apparently bonded with the section in D 225. The wall, and any doorway that may have led into the area, was no longer preserved at the southeast edge.[41]

38 There was evidence that a window or door had been built into the parapet along the south side of the fortification wall near this area; see chapter 3 on the enclosure walls.

39 As shown by a sketch in Hoerth 1964a, no. 1.

40 Thicker wall remnants are also preserved to the east in D 59 and D 68, perhaps also belonging to an earlier structure whose north walls were cut through by the third wall addition.

41 Knudstad 1964a, p. 46.

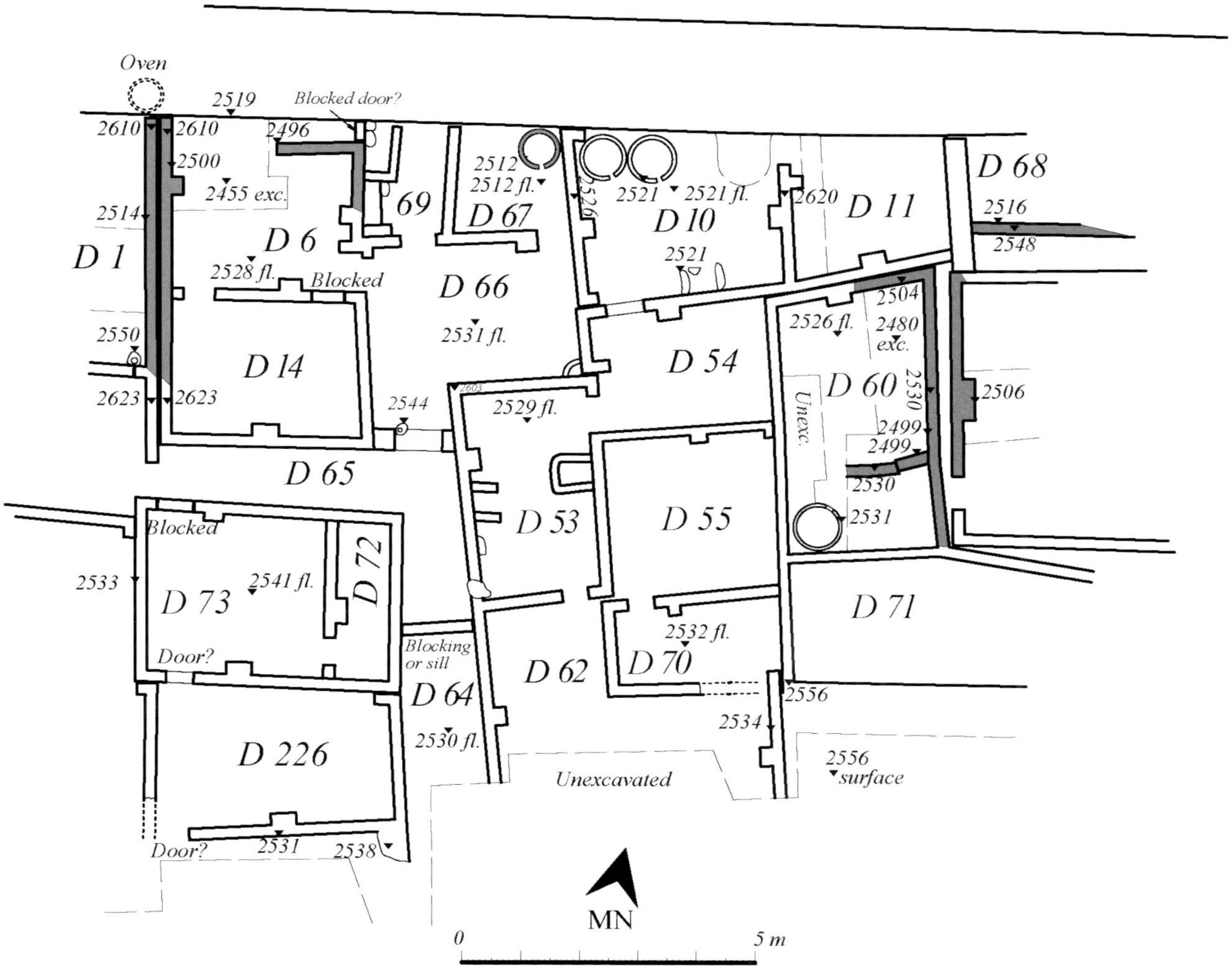

Figure 4.2. Houses along the middle part of the north wall

The oven in D 74 was founded at an elevation of 2551, while the earlier wall to its north was founded at 2527. The wall is barely visible in the photograph of the area, and we can assume it was covered over during the use of room D 74.[42] A wall contemporary with this kitchen area was inserted to close the space on its north. No floor surfaces were found.

A fused faience bead (OIM E50827) was found in the fill of D 74.

House D 1 and Level IV Walls

D 64 and D 65 (figs. 4.1–4.2) were both part of a lane leading back to D 204, an entrance room into house D 1 (see fig. 4.1). The passageway is discussed in more detail below. The area could have been covered over by the staircase in D 207 at a later phase, which would have rendered both entry room D 204 and house D 1 unusable. But if the staircase was L- or U-shaped and the landing had led southward (and perhaps then westward), some space may have been left by the typical solid stairwell compartment. The entry into the house may also have been accessible if the staircase had incorporated timbers and beams as support instead of solid brick fill.

During the surface clearance over D 1, the walls were traced down through a number of upper strata, including the dark alluvium layer ("floor 2"), which was about 1 m below the original surface.[43] As mentioned in the introduction to this chapter, this darker or "ash" level was found throughout the west side of the fortress. It was not a living surface as initially

42 No wall heights were recorded, although the foundation levels of both walls were noted.

43 Floor 1 was a small, hardened surface in the area of the goat burial and was mentioned earlier in this chapter. See Pierce 1964b, pp. 9–11, for a detailed description of the levels of the excavation, including that of the test pit in the northeast corner of the room.

thought, but rather resulted from the action of repeated water saturations, which caused dark silts and ash deposits to permeate the walls and fill of the buildings. Strata 1 and 2 were later combined into one fill level.

Floor 3, which belonged to the Level III use of D 1, was contemporary with at least some of the walls shown in the plan. There was a quern embedded in it and yet another in the fill above associated with a handstone. The floor was 1.3 m below the natural surface. The walls of the building abutted the fortress wall on the north, which bulged southward slightly in this area from saturation by floods.

A stone door socket was found embedded in the floor in the east doorway of D 1, which had been blocked at some point in time, while the west doorway had also been blocked up with bricks and mud.[44] The west door was only 43 cm wide and was perhaps closed off during the reconstruction in D 207—that is, as a consequence of the building of the stairway.

The east wall of house D 1 ran alongside the wall used by two rooms in house D 66, while its west wall was shared with house D 216. A fragment of a wall and a corner, shown on the plan inside the room, represent earlier Level IV remains in this area. This earlier architecture had been founded on sand 34–38 cm below the wall bases and door socket of house D 1, at elevation 2513, and 10 cm below the base of the inner fortress wall relining. The preserved height of the earlier wall was level with the floor, as represented by the base elevations of the door blocking and stone socket, in D 1. The line of the earlier wall ran north–south before turning to the west and disappearing under the east wall of D 216, which is also an earlier wall that was partly reused in Level III. It had been cut through when the inner wall relining was built. The southern extent of an earlier wall abutted this wall and is also considered a remnant of Level IV that was not reused in the later structures. These walls are not visible in the photographs of the area, although the higher wall between D 1 and D 216 built next to them is visible. The northern extent of the wall contemporary with D 1 and D 216 was cut back to create a doorway between the rooms.

In addition to the objects recovered in the fill of D 1 mentioned above, D 1 (1) yielded multiple pottery disks, one with incised crossed lines (OIM E47415, E47421, E50798), as well as the fragment of a rim from a stone bowl (OIM E50812). The sherds from D 1 (2)—that is, the fill above the floor of the house—included a bread plate (OIM E49583) and a mat-impressed handmade cooking pot. D 1 (3), below the floor level in the room, yielded sherds, crucible fragments (OIM E51021–E51023), and pottery disks, some with incised crossed lines (OIM E47447–E47458).[45]

When the excavators noticed that the floor associated with D 1 ran under the fortress wall relining, they decided to sink a test pit through the floor on the east side of the room, where they proceeded down through layers of fill.[46] The fill of stratum D 1 (4) was found to consist of sterile gray sand, while D 1 (5) was a stratum of burned or discolored alluvium and ash, the bottom of which was 1.47 m below the level of the door socket in D 1. One red-polished handmade silt sherd with a black interior was encountered in D 1 (5), and a bone fragment and a worn sherd were also found (OIM E49012, E49013). This test pit reached approximately 2.60 m below the natural surface (pl. 25b).

A thin test pit was sunk along the east wall of the house, just to the east. It cut 30 cm into the wall relining in the northeast corner and also into the east wall, exposing an earlier oven below the relining.[47] The base of the oven (elevation 2505) lay 45 cm below the door socket and probable floor in D 1, and 18 cm below the base of the inner fortification wall relining. It is significant that the east wall of D 1 and the west wall of D 6 and D 14 were both founded at lower levels, 2514 and 2500, respectively. The lower-lying wall of D 6 and D 14 was thus contemporary with the oven under the wall relining, but it was later rebuilt and used in the rooms to the east.

House D 73

This house was composed of D 72, D 73, and D 226 to the south (fig. 4.2; pl. 26a). An exterior doorway had once led into D 73 from the passageway defined by D 64 and D 65, but this doorway was later sealed off. Another possible doorway may have led into room D 226 at its southwest corner. Although the southern extent of the west wall was found to be missing, the finished edge of the south wall suggested the edge of a doorway.[48] A doorway was built between D 73 and

44 The "door" blocking may also have been a wall patch constructed during the Level III rebuilding and reuse of the earlier walls.

45 Pierce 1964a, January 9. See also January 19 for a discussion of the earlier features noted under, and next to, the wall relining.

46 Pierce 1964b, p. 8, mentions a fourth floor within this test pit, 30 cm below the floor and door socket levels in house D 1. While not shown on the final plans of the site, the field notes depict this possible fourth floor. The lower floor was apparently above walls that were oriented differently from those in D 1.

47 Knudstad 1964a, p. 46.

48 Knudstad 1964a, p. 45.

D 226 in the cross wall dividing the rooms and possibly cut through an existing wall. It was 45 cm in width. The area south of D 226 was not preserved, although a thicker wall abutted, and was perhaps bonded with, the southeast corner and was preserved for 55 cm to the south.

The walls surrounding D 73 were constructed with support pilasters on the interior of the north and south and were perhaps bonded. The east and west walls of D 226 were then built abutting the south wall of D 73 at both corners, with short bent-end sections adjoining the earlier wall. A wall divided D 73 from the narrow room, D 72, with a doorway at the south end. The wall separating D 72 from the main hall of the house also had a brick support construction by the door. The west and south walls of D 73 and D 226 were founded 8–10 cm below the floor level of room D 73.

A stone arrowhead (OIM E24393) was found in D 73 (1) on an ash deposit near its blocked-up doorway and only 10 cm below the excavated surface. In addition, a deposit of two faience beads and one of red quartzite (OIM E50813) were found along the west wall in stratum 1.

Passageway D 64 and D 65

A lane defined by D 64 and D 65 ran north and then west into a number of house units along the north wall (pl. 26a). It gave access to house D 66, the doorway in house D 73 before it was blocked up, and the east door of house D 1 before it was blocked off. Presumably the lane started off from a main road somewhere toward the middle of the West Sector. In D 64, there was a wall or perhaps a high sill blocking the passageway. In the photograph this sill(?) is at the height of the preserved walls to its south. Whether it was meant to prevent animals from passing along the lane into the houses or to block the lane completely remains uncertain. However, at one time the passage had turned west in front of the doorway into D 66, and from there it led back to house D 1, which has been discussed above.

House D 66 and Level IV Walls

This complex of rooms included D 66, D 6, and D 14 on the west, and the smaller areas of D 69 and D 67 along the enclosure wall. A short wall with bent or L-shaped ends was inserted into the southeast corner of D 66 to separate it from D 54 to the east. A mud bin was then inserted in the corner. Two short jambs abutted the walls to the east and west of the door. The doorway had a brick sill and a stone socket on the west side, the latter set atop the sill 13 cm above the floor in the center of the room.[49]

Besides pottery, the objects from the fill above the floor in D 66 (1) included a white faience disk bead and a faience segmented tube (OIM E24388). A pottery disk with an incised cross (OIM E49731), a river cobble of basalt with evidence of use as a pounder, and three querns or quern fragments were also found here. The very worn quern fragment in Chicago (OIM E50772) is made from sandstone. A crucible rim (OIM E51017) also came from D 66 (1).

A doorway at the northwest corner of D 66 led into room D 6 (pl. 26b). Walls in this room seem to have been structural elements that existed before the third wall relining was added and were reused in the Level III house. A test pit was excavated in the northwest corner of D 6, measuring approximately 1.65 m on each side and sunk 73 cm below the floor level associated with the room. No architectural or material remains were found in the test pit, although the base of the west wall was identified.

The west wall of D 6 and D 14 was founded 28 cm (elevation 2500) below the floor level of D 6 (2528). This was 14 cm below the east wall foundation of D 1, which itself was founded below the surface of the Level III walls in D 1. The elevations of many wall bases in house D 66 were not recorded, but the short east–west wall in the northeast corner of D 6 was also at a lower elevation (2496). The photograph on plate 26b shows the preserved tops of the two walls between D 1 and D 6 at the same elevation of 2610 next to the fortification wall and 2623 at their southern extents. The base of the inner fortress relining was 9 cm below the floor level of D 6 but 19–23 cm above the early wall foundation levels.

The east end of the narrow 45 cm space between the wall at the northeast corner of room D 6 and the inner fortress relining was blocked during the Level III reconfiguration of the area. Since the wall was left standing, and perhaps rebuilt, the space was obviously still used. This east–west wall was bonded with the north–south wall between D 6 and D 69. The foundation level of that wall is unknown, but it can be seen standing in photographs of the area.

Two faience Hathor heads (OIM E24394 and Khartoum 14270) with the exterior glaze rubbed away were found on the floor of D 6. In addition, three querns or fragments of querns, one of "basalt" or red granite, stemmed from the fill above the floor in stratum 1. Pounders were also found here (OIM E50774, E50775, E50781), one with traces of a reddish deposit. Four

49 Knudstad 1964a, p. 45 obverse.

plain pottery disks were also uncovered in this room (OIM E47461–E47464). Finally, the pottery included a marl flask, a large wide-mouthed jar, and a jar stand.

There was initially a doorway between D 6 and D 14 at the east end of their dividing wall, but this door had been blocked off and another doorway made at the west end of the wall. The south wall of room D 14 had a brick support pilaster on its south wall. The base of its west wall was left unrecorded, so it is uncertain whether this part of the wall was also founded at a lower level. Two pounders and some plant fiber or bark (OIM E49011) were uncovered in the fill of D 14 (1).

Room D 69 was entered from a door in D 66 (pl. 27a). The small space contained a number of brick features against the west wall, including a rectangular bin measuring 1.03 × 0.57 m. One can see it in the photograph of the area, which also shows the bin walls preserved at varying heights. An adjacent bin or shelf emplacement, 30 cm deep, could have functioned as a workplace for grinding grain or other substances. Grinding and pounding implements were also found in this room, including two querns, one of sandstone (OIM E50773), and a gray sandstone pounder (OIM E50799). Seven pottery disks were also found here (OIM E47403–E47409), some with incised crossed lines. Pottery from the room included a rounded silt jar base (OIM E49793), which had been reused as either a lid or a saucer. The excavators frequently found marl flasks and the common type of "Dorginarti" pot stand together with grinding and pounding equipment, and one of each was uncovered in D 69 (1).[50]

The floor level of D 67 (2512) was 19 cm lower than the floor in D 66, and the small oven in its northeast corner rested upon this lower floor but had been built up for continued use (pl. 27b). The base elevations of the walls dividing it from D 69 and D 66 are not known, but the photographs of the walls and oven show them standing as high as other walls in the area. The difference in elevation between the floor of room D 67 and other floors in this structure suggests that the floor was an earlier Level IV remnant, perhaps with a step down, although no step is visible in the photographs or recorded in the plan. The earlier wall between houses D 216 and D 1 to the west, for example, was founded at 2513, although part of it survived at heights to show that it might have been usable in Level III. The base of the fortress wall relining in D 6 was at elevation 2519, having sloped 4 cm lower than in D 1.

50 See fig. 5.55 in chapter 5.

The objects from D 67 (1) included seven pounders, one of black sandstone (OIM E50778), and a handstone (OIM E50787).

Passageway D 62 and Workspace or Kitchen D 53

D 62 and D 53 were two connecting loci forming a passageway that led back to D 10. They were situated immediately east of the passageway wall along D 64 and D 65 (see pls. 27a and 28a). The north wall of D 53 was shared with D 66. The south end of the long wall on the west was eroded away toward the center of the West Sector. Around its midpoint, north of a support pilaster, was an east–west wall forming a narrow entrance into the kitchen or work area of D 53. The floor level in D 53 was 2 cm above the floor in D 66.[51]

Two bricks were placed perpendicularly against the west wall of D 53, some 67 cm apart. They stood one course high and extended out 45 cm into the passage. This appears to have been a hearth, although there is no mention in the record of ash or burning in the area. Immediately to the south of this emplacement, a wall chink had been repaired with a stone plug. A bin with two courses of brick was built against the east wall and projected out into the space opposite the hearth. The bin measured 65 × 70 cm. Just beyond this to the east was the doorway into D 54.

The objects from D 53 included two reworked pottery disks (OIM E47497, E49726), a faience bead (OIM E24384), and a quern (left in the field).

House D 54

This house included rooms D 54, D 10, D 11, and probably also the previously discussed courtyard or room D 53.[52] This area was located partially below the later Level III or Level II wall fragment built of 40 × 20 cm bricks, which ran over the west wall in D 11 and to the south into D 71.

There were shared walls throughout this complex, with a short cornered wall segment inserted between D 66 and D 54. D 10 was entered through the west end of a long wall running to the south of D 10 and D 11, through a doorway that had been cut into the

51 This area is also shown in Knudstad 1964a, p. 45.

52 The photographs of the area appear to show a separate wall built at the north side of D 10 and D 11; since the plans do not show this, it was perhaps an eroded part of the inner fortress relining.

wall.[53] The walls had support pilasters along their lengths. None of the wall base elevations in this area were noted.

There was a segment of wall inserted between room D 10 and rooms D 66 and D 67 to the west (pl. 28b). The wall shows the familiar L-shaped segments at both ends, abutting both the main fortress wall and the wall on the south, and it too had a support pilaster.

The two contemporary ovens shown on the plan of D 10 in the northwest corner of the room, along the fortress wall, were placed at level 2521, which was also the level of the floor of D 10. This was 8 cm below that of D 53 and 10 cm below the floor of D 66. The ovens each measured about 80 cm in diameter, with stoke or ventilation holes oriented south. The area below the base of a later oven at the east of the room was left unexcavated, but it was not contemporary with the room, instead belonging to an upper level in the West Sector (the later oven is shown on plan 3). The stone hearth on the south side of the room featured a reused quern fragment placed on edge to form its west support.

A narrow doorway between D 10 and D 11 was found at the north end of a dividing wall that bonded with the long east–west south wall. Part of the area within D 11 was left unexcavated because the later wall ran over the westernmost side of the room. Support pilasters were built on the west and south walls. The east side of D 11 was bounded by a thicker wall fragment 37 cm wide and was shared with D 68. The foundation levels of this wall are uncertain, and it cannot be determined which walls in this area may have originally belonged to buildings located beneath the inner relining wall. The insertion of short walls as boundaries between rooms or courtyards was a common way of reconfiguring areas after the addition of the relining in Level III.

Potsherds were found in all these rooms, with a large quantity of different types of both wheel-made and handmade vessels coming from D 10 (1). These included a marl wide-mouthed jar and a pot-stand type commonly found at the fortress. Two pottery disks with crossed lines incised on them (OIM E49762, E49763) were excavated in D 10 (1), and a nodule of rock with a cavity was found on the floor of D 10. A silt bowl with red rim and a quern also came from D 10 (1) (OIM E24348).

53 Knudstad 1964a, p. 45.

House D 70

House D 70 comprised rooms D 70 and D 55 (fig. 4.2). It was perhaps entered through a doorway along the ruined east end of its south wall, since there was no other evidence for a door into the two-roomed complex (see also pl. 27a). The two rooms were mostly bounded by wall segments displaying the typical cornered ends, not bonded with adjoining walls, and with occasional pilasters erected along them for support. In addition to abutting one another, they also abutted the long north–south wall forming the east limit of the house, which was shared with the house next door. The narrow doorway between D 70 and D 55 was located at the west end of the dividing wall.

The end of a wall was found to be jutting into house D 70 from the south at its southeast corner, suggesting another room to the south, although the architectural remains were not preserved here.

The level of the excavated floor in room D 70 was about the same as the floor in D 53. It was uncovered 24 cm below the unexcavated sand surface to the south of the building. The preserved corner of the adjacent area, D 71, was level with this unexcavated sand surface and is barely noticeable in the photograph.

The fill within D 70 (1) yielded a faience bead (OIM E50828) and two gray soapstone whetstones(?) (OIM E24390A), one of which showed nicks at one end and had been worn down in the middle of one side. Also uncovered here was a corroded copper blade that had been broken into four fragments (OIM E24390B). The latter two objects were found near the possible door into the room and below the "ash" layer.

Area D 71

The west wall of area D 71 was shared with house D 70 and ran north–south along the entire area. Another wall fragment was preserved on the south side, and a short stretch ran out diagonally from the southeast corner with D 60 (pl. 29a). All the east–west walls disappeared toward the east. No analysis can be made of the function or structural association of this area.

Passageway D 56 and D 58

The southern extent of this passageway had disappeared, along with all the walls toward the center of the West Sector. Only the stretch leading into house D 59 and the stairway of D 57 were preserved (pl. 29a). The passageway continued north before it apparently turned eastward into a group of rooms along the north wall, D 44 and D 42–D 43, perhaps all remnants of the earlier Level IV.

House D 59 and Level IV Walls

This house included rooms D 59, D 68, and D 60 (fig. 4.3; pl. 29a–b).[54] The designation D 59 was assigned to both this room and a poorly preserved "building D 59" at the northeast corner in the West Sector, which will be discussed later.

The doorway leading into house D 59, along the west side of the corridor mentioned above, was 60 cm wide. The east wall of the house was 37 cm wide along most of its preserved extent, matching the width of the walls at the east and west sides of room D 68 to the north, both of them much thicker than the walls in this residential quarter, and indicating the remnants of a sturdier building that stood here at an earlier time. None of the wall base elevations were recorded. The thinner walls of the building did not usually exhibit support pilasters to strengthen them. The west wall was shared with room D 60. The base of this wall was founded 18 cm lower than the floor level in the room (elevation 2524), so some reuse of earlier walls occurred here. The south wall was, however, built only 2 cm below the floor level in the room.

A fireplace rimmed with bricks, 74 cm in diameter, was placed off-center in the main hall. The brick edges were flush with the floor level of the room, but evidence of burning within the fireplace ring had been preserved.

The entrance to the back room of the house, D 68, was to the right upon entering the house. There was a large stone on the east side of the doorway in D 68, although no wear pattern was found on it to suggest it had supported a doorpost. Two wall partitions are shown on the final plan of the room, which were noted as earlier walls in the field drawing. A test trench was sunk down to the top of a level of alluvium to clarify the nature of the lower walls, reaching bottom at 2424. The north–south wall was founded at 2495 and was cut by the relining wall of the fortress enclosure (2503), while the foundation of the north–south wall was 29 cm below the floor found in D 59. The east–west wall it adjoined was founded higher up, at 2516. The latter wall can be seen in the photographs of the area but was undoubtedly covered over by debris and a floor when house D 59 was in use (pl. 30a). But it was still preserved standing at a height 24 cm above

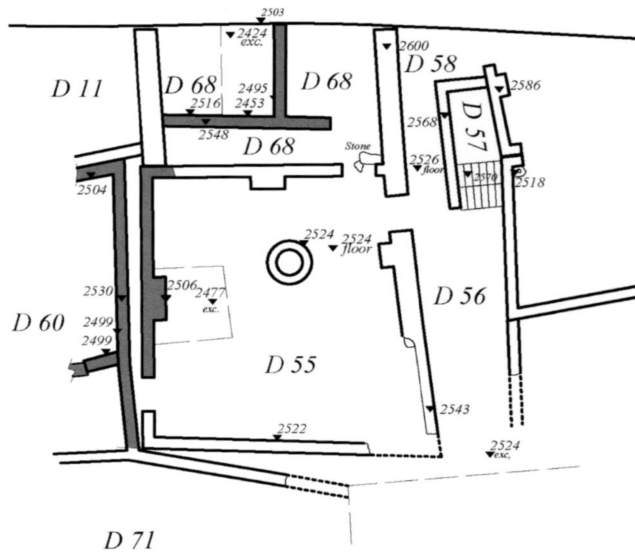

Figure 4.3. House D 59 and area (earliest walls are gray)

the floor in D 59, so the back room was situated on a higher level.

The field and pottery notes for the area indicate that four strata were excavated in D 68, with a lower floor at D 68 (4) being noted as on the same level as the floors in D 40 and D 41 to the east (see below) and D 68 (3) being noted as on the same level as the floor in D 60 (2).[55] These floors are not noted on the final plans. The floor in D 68 (3) was associated with the cross walls in the south half of the Level III room. There are no elevations on the final plans for this test trench, which was excavated at the end of the season.

The back room, D 68, may thus have had a higher floor than room D 59 to its south, having been built rather haphazardly on the Level IV architectural remains below it.

A doorway connected D 59 with D 60 to the west, where a contemporary floor level was about even with that in D 59 (2526 and 2524) (fig. 4.2). The foundation of the north wall of room D 60 was found to be 22 cm below the level of the Level III floor in the room, while the south part of the east wall and another lower partition below the floor were founded 27 cm below the Level III floor. The wall shared with D 59 was also resting on a level that was 18 cm lower than the floor in D 59, so it too may have originated in the earlier Level IV. The wall heights noted for the lower east wall and the "partition" wall show that they had been preserved 4 cm above the floor level in the room, but they

54 The most distinctive sherds from the various stratigraphic levels at D 68 are presented in fig. A.I.17 in appendix I.

55 Hoerth 1964a, no. 1.

are not visible in the photographs (pl. 29a–b). The photographs appear to show other wall fragments in the northeast (and northwest?) corner in D 60, which had been left standing during the Level III use of the room. The base of the oven in the southwest corner, however, was on the same approximate elevation as the floor to the north of it and was contemporary with the Level III use of the room. It appears on plate 29b below the later long wall of an upper level in the West Sector.

A test pit was excavated below floor 1 in D 60 (2526) and eventually was expanded northward.[56] This test pit ended 46 cm below the Level III floor. A floor level that is mentioned in the field notebook but not noted in the final plan seems to have been found 40 cm below the upper floor.[57] This floor would have been below the levels of the earlier wall foundations. Since this detail is not evident in the final plans, subsequent examination may have led to a rejection of this feature.

The excavations in the area of house D 59 offer strong evidence of the complexity of the site, with the field records pointing to a Level IV or earlier occupation along the north enclosure wall.

Other than potsherds, the few objects uncovered in this complex of rooms included a pottery disk with incised crossed lines (OIM E49751) and a bread plate with incised decoration on its upper surface. These were found in D 60 (2) north, which was even with the top of the earlier east–west cross wall shown on the plan and thus is associated with the Level III floor in stratum 1. A stone net weight was found in "D 59, by the west wall at the north" (OIM E24380) and is assumed to come from here rather than from building D 59, which is discussed below. Two querns were found in D 59 (1), and three more were uncovered in D 60 (1); all were left in the field. The sherds of a broken but complete marl jar (MJ 1) with multiple potmarks were found in D 59 (1) (OIM E49527). Finally, a marl flask and silt "Dorginarti" pot stand were also found in D 59 (1), and a wide-mouthed marl jar came from D 60 (1).

Stairway D 57

Passageway D 56 led north to the stairway preserved along its east side, D 57 (pl. 30b).[58] Only the two lower steps remained. These steps had a depth of approximately 39 cm, and the topmost tread was situated 44 cm above the surrounding floor level near the base of the stairs. The west wall supporting the stairs turned eastward to abut the wall that was shared with room D 44. The width of the staircase was 83 cm.

This flight of steps was built outside any of the structures, as was the case with the stairway preserved in D 207. Thus, both offered public access to rooftops in the area or to the top of the main fortification wall. It was atop this area that an intermediate-level wall had been built, as mentioned above. Its base was measured at an elevation of 2568, which was the same level as the preserved wall top along the west side of the Level III stairway, although the east wall was left standing 18 cm higher.

The only objects recovered from this area were two pottery disks from D 58 (1), one with incised crossed lines (OIM E49733–E49734).

LEVEL III AND LEVEL IV WALL REMAINS BY THE NORTH WALL BREACH

House D 43 and Area D 44–D 45

The floor levels in the group of buildings immediately to the west of the north wall breach had elevations that differed by as much as 27 cm, and the lowest floors were therefore on par with some of the earlier wall foundations and floors to the west, as described above (fig. 4.4). The floors in D 43 and D 42 were 8–14 cm lower than the floor of D 44 (2523), and they were also lower than the floors in rooms D 228 (2533) and D 232 (2523) under the wall relining at the north breach. The lowest elevations in this area are therefore associated with Level IV walls and floors, of which some of the former were reused during Level III. The damage in this area removed much of the architecture, with many of the later floor levels and walls in the area having been eroded away. The higher levels associated with the structures under the wall relining suggest that the walls and buildings to the south were cleaned out to a deeper floor level to be reused after the wall relining was built.[59]

Although the southern portions of the walls in this area were not preserved, the complexes here were smaller—and also less regular—than the structures to the west.

House D 43 includes area D 44 to its north and a small side room in D 42 (pl. 31a–b). D 45 was perhaps

56 Hoerth 1964a, no. 1.

57 Hoerth 1964a, no. 1: "A cross wall was found in this test pit, and a trench was then run north in the room on the other side of the wall [and] ca. 40 cm down (which equals level 3 in D 68 and the fill is the same). D 60 North (3) [indicates] the sherds from ca. 20 cm beneath this second floor (only tested in the northeast meter of the trench)."

58 Knudstad 1964a, p. 44.

59 The lowest floor surfaces under the wall relining are similar to those in D 42 and D 43 (elevation 2516 in the space west of D 230, 2509 in D 231).

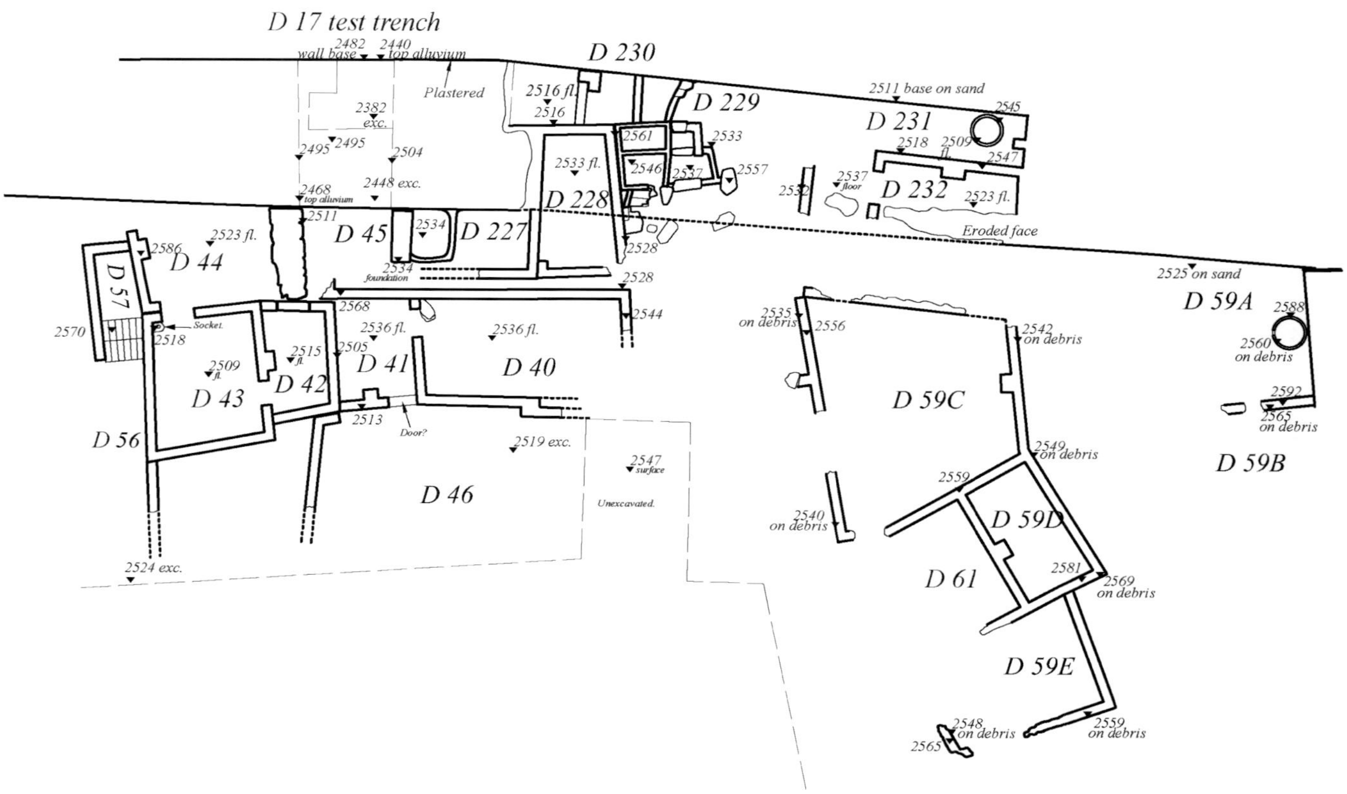

Figure 4.4. Houses and building D 59 at the north wall breach

also part of the unit before a wall was built at some point in time to seal it off. A stone socket was found on the west side of the doorway leading into D 43. The front room was linked to a slightly higher floor in D 42 through a doorway in their shared wall, which had a support pilaster next to the entrance.

Another doorway led from D 42 into D 44, but it had been blocked off by a thick wall of mud and stone that had been constructed to the north of it. This wall of mud and stone was 60 cm wide and was founded at a lower elevation (2511), and perhaps cut through by the wall relining. The test trench D 17 was sunk beneath the inner relining to the east of this wall and north of D 45.

The south and west walls of D 43 were apparently bonded, and the walls built to enclose D 42 abutted them. The base of the east wall of D 42 was at an elevation of 2505 and therefore can be associated with the lower floor levels in D 43 (2509) and D 42 (2515). The floor level in D 41 to the east, however, was 21–27 cm higher. The elevation of the fortress relining to the north, where an exploratory excavation of the wall had been undertaken as D 17, was 2504 on the east side and 2495 on the west (see below).[60]

60 Knudstad 1964a, p. 44. The inner wall of the main enclosure, built before the inner relining, was founded at 2482. No building remains were uncovered during this wall base exploration.

The other walls in this area included a wide brick wall stub to the east, which partially closed off the area between D 45 and D 227 and against which a mud basin had been built. Both were at a level of 2534 and were comparable to the floor in D 41. The fragmented north wall of D 41 had a corner segment that had once continued to the north into the area of D 45. The latter wall had a foundation level of 2528 at the east. As to what chronological phase all of these structural remains may have belonged, we simply cannot determine with the evidence recorded.

House D 41 and Area D 46

The south portion of the area just described had been destroyed by floodwaters. The area south of house D 43 was not assigned a separate locus number, but there were two separate wall fragments that ran in a southerly direction before disappearing. The west wall abutted the southwest corner of house D 43 and was aligned with the house wall. The cornered end of the east wall rested up against the southeast corner of the house and extended out diagonally to the southwest. There was nothing of note preserved in area D 46 except a possible doorway leading into house D 41 and an irregular pit on the west side of the unit.[61]

61 Hoerth 1964a, no. 1.

House D 41 was preserved as a small two-room unit. There was a stone socket in D 40 on the north side of the doorway between D 41 and D 40. As mentioned already, the two rooms had the same floor level, which were 21–27 cm higher than those to the west. The elevation of 2536 was comparable to a few floor levels found in structures under the wall at the north breach (D 228, D 229, and D 232).

Objects from this area include a copper kohl stick (OIM E24415) and a bone point or awl from the same area (OIM E24416). Both were found east of D 40 and below the surface. D 41 (1) produced a black sandstone quern and two stone pounders (one of which is OIM E50776). A faience bead (OIM E24378) was found on the northwest corner of D 42, and one fragment of a white stone quern came from the fill of D 42 (1). Two pottery disks (OIM E49727, E40728) were found in D 44.

NORTH WALL BREACH AND LEVEL IV REMAINS

As mentioned previously, the excavators cleared the remains of the inner wall relining at the north breach.[62] The relining was found to be resting on layers of ashy flooring that ran under it and then abutted the base of the earlier core wall. These layers were also observed in the test excavations in D 1.[63] A plastered finish was found on the inner face of the double-walled enclosure in this area, with the buildings found here being constructed against the innermost face of the inner core wall, or cut by it.

Also as mentioned previously, the removal of part of the inner relining at house D 1 revealed a similar situation. Here, the floor of the house continued under the wall as well. As mentioned previously, the test trench D 17 was cut through the inner relining to the west of the breach in order to explore the base of the inner fortress wall (pl. 31b). The probe under the inner wall exposed wind-laid layers of sand, ash, and a few sherds above a surface of natural alluvium, but no structural remains. It was noted that

> sand levels deeper than any of the outer wall foundations were also tested and found to contain ash lenses and a few nondescript sherds. Some form of light occupation prior to the erection of the fortress is thus supposed for the spot tested, but as our exploration did no more than establish this fact, little can be said about the occupation itself.[64]

Two test pits, on the interior and exterior sides of the exterior core wall of the enclosure, were located at the eastern extent of the north breach and are described in chapter 3.

Rooms and Bins D 227–D 230

The wall dividing D 227 and D 228 extended out from beneath the inner wall addition and was probably not used after the relining was built. The wall was founded at elevation 2528, as was the wall at the northeast corner of D 40. The south wall of D 228 abutted the corner of room D 227 with an L-shaped end, as found throughout the various structures of the fortress. The floor in D 228 had an elevation of 2533, just 3 cm below the floor in D 40 to the south.

Next to the east wall of D 228 was a group of small bins or compartments that may have continued under the innermost of the two main fortification walls (pl. 32a).[65] The preserved brick showed that each bin ranged between 48 and 58 cm from north to south and between 65 and 80 cm from east to west. One bin wall rested on level 2533. This was 17–24 cm higher than some of the floors under the relining but also similar to other floors found in D 228 and D 232 and south of the wall.

In the northwest corner of this exposure was D 230, an area defined by wall fragments. The floor level in the space to the west was at an elevation of 2516, which is comparable to the low floors in D 42 and D 43 but 17–20 cm below the floor levels in D 228, D 40, and D 41.

Rooms D 231–D 232

A fragmentary north–south wall was uncovered between the bins and D 232. Just to the east, the floor level outside the west end of D 232 was measured at elevation 2537. The longest stretch of wall in this area

62 The sherds from deposits found in the rooms under the wall relining are presented in appendix I. At what point in time the wall breach occurred is uncertain, and some of the sherds excavated here undoubtedly fell into the area during floods or other later disturbances. However, Knudstad notes that remnants of the inner lamination covered the area, but with its inner face gone.

63 Knudstad 1966, p. 184. The oven under the inner lamination at D 1 was resting on an elevation of 2505.

64 Knudstad 1966, p. 184; see also Knudstad 1964a, p. 44. Pierce (1964b, p. 62) describes the various strata shown in the photographs (one shown on pl. 31b): "a wooden beam between the two topmost tags, shown next to the meter stick; 1.0 meter below the beam is the bottom of the enclosure wall; a 4-cm layer of ash below the wall; 6 cm of light gray soil; 11 cm of darker gray soil; ca. 23 cm of gray-black soil; 6 cm of burned alluvium; and then sterile brown river silt."

65 Knudstad 1963–64, p. 330.

was on the north side of room D 232. It was found to have a support pilaster on the interior side, and its foundation was at elevation 2518. The floor level within D 232 was recorded as 2523, and the preserved patch of floor to the west of the doorway was 14 cm higher. To the north in D 231, the excavators uncovered an oven resting on a floor level 14 cm below the floor level in D 232 and more comparable to the floors in D 230 and in the rooms to the west along the relining. The natural topography of the site in antiquity is unknown, but it is clear that there were various phases of occupation that had been obscured by the extremely poor preservation and in fact were altogether unknowable due to incomplete excavation of the site. But it seems that even the area under the wall relining had gone through a renovation before the building of the inner wall.

Two pottery disks were found in D 227 (1) (OIM E47470, E47471). The types of pottery from the structures under the inner wall addition here were the same types as those found in the structures against the inner relining in the West Sector. No chronological differences can be discerned.

Considering the Structures under the North Wall Breach

The floors and wall foundations of the buildings under the inner wall relining were found to be at a similar height as, or higher than, those uncovered farther west. The floor levels of the north breach buildings were higher on average than the floor level, for instance, in D 67 (elevation 2512), although the floor level next to D 230 (2516) was close. Floor levels comparable to those of the upper levels under the relining were found in D 6 (2531), D 60 (2526), and D 70 (2532), while the lowest wall foundations along the north fortress wall were quite a bit lower than those under the relining. This would include the wall base levels in D 220 (2502), D 1 (2513), D 6 (2496 and 2500), D 60 (2504), and D 68 (2453) and that between D 41 and D 42 (2505). Therefore, it is assumed that the buildings excavated under the third wall addition also show two phases of rebuilding before the addition of the inner relining. Evidence of this earlier phase is thus found both under the north wall and to the south of the later fortress wall relining.[66]

In some cases, the occupants of the fortress during the Level III phase of its existence were digging out and reusing the earliest walls and floor levels that remained south of the enclosure wall relining.

The area south of the north wall breach was too eroded to preserve any walls belonging to the later Level III and Level II occupations. They had been washed away by floodwaters.

Building D 59A–E and D 61

Locus number D 59 was used to designate both a house, which was discussed above, and a building, which is discussed here. D 59A was the locus designation assigned to a space east of this building complex (see fig. 4.4). It was north of a wall extension abutting and perpendicular to the dividing wall between the West and Central Sectors. Locus D 59B was to the south of it. The wall between D 59A and D 59B was founded at an elevation of 2565, while the oven in D 59A rested 5 cm below it. No floor levels were associated with the wall or oven. The base of the inner relining to the north (elevation 2525) was 35–40 cm below the foundation levels of these two architectural features and 17 cm below the nearby wall foundation of D 59C.

The walls of D 59D and D 61 were apparently bonded to one another, while the skewed east wall of D 59E abutted the wall of D 59D. They were not preserved to a great height due to their erosion, but two of the walls still showed evidence of interior support pilasters. (The best photograph is shown on plate 35a, where the low walls of the area—at the upper left of the photograph—are partly obscured because the excavators used this place to process pottery.) The elevations of the bases of the walls in the south and middle of the complex, in D 59E and D 59D, closely matched the elevations of the higher standing oven and wall fragment along the dividing wall between the West and Central Sectors.[67] But the north half of the complex was 17–25 cm lower. The exterior northwest corner of D 59C was at about the same level as the floor and wall foundation in D 228.

The central part of the complex was composed of two areas, room D 59D and room D 61, with no apparent connecting doorway preserved. The walls were left standing only 10–12 cm high. Their walls were aligned differently than in the adjoining rooms, running on a northeast–southwest axis.

The fragment of a wall was found on the south side of D 59E, while the finished west edge of the north wall of D 59C abutted the finished edge of the west wall of D 59C in a clumsy manner.

66 See appendix I, fig. A.I.20, for the pottery types found under the breach.

67 Knudstad 1964a, p. 43 obverse.

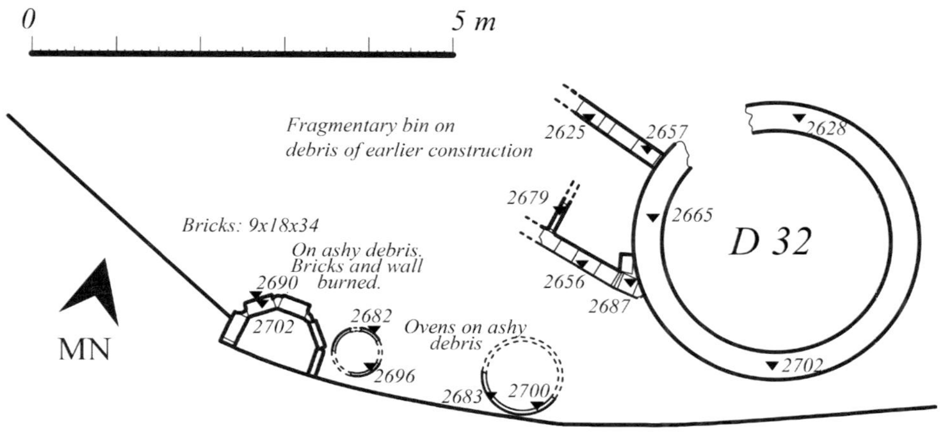

Figure 4.5. Silo D 32 in the top level

The building certainly could have been part of the Level IV settlement found under the walls, but it is uncertain whether part or all of it was reused in Level III. The skewed and awkward character of the surviving walls found near the north wall breach differs from that of the buildings farther west. As noted above, however, the loss of coherent structural remains and the possible shifting of the walls through floods heighten this effect.

Some slag pellets were found together below the "ash" layer of D 59B (OIM E51029), and a stone blade preserved to a length of 5.5 cm came from D 59C (1) (OIM E24417). The other objects that received the designation D 59 are assumed to have stemmed from the room of that number along the north wall, since no letters were appended to the locus number to identify rooms of the complex under consideration here. A faience bead and a globular-bodied marl flask were found just below the surface in D 61 (OIM E24373–E24374).

CLEARANCE ALONG THE SOUTH WALL IN THE WEST SECTOR

The multiple levels of architectural remains along the dilapidated south enclosure wall could not be securely linked to the various phases on the north side of the West Sector, or even to other structures nearby. Two large buildings were placed strategically here, one at the entrance into the fortress from the West Gate, and the other near the southeast corner of the West Sector. The structural remains and associated finds were of little help in determining their purpose, although one would expect the fortress to have had larger administrative buildings and a temple.

The discussion below will begin with a description of the multiple levels of structures and the large administrative building ("bedrock building") at the southwest corner and finish with the multiple levels and large administrative building found at the southeast corner.

UPPER LEVEL IN THE SOUTHWEST CORNER

Silo D 32 and Area to the Southwest

The topmost level along the south enclosure wall consisted of a reused earlier silo, D 32, with the fragmentary walls of an attached bin on its west side and a possible opening on its northwest side (plan 2, fig. 4.5).[68] The bin construction to its west rested atop the walls of another bin associated with the earlier use of the silo. The exterior diameter of the silo was approximately 3.5 m. This silo and its associated walls represent the first level of more than four strata excavated in a location that was a jumble of ovens and silos built atop more of the same (pl. 32b).[69] Along the wall to the west were the remains of two ovens and a hearth on ashy debris. The easternmost oven measured 86 cm in diameter, and the west one measured 52 cm. The fortress wall relining next to the hearth was burned. The bricks used in its construction were 34 cm in length. Other brick sizes were not recorded, but in any case all the bricks likely had been reused from earlier structures.

Although the elevations cannot prove contemporaneity across such wide areas, it should be noted that the foundations of the upper-level walls at the southwest corner were somewhat equivalent to those of the upper-level silo, D 4, in the northwest corner. The walls of the bin up against D 32 rested on elevations ranging from 2625 to 2656, while the ovens were

68 Knudstad 1964a, p. 38.

69 Knudstad 1964a, p. 38.

at elevations 2682 or 2683, and the hearth sat atop a level of 2690. D 4, on the other hand, was founded at elevations ranging from 2675 to 2693. The variety of elevations on the south side of the fortress was partly due to terrain (since bedrock was encountered at the southwest corner) and accumulations of earlier debris that was not leveled. The congestion of the area shows that there had been multiple rebuildings and repairs over time.

Clearance of surface debris around D 32 yielded a stone net sinker (OIM E24356), a flint blade fragment (OIM E24363), a faience bead (OIM E24364), a glass rod (OIM E24365), twenty-eight assorted Meroitic beads (some on the original string) (OIM E24366), and a fragment of a blue faience New Year's flask (OIM E24350). The fill within the silo contained a stone arrowhead (OIM E24353). The faience New Year's flask fragment should be contemporary with the Level II use of the Central Sector, since such flasks are typically considered Saite (see chapter 7). The area was located below three Meroitic burials that had been cut into the top of the enclosure wall and had been looted. In addition, Meroitic beads and sherds were found in the area between D 32 and the south wall.[70] East of D 32, at the surface, was a crucible fragment (OIM E51024). Christian sherds were also scattered across this area, particularly to the west of the silo.

INTERMEDIATE LEVEL IN THE SOUTHWEST CORNER

D 32 and Area

This area displayed a number of architectural fragments representing at least two levels of reconfiguration and construction (levels 2 and 3) at the southwest corner (fig. 4.6; plan 3).[71]

A number of flimsy bins were found between silo D 32 and the silo to the west, D 32 W, and a few of the walls were contemporary with the two silos and abut their walls. D 32 W was actually contemporary with this phase of D 32, and both sat atop the earlier area of small workrooms, D 117 through D 122, while D 32 W was built atop intermediary wall fragments between it and the earlier bedrock silo. The exterior diameter of D 32 W would have been more than 4.5 m and, like D 32, its walls were 40 cm thick. The thin walls of the bin attached to silo D 32 extended out to the west and were partially covered over during the building of the west silo.[72] However, the northern extent of the wall of D 32 W was no longer preserved, since it was dismantled in the later reuse of silo D 32 and its adjoining bins. A third silo, D 32 E, with thinner walls, was connected to the east side of D 32 by a short wall.

D 115 was an oven located to the south of these silos and next to the fortress wall, founded on a level very similar to the nearby floor and wall of D 32 W.

The constructions were all built of bricks measuring 35–37 × 20 × 8 cm or 37–38 × 19 × 7 cm, and there was evidence for the plastering of walls and floors. All structures rested on ashy debris.

Fragmentary walls and an oven farther east along the south wall were assumed to have been contemporary with this group of silos; although no walls connected the areas, their wall foundation elevations are similar. These are not shown in the detailed plan in figure 4.6.[73]

Staircase D 113

A staircase was found along the south enclosure wall to the west of these silos, built atop the Level III silo in D 124.[74] One of its walls was preserved just below the surface (pl. 33a). The north wall of the staircase was built atop the east–west wall of a space, also designated as D 113, in the stratum below (see plans 3 and 4). The earlier walls are not included in the plan of the intermediate level in the West Sector, but they may have served as part of a foundation built for the staircase.

A doorway leading into the staircase from D 114 was located at the west side and led into a narrow, 42 cm wide hallway. The doorway was blocked off at some point in time.[75] The steps were no longer preserved. It undoubtedly turned direction at a landing, from which point steps led to the top of the wall. Underneath the stairs was a small storage area entered through a door off the hallway. A fragment of floor preserved at a comparatively high elevation (2599)

70 Pierce 1964b, p. 29.

71 Knudstad 1964a, p. 39.

72 The bin walls of the intermediate level of D 32 were contemporary with the wall fragment below D 32 W.

73 These latter structures are above the area of the Level III locus D 110.

74 Knudstad 1964a, p. 48, where he questions whether the silo walls and the cornered walls built atop the silo are Level IVA or IVB. No Level II pottery was found in the area that might indicate the stairway was later, but this date also remains a possibility.

75 Knudstad 1964a, p. 48.

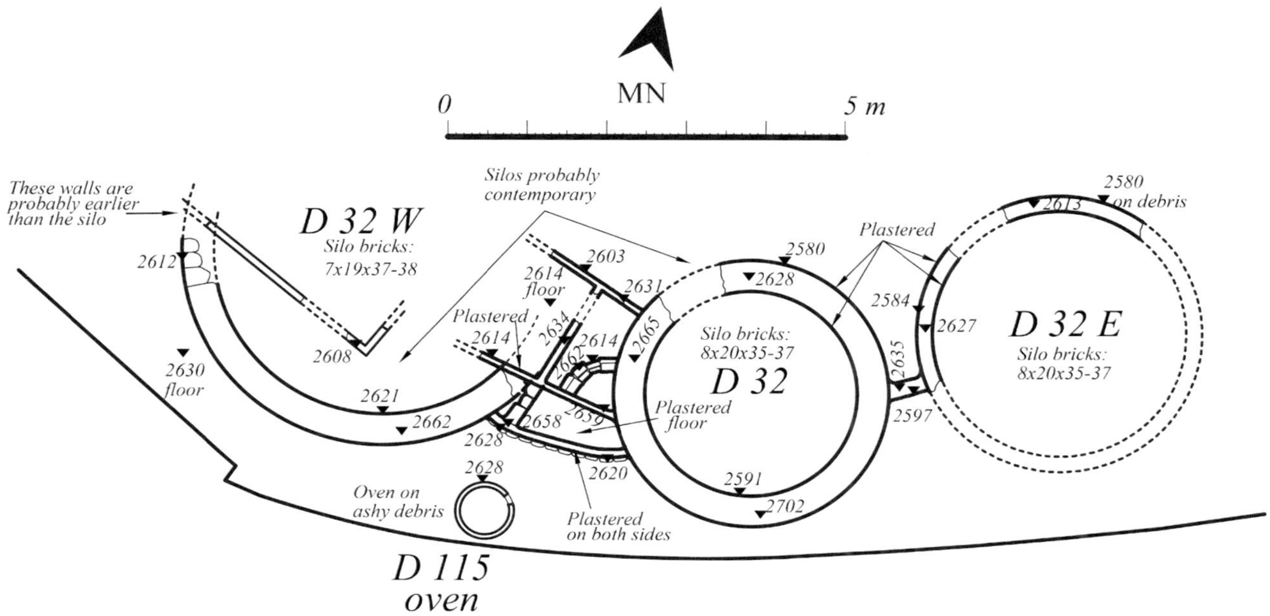

Figure 4.6. Silos D 32, D 32 W, and D 32 E in the intermediate level

within the earlier silo D 124, below the staircase, undoubtedly belonged to this later phase and not to the silo.

The north wall of the staircase was thick, some 40 cm wide, and had a corner wall fragment running north. The east wall of the staircase structure appears also to have continued farther north.

The objects stemming from this intermediate level consisted of some copper or bronze fragments of uncertain form (OIM E24403) from the fill of D 113 (1), a sherd of a Nubian cooking pot (OIM E49605), and a pottery disk (OIM E49758). Two hundred seventy-five faience beads were also found on the floor along the west wall of D 113 (1) (OIM E24404). In the rubble of D 114 (1), just to the west of the staircase, were found three thick, reworked sherds (OIM E24407) (see fig. 7.6a in chapter 7). The ends of these sherds had been ground down to a smooth, rounded rim, beneath which a groove set it off from the rest of the sherd, with surfaces showing fine scratches running perpendicularly to the long side. The largest was 10.5 × 11.0 cm. A similar sherd was found in D 113 (1). A seal-impressed sherd was found south of D 113 and D 114, on the enclosure wall relining, with the mirrored image of a cluster of three lotuses (OIM E24406).[76]

76 The form mirrors those shown in fig. 5.30f in chapter 5.

SOUTHERN BUILDINGS OF LEVEL III AND LEVEL IV

Earlier Silo D 124 and Walls

Three more levels of rebuilding below the topmost phase of the staircase in D 113 are indicated in the architectural drawings of this area on plans 1 and 4 (fig. 4.7). The thin walls of the space below the stairs, also called D 113, were found to be preserved at a height of 33 cm and were located just below the staircase construction and atop silo D 124. The north wall of the room perhaps continued west under the later staircase, the remains of which were left unexcavated. A north–south wall abutting the room wall on the north may have continued on over the top of the silo wall.[77] The neck fragment of a highly burnished wheelmade jar or bottle, with a red-slipped exterior, came from lower fill in D 113 (1) and the fill of D 124 (1) (OIM E50074, E50078).[78]

Although the walls surrounding locus D 113 were technically built earlier than the staircase, their construction phase may have been associated with the building of the stairway's foundation in the intermediate level at the southwest corner.

Silo D 124 lay below D 113 and stood atop another construction composed of two thin walls joined at

77 Knudstad 1964a, p. 48, with a question mark at the preserved end of this wall.

78 Shown in fig. 5.63a in chapter 5.

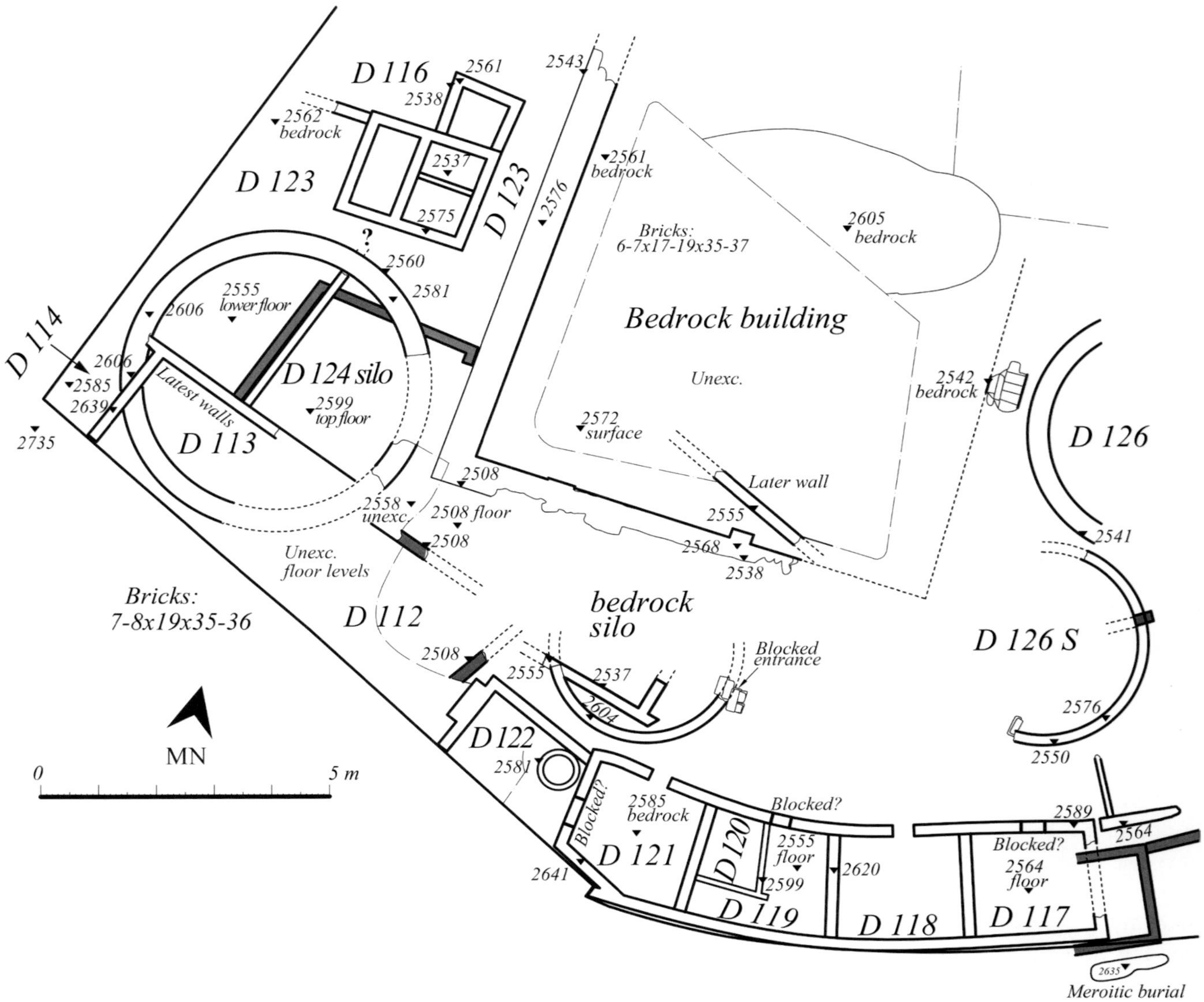

Figure 4.7. Buildings in the southwest corner (earliest walls are gray)

a corner. D 124 measured more than 5 m in diameter, and its wall was preserved to a maximum height of 21 cm on the north side. The top floor level (2599) within the silo was 39 cm above its wall foundations. This surface was contemporary with the staircase and the foundations of D 113. Another floor surface within the west part of the silo was measured at an elevation of 2555, or 5 cm below the silo foundations at the northeast edge.

The east–west wall defining the north edge of the earliest structure excavated, which has no locus number or recorded elevations, abutted the "bedrock building" to its east with an L-shaped, cornered end. This earlier Level IV remnant was perhaps contemporary with other wall fragments to the east and the original bin construction D 116 to its north.

Bins D 116 and Area D 123 of Level IV

To the west of the large bedrock building was a group of bins in D 116. They measured between 0.80 and 1.50 m from north to south, and between 0.75 and 1.00 m from east to west. The southern compartments were built as a unit, with an extension added to the north and another wall stub placed atop a bedrock outcrop to the west. The bins are clearly seen on plate 33b.

The levels of the floor and foundations of the bin walls were approximately equivalent to the level of the wall of the bedrock building to the east. The bins may have been left intact after the construction of the later silo, D 124, since the height of the south wall of the bins remained 15 cm above the foundation level for the silo. The four compartments were filled with ash and charcoal to an uncertain depth.[79] The only object found was among the surface sherds north and west of the bins. The find in question was a plain pottery disk (OIM E49743).

D 123 was a passageway between these bins and the bedrock building and between the bins and silo D 124 to the south. The area was filled with ash to a depth of about 20 cm, and a fine white ash layer divided the layer in two. The fill level, D 123 (1), contained a bone awl (OIM E24416), a pottery disk (OIM E49742), and a miniature handmade bowl with incised decoration (OIM E50715).

BEDROCK BUILDING AND ENVIRONS

The excavators encountered the bedrock of the island in the southwest corner of the enclosure wall. Therefore, the building in the southwest corner is here christened the "bedrock building" (pl. 33a–b). The outcrop rose about 63 cm above the foundations of the east and west walls of the structure. The erection of the west wall of the building would have necessitated cutting a trench into the bedrock, although there is no mention of this. The substantial walls surrounding the natural rock formation would have required a curious interior arrangement to allow movement within, unless the floor was raised over the bedrock. The bedrock could have been removed if it was of concern, unless the rock feature was central to the function of the building that incorporated it. The central part of the building was left unexcavated, mainly because there were no indications that interior walls had survived except for a later, intrusive wall on the south side.

The bedrock outcrop that was exposed within the building dropped off drastically toward the east by 63 cm in elevation but less dramatically to the west by 44 cm. It formed a sloping ridge from its north exposure in the building to the south wall area, dropping 20 cm in height within D 121 and down 43 cm west of D 116. It was apparently absent in the southwest corner. There is no indication that bedrock was exposed during the excavation of the glacis trench to the south of bastion N just on the other side of the enclosure walls.

The walls of the bedrock building were 58–60 cm wide and were constructed of bricks 35–37 cm long, the same size as those used in most structures in the area. The preserved extent of the walls measured more than 8.5 m from east to west, if one takes the east wall fragment found next to silo D 126 as the eastern limit of the structure. The preserved wall on the west side was about 8 m from north to south. The south wall was badly eroded on its exterior and had a support pilaster built onto the interior side.

A wall fragment along the south side, placed on a diagonal to the orientation of the bedrock building, was founded at a level (elevation 2555) 17 cm above the foundation of the building's south wall (2538). The preserved height of this wall was not noted, but it appears to stand higher than those of the bedrock building in the photograph shown on plate 34a. On the architectural plans, Knudstad labeled this wall and the silo south of it as later features.[80] The foundation of this "bedrock silo" to the south of the building was on par with that of the diagonal wall and the silo in D 124 (2555 and 2560, respectively).

This one diagonal wall fragment suggests that the bedrock building went out of use at some point in Level III, unless it was part of an interior construction of some sort. The ruinous state of the architectural features around and atop this area of elevated bedrock hinders precise interpretation.

The bedrock silo to the south of the building, as just mentioned, was founded at the same level as that of the diagonal wall just described. The preserved wall height of this silo was 49 cm (2604), and it would have been about 3.25 m in diameter. There was evidence of its door having been blocked. Below the silo was a thin cornered wall fragment founded at the same elevation as the foundations of the large building (2537). But it had a slightly different orientation than that of the large building. Below this wall fragment was yet another, earlier wall to its west, on a level of 2508, which matches the bottom elevation of a wall fragment and a floor level even farther west. The southwest corner of the bedrock building was founded at 2508 as well, having perhaps been built upon bedrock and thus founded 30 cm below the same wall base to the east.[81]

The walls of this large building were wide enough to have supported more than one story, but no evidence was found to support such an assumption. The size of the building suggests an official function, but there is no clear evidence to prove this since the area was scoured clean of architectural remains and objects that might clarify the situation. The many silos and small rooms in the southwest corner suggest that the area was an important storage and crafts area surrounding an administrative or religious area.

79 Hoerth 1964a, no. 1.

80 Knudstad 1964a, p. 48.

81 "D 112 and area," to the east of the staircase D 113, was also noted as having unexcavated floors; see Knudstad 1964a, p. 48.

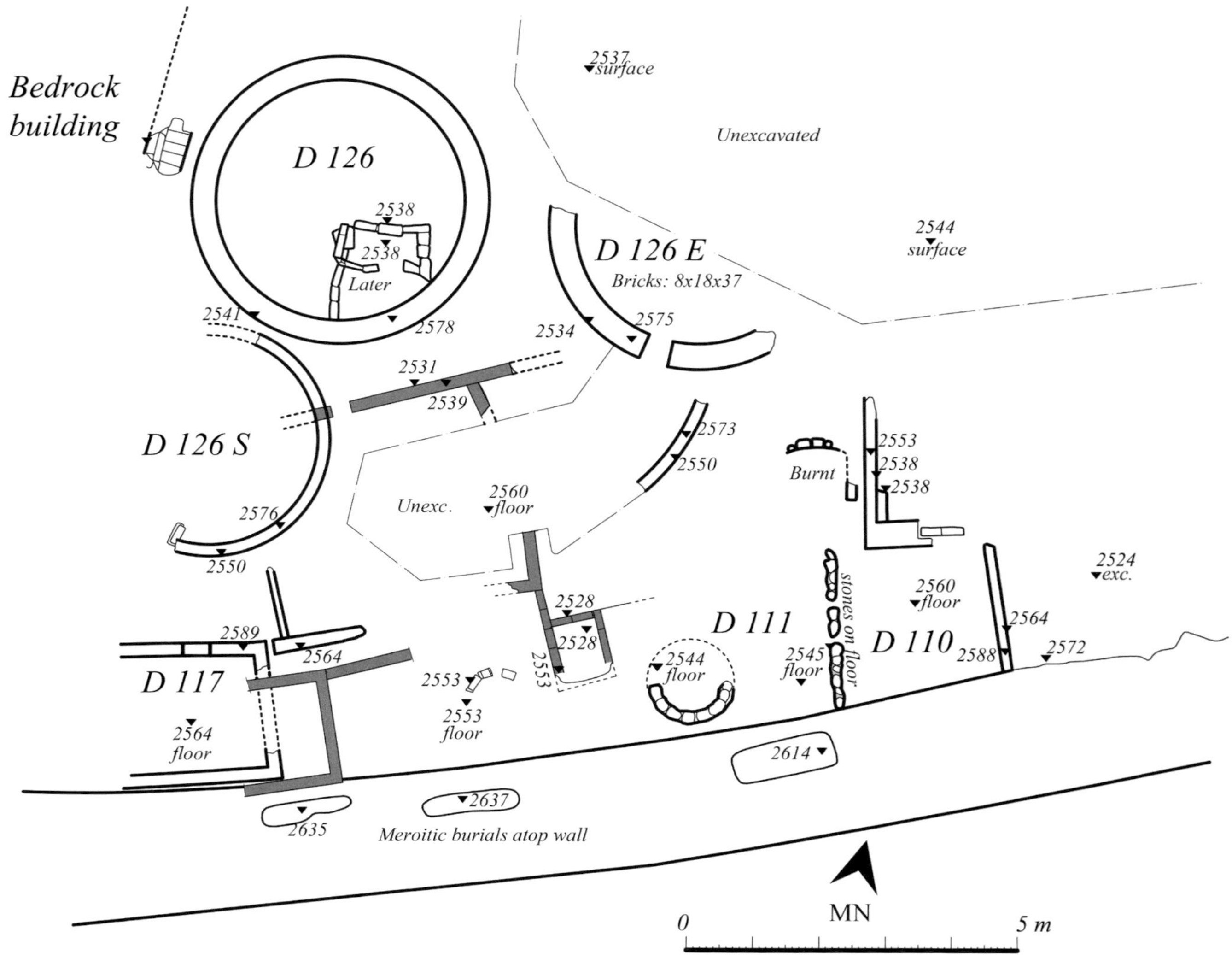

Figure 4.8. Silo D 126 and area (earliest walls are gray)

Silo D 126 and Its Neighbors

At least three large silos, one of which was assigned the locus number D 126, were located to the east of the bedrock building (fig. 4.8; pl. 34a–b). All the silos were founded at slightly different elevations. Despite the uneven terrain of the site, these silos were at least partially or entirely contemporaneous with each other, with silo D 124 in the southwest corner, and also with the bedrock silo.[82]

D 126 was 4.5 m in diameter, with walls 37 cm wide, and was preserved at a height of 37 cm. Its foundation was level with the base of the adjacent wall fragment of the bedrock building, 7 cm above the foundation of the silo to the east, D 126 E, and 9 cm lower than silo D 126 S. The preserved height of its walls is comparable to those of the other nearby silos. A later group of bins was haphazardly constructed within this silo.[83] A stone arrowhead (OIM E24419) and a length of corroded copper tubing (OIM E24420) were found when the wall of D 126 was cleaned.

The silo south of it was slightly smaller, about 3.5 m in diameter; it was given no locus number but is here referred to as D 126 S. It was less well preserved and had thinner walls. To the east of D 126 were the partially preserved walls of another silo, D 126 E. The latter was more than 4 m in diameter, with walls that were 40 cm thick. The bricks used in its construction were 37 cm long, which was probably true of all three silos in this area, although this was not noted in the excavation records. A narrow doorway was preserved on the southwest side.

82 Silo D 124, in the southwest corner, had a wall foundation at elevation 2560, while the bedrock silo was founded at 2555, silo D 126 at 2541, the silo south of it at 2550, and the easternmost fragmentary silo at 2534.

83 Knudstad 1964, p. 48 obverse. Recorded as "later."

A thin curved wall would have abutted the last-mentioned silo on its south and was situated 16 cm above the silo's foundation. It was only partially preserved. An earlier wall stretching between this silo and D 126 S was not contemporary and ran beneath the latter. These wall bases, however, match up with some of the earlier wall fragments farther west. Adjoining this wall was another one running south, toward a bin-like construction that was contemporary with the earlier wall remains both here and along the fortification wall to the southwest (see below).

A hardened floor surface was preserved next to D 126 E at an elevation of 2560, extending westward from the adjoining curved wall. It would have run over the top of the earlier walls to the north and south, which were 7–21 cm below it, and would have covered the square bin construction and earlier architectural remnants near the south fortress wall. Unfortunately, the mud floor was not preserved anywhere else.

No locus numbers were assigned to the brick oven remnants and rounded bin flanking the rectangular bin along the fortification wall. Both of the former features rested on surfaces above that of the rectangular bin, although still 7–16 cm below the hardened floor surface to the north.

Workrooms, Storerooms, or Granaries in D 117–D 122

These single-roomed, thin-walled buildings were mostly built with a south wall extending along the south relining (fig. 4.7; pl. 34b). The relining was built of a mix of bricks 38–40 cm long. The room walls, however, were built of bricks 35–36 cm long and were sometimes bonded, and the complex was mostly built as a single construction project. The westernmost room was an addition or reuse and was not well preserved. This group of rooms was kept in place during the later Level III relining project, or it was possibly built after the relining had been completed.

The buildings were placed on a higher level than those to their immediate north, perhaps due to the presence of bedrock in the area or earlier structures that were never uncovered. The bedrock was exposed within D 121 at elevation 2585. The base of the oven in D 122 (2581) was 4 cm below the bedrock level next door, and it was 26 cm higher than the base of the silo wall to its immediate north (2555).

D 122 used the fortress wall for its southern limit, and its north wall abutted the wall of the room to its east. Its south wall was perhaps cut down or incorporated when the relining was constructed. An oven or circular bin was placed in its northeast corner. The room was connected to D 121 by a doorway that was later blocked. The room was filled with fine black ash mixed with charred wood, and a fine white ash layer divided the deposit in half.[84]

To the west of D 122 was an area designated as D 112, where an unexcavated block of floors was left standing. At the north end of this area was found a quartzite river cobble used as a pounder (OIM E50780), along with a quern, a round stone bead (OIM E50817), and a faience Hathor head pendant (Khartoum 14271). A marl flask fragment (OIM E50038) and the rim of a pot stand were also among the potsherds stemming from this area. A fragment of an unidentified clay object was found in the surface debris, looking very much like a net weight, with a rounded end and restricted area where a string may have been tied (OIM E24410).

The objects from D 122 (1) included a small burnished double flask (Khartoum 14267), found lying next to the oven, and a quern found next to the north wall.[85] An unbaked clay stopper (OIM E24413) was also found, along with sherds of a thin-walled shallow bowl with red slip and a fine continuous burnish (OIM E50076), a marl wide-mouthed jar, and a fragment of a later Christian bowl.[86] D 121 (1) yielded numerous sherds. It also yielded a ground-down sherd disk with a perforation, another sherd worn down into a cube-shaped ball, and a small stone ball. The latter two perhaps functioned as weights.

A centrally placed, narrow doorway in the north wall of D 121 led into the room. The floor level was at bedrock, as mentioned previously. Next to it was the single room, D 119, where two floor levels were noted in the architectural drawings. The floor on the west side of D 119 by the bin D 120 was situated 44 cm above the floor level taken in the east part of the room (2555). This higher floor level was also the foundation elevation of the high bin. The 32 cm wide doorway into D 119 was blocked at some point. Both D 119 and D 120 were filled with ash and debris. A ceramic lid, its interior coated with resin (OIM E49581), was found within the bin (fig. 5.57c in chapter 5).

To the east along the row of rooms were D 117 and D 118, both single rooms with centrally placed, narrow doorways on the north. The doorway into

84 Hoerth 1964a, no. 1.

85 The double flask is shown in fig. 5.53c in chapter 5 and on pl. 69a.

86 Meroitic graves were dug into the wall atop the area to the east, but it is clear that the context was an open one even during the Christian period, when a house was situated atop the Central Sector Level II platform.

D 117 had been closed off at some later time. The floor level in D 117 was 9 cm above the floor in D 119 to the west, while no floor or excavated level was noted for D 118. The walls to the east have no locus designations and were very fragmentary. Like D 122 to the west, the inner fortification wall would have formed the southern boundary of any rooms that may have continued to the east. The fragment of an east–west wall along the north side of the unnamed area suggests that the line of small rooms did in fact continue toward the east.

An earlier structure was located under D 117 and the area to its east, but the wall elevations were not recorded. The relining of the enclosure wall covered over the southern wall fragment.

As mentioned previously, these rooms may have been workshops or used for the storage of food or other commodities.[87]

Area D 111 and D 110

D 111 was an open area to the west, containing the round bin mentioned earlier (fig. 4.8; pl. 34a).[88] A thin wall defined the east side of D 110 and rested 4 cm above the floor (2560). Fragmentary walls along the north side of the area, however, were founded at elevation 2538, while a line of stones to the west rested on a floor level of 2545. The earlier line of stones was set 15 cm below the floor level of D 110, while the wall footing to the north was 22 cm below the same floor. The D 110 wall and floor undoubtedly belonged to an architectural reconfiguration of the area that covered over these earlier features, including the area along the enclosure wall to the west. The wall and floor of D 110 were thus contemporary with the patch of floor to the northwest and by silo D 126 E.

The elevations of the east wall and the floor of D 110 indicate that this structure was perhaps an eastern extension of the row of small rooms or granaries discussed above. No objects and few potsherds were found in these two areas.

South Wall Breach

The 11 m stretch of the fortification wall to the east, between D 110 and D 109 (figs. 4.8–4.9), had been eroded away by floodwaters running through the fortress interior. The wall was severely eroded near D 110 and then almost disappeared between D 109 and the southeast corner (pl. 35a). Two breaches were noted on the plans of the walls; the smaller breach to the west had few architectural remains by the wall, while the larger had completely disintegrated the walls and most of the interior structural remains.

A stretch of river sand was excavated to a depth of 48 cm from D 110 to the east, exposing the relining's foundation at elevation 2572 and lying 12 cm above the floor of D 110. The inner face of the wall was eroded, and although the fortress wall had slumped, part of it had also been lifted up by the force of the water and rested atop a mound of sand at an elevation of 2588. The sand under the wall was composed of dirt, wind-laid ash, and sherds for this entire stretch. The waterlogged brick of the wall sloped down both to the east and west at the breaches along the south wall.[89] The area between D 109 and D 110 was barren of any architectural remains.[90]

Cataract stones were piled up beneath the inner face of the fortification wall toward D 109, where a mud or pottery basin, filled with stones, rested against the remaining base of the relining wall. The first architectural remains encountered here were those of D 109 and D 108, the latter a staircase built along the fortress wall. How these structures relate to the phasing at the southwest corner is uncertain, but both staircases here were built on higher elevations (as discussed below).

All the bricks used in the area, including in the buildings farther east, were 33–35 cm long. (It will be noted where brick sizes differ.) One noticeable difference in the construction technique used in this area is the lack of support buttresses or pilasters built in the walls of the rooms along the south enclosure wall, which were typically 18–20 cm wide. However, this support feature was common in the rooms at the southeast corner of the sector, as it also was in the rooms along the north wall.

The structures along the south wall also included more stairways built alongside the wall, which differs from the layout along the north wall.

The topmost surface clearance (D 35 and D 36) exposed a black deposit, left by flooding, which covered the walls of the area and also appeared below the floor levels associated with some of the walls.[91] Numerous walls were left standing only 10 cm in height, disappearing toward the center of the fort, which was also typical of the structures along the north wall.

87 See appendix II.

88 Knudstad 1964a, p. 45 reverse.

89 Knudstad 1963–64, p. 307.

90 Hoerth 1964a, no. 2.

91 Hoerth 1964a, no. 2.

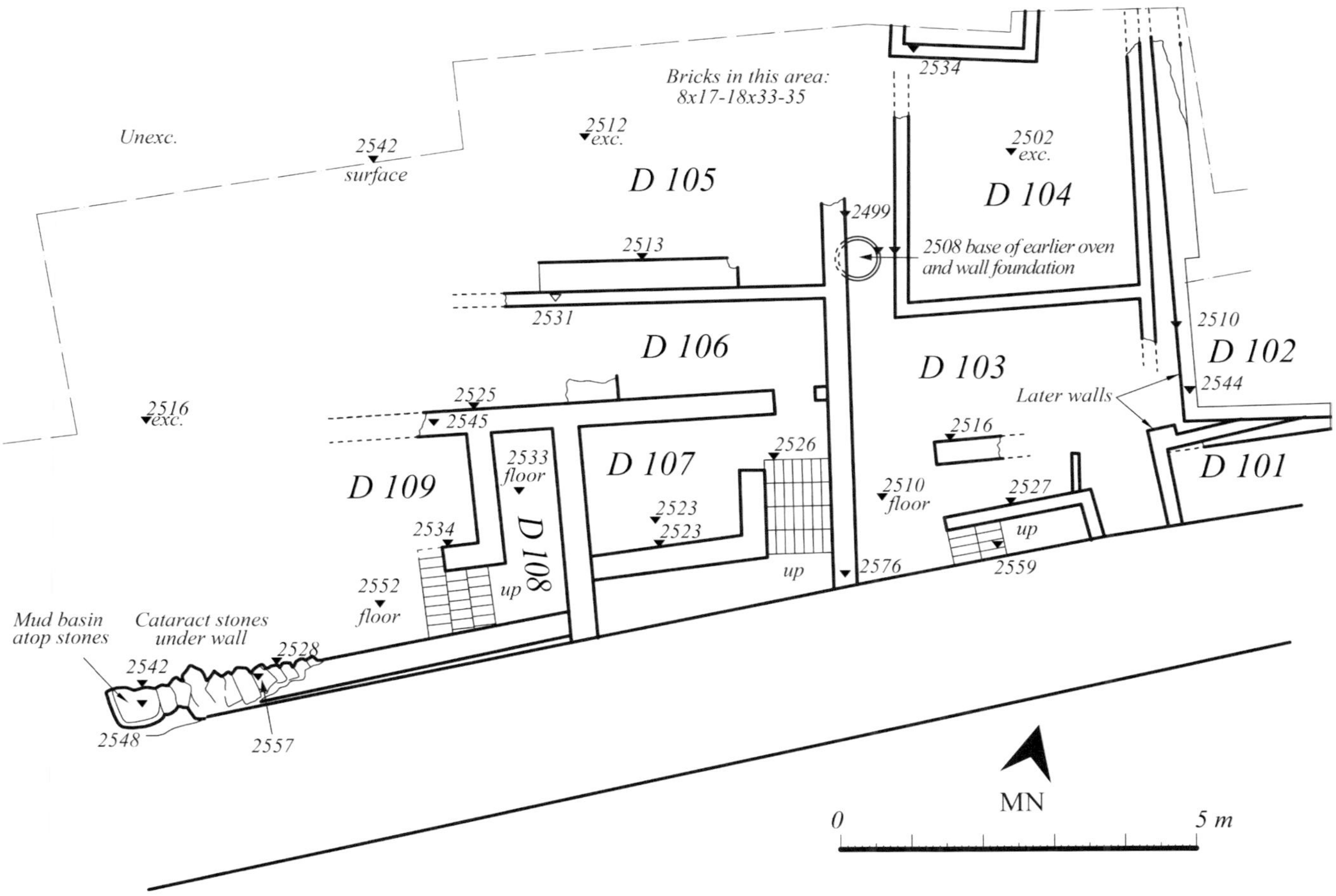

Figure 4.9. Area along the south wall at D 108 and D 109

Stairways D 108 and D 107 and Area

Locus D 109 was situated west of the entrance to staircase D 108, which stood along the fortification wall (fig. 4.9). The areas and rooms here were D 106–D 109.[92] A number of thicker walls 35 cm wide were shared by the two staircase structures in this area. The bases of these walls were more deeply founded than the associated patches of floor, but clearly they were built sufficiently thick to support the staircases and perhaps upper rooms. The recorded elevation for the base of the fortress wall relining in D 101, farther to the east, was 2512. The foundation of the wide north–south wall at the east side of D 106–D 107 was recorded at 2499. This wall ran over, or cut through, an earlier oven, which still stood 9 cm higher than the wall foundation. The thicker, shared wall to the north of both staircases was based at an elevation of 2525.

The two rooms containing staircases D 108 and D 107 shared a wall. In addition to the interior fortress relining, there was a separate wall on the south side of D 108 and D 109. Both stairways had three or four preserved steps composed of bricks laid on edge, their depths being about 35 cm. Staircase D 108 was 1.10 m wide, while D 107 was 1.42 m wide. They both possibly had landings at right angles before another flight of steps would have led up to the upper floors, rooftops, or wall walk.

The space below the stairs in D 108 was excavated down through fill to a floor level 19 cm (2533) below the preserved floor at the bottom of the staircase in D 109. However, the staircase walls were founded atop this particular surface.

The area below the stairs in D 107 was a narrow gap between the south wall of the structure and the fortress wall. It was filled with debris that was left unexcavated. The stairs here would have turned westward and proceeded up. The wall supporting the stairs in D 107 was founded 11 cm (2523) lower than the wall supporting the stairs in D 108, but it is possible that both sets of stairs were contemporary. The only clear floor levels recorded here were those before the stairs in D 108 (2552) and under the fill of those stairs (2533), although the thicker walls around the two structures were all founded at elevations ranging between 2523

92 Knudstad 1964a, p. 42.

and 2534. A much lower floor level was excavated in D 103 to the east, where its elevation was recorded as 2510.

Areas D 106 and D 105

North of the two staircases was a room or corridor, D 106.[93] The remaining stub of a 75 cm wide wall(?) divided the space. Its foundation level, and that of the thinner wall to the north of this space, was not recorded. A deeply founded wall to the north, running alongside the thin wall, was sitting upon an elevation of 2513. Its lower foundation may indicate an earlier wall, but in any case it was probably associated with the longer north–south wall to its east, which sat upon a slightly lower level (2499).

Four pottery disks were found in D 106 (1), two with incised crossed lines (OIM E49766–E49769).

D 105 was the area to the north of D 106. It was defined only by a fragment of the east wall and the walls to the south. The east wall, as mentioned above, cut through an earlier oven that sat at about the same elevation as the floor to its south in D 103 (2508).

D 101–D 104

The southwest corner of the walls enclosing D 104 defined the passageway into D 103. The elevation in this corner was measured at 2508—that is, at the same level as the earlier oven to its west. The base of the wall was almost level with a floor found in D 103 (2510), and it was 9 cm above that of the thicker west wall at D 106 and D 107.[94]

The stairway construction in D 103 was built less solidly than the two stairways to the west. It was also quite a bit narrower. Its north wall rested 17 cm above what had been an earlier floor level in the room. A thick wall fragment to the north of this stairway was only 6 cm (2516) above that floor surface and belonged to earlier architecture in the area. It seems to have been partly dismantled during the later construction of the staircase, although its preserved elevation was not recorded and its photograph on plate 35b leaves this suggestion ambiguous. The elevation of the short, north–south wall abutting the staircase to the east, or perhaps running under it, was not recorded. Two steps of the staircase remained, with the elevation of the topmost step rising 32 cm above the wall foundation. The top of the ruinous fortification wall at this spot was only 3 cm higher than the remaining top step.

A pottery disk with incised crossed lines was found in D 103 (1) (OIM E47459).

One thin wall of D 104 was situated on a level that was equivalent to the floor level within room D 103. Remains of walls were also found at the northwest extreme of the excavated area here, but how they are associated is uncertain. No floor level or doorways were found in this structure. The east wall of this room had once continued south. A doorjamb(?) was built onto one of the east–west walls along the north side of D 101. This wall was built over an earlier wall, and then it too was built over by another wall, which would then have blocked the doorway.

The area yielded numerous potsherds, many probably washed in by floods, although in D 104 (1) a small clay plug or ear stud was found, some 2 cm long and 1 cm in diameter (OIM E24391).

In addition, two stone querns—one used on both of its sides and of white stone, the other apparently of basalt—were found on or near the surface of D 104 along with potsherds, including a marl flask and wide-mouthed marl jar from D 104 (1). But the flood deposits in this area make the original contexts of these objects uncertain.

The west and south walls defining D 102 were built on different levels, with the west one lying on an elevation of 2510 and the south segment lying atop two earlier walls and founded at an elevation of 2535 (fig. 4.10). The earliest wall was the north boundary of D 101, an area cut by the addition of the inner relining and whose preserved wall height was 2502 on the west and a full 17 cm higher on the east.

An almost complete marl jar with a prefired potmark incised on its shoulder was found along the south wall (OIM E24377). The rim and neck of a marl jar (MJ 1) were also found here, and it is assumed that they belonged to the same jar, although no join was found (see fig. 5.37 in chapter 5).

The earliest wall defining the narrow space of D 101 was relatively parallel to the fortification relining and was bonded with the very thin north–south wall that defines the east side of D 102. This north wall and the other walls associated with it belong to a structure underneath the thick inner relining, as mentioned above. The oven lying adjacent to the fortress relining was founded at a level contemporary with other walls in the area (2511) and with the base of the later relining (2512), but still standing above the preserved height of the nearby inner wall lining (pl. 36a). However, its preserved height on the east (2547) suggests that it was built up and reused after the relining was built. The

93 The photographs from the excavation do not include many detailed shots of the rooms between D 109 and D 100, but these are included here when available.

94 This area is discussed in Hoerth 1964a, no. 2.

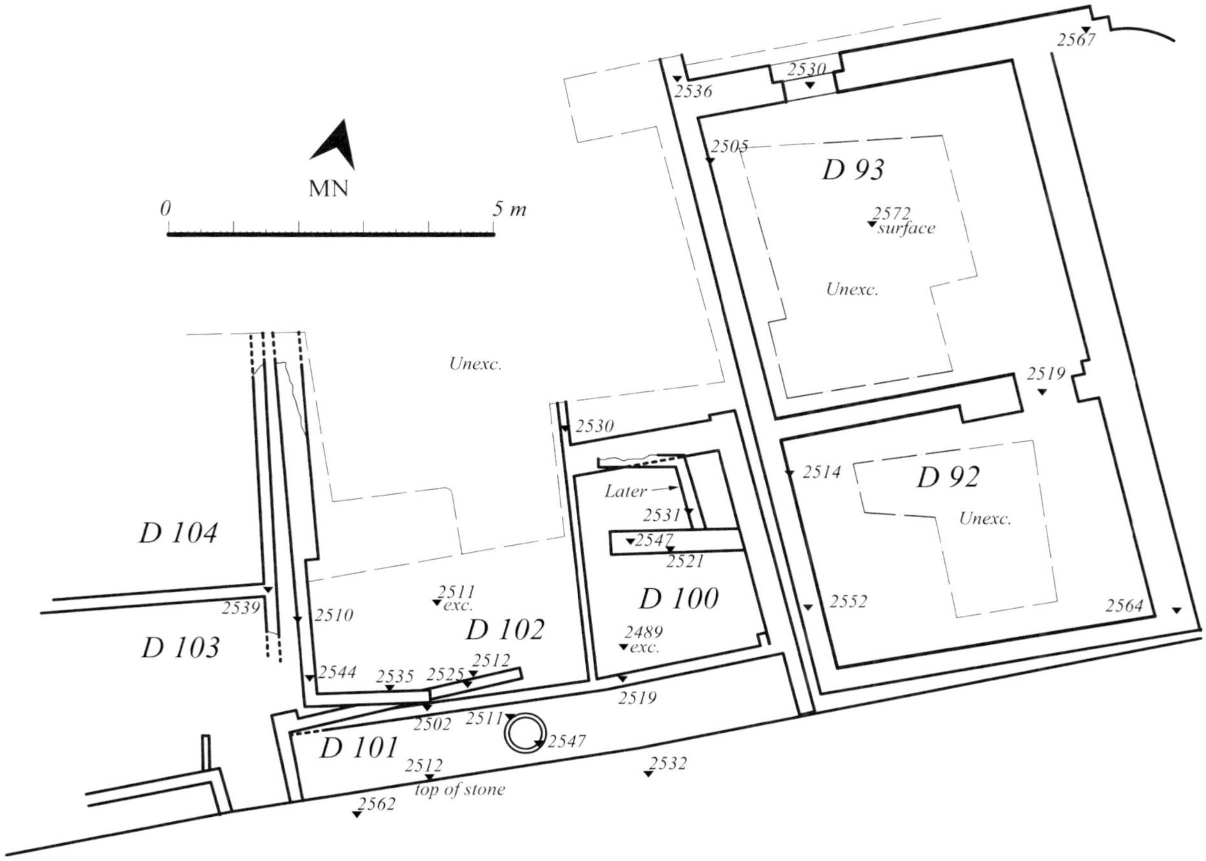

Figure 4.10. Architectural remains at D 100–D 102 and building D 93

width of the room was between 0.80 and 1.15 m. The field notebook indicates that a poorly preserved floor level was found in D 101—undoubtedly the surface upon which the oven was built—which corresponded to the base of the early walls and patches of floors in nearby rooms.[95]

While the topography below the area is unknown, the elevations of the wall bases in the area between D 105 and D 106 (2513), between D 103 and D 104 (2499 and 2508), and in D 102 (west wall 2510) were nearly the same and contemporary. The inner wall relining was built over earlier structures, which necessitated some reconfiguration of the earlier walls remaining north of the wall addition. It is possible that most of the Level III renovations had disappeared in the area except for those preserved along the wall, and that much of the remaining architecture belongs to Level IV. The foundation elevations of the walls in D 108 and D 109 (2534 and 2525), the staircase in D 107 (2526), the staircase in D 103 (2527), and part of the wall between D 101 and D 102 (2535) indicate that they were later additions to the area.

Area D 100

The wall between D 101 and D 102 continued east to form the dividing wall between D 100 and D 101. D 100 shared a very thin wall with D 102 to its west but had wide walls on its north and east (fig. 4.10). The foundation levels of the east and west walls of D 100 are not known. A later wall was preserved in the middle of D 100, with its foundation at 2521, and abutted the east exterior wall. The partition wall on the north side of D 100 was higher (2531) and lay atop an earlier north wall.

The thicker walls surrounding this locus were probably part of a building complex associated with the large building in D 92–D 93, although the lack of elevations from the area provides no certainty.

Building D 93

The large building at the southeast edge of the West Sector, comprising D 93 and D 92, was better preserved than its companion near the West Gate (fig. 4.10).[96] Its

95 Hoerth 1964a, no. 2.

96 Unfortunately, the photographs of this building were in very poor condition and are not included in this publication.

position near the dividing wall and the earlier entryways into the Central Sector suggests that it had an administrative function, although nothing was found within it to verify this.[97] Its north and east walls were 60–65 cm wide, while its west wall was 35 cm wide. The south wall of the building measured 37 cm wide and ran along the inner fortification wall relining. The front doorway was off-center toward the west end of the north wall. An inset 1.17 m wide was located before the sill and doorjambs of the doorway. The whole building measured nearly 10 × 6 m, with the interior measurement of D 93 alone being some 5 × 5 m, which made it slightly larger than the preserved bedrock building at the southwest corner. An indirect doorway led into the slightly smaller room, D 92, from the east end of the dividing wall. The doors, in other words, were not aligned.

The thinner west wall of the building continued to the north, where it was not fully preserved, while the north wall extended eastward for 97 cm. A thin curved wall led from the east end of the north wall and abutted an early silo, which had no separate locus number but included parts of D 77–D 78. A northern wall extension to the east would have impinged on the silo construction if they were built at the same time.[98] However, this suggests that the large building was part of a larger complex of rooms or buildings when it was first constructed.

The east wall of the building sat atop a short wall fragment, partly exposed under the east edge of the wall abutting the D-shaped oven or hearth founded at 2500. The wall fragment and hearth represent features that existed before the construction of building D 93 had begun, or perhaps before it had been finished.

The walls of the large building exhibited slightly different foundation levels: the north wall exterior at the east was at a level 13 cm (2518) above the wall base at the northwest corner of room D 93 (2505), while another wall elevation in D 92 was recorded at elevation 2514. No associated floor levels were found in the building, but the surface level of the doorway between the rooms was at an elevation of 2519, and that of the doorsill to the north was 2530.

The brick size used in all the buildings along the south wall and in the southeast corner was 33–35 cm.[99]

The objects from the fill of D 93 (1) included fragments of a crucible spout, rim, and base (OIM E42954–E42956), but these were probably not in situ. A small faience bead was also found (OIM E50824). Two more crucible fragments were found in the fill of D 92 (1) (OIM E51015–E51015), also likely debris thrown into the area at a later time.

INTERMEDIATE LEVEL IN THE SOUTHEAST CORNER AND POSSIBLE LEVEL III ENTRANCE INTO THE CENTRAL SECTOR

The extreme southeast corner had at least three major phases of structural remains, but a fourth bottom stage is discussed below (fig. 4.11).[100] In keeping with the descriptions of other areas with multiple building phases, the uppermost levels will be discussed first. These architectural elements were included in Knudstad's intermediate level in the West Sector, as a "late Level III" building phase.

The remains to the west of this area along the south wall were earlier than the following doorway D 87. Building D 93 may have continued in use, although there is no way this can be proven.

Doorway D 87

The area atop D 79 and D 87–91 was built over by a wide doorway leading into the Central Sector.[101] Since it bears the same designation as an earlier space below it, it is distinguished here by calling it doorway D 87 (plans 4 and 5).

The L-shaped foundation walls for this doorway construction were set atop the earlier silo of D 77/D 78 and house D 79 and will be discussed below in more detail.

The threshold was composed of a recut stone block, with a smaller fragment placed in the area between the brick jamb on the north (pl. 36b). The south end of the sill had a higher-standing stone with a depression in its center, intended to support something other than a doorpost. The inset sill, more than 2 m wide, led to an entrance that was 1.25 m wide. A patch of floor remained on both sides of the sill at an elevation of 2605; this level was 69 cm above the floor of D 90 underneath it. Possible mudbrick paving lay to the east side of the entrance on its north, with the top elevation recorded at 2610.

The doorway was contemporary with an area of mudbrick remaining on the interior side of the dividing wall between the West and Central Sectors and

97 The possible Level IV and Level III entrances into the Central Sector are discussed below.

98 Knudstad 1964a, p. 41.

99 Knudstad 1964a, p. 42.

100 The sherds found around doorway D 87, and from the levels excavated below it, are presented in appendix I.

101 Knudstad 1964a, p. 22.

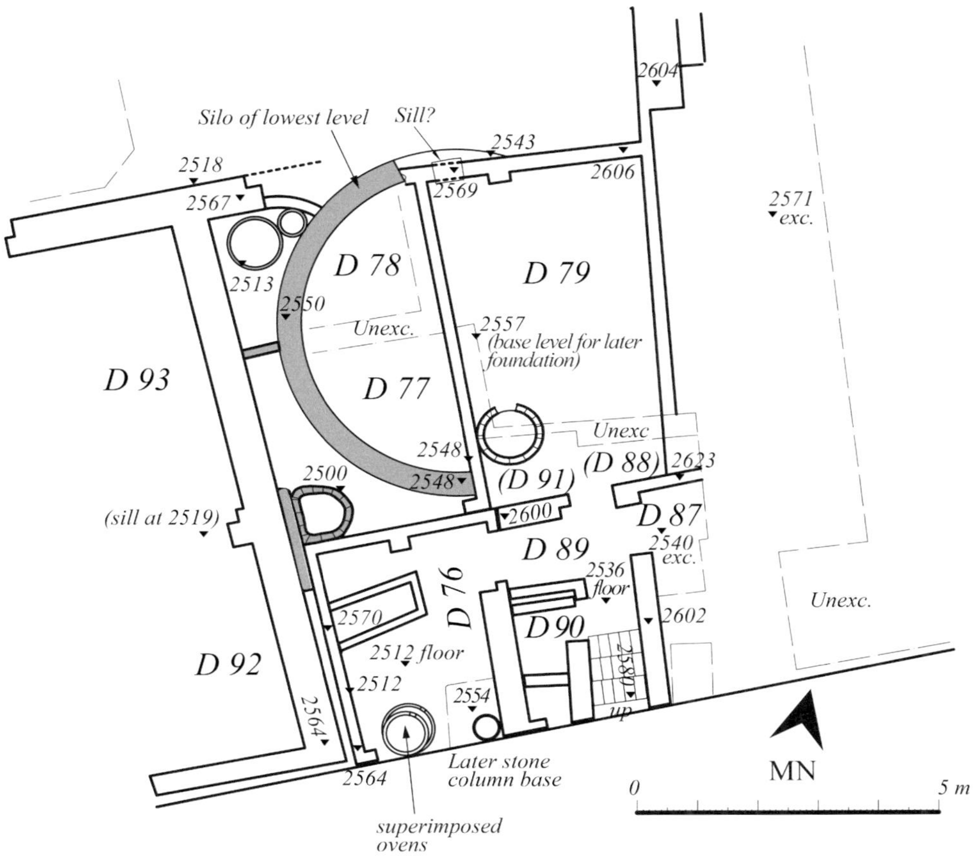

Figure 4.11. Buildings in the southeast corner (earliest walls and features are gray)

north of the brick paving east of the doorway.[102] The first line of bricks, laid on edge, was at an elevation of 2597. The four lines of bricks spanned a length of 1.47 m, with a difference of 37 cm in elevation between the higher south and lower north sides. The remains look like those of a shallow stairway or ramp descending from the doorway toward the north.

Although the notes state that the area was unsolved, the doorway and the possible stairs or ramp would have provided access into the Central Sector during a later Level III phase. The elevation at the foundation of the west stairway of the official residence (2528)—no floor level was found—was 69 cm higher than the lowest level of the conjectural stairway or ramp near doorway D 87, but could have been reached by shallow stairs or a sloping path leading up to the west entrance of the residence.[103] The bedrock in this area was also higher than the surface of the West Sector, although its elevation was not recorded. The possible earlier access into the Central Sector during Level IV will be discussed below in the section on the dividing wall.

The south wall of the later doorway construction was founded at an elevation of 2609, which was 7 cm above the east wall of the earlier staircase in D 90 (pl. 37a). The other walls around house D 79 were incorporated into the fill and walls of the later construction, since they were made level with or were left standing slightly higher than the foundation levels of the doorway's walls.

The preserved heights of the walls in building D 93 were 2536 (northwest), 2552 (southwest), 2564 (southeast), and 2567 (northeast), or roughly equivalent to or higher than the base of the west L-shaped foundation wall (2554). It is assumed that building D 93 was still in use. The later entryway construction, however, need not have infringed on the outline of the building.

The entrance was formal, undoubtedly completely framed in stone, and had an emplacement on its south side that perhaps held a post surmounted by an emblem. Thus, it is assumed that the approach to the doorway was also monumental. Perhaps it was a porch constructed over the earlier buildings alongside the dividing wall between the West and Central Sectors and reached by stairs or a ramp from the north.

102 Knudstad 1964a, p. 41. This brick mass was part of an area that included the possible earlier staircase in the middle of the dividing wall between the West and Central Sectors.

103 The top of the first step into the Level III residence was recorded as 2545.

The only object registered for doorway D 87 (1), the floor associated with the doorway, was a faience bead (OIM E50823).

Foundation Walls

Two wall fragments, without locus designations, were constructed along the interior and exterior sides of the earlier west wall of D 79 and are included in the plan of doorway D 87. Both L-shaped segments remained unexcavated. The width of these walls was 48 cm. Both fragments were designated as "probably Level III walls" in the architectural drawings; they were most likely built as foundations for a porch and stairway construction leading toward the doorway of D 87 in a late Level III renovation of the area.

The earlier wall and the east foundation wall are visible on plate 37a, with the oven of house D 79 shown beneath the west end of the foundation wall at the bottom of the photograph.

LOWER LEVELS IN THE SOUTHEAST CORNER

House D 79 and Level III–Level IV Walls

A level with signs of intense burning begins below doorway D 87 and the associated foundation walls. The tops of the walls in D 76–D 79 and D 88–D 93, however, were burned and discolored for 25 cm down their preserved heights.[104] In D 90 and D 76 the wall tops appeared about 40 cm beneath a hardened crust that was dark gray to black in color, while the upper fill in D 88, D 89, and D 91 also showed signs of intense burning. The wall discoloration can be seen in the photograph on plate 37a. There is no mention of burning on the floors or lower walls of these rooms, and the fill within the lower walls is described as "yellow brown, with no ash."[105] Whatever fire caused the intense burning just described broke out after the rooms had been partially filled in but probably before the doorway into the Central Sector had been constructed.

The walls of the earlier silo in D 77 and D 78 were dismantled during a later renovation of the area and lay immediately below the bases of the walls of D 79, which was a room or house that the excavators thought was a late Level IV, or perhaps early Level III, structure (fig. 4.11).[106] The walls of the house were 18–20 cm thick, but the brick sizes used in the various levels of the area were not noted. No floor levels were discovered in this room. The unit also included rooms D 87, D 89, D 90, and D 76. D 87 was not fully excavated because it lay below doorway D 87.

A narrow doorway leading into house D 79 may have had a sill stone, the top elevation (2569) of which was 26 cm higher than the nearby wall foundation. The wall next to the door was reinforced with a support pilaster on its interior side. The east, north, and west walls were built together, with a slight extension that ran west from the northwest edge of the building. The southwest wall end abutted the north wall of D 76, ending in an L-shaped support. The east wall of D 79 was a thinner wall continuation of the north–south dividing wall between the West and Central Sectors. It seems to have used part of the dividing wall, the width of which decreased from its typical 68–70 cm thickness to 18–20 cm at house D 79. It is unclear whether this dividing wall had completely divided the two sectors, since the architectural remains within the Central Sector at this spot were not well preserved after the later Level III and Level II disturbances in the area.

A stone arrowhead (Khartoum 14274; see pl. 94d) was found 25 cm below the top of the west wall of D 79, while a pottery disk with incised crossed lines was found in D 79 (1).

A door at the south side of the room led into D 89.[107] Room D 87 extended east into the Central Sector just past the door and may well have been a point of access into the sector at some point in time, but it was not completely excavated because it lay under the later walls of doorway D 87. The wall on the north side of D 87 was perhaps also used in the fill of the later doorway construction, since it remained standing at a height of some 17–23 cm above the other walls associated with house D 79. Its foundation level was not recorded.

The staircase in D 90 had thick support walls, its west wall, for example, being 47 cm wide. A thin wall blocked off a possible storage space at its southwest edge, although this may have been an earlier wall since no elevations were recorded. Another possible earlier wall backed up against the north wall of the area. Again no foundation levels or wall heights were recorded here, and there is no clear photograph of these walls. Three steps rose 44 cm from the floor level at the base of the stairs (2536). They were built of bricks

104 Hoerth 1964a, no. 1.

105 Hoerth 1964a, no. 1. The same fill is within both D 76 and D 90.

106 Knudstad 1964a, p. 41.

107 Loci D 88 and D 91 consisted of the fill excavated from the south side of D 79, which was separated from the main room by the unexcavated wall belonging to the foundations for the later doorway of D 87.

placed on edge, the risers were about 14–15 cm high, and the treads were 40 cm deep. The fourth step was missing, and the fill for the construction was apparent in the gap. But the stairs should have turned either east or west at a landing to reach the top of a roof or the fortification wall.

A number of handmade bowls were found in D 90 (1). One pot showed a plain brown fabric (OIM E24414), and the other sherds were those of a black bowl with incised pendant triangles and diagonal slashes incised on the exterior and atop the rim (OIM E36383) (see pl. 71a–b and fig. 5.75a in chapter 5). Three unregistered stone pounders were also found but seem to have been left in the field.

The kitchen in room D 76 had a bin in its northwest corner and two superimposed ovens in the southwest corner (2512 and 2533), with the higher oven being contemporary with the use of house D 79 (pl. 18a). The earlier oven rested on an associated floor, which was at the same elevation as the base of the west wall of the room (2512). This level was 24 cm below the floor level in room D 90 to the east and was part of the Level IV remains, before the addition of the wall relining. It is uncertain in which phase the rectangular bin was used, since no good photographs exist, nor are there any notes or elevations on the architectural drawings to clarify the dating. A column base resting in the southeast corner of the room was also identified as a later addition, perhaps reused as a stool or for some other purpose appropriate to a kitchen. It was not identified as being in situ or on a floor. The column base was left in place on an unexcavated pedestal standing 21 cm above the base of the later oven to its west (2533).

The foundation of the west wall in room D 76 (2512) can be associated with the similar footings of three of the walls in building D 93, with the earlier oven in D 76, and with the oven next to the silo at D 78. The base of the D-shaped bin or oven near the silo at D 77 was, however, set 12 cm lower.

The objects found in D 76 came mostly from "below the ash layer," or the fill below the darkened level associated with fire and numerous flood deposits, and from above a possible floor level at an elevation of about 2533 associated with the latest oven. They included a talc net weight (OIM E24379), a pottery spool with an incised line around its circumference (OIM E24381), an unregistered sandstone "game" piece or weight, and a faience plaque with a recumbent sphinx and hieroglyphs (OIM E24386) (see pl. 89c and fig. 7.2c in chapter 7). A slightly reworked marl body sherd with black-painted bands was also found here (OIM E50626).

Level IV Silo D 77/D 78

One of the earliest constructions uncovered in the southeast corner, to the east of building D 93, was the large silo D 77 (including part of D 78), measuring about 5.5 m in diameter and with walls 40 cm wide. No elevations were noted for the foundations of the silo, but the north wall of house D 79 was built atop it at an elevation of 2543, its west wall was preserved to a height of 2550, and its south wall was below the wall of D 79 at 2548, indicating that the silo had been dismantled during a later reconfiguration of this area.

The excavators cleared the area between the east wall of D 93 and the large silo when exploring the silo wall. The space contained two ovens and a D-shaped hearth or bin farther south, both sets of features mentioned above. It would have been difficult to use these ovens and the hearth/bin during the existence of both the silo and the large building, since the open space near them would have been very limited. The larger of the two ovens was built on a surface that was 5 cm below the nearby foundation elevation of D 93 (2513 and 2518, respectively). The hearth/bin sat on a level below the north wall foundation of D 93, as mentioned above. All these features represent a phase that was at least slightly earlier than that of the silo. The thin east–west wall between the large building and the silo may also have been part of an earlier structure, although no elevations were noted for it.

The objects from the fill under the "ash" layer of D 78 included a copper spatula (OIM E24387) and a reworked rectangular potsherd (OIM E49795). In addition, the records mention a bin in D 78—it is unclear whether this was the D-shaped feature that was said to have contained charcoal (OIM E45999) and pottery disks, one with incised crossed lines (OIM E47443, E47444).

THE DIVIDING WALL

The partition wall dividing the West and Central Sectors was built at some point in the Level IV occupation of the site and continued through Level III (fig. 4.12).[108] It appears to abut the north interior of the fortress wall relining, but it was probably cut back when the relining was added. Because this wall was situated west of the later Level II platform, it was possible to excavate it, although parts of it were degraded due to later construction activity (pl. 37b).

Toward its north edge, the dividing wall rested on an elevation of 2556, or 3 cm above the floor surface

108 Knudstad 1964a, p. 40.

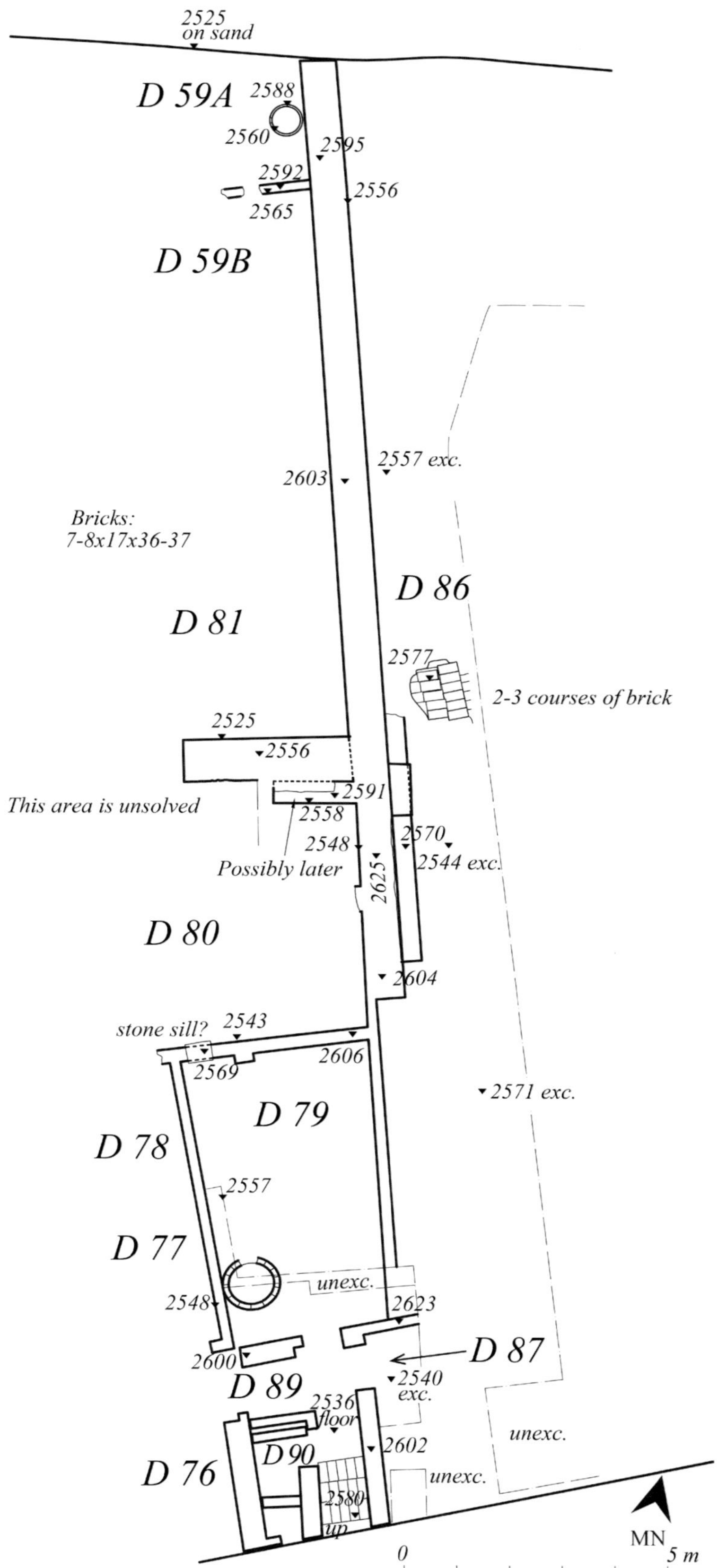

Figure 4.12. Dividing wall between the West and Central Sectors

in D 39, east of the dividing wall. Its base was 31 cm above the deeper foundation of the innermost relining of the enclosure wall (2525) nearby. It continued south, with the foundation level rising slightly to 2564, before reaching a possible stairway or ramp construction at its center portion. A pile of bricks was located just to the north of this, within the Central Sector. These bricks were not part of an intact structure.

The original height of the dividing wall is uncertain. But it did not have to be tall to serve as a boundary segregating the two areas and thereby restricting access to soldiers, workers, and visitors from the West Sector to the Central Sector.

Level IV(?) Entrance into the Central Sector

The walls perpendicular to the dividing wall at its midpoint were not well preserved. The north wall fragment was 82 cm in width and ran parallel to a shorter fragment that was founded 33 cm (2558) above the larger wall (2525). The end fragment of a parallel wall was found about 2 m farther south. On the interior side was a wall or walls strengthening the dividing wall in this location. These walls were probably all built at the same time as part of an entranceway into the official area. A note on the architectural drawing refers to the area as "unsolved" but observes that a Level IV doorway between the West and Central Sectors was needed.[109] These remains look like a good candidate for such a construction.

The dividing wall disappeared toward the south fortification wall. As mentioned previously, it may have been partially dismantled during the later Level III reconfiguration at the southeast corner. The line of the wall continued on at D 79, but only as the thin east wall of the house. No foundation levels or wall heights were recorded here. The upper remains of doorway D 87 represent an even later reconfiguration for access into the Central Sector, while the room below it, D 87, may also have allowed entry into the sector at one time. There were few architectural remains on the east side of the dividing wall, which appears to have been cleared and the bricks reused during the construction of the Level II platform.

Passageways and Streets in the West Sector

The absence of structural remains throughout the central part of the West Sector limits one to a purely hypothetical reconstruction of how people moved around within the fort. A hypothetical reconstruction of pathways in the fortress is shown in figure A.II.2 in appendix II. A good comparison for the typical lanes one would have found in a walled settlement of this general period are those discovered in the residential area at Medinet Habu dating to the Twenty-Second through Twenty-Fourth Dynasties.[110] The passageways were located between the walls of the houses and wended around based on the way the exterior walls of buildings had been constructed. When the ground level required it, there were steps or slopes facilitating movement up or down. The lanes in the residential areas at Dorginarti were undoubtedly similar.

A major road in the center of the area would have provided access to all areas along the walls through connecting passages, which have been mentioned in the discussions of the residential remains along the north wall and are evident to a much lesser extent along the south wall. Access into the two large buildings of the West Sector would have also branched off from this main thoroughfare.

One means of entering the Level III official building in the Central Sector was the later doorway, D 87. The use of stone architectural elements in both structures indicates a similar formal aspect, which supports an assumption that they were at least partly contemporary. In the levels below doorway D 87, and perhaps contemporary with both Level IV and early Level III, access into the Central Sector may have been via building D 79 and the lower-level remains in the area of room D 87. However, the remains in the latter area were not defined well enough to provide supporting evidence. Level IV access into the Central Sector may have been via a centrally placed stairway or ramp at the middle of the dividing wall. Whether this conjectural stairway or ramp was used throughout the Level III and Level IV phases of the fortress remains uncertain, although the latest access is through doorway D 87.

There may never have been much traffic across the Central Sector to reach the east storage areas from the west settlement, but it seems likely that access to the main administrative center of the fortress was necessary for daily communication between the officials, service personnel, and their troops.

The house of the commander was set apart from other residential and work buildings. Moreover, there was a stark division at Dorginarti between the area of the official building and east storage areas and the area of workshops and residences. Such divisions were not typical of earlier Egyptian fortresses in Nubia. Since the person in control of the fortification was undoubtedly an officer of high esteem, if not someone of royal

109 Knudstad 1964a, p. 40.

110 Hölscher 1954, p. 8, fig. 7.

lineage, the stark division between officials on the one hand and ordinary soldiers, their families, and service people on the other is understandable.

Extramural Buildings

Floodwaters obliterated much of the evidence for extramural structures around the fortress, at least by the West Gate and along the north wall (see plan 1). The structures that survived were mainly situated in protected areas along the west and southwest sides, although ovens and built-up areas remained on the exterior of the north side between bastions C and D and between the crenelated bastion and bastion A, where the wall bricks were the larger 40 × 20 cm size. All exterior walls and ovens were built atop the stone glacis constructed around the fortress walls. In addition, the exterior enclosure wall relining between bastions C and D was either eroded or cut back 45 cm to accommodate a shelter or guard post at some later occasion.[111]

Because the excavation team was on a tight schedule, the degree of attention directed to these areas was often determined by the presence of potsherds in the bays between the external bastions. Sherds were especially evident outside the West Gate, in the bays along the west end of the south wall, and also between the bastion along the north wall past the breach and near the North Gate.[112]

West Exterior

Just outside the West Gate, and between the north bastion of the gate and the northwest corner, a number of structures survived (plan 1; fig. 4.13).[113] They were not given locus designations. They may have been storage rooms given the limited space between the surviving walls, which would not have easily accommodated human movement. The walls here were mostly one brick in width. One fragmentary wall closer to the gate was 38 cm wide, but its remains consisted only of a section 1.42 m long joined at right angles to a thinner wall that stretched east and west for 1.20 m. The walls mainly rested on cataract rock.

The remains to the north of the first structure consisted of an area 1.47 m wide between three wall fragments. A small space was located within the walls at the southeast edge of the preserved wall. It was more or less square and measured 88 × 95 cm from the exterior edges. The walls also rested mainly on cataract stones.

North and west of the latter building was another area of walls next to the wall relining. This area showed at least three stages of walls; the earliest fragment turned a corner heading south, and a three-pronged fragmentary construction was superimposed on it. A later wall ran along the same line as the earlier walls, and its north wall fragment was partly covered by the later 40 × 20 cm brick relining. Another east–west wall ran alongside that wall against the relining, but 73 cm higher up and placed atop a floor level of 2624.

The size of bricks used in these buildings was not noted on the architectural plans.

South Exterior

The extramural buildings along the south wall between the southwest corner bastion and bastion N were perhaps dwellings, workshops, or both. These structures, running from west to east, were D 18, D 19, D 21, and D 25.[114] A photograph of the bastion shows the walls of D 19 and the area of the column base in D 25 (see pl. 15b; fig. 4.14).

The walls of D 19 and the oven in D 21 were built against the exterior face of the fortification wall. The later relining, with smaller-sized bricks (i.e., 34–35 cm in length) disappeared halfway along the wall face to the east. It had clearly eroded away or been dismantled before this area was built up. The bay was covered with a second layer of glacis stones at some stage, and the buildings and oven all rested on this upper glacis surface. The top of the relining wall remained standing at an elevation of 2649, with three walls in D 18 and D 19 preserved higher.

The structures' walls were thin and enclosed very small spaces. D 19 was the centrally placed building, with a packed-mud floor lying atop the rocks (2596). Its doorway was probably on the east side, although it was not preserved. The walls of D 18 and D 19 were placed atop heights that ranged between 2615 and 2591, while the east wall of D 19 was preserved at a height of more than 1 m above the floor level. The oven had an opening to the southwest and abutted both the fortress wall and a relining of 36–37 cm bricks on the side of bastion N. The west relinings on this bastion were not entirely preserved, having been either dismantled or eroded away.

To the east of bastion N was locus D 25, which consisted of a single column base of stone placed atop a lower glacis layer at elevation 2538, and perhaps in situ.[115]

111 Knudstad 1964a, p. 35.

112 Pierce 1964a, January 22.

113 Knudstad 1964a, p. 56.

114 Pierce 1964b, pp. 63–66, 78–79, 92, 99.

115 Knudstad 1964a, p. 52.

Figure 4.13. Extramural structures outside the West Gate

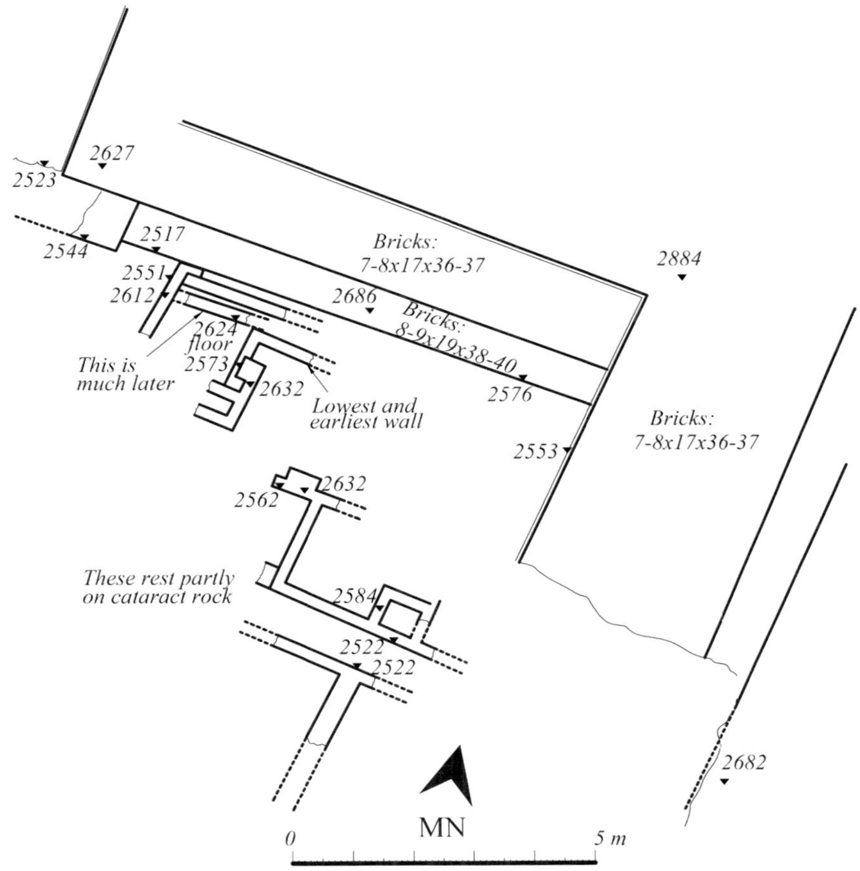

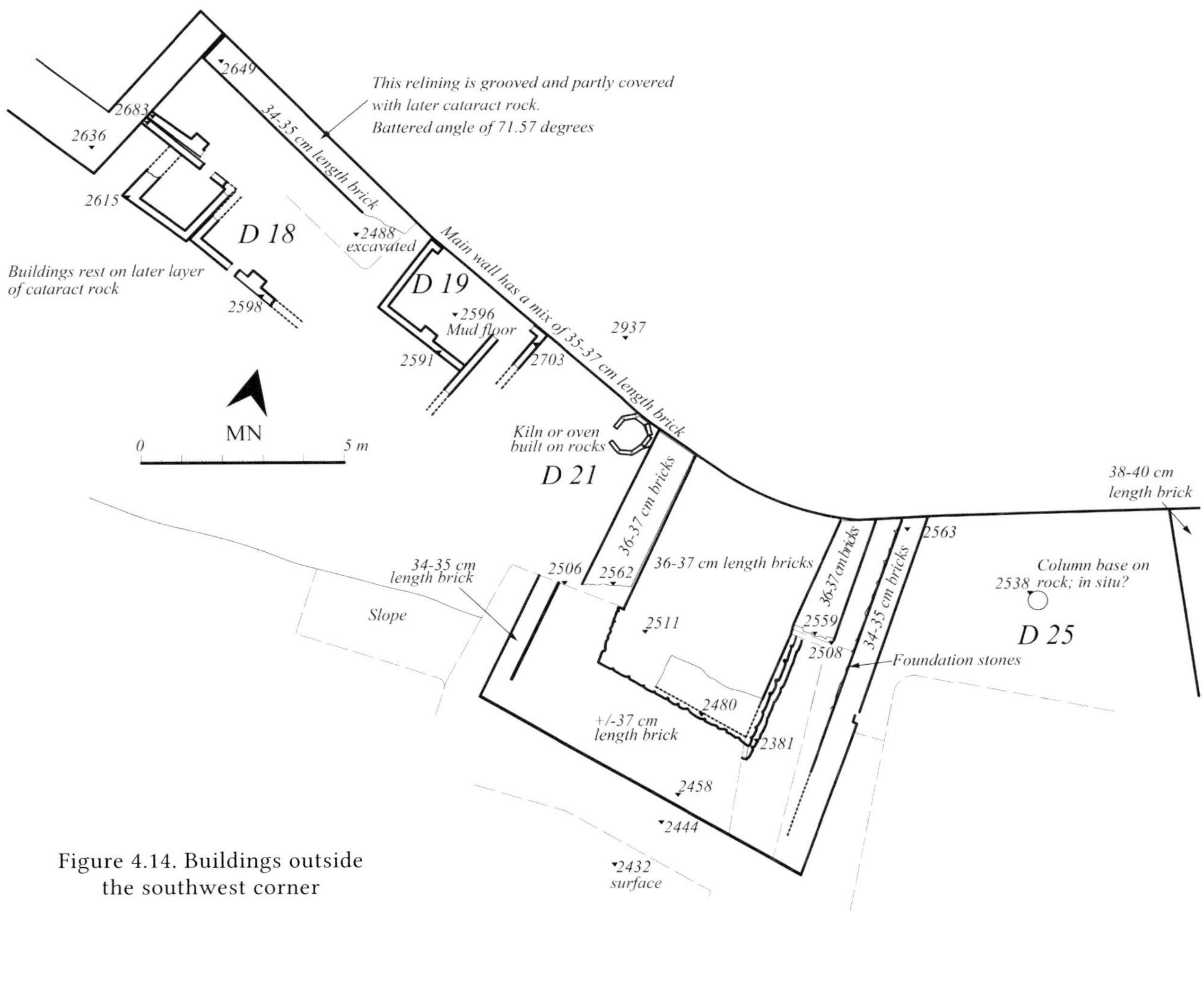

Figure 4.14. Buildings outside the southwest corner

This group of structures yielded a number of interesting objects and sherds of both wheel-made and handmade vessels. An early Iron Age flask import from Phoenicia (OIM E24327) stemmed from D 19 (1), a faience ring plaque with "lord of heaven" inscribed on it (OIM E24325) was found in D 18 (1), what may have been two flattened stone weights with drilled holes and a handstone (OIM E50785, E50786, E50789) came from D 19 (1), and stone fishing-net weights (OIM E24328, E24329) were recovered in the same area. Two miniature handmade bowls were also uncovered here (OIM E24399). In addition, fragments of crucibles, some with interior deposits (OIM E42942–E42944, E51013, E51026, E51031) were found scattered in D 19 and D 21, indicating perhaps small-scale melting of metal, if they were not transported here from elsewhere at the site. Finally, sherd disks, one with incised crossed lines, were found near the column base in D 25 (OIM E49761, E50799), as were fragments of a stone bowl (OIM E50811).[116]

HOUSE AND SETTLEMENT PARALLELS[117]

Domestic buildings from Third Intermediate Period and Late Period levels at other sites show some similarities, especially in the urban arrangement of the small- and medium-sized houses inside the great girdle wall of Medinet Habu at Thebes.[118] The passageways and blind alleys, which allowed access to the residences, reflect the arrangement at Dorginarti. Hölscher noted that

> the houses were laid out arbitrarily without planning. The streets were usually no larger than 1.50 m wide, often even narrower, and wandered crookedly uphill and downhill, with steps here and there, over existing rubbish heaps.[119]

The Medinet Habu houses attributed to the Twenty-Fifth and Twenty-Sixth Dynasties show a change in the settlement pattern at the site, with the village houses still there but more houses displaying a formal arrangement.[120] The three-room houses, each with one staircase, differ from the Dorginarti dwellings in the West Sector in their larger size, the presence of two columns in some rooms, and the presence of a stairway placed near the wall at the back. They also do not closely resemble the official residences in the Central Sector, both of which resemble the larger houses outside the enclosure wall of the Great Temple. This is discussed in part 2 of this chapter.

Both the houses excavated at Elephantine dating to the first half of the first millennium and those in Nubia dating to the same time period are usually larger buildings and compare better to the Central Sector's official residences. Unfortunately, smaller houses for the less politically and economically powerful are not so prominent in publications.

However, recent excavations at Kawa have exposed numerous smaller residences dating to the Napatan or early Kushite period, the sizes of which compare well with those at Dorginarti.[121] At Kawa there was a dense pattern of houses and alleys, and both larger and smaller structures are built against one another in an agglomerative pattern.[122] Although dating to the New Kingdom, the urban settlement pattern is also reflected in the house remains at Amara West, particularly when the larger houses were subdivided over time into smaller units.[123] A few buildings and residential features in the fortress at Gebel Sahaba may also offer good parallels. Here, although very few of their features have survived, the walls show the inner support pilasters so very common at Dorginarti.[124] The use of these supports is also evident at Kawa and Qasr Ibrim.[125]

116 Most of the objects and pottery are shown in the figures accompanying the discussion in chapters 5 and 7.

117 For the location of sites mentioned in the paragraphs below, see map 1 for the Egyptian sites and map 2 for the Nubian sites.

118 Hölscher 1954, pp. 6–8, especially fig. 7, which represents structures assigned to the Twenty-Second through Twenty-Fourth Dynasties.

119 Hölscher 1954, pp. 6–7. See also the arrangements of various phases at the settlement in Hölscher 1934, pls. 6, 10, 15.

120 Hölscher 1934, p. 14, fig. 19, folio pls. 14–15 (the structures are shaded in light green, although they probably also date to the Twenty-Fifth and Twenty-Sixth Dynasties; see Hölscher 1934, p. 14, n. 52).

121 For example, the smaller houses are mentioned as probable two- or three-roomed structures, but no plans are shown, in Welsby 2002, p. 39 (C4 and C13); see also the description and plans in Welsby 2014, pp. 49–50, figs. 2–3 (C25, C27, and the later and larger unit D21, built over C27).

122 A plan of the upper town is shown in Welsby 1998–2002, p. 185, fig. 1. Note also Welsby 1998, pp. 19–20, and 2014, pp. 67–68, fig. 3.

123 The earlier New Kingdom houses of Amara West are typically larger villas in their initial construction, although the division of these larger units into smaller houses over time, and the pathways between them, show an urban pattern similar to the layout at both Medinet Habu and Dorginarti. See Spencer 2014, pp. 460–64, 470, figs. 1, 7.

124 Säve-Söderbergh and Troy 1991, pp. 321–22, pls. 216, 218–19 (areas 2, 4, 7, and 8).

125 Welsby 2002a, p. 37, fig. 4 (building A3). Welsby notes that the use of thin walls with buttresses is similar to Kerma period buildings at Kawa, and he thus classifies it as a Nubian style of construction. For similar evidence from Qasr Ibrim, see Rose 2008, pp. 201–2, fig. 6.

Part 2. Buildings and Structures in the Central and East Sectors

GENERAL NOTES ON THE RESIDENCES OF LEVEL III AND LEVEL IV

The size of both the earlier and later official residences at Dorginarti was impressive and compares well with a few larger buildings found in similarly dated contexts in both Nubia and Egypt (plans 5 and 6). The suggested dates for the Level IV and Level III residences range between the tenth and eighth centuries BC. The Level III administrative residence was about 14 × 12 m in area, or 168 m², while the Level IV building was slightly smaller, about 13 × 11 m. The stone sills, jambs, lintels, and stone column bases and fragments found in the Level III building were embellishments not found in the earlier residence. The Level IV residence, however, featured brick-paved floors not found in the Level III residence, and in some cases there were several layers of floor replastering atop them.

The stone architectural elements of Level III had clearly been reused. They were adorned with Ramesside-style images as well as fragmentary inscriptions and graffiti, some of which had been turned away from view. They undoubtedly had been transported from nearby Buhen, which had been occupied in the Ramesside period. This is discussed in more detail below.

The distinctive sherds from the various levels in both the earlier and later residences are presented in appendix I, while the locus numbers and descriptions are listed in table A in appendix III.

LEVEL III IN THE CENTRAL SECTOR

GENERAL EXCAVATION NOTES

The plastered brick walls of the Level III official residence and administrative building were encountered immediately below Level II and were encased by the rubble used to fill the later platform (plan 7). Since the east and west retaining walls of the Level II platform were situated to the side of the Levels III–IV buildings, the Dorginarti team was able to excavate these levels without entirely removing the fill and masonry of the Level II platform (pl. 38).[126]

The official residence was located on the highest elevation in the fortress, and its location, large size, and plan indicates that it had been the office and residence of the fortress commander. The archaeological remains found within both the Level III and Level IV residences do not indicate their functions. There were abundant potsherds and some objects, and the stratigraphy was fairly well preserved in Level III, but no official seals, mud sealings, or ostraca were found to prove the building's function. The exception was a small inscribed sandstone block, which is described more fully in chapter 7 (OIM E24326) (see pl. 92a). The stone was found at the bottom of a test pit that had been sunk in D 5 of the Level III residence. This pit had been excavated nearly 1 m down through a floor in Level IV.[127]

The exterior walls of the building were 65–67 cm thick, while the interior walls were the same width or at least 40 cm thick. All the building's walls were built of bricks of 36–37 cm in length. The building was probably two stories in height, but only the lower part of the ground floor walls remained standing after the Level II builders had demolished the upper walls. The relatively even removal of the walls is clearly visible on plates 38–39a. The nearly square plan shows two large rooms on the south side and two smaller rooms on each side of a hall or courtyard to the north. The staircase, however, was now located at the northeast corner of the building instead of at the southwest corner, as we will see in the Level IV residence.

The excavators noted that "advanced erosion" had taken place between the abandonment of Level III and the construction of the Level II terrace and building, although their notes indicate that they were uncertain about the length of the hiatus involved.[128] In addition, the water- and wind-streaked walls of the residence, along with evidence for encrustations of wasps' nests, indicate that the building had been exposed to the open air for some period of time.[129] There was also wind-laid sand and fallen brick immediately atop the Level III floors and below the compacted fill of Level II.

Sand accumulates quickly on the west bank of the Nile in this region, and therefore the chronological gap between Level III and Level II would have been difficult to ascertain.[130] However, the latest pottery from

126 Since the expedition operated under severe time constraints, the Central Sector was not completely excavated; see Knudstad 1966, p. 181.

127 Pierce 1964b, pp. 21–22.

128 Knudstad 1966, pp. 179–80. See also Knudstad 1963–64, p. 245, and Pierce 1964a, January 28, where the possible period of abandonment is discussed.

129 Pierce 1964a, January 28.

130 See Macadam's reference to an account by James Henry Breasted concerning his work at Sesebi: "During the five years between the excavations of 1930–1 and 1935–6 the excavated

the Level II activity indicates as much as 150–200 years separated the later Level II assemblage and that of Level III.

The fill within the platform and around the walls of the lower building consisted of loosely packed sand, brick, and sherd rubble atop densely packed sand and fallen brick debris.[131] The lower compacted fill, which the excavators encountered generally below the loose fill, was also recognized by them as a product of intentional filling from atop the walls, not all of it resulting from simple erosion.

The Level III building contained seven rooms, with D 20 perhaps being a courtyard, and a narrow staircase on the northeast in D 16C.[132] The main doorway into the building was on the west, and there was a smaller door on the north. Both doorways required stairs to access the higher level of a house built atop earlier debris (pls. 39b–40).[133] The height of the new residence was raised about 60 cm above the ground level, which necessitated the stairway entrances.[134] The construction of the Level III building required razing the previous Level IV walls to a height of approximately 40–70 cm above the ground level and using the earlier walls as foundations for the Level III walls, as shown in the north–south section of the Central Sector on plan 7. The newly built walls were laid upon or below the floors of Level IV, with the layer between them filled mostly with brickbats atop 5–10 cm of brown sandy soil lying atop the floors of Level IV.[135] The wall foundations were noted as being slightly wider at the base than above. The plaster on the interior wall surfaces was preserved about 25 cm above floor level in many of the rooms and sometimes ran continuously down the walls and onto the floors with a sloping joint between them. The floors throughout the building were very uneven and pocked by small depressions, perhaps from the activities of the Level II builders. The Level IV walls were exposed soon after the clearance down to the floor level of Level III (pl. 41a).

Six doorways of the Level III residence were adorned with stone jambs, lintels, and sills. The doorway into D 24 had been destroyed by a well cut from above in Level II. The north entrance door into the building had one brick jamb and a socket in place, and broken stone jambs lying on the porch outside the doorway. Since there was no evidence for a substantial structure of an earlier date on the island and the Level IV residence had finished mudbrick jambs, it was supposed that the stones had been transported to Dorginarti from another site such as Buhen, where Ramesside building activity is attested:

> . . . the stones bearing inscription[s] . . . show clear reuse of larger blocks or architectural fragments from somewhere else. An examination of the uninscribed blocks incorporated into the doorways also gave sufficient clues to their having been prepared for and used in a previous context. . . . Some were used in a broken condition (that is, a finished block with broken end), very few seemed to match, and their installation in the brickwork in all cases was clumsy.[136]

Many of the blocks had been discolored by burning and had collapsed after the building was abandoned. Thus, they were found "not on the floors but in a thick layer of . . . mudbrick debris resting on a five to ten centimeter mixed layer of burned mudbrick, ash, and charcoal which rested in turn on floor dust [and sand] above the actual packed mud floors of the rooms."[137] There was evidence of a fire in the southeast part of the building and in the nearby hallway. The burned stratum was found above the floors in D 9, inside the doorway of D 5, in D 20, and along the doorways and stairway of D 16A–C. But in D 5, D 22, and D 24 the evidence was minimal or absent. A burned beam was found lying in the debris of ash and charcoal in the south part of D 9, suggesting a second story, while other roofing materials were found in the room. In addition, burned wooden door sockets were found in D 9, D 16A, and D 20.[138] The ash and charcoal layers were found under many of the architectural stones lying atop the debris layer, but in some cases the stone elements were found lying atop the thin deposit of dust on the floors and not on burned debris.[139]

The reconstruction of, and inscriptions on, the reused stones will be discussed in further detail at the end of the description of the Level III building below.

temples, [which] in places [were] four metres high, were at least half buried [by sand]"; see Macadam 1949, p. 63.

131 See below for a description of the stratified deposits in D 9.

132 The courtyard suggestion is from Knudstad 1966, p. 181. The Level III residence plans are shown in Knudstad 1964a, pp. 12–21, and include plans of the fallen stones.

133 The blocked doorway on the east is actually a part of the Level IV building that was not used in the Level III renovations.

134 Knudstad 1966, p. 181.

135 Hoerth 1964a, no. 2, where he discusses the construction in detail.

136 Knudstad 1966, pp. 182–83; Pierce 1964b, p. 25.

137 Knudstad 1966, pp. 182–83; Knudstad 1963–64, pp. 229–30.

138 Pierce 1964a, February 2, and Pierce 1964b, p. 4, where the beam's measurements were recorded as being 1.6 m long, 15 cm wide, and 6 cm thick.

139 Knudstad 1963–64, pp. 230–31, 236.

OFFICIAL RESIDENCE

The discussion will start with D 9, where the rebuilding and heightening of the east wall included blocking up the east doorway of the Level IV residence. The floor level was very uneven, with an elevation of 2687 at the north side, 2678 along the east wall, and a column base founded on a floor at 2674 in the middle of the room.[140] There were hand- and footprints impressed into the mud.[141] The floor level on the east of the room was 75 cm above the Level IV doorsill pavement immediately below it. The east, north, and west walls of the room were preserved, but the south wall had mostly caved out, exposing the line of the south wall of Level IV, upon which it rested. The wall ended abruptly at the eastern limit of D 5 and was abutted by the 65 cm wide partition wall between D 9 and D 5.

Two burned wooden beams rested in the south half of the room in the sand above a floor level, with burned debris lying atop it. The packed-mud, cataract rock, and sandstone foundation for a column was found in the center of the room, along with an in situ stone column base fragment that proved to be a reused stone with a fragmentary inscription on its west side (field number D-130) (pl. 41b). The column base was flush with the floor level in the center of the room, at elevation 2674 (pl. 42a).

A doorway between D 9 and D 20 was situated at the west of the north wall and featured two reused inscribed stone jambs facing into D 9 (field number D-131 and D-132) (pls. 42b–43a).[142] Both stone jambs were broken, and only their lower parts remained in situ. Each bore two columns of identical but symmetrically opposed text. The upper half of the east jamb was found in D 9 to the south. The top of the doorsill was 18 cm above the floor upon which it had been placed.

The general stratification within D 9, from the top to the floor, included the loose Level II fill, the bottom elevation of which was 2743 (stratum 1); a more compacted level whose base was at 2708 (stratum 2); an uneven burned-debris layer (stratum 3); and finally, a thin deposit of brown soil, sometimes less than 1 cm thick (stratum 4) (pl. 43b). The latter layer covered the gray mud-plastered floor (stratum 5).[143] The base of the north–south partition wall on the west side of the room was recorded at 2622, which was 68 cm below the nearby Level III floor and only 8 cm above the Level IV floor below. Its preserved height stood 1.73 m above the nearby floor associated with it.

The doorway leading into D 5 from D 9 featured a stone door socket on its south side, which was possibly the top part of one of the two inscribed jambs of the doorway in D 5 (see pl. 91a). This stone door socket was also a reused fragment, with a fragmentary offering formula on its underside (field number D-129, OIM E24424). It had been placed on a surface that was level with the D 9 floor leading through the doorway.

Two reused stone jambs were found in situ, one on each side of the doorway leading into D 9 from D 5 (pl. 44a). The stone sill, with graffiti depicting cattle, a boat, and a man (field number D-129) rested 6 cm above the nearby floor level in D 5. There were also fragments of a carved and painted stone lintel west of the doorway (field number D-128), with one fragment found in the center of the room.[144]

D 5 was also a large hall with evidence for a column. The foundation for this column was level with the floor in the room, but its construction went down 71 cm into the doorway between D 312 and D 315 in Level IV (2619).[145] The foundation was made of mud and stones placed off-center toward the south, perhaps to avoid obstructing passage from the main west door to the doorway leading into D 9. A second column may have existed to the north, although no physical evidence of it was found.[146] The fragments of a stone papyrus-bud column capital were found in the center of the room (unregistered and left in the field).

This area was located above rooms D 312, D 315, and D 314 of the Level IV residence, with the north, west, and south walls running atop the earlier walls. The south wall of the room was built along the line of the earlier wall, although it had been shifted farther south and thus enlarged the area within D 5 (pl. 44b). The east end rested against the exterior of the eroded south wall of D 9. A plaster pilaster or molding was placed midway along the wide partition wall shared by D 9 and D 5, with its base 22 cm above the nearby floor elevation of 2682. The molding stood opposite the

140 Knudstad 1964a, p. 13.

141 Pierce 1964b, p. 40; Knudstad 1963–64, p. 230.

142 One of the jambs was given to the Antiquities Service (field number D-131) and one to the Sudan National Museum in Khartoum (field number D-132). Both were cut from pink sandstone, and their texts are discussed below.

143 Pierce 1964a, February 5, describes the layers in more detail: a loose intentional fill (stratum 1a, 60–65 cm); then a layer of more compact intentional fill (stratum 1b, 70–80 cm); stratified sand layers and brick fragments (stratum 2, 40 cm); an ash, charcoal, and burned-brick layer (stratum 3, 5 cm); and a thin layer of brown soil (stratum 4, 1 cm), which lay upon a gray mud-plastered floor (stratum 5). See also Knudstad, 1963–64, p. 230, and elevations noted in Knudstad 1964a, p. 13.

144 The lintel fragment was left in the field but registered in the field object register as D-128.

145 Knudstad 1964a, p. 21.

146 Pierce 1964b, p. 73, sketch of D 5.

column base. It is unclear whether it was merely decorative or served a functional purpose.

The west wall was mostly collapsed, but it too ran atop the Level IV wall below it, with the main entrance into the residence at its north end.

The floor was slightly undulating and incorporated the tops of the walls of the Level IV building.[147] The excavation diary mentions fish bones embedded in the mud floor of D 5.[148] The wall plaster along the north side of the room continued down onto the patch of plaster forming the floor, which stood at an elevation of 2675 next to the exterior doorway. The level of this floor was actually fairly even except in the northwest corner just mentioned, where the level was 6–9 cm lower than the levels in the corners of the room (2684, 2682, and 2681).

The elevations of the fill strata within the room, and at its northwest corner, were recorded as being 2762 between strata 1 and 2 of the Level II fill and 2694 between strata 2 and 3, while the nearby floor level of stratum 4 was at elevation 2675, as mentioned previously.[149] The bottom of the loose Level II fill was thus 19 cm higher in this room than in D 9.

A test pit was sunk through the floor along the south wall of the room, although the test is not noted on the final plan.[150] It was excavated down nearly 1 m, through the Level IV layers below it and the earlier floor levels. It was in this test pit that a stone inscribed with the name of the fortress was found (OIM E24326; see pl. 92a). It could possibly be a foundation deposit belonging to the Level IV residence.[151]

D 5 was clearly the main entrance hall into the Level III residence, with a wide porch and stairway leading up into the residence from the west doorway. Visitors entering this room would have seen the lintel and jambs framing the doorway leading into D 9, at the back of the room. The doorway into the room was not well preserved, although a brick jamb that was preserved on the north side of the room seems to indicate the door was offset from the main axis of the staircase. No stone elements were found here, and the south wall and jamb were gone. The raised brick sill, preserved at the north, stood a full 17 cm above the nearby floor level.

The stairway landing was packed with debris, which was later excavated as D 302, but no prepared surface was detected atop the fill.[152] At the top of the stairs, however, four bricks had been laid on edge atop a surface level with the top step and perhaps represent the remnants of the porch floor. But the top level of the nearby doorsill was 62 cm higher, and more steps would have been necessary to reach it. The staircase construction was founded on sand and debris at various elevations around its perimeter, as well as in the area extending westward beyond the range of the Level IV walls below. The foundations of the staircase platform were 92 cm below the preserved height of the north wall of the staircase. The first, westernmost, step was built 1 cm below (2528) the foundation level of the west wall of Level IV (2529).

The bottom step rose about 17 cm from its base, and each subsequent step rose 13–17 cm. They were built of headers laid flat, their depth ranging between 30 and 40 cm. The staircase was 2 m wide, excluding its sidewalls. Its enclosing walls were not symmetrical, with a 90 cm wall running along the southwest residence wall and abutting a short, 63 cm wide east–west wall at the south edge of the porch. In turn, this wall was abutted by another east–west wall that ran along the south side of the staircase.

The north side of the stair construction had a 63 cm wide, L-shaped wall configuration enclosing the landing and abutting a short L-shaped wall that ran along the north side of the stairs. The latter wall had a circular hole sunk into its east end, and although the drawings are not specific about the feature and the depth was not recorded, it represented some type of emplacement. Another stub wall filled in the space created by the two walls.

A sandstone water basin rested on the bottom two steps. It measured 63 × 53 cm, and it was 14 cm in depth and 21 cm in height.[153] This basin was undoubtedly intended for visitors to cleanse themselves before entering the official residence, much like the basin found inside the east doorway of the Level IV residence (see below).

There was an entrance corridor or courtyard, D 20, leading from the north stairs and doorway into the residence.[154] Four smaller rooms and a stairway

147 Pierce 1964b, p. 21.

148 Pierce 1964a, January 11 and February 8.

149 Knudstad 1964a, p. 13. Various designations were used for the strata, with the fill levels of Level II being called variously 1a or 1b, depending on whether it was a looser fill on top or the more compacted fill below.

150 Pierce 1964a, January 14; Pierce 1964b, pp. 21–22.

151 The resulting test pit, if it was 1 m in depth as recorded, would have gone down to an elevation of about 2582.

152 Knudstad 1964a, p. 14.

153 Knudstad 1964a, p. 14. This object was not registered and was probably left on site.

154 See Pierce 1964b, p. 68, who suggested it may have been a courtyard. The plan is shown in Knudstad 1964a, p. 12.

opened onto the hall, D 16A–C, D 22, and D 24, and reused inscribed stone jambs were set into the brick moldings of D 22, D 16A, and D 16B (pl. 45). There were also stone sills with incised graffiti set into many of the thresholds, including at the north entrance into D 20 (pl. 46a). The floor was covered by debris, upon which rested fallen lintel fragments. These either had fallen neatly into place or had been removed and lined up (pl. 46b).[155] The lintels had perhaps not even been installed yet.[156]

The Level II well construction cut through walls in this area, even to a depth below that of the Level IV residence. Two Christian sherds that were discovered at a level equal to that of the Level III building joined a sherd from the same vessel that had been found at the surface in Level I.[157] These later sherds were uncovered below the sand that had filled the well.

The top of the stone sill from the D 9 doorway, which led into D 20, was situated 8 cm above the brick sill set into the same doorway at its north side. The floor level within D 20 sloped upward 20 cm toward the north part of the room. There is no mention that two floor levels were uncovered here. The clear imprint of a child's foot was found preserved in the mud plaster of the floor.[158]

The east rooms of D 16 were separated from each other by 42 cm wide dividing walls that abutted the east residence wall, which had been plastered before the partitions were built up against it (pl. 47a).[159] In D 16A, which was 1.90 m wide, the excavators discovered stone doorjambs, a brick sill, and a wooden door socket on its south side, which had been burned. The floor level sloped downward 5 cm from west to east (pl. 47b). D 16B was 1.75 m wide, and it also featured a brick sill with a wooden door socket at the south, along with two stone jambs. The one floor level recorded within D 16B was 13–18 cm higher than the floor levels in D 16A. Both the inner and outer doorsills of D 16A and D 16B rose above the floor levels in the rooms and in the hallway or courtyard of D 20.

The narrowest space was occupied by the stairway in D 16C, whose width from north to south was 70 cm (pl. 48a). The stairs rose toward the east for 2.9 m and were constructed of headers laid on end. The rise for each step was approximately 17 cm, totaling 96 cm in elevation difference between the top levels of the bottom tread and the seventh tread that was preserved. The stair depths ranged between 38 and 42 cm. The construction of the bottom step was slightly different, as its bricks appear to have been laid flat in a disorganized fashion. In addition, it rose 21 cm above the nearby floor level. The top of the stairway had been destroyed, exposing the brick and dirt that composed its fill.

The precise plan of the staircase is uncertain, although a landing and second flight of stairs would have been necessary to reach any roof or second story that may have existed.

The rooms of D 16A–B were packed with loose fill (stratum 1a) from the preparation for the Level II platform, a more compact fill layer that averaged 60–70 cm in depth (stratum 1b), and another layer near the doorways consisting of charcoal, ash, and burned brick that was 3–6 cm deep (stratum 2). A thin deposit of sand (stratum 3) lay atop the floor of stratum 4, which again was an uneven, mud-plastered surface with the imprint of hands and feet.[160] As elsewhere in the building, the destruction of the roof or second story left these rooms open to the sky, leaving streaks on the walls resulting from wind and rain erosion, along with the remains of wasp nests.

D 22 and D 24 were located to the west of D 20. As mentioned previously, the Level II well that had been sunk into the Central Sector in this area had destroyed much of the eastern part of D 24. It was assumed that a doorway led into this room from D 20. The south and west walls of D 24 were partially rebuilt sections of the Level IV walls below them, with foundation elevations varying from 2639 to 2678—that is, the elevations varied depending on whether the rebuilding of the earlier walls had either used the earlier wall top or cut into the earlier wall. The floor level at the west of the room was level with the south floor elevation in D 20 (2687).

The doorway leading into D 22 still had an arrangement for a jamb at its north side, with an inscribed stone block lying face up as a plinth for the jamb (see pl. 46a). The south side had been destroyed by the later well. The floor surface was at 2676, which was a bit lower than the floors in the nearby rooms. The brick sill of the door was level with the southern floor surface of D 20 and was 10 cm above the floor level within the room. The north wall of the room was founded atop the Level IV doorjambs below it (2691 and 2771), and the rebuilding levels were evident along the entire length of the residence wall. The west wall of the house had collapsed, which is why the photographs of

155 Pierce 1964b, p. 72.

156 Knudstad (1966, p. 182) remarks that the lintels "in all but one case lay broken but nearly complete just outside their doorways."

157 Pierce 1964b, p. 69.

158 Pierce 1964b, p. 70.

159 Pierce 1964b, p. 57.

160 Pierce 1964b, pp. 59–60.

this area show the interior of these rooms so clearly (see pl. 45).

The north entrance indirectly approached the building from a staircase, the top level of which was not preserved, thus exposing the broken bricks and debris in its fill (pl. 48b). The stairs rose from the east. The floor level below the first step was at elevation 2646, and the first step rose 17 cm from this point. Each of the preserved steps was between 38 and 48 cm in depth, and their height varied from 9 to 17 cm. The last step turned an angle to enter the building over a brick sill 10 cm high, before stepping down to a lower sill at 2712 and entering D 20, where the nearby floor level was preserved an additional 6 cm lower.

The north staircase was dismantled to expose an uneven, corrugated floor with depressions below. The floor was covered in white plaster.[161] The floor that was associated with the north stairs, however, was some 15 cm above this corrugated flooring. The staircase construction also covered over the walls of a lightly built structure north of the Level IV residence (i.e., D 305 and D 306).

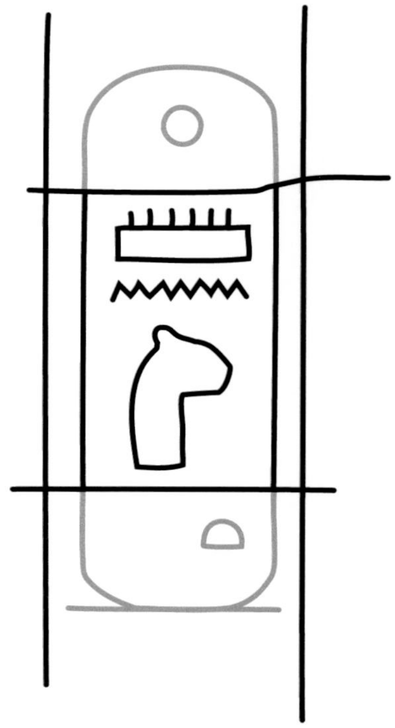

Figure 4.15. Cartouche on field number D-130 (Richard Pierce)

REUSED STONE ARCHITECTURAL ELEMENTS

The column fragment atop the foundation in D 9 (field number D-130) had been formed from an inscribed block made of a light gray sandstone, and it preserved part of a vertical cartouche that has been reconstructed with the name of Ramesses I, *Mn-pḥty-Rʿ* (fig. 4.15; pl. 41b).[162] The fragment was left in the field at the end of the season. It measured about 50 cm in diameter, varied from 49.9 to 53.0 cm across, and was 25 cm thick. Yellow paint was preserved on the fragment.

The two jamb fragments on the doorway from D 9 into D 20 were found in situ and were inscribed with two columns of identical hieroglyphic text (field numbers D-131, Antiquities Service and D-132, Khartoum). The entire doorway is shown on plate 43a and illustrated in figures 4.16 and 4.17. The upper fragment of the east jamb was found south of the door, in D 9 (2). Both jambs were made of pink sandstone and were either 54 cm high (without the upper fragment) or 88 cm high (with an upper fragment), 31–32 cm wide, and 16–17 cm thick. Chips of red and blue paint were found within the hieroglyphs on both jambs.

The inscription would have read the same on each jamb when the blocks were joined. The east jamb reads: ". . . Buhen, *Pn-tꜣ-wr.t*, triumphant" (pl. 49a). The upper portion on the west jamb preserves more hieroglyphs, with the inscription reading: ". . . for the *kꜣ* of the *ḥꜣty-ʿ* of Buhen, *Pn-tꜣ-wr.t*, triumphant" (pl. 49b). Richard Pierce copied the jambs in his field notes (fig. 4.17).

A number of the blocks in the house also bore graffiti. Most of them were on sills and were partially covered by jambs or had been turned away from view.[163] What is possibly a horse is depicted on the sill between D 9 and D 20, with its neck bending toward the ground (pl. 50). The fact that the head is missing, replaced by two lines set perpendicularly to each other, is unfortunate. The space atop the block perhaps restricted a more faithful rendition.

The stone lintel at the doorway in D 5 showed similar scenes on both sides, divided by a bouquet of lotuses (field number D-128) (pl. 51).[164] This doorway was important because it was highly visible in the first room inside the residence (no pictorial elements were found at the west entrance). Light-gray sandstone fragments of the sill were found in D 5, above the floor, and one fragment came from the center of the room. The reassembled fragments measured 152 × 53 cm in length and height, respectively, and were 18 cm thick. The

161 Hoerth 1964a, no. 2.

162 Pierce 1964b, p. 43.

163 Pierce 1964a, February 2.

164 The block was 1.53 m long, 0.53 m high, and 0.18 m thick and was left in the field.

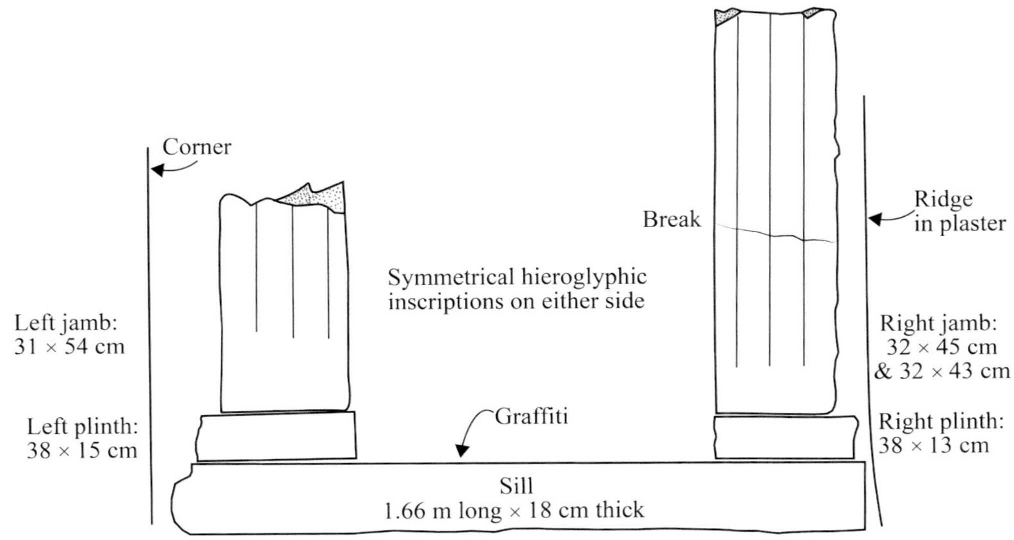

Figure 4.16. Construction of doorway from D 9 into D 20 (James Knudstad)

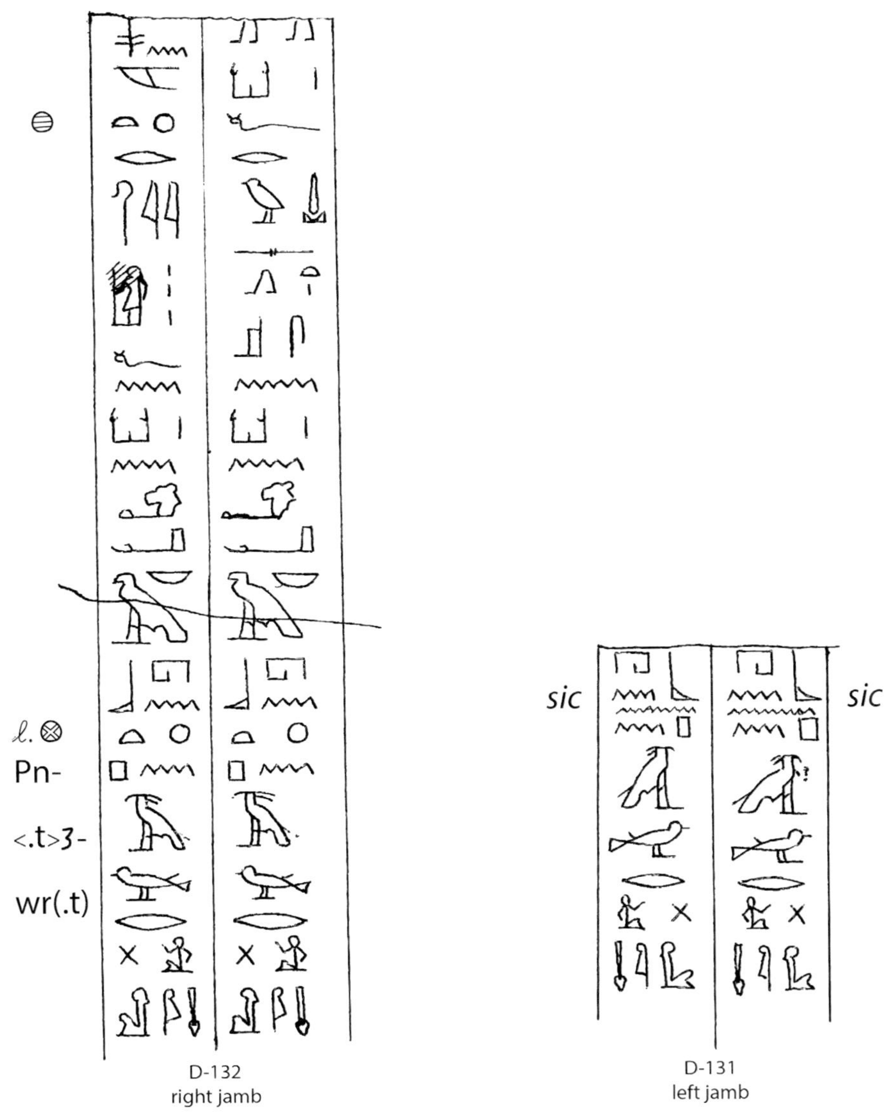

Figure 4.17. Jamb inscriptions from door into D 20, field numbers D-131 and D-132 (Richard Pierce)

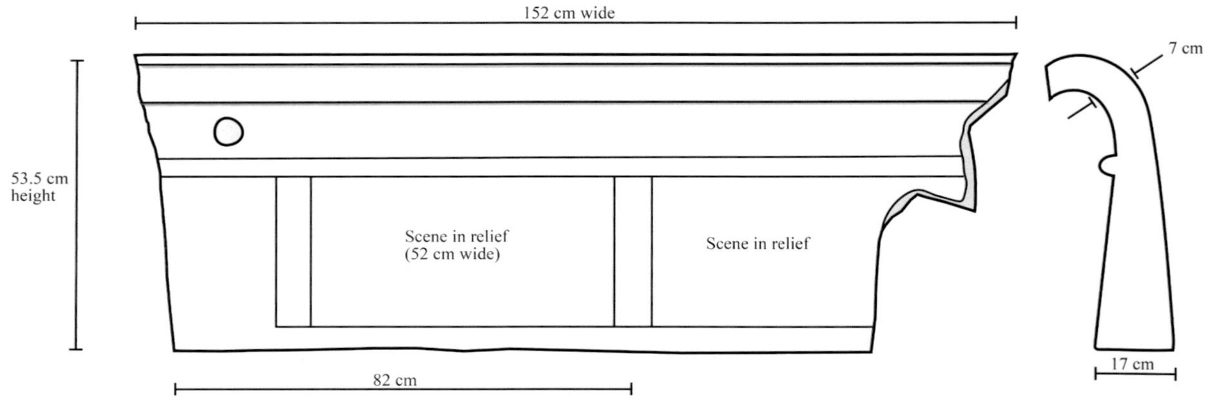

Figure 4.18. Drawing of lintel in D 5, field number D-128 (James Knudstad)

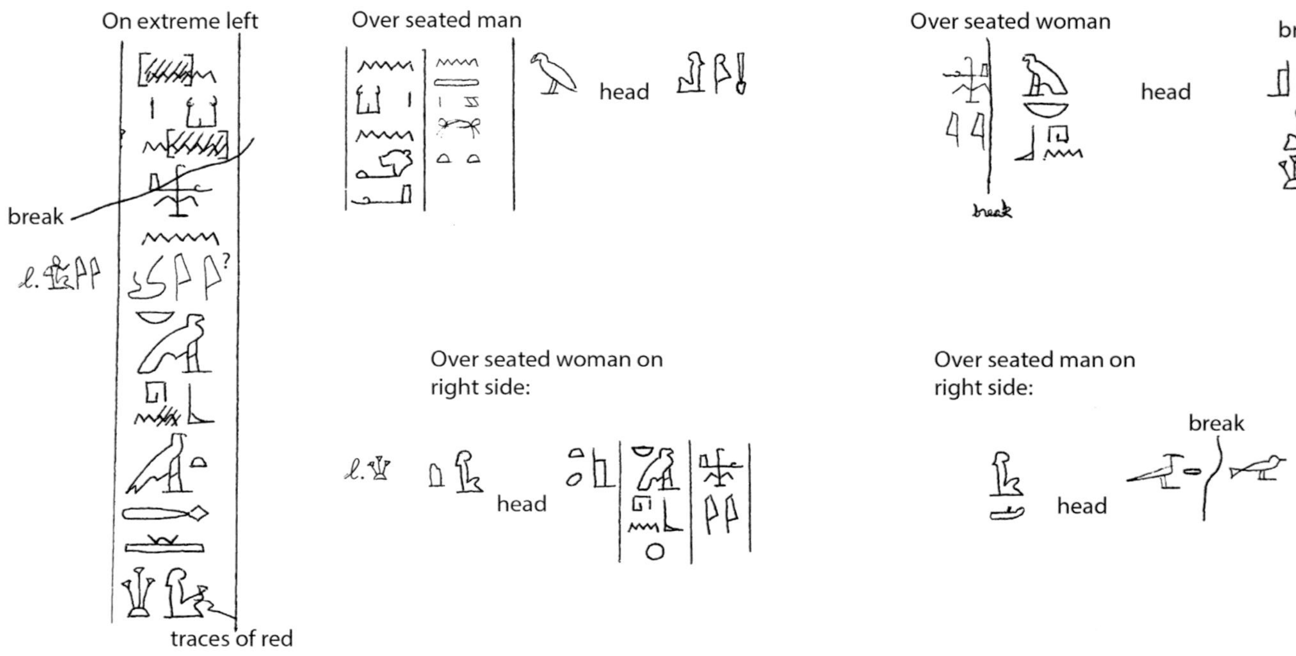

Figure 4.19. Text on lintel from D 5, field number D-128 (Richard Pierce)

profile of the lintel showed a cavetto molding placed over frieze panels (fig. 4.18). The scene showed figures in low relief, with hieroglyphic inscriptions, traces of yellow paint on the background, and red paint on the relief. The molding had alternating red-blue-white-blue-red vertical bands. The inscription was partially preserved on both the left and right sides.

The two sides, one of which was complete, showed a man in a long garment with lotuses over his head and holding a cup and lotus in his hands, with his wife seated behind him in the same manner (pls. 51–52a). Standing before the two of them was a servant woman in a flowing robe holding a cup and lotuses over a small serving table. The inscription over the man's head in both scenes reads: (1) "For the *kꜣ* of the *ḥꜣty-ꜥ* (2) . . . (3) . . . , triumphant" (fig. 4.19). Over the wife's head it reads: (1) "The chantress of (2) Horus, Lord of Buhen, (3) Isis, triumphant." The column of text behind the servant woman reads: "For the *kꜣ* of the chantress of Horus, Lord of Buhen, *Tꜣ-ꜥꜣ*, triumphant." The name of the man was not preserved.

Two broken jambs were found in situ at the door within D 5 (unregistered). No records preserved the dimensions, nor is there a photograph, so it is presumed these blocks were uninscribed.

A small block, perhaps from the upper part of one of the jambs of the door between D 9 and D 20, was reused as a door socket. It was found in situ at the south side of the doorway within D 9 and leading into D 5 (field number D-129, OIM E24424). It was made from light-gray sandstone, with the face measuring 19 × 27 cm and with a thickness of 16 cm. It bears the

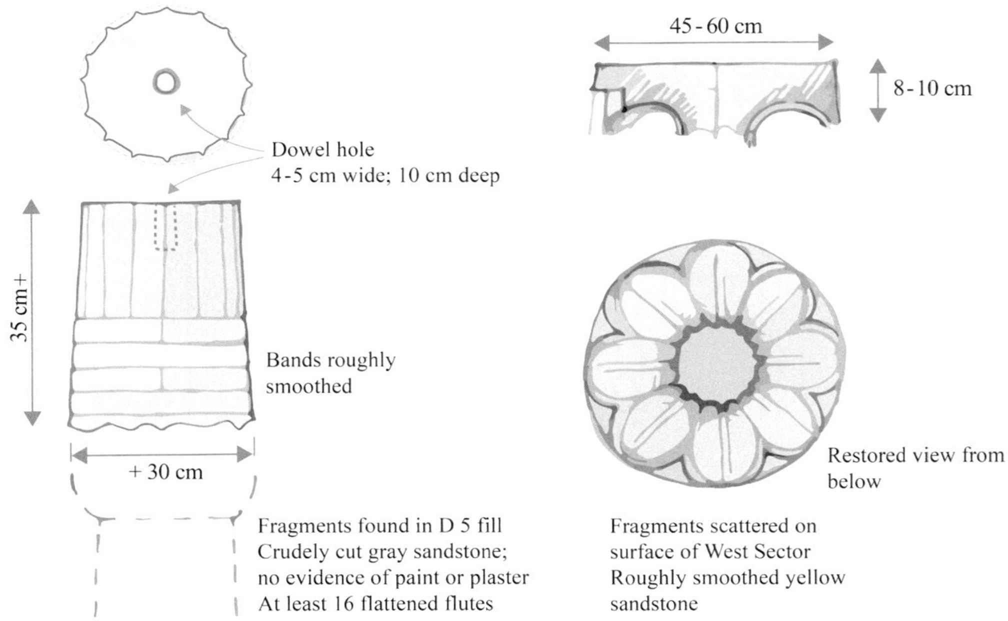

Figure 4.20. Level III columns (James Knudstad)

beginning of the *ḥtp-di-nsw* offering formula and was discovered lying face down with an abrasion from the door socket evident on the side opposite the hieroglyphs.[165] The back shows herringbone-shaped quarrying marks. (This object is discussed in chapter 7 and shown on plate 91a.)

The fragments of a gray sandstone papyrus-bundle column were found toward the center of the room in D 5 (pl. 52b). A reconstruction was presented in the architectural drawings, along with the fragments of a yellow sandstone palm-leaf capital found on the surface of the West Sector and left in the field (fig. 4.20; pl. 53a).[166] None of these fragments were registered, and they were left in the field at the end of the season.

A doorsill between D 5 and D 9, which was not registered, bore graffiti depicting cattle, a boat, and a man whose head was no longer distinct because it had been hidden below the doorjamb. The photograph of the sill is not clear enough to reproduce here. The decorated lintel fragments found in D 20 were mentioned above, but other lintel fragments were also found in D 20, near D 16B and D 22 (pl. 53b). There were no hieroglyphs or decorations preserved on these lintels, which were made of white, pinkish, and yellow sandstone.[167] The fragments clearly show the cornices, moldings, and holes for mounting the lintel (pl. 54a). Illustrations of these stone lintel fragments show the torus moldings more clearly (fig. 4.21).

A reused block of light-gray sandstone was used on the north side of the doorsill between D 20 and D 22 (field number D-133, Antiquities Service), with the inscribed part lying face up (pl. 54b). The block measured 31 × 38 cm and was 16 cm thick. It served as a plinth under the stone jamb. Its fragmentary inscription reads:

> (1) ". . . Amun-Re, Lord of the Thrones of the Two Lands, and(?) the royal *kꜣ* of the Lord of the Two Lands *Wsr-mꜣꜥ.t-Rꜥ-[stp.n-]Ỉmn* . . . (2) . . . mother of the god, lady of heaven and(?) the royal *kꜣ* of the Lord of Diadems *Rꜥ-[ms-sw ḥḳꜣ-mꜣꜥ.t]-Rꜥ mry-Ỉmn*. . . ."[168]

The cartouches were not entirely preserved, but they were reconstructed as those of Ramesses IV (fig. 4.22).

A stone sill located at the doorway between D 16A and D 20 preserved some graffiti, as did the sill between D 16B and D 20, one with a sketch of a boat and the other with a fish (pl. 55a).[169] The other stone architectural elements, however, were obviously uninscribed

165 Pierce 1964b, pp. 44–45.

166 Knudstad 1964a, p. 20. The column capital may well have come from the Level III residence, although D 205, along the north wall, had a column foundation in place at the center of the room. It may also have come from the Level II superstructure, of which little remained.

167 Knudstad 1964a, p. 19.

168 Pierce 1964b, p. 77.

169 It remains uncertain whether the ostrich referred to by Knudstad (1964a, p. 19) is actually this graffito.

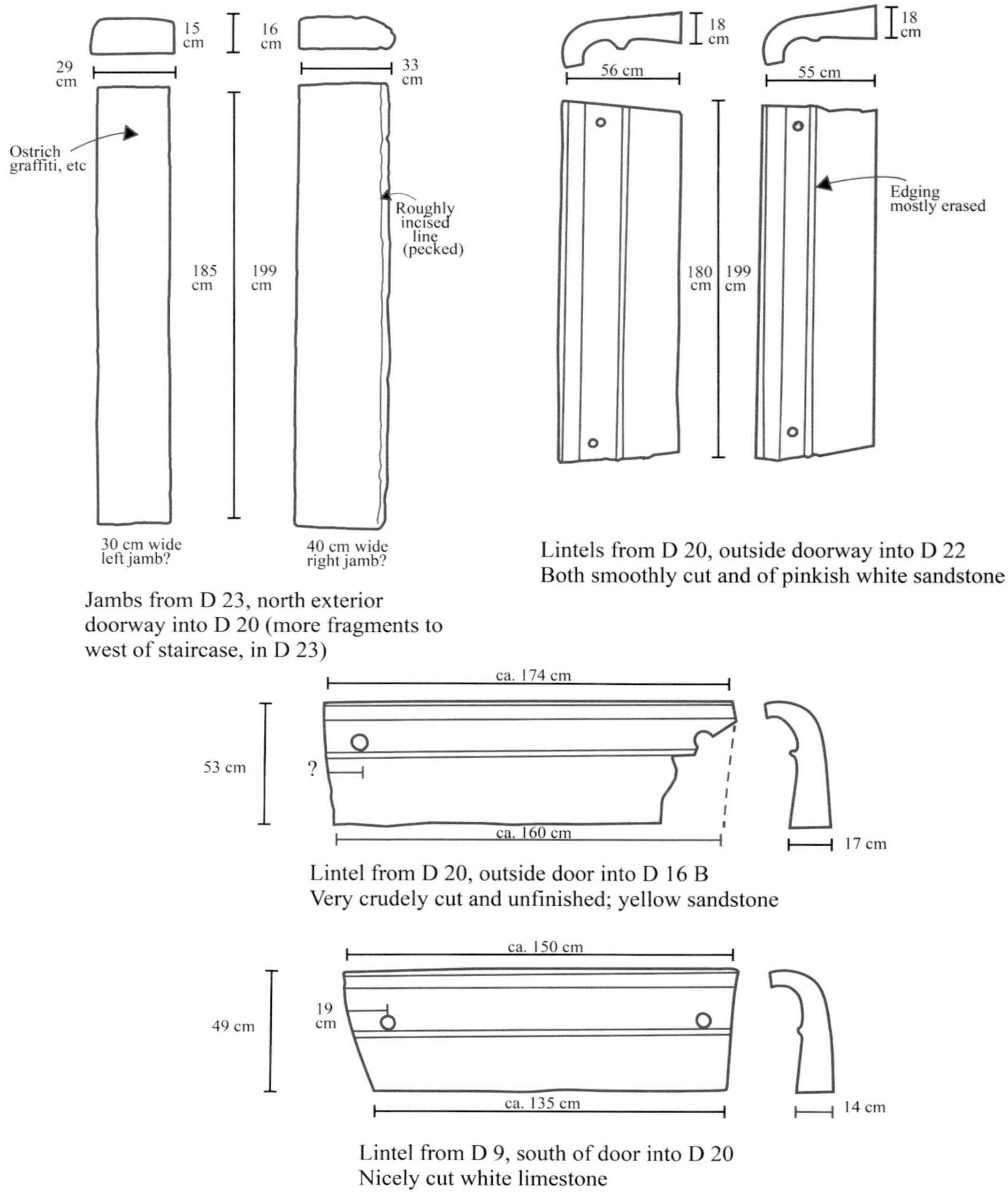

Figure 4.21. Lintel fragments in D 20 (James Knudstad)

and bore no graffiti, such as the jambs of the doorways to D 16A and D 16B.

The north stairway outside the building from D 20 was covered over by fragments of fallen architectural elements as well, perhaps all from jambs, but these were uninscribed and apparently bore no graffiti (see pl. 48b).

The excavators also discovered here a stray stone with three triangles, a square shape with one rounded corner, and a quadruped with a triangular head (pl. 55b).

The stone architectural elements had clearly been robbed from a site close to Dorginarti for reuse in embellishing the Level III official residence. The style of decoration and execution, along with likely cartouches of Ramesses I and perhaps Ramesses IV, point to the materials' coming from buildings at Buhen.[170] The

170 For evidence of Ramesses IV's reign, see Caminos 1974, vol. 1, pp. 23–24. For evidence of Ramesses I's reign, see Caminos 1974, vol. 2, p. 106.

Figure 4.22. Text on field number D-133, plinth under north jamb into D 22 (Richard Pierce)

south temple at Buhen yielded inscriptions of Ramesside kings down into the reign of Ramesses XI.

The male personal name on the lintel in D 5 was missing, but the name Pentaweret is found on the jambs between D 9 and D 20. This name is associated with the Ramesside era, most famously with a son of Ramesses III,[171] although it is also found on the triumphal stele of Piye from Gebel Barkal, where the individual is identified as one of the chiefs of the Ma and a follower of Tefnakht.[172] However, the fragments were Ramesside in date.

The excavators identified the graffiti on the jambs and sills as C-Group carvings and surmised that they were drawn on the blocks while in their original positions somewhere around Buhen. On the other hand, the stones that were turned upright may have had these images incised into them during the life of the Level III residence or before the Level II builders covered them with debris.[173]

OBJECTS FROM THE LEVEL III RESIDENCE

In addition to the items mentioned above, the objects from the Level III residence included a broken faience figurine (OIM E24332) lying atop a fragment of the south wall in stratum 3 of D 9 and a soapstone net weight (OIM E24411) from the doorway between D 9 and D 20. Talc and soapstone net weights were also found in strata 3 and 4 in D 22 (OIM E24330, E24331)—that is, in the fill of the room and on the floor. The fill of the platform at D 9 (1) yielded two faience beads (OIM E24358). Charcoal samples from the burned layer were brought back to Chicago but remain unanalyzed (OIM E49001, E49003, E49005, E45996). A folded copper or bronze fragment was found in D 5 (2) fill (OIM E24357), while a pottery disk was found in the test pit below the floor in D 5 (OIM E49741). Great quantities of wheel- and handmade sherds were uncovered in both the loose and more compacted fill, and in test trenches outside the building. At the northwest corner of the Level III residence, a travertine stone bowl rim (OIM E24402) and a stone arrowhead (OIM E24401) were recovered as well.

Christian sherds were found in the fill strata of the well cut from the top of the Level II platform, mostly above the hallway, D 20, and within D 24, while the latter room's stratum 4 contained a number of imported amphorae sherds that had intruded from the Level II use of the site.

OTHER STRUCTURES AND REMAINS

The area surrounding the official residence during both phases, Level III and Level IV, was incompletely excavated due to the presence of the Level II foundations on the east and west sides, and therefore any earlier architectural and archaeological materials that may have existed in these areas remained buried.[174] The elevation of the passageway surface by the north stairs was 2646, and it was recorded as the top floor—that is, it was associated with the use of the Level III residence. There was an earlier Level IV floor to the east at an elevation of 2603.

West of the Level III building, the later Level II brick fill was cleared back, exposing yellow sand or floors with some small features running under the relining of the north wall interior.[175] Most of these features are associated with either Level IV or Level III phases that were subsequently covered over by the later Level III interior relining walls. These will be discussed below.

Two silos were excavated in the northwest corner of the Central Sector, an area with the locus numbers D 37–D 39 (plan 5).[176] The base of the nearby relining stood higher than the preserved elevations of the floors in D 38 and may have been built over or abutted

171 See, for example, Dodson and Hilton 2004, p. 193.

172 Eide et al. 1994, p. 105, 1.116, from the triumphal stele of Piye from the Amun temple at Gebel Barkal. See also Grimal 1981.

173 Knudstad 1966, p. 183.

174 Knudstad 1966, p. 181.

175 Knudstad, 1963–64, p. 305.

176 Knudstad 1964a, p. 40. There are no good photographs of this area, but see the wall of the silo south of D 38 on pl. 37b.

the wall of silo D 38. Both silos' walls were composed of the later, 40 cm long bricks and were founded at elevations 2559 and 2567. The main enclosure wall had slumped inward at some point in time, partially covering over the north wall of silo D 38. The size of the bricks used in the building of the relining at this spot is not mentioned.

The mudbrick paving in D 37, to the east of D 38, extended into the south part of the silo at an elevation of 2584. The floor patch within D 38 lay atop thin walls that were perhaps inner compartments covered over at a later time either by a new floor or by a structure associated with the lowest floor found within the silo (2568).[177] The elevations of the footings of these thin walls were not recorded. The floor found in D 39, in the northwest corner between the silo and the dividing wall, was a further 15 cm below the earliest floor in the silo area. The dividing wall between the West and Central Sectors, and the wall of D 38, were built 3–6 cm above the lowest floor level recorded in D 39. Finally, the oven and wall fragments west of the dividing wall were all founded at about the same levels.

The interior relining that was built along the north wall and continued eastward to the vicinity of the North Gate was approximately 1 m thick. It was laid against the 6.3 degree batter of the enclosure wall, and its east end was curved. The curved end of the relining segment was founded 45 cm lower than the short segment abutting it to the east, which rested at 2667 on brick debris.

The test excavations that were undertaken along the south wall relining of the Central Sector exposed only the face of the later relinings of the sector and a stone slab to the south of the residence. It is unclear whether this block rested on a living surface. Its elevation was 2551. The base of the relining on the south was below the north wall relining just described.

Doorway D 87 allowed passage between the West and Central Sectors at the south edge of the dividing wall line. This doorway was monumental in proportion to anything preserved to its west, with a 1 m wide doorway, a stone sill, and an emplacement for a pole, making it an appropriately imposing entryway.

While the renovations and reuse at the northwest corner of the Central Sector are evident, nothing in the areas around the Level III residence shows a cohesive set of outbuildings that serviced the official residence. If we can determine a date by the presence of bricks 40 cm in length, both silos were part of a later Level III building program, but whether before or after the building of the relining is uncertain. Did the relining, built high on a pile of debris, abruptly stop at silo D 38, or was it dismantled to make way for a deep-seated silo? The longer brick size, mentioned throughout these chapters, was also associated with the interior fortification wall relinings throughout the fortress in all sectors, including to the east along the north wall, and was also used for the large silos in the East Sector. It may be assumed, therefore, that the 40 cm long bricks were also used to build the relining wall next to D 38, although the field notes are silent on this detail.

LEVEL IV IN THE CENTRAL SECTOR

GENERAL EXCAVATION NOTES

The excavators traced the tops of the walls of the Level IV building while removing the floors of Level III, and, as mentioned above, they discovered that the builders of the later residence had torn down the early walls to a height varying between 40 and 70 cm above the earlier floors (plan 6; pl. 56). Cuts had been made into several walls of the Level IV building to accommodate the later constructions, while a few of the earlier walls had been used for the foundations of upper walls. The size of the residence was slightly smaller than its successor, about 13 × 11 m, and had seven rooms built on a tripartite plan with the two largest rooms in the middle. The north entrance hall of D 310 and the accompanying main room of D 311 reflected the east entrance room of D 313 and its associated main room, D 312. Smaller rooms filled in the east and west portions, including a staircase at the southwest corner.

The exterior walls of the residence were all approximately 75 cm thick and undoubtedly supported a second story, with a stairway in D 315 at the southwest corner providing access to the roof or second floor. The interior walls mostly ranged between 67 and 70 cm wide, with a few that were 42 cm wide, while a 1 m wide wall supported the stairway construction in D 315. The bricks used throughout, when noted, were 36–37 cm in length.

A deep test pit was sunk below the pavement in D 311, the floor elevation of which was 2635, extending through a leveling deposit of soft, crumbly sandstone chunks of different colors mixed with yellow sand and ending 54 cm below the initial floor surface. The base of the room's partition wall was even with the bottom of a yellow sand layer at elevation 2557. A stratum of sterile gray sand ended 1.20 m below the floor level. This gray sand layer rested atop alluvium, and the excavation continued down through the alluvium for 2.48 m below the floor, ending at an elevation of 2387,

177 Knudstad 1964a, pp. 40, 43.

although it had not yet reached the bottom of the cut for the Level II well.[178] To the west of the residence, a deep test pit reached the top of the alluvium at 2441. No structural or cultural remains were encountered in these test pits.

Beneath the Level III floor on the south side of D 312 was a 60 cm layer of broken brick and soil, then about 5 cm of sand resting atop the brick paving of the Level IV room. The pavement was 13 cm thick and rested atop leveling stones mixed with yellow sand. At the north side of the room, the test pit below the floor level of 2616 revealed the same multicolored stone floor foundation at elevation 2606, with the bottom of the stones at 2575 (pl. 57a).

The leveling stones were also found below the floors of D 311 and D 315 but not in D 310, D 313, D 314, or D 316. Two brick pavement levels were found in D 314 below an upper surface of tamped brown mud, and multiple layers of floor replastering were discovered in D 316, although they were not traced throughout the room and their elevations were not all recorded or noted on the plan.[179]

The pavements found in D 312, D 313, D 314, and D 315 were constructed of broken bricks. The paving found in D 312 and D 315, and the mud floor in D 311, all rested atop this deposit of soft, crumbly sandstone chunks of different colors mixed with yellow sand. The layers below the bottom pavement in D 314 are unknown. The compacted yellow sand underpinning the entire building was level with the foundations of the walls.

The structural remains were difficult to excavate because water saturation had destroyed the textures and densities of the bricks and floors.[180] Excavation notes do not mention the presence of any bedrock surface encountered during operations either in this building or in its surroundings.

The floors of the residence displayed an untidy range of elevations, with the floors of D 311 and D 316 being much higher than the floors elsewhere.[181] One suspects that more than one level of floor was recorded in different areas. Even the floor levels recorded in D 311 differed by as much as 21 cm. The mention of two floor pavements in D 314 indicates stages of renovation that were not discernible in the Level III residence. The topmost plaster floor in the room was still 10–13 cm lower than the closest floor level in nearby D 311 (2635), while its lowest floor was 28 cm below that floor level in D 311.

Two floor levels were recorded for D 313, with the north floor elevation being 12 cm above the elevation taken on the south (2607). The entrance paving at 2603 was close to the south floor elevation. The brick paving was 34 cm thick at the south end and 22 cm thick at the north end.

A great variety of floor levels were recorded across the official residence, even within the same rooms. Thus, the multiple floors exposed during the excavation, and recorded on the plan for D 311–D 314, may not all belong to the same floor phases. The Level IV residence, as mentioned above, perhaps went through a longer period of use than did the residence in Level III.

OFFICIAL RESIDENCE

There were exterior entrances on the east and north sides of the building, but because the floor was mostly level with the outside ground surface, the only staircase present in the building was located in D 315, leading to the roof or second story. The five interior doorways of the Level IV residence exhibited mud-plastered jambs that showed no evidence of disturbance, so it was presumed that the inscribed stones used as doorway fixtures in Level III did not originate in this earlier structure.[182] The interior doorway between D 311 and D 314, as well as the south jamb of the door between D 311 and D 310, were missing because the area had been broken through by the Level II well, which had been excavated to a depth of more than 2.5 m below the floors of this house. Clearly, if bedrock had been present, the digging of this feature was a massive undertaking.

The exterior side of the east doorway into the residence was paved with two courses of worn brick laid on end and was 2 m wide (pl. 57b). The doorway was inset, its wider opening narrowing toward the interior, where the interior jambs provided more restricted access into the room. They appear to have been later additions to the door. The Level III builders later shaved them off when blocking this entrance and incorporating it into their house wall. An oblong mud feature abutted the jamb on the south, measuring 1 m from

178 Knudstad 1964a, p. 23. This would mean that the depth from the top preserved height of the well to the bottom of the excavated test pit was 4.64 m.

179 Hoerth 1964a, pp. 6–7.

180 Knudstad 1963–64, p. 304.

181 Measuring included the two floor levels in D 316 and the three east levels in D 311.

182 Knudstad noted that there was "no real evidence . . . for the original position of the Level III stones in doorways of Level IV. All floors and plastered jamb niches [of the Level IV building] look intact and built in mudbrick only"; see Knudstad 1963–64, p. 326.

east to west (pl. 58a). The floor level under the mud construction was at elevation 2613, or 6 cm below the nearby floor at the north. The brick floor paving was mentioned above, but below it in the northwest corner a test pit was sunk to a depth of 62 cm.[183] The wall foundation at that spot was 56 cm below the floor and rested on a layer of yellow sand.

The thick partition wall of 73 cm in width ran between D 312 and D 313, with a 97 cm wide doorway at its south end. The middle portion of this larger hall was largely unexcavated due to the Level III remains and fill left in place, although a cut made toward the center from the south was probably undertaken to locate a potential column base. The floor levels here also varied in elevation, with the elevation in the northeast corner being at 2611, the level before the doorway into D 311 being at 2616, and the level at the southwest of the room being at 2610. Three exploratory pits were dug below the floors, showing that the paving was 10–25 cm thick in various spots around the room and that it lay atop stone foundation layers that were 31–36 cm deep.

The south wall of the room had been cut down so that it was even with the Level III floor in D 5 (2680), but the wall still stood 45 cm in height above the nearby Level IV floor.[184] The north wall of the room, east of the doorway into D 311, was also partially removed to accommodate the Level III door between D 9 and D 20. The west side of this wall, however, remained at a height of 85 cm above the nearby associated floor level and 1.44 m above its associated foundation level. As mentioned above, the west end of the same wall had been destroyed by the later well.

At the southwest corner of D 312 was a doorway that led into the antechamber of D 315 and onto a stairway. Off to the north was a narrow storage area with remains of a floor at elevation 2610. The stairs led directly west from the door. The construction of the steps varied from the steps found in the smaller stairways of the West Sector and in the Level III residence (pl. 58b). The first step displayed a course of headers at the foundation that had been sunk 28 cm below the floor, then a row of bricks laid on edge, and finally its top featured a wooden nosing and a row of headers at the back atop the foundation fill (fig. 4.23).[185] The two upper steps were constructed in the same way, with the wooden nosing and headers laid atop a course of flat bricks. These wooden stair moldings were sunk about 3–4 cm into the south sidewall. The topmost plank extended 75 cm beyond the stairs and under the area of a landing.

The steps rose 58 cm in height to the west before reaching the landing, from which another flight would have proceeded upward to the north and then east. The treads were either 45 cm or 55 cm in depth. The rise for the first step was 13 cm above the floor level (2610), but the next step was 23 cm high. Much of the evidence for the upper staircase was lost due to Level III construction activities around the west wall.

Room D 312 led into room D 311 through a doorway in the center of the dividing wall. The sill at the entrance into the north hall was 17 cm below the nearby floor level within D 311 (2635), although it was nearly level with the floor in D 312. Thus, a rather steep step (or set of steps) or slope led into D 311, at least as suggested on the final plan of the house.[186]

The only item that was uncovered in room D 311 was an off-center brick basin around 80 cm in diameter, although its south side had disappeared (pl. 59a). Its 38 cm high wall was built of three courses of bricks laid in a rough circle and plastered inside and out. Its center was filled with rubble, the top part of which had been hollowed out, perhaps to serve as a stand for a pot or as a pottery hearth.

As mentioned previously, the floor levels across D 311 varied by as much as 21 cm (e.g., 2626 vs. 2647), with those along the north wall being higher than elsewhere, reflecting either floors that had been laid at different times or a rise in level along the north wall. The rebuilding of the official residence during Level III may have destroyed conclusive evidence for multiple floor levels in the room.

The Level II well had damaged the extreme southwest edge of the hall, and therefore the door into D 314 was gone. This room had three floor levels. The bottom two floors had pavements composed of bricks, while the upper floor was composed of brown silt or mud. The elevation of the lowest floor was 2607. It was separated from the uppermost paving by 15 cm, while the top of the tamped-silt floor was 3 cm above the intermediate floor level.

Room D 316 was located in the northeast corner of the residence and was entered through a doorway at the north end of the wall shared with D 311 (pl. 59b). The size of this room mirrored the size of D 313 to its south, each being about 5 m from north to south. The

183 Knudstad 1964a, p. 24 obverse.

184 The base of the Level III house wall in this area was at elevation 2555, or some 15–21 cm above the foundations of Level IV's south wall. The Level III floor level was around 2680. The remaining height of the Level IV wall above its foundation level would have been 1.40 m.

185 Knudstad 1964a, p. 24 reverse. A similarly constructed stairway was found in a fourth- or third-century BC house at Elephantine; see Krekeler 1996, pp. 111–13, B 6.

186 Knudstad 1964a, p. 23.

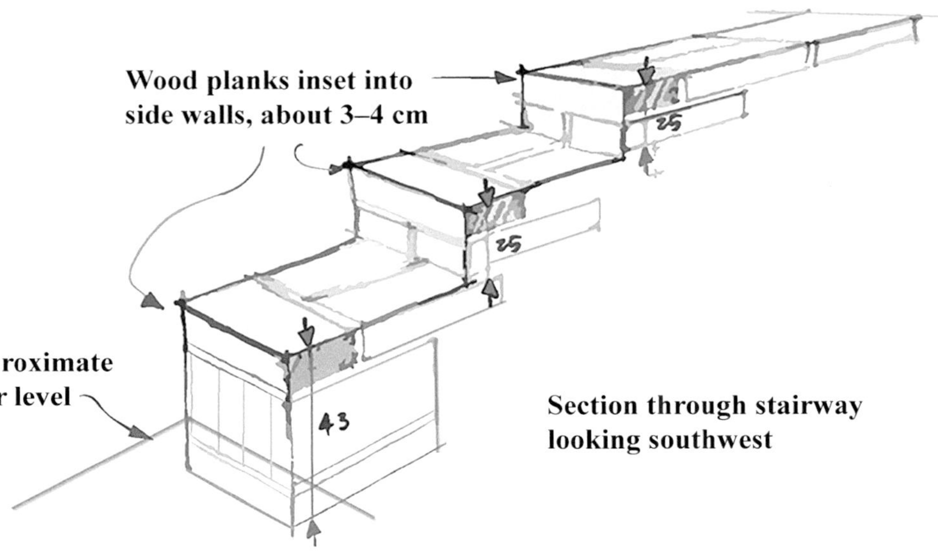

Figure 4.23. Stairway construction in D 315 (field sketch by James Knudstad)

floor levels within D 316 sloped down 8 cm from north to south. The multiple layers of plastered floors found in the room were mentioned previously but are not noted on the final drawings of the room.[187]

D 310 represents a vestibule into the house from the exterior doorway on the north. The floor level of the entrance room, which perhaps had a mud-plastered surface, was preserved in the southwest corner of the room at an elevation of 2618. The rest of the room was excavated 33 cm below this level, where the foundation of the north jamb of the doorway was exposed, but no other floor levels. The exterior door may have had a brick sill, but it was worn and indistinct.

There was no evidence of earlier walls in any of the test pits sunk in these rooms.

OBJECTS FROM THE LEVEL IV RESIDENCE

The area had obviously been cleaned out before construction of the Level III building, and therefore little besides pottery was found in the earlier residence. A spout fragment from a crucible in D 316 was found in the fill of a trench between strata 1 and 2 (OIM E51012). Pottery disks, one with an incised cross and one with a star, were found in D 310 (2) and D 311 (1) (OIM E47498, E47499, E49744). Finally, a highly burnished wheel-made beaker, with a round to pointed base, was uncovered from the fill in D 310 (1) (OIM E49523) (see pl. 70b and fig. 5.63b in chapter 5). The latter vessel was unusual, featuring a bright-red slip and overall vertical burnishing, and with a small knob protruding from its midsection.

OTHER STRUCTURES AND REMAINS

The excavators encountered the Level II fill in trenches outside the residence, along with a varying depth of brown and sandy soil atop the yellow sand upon which the early residence was built.[188]

To the north of the Level IV residence, the thin walls of D 305 and D 306 were uncovered beneath the landing of the north stairs leading into the Level III residence. The preserved tops of these walls were as much as 13 cm below the base of the later staircase, and their foundations stood atop debris at a level nearly even with a patch of floor over to the east (2603) and the foundations of the north residence wall. The walls may have abutted, or been cut through by, the north wall of the Level IV residence. The surface upon which they were built was uneven and filled with indentations, but it was still covered with white plaster.[189]

The excavators suspected that these walls were earlier than the residence, although they found no earlier walls in the deep test pit they sunk within D 311 to the south.[190] Another wall adjoining this group at the

187 These thin layers of mud floors are not reflected in the final drawing of the room and thus may not have been solid evidence for multiple restorations of the room's floor.

188 Hoerth 1964a, no. 2.

189 Hoerth 1964a, no. 2.

190 See Knudstad 1963–64, p. 327; Knudstad 1964a, p. 12 reverse. He noted that the east wall of D 306 "probably abutted the Level IV wall (?)" and also noted that these were "Level IV (or earlier) walls under the Level III northern stairs."

northeast had been cut through by the later relining of the fortification wall, the base of the relining being 12 cm below the preserved top of the wall.

To the east of D 306, in D 303, a line of stones was found resting 5–15 cm above the yellow sand level and was sealed by the north landing of the Level III residence.[191] Close to the northeast corner of the residence was a patch of mud floor placed atop a 12 cm layer of yellow sand. The elevation of this floor (2603) was level with the east doorway paving, while the northeast corner wall of the residence was founded at only a slightly lower elevation, at 2601.

To the west of the Level IV residence was another short wall fragment running under the later fortress relining, with both the relining and the wall founded at an elevation of 2591.[192] An oven to the east of the latter wall rested 5 cm higher, but it was also covered over by the relining. A second oven to the west rested at an elevation of 2576. Since the remains were so disparate, it is hard to determine which walls and ovens were contemporary with the earlier levels in the Central Sector, and how they may be associated with one another.

PARALLELS TO THE OFFICIAL RESIDENCES

Comparisons to the large residential buildings at Dorginarti are found in both Upper Egypt and the Sudan (see maps 1 and 2). A few residences at Thebes—most notably Medinet Habu—and Elephantine are examined here, while in Nubia there are a number of similar buildings at Amara West, Kerma, and Kawa. The large size and almost square plan of both residences are the most useful features for comparison, in addition to the presence of stairways, but as we will see, the arrangements of the rooms and the features within them are different from their counterparts in Egypt and elsewhere in the Sudan. Neither of the structures at Dorginarti was preserved at a great enough height to offer evidence of windows, and there is also no evidence of any fine details other than the stone architectural embellishments in Level III.

The size and shape of the houses of the Twenty-Fifth and Twenty-Sixth Dynasties that were located to the south and southwest of the Great Temple at Medinet Habu are similar to the official residences at Dorginarti, although the room arrangements are not.[193] The superstructures, however, were largely destroyed.[194] The homes included storage areas and transverse courtyards, which are not found in either residence at Dorginarti. House 1 at Medinet Habu was mostly arranged in a tripartite manner, with two main halls running east to west and small rooms on each side, but with a long additional "court" running the width of the building on the west. Its rooms contained a well, underground cellars, and a stairway at the northeast corner, and were backed by a long courtyard along the west side. The stairway is reconstructible since the walls around it stood 2.5 m high.

At Elephantine, houses from the Twenty-First through Twenty-Sixth Dynasties were excavated to the northwest of the temple of Khnum.[195] The nature of the site, with layers of urban remains from different periods, makes it difficult to find a nearly complete house to compare with the buildings at Dorginarti. House AB, attributed to the Twenty-Fifth or Twenty-Sixth Dynasty, is smaller and linear. But the walls of the earlier or contemporary buildings (e.g., BC, G, H) are mostly too fragmentary and appear to have been smaller in size.[196] No near parallels emerge from their plans.

At Amara West, the Level One building above the governor's house, located west of the temple, may date to the Napatan period.[197] The house is much larger than the Level III and Level IV official residences at Dorginarti, and the room arrangements are different. The stone arrowheads found on the surface above the house, and a handmade sherd below the floor of Level One, may indicate a first-millennium BC date based on similar artifacts elsewhere in Nubia from the Kushite

191 Hoerth 1964a, no. 2. This line of stones is not shown on the final plan; perhaps they were builders' guides.

192 See Knudstad 1964a, p. 25.

193 Hölscher 1954, pp. 14–16 and fig. 20. The pottery from House 1, for example, was mostly later in date, when the building site was reused for pottery kilns. Only type A5 was contemporary with the house, which is similar to a few red-slipped examples found in Level II at Dorginarti. See also Hölscher 1934, folio pls. 14–15. (Some of the structures are shaded in light green, although they probably also date to the Twenty-Fifth to Twenty-Sixth Dynasties, as do the darker green buildings; see Hölscher 1954, p. 14, n. 52.)

194 The lack of preserved doorways in most of these houses, the use of concave courses of bricks, and the thickness of their walls led Barry Kemp to suggest that they represent platformed foundations upon which the houses were built; see Kemp 2006, pp. 353–55.

195 Kaiser et al. 1988, pp. 167–74 and especially figs. 11–12; Kaiser et al. 1990, pp. 209–17 and figs. 7–8.

196 Kaiser et al. 1990, fig. 7.

197 Fairman 1948, pp. 6–8 (E.13.2), pl. II, and shown more clearly in Spencer 1997, pl. 116 (Level I walls).

and Napatan periods, but this is tentative.[198] House 9 is south of the temple, and is very similar in size and plan to the Dorginarti residences, but it is not described in detail.[199] A west stairway and porch accessed the interior of the house, which had at least nine rooms arranged around the east, south, and west sides of a hall or courtyard.

The results of recent excavations at Amara West, which exposed New Kingdom villas, demonstrate how the smaller structures at Dorginarti lacked the working and storage spaces evident within the walls of the Amara West houses.[200] It is evident, then, that the space around the Level III and Level IV residences in the Central Sector must have contained the kitchens, storage areas, and other service buildings for the official residence. This left the large residences to serve for official functions, and for the private quarters and offices of the fortress commander.

At Kerma, the Napatan residential building (*bâtiment napatéen*) I, in its Phases I and II, is similar in both size and outline to the residences at Dorginarti. The interior arrangements, however, differ. During each rebuilding stage, the structure became larger and its walls thicker.[201] The bricks in all three phases were 0.4 m in length. The Phase II residence was built atop the razed walls of the preceding house, and the Phase III house was built on much the same plan as that in the previous phase, but it was larger in size and no doorways were preserved. The excavated remains of the residence in Phases II and III represent only the platformed foundations for first-floor living quarters, with the foundation rooms used for storage.[202] The ceramic material uncovered in the three phases of the house, as well as that recovered during excavation of the earliest walls nearby, appears to be later in date than the pottery from Dorginarti.[203]

At Kawa, before recent work there was undertaken, the closest comparison to the official residences at Dorginarti was building F1 in the area around the temple of Taharqo (Site I), but this building did not offer a close comparison due to the different arrangement of its rooms and its larger size, measuring 26.5 × 15.5 m.[204] Recent excavations, however, have exposed the foundations, and sometimes the walls, of a number of buildings that mirror the size and arrangement of the Level III and Level IV residences at Dorginarti.[205]

For example, in the published reports of 2002 at Kawa, building F2 is 16.5 × 11.9 m in size and perhaps arranged on a tripartite plan much like the Level IV residence at Dorginarti, with a staircase foundation similarly located in the southwest corner.[206] The interior of F2 was paved with fragmentary and complete bricks. The square buildings C2 and B5 also had tripartite layouts, with central axes and side rooms. Staircase foundations were found in C2 and in the smaller buildings, B2 and C3, and always in the southwest corner of their interiors. Buildings B2 and C3, however, were divided into two interior sections along their lengths, which was not the case with either the Level III or the Level IV residence at Dorginarti.[207]

The external walls of D1 at Kawa were more than 1 m and up to 1.5 m thick, and the structure measured 15.0 × 17.5 m. It may have mirrored either of the Dorginarti residences, but it was not well enough preserved to say so definitively.[208] It may also represent a foundation platform like those discussed in chapter 6.

Therefore, the closest similarities to the official residences at Dorginarti were the larger houses at Medinet Habu and those from Kawa and Kerma. The

198 However, only a few of the handmade Nubian sherds are shown in Spencer 2002, p. 22 (type D), 32–36, color pls. 1–3. See also Spencer 1997, pp. 15, 157.

199 Fairman 1948, pl. II; Spencer 1997, pl. 78 (also a Level One construction).

200 See, for example, Spencer, Stevens, and Binder 2014, p. 36. They note that the service and storage areas were in the compounds surrounding the walled estates at Tell el-Amarna, while the E12.10 building combined these areas within its house walls.

201 Mohamed Ahmed 1991.

202 Entrance was gained into the Phase II house through exterior doors and via stairs and a ramp to the west, while the Phase III residence was accessed by a south side ramp and the north exterior entrance and associated stairway; see Mohamed Ahmed 1991, pp. 40–42, fig. 4. See the discussion of parallels to the Level II platformed fortress in chapter 6.

203 The marl products from the Theban region, with highly corrugated jar bodies, and the distinctive channeled-rim store jars are mostly later than the material from Dorginarti, and such sherds were found in the "*phase primitive*"—that is, the phase before the building of Napatan building I.

204 Kirwan 1955, pp. 209–13, pl. 18 (Site I). The pottery from the early levels is later than the assemblage of Levels III–IV at Dorginarti and is more representative of vessels found toward the end of the Twenty-Fifth Dynasty, i.e., the channeled-rim jar forms and the finger-impressed jar of type 1g in Macadam 1955, pl. XXXII. The house is discussed in Welsby 2002b, p. 32. The measurements are listed in Welsby 2002a, p. 44, table 1.

205 Welsby 2002b.

206 Welsby 2002b, p. 32, fig. 1, and with the size taken from Welsby 2002a, p. 44, table 1.

207 Welsby 2002a, pp. 37–38, fig. 5.

208 Welsby 2002a, pp. 40–41, fig. 8.

building would be fitting for the residence and administrative headquarters of the garrison's commander.[209]

THE EAST SECTOR

A number of storage installations were found in the East Sector of Dorginarti, along with some possible administrative rooms at its easternmost end in D 33 and D 34 (plan 8). The unexcavated surface in this sector was level with the eroded tops of the main fortification walls, and as with the Level II remains in the Central Sector, there were a number of Meroitic burials dug into the fill (pl. 60a). A layer of rubble 82 cm thick, consisting of brickbats and soil with very few sherds, was encountered lying atop 7 cm of a compacted, blackish rubble fill. This stain or "ash" level was the same as that encountered throughout the fort, and represented a deposit left by floodwater permeation.

LEVEL II LIME PLASTER PREPARATION AREA

Southeast of the gate and level with the bottom of the stairs leading to the top of the Level II platform was a circular (2.25 × 2.30 m) slab of burned and crumbled limestone lying 2–3 cm below the unexcavated surface, and 5.35 m to the southeast of the Level II staircase.[210] The 8–15 cm thick deposit undoubtedly represented the material used for plastering the walls and floors when building or renovating the Level II platform. It rested 55–65 cm above the south part of the easternmost silo along the north wall, D 351.

LEVEL III SILOS

Two circular brick silos were found just inside the North Gate, each measuring 4.4 m in diameter. Their walls were constructed using 38 cm long bricks. The wall of D 350 rested atop the demolished end of the east wall flanking the gateway, as well as being atop the walls of a rectilinear structure just inside the gate (pl. 60b). A floor level found within the silo sat at an elevation of 2600 and was likely associated with the earlier structure that it sat upon. The silo wall itself was founded 14–17 cm above that floor. D 351 also rested atop the thin wall at D 353. The floor elevation within this silo was 2602, a similar depth as the wall foundations in D 353, and like those of the earlier walls at D 350.[211]

The foundations of the three silos along the south wall, D 361–D 363, were built using stone, although bricks remained atop them in some places (pl. 61). The bricks used in the construction of D 361 and D 363 measured 40 × 18–20 × 7–8 cm, and it is assumed the same brick size was used for D 362. Since both the brick size used and the building technique employed in these silos differed from the other two found in the East Sector, they may have been constructed in a slightly different time period, although they may also have been partly or wholly contemporaneous. The stone foundations had been sunk some 19–33 cm below the foundations of D 350 and D 351, while the walls of the south silos were preserved at a lower height. The south silo walls remained standing 23–42 cm above their foundation levels. They ranged between about 4 and 4.5 m in diameter. The stones varied in size but were all rectangular. No floor levels were apparent.

At some point in time, a wall had apparently already blocked off the easternmost rooms, D 33 and D 34. A higher additional wall adjoined the earlier blocking wall, measuring 73–77 cm in width. The foundation level of the inner blocking wall was recorded as being at elevation 2577 at its south end, while the thinner relining stood higher, at 2603. This later relining was composed of the same 38–40 cm long bricks as the south silos, and it ended in a finished or cut edge at the north side of D 361. The construction of D 361 had perhaps necessitated the partial dismantling of the relining, and possible fragments of it were also found at the north and south corners of the area, in D 359 and D 364.[212]

No floor levels were found in any of the silos along the south wall or in D 366 to the west. D 362, however, was excavated down to an elevation of 2580, where cataract stones were encountered. The base of the third inner relining wall in this sector was measured at elevation 2564–2465 along the south wall and at 2577–2576 along the north. The south wall had been placed atop cataract stones that seem to have underpinned at least the south side of the East Sector.[213] The excavators mentioned finding a number of floor levels in D 360, in the center of the sector, but they did not

209 For example, the administrative headquarters of Egyptian officials in the Levant, during the New Kingdom, are typically set on the highest part of the mound; see Oren 1984a, p. 50. They are square in plan and usually symmetrical in their room arrangements, and staircases are often located at one of the corners of the building.

210 Pierce 1964b, East Sector, pp. 1–2.

211 Knudstad 1964a, pp. 29, 31.

212 Knudstad 1964a, p. 30.

213 The West Gate walls were also placed atop this cataract rock, but the rock foundation was typically found only under the outer enclosure wall—that is, in areas where test pits were excavated.

mention encountering cataract rocks below them. Few elevations, however, were recorded in this area.[214]

D 363 was placed next to a rectilinear structure, D 366, the wall of which rested on a level that was 33 cm below the foundation of the silo. The walls of the structure remained standing at a higher level than those of the silo next to it, and it may have been an earlier structure that was reused. D 366 may be considered part of the original plan of the East Sector in Level IV.

LEVEL III–LEVEL IV GRANARIES

The earliest features in the East Sector are represented by the rectilinear structure in D 366 at the south, the walls under D 350 and D 351, and the line of magazines or granaries along the north wall, D 354–D 358 (pl. 62). The thin-walled rooms here had no apparent entrances and were possibly accessed through their roofs. The base of their walls was approximately on a level that was even with those of the south silos, but below that of the two silos by the North Gate. The rooms in question may have continued in use through Level III; their walls mostly remained standing at the same heights as the silos in the sector. Knudstad noted that the south wall of D 355 and D 356 was possibly a later wall.[215]

D 33, D 34, and the East End

D 33 was the westernmost of two rooms found at the east end of the sector (pl. 63). A 97 cm wide doorway had once led into it from D 360, before it was closed off. The vestibule before the southeast stairway, leading up onto the wall, was located on the right-hand side after entering.[216] The doorway into D 34 was narrower and was placed slightly south of the door to D 33. All walls in this area were built directly atop cataract rocks, or atop a thick brick foundation built atop these rocks, and their floor levels varied: 2548 before the stairway in the vestibule, 2571 inside D 33 near the door, and 2562 in the middle of D 34.[217] The top of the stones in the foundation in the vestibule was, however, 21 cm above the floor level recorded, and the cataract rocks were also recorded as being higher than the floor patch and brick foundation(?) in D 33. The stones had been displaced. The architectural drawing notes specifically that the cataract rocks were also found under the floors, and the photographs appear to show the floor sitting atop rocks in D 33 (pl. 64).

The packed-mud floor of D 33 was found 1.35–1.68 m below the top of the nearby preserved walls, with the last 47 cm of fill above the floor consisting of compacted rubble with a series of thin ashy or dark lenses. Below the north part of the floor was cataract rock, while on the west side of the room a wide brick foundation was uncovered, with its top surface at an elevation of 2580 and sloping down 8 cm to the south by the doorway leading in from the East Sector. This brick foundation ran at a different angle from that of the room walls but matched that of the fortified east end. It was founded on sand.

D 34 was entered through a door from the stair vestibule. The floor level and top of the cataract rocks in the room were 9 cm below the floor level in D 33. Two wooden planks had been set into, and laid flat onto, the wall at the easternmost interior. The foundation of the east wall of D 34 was recorded as being at 2639, 77 cm above the floor level of 2562. Another wooden plank was found lying flat against the wall at the juncture of, and the jog between, the two main fortification walls to the northwest of D 33, just beyond the west wall of the room.

The rebuilding joints between the wall blocking off the easternmost rooms and the west wall of D 33 were untraceable, but together the two walls formed a 1.5 m thick barricade. Beyond D 34 was a space at the east end of the fortification. The interior surface of the enclosure wall in this area had been plastered, perhaps indicating that this had initially been a functional space. But sometime afterward it was filled with rubble that was subsequently removed during the excavation (pl. 64). The blocking off of the two rooms and the southwest wall stairs may have been part of a later project intended to renovate and strengthen an upper surveillance platform at the east end, perhaps in Level III. The thick mudbrick foundation found within D 33 may indicate that there was an upper-level observation platform above D 33 and D 34 at all times, which enlarged the area at the east end of the fortress out to the west.[218] This is speculation, but it seems reasonable.

All bricks in the rooms just described and in the wall stairs were 34–36 cm long, when they are identified, as were those of the exterior wall relining to the south and the innermost blocking wall at the west. The latest blocking wall on the west was constructed of bricks measuring 38–40 cm in length.[219]

214 Knudstad 1964a, p. 29, where the excavated floors are noted.

215 Knudstad 1964a, p. 29.

216 The stairway is discussed in chapter 3.

217 Knudstad 1964a, p. 28.

218 This is discussed in chapter 3.

219 Knudstad 1964a, p. 28.

Very few sherds were found in these rooms, but a collection of forty-seven small cubes of galena (OIM E24355) was found atop the floor in the northeast corner of D 33.[220] Ten stone arrowheads were uncovered resting atop the charcoal-covered floor in D 34 (OIM E24349), and one was found in D 33 atop the foundation rocks below the mud floor at the northeast edge of the room (OIM E24421).

In D 33 and D 34 the stratigraphy was recorded as (1) the top of the wall, as well as the unexcavated ground level (surface clearance); (2–3) 82 cm of fill above 6–8 cm of "ash" or a "dark stain" level; (4) a 47 cm thick deposit of mudbrick and rubble, with very few sherds and a series of thin ash lenses; and (5) a mud floor on top of a cataract rock foundation.[221]

220 Pierce 1964b, p. 4. Pierce correctly identified the cubes as possibly for kohl. The structure of galena breaks naturally into cubes.

221 Pierce 1964b, p. 4.

5

The Pottery from Level III and Level IV

Regional pottery differences, varying site functions, and varying rates of pottery development for different forms greatly complicate the task of dating the pottery assemblage from the Level III and Level IV contexts at Dorginarti.[1] I have tentatively dated the Level IV forms to the tenth and earlier ninth centuries, while the Level III forms are thought to be entirely within the early half of the eighth century BC. The pottery is presented here together, because the sherds were often found together in mixed archaeological deposits and it remains unclear whether the lifespan of any of the earlier forms lasted until the eighth century BC. Future ceramic seriation and independent dating evidence will, of course, lead to greater precision.

The bowl sherds from the deeper test pits are usually very worn or were handmade, and later builders had largely emptied the Central Sector Level IV residence of all debris. Some sherds were noted as coming from primary contexts atop floors in the West Sector or had been sealed in fill between two floor levels. But many of them stem from Level III contexts that were not entirely sealed from Level IV.

Sherds from areas with more than one stratigraphic level have been compiled in appendix I at the back of this book, using the most distinctive sherds in each locus. The appendix includes information from the various strata in D 4, D 60–68 and its surrounding area, doorway D 87 and the levels below it, contexts in the Level III and Level IV residences, and the deposits under the wall relining at the north breach—although some of the latter deposits were no longer sealed beneath the wall relining when excavated.

The fortress of Buhen and its surroundings, along with the nearby fort of Mirgissa, were active in the New Kingdom, and the island was undoubtedly used for its vantage point over the river and the west bank, but there is no architecture to indicate extensive use of the island during the New Kingdom. The few New Kingdom sherds found are discussed at the end of this chapter.

There are ceramic forms from the site that reflect earlier New Kingdom traditions, such as the marl jar form with a swollen or bowed neck (MJ 1 below), the life span of which probably ended within the first century of the first millennium BC. The two most complete jars of this form were found in D 102 and D 59, areas where the upper levels had been eroded away by floods. However, fragments of the rims were also found in contexts within the Level III structures along the north wall.

Silt food bowls with red rim bands, marl pilgrim flasks, silt jars with brushed-on cream slip, marl jars with black-painted bands, and bowed-neck marl jars are typically considered earlier forms. But other ceramic forms at the site, the use of marl fabrics, and the marl industry's distinctive manufacturing techniques represent eighth-century BC trends.[2] The different loci at the fortress often contained both earlier and later pottery types.

The rarity of white, spiral-banded silt jar sherds at Dorginarti is another chronological or regional indicator, and there is no evidence of the similarly decorated

1 For a discussion of the discontinuity of pottery traditions between the Third Intermediate Period and the Twenty-Fifth/Twenty-Sixth Dynasties, see Aston 1996a, pp. 71–72, 87–89. For the northern perspective, see French 1992, p. 84. Newer publications confirm a similar trend, which continues during the Twenty-Sixth Dynasty; see Boulet and Defernez 2014.

2 A comprehensive analysis of this new technology, with references to earlier publications, is found in Boulet 2017. These imports are found at numerous sites in Nubia, including Qasr Ibrim, Kawa, and Tombos. They are also found at el-Kurru; see Heidorn 2018a.

silt bottles.[3] The white-banded jars are believed to be a hallmark of the Twenty-Fifth Dynasty, although at Buto they are very common in the early Saite period and absent from later levels.[4] The few examples from Dorginarti stem from insecure contexts such as D 13 (1), which was a debris layer over the surface around and on the west fortification wall, and from D 35, a surface to "ash" layer.

The use of white slips on silts, and sometimes around the rim of red-slipped jars, is found in the later Third Intermediate Period materials from both Elephantine and the later deposits around the South Tombs at Amarna.[5] The silt jars from Dorginarti commonly have red slips or are unslipped, but with no white-coated rims.

The almost complete absence of parallels to the types of marl jars found with the Twenty-Fifth Dynasty royal burials at el-Kurru is more indicative of a chronological than a functional difference.[6] There are similarities to pottery types from Dorginarti in the much-disturbed debris around the ancestral tombs at el-Kurru, including marl jar rims from Kurru tomb 19 and marl flasks from Tumulus 6, both attributed to Generation B.[7] Some of this pottery from around the tombs stems from the original burial goods, but the nature of the later activity causes some concern, particularly considering the presence around the ancestral tombs of marl channeled-rim store jars that are more securely dated to the late eighth and seventh centuries BC, or perhaps initially in the reign of Taharqo and thereafter.[8] Later pilgrimages to the tomb sites undoubtedly brought in their wake looting and other disturbances of the archaeological contexts that must not be underestimated.

The contexts in and around the temple of Taharqo at Qasr Ibrim have yielded pottery forms that parallel those found in the royal tombs at el-Kurru and in the early tombs at Nuri, but earlier Upper Egyptian marl jar forms were excavated in levels below the temple foundations and around the gateway.[9] The pottery from these contexts shows that Qasr Ibrim was partly contemporaneous with Dorginarti, although activity at Qasr Ibrim was both earlier and later than that at Dorginarti. It is assumed that Dorginarti was abandoned before the last third of the eighth century.

The wheel-made pottery from Dorginarti reflects the transition stage between the earlier ceramic traditions of the first millennium BC—with a morphological development from the Ramesside and early first-millennium BC forms—and the new pottery forms, clays, and manufacturing techniques that originated in the eighth century. Thus, while some vessels at Dorginarti closely resemble those of Phase IIb at Elephantine, other forms and fabrics at the fortress resemble those of the Phase III materials at Elephantine.[10]

Handmade pottery was found throughout Level III and Level IV contexts and is typical of the long-lived ceramic traditions of Nubia. It is not closely datable, although the Napatan types can be quite distinct in rim form and incised decoration from the earlier second-millennium traditions of handmade vessels. The sherds consist mainly of bowls and a few closed jars. Comparable forms, with similar streak or pattern burnishing as on the wheel-made vessels, were found at sites along the Nubian Nile, at the Letti Basin and Debba Bend, and in the Wadi Howar at Gala Abu Ahmed.[11] There is a wide range of forms and incised decoration, particularly incised rims, mat impressions on the larger bowls, black-topped vessels, and burnished black-topped red ware bowls.[12] The handmade types from Sanam also exhibit similarities, although only a few of the wheel-made vessels are similar to those from the fortress at Dorginarti.[13]

3 Silt bottles with white spiral decoration are found in layers attributed to the Twenty-First Dynasty at Karnak North, but other examples come from Twenty-First through Twenty-Fifth Dynasty layers; see Jacquet-Gordon 2012, figs. 88aa–bb, 95c–e. Aston notes they are known in the late eighth and early seventh centuries BC, citing one found at Buto dated to the second half of the eighth century; see Aston 1996b, pp. 9–10, n. 35.

4 Aston 1996a, p. 52, n. 329, and p. 81 for Phase III.

5 For examples, see Aston 1999, pp. 175–78, nos. 1620–25, 1627–30. All are dated to Phase III. The Amarna example was published by French 1986, pp. 156, 174, figs. 9.8, SJ1.1.1, and 9.9, SJ2.6.1.

6 Heidorn 2018a. However, in D 24 (2), a locus containing disturbed fill from the Level II platform construction, there is a silt copy of a marl channeled-rim jar prototype that should be dated to the Level II activity at the fortress; see chapter 6.

7 Dunham 1950, pp. 72–75, and especially fig. 24c, 19-3-844 (from debris of the superstructure). The marl flask sherds were not published by Dunham. No registration numbers were associated with them, although they were in a storage tray with other Tumulus 6 materials. See also Heidorn 1994, pp. 121–23, nn. 31, 34, and figs. 2k, 3i–j.

8 Heidorn 2018a, pp. 329–30.

9 Personal observation 1998. French notes a number of marl jar and bowl forms corresponding to examples from Qasr Ibrim; see French 1986, p. 166.

10 Aston 1999, pp. 68–207.

11 For example, the Nubian vessels at Kawa, the Soniyat temple at the Debba Bend, and Gala Abu Ahmed in the Wadi Howar, all of which are mentioned below in the pertinent footnotes.

12 Żurawski 2003. See also Eger, Helmbold-Doyé, and Karberg 2010 and Jesse 2016. The precise parallels are mentioned in the handmade pottery section below.

13 The wheel-made Egyptian pottery at the cemetery of Sanam is mostly later in date than that from Dorginarti, resembling

A NORTHERN ORIENTATION

The preponderance of Egyptian wheel-made vessels from Level III and Level IV at Dorginarti, and the imports and Egyptian pottery from the later Level II contexts at the fortress, reflect the northern orientation of the fortress and its lines of supply. The fortress may have been built and occupied by Nubians—before the Kushites rose to power as the Egyptian Twenty-Fifth Dynasty—who used wheel-made pottery for food transport and storage, and wheel- and handmade pottery in their everyday activities. The pottery suggests that a close interaction had been established with Upper Egypt before the rise of the earliest Kushite dynasts. It might also be assumed that the Second Cataract was an outpost of Egypt in a period before the Kushites controlled Lower Nubia, and that the army and its staff were composed of a mix of people from both north and south. This could have occurred sometime in the Twenty-Second Dynasty. Whoever held sway over the fortification, it remains clear that the bulk of their provisions was packaged in Egypt and sent to the Second Cataract.

The significant corpus of handmade pottery at Dorginarti was found in direct association with wheel-made pottery throughout the fortress. Cultural indicators such as handmade pottery and stone arrowheads suggest people who originated from areas along the Nile in Nubia or in the deserts, although their identity is not entirely visible in the archaeological record, unless we assume that only northerners used wheel-made vessels and only southerners used handmade pots. People tend to retain their familiar cooking traditions and diets, but a familiarity with northern customs might be presumed.

Women are usually identified as the manufacturers of handmade vessels, so we may suppose that women from local cultures were present. It is probable that all types of cooking—baking, stewing, grilling, simmering—were done at the fortress and that both handmade and wheel-made vessels were used for storing, preparing, and serving food and drink for all residents.[14] Numerous ovens were found throughout the West Sector, indicating that bread was baked, and bread plate (*dokka*) sherds were uncovered. The precise cultural identity of the fortress's builders and residents cannot easily be determined through a simple analysis of the ceramics or small objects, but the handmade pottery and stone arrowheads indicate that at least some Nubian or other Sudanic peoples were present.

Dorginarti exhibits a pottery assemblage that is mirrored in the pottery found in the late dynastic deposits around the South Tombs at Amarna and from the surface survey at the fortress of Abu ʿId in Upper Egypt (see references below), both of which are dated to single periods.

THE WHEEL-MADE CORPUS

Not all the excavated sherds were brought back to the Oriental Institute Museum, since shipping the copious number that had been excavated would not have been viable. The potsherds and vessels available for study therefore could not be statistically analyzed, and not all vessels could be reconstructed.[15] A short section toward the end of this chapter will cover the earlier sherds of New Kingdom date that were found in various locations around the fortress, and in one case under a Level III–IV wall, but that were not associated with any known architectural unit. Plates 65–74 present photographs of selected sherds to indicate fabric color, surface finish, and decoration more clearly.

It is certain that the Upper Egyptian marl vessels, and perhaps some of the silts, were imports into the Second Cataract region. But some of the silt vessels could easily have been manufactured locally.[16]

FABRICS OF THE WHEEL-MADE VESSELS

The vessels were manufactured from both alluvial and marl clays, with the marl vessels (Marl A4) coming from a source typically identified with Upper Egypt

types found in the royal tombs of the Twenty-Fifth Dynasty at the end of the eighth and beginning of the seventh centuries BC; see Griffith 1923. The handmade pottery shows some similarities, especially the black-topped vessels and the use of a streak burnish on the vessels' surfaces. Only the marl rim profile on IIIe, the single deposit from tomb 673, and the flask VIIh from the fill of grave 1009 (which contained a bowl and marl jar rim IIIf, similar in form to the rim of IIIe) are found among the materials from Dorginarti. Only two examples of the jar form were found at the fortress, but there were numerous examples of the globular wide-rimmed flask. For some of the forms, see Heidorn 2018a.

14 For the distinction between Near Eastern consumers of bread and African tendencies to eat porridge, see Haaland 2007; Edwards 1996; and Pope 2013.

15 The ceramic terminology employed here largely follows that used by Rice 1987, but see also Bourriau 1985. Munsell soil-color charts were used to describe the color of some of the pottery fabrics or the color of coatings found on the vessels (Munsell Color 1988). These readings are given in the text or in the figures.

16 The latest research on silt clays is published by Carrano, Girty, and Carrano 2009. See also Carrano et al. 2008.

around Qena/Ballas or Thebes.[17] Both Marl A4 variants 1 and 2 are considered in the material here, although there is little variation in temper or fired color of the majority of sherds. It seems the difference is more in the manufacturing technology used to produce thinner walls and, during the later part of the eighth and seventh centuries BC, the prevalence of ribbed or corrugated exteriors. The corrugation is not so prevalent on sherds in the Oriental Institute Museum collection, undoubtedly because not all body sherds were salvaged. But this fact is also chronological. The sherds are fired a greenish color in only a handful of cases, and are typically fired with pinkish layers in the break, sometimes with pale reddish-brown or yellowish cores and with pale-brown to pink surfaces. The same jar or bowl forms, however, were also fired with gray-brown cores and pale-brown surfaces. Their inclusions always consist of myriad small black particles, sands—some of which are clear and shiny—and small limestone particles. Occasionally there are subangular red particles or small lumps of pink unmixed clay. The few tempers that differ are mentioned in the descriptions below.

Although there are wheel-made silt jars that appear to imitate marl Egyptian forms, they are not made of the distinct sandy fabric that was used for Nubian copies of the Upper Egyptian marl jars during the Twenty-Fifth Dynasty and later.[18] The one exception is a silt copy of the Upper Egyptian channeled-rim marl jar found in Level II contexts, although the fabric is not very micaceous.[19]

Jars of Dorginarti's Nile D, a silt fabric with some limestone temper and a marl-based slip on the exterior, were perhaps produced in Egypt. Helen Jacquet-Gordon prefers to place this fabric within a class of Nile B2 clays, while it is David Aston's category Nile D, or perhaps more precisely his Mixed-Clay Fabric 1, in his work on late New Kingdom through Ptolemaic period pottery at Elephantine.[20] The related jars, made of typical Nile B, usually without limestone temper, have the same shapes as Dorginarti's Nile D jars, and light-colored slips are also brushed onto the exterior surfaces.[21]

The Nile B2 or Nile C silt vessels could have been manufactured in Lower Nubia or near the site. The details of ceramic fabrics will be discussed in each separate shape category. However, a number of wheel-made silt sherds—two shallow bowls, a beaker, and a jug—are distinctive by reason of their temper and highly burnished and polished surfaces, and these are discussed separately below.

SILT BOWLS AND BEAKER

The most common wheel-made forms at Dorginarti, encountered in every locus and stratum, are the hemispherical or slightly elliptical silt food bowls with rounded bases (SB 1). The bases of the sherds show evidence of a scraped trimming on the exterior. This form is assumed to represent the usual bowl for eating and drinking, but it undoubtedly served other purposes, and the sherds are sometimes blackened from fire or use near a fire. The sherds of these bowls are very common in the materials brought back to the Oriental Institute Museum.

There is a single silt beaker similar to a form that is common in later Twenty-Fifth Dynasty contexts in Nubia and Egypt, where it is associated with burials of Kushites. It is taller than its counterparts, however, and different from the typical hand- and wheel-made beakers found in the later eighth century BC. It was found in the Level IV residence and is presented below in the Unusual Vessels category (fig. 5.63b; pl. 70b).[22] The base of a possible beaker, either hand- or wheel-made, is also discussed below (fig. 5.13c). There is no clear evidence for handmade silt bowls that imitated the wheel-made SB 1 examples discussed below. There is also little evidence for any of the bowl forms containing a large amount of mica temper, a feature sometimes found at Nubian sites of different periods.

All bowls were manufactured of Nile B2 or chaffy Nile C. The cups and some very thin-walled SB 1s were

17 The general fabric categories used here loosely follow those first identified in the Vienna System, although they have since been refined for use at various sites and for various periods; see Bourriau and Nordström 1990.

18 Pinkish to reddish-brown marl and silt mix, or silt, was used for the local production of Egyptian forms at sites in Upper Nubia, such as those found in the royal tombs at Napata (personal observation) and at Qasr Ibrim (personal observation). The fabric is typically sandy and micaceous, with chaff and occasional limestone temper. See, for instance, Rose 2008, p. 205, and Sjöström and Thomas 2011, p. 64.

19 See chapter 6 and fig. 6.19a.

20 Aston 1999, pp. 3–4, 6.

21 Jacquet-Gordon 2012, p. 5. The fabric was used for large jars, dated from the late New Kingdom through the Third Intermediate Period and up to the Ptolemaic period. See also Aston 1999, pp. 3–4, 6.

22 Beakers are shown, for example, in Williams 1990, p. 54, fig. 2b–c, and Vila 1980, p. 156, fig. 169, types II-1A–1B, from tomb 46. These beakers are also found at Sanam and Kawa, and in the tombs of Nubian individuals around Thebes. The beakers found in a storage area of the chapel of Amenirdis are shown in Hölscher 1954, p. 74, pl. 47, types X4–X5; see also Budka 2010a, p. 507, fig. 3, with parallels from Nubia.

made from a finer Nile B2 with mica and small rounded sands, which may be natural inclusions in the clay source, but fine chaff bits are sometimes noticeable. The thicker-walled varieties usually show a coarser texture and were made of a chaffy Nile C, with larger temper particles and organic additions.

The most frequently occurring bowl types are placed in numbered categories below, but other bowl forms are also discussed. The number of size and form variants within each group indicates production by various individuals or pottery workshops, perhaps consistent with the idea of local production facilities. While many varieties of bowls are not very common among the sherd material studied, none of the silt bowls from the excavation of Level III and Level IV contexts show clear differences for dating purposes.

Silt Bowl 1

The SB 1a and SB 1b types are the commonest of the various types found. It is a deeper bowl form that often shows evidence of wheel ribbing on the exterior surfaces and a lathed or scraped finishing on the base exteriors. The manufacturing method is sometimes obscured by the pattern burnishing on the interior and exterior surfaces of SB 1b, but there are only a handful of sherds where the wheel or hand fabrication is not at all apparent upon visual or tactile examination. The shape of this deep, hemispherical to ovoid bowl is shown in the illustrated vessels, although shallower stances are also present. A number of vessels were found (pl. 65a, c).

SB 1a bowls most often display a red rim (RR) on both the interior and exterior surfaces but can also have a thin red slip (RS) on the interior surface (figs. 5.1–5.2). There are also examples with no traces of slip or rim bands. Many of the red-rimmed bowls were clearly painted with a red clay solution that ran down the sides of the vessel before the paint was set. The banding is typically not at the same latitude on the interior and exterior surfaces. The rim diameter averages 10–18 cm but can range from about 8 to 23 cm.

A few examples of the smaller SB 1a type are like a marl cup found rarely in the sherd collection (figs. 5.1a–5.2a). One or two of the small silt bowl sherds show traces of red slip or a red rim band.

The surface finish is what distinguishes SB 1b from the previous type, since the forms and their sizes are comparable, but streak burnishing—or pattern burnishing—is found on all sherds in this category (fig. 5.3). Streak burnishing decorates both the wheel-made and

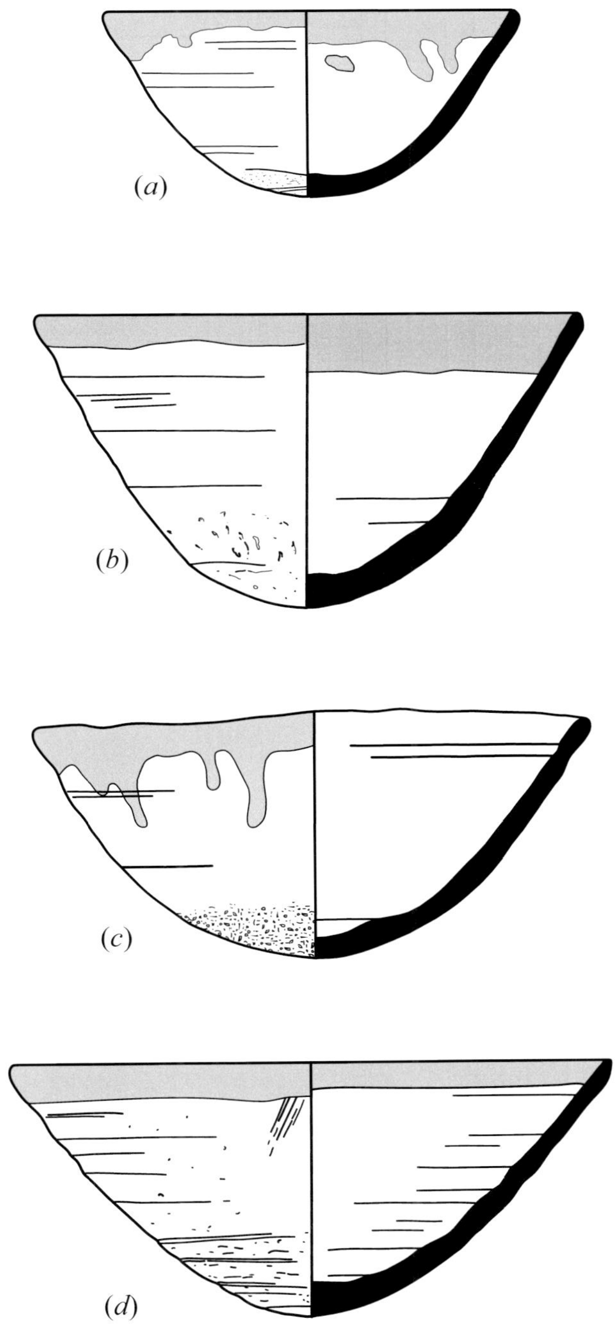

Figure 5.1. Silt bowls 1a (2:5)

(*a*) D 116, west bin; RR interior/exterior
(*b*) D 10 (1); OIM E24347; RR interior/exterior
(*c*) D 15 (1); OIM E24408; RR exterior, RS interior(?)
(*d*) D 7 (1); OIM E24343; RR interior/exterior

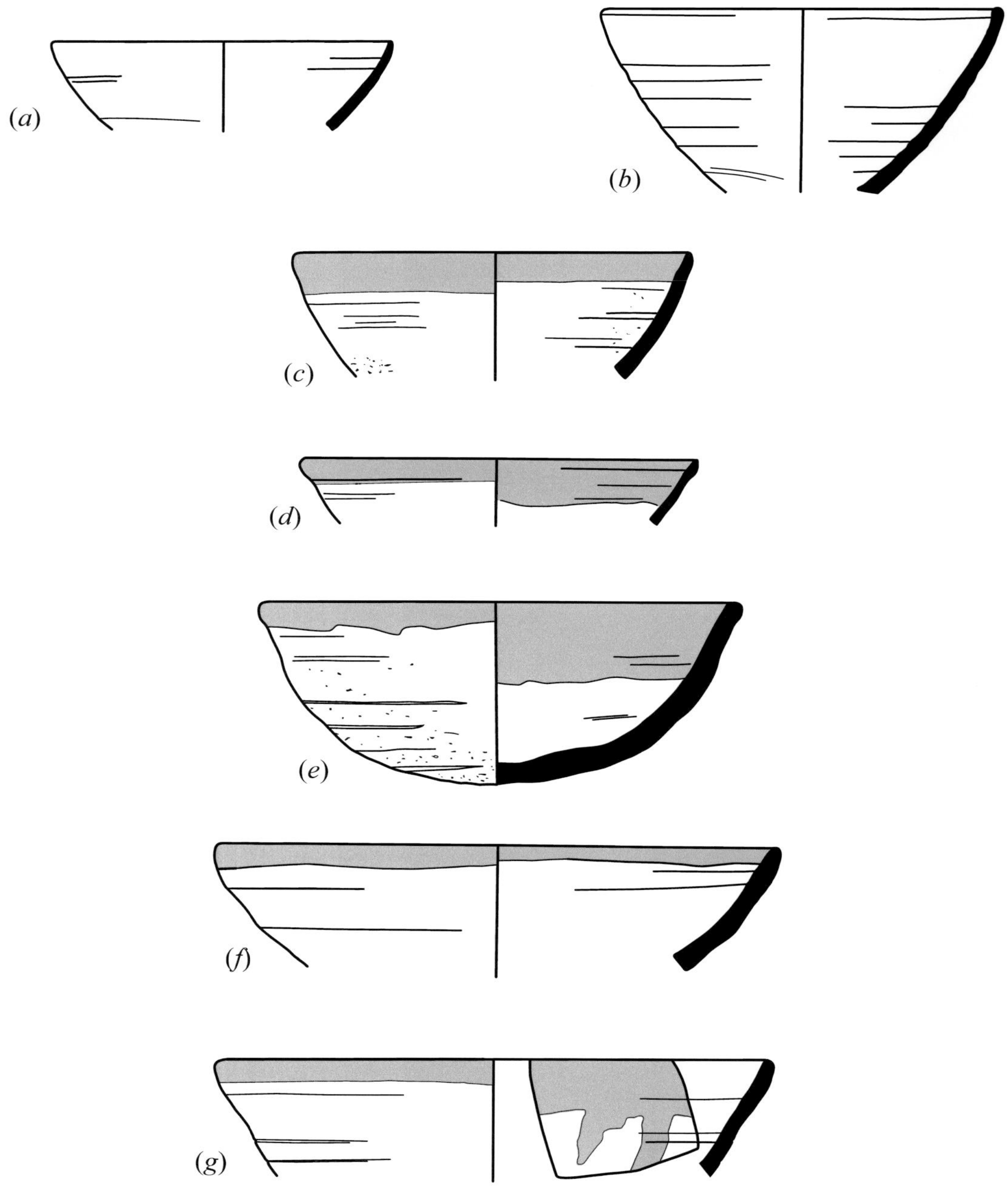

Figure 5.2. Silt bowls 1a (2:5)

(*a*) D 303 (1)
(*b*) D 12 (1)
(*c*) Structure atop north wall (crenelated buttress); RR interior/exterior
(*d*) D 310 (2)
(*e*) D 230 (1); RR interior/exterior
(*f*) Structure atop north wall (crenelated buttress); RR interior/exterior
(*g*) D 310 (2); RR interior/exterior

the handmade silt vessels.[23] The burnishing is horizontal or vertical on different parts of the pot, or is entirely random. The vertical movement used by the potter to burnish the interior and exterior surfaces of the bowls created strokes that meet toward the rim in an inverted V–shaped design. The manner of burnishing was haphazard and not an attempt to produce a continuous burnish.

Most parallels to the SB 1 types are found within the Third Intermediate Period. The hand- or wheel-manufacturing technique, the presence of red rims or slips, and the streak-burnished decoration are found throughout the Nubian region at Qasr Ibrim,[24] Qustul, Missiminia, Amara West, Kerma (Napatan residence), Soniyat, Tombos, Sanam, and Hillat el-Arab.[25]

The form is found at many Egyptian sites, but usually not with the red rim band or with burnishing. For example, numerous silt bowls of the same general forms as those from Dorginarti were uncovered in the layers dating to the Twenty-First through Twenty-Fifth Dynasties at Karnak North.[26] The pottery from the Twenty-Sixth and Thirtieth Dynasty levels at Karnak North, as well as elsewhere within Egypt, shows that marl bowls replaced these silt forms, which is not the case at Dorginarti, where the silt and marl bowls were frequently found in association with each other.

The ceramic deposits in Amarna's South Tombs and in the fort at Abu ʿId include the same SB 1 forms in uncoated or red-slipped silt fabric, but not burnished in any way.[27] Elephantine also has similar unburnished bowls in the late New Kingdom contexts down through Phase III, but the red rim decoration diminishes during the Third Intermediate Period.[28] The Phase III bowls from Elephantine, however, are very close in shape to the hemispherical bowls found at Dorginarti.[29] These forms, with the scraped bases apparent, were found at the temple of Mut at Karnak, where red rim bands, and rarely burnishing, were found on bowls stemming from Stratum 3, which is attributed to the Twenty-Second and Twenty-Fourth Dynasties, but also later in Stratum 2.[30] A slightly similar form also stems from the town site of Ashmunein, where examples were dated to Level 2, attributed to circa 850–750 BC, and also not burnished.[31]

23 Similar pebble burnishing on vessels dated to the New Kingdom is also found at Sesebi and Kerma, but only references to the examples dated to the Napatan period or Twenty-Fifth Dynasty in Nubia are given here. For a good parallel to Dorginarti's burnished silt bowls, see Williams 1990, p. 71, fig. 22b. Streak-burnished wheel-made beakers are also common; see, for example, Williams 1990, p. 68, fig. 20b, and p. 70, fig. 21b. Another wheel-made, streak-burnished bowl is shown in Vila 1980, pp. 156–57, fig. 169, 200/4. Note the bowls found in the Napatan reuse of tomb S 515 (Room A) at Semna, in Dunham and Janssen 1960, pp. 78–79, fig. 36. Two pebble-burnished bottles also come from the tomb; see Dunham and Janssen 1960, fig. 39. They are also found at the cemetery of Sanam, where beakers, but also other forms, are described as polished or having vertical or horizontal burnish; see Griffith 1923, p. 100, pl. XVIII, types XIIc, h.

24 Examples are shown in the recently published pottery from Qasr Ibrim; see Rose 2019, figs. 8–15 (both wheel- and handmade).

25 For example, see Williams 1990, pp. 7–8 for Qustul; Vila 1980, pp. 155–57; Griffith 1923, pp. 100, 101–2, pl. XVIII, types XIVa, d–f; and Vincentelli 2006, p. 62, figs. 2.35–2.36, from tomb 10. These types of bowls are variously dated to the Kushite or Napatan periods or to the Twenty-Fifth Dynasty. There were wheel- and handmade beakers and bowls in Nubia at Sanam and in the Napatan burials at Missiminia and Qustul. See also the red-rimmed and streak-burnished examples from ARA 10 in Vincentelli 2006, pp. 57–60, 62–64, figs. 2.35–2.37. This tomb's remains come from multiple uses and are dated from the late New Kingdom to the Twenty-Fifth Dynasty, with evidence of Meroitic period reuse. The bowls from Amara West are shown in Binder 2011, pp. 46–47, pl. 14 and fig. 9. Binder dates these types to the eleventh and tenth centuries BC; see Binder 2011, p. 50.

26 Jacquet-Gordon 2012, pp. 228, 238, and figs. 88a–f, n (Twenty-First Dynasty?), 92 (some with red slip exteriors and red rims on the interior).

27 French 1986, pp. 157, 181, SB1.3, SB 2s, SB 4s (specifically SB4.3.1), SB 6; Aston 1996c, p. 22, pl. I, nos. 2–6, and p. 26, pl. III, nos. 38–39 (red-slipped). The material from Amarna was tentatively dated to in or about the Twenty-Fifth Dynasty, while that from the fortress at Abu ʿId was attributed to the reign of Piye, although there was no independent dating evidence at either site.

28 Aston 1999, pp. 25–26, no. 51; pp. 73–74, no. 533; pp. 75–76, no. 555; pp. 158, 163, no. 1497. This deeper shape of bowl developed in Phase IIa, the late New Kingdom, and continued through Phase IIb, ca. 1000/950–750/600 BC. Aston notes that red rims on bowls persisted into the later eighth century and beyond, but this decoration lessened in significance during the Third Intermediate Period in Egypt. See also the observations in Aston 1996a, p. 80. In the north of Egypt, a similar observation is made in Laemmel 2008, p. 182, although here the bowls are shallower, with outturned rims. The particular examples stem from Stratum Ba, tentatively attributed to the late Twentieth and Twenty-First Dynasties.

29 For some examples, see Aston 1999, pp. 166–67 (bowls), all uncoated silts, and pp. 174–75 (bowls), with red slips.

30 Sullivan 2013, pp. 82, 172–74, 203, mostly types D-1 through D-13. Sullivan cites parallels with materials from Tell el-Balamun and Ashmunein. Also see the smaller cup with a slightly pointed base in Sullivan 2013, pp. 175, 205, type E-10, where the author cites Third Intermediate Period parallels to examples from Balamun and Amara West.

31 Spencer 1993, p. 44, pl. 44, type A.2.10. See also the materials collected in Aston 1996a, pp. 41–43, figs. 91–109. Aston dates most of the pieces illustrated in Spencer's publication as contemporary with the Amarna South Tombs' late dynastic material, ca. 750–600 BC, although he notes that some pottery types begin earlier in the first millennium BC.

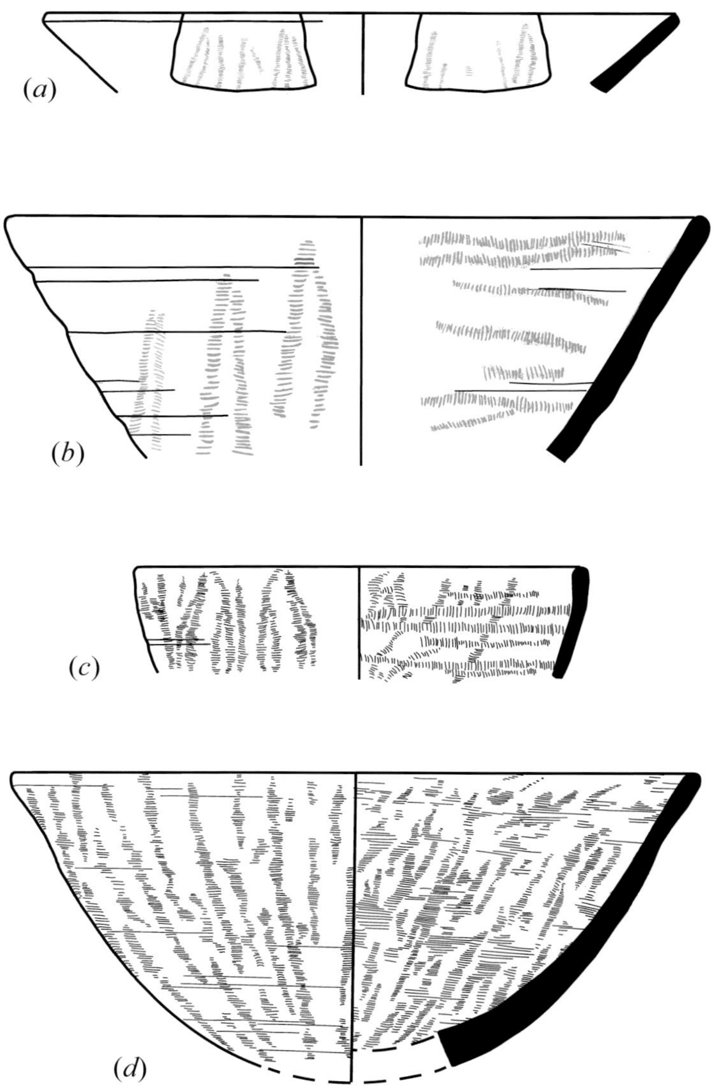

Figure 5.3. Silt bowls 1b (2:5)

(*a*) D 90 (1); OIM E50077; RS and burnished interior/exterior

(*b*) D 113 and D 114, surface

(*c*) D 19 (1); brick-red fabric; burnish strokes atop rim

(*d*) D 303 (1); RS interior/exterior(?); streak-burnished interior/exterior

Comparable bowls come from Third Intermediate Period deposits at the site of Tell el-Retaba in the Delta, where they are compared to vessels found at Tanis and Memphis.[32] Late Twentieth Dynasty levels at Tell el-Retaba yielded bowl forms close to the SB 1 type at Dorginarti, and these bowls then continue on into later levels.[33]

Shallower bowls, which are not as numerous at Dorginarti, are included here with the more typical deeper SB 1 examples and also exhibit evidence of red rim bands and streak burnishing. Their dating can range from the New Kingdom to the Third Intermediate Period.

Silt Bowl 2

Bowl forms with knobbed lips set atop vertically placed rims are fairly common and are categorized as SB 2a (figs. 5.4–5.5a). There are also examples of this form from the late dynastic assemblage at Amarna, dated to in or about the Twenty-Fifth Dynasty, and from the fortress at Abu ʿId of the same time period.[34] The bowl sherds at Dorginarti mostly show traces of an interior red slip and a red-rimmed exterior. Burnishing strokes are often found on the interior surface, but both surfaces could display this decoration. One example has a continuously burnished interior surface and burnish strokes on its exterior. The rim diameters of SB 2a are typically 16–20 cm but can measure 12–32 cm wide. The base areas show a roughened scraping of the surface.

A less common bowl form, SB 2b—in this case exhibiting a morphological difference rather than a surface finish variation—has an incurved upper rim that occasionally ends in a small knob, sometimes set off by a lightly incised line on the exterior, and the form can be quite deep (figs. 5.5b–c, 5.6a). The bowls in this category range between 20 and 24 cm in diameter at the rim. A base with a small knob or button that was found in the same locus as one of these bowl rims may belong to it (fig. 5.5b). There are parallels in shape and fabric to this form and its varieties in Phase IIb at Elephantine—attributed to the Twenty-First or early Twenty-Second Dynasty—in addition to parallels in marl and silt in the late dynastic deposits at Amarna.[35] The complete profile of this bowl with a round to pointed base is shown in material from Tell el-Retaba, and although it is shorter in height, its shape is similar to these bowls.[36] The forms also resemble a silt bowl type, with a low ring base but no slip, from Tanis.[37] The Tanis example, which has a wider mouth diameter, is dated to the Twenty-First or Twenty-Second Dynasty.

Incurved rims are also found on shallower bowls such as those shown in figure 5.6b–d, which are not morphologically the same as the preceding shapes except for their rim stance. A thin line is usually incised below the rim on the exterior. Traces of a thin, bright-red slip are found on the exteriors and sometimes on the interiors of this bowl type. The rim diameters measure 20–25 cm. Parallels are found among the late dynastic deposits at Amarna.[38]

An uncommon form among the materials from Levels III and IV at Dorginarti are the bowls with rounded shoulders and incurved rims shown in figure 5.7a–b, and the form can be quite deep. They too had traces of either a red slip or a red rim, sometimes with streak-burnished decoration on the interior surface. The entire profile of the form is unknown, and except for the small cup, the sherds attributed to it actually could have belonged instead to vessels resembling the preceding types (e.g., fig. 5.5b). Qantir and Tanis yielded somewhat similar bowl rims in Twenty-First through Twenty-Second Dynasty levels, where they are noted as continuing on from the Ramesside period through the Third Intermediate Period.[39]

Silt Bowls with Everted Rims

These bowls have been separated from the next group because of their everted rims (fig. 5.7c–e). They are not common at Dorginarti. Like the rest of the bowl types discussed here, they are made from Nile B2, sometimes with evidence of red-slipped surfaces, but with no evidence of red rim bands. One of the bowl sherds shows burnishing strokes on both the interior and exterior surfaces.

---

32 Wodzińska 2011a, pp. 160–61, figs. 52–53, nos. 1–2, and Wodzińska 2015, p. 111, fig. 117, nos. 2–3, 7. For vessel parallels with examples from Dorginarti, see also Wodzińska 2011b, pp. 1021–22, 1034, fig. 9.

33 Wodzińska 2015, pp. 107–8, fig. 114, no. 2.

34 French 1986, p. 181, SB4.4.1. The parallels from Abu ʿId are in Aston 1996c, p. 23, pl. I, nos. 13–14.

35 See the bowl from Tanis in Aston 1996a, pp. 24, 67, fig. 206d, Group 2, dated to Phase II (tenth to eighth centuries BC). A marl parallel from Amarna is shown in French 1986, p. 184, MB3.1.1.

36 Wodzińska 2011b, p. 1034, fig. 9, no. 6.

37 Laemmel 2011, p. 28, pl. 18, no. 104.

38 French 1986, p. 181, fig. 9.15, SB4.4.1.

39 Laemmel 2008, p. 183, pl. 11, nos. 4–5. Laemmel notes (p. 181) that many of the forms from Strata Ba–Ac began already in the earlier Ramesside period but continued through the Twenty-First Dynasty into the Third Intermediate Period. See also Laemmel 2011, p. 27, pl. 17, no. 100, with light ridges below the rim and at the widest diameter.

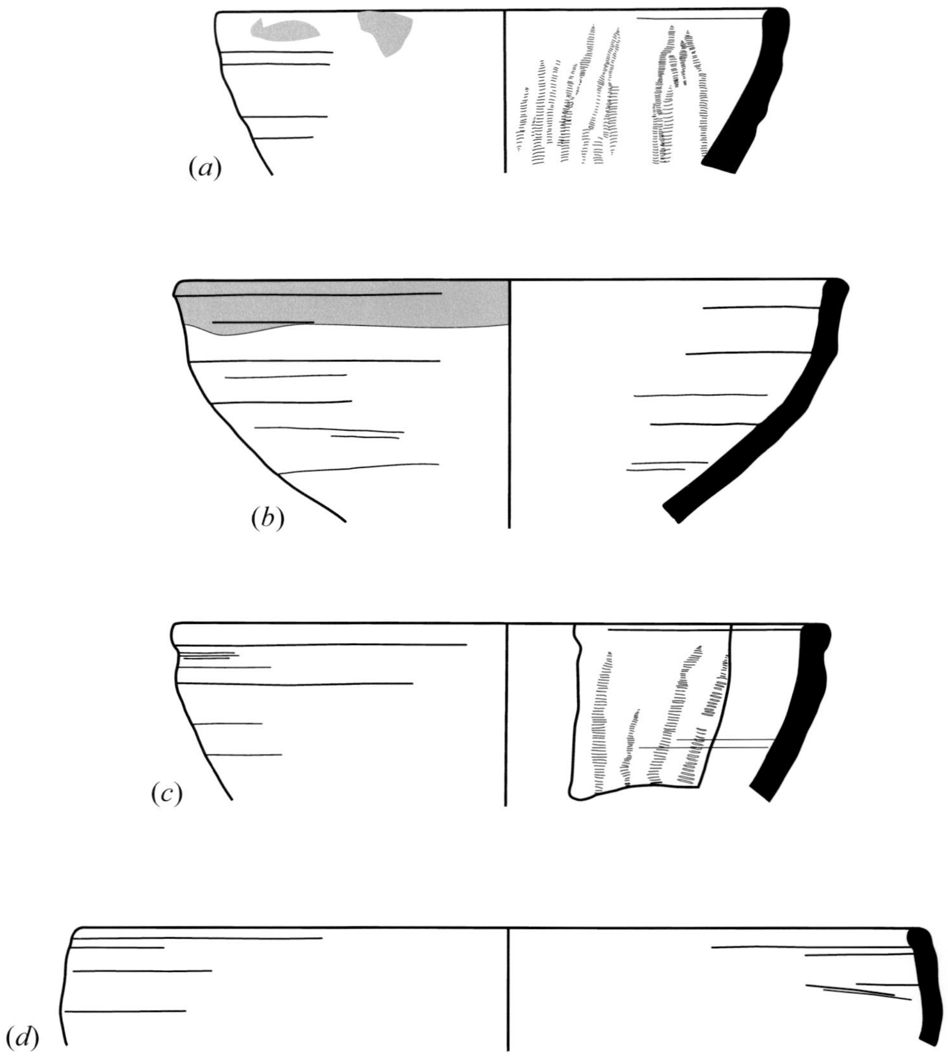

Figure 5.4. Silt bowls 2a (2:5)

(*a*) D 305 (1); RR exterior, RS and burnished interior

(*b*) D 122 (1); RR exterior, RS interior

(*c*) D 48 (2)

(*d*) D 67 (1)

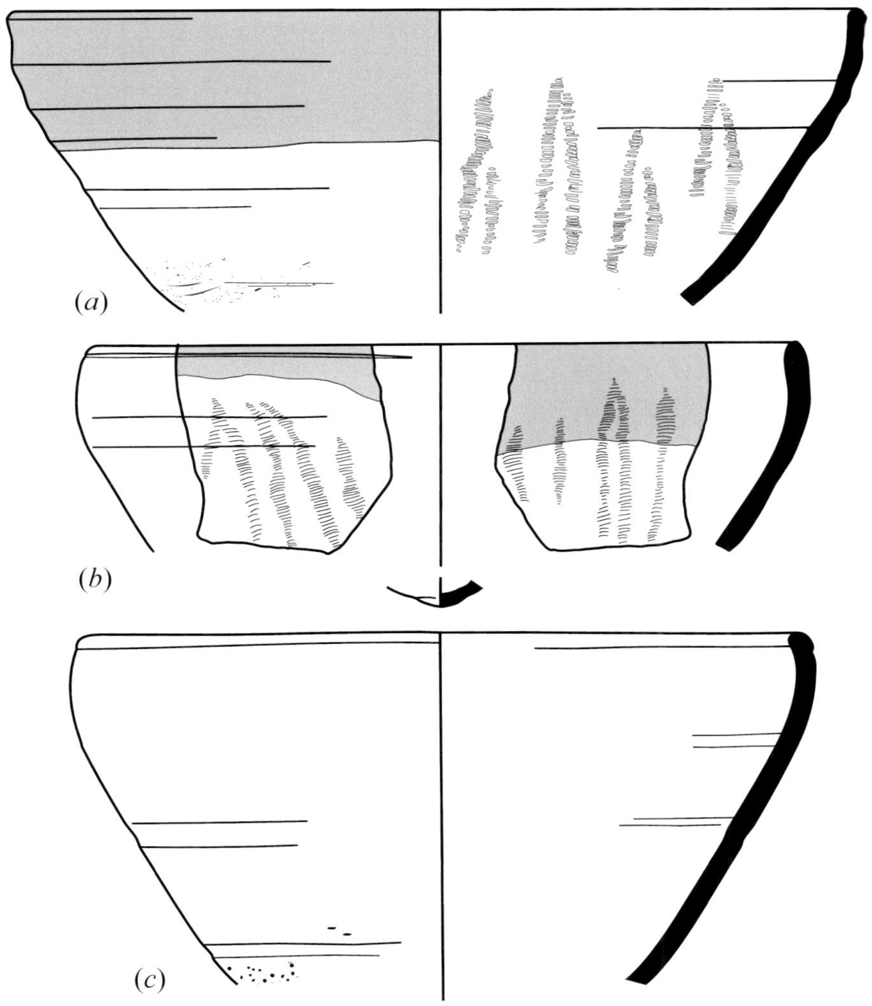

Figure 5.5. Silt bowls 2a–2b (2:5)

(*a*) D 303, trench at northwest corner Level III residence; RR exterior, RS interior

(*b*) D 115 (1), fill under ovens, RR interior/exterior; D 115, fill under ovens, burnished base interior, RS exterior (two sherds)

(*c*) D 6 (1)

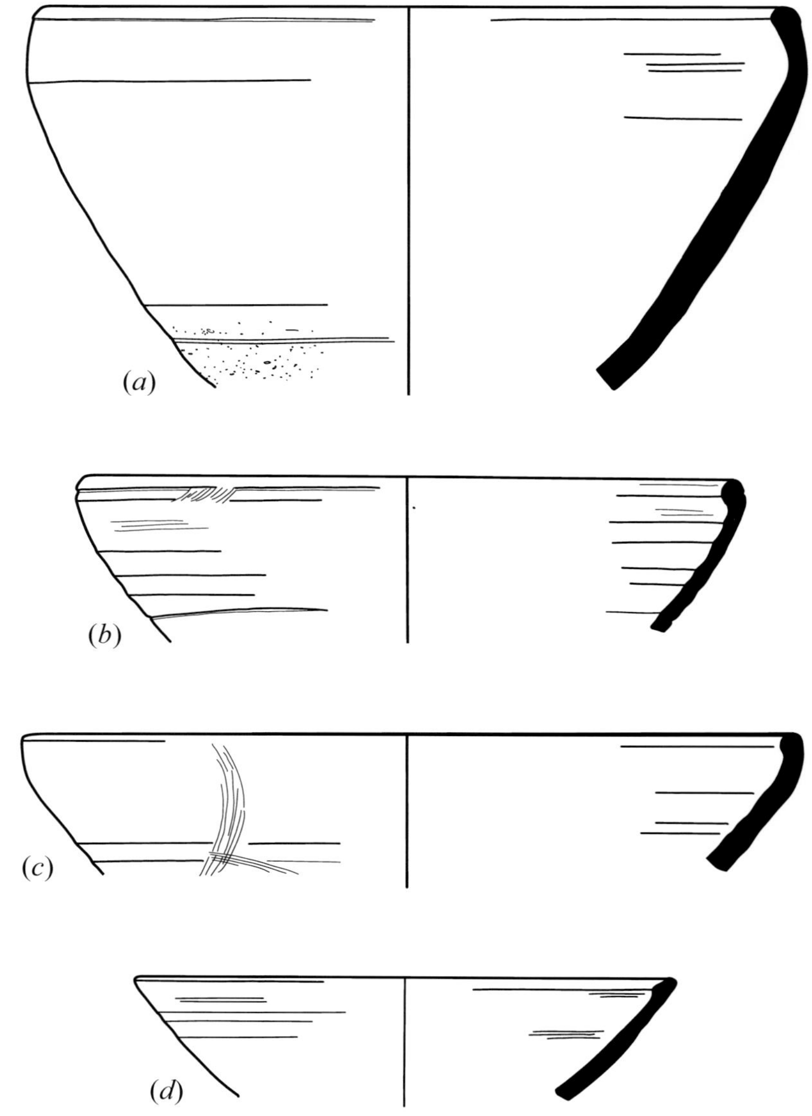

Figure 5.6. Silt bowls 2b (2:5)

(*a*) D 19 (1); coarse finish on base exterior, red-brown fabric with black core; fire blooms on exterior

(*b*) D 114 (1); RS traces interior/exterior

(*c*) D 113 (1)

(*d*) D 90 (1)

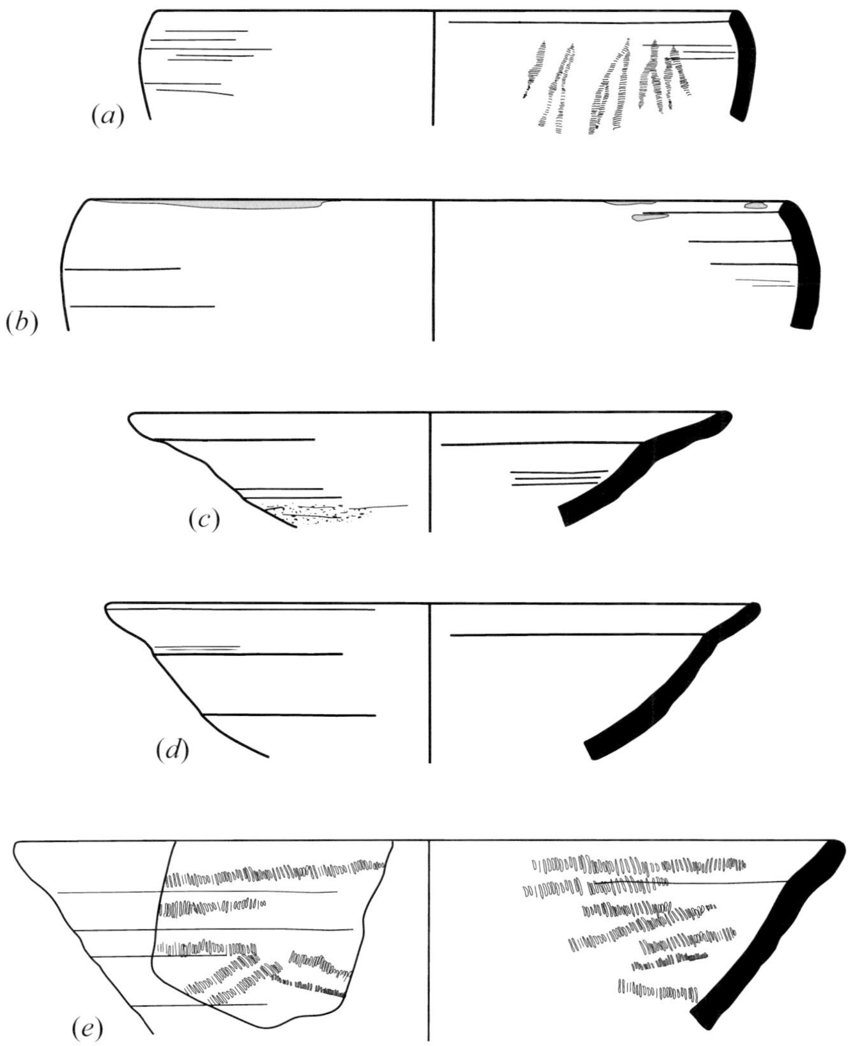

Figure 5.7. Silt bowls 2b and everted rim bowls (2:5)

(*a*) D 80 (1); RS traces interior/exterior

(*b*) D 4, D 47, D 48, tearing down walls in northwest corner of West Sector; RR or RS traces interior/exterior, bright-red slip

(*c*) D 49

(*d*) D 10 (1)

(*e*) D 307 (1)

Shallow Silt Bowls

The shallower bowls have been grouped in a single category, although there are variations in their wall profiles and rims. Some display deeper profiles but have similar rim forms. There is a pronounced ledge on the interior of certain bowls, which may indicate that they are lids, and the rims can be either everted or straight. Others have straighter walls and simple rims, but there are also examples with rims that end in a lip with a beaded or ledged interior (figs. 5.8, 5.9a–b). Red rim bands or traces of a red slip and streak burnishing remain on some sherds. The examples of shallow bowl types such as figure 5.8f are somewhat common at the site, but the others are rare.

Many of the bowl forms are poor chronological indicators since there are parallels in many earlier contexts in Egypt. But once again, similar varieties are found in Third Intermediate Period contexts at Tanis, in the late dynastic deposits around the South Tombs at Amarna, and at Abu ʿId, Karnak North, and Elephantine.[40]

Silt Bowl Variations

Other varieties of silt bowls were uncovered at Dorginarti, but the types of bowls discussed above were more prevalent in the collection. The bowl variants below are illustrated, but not all are discussed (figs. 5.9c–f, 5.10a–c).

The carinated cup shown in figure 5.9d is a unique example, although it could be a sherd from a vessel like the carinated bowls shown in figure 5.11, if the stance and diameter determined for the sherd are incorrect. It has traces of a red slip on both surfaces, and burnishing is preserved on the exterior side.

The gracefully flaring rim of the example shown in figure 5.9f is exceptional, having a red slip on both surfaces and a continuous light burnish—perhaps a polish—on the interior. The fabric is compact silt with no visible organic temper, it was well fired, and the walls are thin.

A closed bowl form, with an upper wall and rim slanting inward, is shown in figure 5.10a–c. Sherds of this type were found more frequently than some of the other variants. There are traces of a thin, bright-red slip on the exterior of some examples, and burnishing strokes are found on one sherd.

40 For example, Laemmel 2011, pp. 26, 30–31, pl. 16, no. 95, pl. 21, no. 117, and pl. 22; French 1986, p. 181, fig. 9.15 (SB 1s) and fig. 9.16 (SB 10); and Aston 1996c, pl. III, no. 42 (red-slipped silt).

Handled and Carinated Silt Bowls

A silt form that clearly mirrors a more common marl form at the fortress is the handled silt bowl, but it is a rare occurrence in silt (fig. 5.10d). The body has a rounded shoulder below a slightly everted rim. The evidence for one or two handles is not usually present within the same sherd bags. The constricted neck and slightly everted rim of the form is distinct. There is one complete profile of the marl form (MB handled), with a ring base, but there is no evidence for the second handle. No bases can be conclusively associated with the silt form. The silt examples both have traces of streak burnishing. The sherd in figure 5.11a, which has no indications of a handle, may also be an example. It has a red slip on its exterior and perhaps also on the rim interior.

A number of rim sherds may belong to a handled bowl or jar form. One example was found in a context along the west wall of the Level II platform and may represent a later form (fig. 5.10e).[41]

Carinated bowl sherds, some with necks up to 3–4 cm in height, are uncommon and may stem from the earlier Level IV phase at the site (fig. 5.11c–e). The shorter-necked vessel in figure 5.11e may also belong to this category. The sherds in figure 5.11c–d show evidence of a red slip on their upper exterior surfaces, another sherd of this type had a slip also on its interior surface, while the other sherds have no surface finish preserved. The rims, when they are present, have diameters between 20 and 24 cm. One carinated shoulder sherd had a red-slipped and streak-burnished exterior surface, and possible traces of a red slip on the interior (fig. 5.11b).

The form is found at Elephantine from the Ramesside period onward and was very popular at the site in Phases IIa–IIb. The fort at Abu ʿId also yielded examples that are dated to circa 750–650 BC.[42] The bowls from Elephantine have a red slip on the exterior over onto the interior rim and show evidence of use as cooking pots. The Dorginarti examples show no such evidence.

41 For a later form, see Oren 1984, p. 15, fig. 15. There are close parallels from Thebes that are attributed to the later Third Intermediate Period; see Schreiber 2008, p. 62, pl. LVII, fig. 2.1.7, 8, 12, 13. These are from a reuse of Theban Tomb 32.

42 Aston 1999, pp. 29–31, nos. 95–99 (late New Kingdom), pp. 68, 73–75, nos. 538–44, p. 122, no. 1086, and pp. 167–68, no. 1561. Note his comments (p. 168) about these bowls in Phase III; however, the form shown is not very similar to the Dorginarti examples. Aston also found these in contexts attributed to ca. 750–650 BC; see Aston 1996c, p. 26, pl. III, no. 36, where there is an example that is the same shape as the more complete form from Dorginarti.

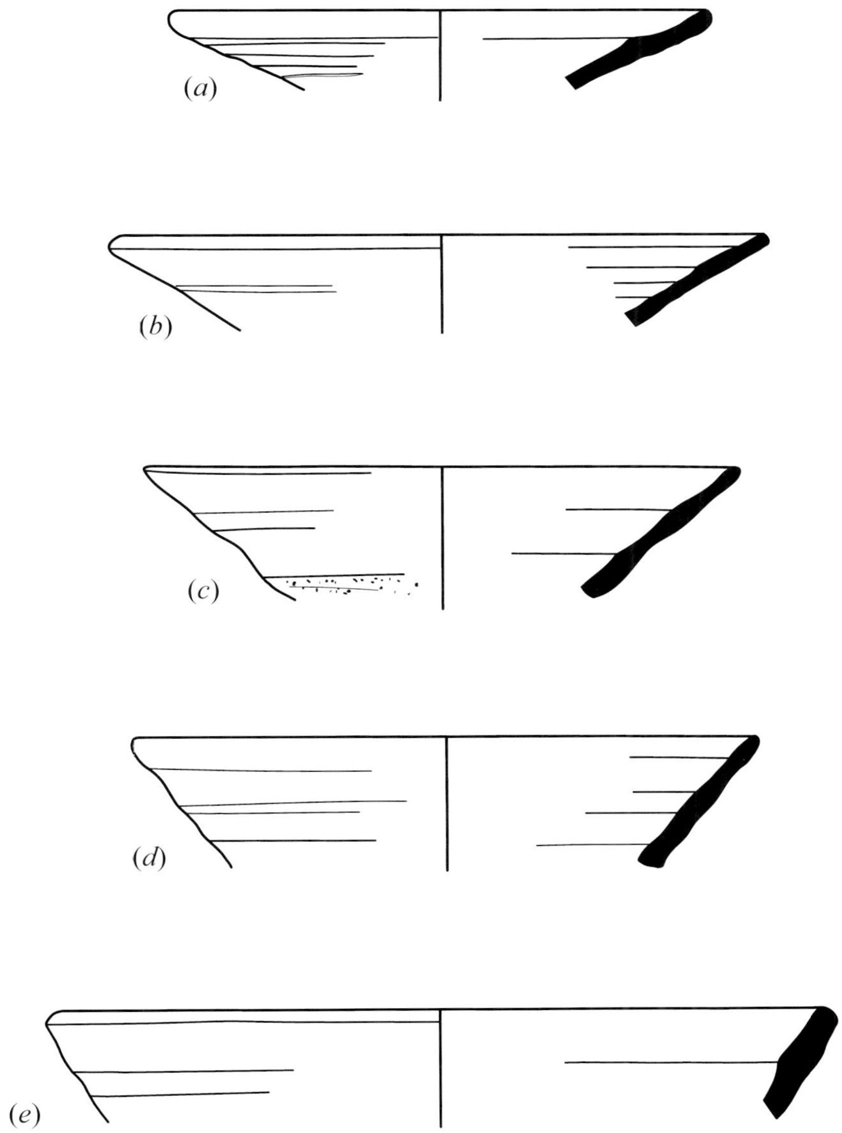

Figure 5.8. Shallow silt bowls (2:5)

(*a*) D 39; shallow ledged

(*b*) D 308 (1) and D 308 (2)

(*c*) D 14

(*d*) D 302 (1)

(*e*) D 305 (1); RS interior/exterior, horizontal burnishing on interior

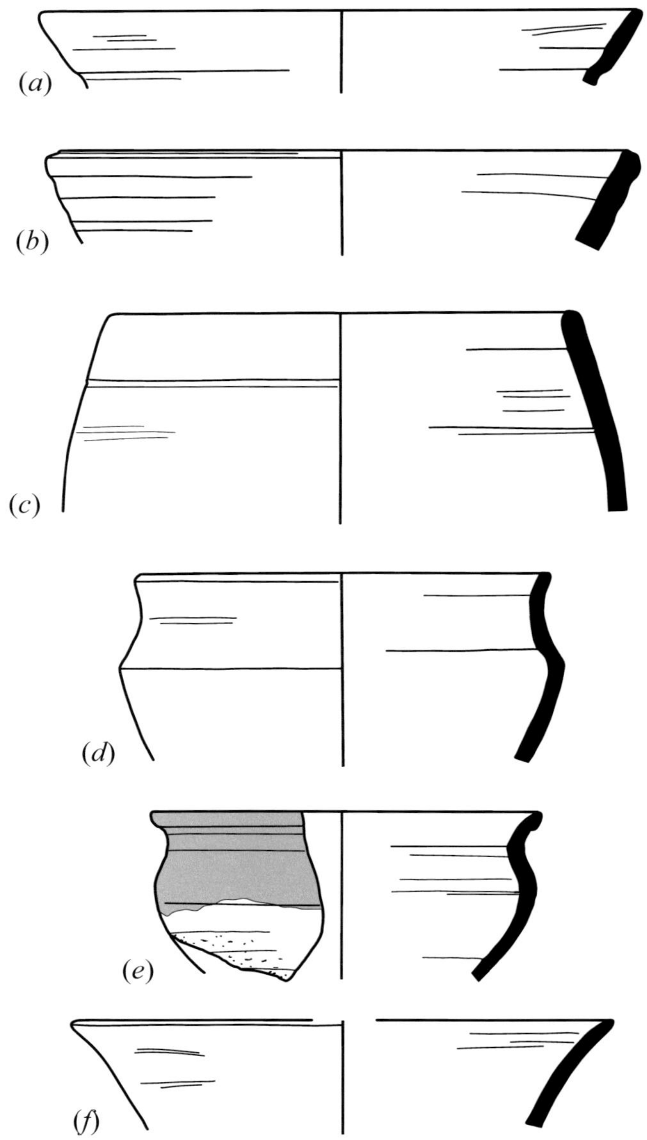

Figure 5.9. Silt bowl variants (2:5)

(*a*) D 10 (1), middle oven; RS exterior/interior(?), lid(?)

(*b*) D 17, test trench under enclosure wall

(*c*) D 103 (2); mixed clay

(*d*) D 86 (1); RS traces interior/exterior, burnish traces on exterior

(*e*) D 60 (1); RS interior/upper exterior

(*f*) D 122 (1); RS exterior, RR interior, 5YR 5/4, polish or burnish on interior

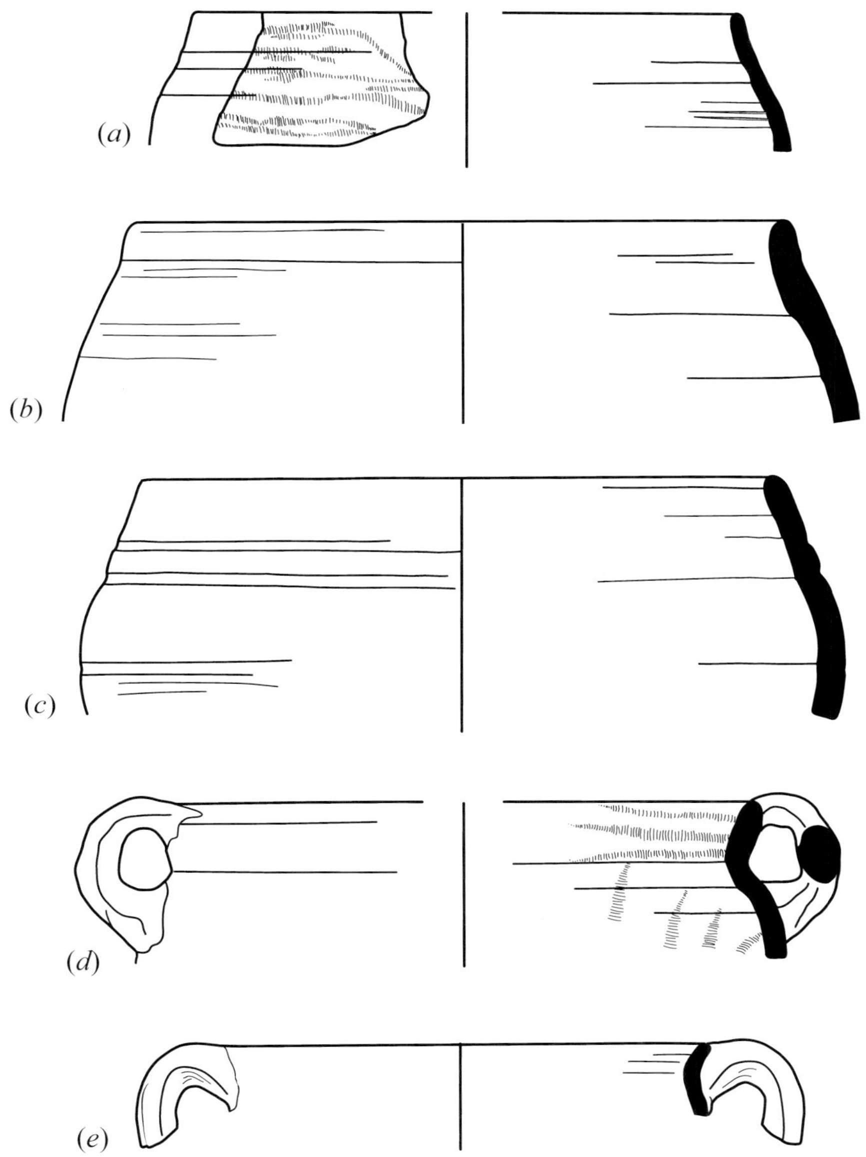

Figure 5.10. Silt bowl variant and handled bowls (2:5)

(*a*) West of D 32; streak-burnished exterior, burned exterior

(*b*) D 208 (2), area of ovens

(*c*) D 6 (1); 10R 5/6 RS exterior

(*d*) D 114 (1); coarse silt, RS interior/exterior(?), streak-burnished interior

(*e*) D 75

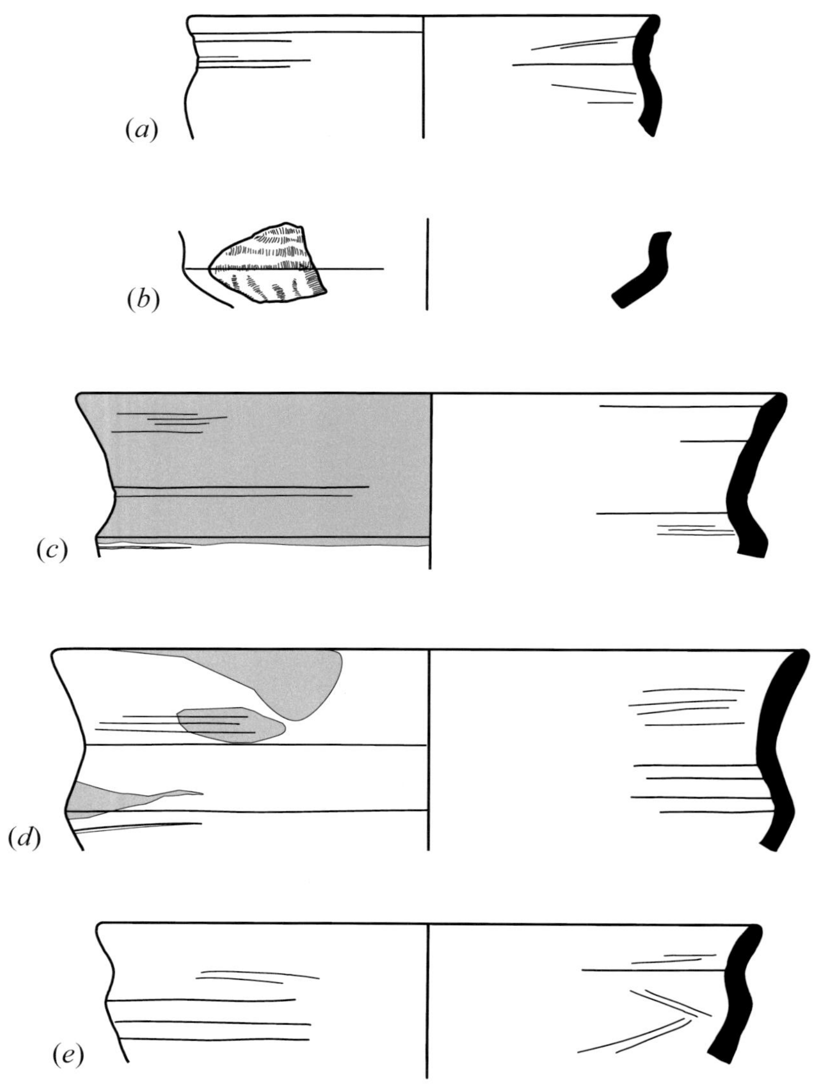

Figure 5.11. Handled(?) and carinated bowls (2:5)

(*a*) Structure atop north wall, crenelated buttress; RS exterior, RR interior(?), jar(?)

(*b*) D 35 and D 36 mixed; RS exterior(?)

(*c*) D 55 (1); RS exterior, very micaceous

(*d*) D 68 (3); RS traces exterior

(*e*) D 12 (1); RS traces interior/exterior

Large Silt Bowls

This class of large platters or bowls is known from earlier periods, but the form continues into the Third Intermediate Period (fig. 5.12). At Dorginarti, these vessels usually have a triangular rim profile and always exhibit string marks on the exterior, resulting from twine being wrapped around the vessels to hold them together while drying. The types of bases associated with these sherds are uncertain in some cases, but there are ring bases present with rim sherds, including one each from D 44 and D 10 (1) (fig. 5.12d). Parallels from Nubia and Egypt have round, ring, or flat bases.[43] Fragments of this form are fairly common at Dorginarti. The form is not a good dating indicator on its own, since it is found in New Kingdom and Third Intermediate Period levels throughout Upper Egypt.[44]

These wide bowls may have been used to hold or transport coals and hot ash for cooking, serving, or other utilitarian activities. There is little evidence on the sherds, however, for burning or smoked areas on the surfaces. Many of them were darkened either by use or by water saturation.

A unique vessel is shown in figure 5.13a. The rounded rim has a channeled exterior, along with cord marks around the exterior surface and possible red slip traces, in addition to worn burnishing strokes running horizontally along the bottom of the interior. A ring base fragment, which may belong to the rim and body sherd just discussed, also has burnishing strokes, although they are randomly produced. The depth and width of this bowl are similar to the depths and widths of some of the larger vats or basins from the site, which are mentioned below.

Silt Rounded and Flat Bases

Base sherds found in the same contexts as the bowls either are rounded (sometimes with small protrusions in the center) or are smaller ring bases. The silt bases are shown with the vessels, or in figures 5.13–5.15. More rarely occurring are silt bases with the center hanging below the edge, like those of the marl bowls (fig. 5.14b). A few flat-based sherds, cut from the wheel with a string, were also found (fig. 5.14d–e). A handful of flat-based sherds, two of them having pierced holes in their bottoms, usually show finger imprints at the base; they belong to a form referred to in the literature as *beer jars* but could also be construed as beaker forms (fig. 5.15a–d).[45] Their base diameters range between 7 and about 12 cm. Their function is uncertain, but they may be bread molds.[46]

MARL BOWLS

The marl bowls showed much less variety than the silt bowls and were not as common. They were found in the same contexts as the silt bowls and were unhelpful in offering chronological distinctions. They are all wheel-made and exhibit Marl A4 fabric. The majority of base sherds brought back to the Oriental Institute do not have the round or pointed shape of other marl bowls of the Third Intermediate Period or Saite era, but they were found in association with ring bases, often with the center protruding below the edge of the base. The rim diameter of the most common marl bowls is about 25 cm, although it ranges between 16 and 30 cm.

The examples with wider diameters and ring bases would suggest they were used to serve food. A number of marl forms are too shallow and wide to be bowls and therefore are classified as marl lids (see below).

Marl Cups

A small marl bowl has a slightly incurved rim and parallels the silt cups mentioned above. One complete profile of this marl form has an incurved rim and a pointed base that is quite distinctive (fig. 5.16a–b). One sherd shows the coloration differences from stacking during firing; the top part has a cream color, and the areas below the middle are pink. The form occurs infrequently, but it is counted in at least four different contexts.

43 Flat-, ring-, and round-based platters and bowls were found in levels dated to the late Twentieth Dynasty and Napatan period at Amara West; see Spencer 2002, pp. 16–17, pls. 19–20 (B6, B15–B17), pp. 22–23, pls. 25–26 (E). Spencer notes (p. 22) that E7, with a slight ring base, and E8–E9 are typical Third Intermediate Period bowls in Egypt.

44 Similar types are also present in Spencer 1993, pl. 49, types A3.65–A3.69. The types were found in all three strata, which were dated by the excavator to ca. 950–850 BC for the earliest levels and ca. 750–650 BC for the latest level. See also Aston 1996c, pp. 23, 26, pl. I, nos. 10–11, pl. III, nos. 41–42 (red-slipped vessels), and the slightly similar examples in French 1986, p. 181, SB1.2.1, and p. 182, SB10.1.1.

45 This type of vessel has a long history, but for first-millennium parallels see Jacquet-Gordon 2012, fig. 94m, q, s (all red-slipped). The form is evident at Heracleopolis Magna, where they are called *funerary beakers*; see López Grande and Quesada 1993. Flat-based examples, called *beakers*, were found in Nubia, for instance at site E10 in the Northern Dongola Reach survey; see Welsby and Welsby Sjöström 2006–7, fig. 2, types BB1–BB2. The site is dated to the New Kingdom but may run down into the pre-Twenty-Fifth Dynasty period.

46 Smoláriková 2008a, p. 61, 153, fig. 46, C048.

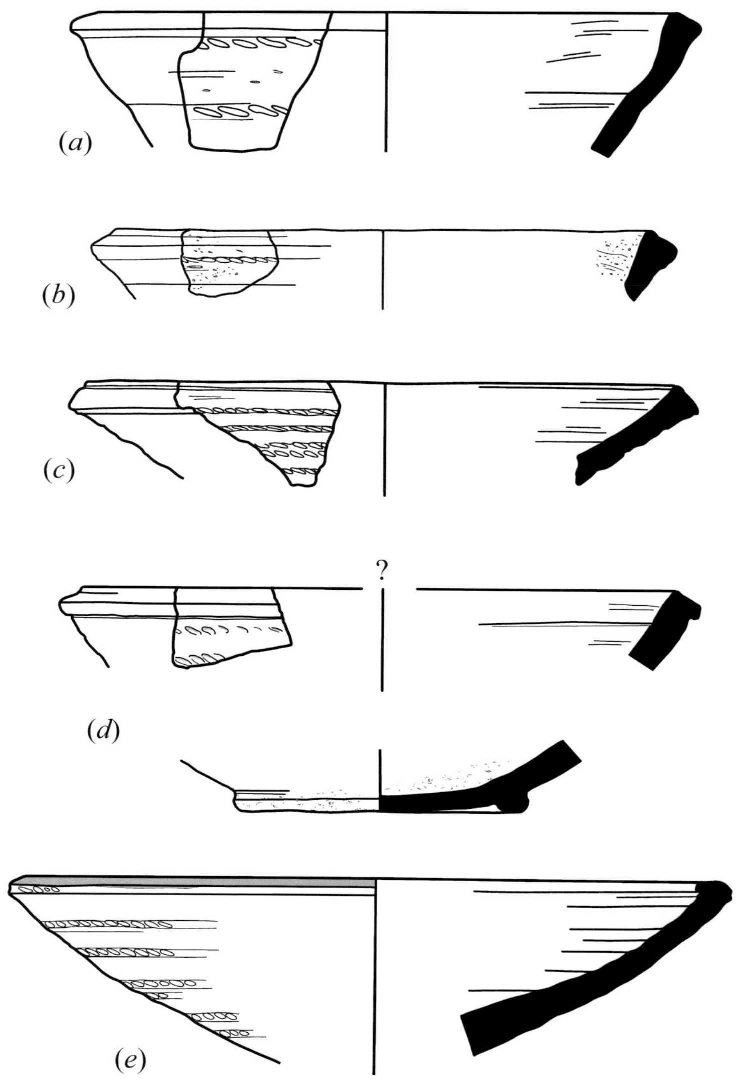

Figure 5.12. Large silt bowls (2:5)

(*a*) Sealed under D 4; shallower(?)
(*b*) D 302 (1), also sherd from D 303 (1); shallower(?)
(*c*) D 200–D 202, surface and (1); same as D 4, D 47, and D 48 sherds
(*d*) D 44; RS interior/RR exterior, similar rim from D 10 (1)
(*e*) D 104 (1); RS splashes on interior, RR and red splashes on exterior

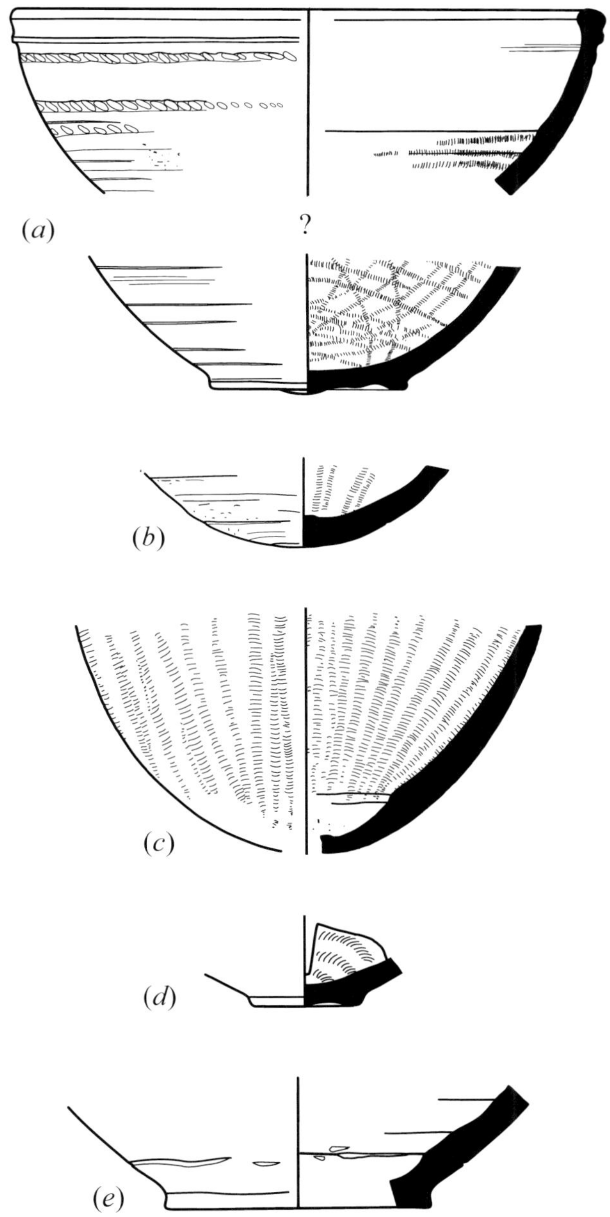

Figure 5.13. Large silt bowl and bases (*a*, 1:5; *b–e*, 2:5)

(*a*) Both sherds D 68 (2); thin RS exterior, streak burnish on interior

(*b*) D 111; RS traces interior/exterior, burnish on interior, traces on exterior, edges very worn, reused as dish(?)

(*c*) D 112 and area; stance and diameter uncertain, RS exterior(?), handmade(?)

(*d*) D 9 B (1); RS interior/exterior, 10R 5/8, 7.5YR 6/4 fabric, concentric burnishing on interior

(*e*) D 112 and area; vertical burnishing strokes on exterior, RS interior

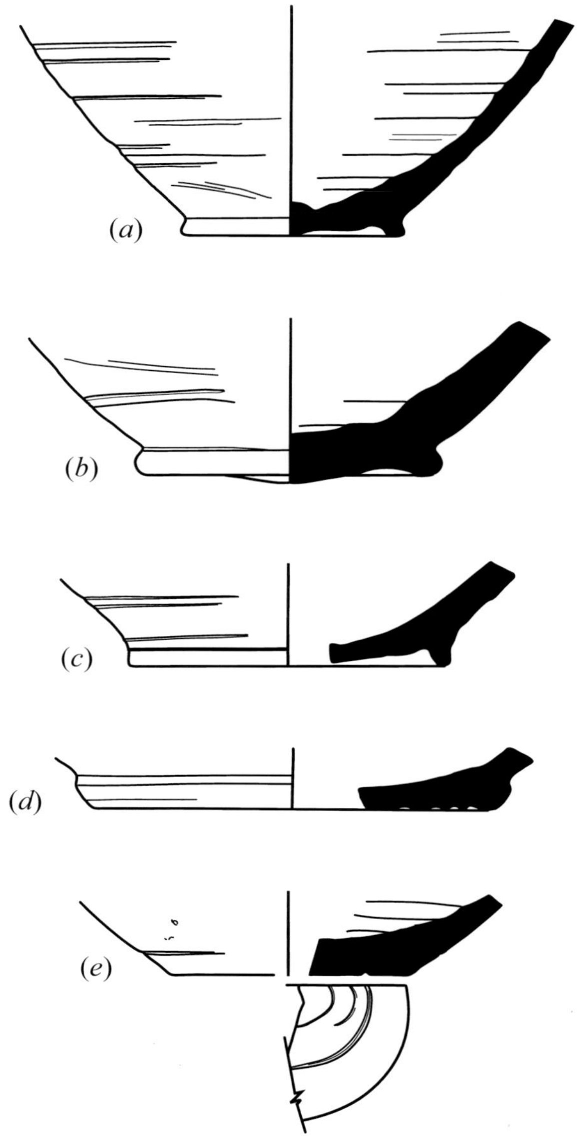

Figure 5.14. Silt bowl and bases (2:5)

(*a*) D 17 (1); note bowl from D 17, under inner relining
(*b*) D 73 (1)
(*c*) D 112 and area
(*d*) D 215 (1); string-cut
(*e*) D 10 (1); string-cut

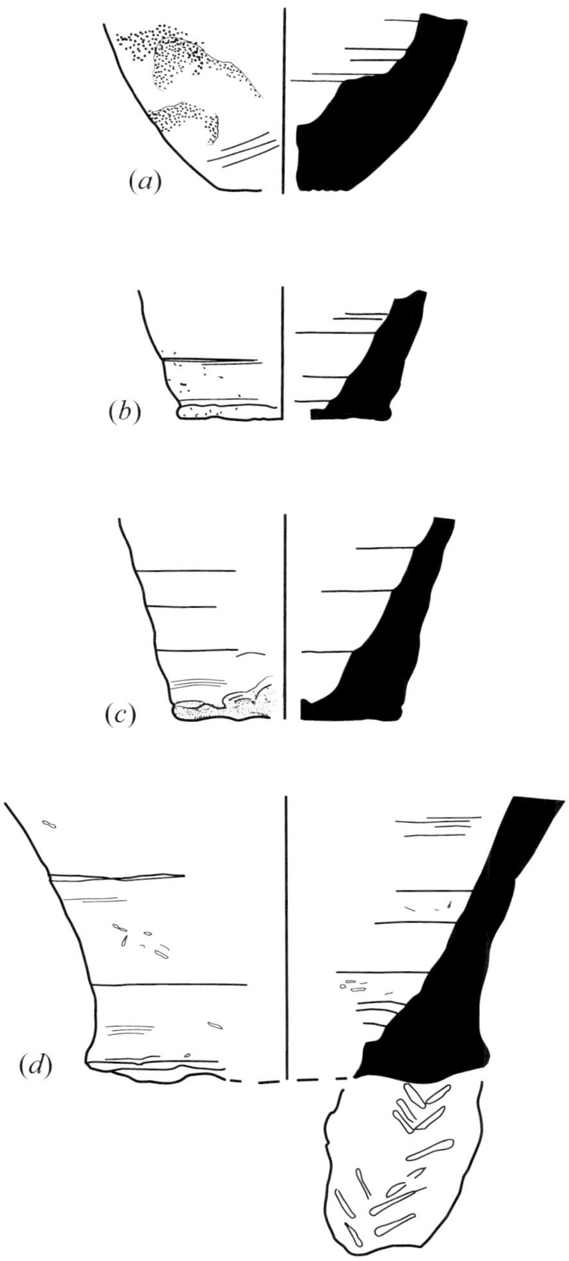

Figure 5.15. Flat silt cup or beaker bases (2:5)

(*a*) From top of north wall at northwest buttress; string-cut base, finger-tip(?) impressions on exterior

(*b*) D 3 (1); pierced base

(*c*) D 200 (1) and D 202 (1); pierced base, finger-tip impressions on exterior, another example from D 215 (1) with pierced base

(*d*) D 23, west of and from lower part of rubble; mat-impressed on base exterior

Carinated Marl Bowl

A sherd from a carinated bowl shoulder parallels an example from the pottery magazines in the tomb chapel of Amenirdis at Thebes and is shown in figure 5.16c.[47] The Dorginarti sherd measures about 18 cm at the carination, although the stance and diameter are uncertain. Other parallels stem from Egypt, including Karnak North, where they were found in the Twenty-Sixth through Thirtieth Dynasty levels.[48]

Marl Bowls

These bowls were initially categorized by rim type, but the rim profiles are closely related from one variant to another and this division proved unuseful as a chronological indicator. They are grouped together here. A type with the rim flap thickened or folded over onto the interior is very common (figs. 5.16d–e, 5.17a). An incised line below the rim on the exterior matches the finishing of the rim on the interior. A second distinct type has the rim clearly folded over toward the exterior, with its profile enhanced by the channel below the rim edge and a more strongly carinated shoulder (figs. 5.17b–c, 5.18a). The channel in two cases shows chattering from the finishing touches of the leather-hard vessel on the wheel. The bowl in figure 5.18b could be either one of the types just described, since the manufacturing lines are indistinct. The third rim type is less common. Its rim is thinner and slightly everted, but it also has a channel below the rim on the exterior, thus rendering a slight shoulder carination (fig. 5.18c). These rims are associated with marl ring bases, as shown in some of those illustrated, and often with the central part of the base hanging below the edge of the base rim. No other types of bases are found in association with these rim sherds, as mentioned previously.

Marl bowls largely replaced the silt food bowls at Karnak North during the Twenty-Fifth Dynasty.[49] But at Dorginarti, the marl bowls were found in the same contexts as the silt bowls discussed above. The marl bowls generally have rim diameters of 20–26 cm and would not supplant the smaller silt bowls for eating or drinking.

Similar marl bowls are found at Ashmunein, where they are present in all levels, but particularly in the upper strata Levels 1–2. The closest rim parallels to the Dorginarti types are associated with round to pointed bases at Ashmunein.[50] Amarna and Abu ʿId also yielded similar marl bowls, although they do not have the articulated rims found on most of the Dorginarti examples.[51] Parallels are also found at the Seti I temple at Qurna, the temple of Mut at Karnak, and in Phase IIb–III materials at Elephantine.[52] The marl bowls appear first in the materials from Heracleopolis Magna in stratum 8, attributed to circa 760–740 BC, and last until the end of the seventh century BC.[53] At Heracleopolis Magna, they are also found in contexts together with silt bowls.

Marl Handled Bowls and Related Vessels

Marl handled bowl fragments are common. They were uncovered in every area of the fortress (figs. 5.18d, 5.19; pl. 66a). A complete profile of the form can be reconstructed, including a low ring base, fragments of which are often found in association with these rim sherds (fig. 5.19b), although the bases also resemble those of the marl bowls described above. It is easy to spot the rim among the small sherds because of its distinctive rounded lip and the everted rim stance atop a rounded shoulder. The presence of a second handle remains conjectural since there were no sherds with evidence of multiple handles, although other marl handles were sometimes present in the same sherd bags. The rim diameters are most often 20–22 cm.

47 Hölscher 1954, p. 74, pl. 47, type W6. It was described as "finer red" fabric. It was found together with W4–W5 in the same deposit, and these bowls are like the MB 1 examples described below.

48 Jacquet-Gordon 2012, p. 253, fig. 99a–e. These examples are also are associated with many marl bowl types found at Dorginarti.

49 Jacquet-Gordon 2012, p. 254, fig. 99j–l, versus the silt bowls similar to Dorginarti's examples in the Twenty-First through Twenty-Fifth Dynasty layers in fig. 92 (pp. 236–39).

50 Spencer 1993, pp. 44–45, pl. 51, types A4.14, 17; pl. 54, types A4.40, 44; and pl. 55, types A4.54.2–54.3. Most common at the site were A.4.13–21, less frequently A.4.40–54. Some of these bowls measured as much as 43 cm in diameter.

51 French 1986, p. 184, fig. 9.18, MB 2s; and Aston 1996c, p. 30, pl. IV, nos. 56–58.

52 Myśliwiec 1987, pp. 69–70, nos. 600, 620. For the Mut Temple parallels, see Sullivan 2013, p. 118 (for their probable use) and pp. 175–76, 207–8, type G. Most of these are marls and are found in either Stratum 1 or 2, thus attributed to a period from the Twenty-Fifth Dynasty into the Twenty-Sixth Dynasty. Elephantine's Phase III examples are shown in Aston 1999, pp. 181–83, nos. 1672, 1674–78. There are a number of parallels with Dorginarti's marl bowls in Phase IIb as well; for example, see Aston 1999, pp. 134–35, no. 1234.

53 Aston 2010, pp. 80–82, pl. 27, nos. 515, 531 (Stratum 7–8, ca. 760–740 BC), pl. 28, nos. 560–62 (Stratum 6, ca. 740–720 BC), pl. 29, nos. 575–76 (ca. 720–700 BC), and down into Stratum 1 of the late seventh century.

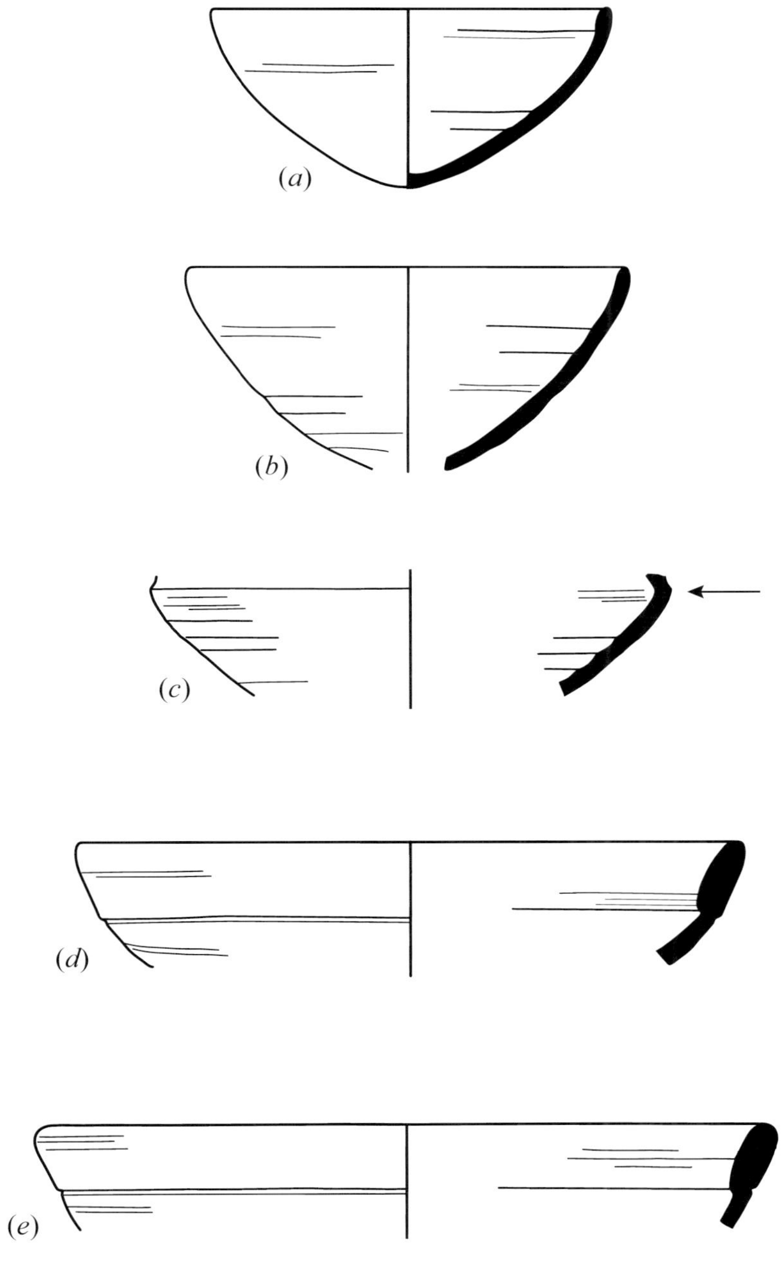

Figure 5.16. Marl cups and bowls (2:5)

(*a*) D 69 (1), examples from D 80 (1) and D 112 and area
(*b*) D 92 (1); marl
(*c*) D 200–D 202 (1); stance uncertain, 17–18 cm diameter at carination
(*d*) D 66 (1)
(*e*) D 60 (1)

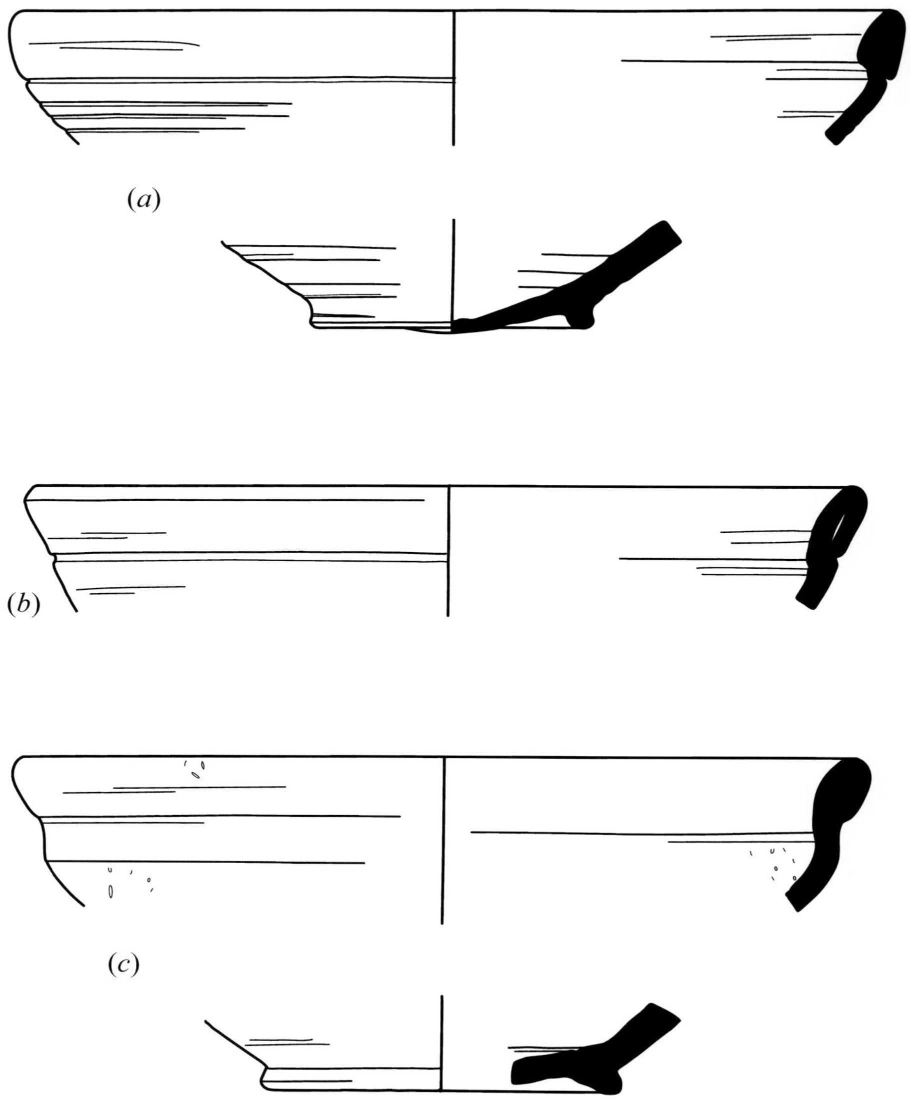

Figure 5.17. Marl bowls (2:5)

(*a*) D 12 (1); rim and base

(*b*) D 104 (1)

(*c*) D 45; rim and base, chaff temper

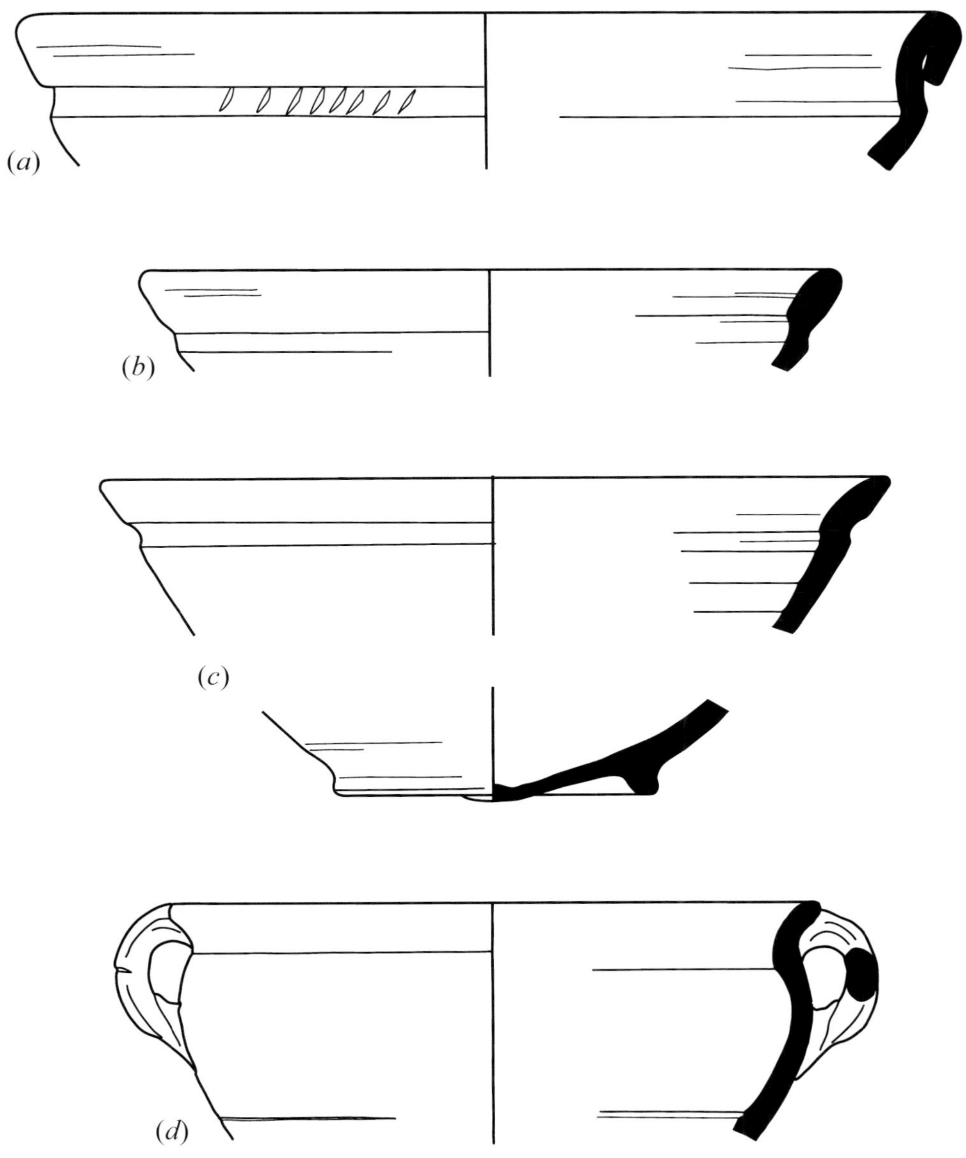

Figure 5.18. Marl and handled bowls (2:5)

(*a*) D 117–D 120, fill to north of rooms

(*b*) D 111 (1)

(*c*) D 104 (1); rim and base

(*d*) D 102 (1); second handle reconstructed

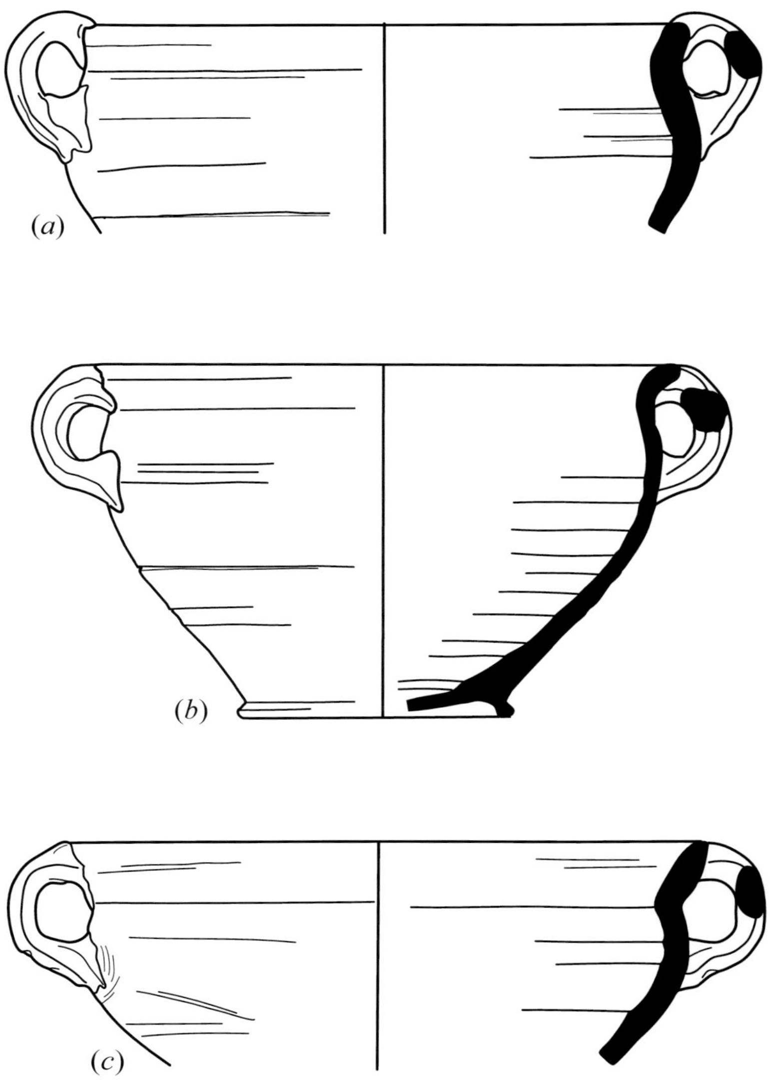

Figure 5.19. Marl handled bowls (2:5)

(*a*) D 35, surface to “ash” layer

(*b*) D 215 (1); OIM E49517, full profile, reconstructed

(*c*) D 200 (2); second handle reconstructed

Not all the marl bowls shown in figure 5.20a–d need belong to this handled variety, and some may belong to a slightly later period,[54] but their profiles were very close to this class of vessel.

These handled bowls are smaller and less well made than the handled krater/bowl from the Level II contexts at the fortress, which is mentioned below in chapter 6 (fig. 6.12c; pl. 86a). Similar shapes from Egypt are generally attributed to dates later than what is assumed for Level III or Level IV. Precise parallels in form, handles, and fabric come from the Ptolemaic strata at Karnak North, although a slightly different silt example stems from the Twenty-Sixth through Thirtieth Dynasty levels.[55] Elephantine has handled bowls in Phase V, although they are dissimilar to the Dorginarti examples.[56] Other similar vessels stem from Tell el-Maskhuta, where the form is a very good parallel, as well as from the Seti I mortuary temple at Qurna.[57] A similar bowl profile, with no evidence for handles, was also found at Karnak South and attributed to a late Third Intermediate Period date.[58]

There are two parallels from Nubia as well, including a marl example from the early levels at Meroe, and another from the reuse of a New Kingdom tomb at Semna.[59]

Similar handled bowls were common in the Iron Age Levant, where they are most often called *kraters*, and these forms were undoubtedly an inspiration for Egyptian potters.[60]

Marl Bowl Bases

In figure 5.21, a number of marl ring bases are shown. As mentioned above, no round or flat bases are associated with the marl bowls. A unique base for a goblet is shown in figure 5.21e. It was made from a coarser fabric that left voids from burned-out limestone, giving the sherd a spongy appearance. The context of D 41 is adjacent to the eroded area at the north wall breach, and the dating remains uncertain. No parallel has been found.

SILT JARS

The silt jar forms from Dorginarti display a great diversity of forms and sizes, which is true of most of the silt forms from the site. The three most distinctive, and common, jar types are SJ 1, 2, and 3, which are discussed below. Wide-mouthed jars were also often found at the site, but they are more difficult to date. The jars are most often made of Nile B2, but the occasional presence of more organic or sand temper indicates coarse Nile B2 or Nile C fabrics. They may have uncoated exteriors or thin, bright-red slips, and very occasionally the exterior shows traces of a thin white coating. Only a few sherds from two eroded contexts show certain evidence of painted white stripes on the exterior surfaces (pl. 65d). Some traces of streak burnishing are apparent on silt jar body sherds, but such traces are uncommon and none of the rim forms illustrated here shows evidence of burnishing. There are a variety of rim types and diameters for the commonly found wide-mouthed jars with rounded and thickened rims, and these are also discussed below. The silt jars may have been transported from distant locales, but it is more likely that they were used for the transport (and storage) of grain or other food supplies from distribution points within or outside the fortress and were manufactured in Nubia.

---

54 For example, the bowls shown in fig. 5.20c–d were in higher strata in the northwest corner of the West Sector (i.e., under the topmost silo D 4 and in the fill of house D 47, an intermediate architectural phase).

55 Jacquet-Gordon 2012, p. 295, fig. 121m–n (of marl). The earlier deposits had silt examples, with a ring base on the example illustrated, but the shape does not compare very well; see also Jacquet-Gordon 2012, p. 270, fig. 107s (red-slipped with a white rim).

56 Aston 1999, pp. 224–25, no. 1991, pp. 228–30, nos. 2024, 2029–31 (silt and marl). These were found in contexts dated to the late Saite and Persian periods.

57 Holladay 1982, pp. 114–15, pl. 19, no. 4 (silt), materials from the blocked-up Persian well, ca. 486 BC. The materials from Qurna are in Petrie 1909, p. 15, pl. XLIX, nos. 768–69 (marls). More recent materials from the Seti I Qurna temple are shown in Myśliwiec 1987, p. 71, nos. 640–51. Both Petrie's and Myśliwiec's materials are generally placed within the seventh to sixth centuries BC; see Aston and Jeffreys 2007, p. 58.

58 Sullivan 2013, pp. 176, 207, types G-3 and G-4 (with ring base). The bowls are from Stratum 2, which is attributed to the late Third Intermediate Period and Twenty-Fifth Dynasty.

59 Shinnie and Bradley 1980, p. 102, fig. 32, no. 53 (J 50 [16]). The marl example from Semna is shown in Dunham and Janssen 1960, p. 79, fig. 36 (Tomb S.515 24-2-537, of marl). A temple of Taharqo was built within the earlier fort. Objects from the area around and perhaps under this temple, and in the reused tombs, include pottery and items of Napatan date. Streak-burnished silt bowls, like the SB 1b form discussed in this chapter, and wide-mouthed marl jars like those at Dorginarti, were also found within the reused tombs and within the fortress confines around and under the temple.

60 Amiran 1969, pp. 216–26. None of the rim forms is similar to these small handled bowls found at Dorginarti. For the northern Levant, see also Lehmann 1996, pl. 11 (one- or two-handled). These are dated to his Assemblages 1–2, both within the eighth century BC, particularly the second half of that century.

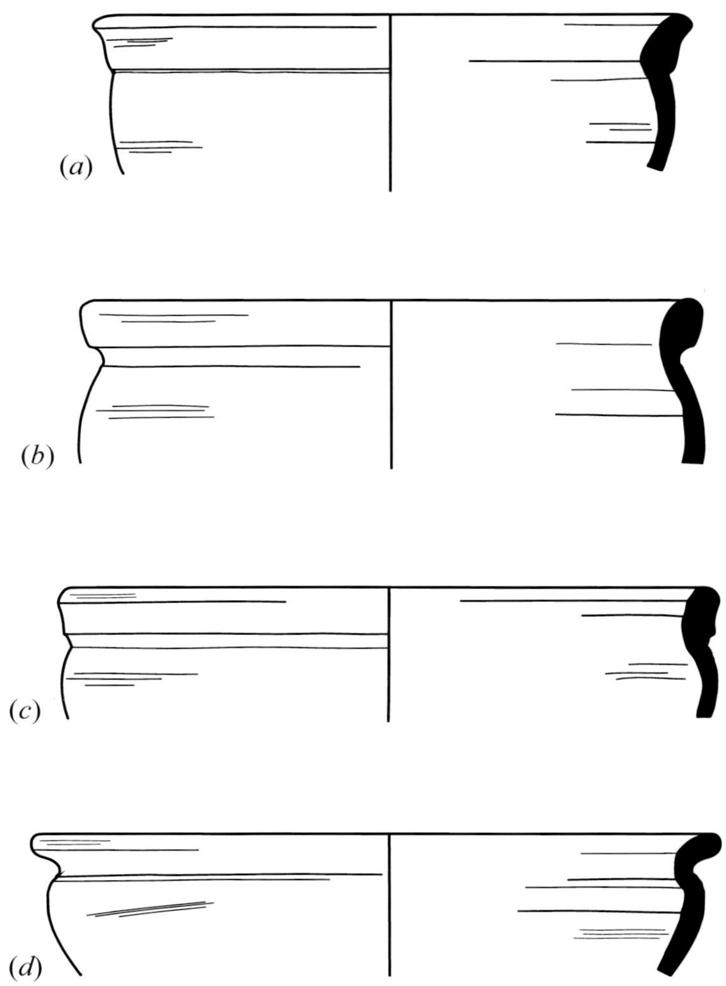

Figure 5.20. Marl (handled?) bowl variations (2:5)

(*a*) D 10 (1)

(*b*) D 208 (1), area of ovens

(*c*) Under silo D 4

(*d*) D 47–D 50, below surface

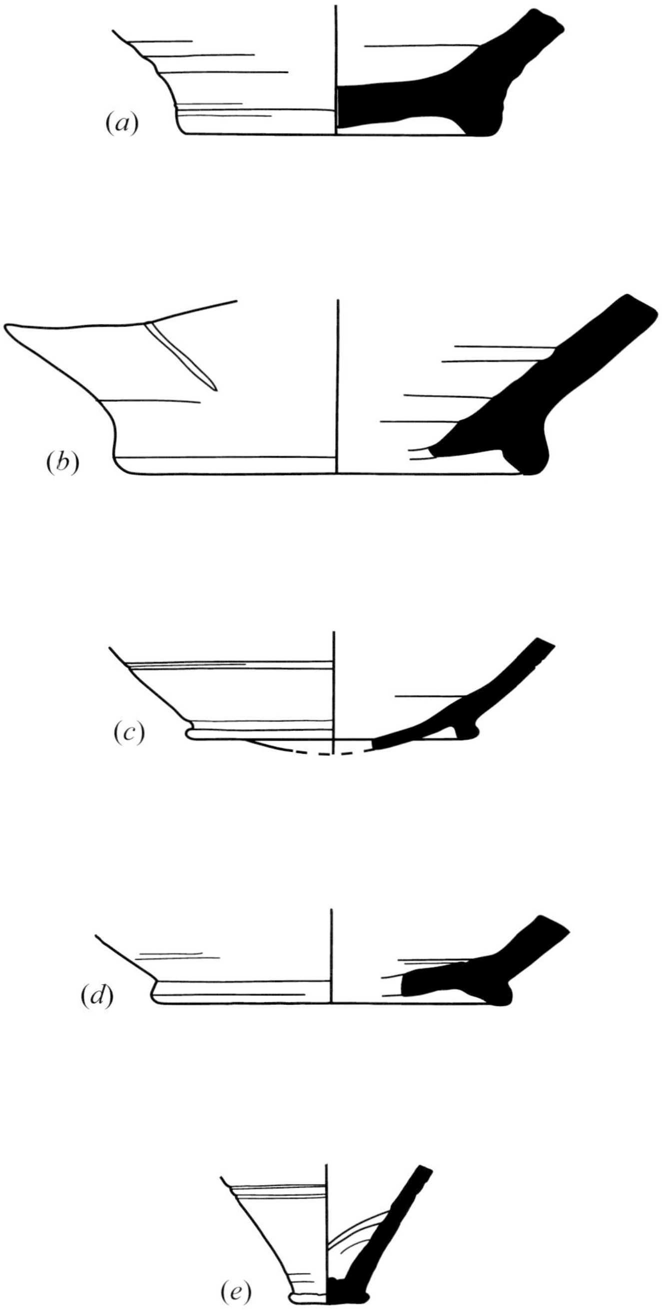

Figure 5.21. Marl bowl and cup bases (2:5)

(*a*) D 4, clearing out silo

(*b*) D 36, surface clearance; incised line intentional(?)

(*c*) D 113 (1)

(*d*) D 45; associated with MB 1 rim, chaff temper

(*e*) D 41; abundant voids in section from burned-out limestone

Silt Jars 1–2

The silt jars presented in this section closely resemble the seventh- through sixth-century jars found in both Egypt and Nubia and are here considered precursors of these later types (figs. 5.22–5.23, 5.24a).[61] The shoulder of the most complete SJ 1 example slopes down from a well-defined neck, and the two bases found in the context with the jar are slightly ovoid—that is, somewhere between rounded and pointed (fig. 5.22a–c). It may be assumed that the shape differs from SJ 2 only in height and body diameter, although the complete body and base shapes of SJ 2 are uncertain. The two types are discussed together here since no chronological distinctions are apparent.

SJ 1 has a taller neck than SJ 2 (about 6 cm versus up to 3.5 cm in height), but they are considered together because the neck height of a small rim fragment could not always be determined. The rim diameters range between 10 and 13 cm for both types. They are often found together in the same contexts. There is no certain evidence of handles, and it seems likely they did not have them, at least to judge from the evidence in the museum collection. The sherds either are uncoated or show a thin, bright-red or red-brown slip on the exterior (pl. 65b). A handful of sherds show traces of a thin cream coating, which may be a degraded red slip or paint (fig. 5.23a).

The closest published Nubian parallel to SJ 1 was found at Kawa during recent excavations in the area of the stone temenos gate.[62] It has been suggested that the gate might be contemporary with the Taharqo temple at the site, but this is uncertain. The earlier debris on the initial pavement of the gate contained a taller-necked jar like SJ 1, although the dating of the type at the site is uncertain.[63]

The Sanam treasury foundation deposits of Taharqo contained a smaller silt jar that is reminiscent of the SJ 1–2 forms from Dorginarti, although the shoulder is high and rounded, and the jar is shorter.[64] It is, of course, later than the Dorginarti types. However, the form reflects, in its subsequent development, the jars

61 For examples from Nubia, see Griffith 1922, pp. 91–92, pl. 21, no. 1 (wheel-made, red-slipped). See also Griffith 1923, p. 95, pl. XVII, types I and II. Clearer illustrations of the type are found in Egypt, as shown in French 2007, pp. 111–12, no. 30, and fig. 17, no. 2.

62 Welsby 2011, p. 57. There is no clear proof for the contemporaneity of the Taharqo temple and the stone temenos gate.

63 The pottery is presented in Sjöström and Thomas 2011, fig. 3.3.

64 Griffith 1922, pp. 91–92, pl. XXI, no. 1 (wheel-made, red-slipped).

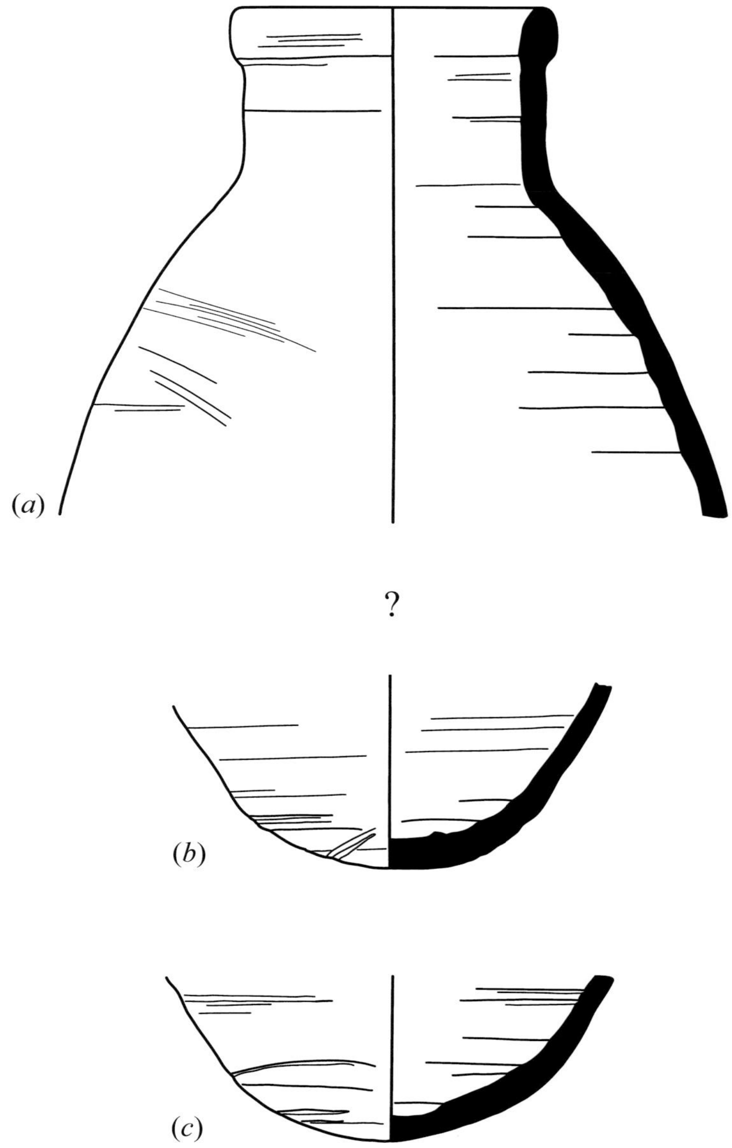

Figure 5.22. Silt jars 1 (2:5)

(*a*) D 6 (1); OIM E49504

(*b*) D 6 (1)

(*c*) D 6 (1)

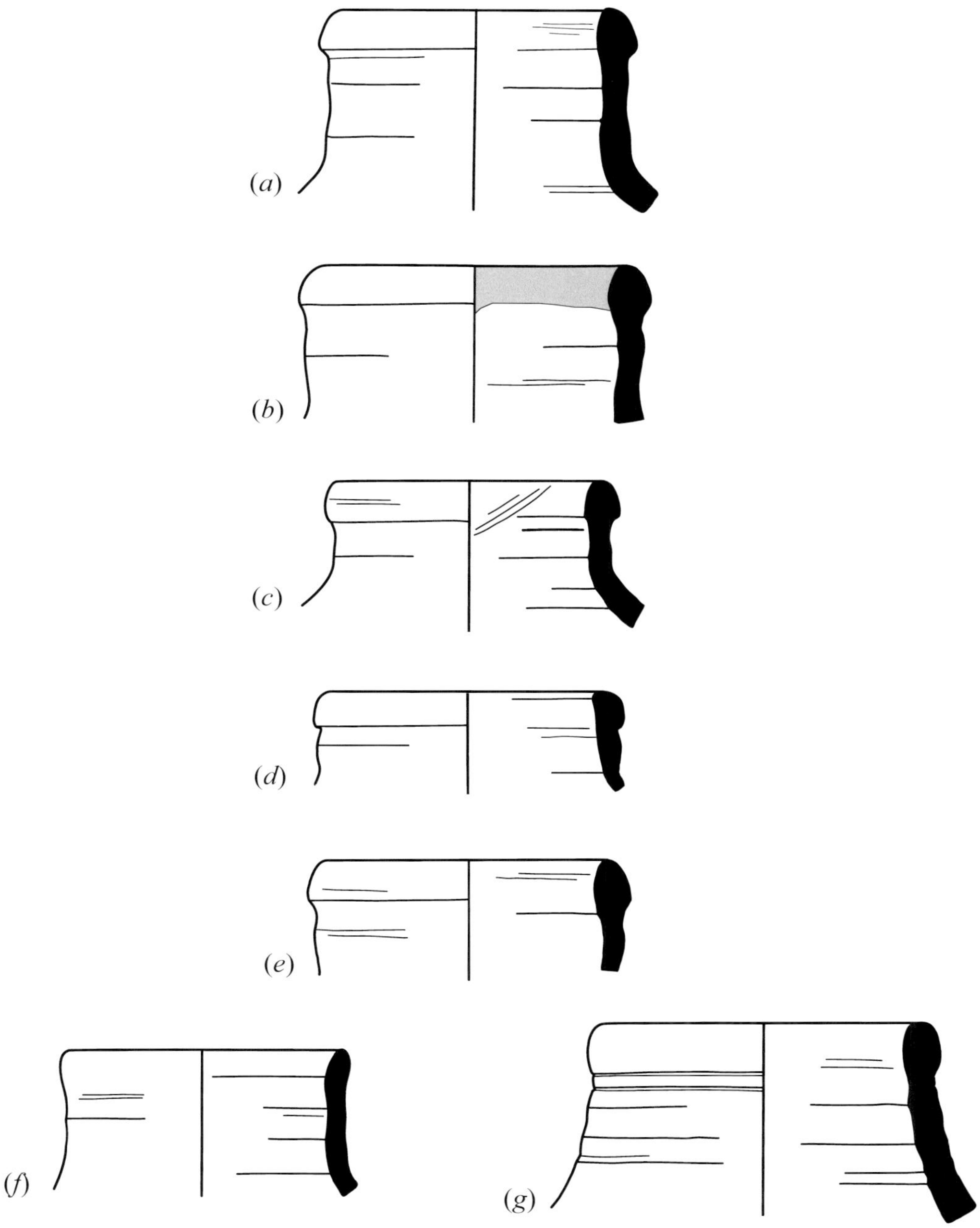

Figure 5.23. Silt jars 1 and 2 (2:5)

(*a*) D 78 (1); thin RS exterior, traces of cream paint(?)

(*b*) D 12 (1); RS traces exterior/RR interior

(*c*) D 36, east end; no trace of slip

(*d*) West Sector rubble against western retaining wall for Level II

(*e*) D 48 (2); RS exterior, abundant sand temper, SJ 1 or SJ 2

(*f*) D 58 (1); variant, traces of slip on interior rim

(*g*) D 14 (1)

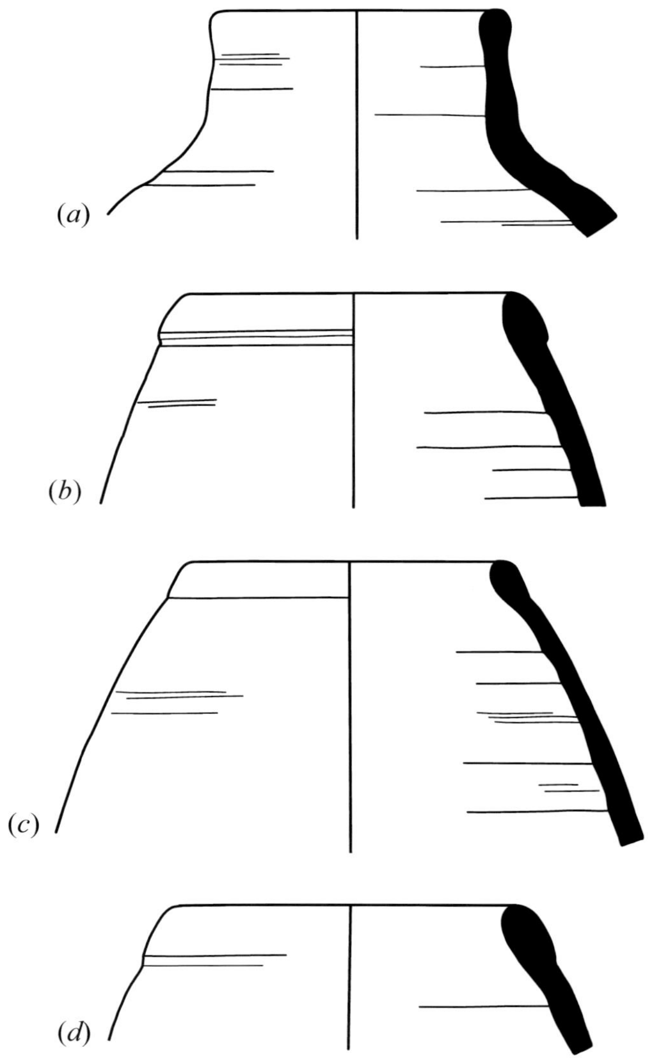

Figure 5.24. Silt jar 1 variant and silt jars 3 (2:5)

(*a*) D 208 (2), south of ovens; RS exterior, 2.5YR 5/8

(*b*) D 309, pit A

(*c*) D 13 (1)

(*d*) D 113 and D 114, surface

from the sixth century, those with longer necks having incised or raised lines and slender bodies.[65]

Close parallels to SJ 1 are found at various excavations around Karnak, where similar rims and necks with thin red slips were uncovered.[66] Karnak North yielded examples from both the Twenty-First Dynasty layer and the Twenty-Sixth to Thirtieth Dynasty layers.[67] The embalmers' deposits on the west bank at Thebes contained a similar jar with a higher, rounded shoulder, as was also the case with an example from the tombs at Dra' Abu el-Naga.[68] The Late Dynastic pottery from the South Tombs at Amarna includes parallels to both the SJ 1 and SJ 2 rims,[69] which David Aston places in his Phase I.[70] The type is also found at the fort of Abu 'Id, although the parallels there are closest to the shorter-necked SJ 2.[71]

There are also a few parallels from the north of Egypt, including from Tell el-Retaba.[72] The vessels from Tell el-Retaba, with the shorter necks like SJ 2, were used for cooking and were covered with soot. There is no clear evidence for their use as cooking vessels at Dorginarti, however, apart from the occasional burned sherd. But it may be that their fragmentary condition or the frequent flooding of the site either obscured or removed such evidence. Similar rims and necks from Tanis are dated to the early to mid-eighth century (Sheshonq V), but there are others attributed to the end of the Third Intermediate Period and Saite era.[73]

Silt Jar 3

The precise body shape and base of this silt jar form is unknown, although the rim sherds in the collection are fairly common (figs. 5.24b–d, 5.25a). The specimens from other sites, dated slightly later than the assemblage at Dorginarti, show a long vessel with a constricted waist, rounded shoulders, and a vertically placed rim.[74] It is assumed that the bases of the form found at the fortress were rounded, although definite evidence is lacking. There are thinner- and thicker-walled variants at the site, but the rim is standardized. The interior and exterior walls display wheel bumps. The jar sometimes has a thin red coating on the exterior, which may run over onto the rim interior, although many sherds show no traces of slip.

Parallels are found in Third Intermediate Period contexts from Egypt—for example, among the Phase IIb material from Elephantine.[75] There is a precise parallel from Tell el-Retaba in Third Intermediate Period

65 See, for example, the discussion and comparisons in Defernez 2001, pp. 116–29, pls. XVIII–XIX. Sanam cemetery's type I and II silt jars are similar in form; see Griffith 1923, p. 95, pl. XVII. See also Lohwasser 2012, pp. 208–9, fig. 61. Note particularly type A2, or Griffith's type Ib, closer to the Dorginarti SJ 1 and SJ 2, and compare them with type AI, with impressed lines around the neck.

66 Defernez and Boulet 2013, p. 6 and fig. 2, ON/P 2081/5173. This parallel is dated to the late Twenty-Fifth or early Twenty-Sixth Dynasty. The Mut Temple excavations produced shorter-necked forms typically coated with white; see Sullivan 2013, pp. 185–86, 223, types 8-3 through 8-6. They were found in Strata 2–4 and date between the late New Kingdom and Third Intermediate Period—that is, before the Twenty-Fifth Dynasty. See also Myśliwiec 1987, at Qurna, no. 834.

67 Jacquet-Gordon 2012, p. 228, fig. 87d. This is a marl vessel with a wider body and carinated shoulders. This pottery group was found under a level dated by bricks stamped with the name of Pinedjem I. The same form is found in the Twenty-First through Twenty-Fourth Dynasty assemblage (p. 246, fig. 96, red slip with white coating was on the upper part). The better parallel for the form is dated to the Twenty-Sixth through Thirtieth Dynasty layers (p. 277, fig. 111, P.497), which however has a red slip and band of white wash on the shoulder and body.

68 Budka 2010b, p. 35, fig. 6, top left. However, the body form is different on this example, which is dated to the Saite and Persian periods. For the Dra' Abu el-Naga example, see Seiler 2003, pp. 366–67, fig. 20, no. 2, where she cites additional parallels. The shoulders are also higher and more rounded than the sloping line of the most complete Dorginarti jar, with the morphological differences attributed to the later chronological development of the Theban jar. The Dra' Abu el-Naga group is dated to the latter part of the seventh century, with a possible extension into the sixth century BC.

69 French 1986, pp. 156, 174, fig. 9.8. The closest parallel to Dorginarti's SJ 1 jar is French's SJ1.7.2, and the closest parallel to Dorginarti's SJ 2 is perhaps SJ1.3.1. French suggests they would have been round-based, and the larger jars may have had handles and stood about 20–30 cm tall. He notes that it was common for the jars in fig. 9.8 to have white wash on the neck and shoulders, and that the spiral-decorated sherds probably derived from these vessels, although practically no evidence was found for this surface treatment on this type of red-slipped jar at Dorginarti.

70 Aston 1996a, pp. 43, 64, figs. 112c, 197c. The exterior of the silt pot is white coated, with a red rim band on the interior and exterior surfaces. Phase I is divided into two phases that span ca. 1200–1100/950 BC.

71 Aston 1996c, p. 25, nos. 24–26 (bottles) and pl. II. These jars are made of Nile B2 and Nile D fabrics at Elephantine and are found in Phases IIa and IIb.

72 Wodzińska 2011b, p. 1034, fig. 10, nos. 1–3.

73 Defernez and Isnard 1998, pp. 166–67, 186–88, 208, pl. X. See also Laemmel 2011, p. 47, pl. 36, nos. 195–96 (unslipped silts).

74 See, for example, the more typical "sausage jar" shapes in Schreiber 2008, p. 73, pl. LXVII. He attributes them to the late eighth through early seventh centuries BC.

75 Aston 1999, p. 128, nn. 82–83; p. 129, no. 1161 may represent the shape of this type, but the evidence for the slightly tapered base found on the Elephantine example is not apparent in the sherd collection at the Oriental Institute Museum.

contexts, and some similarity exists with an example among the late dynastic pottery from the South Tombs at Amarna.[76] There are also parallels with silt jars from Kurru Tumuli 4 and 5, although these jars have slightly rounder shoulders and thinner rims unlike the beaded rim of the Dorginarti examples.[77]

Silt Jar 4

This rim form, with the inner knob at its lip, has many parallels during the late New Kingdom and Third Intermediate Period (fig. 5.25b–d). The rim sherds of this jar type are not very common in the collections from Dorginarti, and there are numerous variants. Most sherds of SJ 4 form have traces of red slip on their exterior surfaces and over onto the rim interior. Their rim diameters range from 10 to 14 cm.

There is, again, no complete body profile for the form, and attempts at reconstructing sherds from the bags in the museum collection were unsuccessful. Parallels from Egypt have been assigned dates as early as the eleventh and tenth centuries BC, but the rim types are also present in contexts attributed to the mid- to late eighth century BC.[78] The Egyptian form is also found in the Levant in similarly dated contexts.[79]

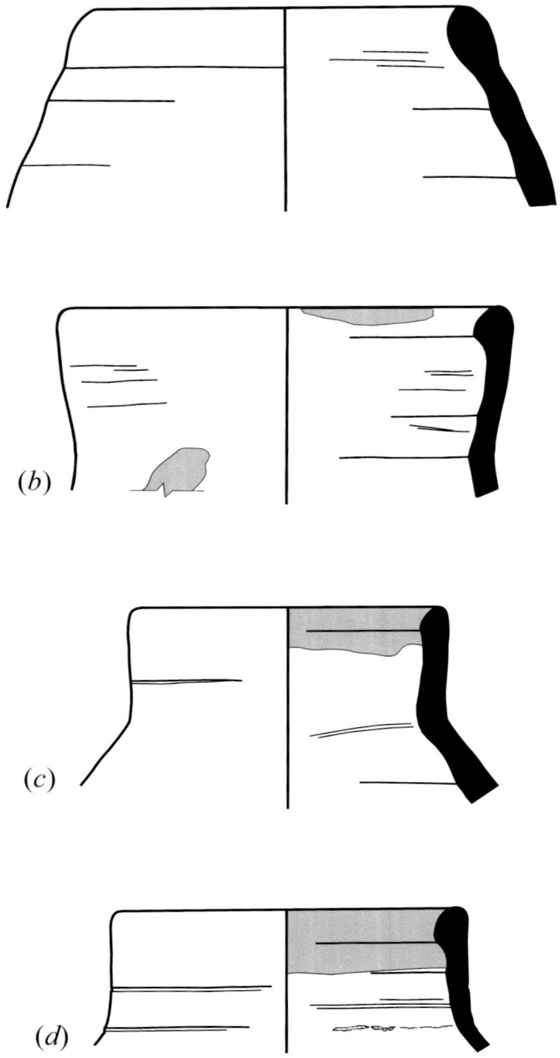

Figure 5.25. Silt jars 3 and 4 (2:5)

(*a*) D 43

(*b*) D 103 (1); very worn and crumbly, 5YR 6/4, light-red thin slip on exterior/RR interior

(*c*) Destruction of walls at D 4, D 47, D 48; RS exterior/RR interior

(*d*) D 75; RS exterior/RR interior

76 Wodzińska 2011b, pp. 1022, 1034, fig. 10, no. 4. It is a common type at the site in the Third Intermediate Period levels. For the example from the Amarna South Tombs, see French 1986, p. 180, fig. 9.14, SJ6.9.1. He dates the material to in or about the Twenty-Fifth Dynasty.

77 Dunham 1950, p. 18, fig. 3a (19-3-409), from Tumulus 4 of Generation A. The brown silt vessel was wheel-made and the fabric is porous, having many voids left by organic matter (personal observation). There is more than one of these jars with the tomb material. The silt example from Tumulus 5 (Generation A) is unpublished but also comes from the debris and siftings of the excavation (19-3-599).

78 For example, French 1986, p. 174, fig. 9.8, SJ1.1s; Jacquet-Gordon 2012, p. 87, fig. 87b, e, f (Twenty-First Dynasty) and p. 95, fig. 95l (Twenty-First through Twenty-Fifth Dynasty levels); Boulet 2015, p. 64, n. 13; and Aston 1996c, p. 24, pl. II, no. 20 (in uncoated Nile B2), pp. 26–27, pl. III, nos. 45, 49 (slipped with red or pink), and p. 32, pl. V, nos. 73–74 (in marl). The same rim and neck profile are found on cream-slipped jars attributed to the eleventh and tenth centuries at Memphis; see Aston and Jeffreys 2007, p. 52, figs. 45–46. See also the parallels in Laemmel 2011, pp. 19–20, pls. 11–12, nos. 62, 67–68. Laemmel notes that the form appears in the Twenty-First Dynasty, either slipped or unslipped, and continues after the Twenty-Second Dynasty—unslipped—until the end of the eighth or beginning of the seventh century BC.

79 For example, Ben-Dor Evian notes the presence of similar jars in Iron Age contexts in the Levant; see Ben-Dor Evian 2011, pp. 103–4, fig. 3. The traditional upper date for this Egyptian type is the eleventh to tenth centuries BC, and variations are found down through the eighth century.

Nile D and Related Silt Jars

There are numerous rim variations of Nile D and related silt jar forms at Dorginarti, but all of them are made from silt, sometimes with a limestone temper typical of a Nile D fabric (figs. 5.26–5.29). Their fabric types are noted in the figures. The technique of firing the fabric quite hard and coating the surface with a light-colored slip is the same for all the jars in this category. Sherds of this jar type are present in small numbers in the collection, although their variety has determined a high rate of illustration for this category. The surfaces are dark brown to reddish, while the break may be almost a dark purple; many of the sherds have a wide grayish-black core. All sherds show evidence of a thin slip that was sloppily applied to the exterior surface with a brush-like implement (pl. 65g–h).[80] The slip is cream, yellowish, or rose colored. Within each band of paint, striated lines are sometimes slightly impressed into the clay from the applicator brush.

Some sherds from D 14 (1) exhibit a thick coat of creamy yellow slip, while the other examples from the same context have a thin, pink slip brushed onto the exterior.

The differences in rims and neck forms gave rise to subdivisions. The first type is a smaller jar rim that mirrors the form current in the late New Kingdom and made from Marl D (fig. 5.26a–b). The rim is pillowed on the interior, while an exterior rim band is set off from the curved neck. The rim diameters are 10–14 cm. Another variant within this category is a wider-rimmed vessel, with the thick rim set above a pronounced bulged neck and sloping shoulders; the rims range between 20 and 24 cm (figs. 5.26c–d, 5.27a). A more complete profile of this type is shown in figure 5.28a. The last type of rim profile was a rarer find among the sherds in the Oriental Institute Museum collection (figs. 5.27b–c). Its angular rim stance and the straightness of its neck sets it apart, although the rim is still pillowed on the interior.

There are a number of loop handles and base fragments made from Nile D, while the related Nile B2 sherds with light-colored slip include one rounded base with a bulbous tip and a number of loop handles, one of which is illustrated (fig. 5.29a–c).

Most rim sherds from Dorginarti are like the small and large jar variants shown in figures 5.26–5.27, although there are several sherds of a wide-mouthed form and one narrower-rimmed jar resembling an Iron Age Levantine amphora (fig. 5.28b–c).[81]

The same forms were made of marl in Upper Egypt or of silt at sites in Lower Egypt, the latter of which are either coated or uncoated. The parallels are dated from the late New Kingdom and into the Third Intermediate Period, with parallels from Memphis, Tanis, and Qantir noted here.[82] The Marl D jars of the earlier Ramesside period, which resemble the smaller jar at Dorginarti, were increasingly made of mixed clay or silt fabrics during the Third Intermediate Period, most often with marl-based slips.[83] Lower Egyptian counterparts were also found at Tell el-Retaba.[84] Jars of this class are found as imports or local imitations in Iron Age IB/IIA contexts in Israel, where they are dated between 1140/1000 and 850 BC or between 1050 and 870 BC.[85]

Nubian parallels to the smaller-mouthed jars come from the tomb of Kurru 5 of Queen Qalhata—Piye's daughter, Shebitqo's wife, and the mother of

80 See the descriptions of the fabric and surface finish in Laemmel 2008, p. 182, where she notes this method of slip application on Twenty-First to early Twenty-Second Dynasty silt and mixed-clay jars. See also Aston and Jeffreys 2007, pp. 23–24. The surface treatment of the Egyptian jars from Dor are the same; see Waiman-Barak, Gilboa, and Goren 2014, p. 331.

81 This rim slightly resembles a narrower Levantine example published in Bourriau and French 2007, pp. 117, 130, fig. 1, no. 1 (ca. 750–675 BC). The fabric and rim profiles vary across this class of amphorae, although none was described as being made of an Egyptian fabric.

82 For Tanis, see Bavay 1998, pp. 323–26, fig. 34, nos. 42–44, fig. 35, nos. 48–50, and fig. 36, no. 66 (transition between the Ramesside period and Third Intermediate Period). See also, for Tanis, Laemmel 2011, pp. 13–16, pls. 6–7 (Twenty-First and Twenty-Second Dynasties). For Memphis, see Aston and Jeffreys 2007, p. 36, fig. 31, no. 253 (tenth to ninth centuries BC), p. 37, fig. 34, nos. 322–23 (eighth to seventh centuries), and p. 55, fig. 51, nos. 593–99, 601 (latter of Marl D) (mainly eleventh to tenth centuries). For Qantir, see Laemmel 2008, p. 177, pl. 3, no. 3 (early Twentieth Dynasty, mixed clay), and pp. 181–82.

83 Aston 2004, p. 200.

84 Wodzińska 2011a, p. 161, fig. 53, no. 7 (Third Intermediate Period).

85 See Waiman-Barak, Gilboa, and Goren 2014. A few are found in later mixed contexts at Dor. The relevant parallels are found on p. 319, pl. 1, nos. 9–10; p. 322, pl. 4, nos. 8–12; p. 323, pl. 5, no. 5, pl. 6, no. 7; p. 324, pl. 7, nos. 4–6; and p. 326, pl. 9, nos. 9–15. Egyptian parallels are listed on pp. 332–38, tables 1–9. The authors propose a chronological development of the surface finish, with thicker slip applications on the earlier examples, then unslipped vessels, and then thinner slips, applied irregularly, at the end of the development. This development stretches between ca. 1140/1100 and 850 BC; see pp. 316–18. See also Gilboa 2015, p. 253, with mention of similar jars from Ashkelon.

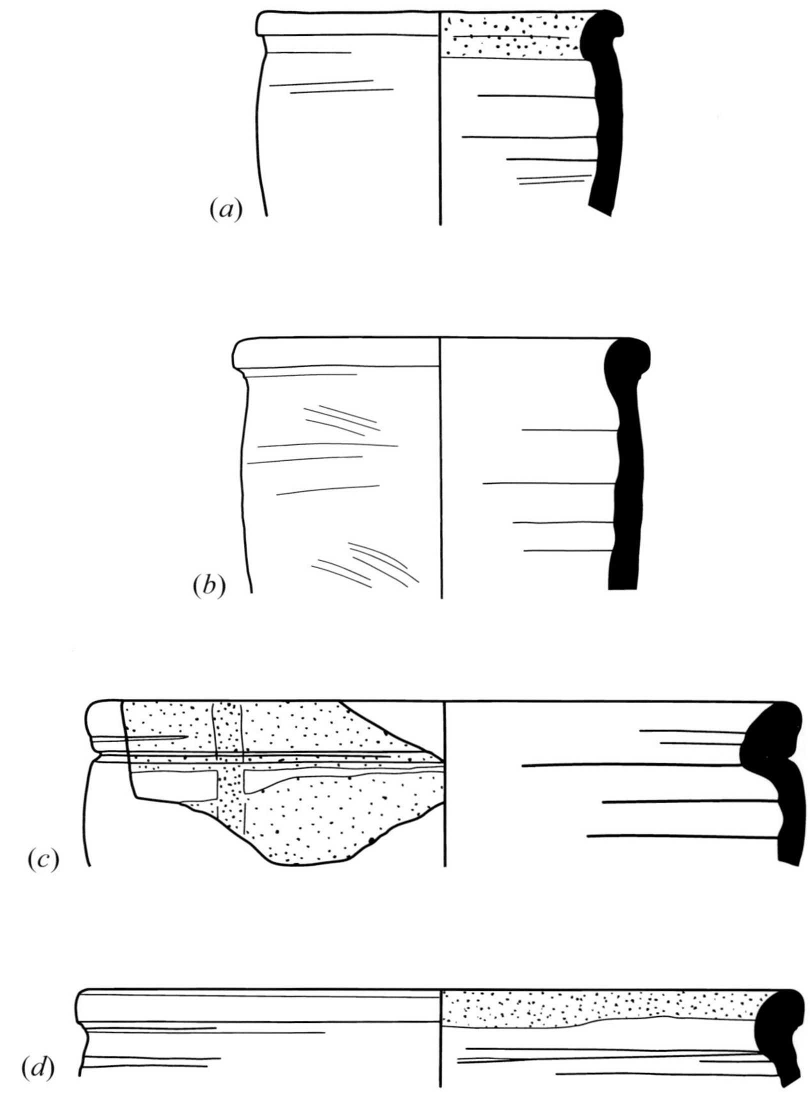

Figure 5.26. Nile D and related jars (2:5)

(*a*) D 116 and area; OIM E49706, reddish-yellow slip on exterior and on rim interior, Nile D

(*b*) D 206 (1); Nile B2, no slip

(*c*) D 1 (1); OIM E49567, Nile D fabric, cream-yellow slip on exterior and atop rim

(*d*) West of D 32 (1); Nile B2, whitish-cream slip on exterior and rim interior

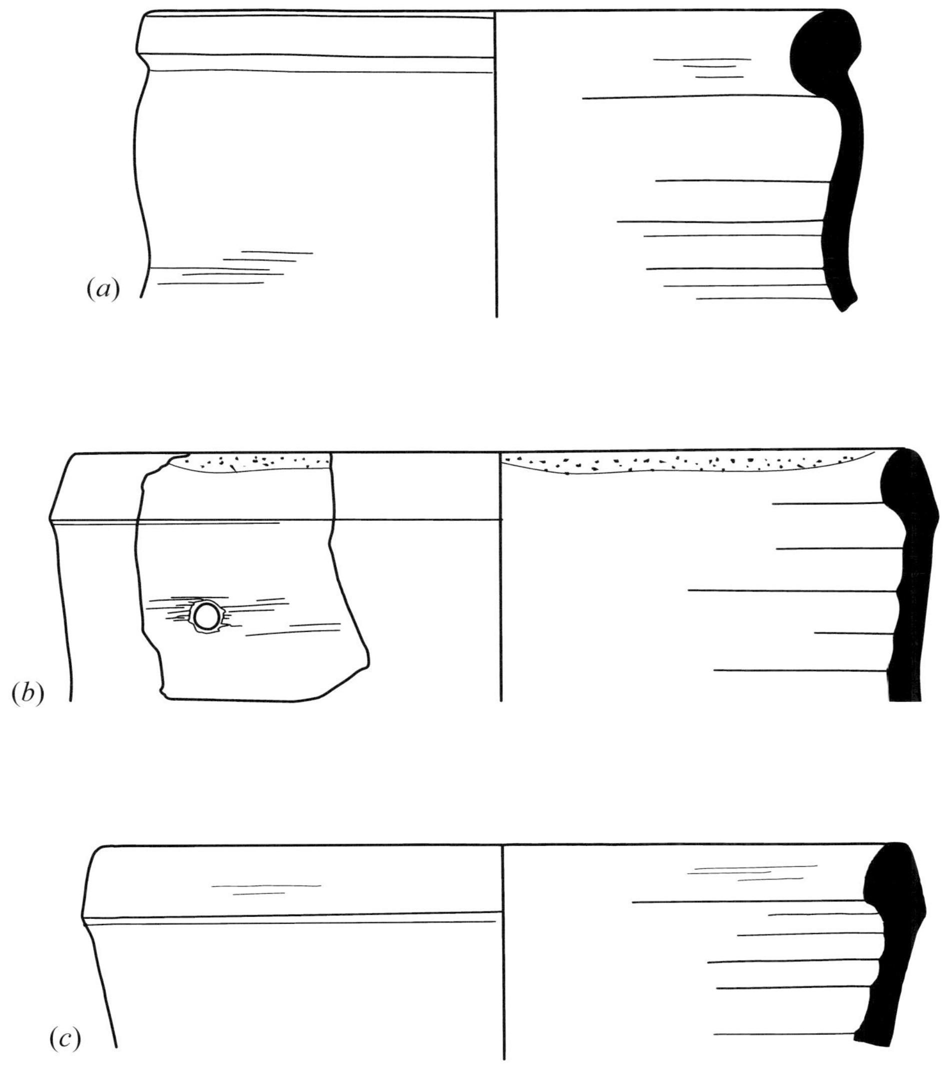

Figure 5.27. Nile D and related jars (2:5)

(*a*) D 35 and D 36 mixed, surface to below "ash" layer; Nile D, no slip present

(*b*) D 1 (1); OIM E49721, Nile D fabric, creamy-pink slip traces on rim interior/exterior

(*c*) D 4, D 47, D 48, dismantling of walls in northwest corner; Nile D, pink slip on exterior

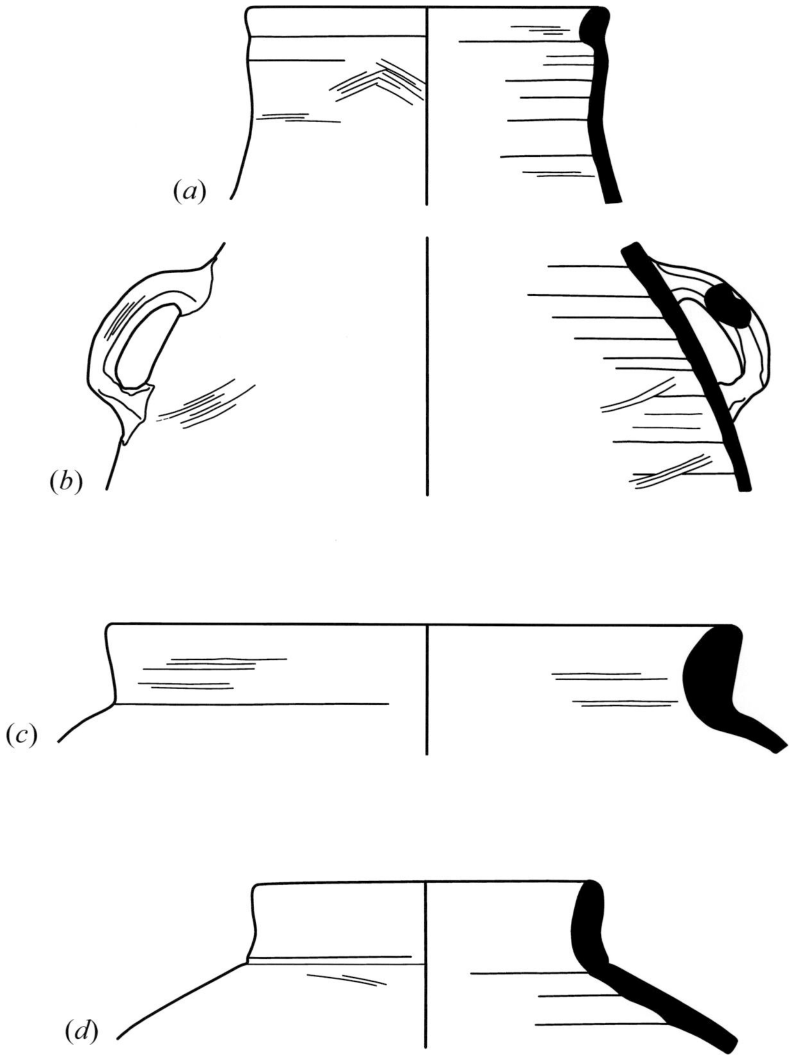

Figure 5.28. Nile D and related jars (*a–b*, 1:5; *c–d*, 2:5)

(*a*) D 14 (1); OIM E49716, Nile D, brush strokes on exterior, pinkish-white slip on exterior

(*b*) D 14 (1): OIM E50741, Nile D, pink coating on exterior

(*c*) D 200 (1) and D 202 (1); OIM E49719, Nile D, some chaff, cream slip swiped horizontally on exterior

(*d*) D 35, surface to "ash" layer; yellow-cream coating on exterior in sloppy white bands, Nile B2, Levantine SJ imitation(?)

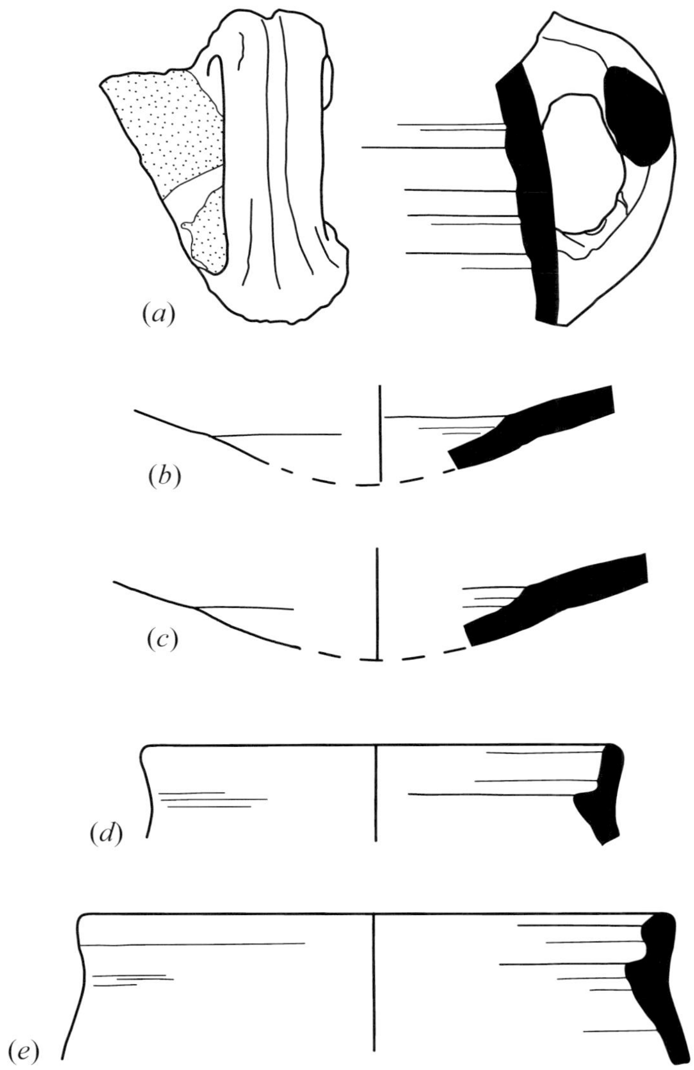

Figure 5.29. White-slipped silt and ledge-rimmed jars (2:5)

(*a*) D 46; Nile B2, thin cream-colored 10YR 7/3 slip on exterior

(*b*) D 46; Nile B2, yellowish-brown slip painted on exterior

(*c*) D 212 (1); Nile D, thin whitish slip, 13 cm maximum

(*d*) D 200–D 202, surface and (1); joins with sherd from D 48 (2)

(*e*) D 6 (1); white slip(?) on exterior

Tanwetamani.[86] Gala Abu Ahmed, located in the Wadi Howar, yielded examples among its Napatan materials.[87] The slightly similar jars in the late dynastic deposits around the South Tombs at Amarna are either uncoated or made with white-slipped silt, and at the fort of Abu ʿId these jar types are manufactured of marl.[88] The slipped Nile D fabric was used at Abu ʿId for the wide-mouthed form and for Dorginarti's SJ 4 form.[89] Parallels were also found at Karnak North in the Twenty-First through Twenty-Fifth Dynasty levels, and at the temple of Mut at Karnak.[90] They were found in the area around and below the chapel of Osiris Wennefer Neb-Djefau at Karnak.[91] At Elephantine the smaller amphora form was uncovered in contexts dating from the late New Kingdom down into Phases IIA–IIB and was made of varying fabrics and finishes, while the slipped Nile D or Nile B jars from Phase III are mostly not similar to the Dorginarti vessels.[92]

Silt Jar with Ledged Rim

This form, the sherds of which are fairly common in the pottery from the site, shows a pronounced ledge below the rim on the interior (fig. 5.29d–e). The jars are made of Nile B or Nile C, mostly with the visible addition of organic materials. Mica and sand were apparently added as tempering agents in some cases. Sherds of this type occasionally have red slips on the exterior, while one example may have a white slip (or deposit) on the exterior. Their rims range in diameter between 16 and 22 cm. The pronounced ledge on this jar type was presumably meant to hold a lid. The form has no parallels in the literature. A wide-rimmed variety of jars with much less distinct interior ledges was also found but has been classified with the wide-mouthed jars below.

Wide-Mouthed Silt Jars

Numerous rim profiles and sizes are included in this category. Almost all contexts in the West Sector yielded sherds of different vessels belonging to storage vessels made with either a Nile B2 or a Nile C fabric (figs. 5.30–5.32). The most common rim types are rounded and set more or less vertically atop either a vertically placed neck or a rounded shoulder and body. The rims can also be slightly everted. The jars with the widest diameters are described below, along with other items for kitchen use, since they were considered either vats or basins, and thus not easily transportable. Two seal impressions were found on the upper shoulder of a jar sherd found atop the wall relining south of D 113 and D 114 (OIM E24406; fig. 5.30g). The form was similar to the jar shown in figure 5.30f. The seal impressions show a mirrored image of clusters of lotuses.

The most complete examples are shown in figure 5.32. Two of the jars, represented by the example shown in figure 5.32b, came from the same locus, D 306 (2), a level below the floor of a room north of the Level IV residence. One of these jars, which is represented in the illustration, features a set(?) of loop handles placed

86 Dunham 1950, p. 40, fig. 12c, 19-3-777 (three examples), 785, 786 (both fragments of this jar type and not bowls). These sherds are all from the smaller silt jars, such as the few examples found at Dorginarti. Registration no. 19-3-777 includes about three such jars, all with a creamy-yellow slip painted onto the exterior surface, while the red-brown fired fabric had a black core; whether they are of Nile D or Nile B fabric is uncertain and needs further checking. The rim diameters of these examples range between 14 and 20 cm.

87 Eger, Helmbold-Doyé, and Karberg 2010, pp. 79, 81, 83, fig. 6, ZN 09/13 (Sondage A); Jesse 2016, p. 307, fig. 5, no. 5.

88 French 1986, p. 175, fig. 9.9, with type SJ2.1.1 and the SJ 2 examples being the closest forms to the wide-mouthed examples at Dorginarti, and the smaller necked jars close to p. 182, fig. 9.16, SB9.4.1 (silt with white slip). French dates this group of pottery to in and about the Twenty-Fifth Dynasty. These forms are mirrored in marl in Aston 1996c, p. 32, pl. VI, nos. 89–93, closely resembling all three forms at Dorginarti. Aston dates the marl assemblage at the site to ca. 750–650 BC. The jar rim may also be present at Ashmunein; see Spencer 1993, pl. 50, type A3.75 ("bowl").

89 Shown in Aston 1996c, p. 33, pl. VI, nos. 97–98. They are coated with cream to pink slips on the exterior.

90 See Jacquet-Gordon 2012, p. 243, fig. 94d (Nile D with a red slip and white wash over the upper part and from the Twenty-First through Twenty-Fifth Dynasty levels). The jars are the wider variants, low rimmed and wide mouthed. Most of the jars shown exhibit evidence of a red slip and a white coating around the upper part. The parallel to the smaller diameter is found on p. 247, fig. 97a–b. For similar types of jars from the temple of Mut at Karnak, see Sullivan 2013, pp. 183, 220, types 5-17 through 5-19 (of marl or silt with a whitish coating), and pp. 190, 229, type 16-1 (of marl or a mixed clay). These sherds came from Strata 2–3, which are attributed to the Third Intermediate Period through the Twenty-Fourth Dynasty.

91 Boulet and Defernez 2014, pp. 609–10, fig. 31-1C, where the type is dated to the end of the Twenty-First and the beginning of the Twenty-Second Dynasties. See also Boulet 2015, pp. 66, 73, 78, fig. 2, no. 5 (with a thick white slip, first half of the eighth century BC).

92 Aston 1999, pp. 62–63, no. 448 (white-slipped silt), pp. 65–66, no. 479 (mixed clay with creamy pink slip), from Phase IIa. In Phase IIb the jars continue, in uncoated marl or mixed clay with light-colored slip; for a few examples, not all of them with the pronounced inner pillowing of the rim, see pp. 82–83, nos. 617–21; pp. 98–99, no. 808; and pp. 103–4, nos. 857, 861.

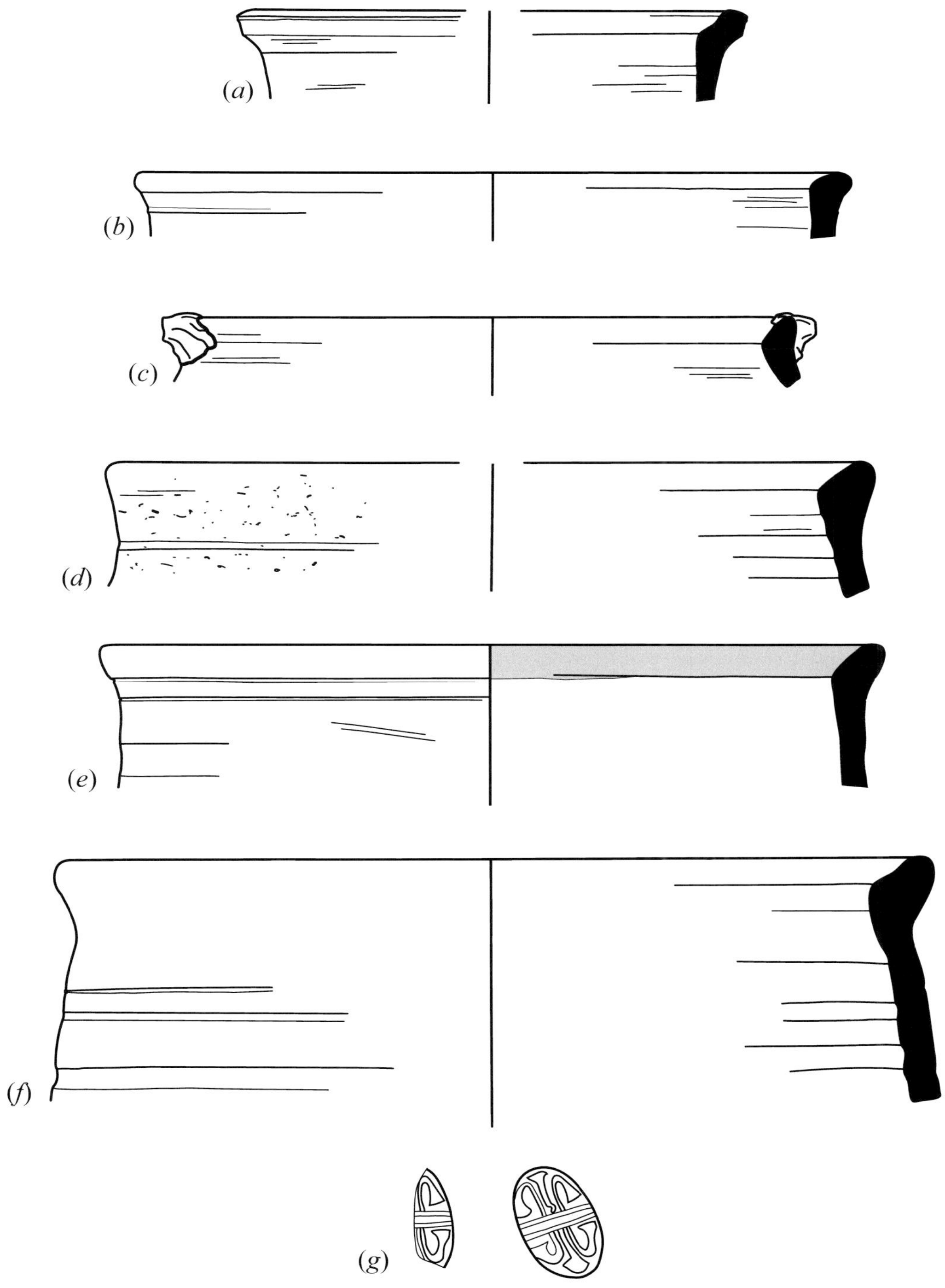

Figure 5.30. Wide-mouthed silt jars (*a–f*, 2:5; *g*, 1:1)

(*a*) D 43, surface to "ash" layer

(*b*) D 103 (1), another example from D 116

(*c*) D 12 (1)

(*d*) D 8, fill above Level III residence

(*e*) D 208 (2), south of ovens in (1); RS exterior/RR interior

(*f*) D 45

(*g*) D 113 and D 114; two seal impressions on upper shoulder of OIM E24406

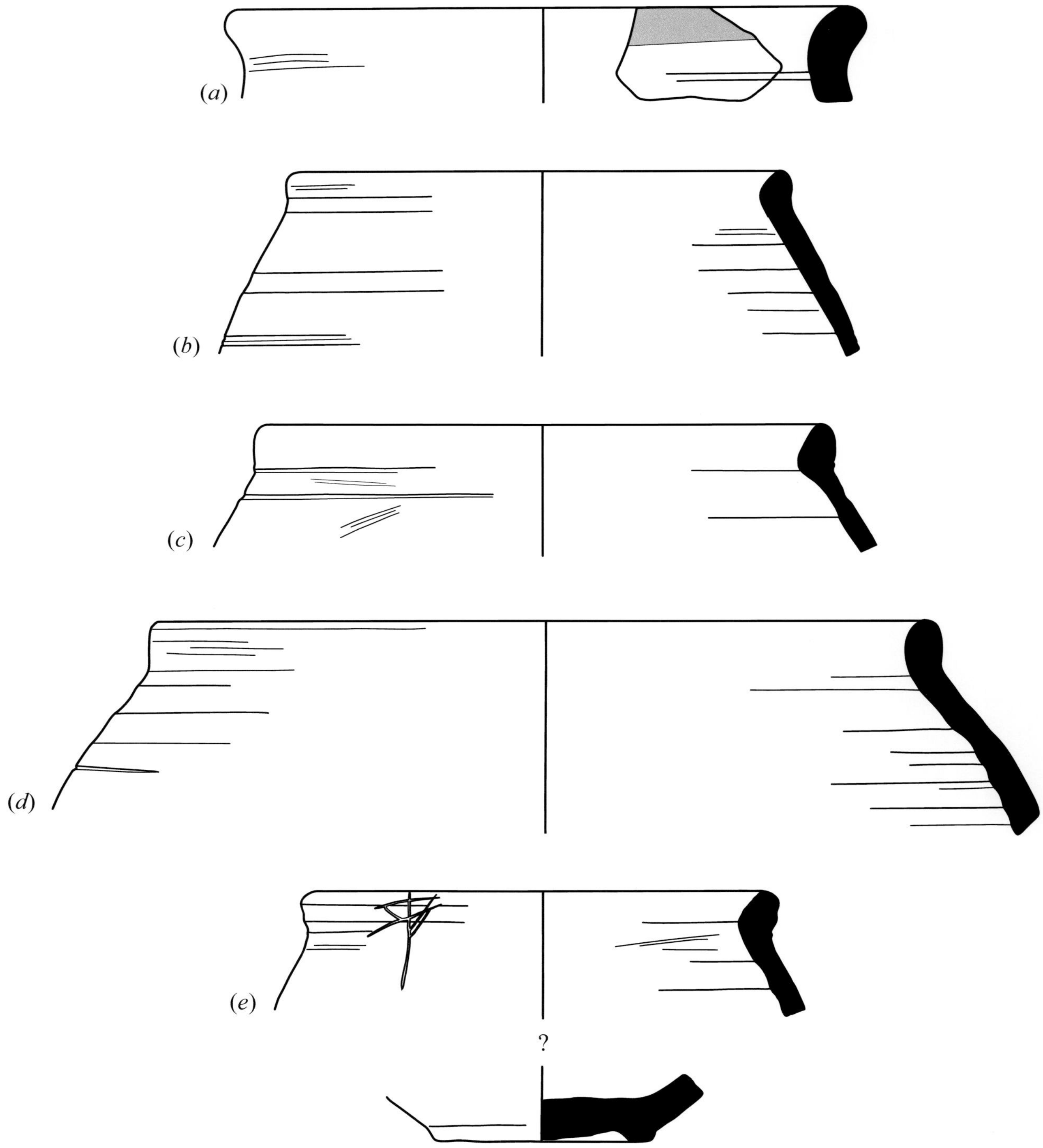

Figure 5.31. Wide-mouthed silt jars (*a*, *d–e*, 2:5; *b–c*, 1:5)

(*a*) D 17, under wall addition; RS exterior/RR interior

(*b*) D 7 (1)

(*c*) D 312 (1), D 313 (1), and test pit D 313 (2)

(*d*) D 230 (1); Nile B2

(*e*) D 10 (1), fill above floor; RS traces exterior, 2.5YR 5/8, prefiring(?) potmark, blackened bottom on base, silt base possibly from this jar

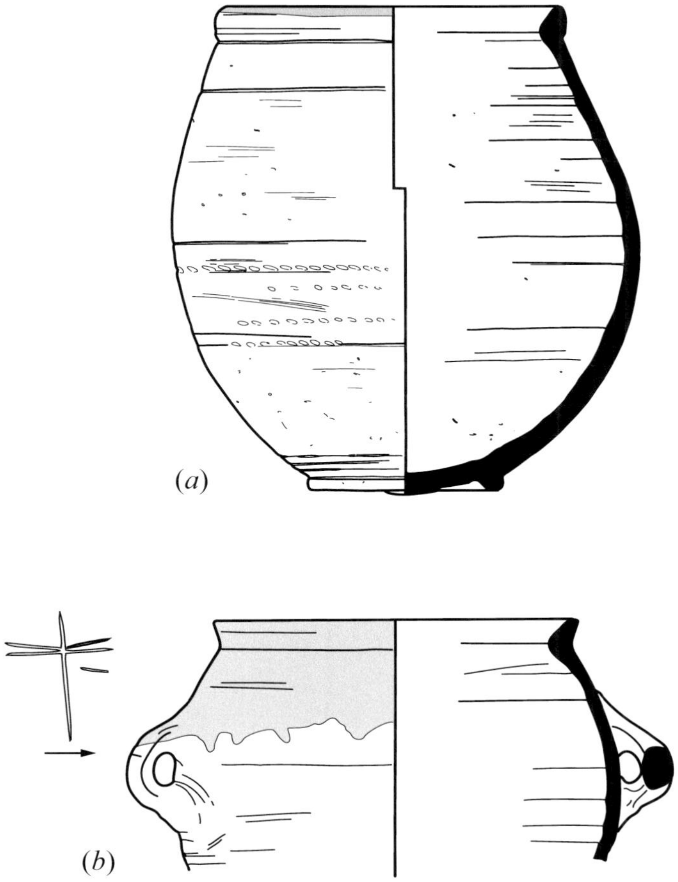

Figure 5.32. Wide-mouthed silt jars (1:5)

(*a*) D 60 (1); OIM E24422, RS rim exterior, very micaceous

(*b*) D 306 (2), one of two examples; OIM E49522, RS exterior top, prefiring potmark (not to scale)

halfway down the body. The upper part of the vessel exterior is covered with red slip. The rim is more everted than most wide-mouthed vessels, and one of the two jars found in this locus has a prefiring potmark on the preserved handle (illustrated). A slightly similar marl vessel was excavated in a Third Intermediate Period context at Karnak North.[93]

The third example, with no handles preserved, has a ring base on which the middle dips below the edge of the base, which would have compromised its stability unless sitting atop soft soil or sand (fig. 5.32a; pl. 65f). The interior base and bottom of the pot area were finished by hand, with finger marks visible on the interior. Cord marks have been preserved around the middle of the body, where a string was wound around the vessel and tied to prevent it from falling apart during its drying stage.

The wide-mouthed forms do not offer good dating evidence on their own, and are also not as widely published as the more distinctive types. But a few forms with comparable rim profiles were found at Amarna South.[94]

Silt Jars with Marl Forms

A number of silt jars resemble marl forms, although they are not very common (fig. 5.33). Some of the jars resemble a marl jar that is common in the materials (MJ 2, see below). Two of the sherds illustrated in the figure stem from mixed contexts in the East Sector and could possibly belong with Level II materials (fig. 5.33b, f). A reddish slip is often present, and streak burnishing is found (fig. 5.33a–b, e). The fabrics sometimes have abundant mineral particles in them. The forms of figure 5.33a–b have a comparable shape in the Late Period materials from Tell el-Retaba.[95]

The jars shown in figure 5.33e–f resemble the MJ 4 form discussed below, and the example from D 47–D 50 has a light-red to pink slip on its exterior, perhaps added so that it would resemble a marl jar. Another jar is either a variant SJ 1–2 or a less bulbous-necked form of the MJ 1 discussed below (fig. 5.33d). It is silt with a red slip and comes from the eroded contexts along the west enclosure wall.

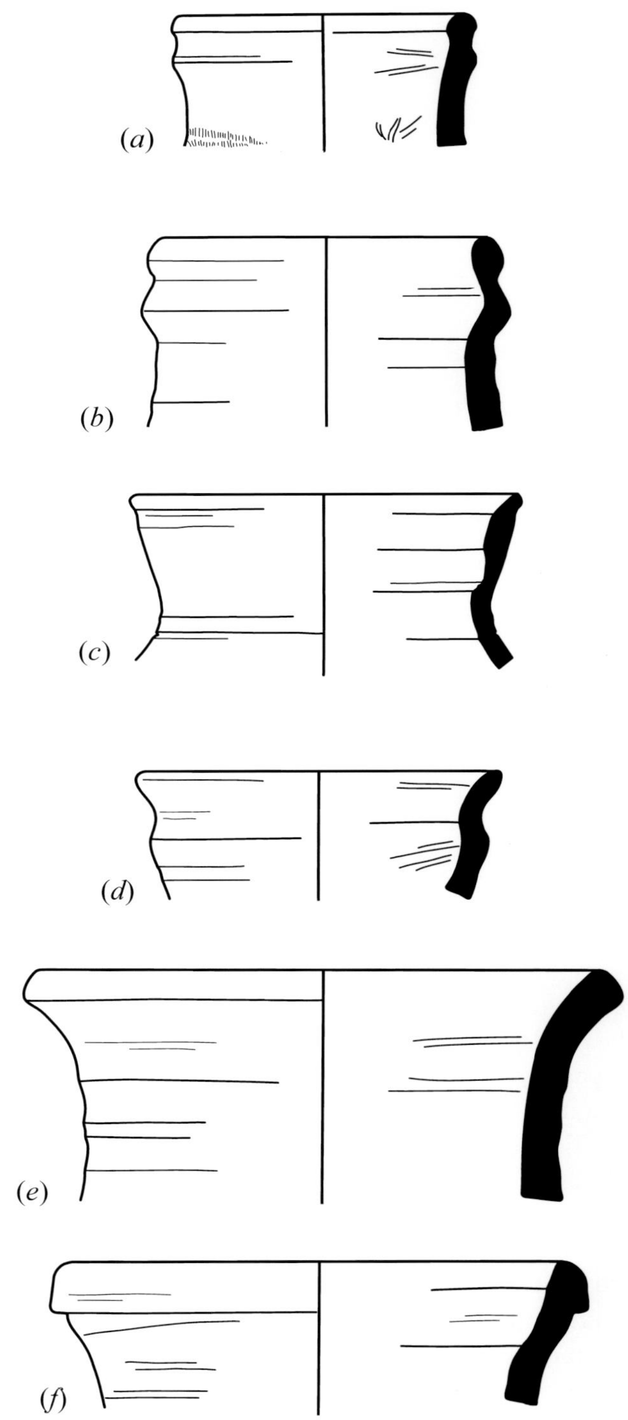

Figure 5.33. Silt copies of marl jars (2:5)

(*a*) D 221 (1); RS exterior and atop rim, horizontal burnish traces on exterior

(*b*) East Sector, rubble over three south silos, sherds also from D 12 (1) west wall and first 40 cm of East Sector; RS exterior, vertical burnishing on exterior, RR interior

(*c*) North of D 32 (1); yellow-brown slip on exterior

(*d*) D 12 (1); RS interior/exterior

(*e*) D 47–D 50, below surface; pinkish-red slip on exterior

(*f*) D 26 (1), first 25 cm

93 There is, however, no evidence for handles; it seems to have had a rounded base, and the rim diameter is narrower; see Jacquet-Gordon 2012, p. 232, fig. 90a.

94 French 1986, p. 176, fig. 9.10 (SJ 3), p. 177, fig. 9.11 (SJ4.2.1), p. 180, fig. 9.14 (SJ6.6.3).

95 Wodzińska 2011b, pp. 1024, 1035, fig. 12, nos. 5, 10.

Silt Jar Bases and Handles

A variety of silt bases are shown with the forms already illustrated, but more are shown in figures 5.34–5.35. The bases could not always be certainly associated with a particular rim.

Silt loop handles are present in small numbers throughout the fortress, usually made in a very cursory fashion, except for the Nile D and related silt jar handles. Two small handles have prefiring horizontal incisions atop their exteriors (one shown in fig. 5.35d). Base fragments from the jars are rounded, slightly pointed at the bottom, ring based, or, very rarely, carinated at the outermost base edge and probably rounded or slightly pointed at the bottom.

MARL JARS

Marl jar sherds were found in nearly every context at the fortress and can be grouped into five major forms, although with small differences apparent in each group. The sherds were not as frequently found as the SJ 1 and SJ 2 sherds in the museum collection, but they are well enough represented. The presence of the MJ 2s and wide-mouthed marl jars, for instance, indicates a date toward or within the eighth century BC, when silt jars no longer dominated assemblages as they are supposed to have done in the earlier first millennium BC. New marl pottery styles and manufacturing techniques are thought to have begun sometime in the mid- to later eighth century BC, when a new corpus of pottery forms mostly parted ways with earlier New Kingdom traditions.[96]

The decorated marl jar with a flaring rim (MJ 4), however, follows in the footsteps of New Kingdom examples, although the rim forms mostly differ from their earlier counterparts. The jar forms from Dorginarti are made from a Marl A4 fabric, fired pink on the interior and creamy white on the exterior, and the temper and color do not visibly differ from the marl bowls. Some of the body sherds found in association with the earlier forms, such as MJ 1 and MJ 4 discussed below, have smooth exteriors. There are, however, body sherds from Dorginarti that are partially or completely corrugated on their exterior surfaces, a trait found among products of the Upper Egyptian marl industry attributed to the mid- to late eighth century BC. Unfortunately, it is not always clear to which rim forms the body sherds belong.

96 See, for example, Jacquet-Gordon 2012, pp. 6–7; French 1992, pp. 84–86; Aston 1999, p. 82.

These marl products all came from Upper Egypt, and perhaps the Theban region specifically, and were filled with particular supplies for the fortress. Many examples of Upper Egyptian marl vessels have been found in Lower Nubia, and their presence has been noted as far south as the earlier tombs at Meroe and as far west as the fortress at Gala Abu Ahmed.[97] The forms found at Meroe and Gala Abu Ahmed all belong to types like the MJ 2 and other forms mentioned below, not the MJ 1 or MJ 4 discussed below. Parallels to the MJ 2s are also found at Tell el-Ghaba in the eastern Delta.[98]

Marl Jar 1

The bowed neck on this marl jar is part of a mostly complete jar body, as shown by the jar in figure 5.36, which has a few wheel striations only around the lower half of the body. Multiple prefiring potmarks have been incised onto the exterior surface. The complete jar body shown in figure 5.37, with a postfiring potmark, has prominent wheel ridges only on its midsection. It presumably also belonged to an MJ 1, because this type of rim and neck sherd was found in the same locus. The shape of these two forms, each with a set of looped handles and rounded base, follows the traditions of New Kingdom pottery manufacture. The base is pressed into a mold as shown by the fingermarks on the interior surfaces, although the base of the jar in figure 5.37 had been finished while being slowly turned, and only the bottom shows hand pressing. Pre- and postfiring potmarks are frequently found on marl body sherds and are presented more fully in chapter 7. A photograph of the jar from building D 59 is shown on plate 66c, although the base portion is not included in the photograph.

The recognizable sherds from MJ 1 are most often the rim and bowed neck fragments, which were found in only moderate numbers at the site, at least as indicated by the museum's collection (fig. 5.38a–c). The form appears in the Twentieth Dynasty in Egypt, but the earlier examples cited typically have a longer neck than those from Dorginarti.[99] Parallels have also been found in later first-millennium contexts at Amarna,

97 Dunham 1963, pp. 337–38, figs. A–B; Eger, Helmbold-Doyé, and Karberg 2010, p. 79, fig. 6 (types ZN 09/6 and 09/12, both made of Marl A4 fabric).

98 Lupo and Cremonte 2011.

99 Hope 1989, p. 123, pl. 6b. See also Aston 2004, pp. 198–99, fig. 13b (of Marl A4, from Thebes). Note the narrow body and rim cited in Aston 1999, pp. 82–83, 87, nos. 616–17, and found in contexts attributed to Phase IIb, which extends to the end of the eighth century BC. See also Aston 2009, pp. 325, 346, fig. 38, type 162.

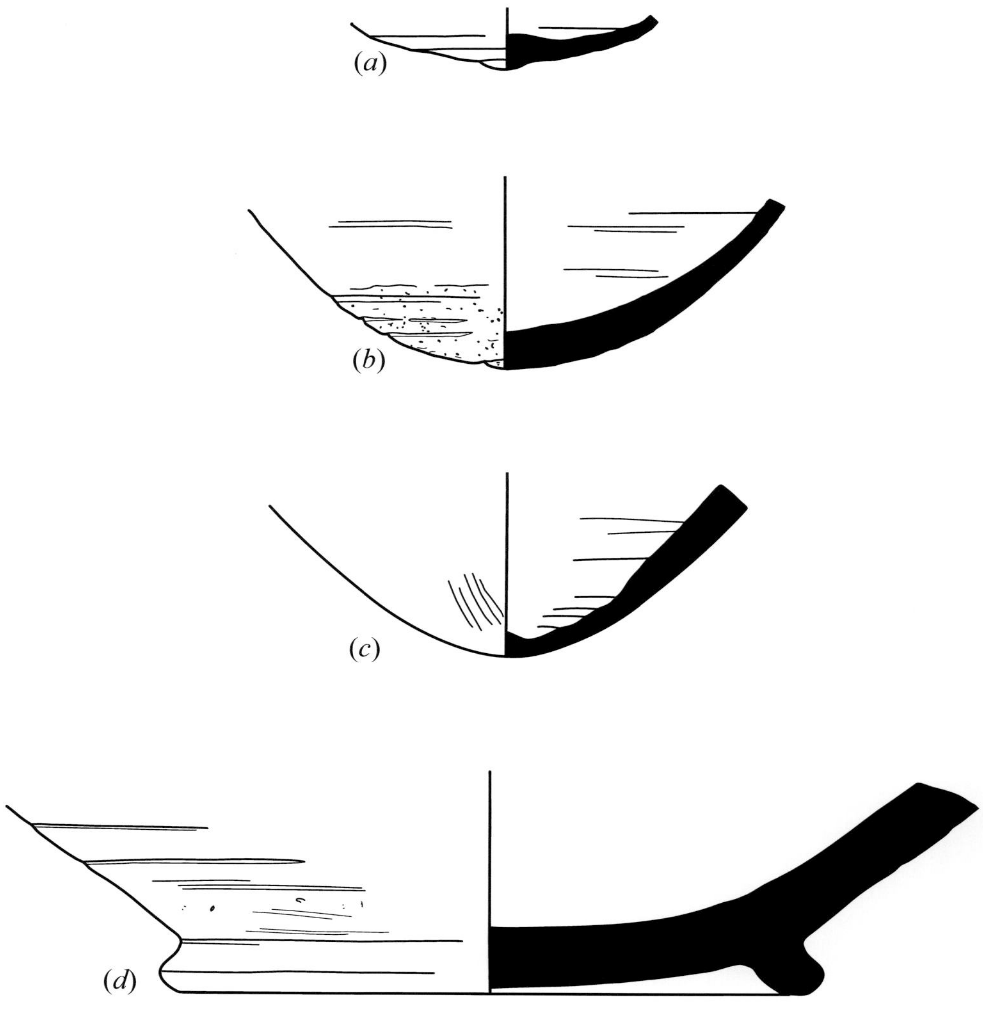

Figure 5.34. Silt jar bases (2:5)

(*a*) D 78 (1), bin; burned on exterior, jar(?) base

(*b*) D 48 (2)

(*c*) D 229 (1); Nile C, stance?

(*d*) D 46, pit; RS exterior and bottom of base

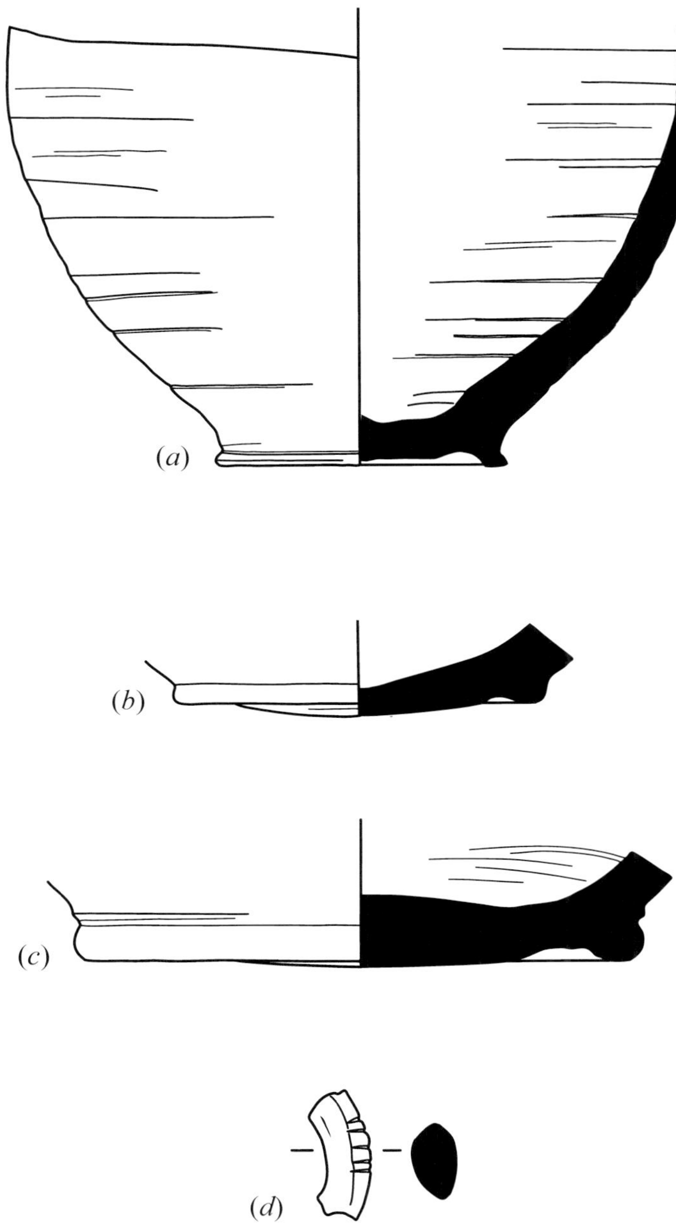

Figure 5.35. Silt jar bases and incised handle (2:5)

(*a*) D 201 (1)
(*b*) D 14; blackened base and exterior
(*c*) D 6 (1)
(*d*) D 53, surface to "ash" layer; prefiring incised lines

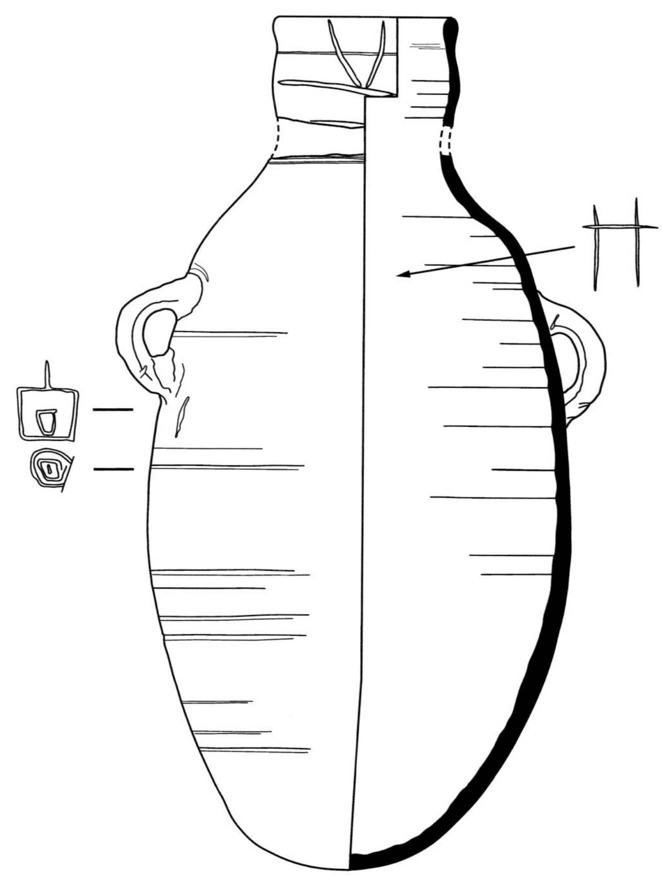

Figure 5.36. Marl jar 1 (1:5)

D 59 (1); OIM E49527, pre- and postfiring potmarks, hand pressed at bottom third of interior, two incised lines at neck break

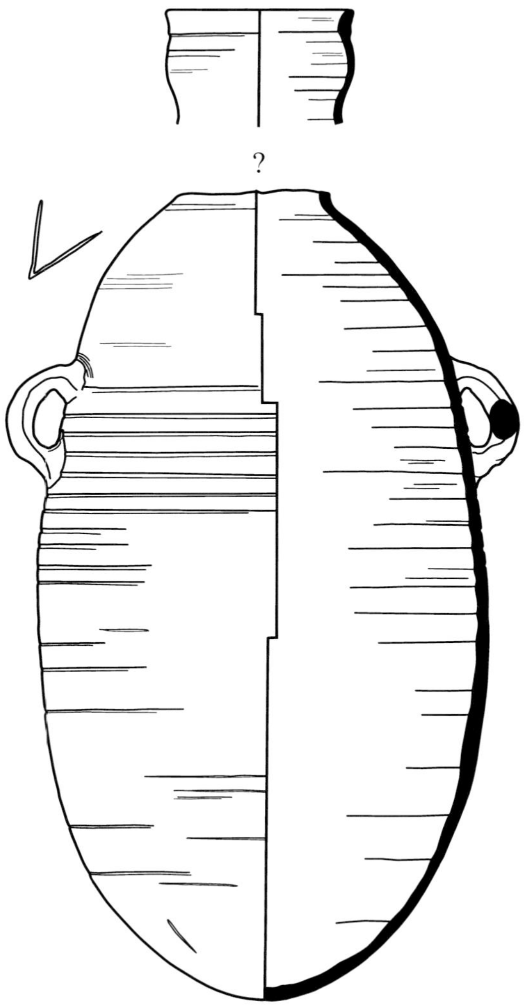

Figure 5.37. Marl jar 1 (1:5)

D 102, along the south wall; OIM E24377, potmark on shoulder, body diameter perhaps wider, hand pressed at bottom interior

around the mortuary temple of Seti I at Qurna, and at Karnak North.[100] A similar form, although with a less bulbous neck, was also found around the chapel of Osiris Wennefer Neb-djefau at Karnak, where the excavators attributed it to the end of the Twenty-First and beginning of the Twenty-Second Dynasties.[101]

Marl Jar 2

This form has many distinct variations, but they are all related by the incised area below the lip followed by a convex fold or bulge at the neck (figs. 5.38d, 5.39a–e; pl. 66d). There is still no clear chronological distinction in the various rim and neck forms, although Aston believes that the grooved rim types as shown are later than those with no external grooves below the rim.[102] However, he notes that they are often found in the same contexts. Sherds from MJ 2 are fairly common in Level III deposits at Dorginarti. It appears that pottery makers were experimenting at this stage of the industry, before the very distinct forms of the late eighth and seventh centuries emerge.[103]

Parallels to the MJ 2 jar are quite common in publications of Upper Egyptian sites with eighth- and seventh-century pottery assemblages. Karnak North has

100 Sherds from both straighter and bulbous rims were found among this material. The one shown in French 1986, pp. 157, 184, fig. 9.18, MJ5.1.1 (in or about the Twenty-Fifth Dynasty) has less of a neck bulge than the Dorginarti examples. A good parallel is shown in Jacquet-Gordon 2012, p. 235, fig. 91d, which is made from Marl A4, variant 2, and is placed in her Twenty-First through Twenty-Fifth Dynasty levels at Karnak North. The jar form is also found in Myśliwiec 1987, pp. 66–67, nos. 522–23, and p. 68, no. 585. He notes that the form may be earlier than most of the materials found in association with it and that the context was unclear.

101 Boulet and Defernez 2014, pp. 609–10, fig. 31-1, type D.

102 Aston 1999, pp. 84, 154–55 (with rim no. 1496), 186. He notes that the rim form of no. 1496, from a deposit attributed to Sheshonq V, has no rim groove and is perhaps the earlier variant. This rim is like that of the MJ 3 discussed below and shown in fig. 5.40. The other parallels from Elephantine are noted in a footnote below.

103 The Egyptian sites yielding these later marl forms are numerous. See the assemblages presented in Boulet and Defernez 2014, pp. 610–15, fig. 31-1, types H–J, and fig. 32-2, types C, I–J. These are dated to the late eighth through the beginning of the sixth centuries BC. There are numerous examples of similar Upper Egyptian marl jars from the temple of Seti I at Qurna; see, for example, Myśliwiec 1987, pp. 60–61, nos. 397–99, and pp. 63–71. Karnak North has examples of marl jars associated with the late eighth and seventh centuries BC; see Jacquet-Gordon 2012, pp. 256–58, and particularly the jars shown in fig. 101. The site of Kerma also has examples of these Upper Egyptian marl imports, as well as locally made silt jar copies. For the marl imports see, for example, Ruffieux 2007, p. 234, pl. 2, nos. 11–13, and p. 236, pl. 4, nos. 28–30. Lastly, note Heidorn 2018a.

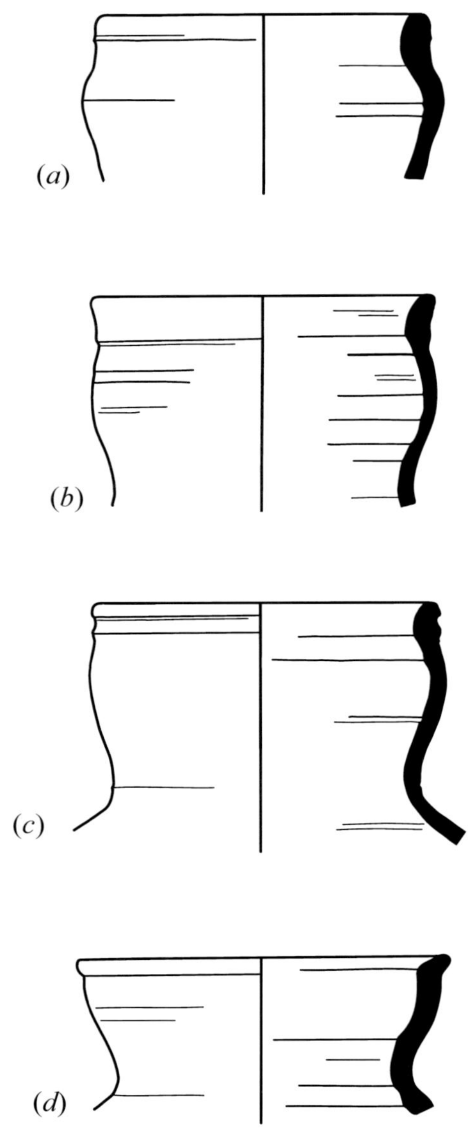

Figure 5.38. Marl jars 1 and 2(?) (2:5)

(*a*) D 310–D 311, trench down to stratum (1); stance?

(*b*) D 4, D 47, D 48, dismantling walls

(*c*) D 60 (1)

(*d*) D 32 (1), south of fireplace

the form both in its Twenty-First through Twenty-Fifth Dynasty deposits and in the Twenty-Sixth through Thirtieth Dynasty layers, and vessels of this type are also found elsewhere around the Karnak complex.[104]

Elephantine has numerous examples that compare well with these jar rims and necks in the Phase III assemblage, with an almost complete body preserved, but there are two earlier examples attributed to Phase IIb.[105]

The last two jar sherds shown on figure 5.39 are not strictly MJ 2 types, but they are related to both this class and the next class of MJ 3. A complete jar profile from Karnak North shows a slightly pointed base and small loop handles, with a rim matching that of figure 5.39g, which has a vertical neck and short triangular rim.[106] This particular type is found around the Seti I mortuary temple at Qurna and at Ashmunein, Amarna, and Abu ʿId.[107] A complete vessel from Amarna also shows a rounded to slightly pointed base and two loop handles.[108]

Two complete jar profiles with rims like that in figure 5.39g come from Elephantine. They are small with tall necks, and have slightly tapering bases.[109] Another complete body profile with a longer vertical neck

104 Jacquet-Gordon 2012, p. 235, fig. 91c (Marl A4 "tardive"), attributed to the Twenty-First through Twenty-Fifth Dynasties, and p. 257, fig. 101e (Marl A4, variant 2), from the Twenty-Sixth to Thirtieth Dynasty layer. Also see Masson 2007, p. 609 and pl. IX, no. 7 (silt). The Mut Temple examples are shown in Sullivan 2013, pp. 182, 218, type 5-1 through 5-6 (marl and silt), pp. 190, 229, type 16-7, and p. 236, types 22-1 to 22-10. All of these vessels are from Strata 2 and 3, or within the Third Intermediate Period.

105 Aston 1999, pp. 84–87, nos. 626–27, from Phase IIb. Phase III examples are presented on pp. 186–87, no. 1698 (almost complete profile); pp. 191, 193, nos. 1736–43; p. 195, nos. 1757–59; and pp. 196–98, nos. 1789–91.

106 Jacquet-Gordon 2012, p. 258, fig. 101j.

107 Myśliwiec 1987, p. 67, nos. 523–25. Nos. 355 and 357 are silt imitations of the marl types; see p. 54, where Myśliwiec mentions a whitish-green slip or white bands running around the exterior surface. See also Spencer 1993, pp. 45–46, pls. 64–65, types E1.5, 8, 10, 18, 19, 22, 25, 26, 30, 33, and 35. Some of the examples shown are like the MJ 3 types discussed below; they are mostly from Level I strata at Ashmunein. For the parallels from Amarna's late dynastic deposits, see French 1986, p. 174, fig. 9.8 (SJ1.2.1 and SJ1.2.2), and p. 175, fig. 9.9 (SJ2.2.4), a silt jar resembling a wider-necked variant. Marl examples are shown on p. 183, fig. 9.17 (MJ 2s). French dates this pottery to in or about the Twenty-Fifth Dynasty. See also Aston 1996c, pp. 31–32, pl. V, nos. 75–84.

108 French 1986, p. 187, fig. 9.21, no. 4. The jar was found at the Workmen's Village, West Street 2.

109 Aston 1999, pp. 184–85, nos. 1694, 1697; see also pp. 204, 207, nos. 1870, 1876. All examples are from Phase III.

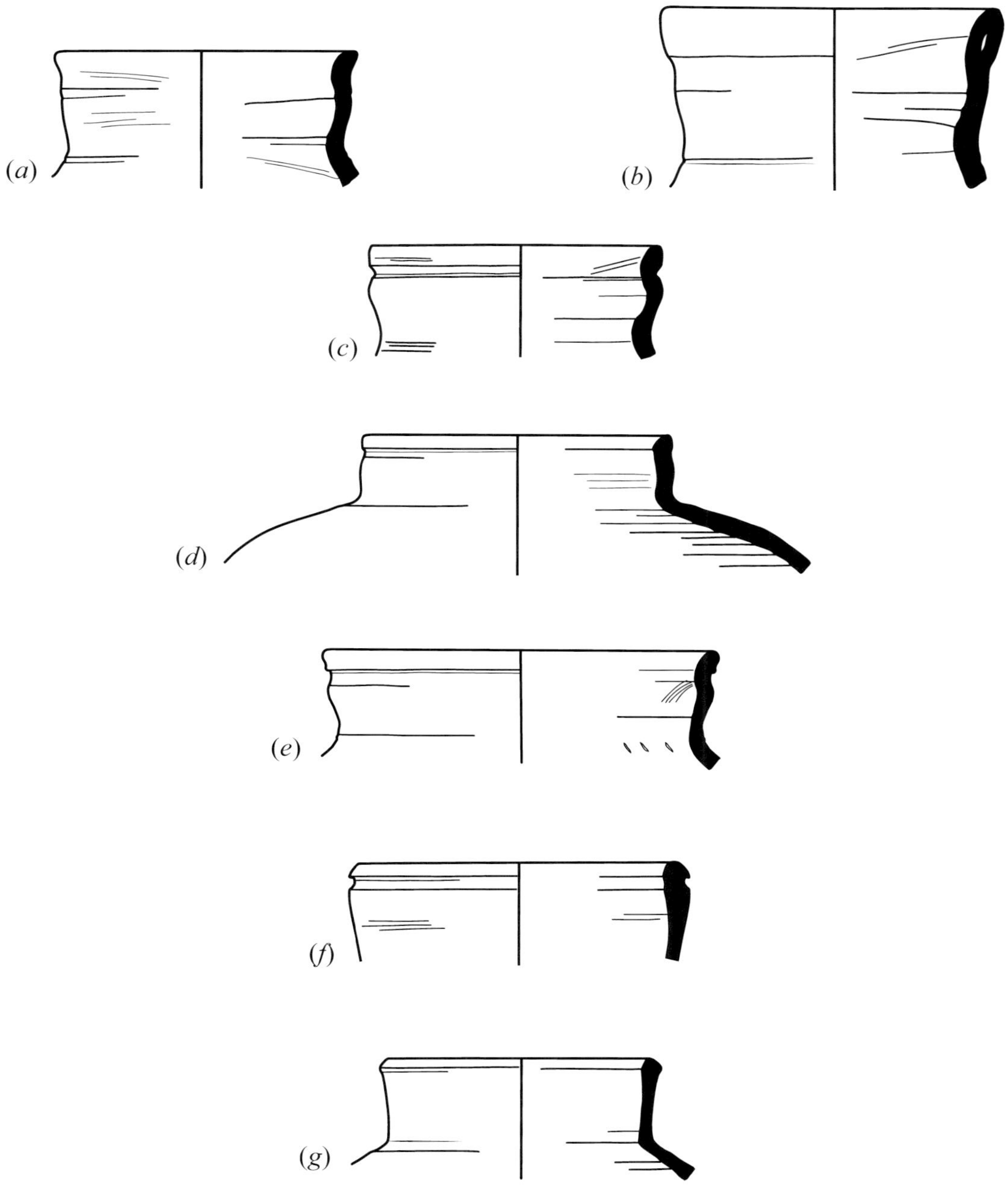

Figure 5.39. Marl jars 2 and variants (2:5)

(*a*) D 113 (1)
(*b*) D 215 (1)
(*c*) D 205 (1)
(*d*) D 7 (1), sherd also from D 32 (1) south of oven
(*e*) D 23 (1), lower part of rubble; chattering on interior neck
(*f*) D 13 (1)
(*g*) D 13 (1)

comes from Karnak North deposits that have been dated to the Twenty-Sixth through Thirtieth Dynasties.[110]

Upper Egyptian marl jars like the examples mentioned above, and the variants shown in figure 5.39f–g, also reached Tell el-Ghaba in the eastern Delta, a site dating to the end of the Third Intermediate Period or early Saite period.[111] They are found at Tanis in late Third Intermediate Period or Saite levels,[112] and at various Napatan sites in Nubia.[113]

A complete jar, with a similar long neck, also came from tomb 1 at Gebel Sahaba in Nubia, and another complete jar was found in the reuse of tomb 8 at Sai.[114] The same rim profile comes from the chamber debris of the tomb of Tanwetamani at el-Kurru, and another complete example from tomb S515 at Semna.[115] The type is also found at Qasr Ibrim.[116]

110 Jacquet-Gordon 2012, p. 258, fig. 101j (Twenty-Sixth to Thirtieth Dynasties).

111 Lupo and Cremonte 2011, pp. 116–18, 121, fig. 2a–d. The destruction layer at the site is dated to the late Third Intermediate Period and early Saite era; see Crivelli, Kohen, and Lupo 2011, p. 205, n. 3.

112 Laemmel 2011, p. 52, pl. 42, no. 219. Laemmel notes that the form is extremely common in levels of the Twenty-Fifth Dynasty and that it continues into the Saite era.

113 Macadam 1955, p. 161, fig. 54, from Kawa. Also note a possibly similar jar from Meroe in Shinnie and Bradley 1980, p. 107, no. 99. An example from Napatan contexts at Kerma is shown in Ruffieux 2007, pp. 231, 237, pl. 5, no. 52, while the reused tombs at Hillat el-Arab also have a number of examples, shown in Vincentelli 2006, pp. 61–62, fig. 2.35, no. 235 (perhaps); pp. 76–77, 82, fig. 2.50, no. 294; pp. 126, 131, fig. 2.82, no. 534; and pp. 139, 149, fig. 2.93, no. 597. French (1986, p. 166) notes parallels between the marl jar examples from the late dynastic deposits around the South Tombs at Amarna and those from Qasr Ibrim, including jars like MJ 2 at Dorginarti.

114 Gardberg 1970, p. 38, pl. 62, nos. 2, 4 (site 100); and in a reused New Kingdom tomb at Sai, which is published in Minault-Gout and Thill 2012, pls. 146, 166, T8 9. The photograph on pl. 166 shows the more precise form of the jar.

115 The unpublished sherd from this jar (19-3-332), from Kurru 16, is shown in the excavation records housed in the Museum of Fine Arts, Boston. I would like to thank Rita Freed and Denise Doxey for permission to study the pottery and records from the royal cemeteries of Kush in 2014. For the publication of the jar from Kurru 16, see Heidorn 2018a. The example from the Semna cemetery is published in Dunham and Janssen 1960, p. 78, fig. 37, 24-2-477. The form has also been found elsewhere in the Sudan; see Geus 1979, pp. 18–19, fig. 8, nos. 8–11. This pottery comes from site K176, perhaps a tomb at Abudiya, near Sai. The deposit is considered early Napatan.

116 Rose 2019, p. 690, figs. 33, 36, 37, for marl jars of types MJ 2 and MJ 3 at Dorginarti.

Marl Jar 3

This jar features a short concave neck and flaring everted rim, or a rim with a more vertical stance (fig. 5.40). There are examples exhibiting variations along the continuum between vertical or everted rim stances, but they typically all have short necks. Parallels to MJ 3 are found at the same sites as those mentioned for MJ 2, such as Qasr Ibrim and the sites mentioned below.

The shorter-necked type is found in association with an MJ 2 at Abudiya, in Upper Nubia.[117] One example was also excavated from a context at Elephantine that is attributed to the time of Sheshonq V.[118] Other sites in Egypt also yielded parallels to this form, where they are dated to the second half of the eighth and the seventh centuries BC.[119]

Marl Jar 4

The jars with flaring rims and concave neck profiles are reminiscent of New Kingdom forms and jar decoration, although some examples are stylistically different (figs. 5.41–5.43; pl. 66e–f).[120] It is supposed that this form originated in the earlier occupation of the fortress in the tenth or early ninth centuries BC, if they were not the remnants of late New Kingdom activity at the site. There are quite a few minor variations in rim form, diameter, and neck height. The necks were sometimes decorated with black bands, and one possible bottle or jar rim has perpendicular spots atop its rim (fig. 5.42e). One shoulder fragment from a marl jar also has black-painted, horizontal bands; its exterior does not display corrugation.[121] There is no complete profile for this jar at the site or elsewhere, so whether it had handles is uncertain, as is the precise shape of the base. It is not very common in Level III and Level IV contexts and is not as popular as the MJ 2s or the wide-mouthed marl jars discussed below.

Although not a parallel for this decorated jar form or fabric, a contemporary decorative style is found at

117 Geus 1979, pp. 18–19, fig. 8, no. 9.

118 Aston 1999, p. 163, no. 1496, from Phase IIb. The deposit was associated with an inscribed jar sealing bearing a reconstruction of the name of Sheshonq V (ca. 767–730 BC).

119 Most references to MJ 2 above also have associated MJ 3 sherds. See, for example, Myśliwiec 1987, pp. 66–67, no. 521, from the magazines; French 1986, p. 183, fig. 9.17, MJ1.1–MJ1.1.4; Aston 1996c, p. 32, pl. IV, no. 68, and pl. V, nos. 70–71.

120 The New Kingdom forms it emulates are like those shown in Jacquet-Gordon 2012, figs. 45a (lower), 48k, 49a, 62a, and 73a, of the Eighteenth Dynasty or Ramesside periods.

121 This sherd was from D 303, the clearance north of the Level IV residence before the walls of D 306 appeared.

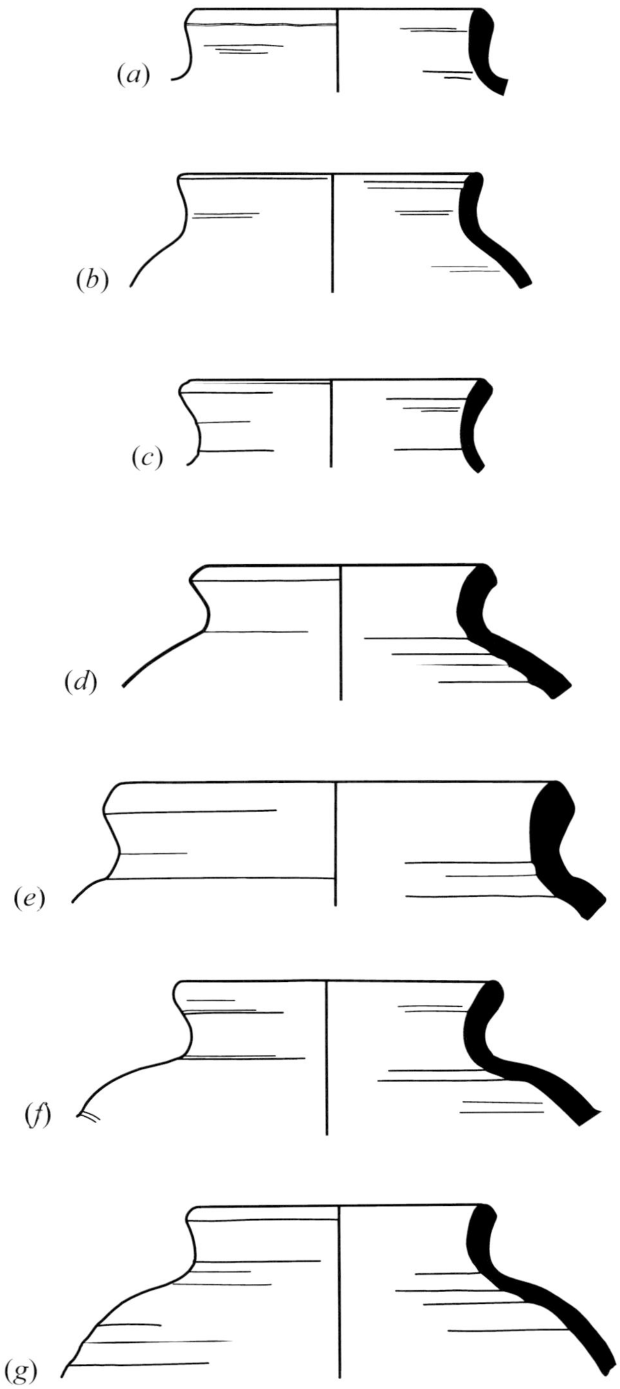

Figure 5.40. Marl jars 3 (2:5)

(*a*) D 23 (1)

(*b*) Removal of brick from stairs north of Level III residence

(*c*) D 13 (1); OIM E49585

(*d*) D 48 (2); burned, Marl A4(?)

(*e*) West of D 32 (1)

(*f*) D 75; handle emplacement visible

(*g*) D 309, pit A

Tell el-Ghaba, a site dating to the Third Intermediate Period in the northern Sinai.[122] The horizontal bands on these sherds are mostly black decoration atop a red- or pink-slipped polished surface, but there is also a black- and red-painted bichrome. In addition, one handmade jar at Tell el-Ghaba has a black-painted band on an uncoated surface. The black bands are around the shoulders of the jars, in the middle of the bodies, and also near the base.

The painted pottery traditions of the New Kingdom are thought to have mostly disappeared in the Third Intermediate Period and Late Period, although such painting appears sporadically on poorly dated, mostly smaller, vessels.[123] Painted decoration, however, including banding on jars and concentric circles on flasks, is found in various regions of the Levant, where it continues from the Bronze Age through the entire span of the Iron Age.[124] Inspiration from farther north obviously appealed to potters, as it did to the residents inhabiting Tell el-Ghaba, who were situated near Levantine, Cypriote, and Phoenician sources of decorated pottery.

The Napatan/Kushite temple at Soniyat in Upper Nubia also yielded examples of painted horizontal bands on silt jars with red or white slip, including wavy-band decoration.[125] They are, however, rarely found outside temple precincts. Site 6-G-9, just north of Dorginarti in the region of Gezira Dabarosa, also has examples of red- and black-banded decoration on wheel-made, loop-handled vessels of local silt fabrics, but all lacking their necks and rims.[126] These were found in association with mostly handmade forms and sherds, some Upper Egyptian marl(?) vessels, blue-painted sherds, and sherds of streak-burnished vessels

---

122 In addition to the jars, juglets, and jugs, there are also two bowls with black bands on the interior; see Fuscaldo 2006, pp. 111–18. The date of the pottery from the site, as mentioned above, is Third Intermediate Period; see Crivelli, Kohen, and Lupo 2011, p. 205, n. 3.

123 Aston 1996a, pp. 80–82.

124 See, for example, the striped banding on amphorae and other forms in Ben-Shlomo, Shai, and Maeir 2004, pp. 1–35 (although not a comprehensive study). The authors discuss decorated pottery with Philistine bichrome, Assyrian, and Judaean influences from the earliest Iron Age through the eighth century BC, when the imported fine wares replaced these types. This decorative tradition was certainly an influence on the painted pots from Tell el-Ghaba.

125 See Phillips 2003, p. 399. These decorated jars are not, however, shown or discussed in the section on the Soniyat temple, where all the painted pottery is dated to the Meroitic period; see Orzechowska 2003, pp. 442–43.

126 Lister 1967, pp. 61–64. There are no illustrations of this pottery, but photographs are published in Lister 1967, p. 62, fig. 23a, and in Adams 2005, pls. 2e–f, 11.

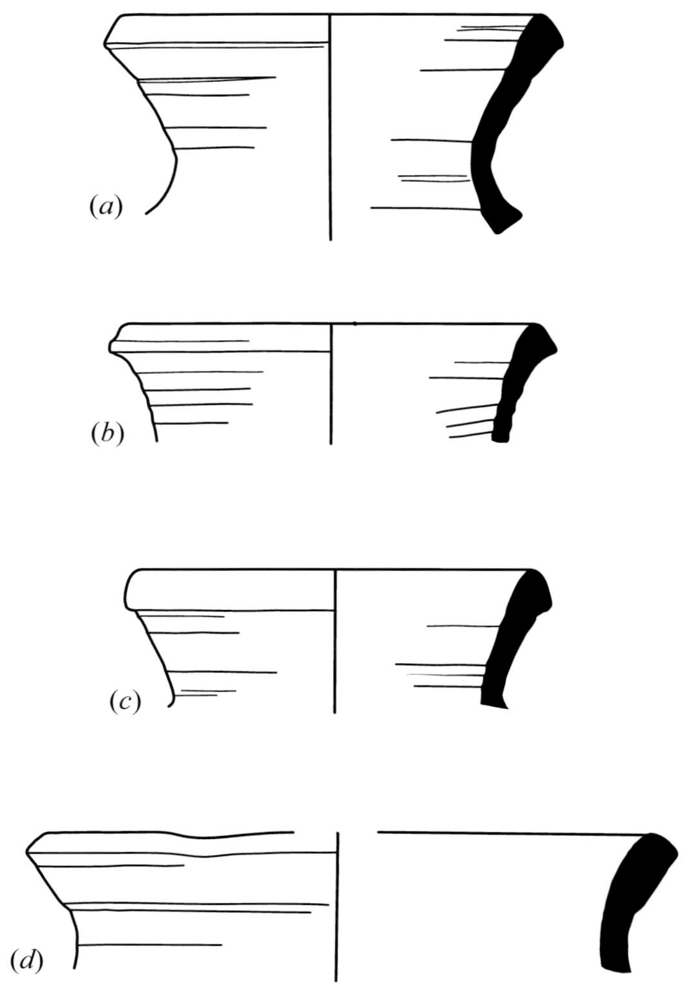

Figure 5.41. Marl jars 4 (2:5)

(*a*) D 7 (1); wider?

(*b*) D 48 (2); burned

(*c*) D 9 (2); worn rim

(*d*) East Sector, first 50 cm above north magazines; warped rim, approximate stance and diameter

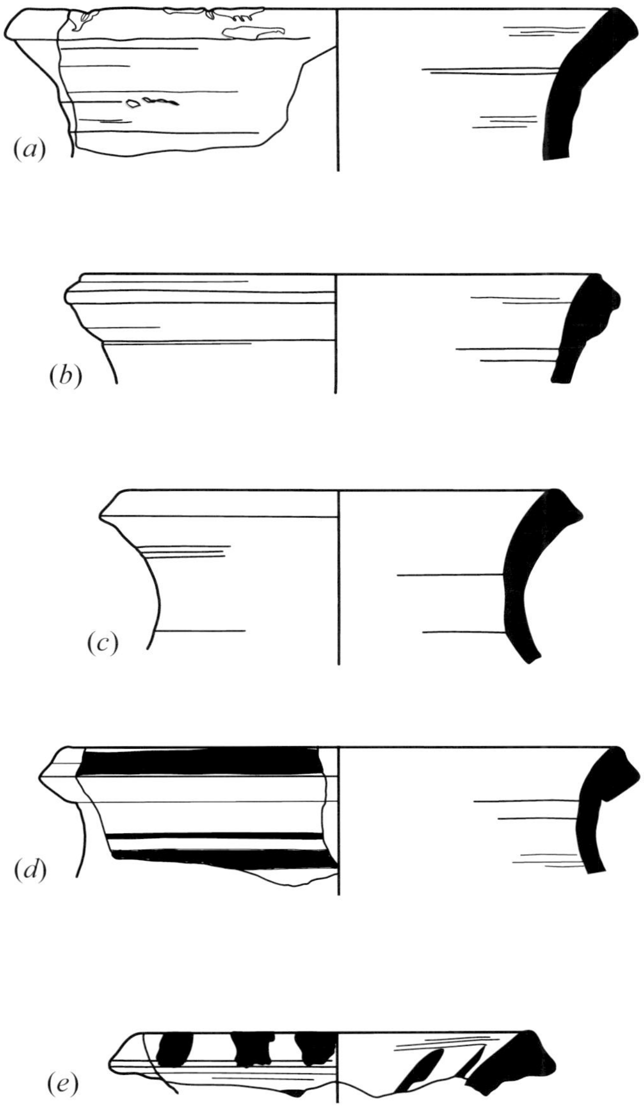

Figure 5.42. Marl jars 4 (2:5)

(*a*) D 32 and area, dismantling walls; same rim grooves on sherd from D 206 (1) and same rim from D 7 (1)

(*b*) D 46, above "ash" layer

(*c*) D 67 (1); similar rims from D 69 (1) and D 68 (3), latter grooved atop rim

(*d*) D 13 (1); brownish-black bands on exterior

(*e*) D 106 (1); black-painted decoration interior/exterior

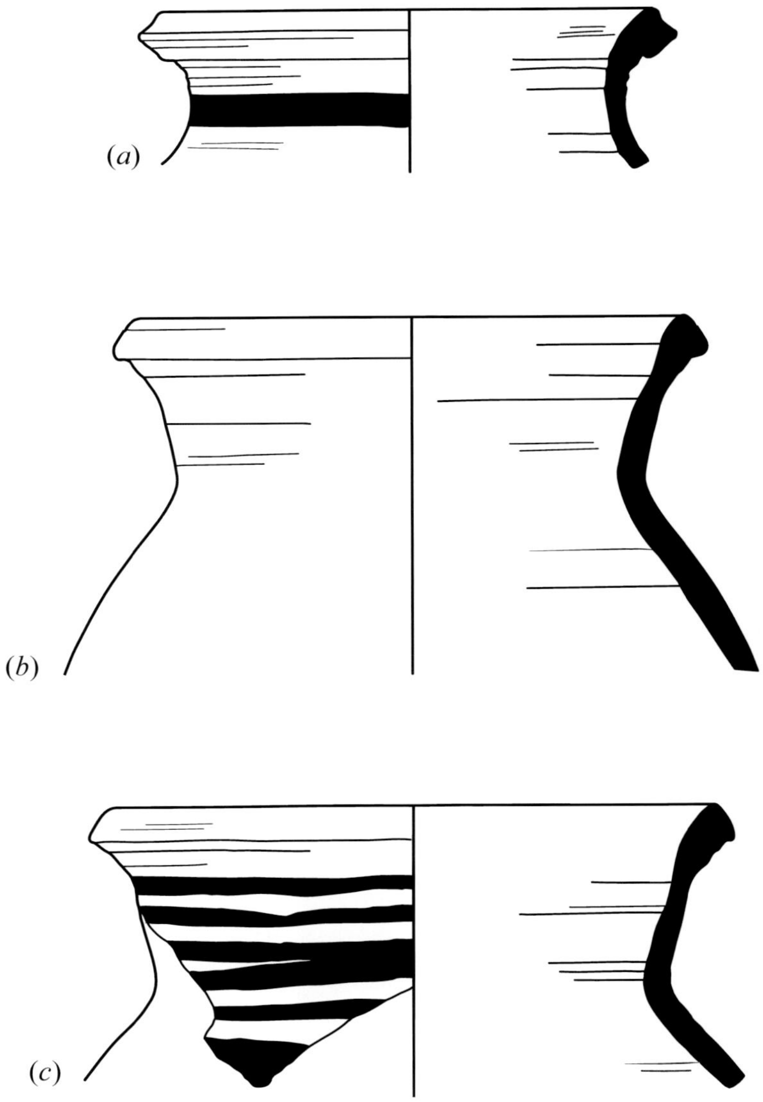

Figure 5.43. Marl jars 4 (2:5)

(*a*) D 18 (1); light-red section and pale-brown surfaces

(*b*) D 208 (2)

(*c*) D 36 and D 36 mixed, surface to below "ash"

of uncertain form. The site yielded materials that were tentatively dated to the late Napatan or early Meroitic periods, but it has been suggested that the seven band-decorated jars were earlier and reused during the Meroitic use of the site for storage, thus predating the rest of the site's early remains.[127]

The black-painted bands found on the flared marl jar form from Dorginarti mirror those on the New Kingdom decorated jars and precede the later dynastic band-decorated vessels of the fourth and early third centuries BC.[128] The band-decorated jars from Dorginarti, and the examples from Tell el-Ghaba and elsewhere, demonstrate a continuity or resurgence in Egyptian pottery decoration that is rarely found in pottery assemblages of the first half of the last millennium BC.

The best parallels to the MJ 4 form are found at sites in Upper Egypt, all of which potentially date from the mid-eighth to perhaps the late seventh century BC.[129] None of these parallels has the black-banded decoration found at Dorginarti. Slightly similar marl and silt forms (*zirs*) were uncovered in the later tomb materials from the Ramesside mortuary complex and tomb of Djehutymes II at Thebes (Theban Tomb 32).[130] A possible silt example was also uncovered from the temple of Mut at Karnak South.[131] There are form parallels from the royal cemeteries of Kush, in tomb Kurru 19 of Generation B and Kurru 14 of Generation C, although the dating of these tombs remains uncertain.[132]

The narrower bottle shown in figure 5.42e, although resembling earlier New Kingdom examples, has a possible Twenty-Sixth to Thirtieth Dynasty parallel from Karnak North, although the Karnak example has no painted decoration.[133]

Marl Jar 5

This type is very rare at Dorginarti, with only three sherds (two joining) having been found in Level III–IV contexts. It is quite distinct, however, and is comparable to a handful of jars found elsewhere (fig. 5.44a–b). Two marl jar rims that precisely mirror those from Dorginarti were found among the late dynastic materials at Amarna.[134] Similar silt types, although of smaller diameters and undoubtedly having narrower body widths, come from the temple of Seti I at Qurna.[135]

Marl Jars with Wide Mouths

Wide-mouthed jar fragments were common at the fortress, much like the sherds of the SB 1, SJ 1, or SJ 2 vessels. There are two varieties, including a finer type with a more sinuous rim and thinner walls, and a heavier-walled and thicker-rimmed version

127 Lister 1967, p. 63. For the dating, see Adams 2005, pp. 90–91. One radiocarbon date from site 6-G-9, which gave a calibrated date of ca. 800 BC, is not cited in the literature (see Adams 2005, pp. 28–29; Adams 2004, p. 67). The sample is, however, published in Trautman and Willis 1966, p. 191. It is to be noted that radiocarbon dates for the early first millennium BC remain somewhat unreliable, particularly considering the Hallstatt plateau flattening between ca. 800 and 400 BC.

128 The linear band decoration is discussed in Schreiber 2003, pp. 43–45, where he mentions the continuity of this tradition from the New Kingdom on, particularly on small vessels (note p. 45, n. 12). See also Schreiber 2002 (particularly pp. 411–15). The vessel type developed beginning in the eighth to seventh centuries, and became popular again starting in the fourth century, but especially during the third to second centuries BC. The decoration on the later vessels consists solely of black lines in its first phase of development, which dates to around the middle of the fourth century.

129 Jacquet-Gordon 2012, fig. 91a (made from "Marl A4 tardive," Twenty-First through Twenty-Fifth Dynasties); also Aston 1996c, p. 32, pl. V, no. 86 (Marl A4, variant 2). A similar rim and neck form (called a *zir*) was found in Phase IIb deposits at Elephantine and is shown in Aston 1999, p. 91, no. 689 (of Marl D). There is a fairly similar marl form from the temple of Seti I, shown in Myśliwiec 1987, p. 59, no. 384, and p. 61, silt with a polished exterior and three incised bands around the neck (Embalmer's Cache 1); marl similarities are shown on pp. 66–67, nos. 530–31, 538. The pottery from this layer of the temple's magazines is dated by the excavator to the Twenty-Fifth and Twenty-Sixth Dynasties. Aston dates all the materials to no earlier than the second half of the seventh century BC, and more likely to the sixth century; see Aston 2003, p. 153.

130 Schreiber 2008, p. 63, pl. LVIII, no. 23, 2.1.7, and p. 73, pl. LXVIII, no. 37 (Tomb B). The reuse of Theban Tomb 32, discussed in Schreiber's section 2.1.7, is thought to be earlier than the intrusive Tomb B, which is dated on the basis of its archaeological remains to the later Third Intermediate Period.

131 Sullivan 2013, pp. 187, 224, type 10-14 (silt, Stratum 2), and p. 238, UN-22 (red-slipped silt, Stratum 2).

132 For the shapes closest to MJ 4, see Heidorn 1994, pp. 121–22, fig. 2k–l (Kurru 19), and p. 126, fig. 5b–c (Kurru 14, marl). See also Schreiber 2008, p. 63, pl. LVIII, no. 23 (perhaps Saite?), and p. 73, pl. LXVIII, no. 37 (beginning around the mid-eighth century BC), which are close in rim and neck form to both the decorated silt and marl jars from the ancestral tombs. Jar 2.2.4, no. 37, has a raised ridge at the join between the neck and shoulder and is a red-slipped silt with a white-painted band spiraling down the body exterior, while jar 2.1.7, no. 23, has a channel with raised ridge just below the neck and is a greenish-gray marl.

133 Jacquet-Gordon 2012, p. 265, fig. 105q. The diameter is 12 cm.

134 French 1986, p. 183, fig. 9.17, types MJ4.1.1 and MJ4.1.2, and the similar types MJ3.3.1 and MJ4.1.3.

135 Myśliwiec 1987, pp. 57–58, nos. 370–71.

(figs. 5.44c–e, 5.45; pl. 66b).[136] Sherds of small and often twisted loop handles and rounded bases—with fingerprint impressions on the interiors where they were pressed into a mold—were found in the same contexts as the rims. The body sherds of the thinner-walled vessels often displayed some corrugation or ribbing, and although the two types were often found in the same contexts, there may be a chronological overlap in the two varieties or a functional difference.

The parallels are far ranging, from sites in both Egypt and Nubia, including the Napatan remains from the Soniyat temple and Napatan/proto-Kushite sites surveyed along the Debba Bend of the Nile.[137] A complete jar, with the more sinuous rim form, was also found in tomb S716 at Semna.[138] Qasr Ibrim also has this form.[139] Early first-millennium examples, ancestors of the New Kingdom prototypical "meat jar," were found at Elephantine, where the thinner-walled types are attributed to the late eighth century BC and perhaps into the seventh century.[140] All the earlier jars at Elephantine display the heavier rim variant. Ashmunein, Amarna, Karnak, and other sites all have examples of these jars in either Third Intermediate Period or seventh-century assemblages.[141]

The thinner-walled type, with a sinuous rim, is also found in small numbers at Tanis in late Third Intermediate Period and Saite contexts, and is made of Upper Egyptian marl.[142]

Other Marl Jars

This general category encompasses a few jars that are unique at Dorginarti (fig. 5.46a–c). The sherds of the everted-rim jar shown in figure 5.46b were found in a couple of neighboring loci in the West Sector and probably belong to the same jar.[143] This context was below the later north–south wall in the clearance of D 7 (1)—that is, the surface to "ash" layer atop D 66 and D 67. A parallel, made of a white-slipped Nile D, was found among the Twenty-First through Twenty-Fifth Dynasty materials at Karnak North.[144]

One example of a decorated, wide-mouthed vessel with a pillowed rim has a handful of parallels at Abu ʿId and Elephantine in Upper Egypt (fig. 5.46c).[145] The sherd was found in D 1 (2). One of the jars from Elephantine was certainly made in the same workshop, given its very close resemblance in shape and decoration; it was found in Phase IIa contexts at that site. But there is another example from a context in Phase IIb. The Dorginarti sherd has a crisscrossed net pattern painted in broad, black strokes on the shoulder, with the tips of the lines running up onto the exterior side of the everted, pillowed rim.[146]

---

136 The two types are also found in Level I at the Mut Temple at Karnak, a level assigned to the Late Period; see Sullivan 2013, pp. 182, 219, types 5-8 and 5-9. They are also shown in French 1986, p. 183, fig. 9.17 (MJ3.1.1, MJ3.2.1, MJ3.2.2).

137 Note Phillips 2003, pp. 397–99, pl. 12. These two thinner-walled vessels are dated to the proto-Kushite period, ca. 1100–900 BC. They stem from the sites of Agri 2 and Abukur 46/47. See also the thinner-walled examples in Orzechowska 2003, pp. 442, 450, pl. 1f, g. Soniyat's assemblage is mirrored at another temple site, Usli 1 (pp. 383–84), and both sites are attributed to the Napatan period (i.e., the Twenty-Fifth Dynasty).

138 For a description of the tomb at Semna and its remains, and for a photograph of the jar, see Dunham and Janssen 1960, pp. 108–9, pl. 116, no. 68 ("green ware" with red interior). The jar, 24-3-743, is illustrated in the excavation's registration records at the Museum of Fine Arts, Boston. The jar appears to be a smaller, thin-walled variant.

139 Rose 2019, fig. 35; first noted in French 1986, p. 166. Here, as for other marl jars like those found at Dorginarti, French cites parallels between the Amarna materials and those found in the materials from Qasr Ibrim.

140 Aston 1999, pp. 191–97, nos. 1733, 1755, 1785–86, all of Phase III; see also the small earlier jars, pp. 184–85, nos. 1689–90. These all have the thinner rim profiles rather than the heavier rim variant. The latter are similar to the wide-mouthed silt and Marl A4, variant 1, jars of Aston's Phases I–II; see pp. 36–37, no. 150 (silt, red-rimmed on white slip), and pp. 39–40, no. 176 (Marl A4, variant 1), both of Phase IIa. Phase IIb examples are shown on pp. 82–83, no. 615 (Marl A4, variant 1).

141 The parallels are too numerous to cite here, and they do not add to any argument for a more precise date. See, for example, Jacquet-Gordon 2012, p. 259, fig. 102e–h (Twenty-Sixth through Thirtieth Dynasty levels; only fig. 102h shows a curving rim), and Sullivan 2013, p. 228, types 14-7 through 14-10 (Strata 1–3 examples; both rim variants). The parallels from the late dynastic deposits around Amarna's South Tombs are shown in French 1986, p. 183, fig. 9.17, MJ3.1.1–MJ3.2.2 (both variants).

142 Laemmel 2011, p. 52, pl. 42, nos. 217–18.

143 The upper levels in this area had later remains atop them, and some of the unusual sherds found may belong to the Level II use of the fortress.

144 Jacquet-Gordon 2012, p. 243, fig. 94g.

145 The similar jar from Elephantine is published in Aston 1999, pp. 39, 41, no. 180. This jar is dated to Phase IIa and is made of Marl A4, variant 1, an earlier fabric that does not continue past Phase IIb. Another fairly similar shape was found in Phase IIb; see pp. 82–83, no. 620, also made from the earlier marl fabric but with red-painted decoration. For the parallel from Abu ʿId, see Aston 1996c, p. 29, pl. IV, no. 55, and attributed to sometime around the reign of Piye.

146 For a detailed analysis of the decorative elements found in the New Kingdom through the Third Intermediate Period, see Aston 1996a, pp. 79–82.

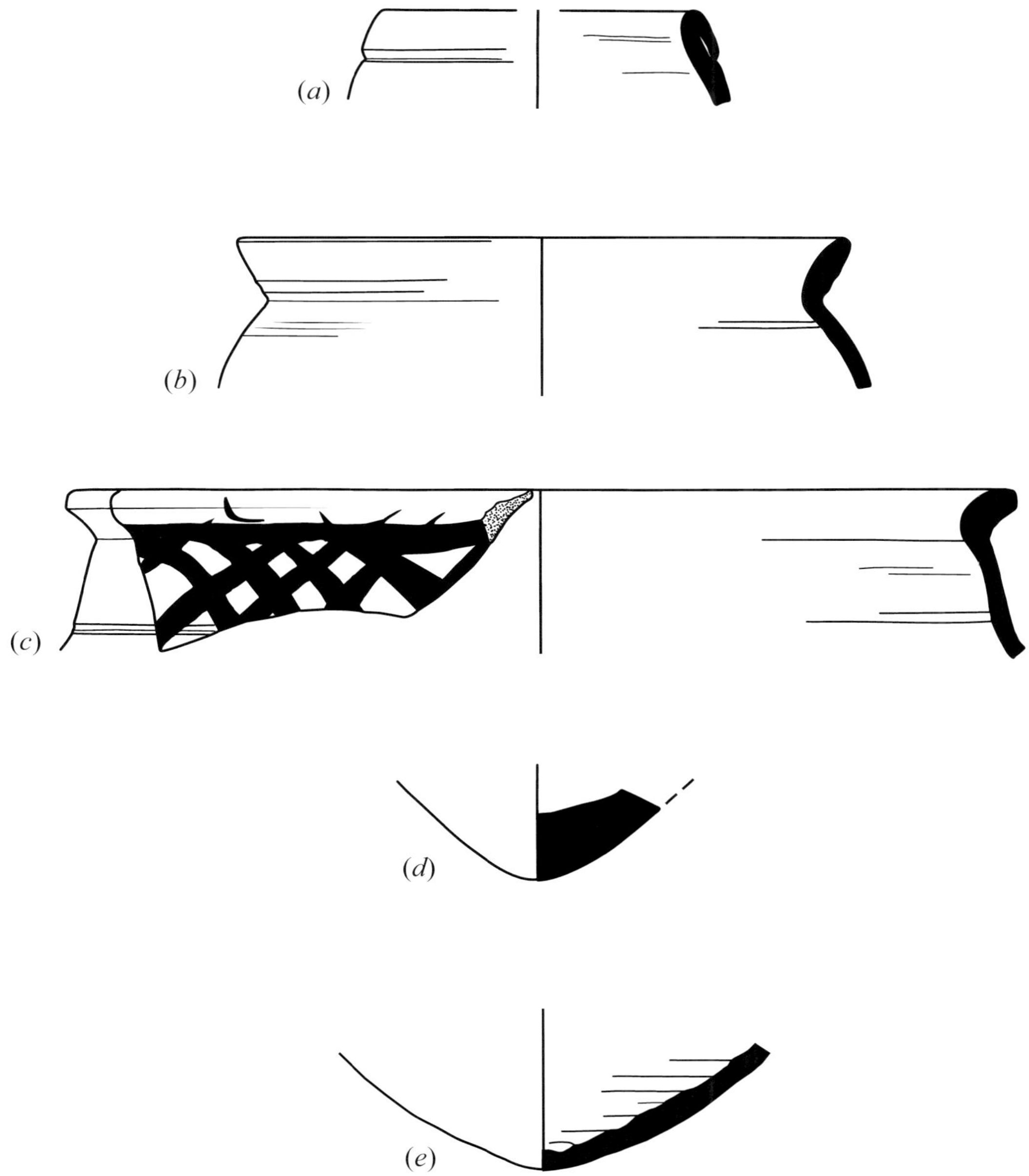

Figure 5.46. Marl jar variants and bases (*c*, 1:5; *a–b*, *d–e*, 2:5)

(*a*) D 104 (1)

(*b*) D 67 (1) and D 66 (1); sherds of same jar in D 7 (1)

(*c*) D 1 (2); OIM E49701

(*d*) D 12 (1)

(*e*) D 69 (1); base

Marl Bases and Handles

The various handles and bases from marl jars are shown in figures 5.46d–e, 5.47. The bases of the MJ 1 presented above are rounded and hand-pressed into a mold, and it remains uncertain whether this base was used for some of the other forms. Those shown in figure 5.46 are tapered. Not many marl jar bases are present in the sherd bags from the site.

Only one of the handles, shown in figure 5.47e, has intentional scored marks on it; only a handful of other loop handles in silt and marl are decorated in the same way.

SILT AND MARL JUGS

There are two reconstructable, one-handled silt vessels with flat or carinated bases (figs. 5.48a, 5.49a; pl. 67a, c). The bodies are ovoid with rounded shoulders that slope up to a vertical neck and simple rim. The example in figure 5.48a has widely spaced, vertical burnish strokes preserved on its exterior surface, with horizontal strokes on its neck and some traces of burnish on the handle. Both jugs have bright-red slips on their exteriors, and the example shown in figure 5.49a possibly also has a red slip on the interior. Only the jar from the intermediate clearance levels below the Level II stratum, in D 8, is technically a jug, since it has a slightly oval-shaped mouth to make pouring easier. These forms were rare at the site, with about three additional sherds identified as possible jug rims.

Since the rims were mostly complete for both vessels, it is clear that these vessels had one handle attached at the rim and top of the shoulder. It is less clear that the examples in figures 5.48b and 5.49b are one-handled pitchers. The handles are attached near or at the rims, but not much of the rims is preserved. The carinated base was found in the same context as figure 5.49b and may be part of the same vessel.

The parallels for these vessels are sparse, with two examples from the Napatan residence at Kerma, a dissimilar example from levels 1–2 at Amara West, and two others found with later burials in reused New Kingdom tombs at Sai.[147] Pitchers or jugs are well known throughout the Iron Age in the Levant, where a number of less precise counterparts may be identified.[148] Like all Near Eastern sites of the late Iron Age, Egyptian and Nubian sites have pottery assemblages that reflect the movement of people and concomitant mix of cultures and material goods at that time. In the Levant they are often called *cooking jugs*, but the examples from Dorginarti show no indication of being used in this way.

FLASKS

The marl flask rims from the site show a variety of forms, but many have very wide rims that could reach a diameter of more than 10 cm (figs. 5.50–5.52). In most cases the sherds found with the marl rim and handle forms did not join, and only two flasks, discussed below, are complete. The wider-mouthed examples that were more common at the site usually exhibit evidence of a handle that was attached under the rim and down to the top of the body. A number of silt and imported variants were also found.

The oasis-ware flask sherds stem almost entirely from mixed deposits containing Level II sherds or from upper levels of fill,[149] and it is assumed that they were associated with the Level II activity at Dorginarti.[150] They are discussed in chapter 6.

Most of the marl body sherds show brownish-black to black concentric bands, but there are a few with worn, red-painted bands.[151] A few sherds show decorative

147 The jugs from Kerma are published in Mohamed Ahmed 1991, p. 54, fig. 18 (I C1a and I C1b). Both date to Phase II of the Napatan Building I, which is dated to the late seventh through the mid-sixth centuries BC. See also the ring-based example from levels 1–2 of House D.14.16 at Amara West, in Spencer 2002, p. 27 (G1), which is silt with red slip on the upper exterior and dated to the Napatan period. The examples from Sai are published in Minault-Gout and Thill 2012, p. 372, pls. 146, 167, T8 54, T20 7. Both are dated to the Napatan period.

148 Very similar types of cooking jugs or jars(?) are presented in Singer-Avitz 2006, p. 200. They are Judaean vessels. The coastal forms are also found in Judah; see Singer-Avitz 2006, pp. 206, 208, the latter being a Late Philistine Decorated Ware vessel with vertical streak burnishing and painted decoration. The examples shown in p. 220, fig. 7, nos. 1a–c have slight ring bases and date to the later eighth century BC. For the longevity of jugs and one-handled jars, see Amiran 1969, pp. 251–62, from Iron I–II, ca. 1200–587 BC.

149 See chapter 6, where the contexts are described, with mention of the one sherd from an oasis-ware bottle stemming from the fill between the Level III and Level IV official residences. It is uncertain at what level the sherd was found, but the fill of the later well dug between the two rooms, D 310 and D 314, was a mixed deposit and contained Christian sherds.

150 Oasis flasks have, however, been found in deposits that are later in date, such as the two from the tomb debris in Kurru 52 (2), 19-3-1162; one found in situ in the burial chamber of Kurru 55 (2), 19-3-1461; another found in situ in the burial chamber in Kurru 72 (4), 19-3-1558B; and so on. See Dunham 1950, pp. 83, 93–94, 105, fig. 35b, pl. XLII.E. They are, for example, also found in Twenty-First to Twenty-Fifth Dynasty contexts at Karnak North; see Jacquet-Gordon 2012, pp. 248, 286–88, figs. 97m, 118.

151 All the sherds with red-painted concentric bands were, however, from mixed contexts in the East Sector.

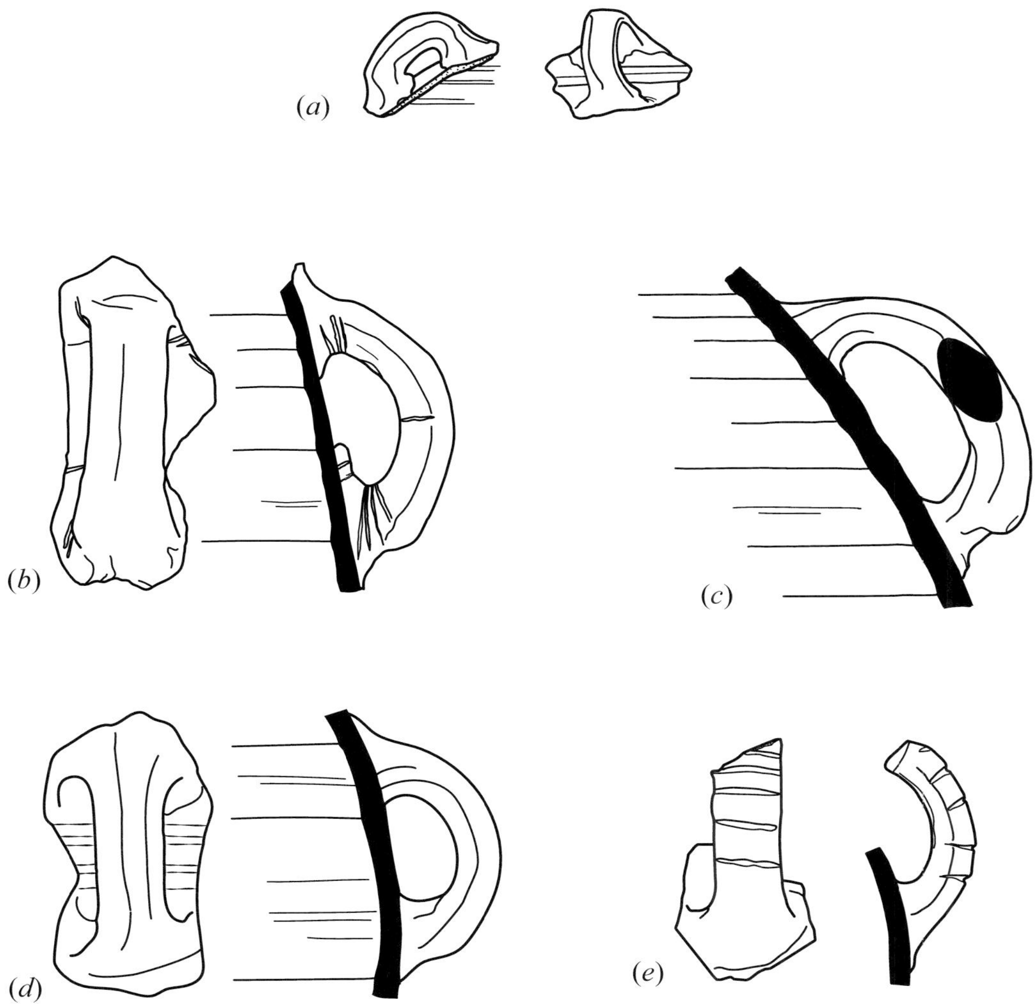

Figure 5.47. Marl jar handles (2:5)

(*a*) D 13 (1); stance and diameter uncertain
(*b*) D 23 (1)
(*c*) D 309, pit A
(*d*) D 9 (1)
(*e*) D 120; prefiring horizontal dashes on exterior handle

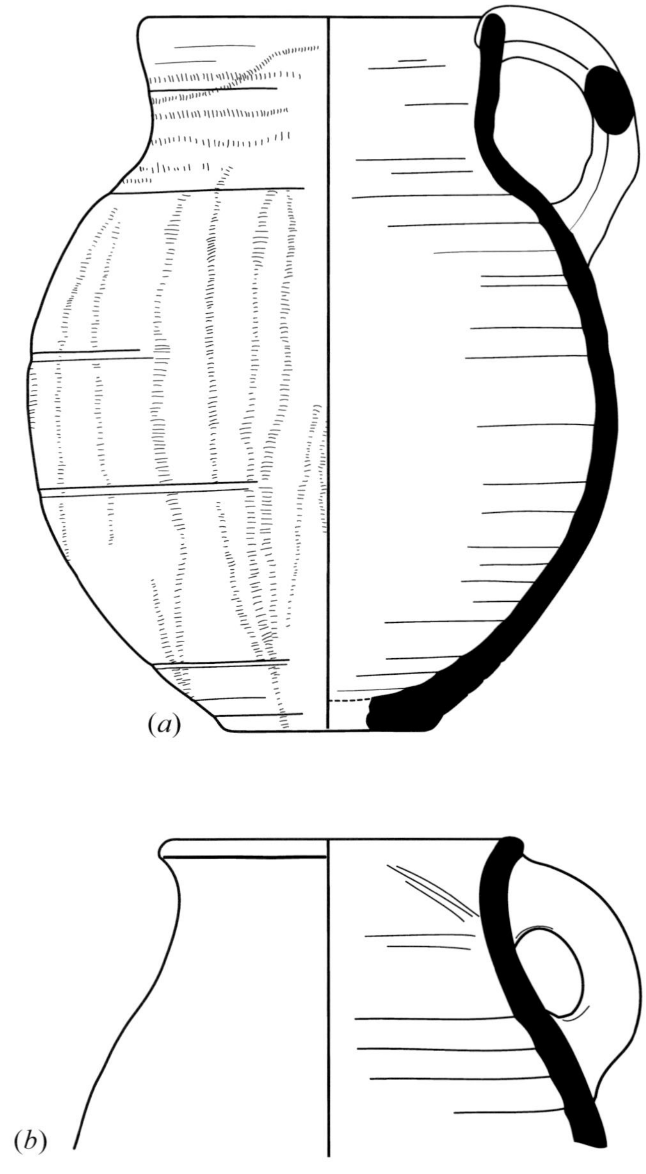

Figure 5.48. Silt jugs (2:5)

(*a*) D 219 (1); OIM E49524, RS exterior, burnishing strokes on exterior

(*b*) D 216 (1); one-handled(?)

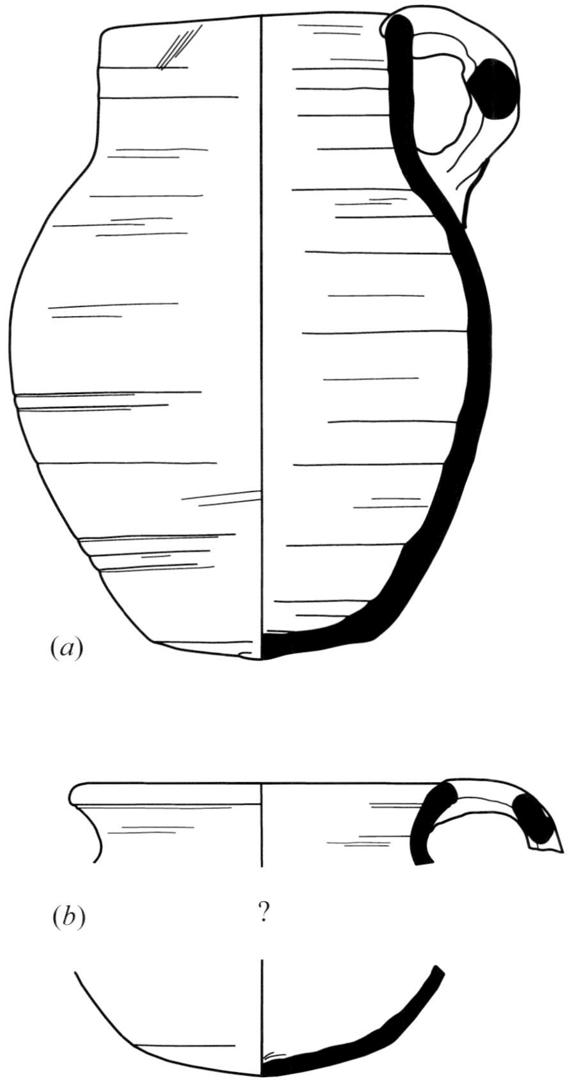

Figure 5.49. Silt and marl jugs (2:5)

(*a*) D 8, west of D 23; OIM E50737, slightly oval mouth, thin RS exterior, perhaps also interior

(*b*) D 112 and area; one-handled(?), marl

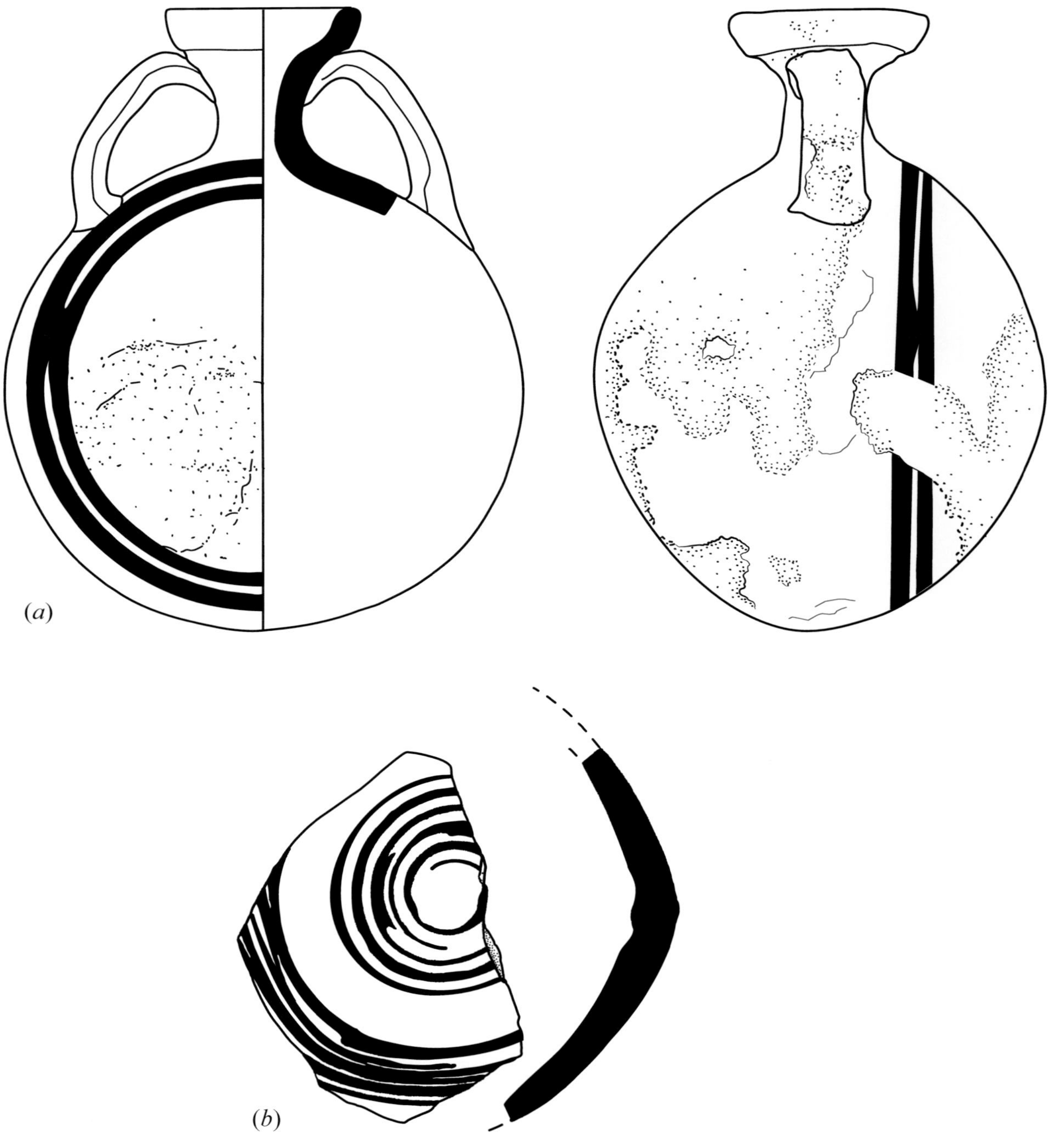

Figure 5.50. Marl flasks (2:5)

(*a*) D 61 just below the surface; OIM E24374, black paint, spalling surface, handle broken and reconstructed here

(*b*) D 102 (1); pilgrim flask body profile, black paint

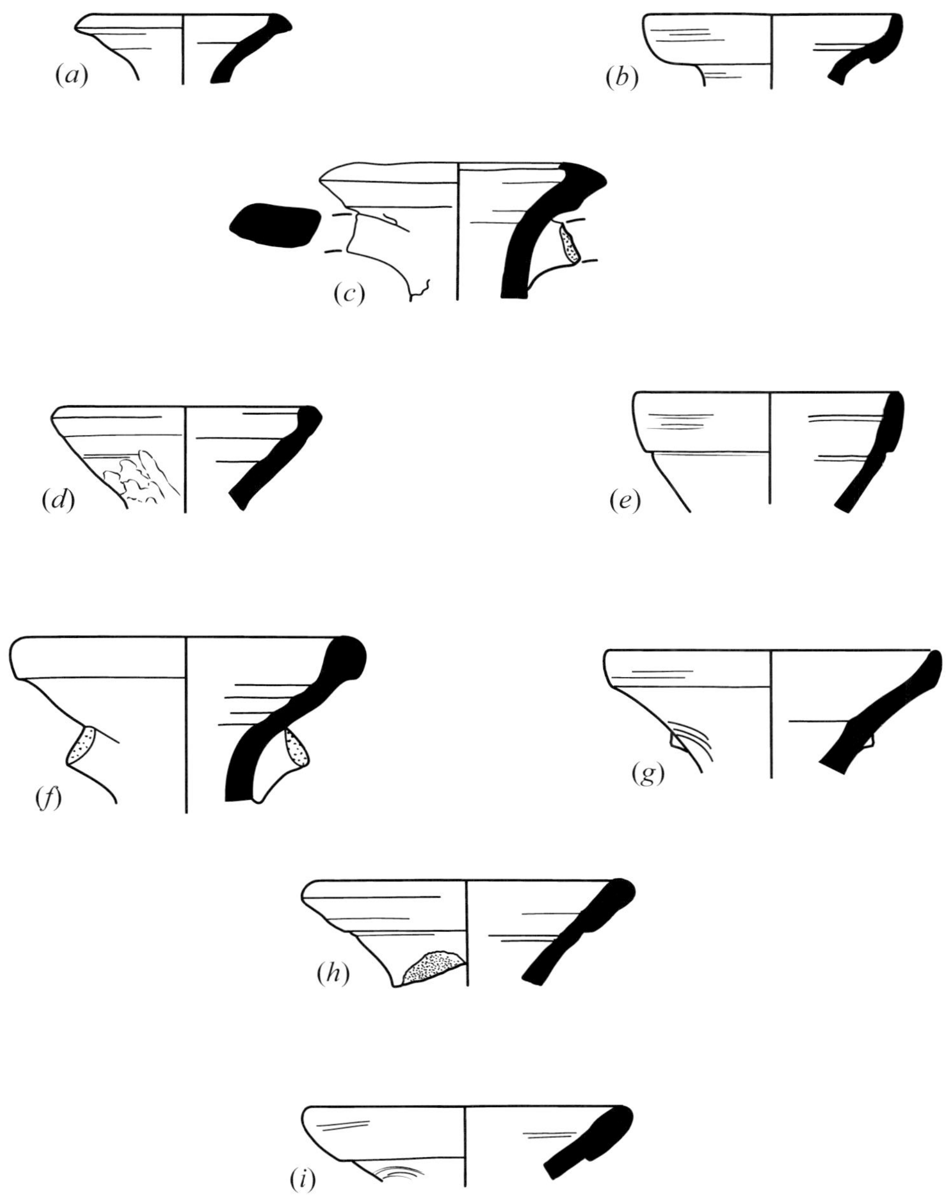

Figure 5.51. Marl flask rims (2:5)

(*a*) D 106 (1)

(*b*) D 55, surface to "ash" layer

(*c*) D 7 (1); OIM E49513, 10YR 7/3 exterior, very pale brown

(*d*) D 41; handle area, associated sherds with concentric black rings

(*e*) D 13 (1); very white surfaces, stance?

(*f*) West Sector surface

(*g*) D 360; handle stamp

(*h*) D 80 (1); handle scar

(*i*) D 79 (1); handle area

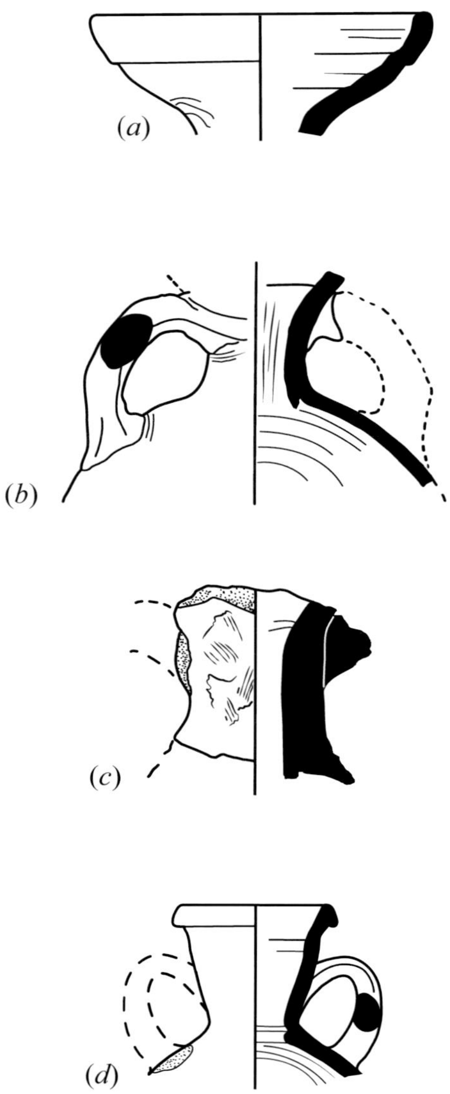

Figure 5.52. Marl flasks and juglet (2:5)

(*a*) D 12 (1); handle scar, same vessel as D 13 (1) sherd in fig. 5.51e(?)

(*b*) D 1 (1); similar sherds from D 5 test pit, D 59 (1), and D 350 (1)

(*c*) D 19 (1); flask neck with handle scars, rough finishing marks

(*d*) D 110 (1)

elements on the body below the handles (pl. 68a). The marl flasks, along with the MJ 4 sherds, represent the majority of decorated pottery in the collection. Sherds from marl flasks are found in moderate numbers in this material. A number of silt flasks with red slip are also present, but they are rarities.

Flasks are believed to have held ointments or condiments and have been found at sites across the entire spectrum from the New Kingdom into the first millennium in Egypt and Nubia, either as imports or as imitations of Levantine and Cypriote vessels.[152] The Levantine and Cypriote flasks of the Late Bronze Age and Iron Age were largely replaced starting in the ninth century by imports or imitations of the one-handled bichrome jugs from the Phoenician sphere. A few of these, particularly the so-called *mushroom-lipped jugs*, are found in Egypt, but they are widespread in the Levant.[153] While these particular imports are not found at Dorginarti, one earlier Iron Age Phoenician globular flask was found at Dorginarti and is discussed below.

The marl flasks represent late New Kingdom traditions with a new first-millennium BC flair for globular bodies and wide rims. There are no examples of the small flasks with tiny pinched handles at the neck that appear in the later eighth century BC and continue into the sixth century in Egypt.[154] They are, however, present at sites in Nubia.[155] Perhaps all the wider-rimmed marl, and most of the silt, flask sherds are from the earlier Level IV occupation at Dorginarti, but this is not certain.

152 The development of the flask during the first millennium BC in Egypt is discussed in Aston 2009, p. 323, and Aston 1996a, p. 89.

153 The mushroom-lipped variant occurs first in the ninth century. For examples from Syria and Lebanon, see Lehmann 1998, pp. 9, 14, and fig. 3, no. 16, which is the predecessor of the true mushroom-lipped jugs shown in his fig. 3, nos. 13–15 (ca. 750–700 BC). For the jugs from Tell el-Retaba, Giza, Shaganbeh, and Thebes, see Aston 1996a, pp. 80–84, fig. 233g–h.

154 For examples from Tell Qedua, see Oren 1984, p. 22, fig. 25, nos. 3 and 5, from the late seventh/early sixth century BC. An example is also shown in Hölscher 1954, p. 73, pl. 47, U2.

155 There is an unpublished sherd from one of these pinch-handled flasks in the materials from the tomb of Aspelta, Nuri 8 (Generation 10) in the Museum of Fine Arts, Boston (personal observation 2015). The flask may be from one of the foundation deposits of the pyramid. However, these flasks appear in tombs attributed to earlier periods at Meroe and el-Kurru; see Dunham 1963, p. 337, fig. A, 31 (23-3-154) from W 503 (Generations 3–4, ca. 747–690 BC); and for examples made of stone, faience, and pottery, see Dunham 1950, p. 80, fig. 27c; p. 89, fig. 29d; p. 92, fig. 30b(?); and p. 105, fig. 35c. The tombs are attributed to ca. 747–690 BC. This sort of small flask also occurs in the materials from Missiminia and Qustul; see Williams 1990, pp. 9, 59, fig. 9b.

One almost complete flask, with a globular body, was recovered from a surface layer in locus D 61 (OIM E24374) (fig. 5.50a; pl. 68c).[156] The rim form has a smaller diameter than many of the other types at the site. This example has splintering surfaces, and the fabric is very weathered, showing plentiful sand and limestone temper. This splintering condition is atypical of the other marl vessels from the site, even when they also show signs of water saturation. A flask from an early first-millennium BC tomb at Amara West is comparable, and more parallels come from the tombs at Sai.[157] A comparison from Medinet Habu was found in contexts assigned to the Twenty-Second Dynasty.[158]

Flasks or bottles with wide rims are the most common type at Dorginarti, as mentioned above (figs. 5.51b–i, 5.52a; pl. 68e). A number of rims are wide and cup-shaped.[159] They were manufactured of a well-fired marl fabric, and the body was made in two halves that were then joined, with the neck inserted at the top. There is no indication that any of the flask bodies were asymmetrical in shape, although only fragments of their bodies accompanied the rims. One body sherd had a slight protrusion or nipple in the center of its side, which is also found on flasks uncovered from a tomb at Hillat el-Arab (fig. 5.50b; pl. 68d), where similar flask rims were also found.[160]

The bodies of these flasks, as indicated by their sherds, are quite unlike the lenticular bodies of earlier flask forms, but they certainly derive from these prototypes. Flasks with spherical bodies are also found in the Levant in Iron Age I.[161] A close parallel from Egypt was uncovered at Medinet Habu in Tomb 19, located near the tombs of the God's Wives of Amun.[162] It has been suggested that it is an earlier vessel that was reused in this Twenty-Fifth or Twenty-Sixth Dynasty tomb. Whatever the original dating, its body profile is not very globular, although its rim is very wide.

Most of the parallels seem to be from sites in Upper Egypt or Nubia, but a precise parallel for one of the rims, with an inward hook at the lip (fig. 5.51c), comes from a context by the gate of Sheshonq III in Tanis, and there is also a white-slipped silt parallel to the wide mouth of figure 5.51g, which also comes from Tanis and has been dated between the Twenty-First and Twenty-Second Dynasties.[163] In addition to the comparable flask from tomb 19 at Medinet Habu in Egypt, the wide-mouthed flasks find parallels in Nubia, including those from early first-millennium BC tombs at Amara West and Hillat el-Arab, as mentioned above, in a Napatan grave at Missiminia, and in the fill of a tomb at Sanam.[164] Examples also stem from the reuse of the New Kingdom tombs at Sai, and there is perhaps one example from Gala Abu Ahmed.[165]

156 The other example was from the Central Sector in D 8 (1), in the Level II fill on the west side.

157 Binder 2011, p. 49, pl. 19 (C9007), which is a marl vessel with red concentric bands from a chamber tomb, Grave 211, with multiple burials and well looted. Chamber tombs are the earliest form of grave in the cemetery, possibly ranging in date from the tenth/ninth centuries BC to no later than the mid-eighth century (Binder 2011, pp. 50–51). For the flasks from Sai, see Minault-Gout and Thill, 2012, pl. 140, particularly T7 30 and T7 42.

158 Hölscher 1954, p. 74, pl. 47, type U5.

159 See, for example, the cup-like mouth on an example in Binder 2011, pp. 41–42, fig. 4, pl. 4, from a chamber tomb, Grave 200, with at least five adults buried in it. The pottery shown with the flask consists mostly of wheel-made beakers and bowls with red rims.

160 For flask rims comparable to those at Dorginarti, see Vincentelli 2006, figs. 2.38–2.39, 2.89–2.92. The examples in Vincentelli's fig. 2.38 are foreign imports that are precise parallels to the contour of the Dorginarti flask sherd. The examples are dated to the late New Kingdom and Third Intermediate Period; see Vincentelli 2006, p. 57. This type of flask body is very common in Iron I in the Levant and Cyprus; see, for example, Gilboa 1999, p. 4, fig. 2.2, and p. 7, fig. 5.9.

161 For more examples, see the flasks in stratum 9c at Tell Keisan, shown in Briend and Humbert 1980, pls. 74–76. These are dated to ca. 980 BC.

162 For the flask, see Hölscher 1954, pp. 31–32, pl. 47, U7. For the suggestion that the flask is perhaps reused and more like Levantine flasks of the Late Bronze IIB, see Aston 1996a, p. 53, with reference to a similar flask from a Ramesside context at Qantir.

163 Bavay 1998, pp. 321–22, fig. 33, no. 17. This example is silt with a red-slipped exterior. See also Laemmel 2011, p. 36, pl. 26, nos. 148–49. These marl forms are dated to the Twentieth to Twenty-First Dynasties. She notes that their appearance in the south (the example in tomb 19 at Medinet Habu) may continue later, into the Twenty-Fifth and Twenty-Sixth Dynasties.

164 Griffith 1923, p. 98, pl. 17, VIIh. This marl flask came from the fill of tomb 1009 and had a rim diameter of 9.1 cm according to the records at the Griffith Institute, Oxford. The tomb itself contained a wheel-made beaker (XIII) that served as a lid for a very ribbed marl jar (IIIf), which is like the vertical-rimmed examples from Dorginarti. The other example is found in Vila 1980, p. 74, fig. 70 (tomb 2-V-6/199), with flask description on p. 160, fig. 180. It was the only item buried with the body.

165 Minault-Gout and Thill 2012, pp. 354–58, pl. 140. The parallels are at the bottom of pl. 140 and stem from tombs 5, 7, and 14. Tombs 3 and 20 contained eleven examples each, and some of the descriptions of the examples not presented point to a cupped rim and spherical body, which would likely indicate a Third Intermediate Period or early Napatan date. See also Jesse 2016, p. 306, fig. 4, no. 12 (marl), and p. 308.

Marl One-Handled Flask

This one-handled marl flask has a precise parallel from site K176, Abudiya, near Sai (fig. 5.52d).[166] The more complete form from site K176 indicates there is one handle. The marl pottery group from this deposit includes MJ 2 forms, wide-mouthed marl jars, marl jars with vertical and everted rims (i.e., MJ 3), and a couple of marl bowl forms, as well as rounded and ring bases that accompanied the rim sherds. The flask, although preserved with one handle at Dorginarti and perhaps also at Abudiya, may find a parallel in a two-handled bottle at the Sanam cemetery.[167]

Silt Flasks

The silt flasks exhibit a variety of shapes, most of them unique in the assemblage (fig. 5.53). Parallels to the silt flask in figure 5.53a were uncovered in Twentieth through Twenty-First Dynasty contexts at Qantir, and there is a later Third Intermediate Period or Saite parallel from Tanis.[168] Comparisons for figure 5.53b were found at Nebesheh, which Aston dates to the twelfth to tenth centuries BC.[169] Another comparable flask was found at Memphis and attributed to the eleventh to tenth centuries BC.[170] The long neck and handles are similar to types from the Syro-Palestinian region in the early Iron Age.[171]

Double Flask

A double flask, an entire vessel, shows a red slip on its exterior that was at one time highly burnished (fig. 5.53c; pl. 69a). The vessel is complete, with no chips, but the fabric should be silt.[172] The surface exhibits a bright-red slip and shows many marks from the hand finishing of the vessel.

Double vessels were popular in New Kingdom Egypt, and a marl flask-pitcher combination dating to the Eighteenth Dynasty was found at Fadrus, in Nubia.[173] There is a double vessel from Third Intermediate Period contexts from Ashmunein in Egypt, and a four-handled double flask from Matmar made from an oasis fabric.[174] Another red-slipped and burnished double flask, published in the *Manual of Egyptian Pottery* and described as Late Period in date, comes from an uncertain location.[175] The best parallel, however, is that from the reuse of tomb 20 at Sai.[176]

Iron Age Phoenician Bichrome Flask

This one-handled vessel is an imported globular jug or flask, but its place of manufacture is uncertain (OIM E24327) (fig. 5.54; pl. 69b). Similar bichrome vessels have been found in Iron Age I–early IIA contexts throughout the coastal Levant and Cyprus. The flask's neck ridge and double-stranded handles are common to these bichrome forms. The thick, red concentric bands between thin black borders are also quite distinctive for flasks described as Cypriote or Phoenician bichrome.[177] Petrographic analysis of other vessels has demonstrated that the production of similar bichrome vessels occurred on Cyprus, along the Lebanese coast

166 Geus 1979, pp. 18–19, fig. 8, no. 2.

167 Griffith 1923, p. 144, pl. XVII, Va, where it is found with types XIIi and marl jar IIIc, in addition to red-rimmed bowls. Also see Lohwasser 2012, vol. 2, pp. 43, 230, fig. 65.

168 Laemmel 2008, pp. 184–85, pl. 12, no. 9 (silt), with references to similarly dated flasks from Mendes, Tanis, and Memphis. See also Aston 1998, pp. 610–11, no. 2494. The later parallel from Tanis is presented in Laemmel 2011, p. 51, pl. 40, no. 214. The find context was disturbed, with mostly Third Intermediate Period pottery, but the area also yielded architecture of the later Third Intermediate Period and Saite era.

169 Aston 1996a, p. 137, fig. 35 (middle row, right-hand side), and p. 295, fig. 193e, for part of the Phase I pottery.

170 Aston and Jeffreys 2007, p. 53, fig. 47, no. 556, from the silo debris.

171 For one example, see Bikai 1987, pl. XXIV, no. 5, from Kouklia tomb 83/83, dated ca. 1050–850(?).

172 The vessel was on display at the National Museum in Khartoum in 1988 and could not be personally handled by me. Nadejda Reshetnikova drew the vessel in Khartoum in 2013, for which I thank her.

173 Bourriau 1981, p. 78, vessel 148 from Saft grave 55(?), dated between the reign of Amenhotep III and the early Nineteenth Dynasty. Bourriau cites the example from Nubia, which is published in Holthoer 1977, pp. 103–4, pls. 23, 56.

174 Spencer 1993, p. 47, and pl. 75, O2, which is a poorly made double bottle of silt (Level 3, Third Intermediate Period). The double flask from Matmar is shown in Aston and Bader 1998, pp. 42–44, fig. 40, no. 7. This particular group of pottery and objects is dated to the ninth and, more likely, eighth centuries BC because of the associated pottery.

175 The flask is shown in Wodzińska 2010, p. 258, Late Period no. 94.

176 Minault-Gout and Thill 2012, pp. 360–61, pls. 141, 161, T20 70.

177 See Gilboa 1999, p. 7, fig. 5 (Iron Age IB) and p. 13, fig. 10, no. 1 (Iron Age IB/IIA at Dor) for the containers that resemble the example from Dorginarti. Gilboa's chronology for the Dor material places Iron IB in the tenth to early ninth centuries BC, but others would date the Iron Age sequence earlier (see below). The bichrome decoration and transition to a more globular flask body appear in Dor's Iron Age IA, but the form and decoration thrive in Iron Age IB and perhaps continue into Iron Age IIA. See also Bikai 1987, pp. 37–41, 48–49, pls. III–V. The shape and decoration of the flask from Dorginarti is like Bikai's Cypriote Kouklia horizon examples, "ca. 1050–850(?)."

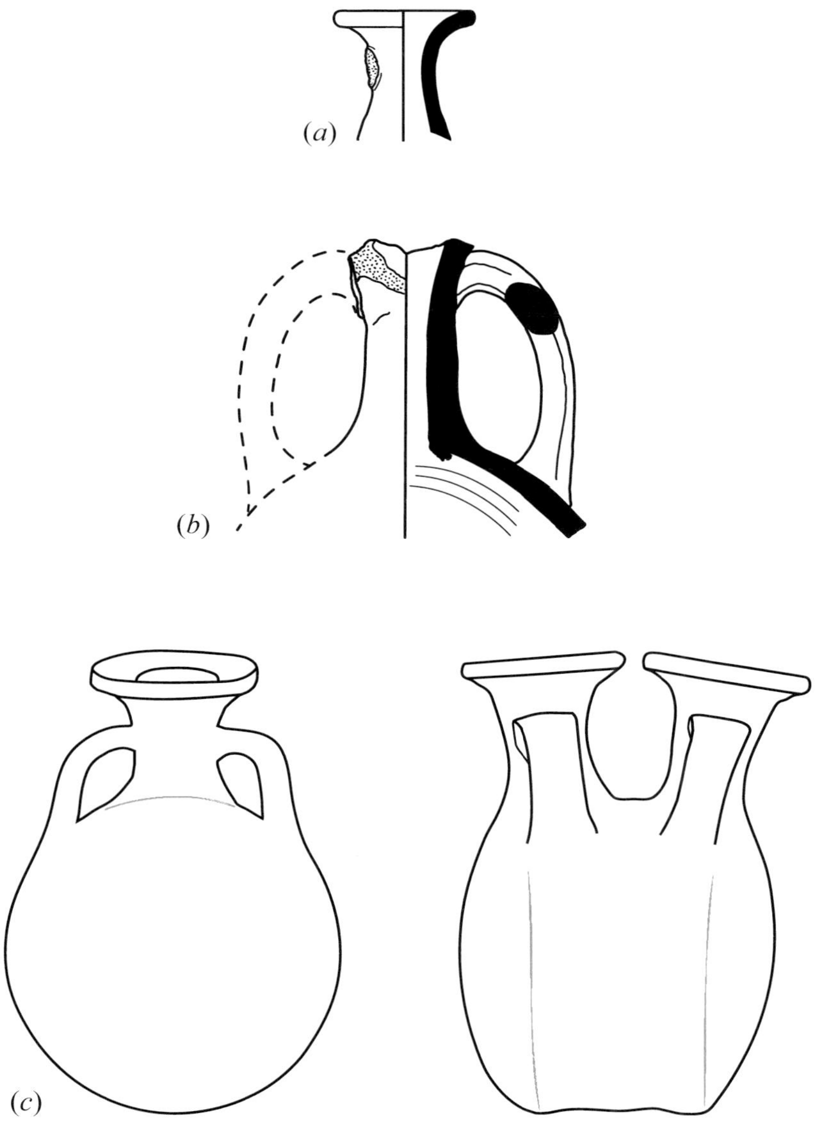

Figure 5.53. Silt flasks (2:5)

(*a*) D 38 and D 39; two-handled(?), thin RS exterior

(*b*) D 90 (1), room east of D 76; thin RS exterior

(*c*) D 122 (1); Khartoum 14267, RS exterior, burnished, drawing by N. Reschetnikova

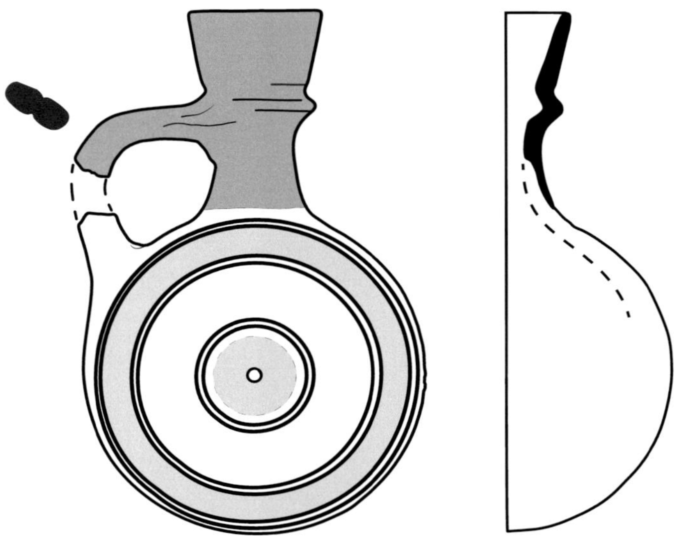

Figure 5.54. Imported flask (2:5)

D 19 (1); OIM E24327, burnished, thin black bands surround either unslipped(?) body surface or thick red bands (gray in drawing), black deposit or slip on neck, rim, and upper handle (drawing by C. Meyer)

(Phoenicia), in the western Galilee, on the Carmel coast, and in the Jezreel Valley.[178]

The fabric is smooth, with no visible signs of temper on the surface. Visual examination of the fabric was made difficult by the position of the handle fracture and the dark staining, with the surface slip and painted bands also being discolored, perhaps from the substance that once filled the vessel. Only small sands are noticeable. The vessel is quite heavy for its size, suggesting either very thick walls or an interior deposit. The double-stranded handle reaches from the ridge on the neck to the top of the shoulder, and the body is round. There is a faint polish left on its surface, and the painted bands are matte. Besides the red and black decoration described above, the topmost part of the vessel was stained, or perhaps slipped, with a dark coating that now appears almost black.

One Egyptian parallel was found at Tell Nebesheh and is thought to date to the eleventh or tenth century BC according to parallels from Megiddo.[179] The dating of Phoenician bichrome is difficult, but the form found here is usually dated to the tenth century BC and should perhaps be no later than around 900 BC.[180] The parallels to the flask from Dor, along the southern coast of the Levant, are sometimes dated to the tenth to mid-ninth centuries BC, according to a lower Iron Age chronology.[181]

The pottery found in the same locus is typical for the mixed assemblages at the site.[182] It was uncovered just below the surface in D 19 (1), a building outside the south wall of the fortress. The architectural remains were resting atop the surface of a second level of glacis stones built to shore up the glacis, after a relining of the exterior fortress wall had been built. The second layer of glacis stones partly covered the relining. This

178 Waiman-Barak 2020. See also Gilboa and Goren 2015.

179 Aston 2009, pp. 318, 326, fig. 18, nos. 2–3. These are grouped under Aston's Palestinian forms. He dates them to the eleventh to tenth centuries BC on the basis of parallels at Megiddo, Stratum VIA, but see Gilboa 1999, p. 19, n. 2, for a dating according to the lower Iron Age chronology (see the following footnotes).

180 Many thanks to Dr. David Schloen, who advised me on the dating of this vessel (personal communication February 10, 2017): "I agree with those who place the Iron IB to Iron IIA transition in the Southern Levant at ca. 950 BC, so I don't think material from Iron IB/IIA would be much later than 900, and could be solidly tenth century."

181 For a discussion of Iron Age chronology at Dor, see Gilboa, Sharon, and Zorn 2004. References to the unresolved chronological debate are cited in Gilboa and Goren 2015, p. 74, n. 8.

182 The associated pottery in D 19 (1) includes marl and silt jar sherds, a marl ring base, handmade bowl sherds (some with burnishing), SB 1s, a marl flask, and a fragment of a crucible. The objects from D 18 and D 19 include a faience cartouche ring fragment (OIM E24325), miniature pottery bowls—one with an incised decoration (all OIM E24399)—and stone net sinkers. The objects are discussed in chapter 7.

vessel may have been a gift or was perhaps acquired by someone as a souvenir, and then somehow arrived at Dorginarti.

STANDS, LIDS, BREAD PLATES, AND VATS

This section includes pot stands, lids, bread trays and plates, and similar kitchen and storage equipment.

Large utilitarian vessels that were used for the storage of food, liquids, and dried goods are very numerous at the fortress, typically in the form of the wide-mouthed silt jars discussed previously. But there is also a variety of even larger vessels, which would not have been easy to transport and therefore most likely served as stationary vats or basins. These forms were made from Nile B2 or—more commonly than was the case for the silt bowls and jars—from Nile C, and most of them showed no evidence of a slip.

Pot Stands

The most common type of pot stand at the fortress was initially referred to as a *Dorginarti stand* because of the quantity of its very distinctive sherds (fig. 5.55). The rim and base portion are very identifiable, and even small fragments can be recognized. The wall of the bottom two-thirds has an undulating profile resulting from wheel-turning bumps, and the top of the everted, pillow-like rim is always worn from use. The rim may be set at different angles, but it is always quite pillowed unless it had been worn down to expose the core and the voids resulting from the use of organic temper.

Examples of the form are found in Napatan contexts at Qasr Ibrim and Gebel Sahaba in Lower Nubia, and from Kawa.[183] A possible rim is found among the later pottery collection around the South Tombs of Amarna.[184] The Third Intermediate Period pottery from Heracleopolis Magna has a similar stand in stratum 25 (ca. 1029–1010 BC).[185]

Very few other silt stand types were found, and these resemble more closely the typical forms known from the second millennium BC onward (fig. 5.56a–b).[186] There are no stands made of marl clay.

Lids

Any shallow bowl could have functioned as a lid, but close examination of the sherds included in this section indicates their dedicated use as lids. They are uncommon in the Dorginarti material and are made of both marl and silt (figs. 5.56c–d, 5.57a–b). These forms are common from the New Kingdom onward, with examples from many sites.[187] The profiles of the comparative examples are not always precisely the same as the lids shown here.

A small ledged lid, made of very fine silt clay, is unique (fig. 5.57c). The interior has been blackened from burning or discolored by a dark residue. Although the top and bottom part of this vessel is missing, parallels for its shape have been found at sites in Egypt.[188] Comparable lids have also been found in Nubia.[189] The lid was from the surface remains at D 120 and may belong to the Level II activity at the site.

---

183 I would like to thank the members of the Scandinavian Joint Expedition who hosted me during my visit to Uppsala in 1988. For the wheel-made stand, see Säve-Söderbergh and Troy 1991, pp. 321–23, pl. 220 (bottom) and pl. 222, nos. 4–5. The wheel-made examples are nos. 143 and 154 in their catalog of finds, from Findplaces 15 and 32 within the fortress. The Qasr Ibrim examples stem from various contexts around the early Napatan gateway and from earlier excavations at the site (personal observation). Many thanks go to directors Mark Horton and Pamela Rose for allowing me to study this material in the 1990s. See also Spencer 2002, pp. 28–29, pls. 44–46, many from E.14.1.

184 French 1986, p. 182, fig. 9.16 (SB11.2.1).

185 Aston 2010, p. 61, pl. 18, no. 333, from Stratum 25. There are also slightly similar, although narrower, versions with a more vertical stance in Strata 7–8 (ca. 760–740 BC), pp. 68–69, 80, pl. 27, nos. 512, 527 (red-slipped and smaller diameters).

186 See, for example, Aston 1999, pp. 25–26, no. 55 (Phase IIa).

187 For examples, see Jacquet-Gordon 2012, p. 114, fig. 50bb–dd (New Kingdom), and p. 264, fig. 105d–f. The latter were found in the Twenty-Sixth through Thirtieth Dynasty layers.

188 Possible parallels—although drawn as bowls—are shown in Aston 1999, pp. 80–81, no. 597, and pp. 136–37, no. 1255 (possible silt lids from Phase IIb), and—drawn as a flat-topped lid—Jacquet-Gordon 2012, p. 279, fig. 112m (silt with red slip), from the Twenty-Sixth through Thirtieth Dynasty levels. A similar lid came from the end of the Eighteenth Dynasty and Ramesside levels (see fig. 83y). See also Sullivan 2013, p. 202, types B-2 and B-3, from Stratum 1 (Saite period), and Myśliwiec 1987, pp. 50–51, nos. 295–96. Examples are also found in the north of Egypt at Tell el-Maskhuta, in Paice 1986–87, p. 102, fig. 8, nos. 1–7, and at Defenneh, in Petrie 1888, pl. 36, nos. 84–85, 93 (Defenneh, Twenty-Sixth Dynasty).

189 A slightly similar parallel from Meroe is shown in Shinnie and Bradley 1980, p. 98, fig. 28, no. 11 (from the earlier levels in J 50c). For the Sanam lid, whose precise details are uncertain, see Griffith 1923, p. 89, pl. 18, no. 4 (marl?). One unpublished example in the Museum of Fine Arts, Boston, comes from the debris of Kurru Tumulus 1 (Generation A) (19-3-339). It is closest in form to the example from Meroe mentioned above. There are only slightly similar lids in Dunham 1950, p. 69, fig. 23c (Kurru 18), and p. 88, fig. 29b (Kurru 53), the latter having a wide diameter. These examples are from tombs dated between Generations 2 and 4.

(*a*)

(*b*)

Figure 5.55. Silt pot stands (2:5)

(*a*) D 92 (1); Nile C, worn rim interior

(*b*) D 41; finger marks on interior, worn rim interior

(*a*)

(*b*)

(*c*)

(*d*)

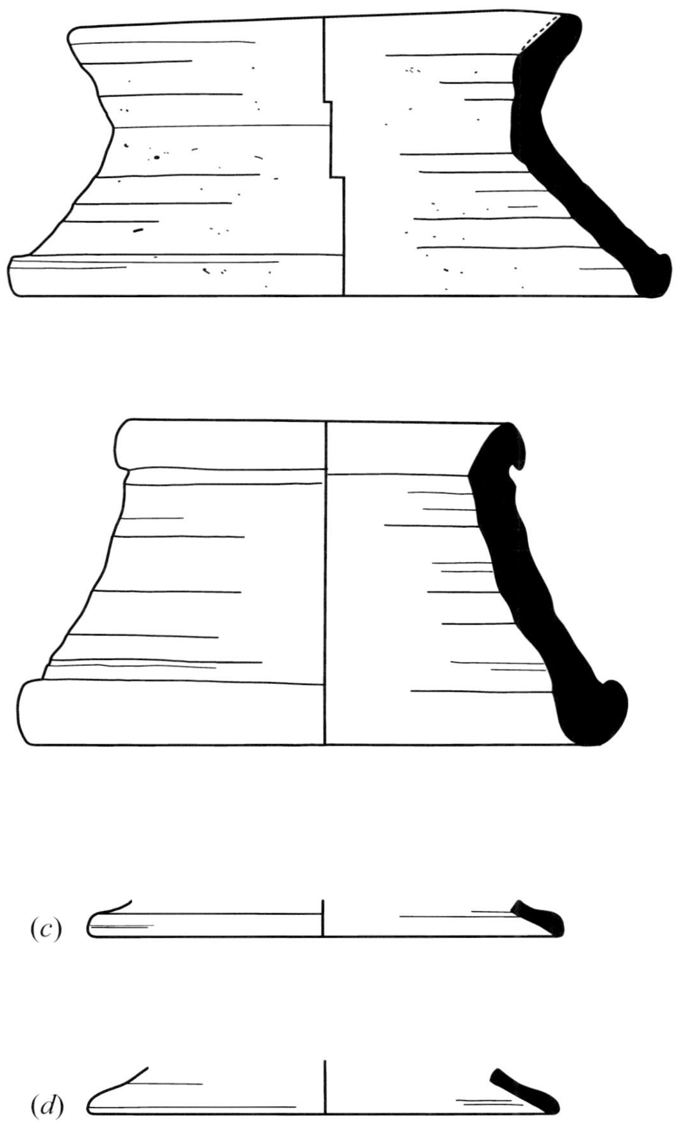

Figure 5.56. Pot stands and lids (2:5)

(*a*) Level II retaining wall, southwest corner on paving; OIM E24338

(*b*) D 36; rim interior very worn

(*c*) D 80 (1); marl

(*d*) D 115, fill under oven; Nile B2, blackened interior/exterior

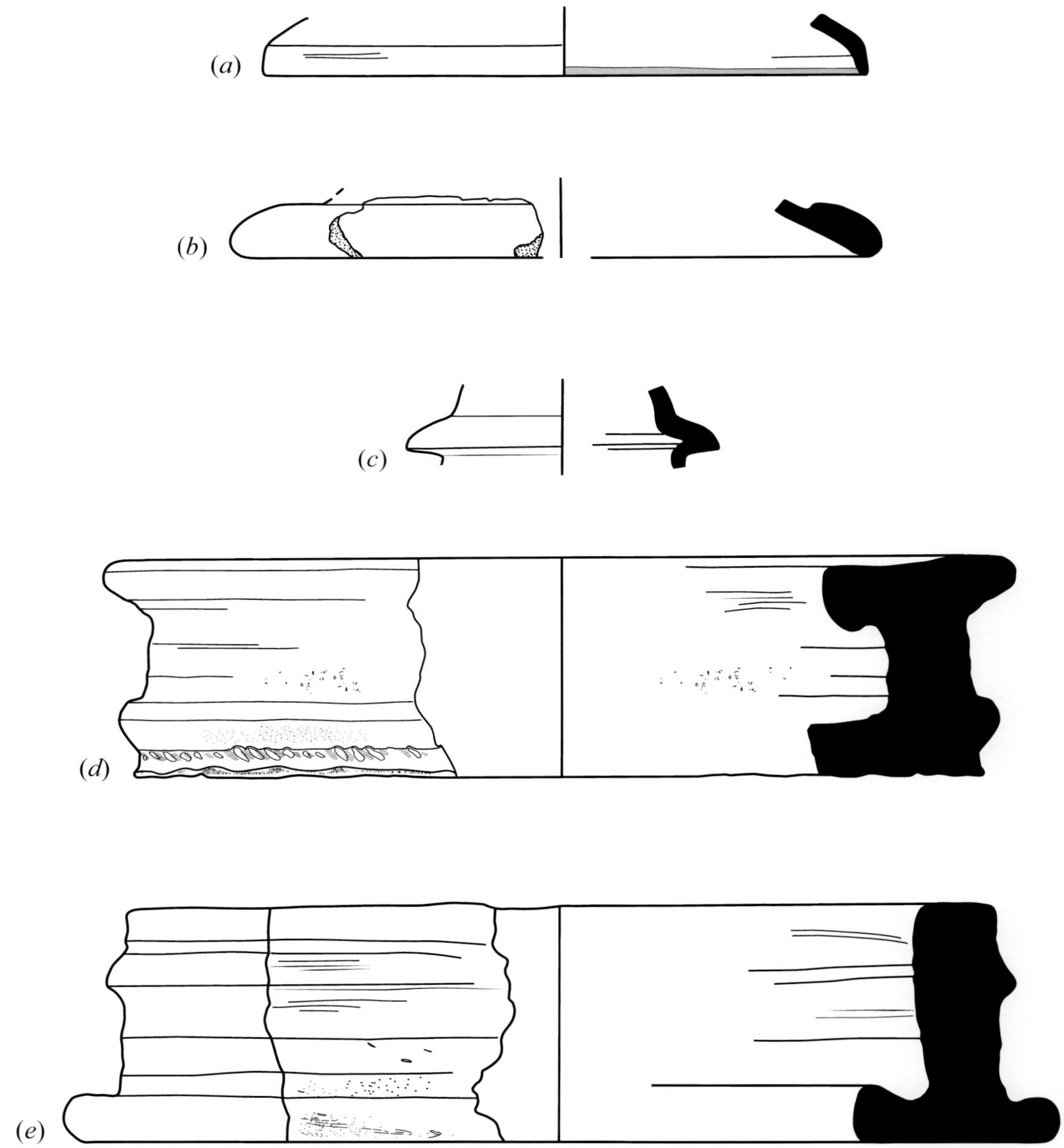

Figure 5.57. Lids and large stands (*a–c*, 2:5; *d–e*, 1:5)

(*a*) D 5 (1); marl, RR interior or deposit(?)

(*b*) D 66 (1); marl, later?

(*c*) D 120; fine Nile B2, darkened surfaces, 10 cm diameter at broken bottom edge

(*d*) D 209 (1); impression of ground on base exterior, 46 cm width

(*e*) D 303 (1); 50 cm base diameter

Large Stands

The large-stand sherds are not usually burned on any of their surfaces, although slightly similar forms are sometimes called *ovens* (figs. 5.57d–e, 5.58).[190] They may, however, have held braziers or large platters of coal. Most of the parallels cited here are not very comparable to the examples from the fortress, since they belong to higher-walled forms with opposing holes cut into the sides, such as on those from Medinet Habu or Karnak North.[191] But many of these display the shelf-like slanting rim and the exterior molding as represented here. There are taller stands from the northeast angle of the enclosure of Psusennes at Tanis, from Phase IIb at Elephantine, and from site 51, the fortress at Gebel Sahaba in Lower Nubia.[192] Kawa also has a few published examples.[193]

The stands sometimes feature ledged rims that tilt down slightly toward the interior. It is not always certain whether they have open or closed bases since the bases are fragmentary, but at least three examples show a finished base rim (figs. 5.57e, 5.58a–b). The manufacturing technique for all these large items is quite crude; the fabric used is very chaffy Nile C; and matting or ground impressions are noticeable on the bottom of the bases, along with evidence of fingerprints. In one case, not illustrated here, there is evidence of a stick or branch impression under the exterior of the rim, where it had been pressed to help form the exterior ledge as on the example shown in figure 5.57d.

The majority of the twenty-four stands in the collection are very wide, reaching as much as 50 cm or more in rim or base diameter, and the more typical examples are 8.5–13.5 cm high. The sherd scatters of these large utilitarian objects were mostly encountered in the West Sector, particularly in contexts of the buildings along the north wall and also in the southeast corner. Sherds representing three different stands also stemmed from deposits serving as fill for the Level III north staircase into the official residence, or were lying upon the floor surface under that staircase.

In two cases this type of stand was uncovered in contexts reflecting possible metallurgical activities. The clearest case is the presence of one in D 218 (1), where three tuyère fragments were found, and the sherds of another type of large stand were found in association with a crucible in the next room, D 209 (1).

Two other large stands of different forms, and a possible basin, are shown in figure 5.59.

Bread Trays and Plates

Bread trays and plates were not common among the sherd materials from Levels III–IV, but this may be due to the types of sherds that were selected to bring back to Chicago. Fragments from completely flat lids or trays, with prefiring incised decorations on one or both surfaces, were recovered in two contexts but find no close parallels (fig. 5.60).[194] The designs are incomprehensible in their fragmentary state. The bottom of the base of figure 5.60a suggests it was laid flat on the ground during manufacture, while the other fragment shows evidence of a finger having been swiped across its base.

The bread plates shown in figure 5.61 appear in the second millennium and continue into the Ptolemaic period.[195] The form is represented in materials from numerous sites, including parallels from Ashmunein and Karnak North.[196] There are also examples pub-

190 Hölscher 1954, pp. 10–11, fig. 11, pl. 6E (Twenty-First to Twenty-Fourth Dynasties). There are two types of supposed ovens at Medinet Habu, although there is no indication in the publication that either type showed evidence for burning. The maximum diameter of the round ovens was 70 cm.

191 See the previous footnote for the reference to the Medinet Habu example. See also Jacquet-Gordon 2012, p. 283, fig. 116c (Twenty-Sixth to Thirtieth Dynasties). Many examples of large pottery stands and basins were uncovered at Karnak North; see pp. 280–83, figs. 113, 114a, d–g, and 115, all from Twenty-Sixth through Thirtieth Dynasty layers. All of these are silt, most with red or white slips, and in one case with red slip and a white wash.

192 Aston 1999, pp. 75–76, 111–12, 125–26, nos. 550, 951–52, 1114. Aston cites the examples from Medinet Habu and Tanis. The unpublished example from Gebel Sahaba is no. 152 in the catalog of finds from the fortress and came from the fill of Findplace 2. I would like to thank Torgny Säve-Söderbergh and his team for making the site registration records available to me in 1991.

193 Spencer 2002, p. 16, pl. 19, A5–A6, from E.14.1 and E.14.22.

194 It is unclear whether the Dorginarti fragments are bread trays, but decorated parallels of a different shape are shown, dating from New Kingdom and Eighteenth Dynasty contexts at Karnak North, in Jacquet-Gordon 2012, pp. 143–44, figs. 64b–c, 65; see also the Twenty-Sixth through Thirtieth Dynasty example in fig. 114b, which is undecorated. Large undecorated lid fragments, with slight rims, are shown in Sullivan 2013, pp. 171, 201, type A-2s (Strata 1–3). Similar lids, bread plates, and platters are shown in Spencer 1993, p. 47, M1.30–74.

195 For their long duration, from the Middle Kingdom until the Ptolemaic period, see Masson 2011, pp. 271, 292, nos. 6–11. Bread trays are present in all three phases of the Late New Kingdom/Third Intermediate Period in Aston 1996a, pp. 72–73, figs. 187a–b, 206a, 216a–c.

196 The Karnak examples are in Jacquet-Gordon 2012, p. 282, fig. 114, from the Twenty-Sixth through Thirtieth Dynasty levels. See also the parallels listed in Spencer 1993, p. 47, pl. 74, M.68–74.

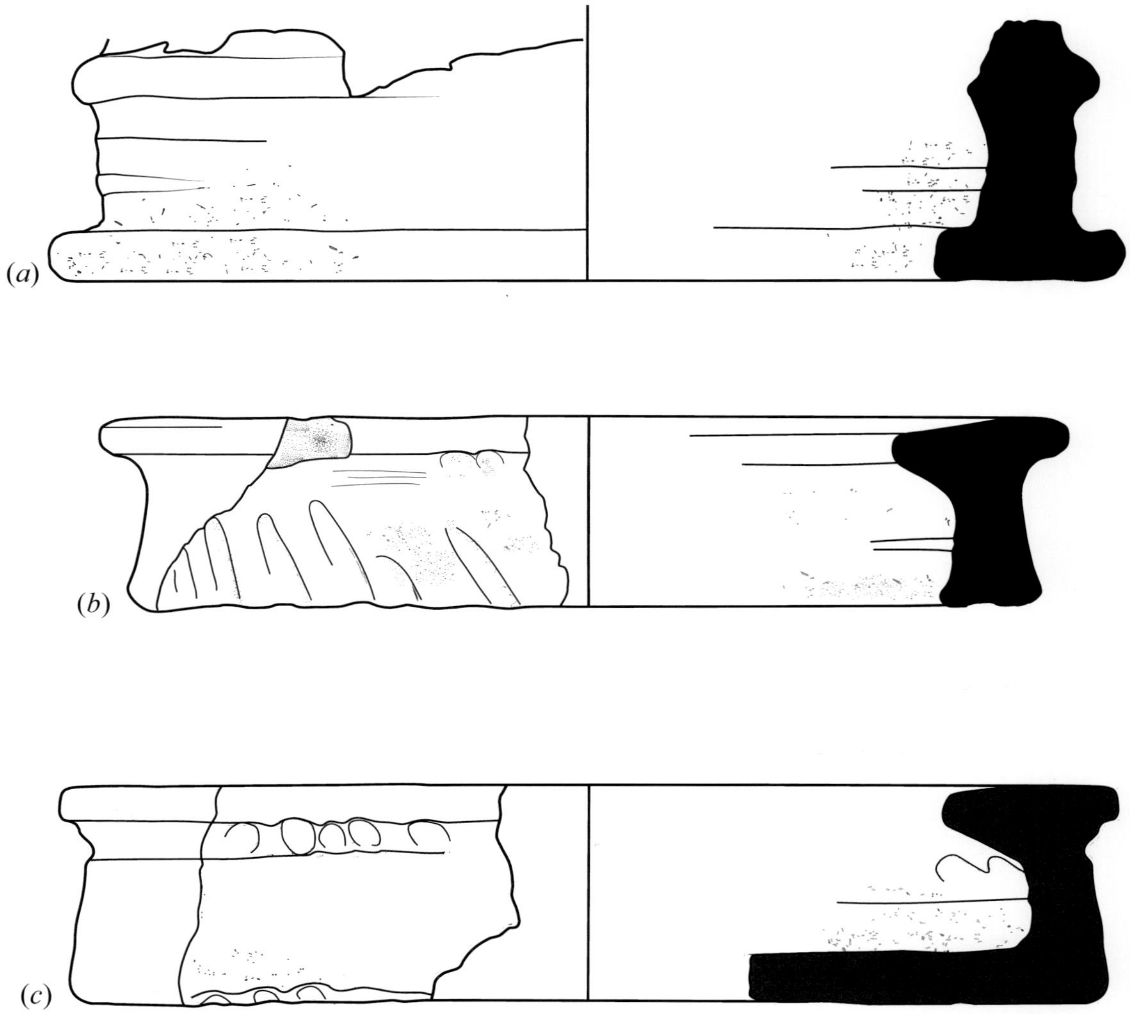

Figure 5.58. Large stands (1:5)

(*a*) D 1 (1)

(*b*) West Sector, north wall at breach, surface; finger impressions on exterior

(*c*) River Stairs, steps 10–17 and steps 24–30; mat impression on base, finger-impressed base on exterior and underneath rim on interior/exterior, base edge broken

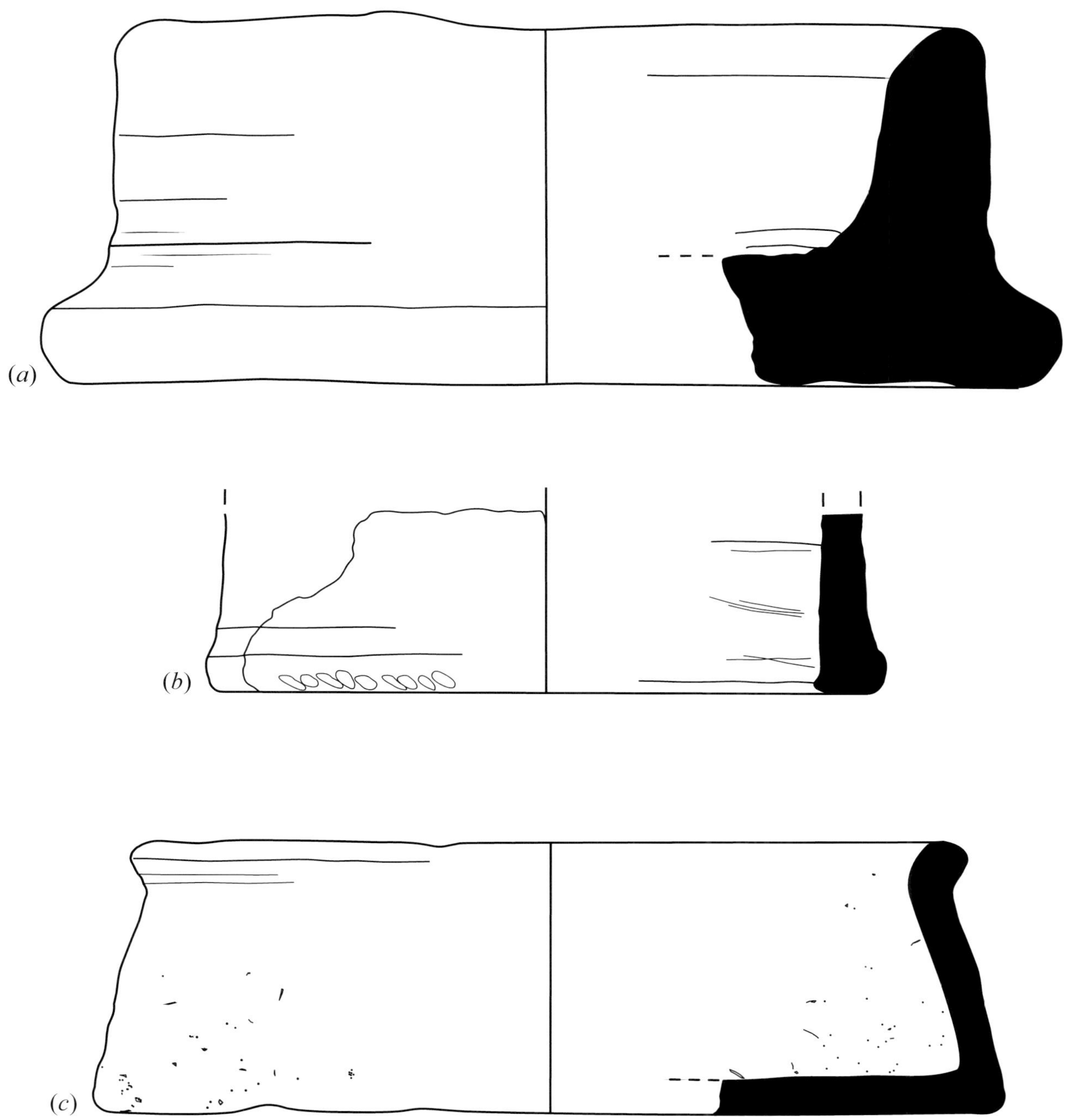

Figure 5.59. Large stands (*a*, 2:5; *b*–*c*, 1:5)

(*a*) D 35 and D 36, surface to below "ash" layer, another example from D 111 (1); fingerprints on base, base diameter ca. 27 cm

(*b*) North wall at breach, surface; base diameter 36–37 cm

(*c*) D 302 (1); finger impressions on base exterior, 46–48 cm diameter

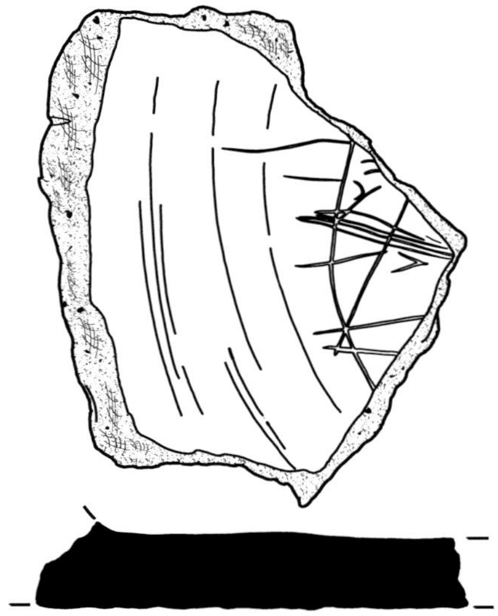

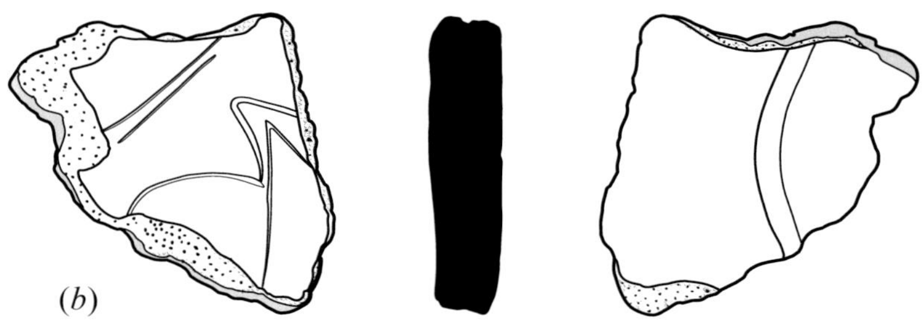

Figure 5.60. Bread plates or trays (2:5)

(*a*) D 60 (2), north

(*b*) D 112 and area; OIM E24410

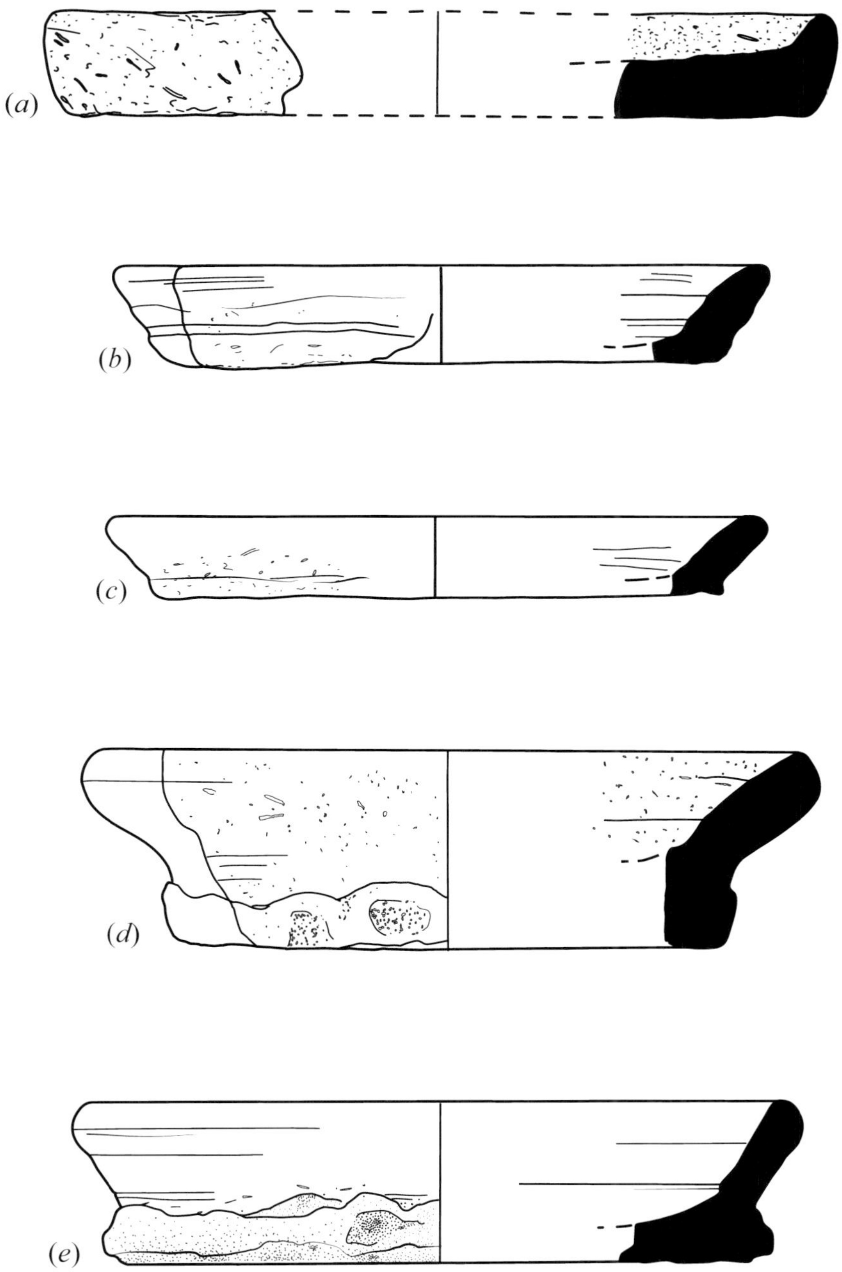

Figure 5.61. Bread plates (2:5)

(*a*) D 118 (1)

(*b*) D 36

(*c*) D 4, northwest of oven

(*d*) D 115, fill between ovens

(*e*) D 114 (1)

lished from Napatan residential and temple contexts in Nubia.[197]

There is not much evidence for burning on the plates from Dorginarti. The exterior of their bases usually indicates they were formed on the ground, with no matting or surface laid down. Many of the base sherds encountered from less complete vessels show finger impressions left behind during their production, such as those in figure 5.61d–e. None of these vessels shows evidence of a slip.

Two possible bread cone base tips were uncovered at the site. One stems from D 111 (1), where it was blackened from fire, and another form was found in the rubble below Level I in the Central Sector, in D 3 (1).[198] No associated rim forms were identified.

Vats and Basins

The wide-mouthed vessels shown in figure 5.62 are even wider in diameter than the wide-mouthed jars discussed above and are very thick walled. They occur infrequently in the collection. The three larger vessels shown at the bottom of the figure are made of coarse silt fabric. The ledge-rimmed jar in figure 5.62a matches the profiles of the smaller ledge-rimmed jars mentioned above. Large, flattened strap handles are found frequently enough to show that the forms in figure 5.62f–g are not unique. The carinated shoulder and tall neck are similar to those of a smaller vessel from Third Intermediate Period contexts at Tanis, and there are also large vessels in Phase IIb at Elephantine that compare well.[199] These vats are sometimes coated with red slips, and in one case with white paint splashed onto it (fig. 5.62e). Heavy ring bases were found in the same contexts with some of these sherds, but no form is reconstructible.

UNUSUAL VESSELS

A handful of wheel-made vessels or sherds from Dorginarti are striking due to their thick red slips and their finely burnished exteriors, in addition to their unusual forms (fig. 5.63; pl. 70a–c).[200] Their fabric is also different, and where they were manufactured is unknown. The burnish strokes are visible on the exterior surfaces of the vessels.

The fabric is different in both cases, seems to combine silt and perhaps desert clay—possibly from a sandstone or kaolin deposit—and fires to a bright red or an orangish red (ca. 2.5YR 6/8 or a 10R red). The fabric has many small dark particles and sand, some finer organics, and an amount of gold mica bits and flakes. The distinct, thick red slip and continuous vertical burnishing on the exterior surface resembles the surface finishing of wheel-made beakers from Napatan tombs in Nubia, although the similarity of the clay and temper is unproven.[201]

A number of similarly decorated bowls—with thick bright-red slip and burnishing or polishing—probably also belong here in view of their unusual finishes and shallow forms.[202]

Flask or Pitcher

Sherds from the neck and handle region of a tall-necked flask or jar, with a rounded neck ridge, represent a different form (fig. 5.63a; pl. 70a). The ware is reddish brown on the interior and has a grayish-black core. A highly burnished red slip covers the exterior, with closely spaced, vertical strokes. Most of the sherds of the vessel are from D 113 (1), but also from a neighboring context in D 124 (1).

There is no parallel for it, although foreign forms may have inspired it.[203]

---

197 Mohamed Ahmed 1991, p. 62, fig. 23 (I F2), where most derive from Phase 2 of Napatan Building I, and p. 81, fig. 28 (III B8). See also Orzechowska 2003, pp. 443, 449, pl. 28a. The bread plates from Kawa are dissimilar and are illustrated in Macadam 1955, p. 215, pl. XXXII, types N.13a–b (2058 and 2062, both burned). No. 2058 comes from House 1, Room 5, phases 1 and 2.

198 The types from D 3 (1) and D 111 (1) are probably comparable to the bases shown in Phillips 2003, p. 401, pl. 20d–e. For the examples from the Soniyat temple, see Orzechowska 2003, p. 455, pl. 22.

199 Laemmel 2011, p. 30, pl. 21, no. 121. The vessel was coated with a whitish-cream slip. For thinner-walled Elephantine examples of Nile B2, whose handles attach below the rim exterior, see Aston 1999, pp. 74–75, nos. 546–47 (flattened strap handle on no. 547).

200 The color of these two wheel-made vessels may be like that of the "cherry" red-slipped, burnished ware at Meroe, which was found only in the lower levels of the excavations; see Shinnie and Bradley 1980, p. 154, and, for example, the beaker and jar on p. 103, fig. 33, no. 7, and p. 106, fig. 36, no. 95.

201 For examples, see Williams 1990, p. 54, fig. 2b–c, pl. 7, and Vila 1980, p. 156, fig. 169, types II-1A from tomb 46.

202 Sherds of a bowl from D 90 (1) (OIM E50077), a similar form to the bowls shown in fig. 5.9f, also have plentiful dark particles and gold mica inclusions. It, too, has a red slip on both surfaces and was finely polished or burnished.

203 The following parallel is a very tentative suggestion. A red-slipped and burnished amphora from Tell Jemmeh, for example, is discussed in Ben-Shlomo and Van Beek 2014, p. 435, fig. 8.43b, of Iron IIB, which is dated to the beginning of the last third of the eighth century BC. The particular jars are examples of Late Philistine Decorated Ware.

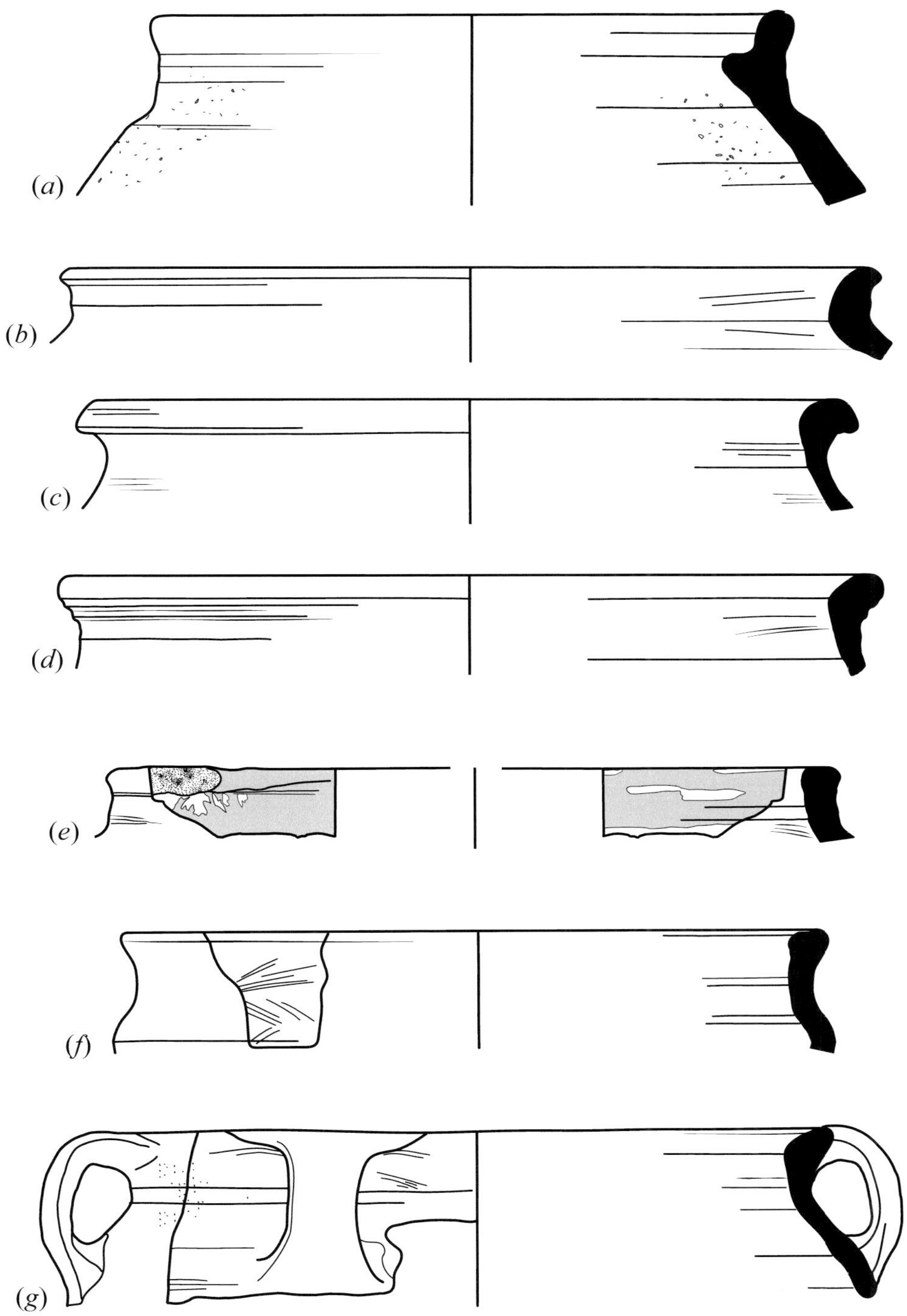

Figure 5.62. Large jars and basins (1:5)

(*a*) D 102 (1)

(*b*) D 104 (1)

(*c*) D 36, surface clearance

(*d*) D 116 and area

(*e*) D 32 and area, dismantling walls; RS exterior, RS applied in horizontal bands on interior, white paint drips on exterior, handle scar

(*f*) D 113 and D 114, surface

(*g*) D 13 (1); 47–48 cm at rim, same sherd types from D 6 (1), D 32 (1), D 78 (1), D 80 (1), D 209 (1), D 218 (1), etc., some with RS exterior/interior rims

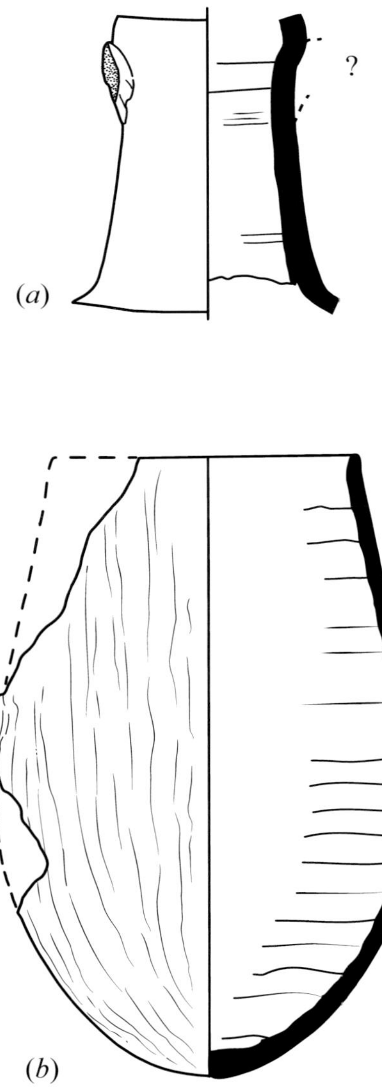

Figure 5.63. Unusual vessels (2:5)

(*a*) D 113 (1) and D 124 (1); OIM E50074 and E50078, RS and burnished exterior, uncertain stance, vertical wheel striations on interior body

(*b*) D 310 (1); OIM E49523, RS and streak burnished exterior, beginning of protrusion at one side, slightly oval mouth

Beaker

A silt beaker was found in the lowest fill above the floor of the Level IV residence in D 310 (1) (fig. 5.63b; pl. 70b).[204] It is very fragmentary and the stance was difficult to reconstruct. The vessel body is slightly flattened, so that the mouth would have been more oval in shape than round. The sides appear to lean in toward the rim, and the base tapers. A small protuberance is found on one of the sides, representing an appliqué or the beginnings of a handle.

It is wheel-made and has a slightly hand-formed base on the interior and at the appliqué. The clay is more micaceous than the previously described vessel. The interior is pale brown, and the exterior has a thick bright-red to orange slip with vertical burnish strokes that are more or less continuous. The facets of the burnishing tool are very evident on the surface.

There are no precise parallels for this creation, but it closely resembles a type of Kushite beaker commonly found during the later eighth century.[205] A number of these wheel- and handmade beakers have also been found in the tombs of Nubians in Egypt.[206]

THE HANDMADE CORPUS

The pottery traditions of Nubian cultural groups, which are part of the Sudanese Sahelian and eastern and western desert spheres, are recognized in the corpus of handmade pottery from Dorginarti. Many of the handmade bowls exhibit simple incised decorations, particularly atop the rim or on the exterior edge and, less commonly, on the surface of the body. Black-topped bowls and jars, sometimes with an evident red slip and always burnished, are found in the corpus. Burnishing is present as a continuous surface finish on the interior and exterior of some vessels, and on a few small red or black bowls, or it is the streak burnishing decoration found also on wheel-made vessels. Kerma and Lower Nubian C-Group, Pan Grave, and perhaps other desert ceramic traditions are represented in the diversity of vessel types and decorative motifs from

204 It appears that the "(1)" level designation was the fill above floor level in this room of the Level IV residence, so the vessel either stemmed from Level IV or may have been part of the rebuilding project in Level III.

205 For examples of other Nubian beakers, see Shinnie and Bradley 1980, p. 103, fig. 33, no. 72, ware Cb, the "cherry-red" in the earliest levels excavated, i.e., M.50 [16]; Vila 1980, p. 156, fig. 169, type II-1A, from tomb 46; and Griffith 1923, pp. 99–100, type XIs, and note the pointed base on XIa), pl. 18. For a better drawing, see Lohwasser 2012, fig. 69, Berlin 7822 (black-topped and wheel-made).

206 Budka 2010a, p. 507, fig. 3, and p. 510.

Dorginarti, suggesting the intercultural aspect of the population living at, or visiting, the fortress.[207] The comparisons indicate that, as in earlier and later periods, the river and desert paths were used as conduits between the regions to the south and to the west and east of the Nile, and that these routes were active and the people on them were mobile.

The process of forming most of these vessels was seemingly undertaken without the aid of a ceramic or basketry mold, although the potter undoubtedly formed the vessel using a pit dug into the sand or silt and then pressed or pounded the coils of clay into shape.[208] Some bowls, perhaps cooking pots, show evidence of the clay resting on matting or in a basket. Not enough bases or large portions of the same vessel were found to indicate the precise manufacturing techniques employed, although the pinching of the clay by hand and the scraping of the vessel walls in the finishing process are evident on sherds.[209]

The handmade beakers so common in the earliest Kushite tombs and settlements are not represented at Dorginarti, and even the many handmade vessels from the cemetery at Sanam and in various Napatan contexts at Kerma show only slight resemblance to the fortress's handmade corpus.[210] The functional differences between cemetery and settlement site assemblages may be operative here, but the lack of beakers is a chronological difference. The wheel- and handmade beakers are found, for instance, in both early and later Napatan domestic contexts at Kawa.[211] It appears that the handmade beakers represent only the pre-Twenty-Fifth Dynasty or proto-Kushite levels.

Most of the handmade forms from Dorginarti find their best parallels at southern sites along the northern Dongola Reach, the Debba Bend, and the fortress of Gala Abu Ahmed in the Wadi Howar, and these will be cited in the relevant footnotes below.[212] There is similar handmade pottery from sites surveyed or excavated at the Fourth Cataract.[213] There are also examples in the Dorginarti handmade corpus that are similar to the rim-decorated bowls and black-topped vessels from site 176 in the region of Debeira and elsewhere north of the Second Cataract.[214]

More comparative material, with incised decorations and mat-impressed surfaces, come from the royal tombs at El-Kurru and Nuri[215] and from Qasr Ibrim.[216] The West Cemetery at Meroe produced one mat-impressed bowl from tomb 641, attributed to

207 The continuation of earlier pottery traditions into the first millennium has been noted frequently, but see Griffith 1923, pp. 102–4, and Phillips 2003, pp. 394–99. See also Ahmed 1985, p. 248, and Adams 1964, p. 161 (burnished black incised).

208 The formation process is discussed, for example, in Williams 1986, pp. 29–36.

209 One possible pottery manufacturing tool is discussed in chapter 7; see pl. 92b.

210 See Griffith 1923, pp. 102–4, pl. XVIII, for similar black-topped vessels, and pl. XXXIII, nos. 10–11, and pl. XXXIV, nos. 1–2, for handmade incised vessels whose forms and decorations are not comparable to the Dorginarti bowls mentioned below. For the closest handmade bowl parallel from the Napatan residential building I at Kerma, see the bowl with an incised rim decoration in Mohamed Ahmed 1991, p. 56, fig. 19, I D6m (Phase III, mid-sixth through beginning fifth century BC). A few other parallels from Kerma are noted, particularly the vessel shown in Ruffieux 2007, pp. 231, 238, pl. 6, no. 66.

211 Wheel-made beakers are found in the early deposits of Kawa Building A2, a domestic building; see Sjöström and Thomas 2011, p. 65, fig. 1. The earlier levels at the site yielded only handmade beakers of somewhat similar form.

212 Note, however, Phillips 2003, pp. 395–401, pls. 8–12. The examples are proto-Kushite and Napatan in date, for which see Klimaszewska-Drabot 2003 and Orzechowska 2003. I would like to thank Friederike Jesse for showing me the pottery from Gala Abu Ahmed housed at the Forschungsstelle Afrika at the University of Cologne in 2011. Some of the Gala Abu Ahmed handmade sherds are published in Eger, Helmbold-Doyé, and Karberg 2010, pp. 77–84, pls. 5–6, and in Jesse 2016. Note, however, other examples shown in an online image database for the site and its objects, at http://arachne.uni-koeln.de/arachne/. This database was consulted during the research on the Dorginarti materials and is mentioned in the footnotes below, citing the Arachne serial number.

213 For example, in Budka 2007, pp. 83–88, and Welsby and Welsby Sjöström 2006–7, pp. 385–97.

214 For site 176, see Säve-Söderbergh 1989, p. 202, pls. 35–37. The incised areas atop and on the exterior of the rim show less variety than the decoration at Dorginarti; see pp. 40–41 (T2, T8, RB2, and RB10). Some of the forms also seem to match the descriptions of handmade vessels from site 6-G-9 at Gezira Dabarosa, although these are not yet fully published; but see Lister 1967, p. 61.

215 The tombs in the cemeteries of Napata also yielded examples of handmade incised pottery and mat-impressed sherds, either within the tombs or from around the tombs. See, for example, Dunham 1950, p. 20, fig. 4b (although from "debris and siftings" of Kurru Tumulus 5 of Generation A); p. 32, fig. 11b, 19-3-541 ("intrusive?" in Kurru 4 of Generation 5); p. 73, fig. 24b, 19-3-876 (Kurru 19 of Generation B); and p. 105, fig. 35c, 19-3-1532 (Kurru 72 of Generation 4). The tombs at Nuri also yielded handmade vessels or sherds from in or around the tomb; see Dunham 1955, p. 29, fig. 16, 18-2-25 (Nuri 74 of Generation 6); p. 50, fig. 32, 17-12-37 (Nuri 71 of Generation 8); pp. 64–66, fig. 42, various (Nuri 21 of Generation 9); p. 70, fig. 46, various (Nuri 22 of Generation 9); p. 114, fig. 86, 18-1-269 (Nuri 40 of Generation 10); and p. 163, fig. 122, 16-11-10, 13, 14 and 17-1-288 (Nuri 7 of Generation 15). Note also the numerous black-ware sherds mentioned but not illustrated, some apparently with mat-impressed exteriors: p. 196, fig. 150, 17-1-28 (Nuri 11 of Generation 19).

216 Adams n.d., pp. 7–9 (Ptolemaic and Roman) and pp. 10–12 (pre-Ptolemaic/Napatan).

Generations 2–5(?), while the earlier levels at the town site of Meroe also yielded similar bowls.[217]

At the fortress of Dorginarti, handmade forms are well represented in the contexts around the Level III–IV residential district along the north fortification wall, while a few stem from the architectural remains under the north wall relining, along with wheel-made vessels. They are represented in lesser numbers—at least in the sherd materials in the Oriental Institute Museum—in the remains at the southwest corner of the West Sector, although they were commonly uncovered in the rooms at the southeast corner. Handmade sherds were less commonly encountered in the layers that can safely be attributed to the Level IV residence in the Central Sector, perhaps due to the clearance of the remains by the Level III builders, but they were found in the Level III residence deposits below the Level II fill. All the structural remains or ovens that were uncovered around the exterior perimeter of the fortification were found to have both wheel- and handmade sherds associated with them. The lack of close dating for these handmade forms does not allow a certain dating of any of these handmade vessels to a later, Level II context. However, they are found in surface deposits or in Level II fill materials.

The following discussion of handmade forms will of course be divided by form, but mostly according to surface finish and decoration. A notable feature of the handmade forms is the common presence of drill holes to mend these frequently broken articles of everyday use.

FABRICS OF THE HANDMADE VESSELS

The clay fabrics used to produce the handmade vessels at the fortress vary by the amount of sand, black particles, mica, and chaff added as temper, as well as by the size of the organics. No tempering particles are evident that are inconsistent with the use of local Nile silt or other resources close at hand. A number of sherds contain small white particles of possible limestone or other rock particles. Although there is little variety in the fabric composition, the shapes of the vessels are not standardized.

The clays used to make the various forms overlap to a considerable extent, although vessels intended for cooking always exhibit coarser fabrics, with more organic, sand, and mineral inclusions. The vessels typically have black cores. The two basic fabric groups are not well defined by differences in temper, only by the larger size and amount of temper in the coarser vessels, which also exhibit a less thoroughly mixed fabric. Details are found below.

UNDECORATED BOWLS AND OTHER FORMS

A few bowls and miscellaneous forms that do not fit with the rest of the handmade corpus are presented in figures 5.64–5.66. They occasionally have a slip or coating, which is present on the burnished hemispherical bowls (fig. 5.66). A reddish-yellow to pink slip covers the interior and exterior surfaces of the small cup in figure 5.64c. Fingerprint impressions from pinching and pressing the clay into shape are found on the smallest bowls. Besides one neck or spout of a bottle, the forms range from small bowls to larger conical or hemispherical bowls. Two slightly carinated bases were also found at the site, one of them with traces of burnishing on both the interior and exterior surfaces (fig. 5.65a–b).[218]

Two bowls exhibit a set of incised lines on the exterior just below the rim (fig. 5.66a–b). They are carefully made and finished with burnishing on their surfaces. One hemispherical bowl has a red slip on its exterior side, with horizontal burnish strokes atop it (fig. 5.66c).

UNDECORATED CLOSED BOWLS

The cooking vessels among these bowls typically show evidence of either completely or partially blackened surfaces (figs. 5.67–5.68; pl. 71a). But this is not true of all the sherds, perhaps due to alternative heating methods or noncooking uses, such as brewing. For the most part, the vessels showing evidence of having been used for cooking do not seem to have been laid directly on the fire, although the incomplete nature of most of the sherds does not allow a thorough analysis of particular cooking methods. The coarse fabric has many fragments of chaff remaining or burned out, leaving voids in the clay, and the temper also contains particles of rocks that are subangular or rounded, and probably additional mica or other shiny particles. The fabric is typically fairly soft and has a rough texture from the amount of temper used.

A feature of many of these jars is the thickened rim, with the top part of the clay turned over onto the interior and finished by rough scraping. The thickened rim is typically found on vessels with inward-slanting or bent rims, but also on vessels with a more rounded,

217 Dunham 1963, p. 337, fig. A, 4 (ca. 747–664 BC). For Meroe, see Bradley 1984, pp. 200–206, and Shinnie and Bradley 1980, pp. 101, 156–62, and fig. 31, nos. 35–36.

218 They are perhaps bases from vessels like that shown in Klimaszewska-Drabot 2003, p. 441, pl. 6b.

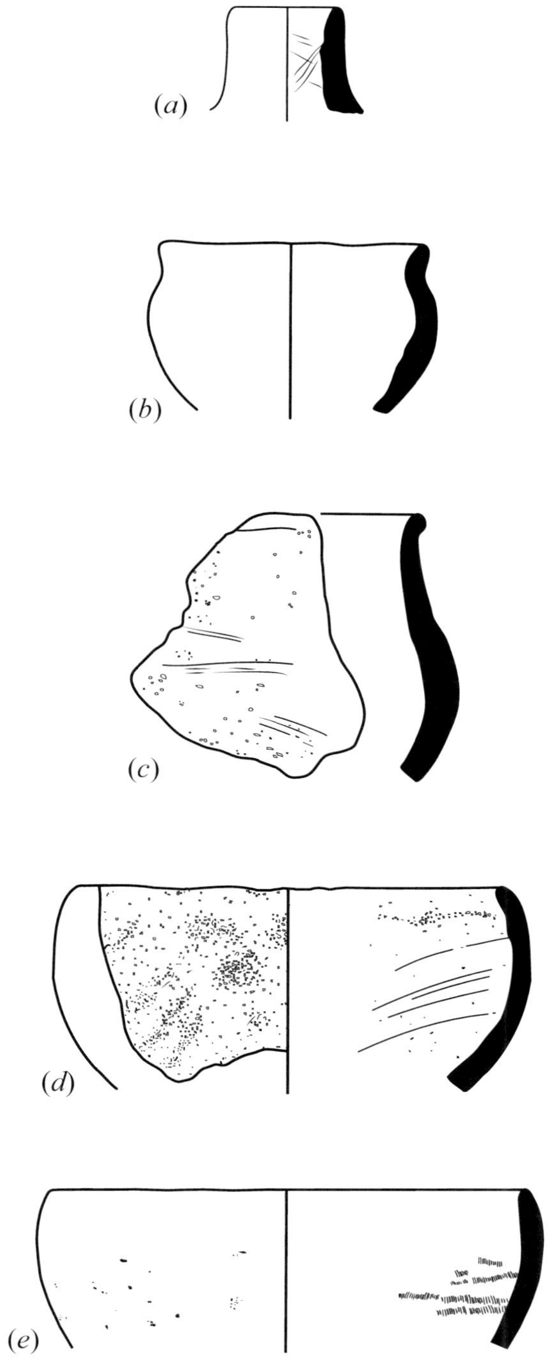

Figure 5.64. Handmade bowls and bottle rim (2:5)

(*a*) D 303 (1); very worn, bottle or spout(?)

(*b*) D 19 (1); red-brown with black core, brown exterior

(*c*) D 60 (1); black base on exterior, black core, reddish-yellow slip interior/exterior, uncertain diameter

(*d*) D 106 (1); brown to red fabric, gray-black core

(*e*) D 13 (1); smoothed lower interior

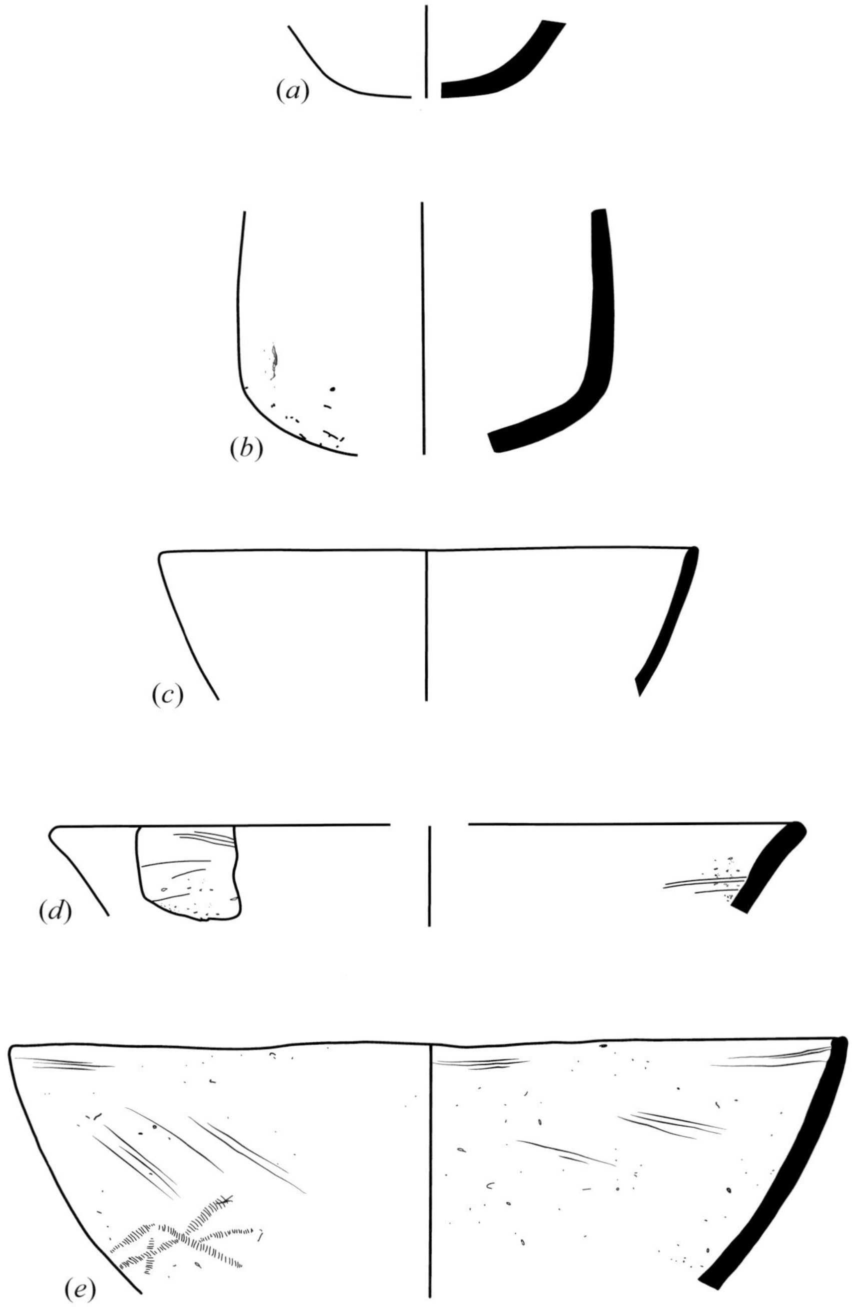

Figure 5.65. Handmade bowls and cups (2:5)

(*a*) D 23 (1); uncertain stance and diameter, medium burnish interior/exterior

(*b*) D 4, above "ash" layer

(*c*) D 2 (1); worn sherd, no surfaces remain

(*d*) D 62 (1); stance uncertain

(*e*) D 227 (1); burnish traces interior/exterior, RS interior/exterior, 10R 5/6

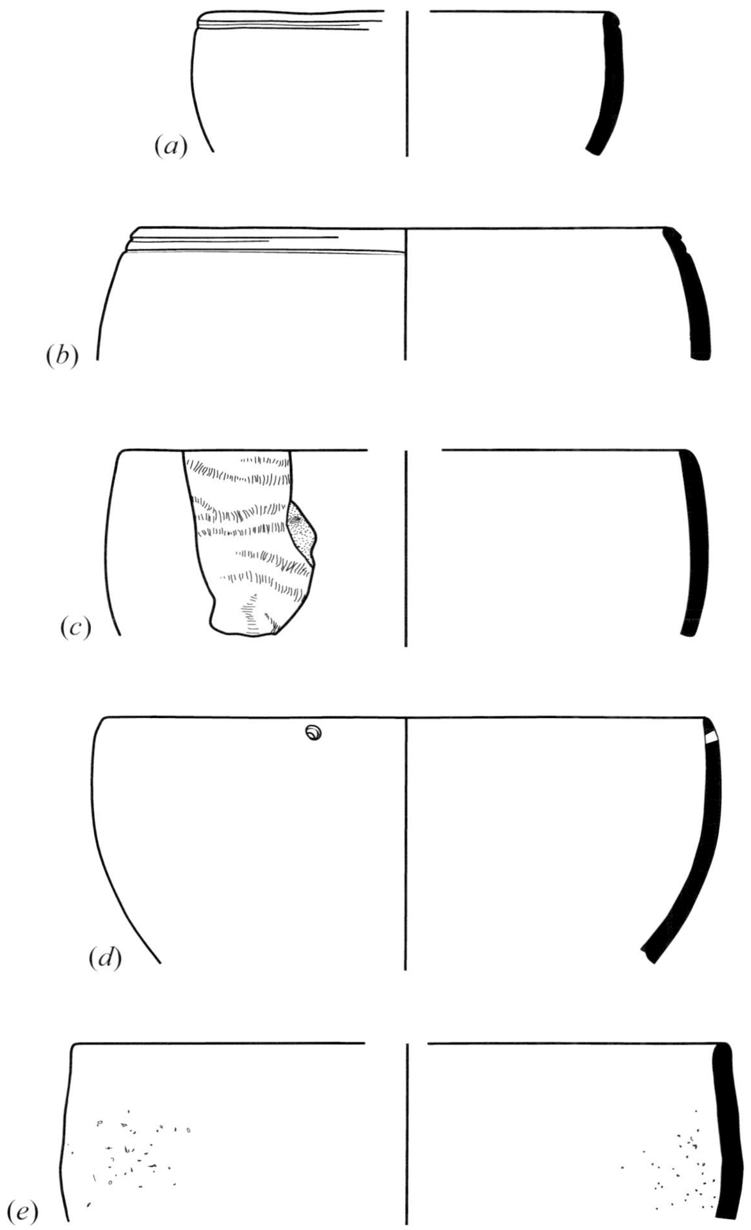

Figure 5.66. Handmade bowls (2:5)

(*a*) D 201 (1); burnished continuously interior/exterior, limestone temper(?), uncertain diameter, drawn as 14 cm

(*b*) D 12 (1); gray-black interior/exterior, burnished interior/exterior

(*c*) D 200 and D 202 (1); burnish strokes atop RS on exterior, blackened areas on exterior, uncertain diameter, drawn as 19 cm

(*d*) D 2 (1); complete burnish interior/exterior, repair hole near rim

(*e*) D 79 (1); black interior, red exterior (slip?), burnish traces interior/exterior, uncertain diameter, drawn as 22 cm

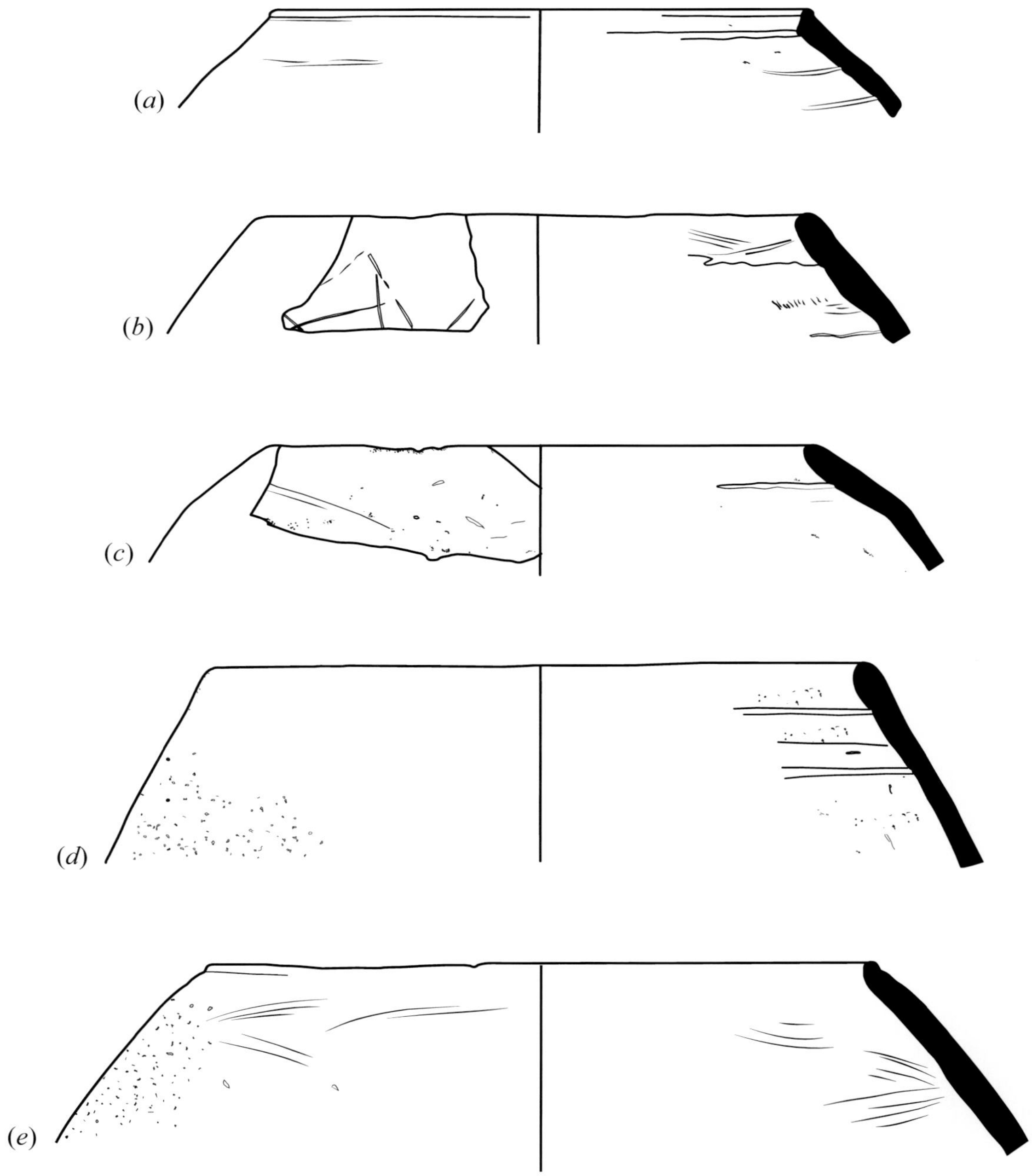

Figure 5.67. Handmade closed bowls (*a–d*, 2:5; *e*, 1:5).

(*a*) D 227 (1); blackened surfaces

(*b*) D 87 (2); potmark(?), light scratches, blackened surfaces, scraped-down interior

(*c*) D 221 (1); chaffy surfaces, blackened exterior, interior 10YR 5/3

(*d*) D 12 (1); RS exterior, burned in areas, chaffy

(*e*) D 200–D 202, surface and (1); OIM E50770, crumbly and sandy fabric, diameter ca. 24 cm

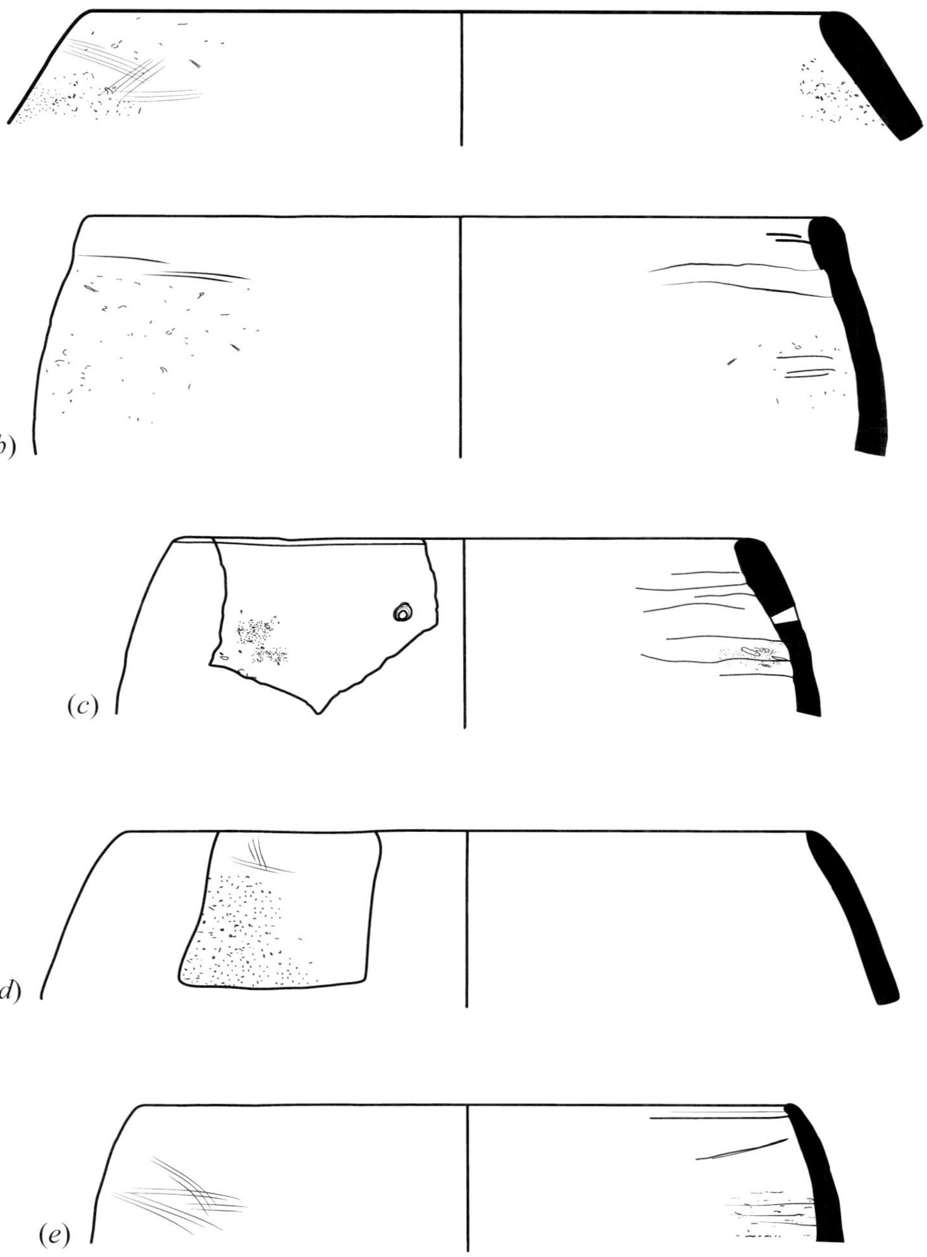

Figure 5.68. Handmade closed bowls (2:5)

(*a*) D 114 (1); blackened brown fabric

(*b*) South of D 208 (1); smoothed exterior rim, burned spot on exterior rim

(*c*) D 33, rubble against west wall

(*d*) D 113 and D 114, surface; similar sherd in D 2 (1) test pit; blackened areas on exterior

(*e*) D 78 (1), bin; similar sherd from oven northwest of D 4; swipes on exterior, bottom interior scraped

hemispherical shape (fig. 5.68). The wall thicknesses are sometimes uneven due to the pressing and shaping of the vessel by an individual not concerned with perfecting the shape. But finishing scrapes also score the interior and exterior surfaces when a more even wall thickness was desired.

Most of these closed pots do not show evidence of burnishing or slip, although some sherds show evidence of a red slip on the exterior (fig. 5.67d). A possible prefiring incised decoration or potmark is found on the exterior surface of the example shown in figure 5.67b.

The pots with hemispherical profiles are typically better formed than the others, with even rim lines and wall thicknesses, and more compacted surfaces. The surfaces of these sherds can also be very chaffy, but the temper and organics are mixed into the clay more thoroughly.

Parallels for many of these vessels are found at sites surveyed along the Debba Bend, and also from the temple site at Soniyat.[219] There, they are dated to the proto-Kushite (i.e., ca. 1100–900 BC) and Napatan periods.[220]

MAT-, STRING-, AND FINGER-IMPRESSED BOWLS

The handmade bowls also show evidence of being formed using matting or baskets, the use of strings to hold them together while drying (or perhaps string rouletting at the neck), and the finishing of the vessels by hand (figs. 5.69–5.71a). The mat impressions could have been part of the decoration of, or perhaps just roughened up, the surfaces of a pot intended for cooking.[221] The variety of rim and body forms indicates that they were produced as needed by various potters. The evidence for finger-impressed mud coating on vessel bases is rare in the materials brought back to the Oriental Institute Museum (e.g., fig. 5.71b), although at least four such sherds, or their outer mud layer, are found among the materials.[222] This technique stretches back at least to the mid-second millennium BC and also continued later. Along with the large amounts of temper in the vessels, it allowed better heat absorption and lessened the thermal shock to the pottery vessel from the heat of the fire.

The mat imprints of the vessels illustrated in figure 5.70b and d are arranged in parallel rows of V-shaped impressions, and this angular woven-mat pattern is the most commonly found pattern on sherds in the collection. More rarely the impressions are those of tight nubs of string running in an oblique direction, perhaps a twined-mat imprint (figs. 5.69a, 5.70a, c).[223] The rim sherd in figure 5.70c is burnished on its interior and on the blackened exterior rim, where the impressions were not present.[224] String rouletting or rocker-stamping was present on the neck of the jar shown in figure 5.71a.

Parallels are found to the south, once again, from Napatan or proto-Kushite (early Kushite) sites surveyed in the Dongola Reach and Debba Bend.[225] The finger-impressed bases of cooking pots are also found at Gala Abu Ahmed, as are mat-impressed handmade bowls.[226]

Black-Topped Bowls and Jars

Along with the mat-impressed bowls, the black-topped examples represent the continuation of a long tradition in Nubia (figs. 5.71c–e through 5.76; pl. 72f, g).

219 Klimaszewska-Drabot 2003, p. 441, pl. 4, and Orzechowska 2003, pp. 453–54, pls. 11–12, 17, 25–26.

220 For the terminology, see Phillips 2003, pp. 397–98.

221 Mat-impressed cooking pots first appear at the end of the Kerma Moyen, or the beginning of Kerma Classique, period; see Gratien 2014, pp. 97–99. See also Phillips 2010. The impressed pottery from Dorginarti is of plaited and tabby weaves, which are dated to the earliest Kerman down to proto-Kushite, and Kerman through Napatan periods, respectively.

222 For a good description of the handmade pottery processes, pebble burnishing, and addition of a finger-impressed mud "stucco" layer on the outer bases of cooking pots, see Lister 1967, pp. 36–37.

223 Soper 1985, pp. 35–39, pls. 5–6.

224 This pot was part of an intentional deposit between the inner wall relining and the innermost enclosure wall, along with a leather bag(?) with straps, the sherds of the vessel shown in fig. 5.80, and a large black-topped bowl with no burnishing present.

225 Klimaszewska-Drabot 2003, pp. 439, 441, pl. 5a–c. See also the mat-impressed bowl from the Northern Dongola Reach Survey in Welsby and Welsby Sjöström 2006–7, p. 386, fig. 2, B17.2. The pottery from site E10 is dated to the post–Kerma Classique, but the illustrated assemblage seems to include a marl jar rim that dates to the later eighth century BC; see p. 386, fig. 2, J44.3. This marl type was found in fairly good contexts in tombs of the royal cemeteries of Kush at Kurru; see Dunham 1950, p. 84, fig. 28c, and pls. LXII.F (19-3-1458), LXIII.A (19-3-1160). This form was also found in Kurru 52 and Kurru 55, which are attributed to a period between the reigns of Piye and Shabaqo.

226 A mat-impressed bowl is shown in Eger, Helmbold-Doyé, and Karberg 2010, p. 79, fig. 6 (ZN 09/15), and p. 83. See also the discussion in Jesse 2016, pp. 308–9, 314, pl. 2, nos. 8–9. For other mat-impressed bowl sherds from the site, see the online photographs at http://arachne.uni-koeln.de/ (e.g., SN 2479422, SN 2480568, SN 2481166, SN 2481691, SN 2481727, SN 2481919, SN 2482025, SN 2482235, and SN 2482256). Not all the mat impressions from the site are similar to those from Dorginarti. For finger-impressed base sherds, see the online photographs for SN 2481008, SN 2481100, and SN 2481369.

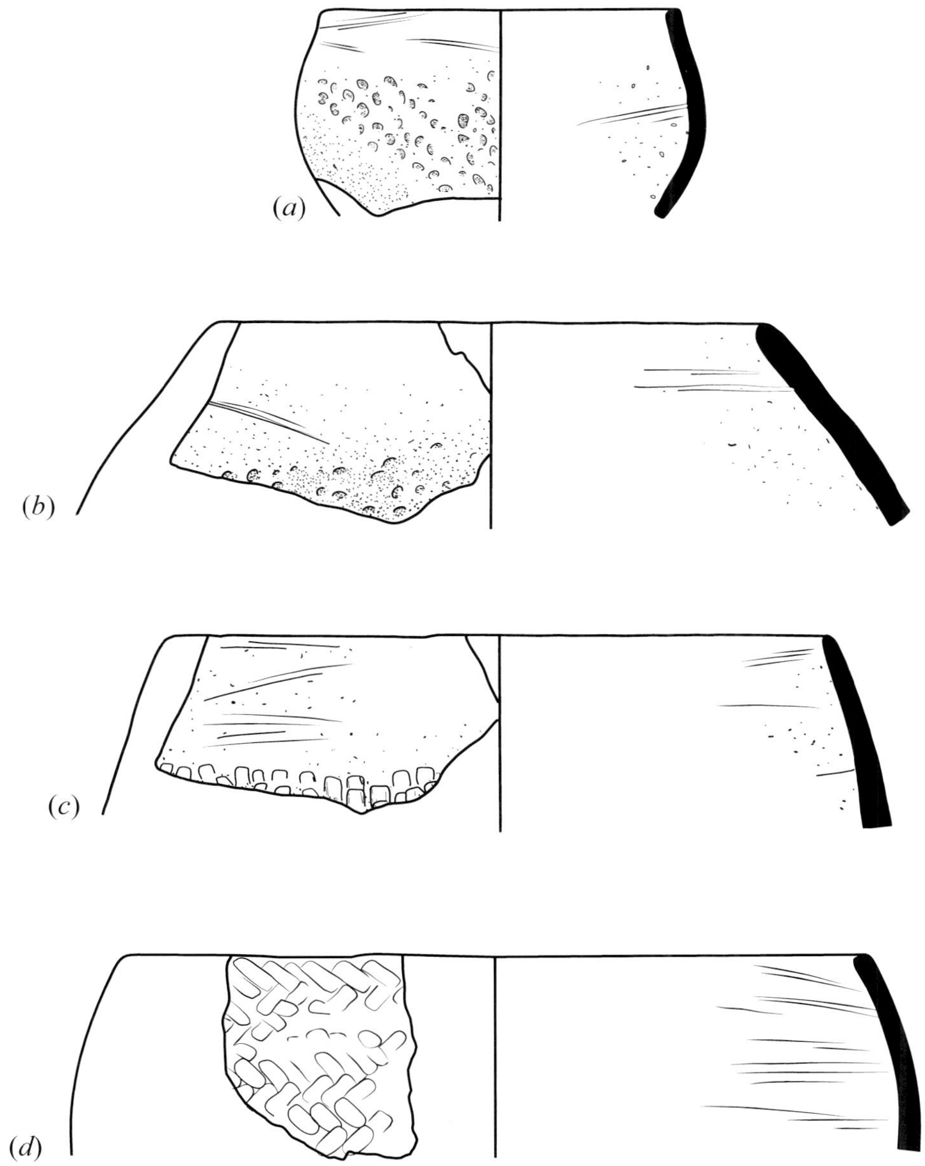

Figure 5.69. Handmade impressed and closed and open bowls (2:5)

(*a*) D 92 (1); mat- or cord-impressed exterior, reddish-brown with gray spots on exterior and at rim

(*b*) D 13 (1); burned on exterior, indistinct mat impressions on exterior

(*c*) East Sector, rubble over three south silos; mat-impressed(?) exterior, roughened interior, gray-black marks on exterior

(*d*) West Sector surface; mat-impressed exterior

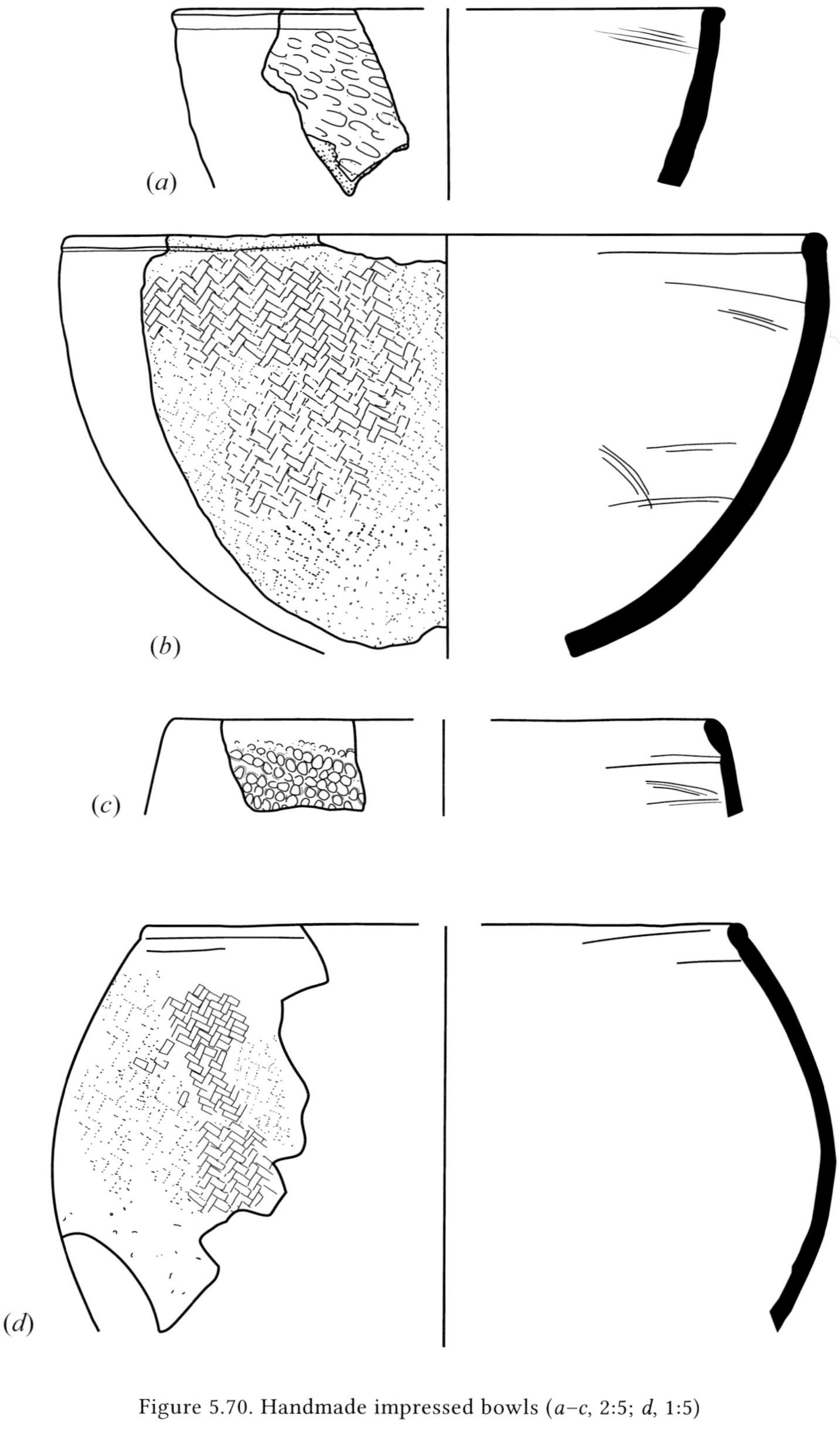

Figure 5.70. Handmade impressed bowls (*a–c*, 2:5; *d*, 1:5)

(*a*) West side of dividing wall between Central and West Sectors; twined mat- or cord-impressed exterior, diameter drawn as 19–20 cm

(*b*) D 76 (1), northwest corner, similar sherds from D 9, D 12, D 13, and West Sector surface; OIM E49558, mat-impressed exterior

(*c*) D 125 (1); light burnishing traces interior/exterior rim, black exterior, reddish-brown interior, roulette or mat-impressed exterior, diameter ca. 20 cm

(*d*) D 9; indistinct mat impressions

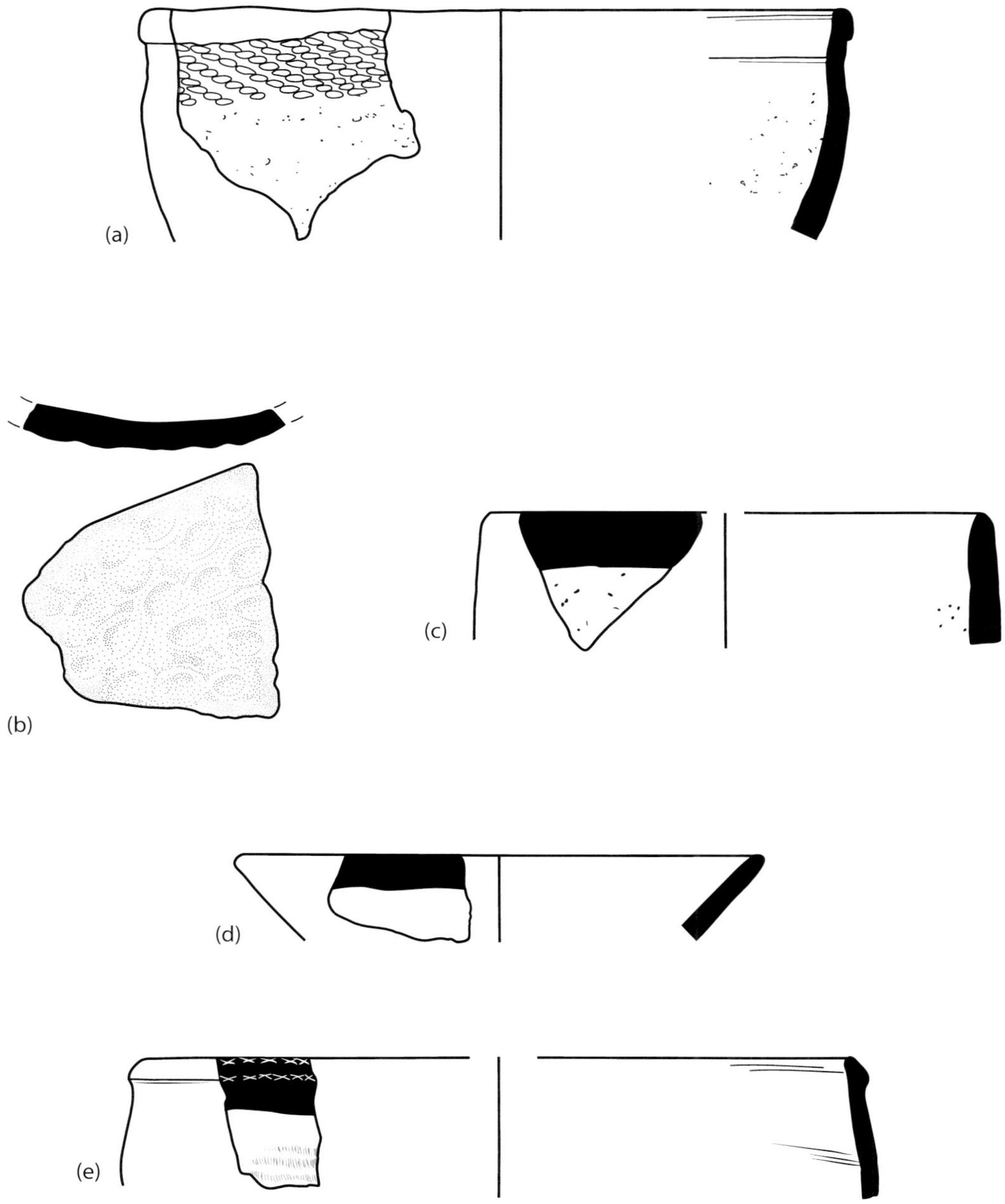

Figure 5.71. Handmade decorated, impressed, and black-topped bowls (2:5)

(*a*) D 23 (1); string or string-roulette impressed exterior, diameter ca. 24 cm

(*b*) D 62 (1); mud layer on exterior with fingerprints, sherd interior black

(*c*) North–south trench, West Sector; black-topped, traces of burnishing on interior, diameter uncertain

(*d*) D 68 (4); black-topped, black interior, RS(?) exterior, continuous burnishing interior/exterior

(*e*) D 10 (1); black-topped exterior rim, brownish interior, streak-burnished exterior, burnished interior, crosshatched incisions on exterior rim, diameter drawn as 25 cm

The type with thickened rim and black mouth is fairly common in all the Lower and Upper Nubian traditions. Evidence of a red slip is not always present. The bowl and jar forms sometimes have an incised decoration atop or on the exterior of the rim, and some sherds display streak burnishing. The interiors are sometimes a burnished black, but they could also be a matte brown or reddish-yellow.

It is unclear whether the diameter reading of the beaker-like rim shown in figure 5.71c is correct. The jar form in figure 5.72a was found in other contexts around the fortress, where it does not always have a blackened rim but always has an incised rim decoration (pl. 72a; compare fig. 5.78a–b). Similar rims were also found at Gala Abu Ahmed.[227]

The black-topped bowls shown in figure 5.73 could be confused with earlier Nubian products if they were found on their own.[228] Their interiors show the fine, haphazard lines of their finishing, except in cases where burnishing strokes have obscured the scraping marks. There are always traces of burnish strokes or continuous burnishing on the interiors of this form.

Parallels for some of these forms have also been found at other sites in Nubia.[229]

Bowls and Jars with Incised or Punctate Decoration

The handmade bowls and small jars in this category exhibit a wide variety of shapes and designs, with the motifs shown in the examples illustrated in figures 5.74–5.81 and plate 72a–d. The sherds shown in figure 5.74 have no precise parallels, at least not in first-millennium BC contexts, while the sherd shown in figure 5.76c may be an Abkan sherd from the Neolithic period. The forms shown in figures 5.75 and 5.76a–c are also unique or rare and show affinities with C-Group or Pan Grave precedents, or even with traditions from farther afield (pl. 71b–d). The bowl shown in figure 5.75d has the unique attribute of having both surfaces decorated with incised oblique lines, oval punctates, and crisscrossed lines as outer rim decoration (pl. 71c).[230] No clear parallels emerge at this time.

The earlier traditions inspiring the bowl with pendant triangles, and a chevron pattern atop and on the exterior of the rim, is clear, although it is not here considered an earlier vessel (fig. 5.75a; pl. 71b). The punctates on the body are round and punched deeply into the clay, and the interior is burnished, unlike C-Group prototypes.[231] There is no evidence that the incisions were filled with a colored substance. The bowl is made of light-brown silt, filled with natural inclusions of sand and mica, with a few small organics added, thus perhaps representing a dung temper.

The two sherds with triangular and horizontal incised lines (fig. 5.75b–c) also reflect a continuity of C-Group and Pan Grave traditions, and may in fact be earlier in date. Both sherds have black interiors and either black or brown-colored exteriors. One is burnished on both of its surfaces, while the other shows no traces of burnish. The stances and diameters of both sherds are uncertain.

The most typical sherds found are those of bowl and jar rims ending in a mostly tapered, narrow lip atop a vertical or slightly everted body profile (figs. 5.76d–e through 5.78; pl. 72b–e). As is the case with most handmade sherds, a few of the stances and diameters are unreadable. They all have incised decoration atop the rim and sometimes reaching onto the exterior. Oblique lines, crosshatches, and chevron patterns are very frequent, followed by a lattice design of long, crossed lines, which have all been incised by a fingernail, thin bone utensil, or stick (fig. 5.81). There are typically no triangular imprints in the tradition of earlier pots, except for the rim-top decoration on one vessel that does not resemble its earlier counterparts (figs. 5.74b, 5.81l). Generally there is evidence of continuous burnishing on the interior and exterior surfaces. The interiors may be black or gray, but they may also be the yellowish- or reddish-brown color of the exterior. The cores are almost always black. One bowl has an impressed wedge pattern and oblique lines on its rim and was fired black on both surfaces and highly burnished (fig. 5.81m; pl. 72e).

This type of rim-decorated handmade bowl is fairly common at sites throughout northeast Africa at various periods, but these wide-ranging parallels will not be explored here. The form and rim decoration have

227 Jesse 2016, p. 307, fig. 5, no. 8, and p. 314, pl. 2, no. 14.

228 See, for example, the black-topped bowls from earlier periods, and those of probable early first-millennium BC date, in Säve-Söderbergh 1989, pls. 16–19, 21–22, versus those on pl. 35 (site 176).

229 For example, a similar example was found in Tomb 14 at Hillat el-Arab, the remains of which were dated between the New Kingdom and Twenty-Fifth Dynasty; see Vincentelli 2006, pp. 89, 93, fig. 2.58, no. 347.

230 While the decoration and form are not the same, examples from the Handessi Horizon B are sometimes decorated on both surfaces; see Jesse 2007, pp. 992–93, 995, fig. 4, no. 3.

231 I would like to thank Bruce Williams for reviewing these sherds with me and providing his expertise.

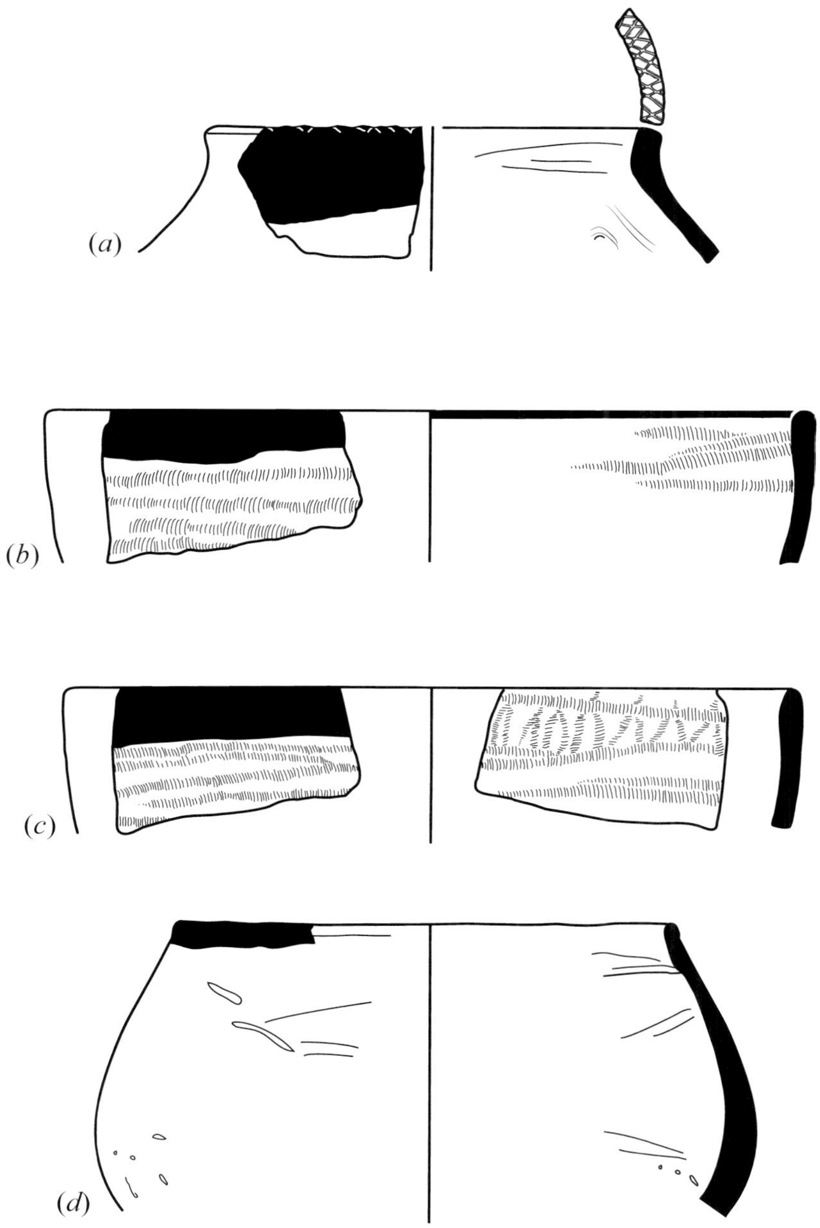

Figure 5.72. Handmade black-topped jars and bowls (2:5)

(*a*) Surface; OIM E36384, reddish-yellow scraped interior, burnish strokes interior/exterior, lattice decoration atop rim, diameter uncertain, drawn as 15 cm

(*b*) D 303, trench at northeast corner of Level III/IV residence; black interior, RS exterior(?), horizontal burnishing interior/exterior

(*c*) D 10 (1); black interior, burnish strokes interior/exterior

(*d*) D 215 (1); OIM E49568, partly smoothed interior, traces of burnishing on exterior, RS exterior(?), uncertain diameter, drawn as 18 cm

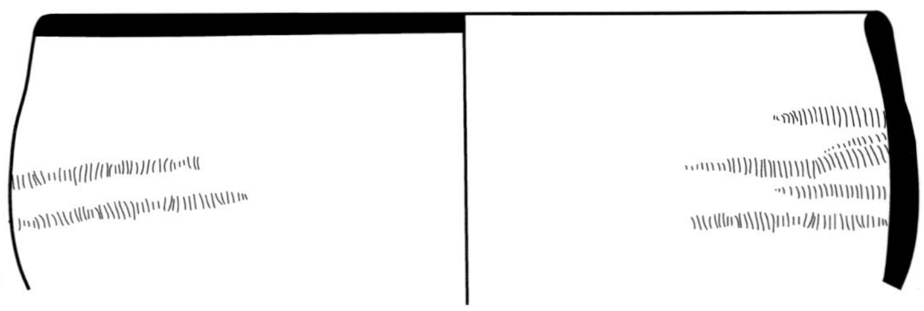
(*a*)

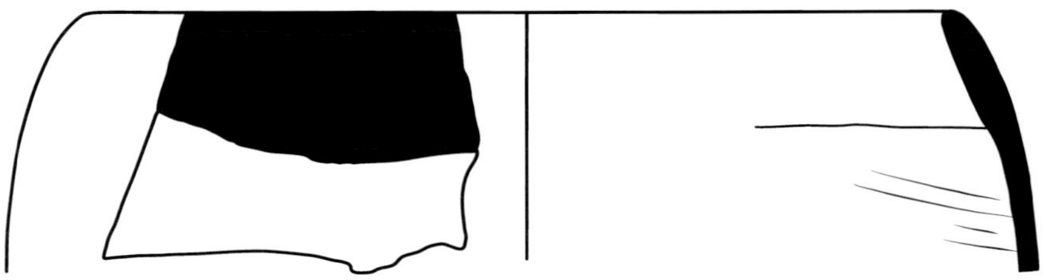
(*b*)

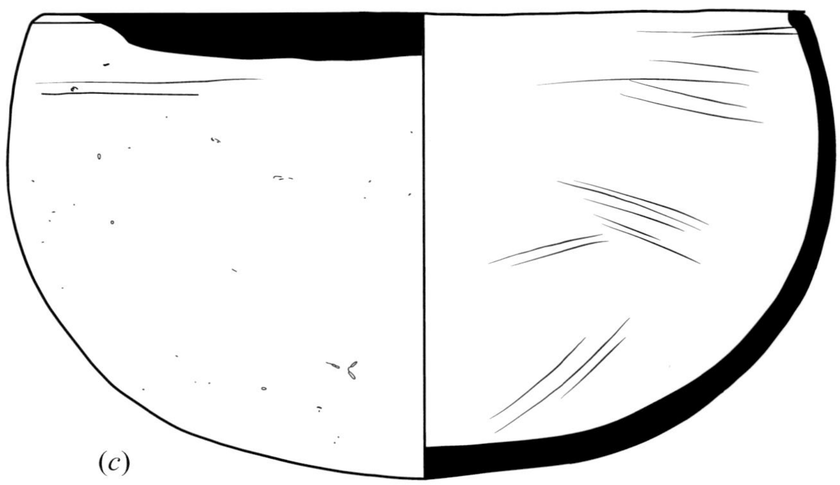
(*c*)

Figure 5.73. Handmade black-topped bowls (2:5)

(*a*) D 23 (1); black-topped, red exterior bottom, 2.5YR 4/6, horizontal burnishing strokes interior/exterior

(*b*) D 200–D 202, surface and (1); another sherd of same, with repair hole; OIM E49533, black interior, mottled black-topped rim, continuous burnish strokes interior/exterior

(*c*) D 44; OIM E24375, burnish traces on interior, 10R 5/8, slip on exterior, black interior, fire blooms on exterior

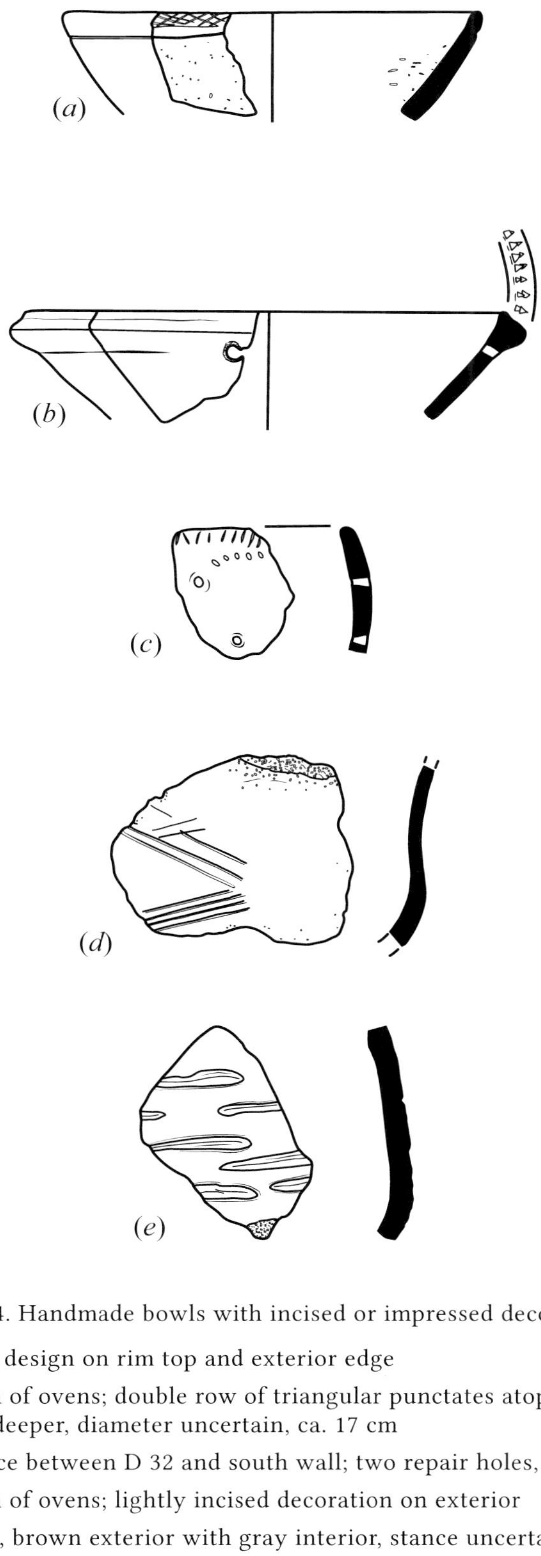

Figure 5.74. Handmade bowls with incised or impressed decoration (2:5)

(*a*) D 12 (1); lattice design on rim top and exterior edge

(*b*) D 208 (2), south of ovens; double row of triangular punctates atop rim, repair hole, stance may be deeper, diameter uncertain, ca. 17 cm

(*c*) Rubble at surface between D 32 and south wall; two repair holes, incised exterior rim

(*d*) D 208 (2), south of ovens; lightly incised decoration on exterior

(*e*) D 62 (1); coarse, brown exterior with gray interior, stance uncertain

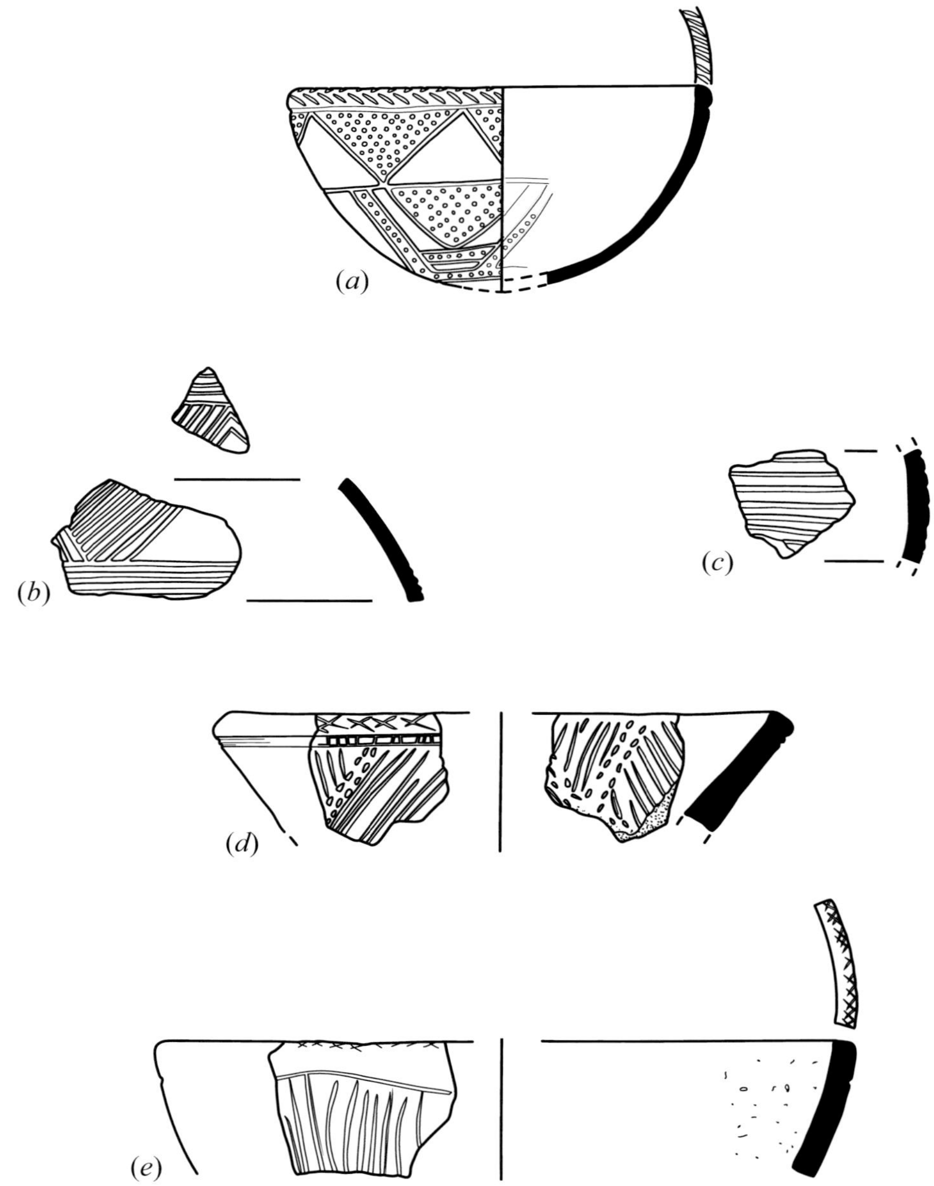

Figure 5.75. Handmade bowls with incised decoration (2:5)

(*a*) D 90 (1); OIM E36383, incised exterior, leaf-shaped chevron at rim and rim exterior, black-gray surfaces interior/exterior, burnished interior/exterior

(*b*) D 113 (1); two sherds, stance uncertain, incised decoration on exterior, black surfaces interior/exterior, burnished interior/exterior

(*c*) Central Sector, 10 cm below Level II walls; black interior, brown exterior, parallel incised lines on exterior

(*d*) D 10 (1); gray-black to reddish-yellow fabric, angular rock(?) particles, uncertain diameter, drawn as 19–19.5 cm

(*e*) D 18 (1); burnished interior/exterior rim, reddish-brown fabric, incised crosshatched decoration on exterior and atop rim, uncertain diameter, drawn as 23 cm

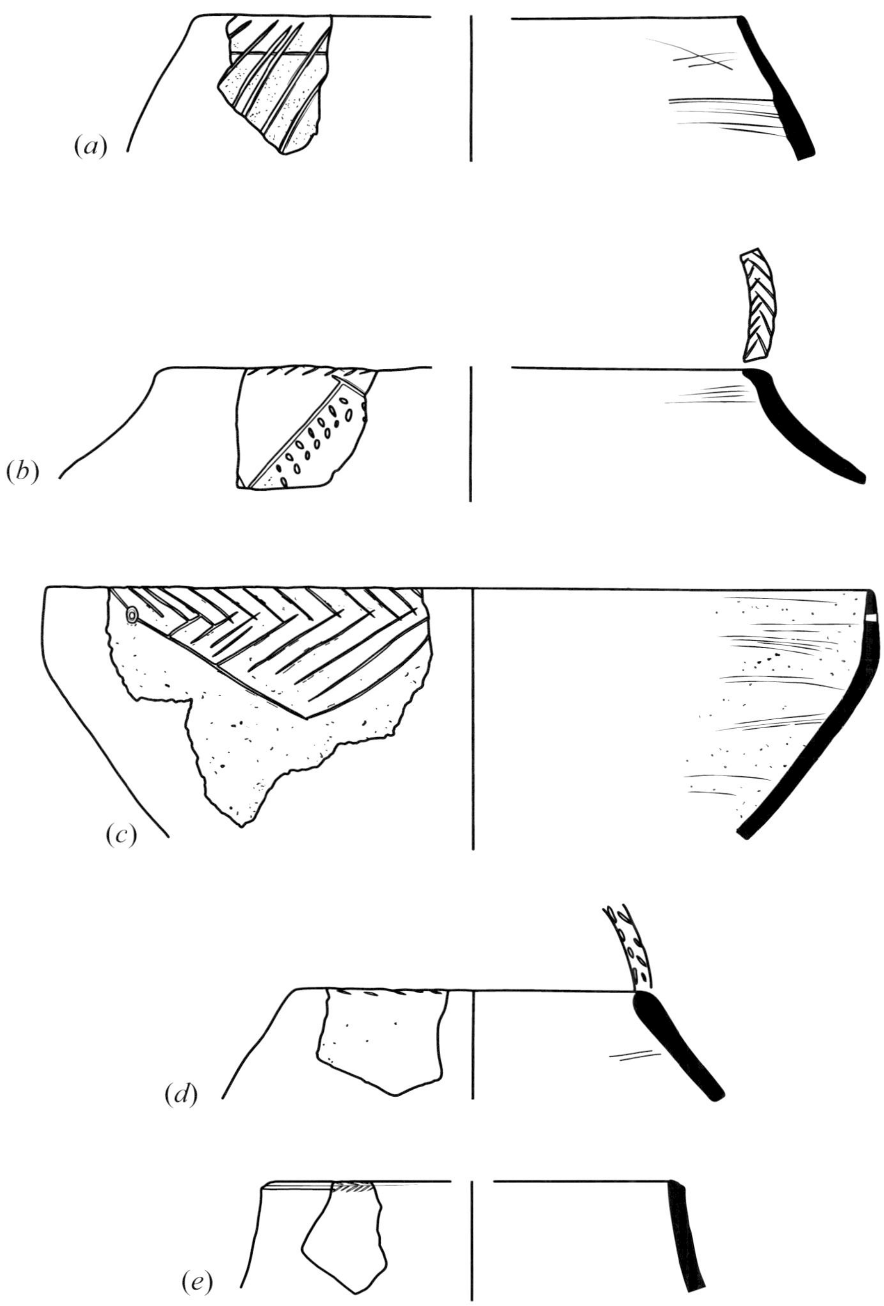

Figure 5.76. Handmade bowls and jars with incised decoration (2:5)

(*a*) D 13 (1); incised oblique lines on exterior, diameter uncertain, drawn as 18 cm

(*b*) West Sector surface; V-shaped lines atop rim, triangles and dots incised on exterior (note start of triangle at bottom left of sherd), diameter uncertain, drawn as 20 cm

(*c*) North–south trench in West Sector; OIM E49525, very sandy and crumbly, repair hole, RS exterior and incised decoration

(*d*) West Sector surface; double row of dashes atop rim

(*e*) D 14; chevron leaf pattern on exterior rim, light-brown exterior, blackish-brown interior, stance and diameter uncertain, drawn as 14 cm

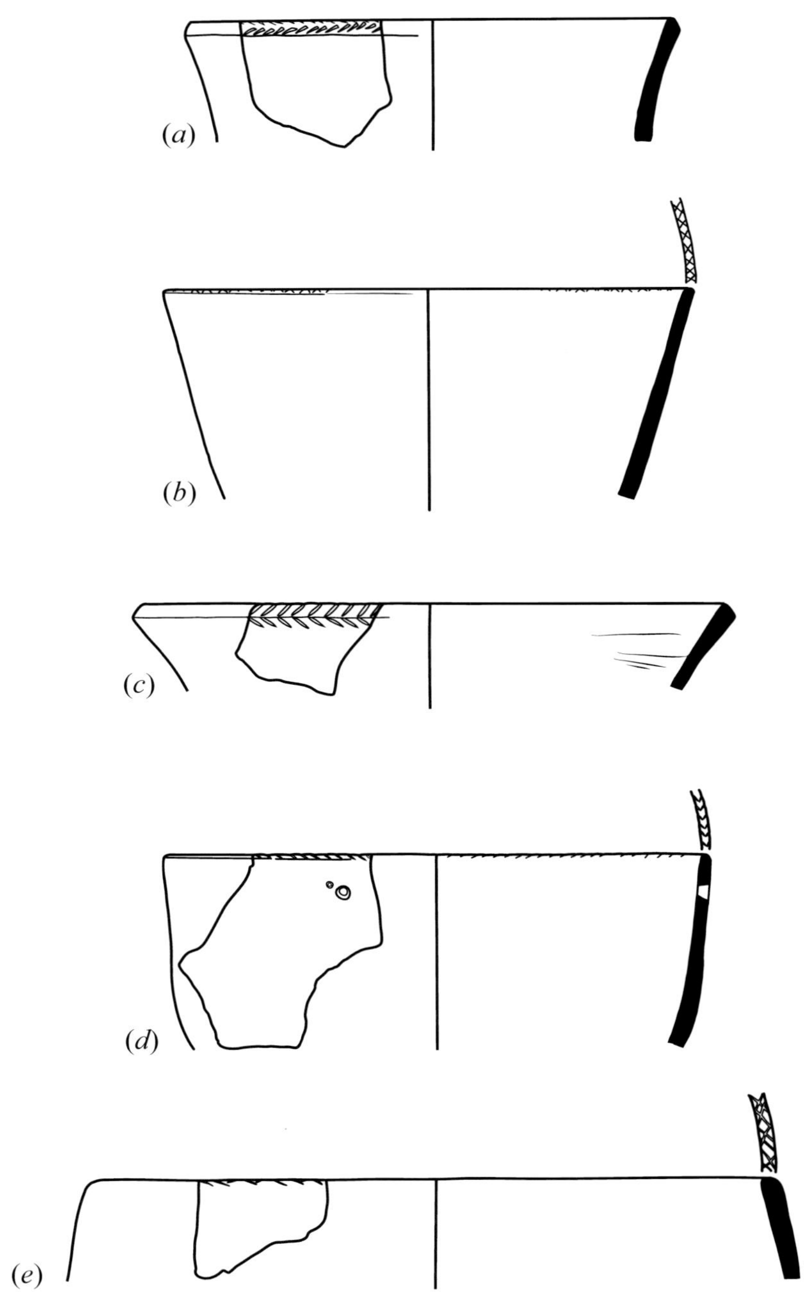

Figure 5.77. Handmade bowls with incised rims (2:5)

(*a*) D 123 (1); highly polished interior/exterior, triangular punctate chevrons atop exterior rim

(*b*) D 1 (3); OIM E49534, burnished interior/exterior, crosshatched incisions atop rim

(*c*) D 92 (1); OIM E49532, black interior, red to gray exterior, burnished interior/exterior, chevrons on exterior rim

(*d*) D 13 (1); OIM E49535, black interior/exterior, burnished interior/exterior, chevron incision atop rim, repair hole, secondary dent next to drill hole

(*e*) East end of West Sector, surface; incised lattice decoration atop rim

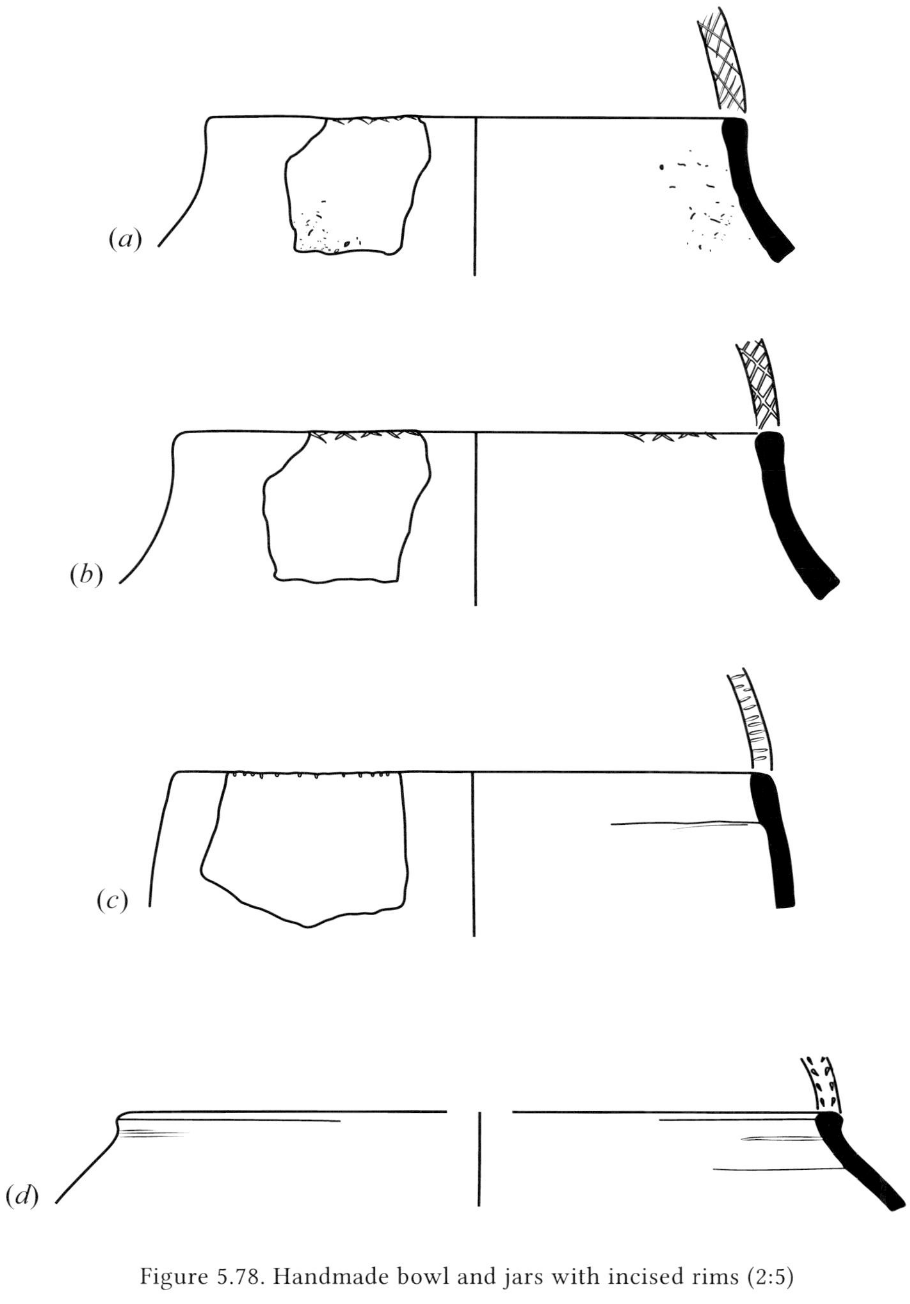

Figure 5.78. Handmade bowl and jars with incised rims (2:5)

(*a*) D 3 (1), lower rubble; incised rim with lattice decoration

(*b*) D 125 (1); black surfaces, burnished interior/exterior, incised lattice decoration atop rim

(*c*) Along west wall of fort at northwest corner; black exterior and black-topped interior

(*d*) D 66 (1); blackish-brown surfaces, burnished exterior and traces on interior, two rows of teardrop punctates atop rim, diameter ca. 24 cm

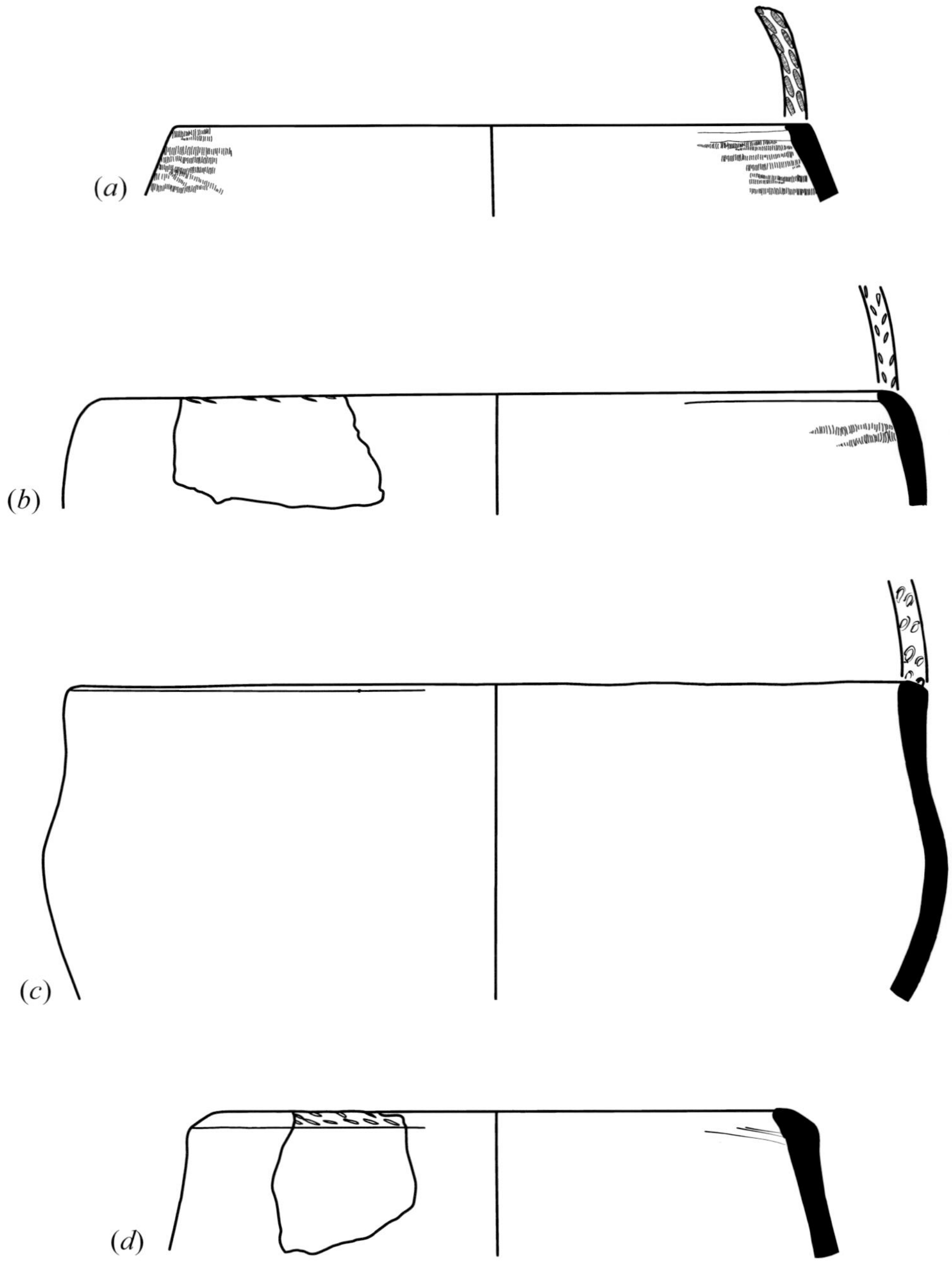

Figure 5.79. Handmade bowls with incised rims (2:5)

(*a*) Below D 4, D 47, and D 48; impressed twine atop rim, burnished strokes interior/exterior, black sherd, interior channel below rim

(*b*) D 19 (1); black exterior, burnish traces interior/exterior, teardrop punctates atop rim, traces of streak burnish on interior

(*c*) D 4, D 47, and D 48, dismantling walls; double row of ovals atop rim, burnish streaks interior/exterior

(*d*) D 1 (1); burnished interior/exterior, black interior, reddish-brown exterior, two rows of teardrops atop exterior rim

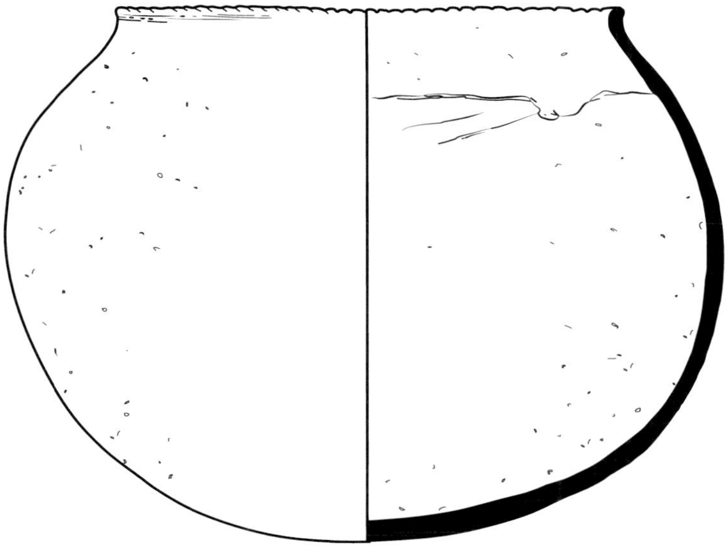

Figure 5.80. Bowl with incised rim decoration (2:5)

D 125 (1); OIM E36380, mottled surfaces, vertical burnish strokes on interior, horizontal and slanted burnish strokes on exterior, rim incised with cross-hatches, diameter ca. 17 cm

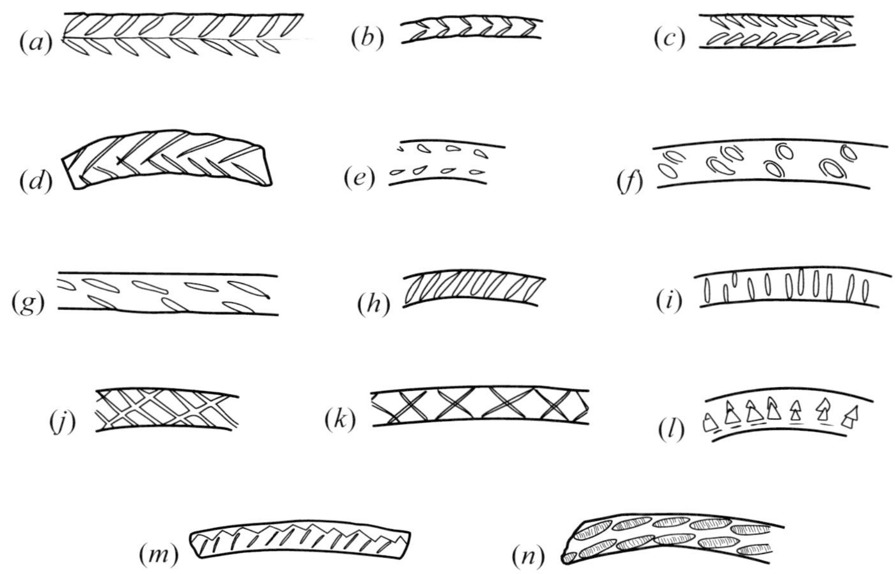

Figure 5.81. Incised rim decorations (1:2, unless noted)

(*a*) Chevron, line in middle is rim edge, not an incision
(*b*) Chevron
(*c*) Chevron on exterior and top of rim
(*d*) Chevron
(*e*) Teardrop punctates
(*f*) Oval punctates
(*g*) Dashes
(*h*) Oblique lines
(*i*) Ladder
(*j*) Lattice
(*k*) Crosshatches, 1:1
(*l*) Triangular punctates, top or exterior of rim (if more vertical stance)
(*m*) Impressed wedge with oblique lines
(*n*) Double row of impressed twine

also been frequently found at sites in southern Nubia.[232] For example, the jar profiles in figure 5.78a–b are similar to the black-topped sherd shown in figure 5.72a, the form and decoration of which are comparable to examples from Gala Abu Ahmed.[233] Similar decorated rim forms were found in the Debba Bend surveys and are similar to the bowls, either open or closed, shown in figures 5.67–5.79.[234]

The sherds of a number of vessels were left in a small space created when the major interior relining of the fortification was built against the two core walls, in D 125 (1), including that shown in figure 5.80 (pl. 73) (see also fig. 5.70c).[235] The bowl is more complete after reconstruction than the other vessels mentioned above, and it also exhibits a different form, with a slightly constricted neck and a short, everted rim knob at its tip. The clay at the rim has been folded over onto the interior of the vessel, where the finishing marks are evident. The surfaces were pitted, though highly burnished in all areas except where it had worn off at the base. Fire clouds of various colors and blackened areas mark the exterior. A blackened area toward the base was possibly where the pot had been set near the flames when used for cooking. The interior was black, and the surface showed the grooves made by a burnishing tool. This cooking pot also has crosshatch incisions atop the rim.

232 All examples cited are roughly contemporary with the materials from Dorginarti. An example from the Napatan Building I at Kerma is shown in Mohamed Ahmed 1991, p. 56, fig. 19, I D6m. The site of Gala Abu Ahmed also yielded similar bowls with lightly incised rims; see Eger, Helmbold-Doyé, and Karberg 2010, p. 78, fig. 5, types ZN 09/1 and ZN 09/14. The online images from the site, at http://arachne.uni-koeln.de/, also show rim-ticks on bowls and small jars, some of them incised more deeply, like gouges, and not good parallels (e.g., SN 2482073, SN 2481231, and SN 2482081). An infant's tomb next to the superstructure of tomb 16 at Sai contained a handmade bowl with incised hatching atop its rim; see Minault-Gout and Thill 2012, p. 89, pl. 165, T16 1.

233 The parallels were cited above; note Jesse 2016, p. 307, fig. 5, nos. 8 and 10, and pp. 311, 314, pl. 2, no. 14. Examples are shown in the online database: http://arachne.uni-koeln.de/. See Gala Abu Ahmed SN 2481444, SN 2481143, and SN 2481148, but note also SN 2480650, SN 2481144, SN 2481145, and SN 2481516, which also seem to be from similar vessels.

234 See, for example, the parallels from the Soniyat temple shown in Orzechowska 2003, pp. 442, 453, pls. 11–13.

235 A rolled-up leather bag(?) and other handmade-vessel sherds were found along with the sherds of three handmade pots in the triangular space left between the relining and the inner core wall, and may be an intentional deposit rather than garbage.

SUMMARY OF THE HANDMADE ASSEMBLAGE

The handmade pottery alone does not help fine-tune the dating of the fortress or identify the cultural groups involved. The handmade sherds represented at the site, however, indicate the continuity of the southern pottery traditions from the second millennium BC into the first millennium. Some residents of the fortress used handmade vessels identified with Nubian cultural traditions for food preparation and eating, despite a reliance on a northern assemblage for the majority of their storage and transport jars.

There is little evidence of wheel-made jars or bowls being used for cooking, and no single wheel-made type could be identified as a typical cooking vessel—that is, there is no consistent evidence for sooting or burning. There are, however, numerous handmade bowl and jar sherds exhibiting evidence of having been used in cooking, but we cannot assume that they were the only cooking vessels on-site. But again, not all sherds were included in the materials brought back to the Oriental Institute.

Individuals whose kit consisted of handmade vessels undoubtedly also utilized wheel-made pottery for various functions, since the handmade types are always found in contexts associated with wheel-made vessels. Ovens were found near the living quarters, but not usually in them, although there is less evidence for hearths and no record of fire pits.[236] Bread plates were discovered at the site, but a fairly small number of bread plate sherds were brought back to Chicago. In any case, bread can be made in other ways. It is unclear to what extent the inhabitants preferred a porridge- or bread-based cuisine, which varied according to Sudanic or Near Eastern traditions, or whether typical southern foods were eaten alongside those of a northern cuisine.[237]

Since it is typical in the Sudanic and African regions to identify domestic potters and cooks as female, we can speculate on a number of different scenarios. Some soldiers or mercenaries may have brought their wives and families to live at the fortress, or they may have lived with members of a local group residing nearby or temporarily visiting the Second Cataract region. Outsiders may have also been employed to cook and work for the residents at the fortress, or men themselves may have done at least some of the pottery

236 It should be pointed out that cooking and baking could easily have been accomplished elsewhere on the island or on the west bank. See, for example, Nordström 2014, pp. 121–23.

237 The variance between the Sudanic or Near Eastern dietary traditions are discussed, for example, in Haaland 2007; Edwards 1996; and Pope 2013.

making and cooking. There is no clear evidence indicating where or by whom the handmade kitchen and serving vessels were made, although the lack of standardized forms indicates that individuals produced vessels for themselves or others when needed.

At least some soldiers, family members, or service personnel at the fortress were indigenous to the regions along the Nile or the deserts bordering the river in Nubia. Unfortunately, the ceramic assemblage and objects from the fortress may not accurately reveal details about the social composition of its inhabitants or the political and economic considerations surrounding their presence at the fortress.

NEW KINGDOM SHERDS

A number of sherds among the Dorginarti materials are clearly, or probably, New Kingdom in date. No Middle Kingdom material was found. The fortresses of Buhen and Mirgissa were active throughout the New Kingdom, and the island was undoubtedly visited during these periods of high activity. The Dorginarti material that can be identified as earlier is sparse, although very distinct. Most of the more diagnostic sherds are of Eighteenth Dynasty types, while a few may mirror materials running on into the Ramesside period.

The mixed locus D 75, which represents a layer under the surface clearance along the west wall of the Level II platform, contained two identifiable New Kingdom sherds. The deposit consisted of rubble collected by the construction workers to fill the later platform. The first is a flattened jar base with a prefired potmark (fig. 5.82a). It is a mixed fabric of silt and marl(?), and fired to a reddish-yellow color with a brownish-purple core. The fabric contains medium-sized chaff, small angular sands, and limestone chunks. The potmark is not well enough preserved to identify it, consisting only of a rounded, raised "arm." The base may be either flat or slightly pointed and thus is comparable to either late Eighteenth Dynasty or Ramesside amphora bases, in imitation of Canaanite amphora bases.[238]

The second sherd from D 75 is a silt jar with a whitish slip preserved as a thicker band on the rim exterior (10YR 8/3). Below this rim band are painted neckbands consisting of a thin brownish-black line above a thicker red band (2.5YR 6/8) (fig. 5.82b). The fabric has subangular to rounded sands of different colors, many organics, and much mica, but also exhibits many limestone or white particles. It is either a coarse Nile B or Nile C. The surfaces are worn and pitted. The amphora is comparable to Eighteenth Dynasty profiles, although no similarly decorated comparisons are known.[239]

Various contexts in or near the East Sector also yielded New Kingdom sherds. A small body sherd of a pilgrim flask is made of a soft marl fabric with an exterior pink coating and some traces of black paint preserved on the exterior (not illustrated). The rubble over the three southern silos in the East Sector contained the sherd of a large silt jar, with lots of organics and sand temper (pl. 74b). The exterior exhibits a thin yellowish-cream slip. The rim is painted with a red band, and a wide red band on the neck is bordered at the top and bottom by thinner black-painted lines. This sherd also represents an Eighteenth Dynasty jar.[240] It may be part of, or similar to, the sherd shown in figure 5.82b.

A silt jar sherd, decorated with upright red triangles, having a center black stroke in each, came from D 10 (1) (pl. 74a). This shark's-tooth design is situated just below the rim on the neck and above a band of multiple black lines in a lattice pattern atop a creamy white slip; below it is another red-painted band. The rim is red with a thin black-painted band at its base. Similar jars have been uncovered in both Egypt and Nubia in contexts dated to the Eighteenth Dynasty.[241]

Under the west enclosure wall relining at the northwest corner of Dorginarti, a silt jar rim was uncovered that resembles examples from Ramesside or early Third Intermediate Period contexts elsewhere (fig. 5.82c). It is made of a Nile B2 fabric with no evidence of slip. Parallels for this form come from Egypt.[242]

A sherd from a carinated jar, or one-handled pitcher, is more clearly datable, and the type is found throughout Nubia in Eighteenth Dynasty contexts (fig. 5.82d).[243]

238 For a carinated Canaanite amphora with a base having a slight central point, see Bourriau 2010, p. 225, fig. 59d (Fabric P31), and p. 285, fig. 74a–c, e (early to mid-Nineteenth Dynasty).

239 See again Bourriau 2010, p. 171, fig. 40, no. 10.16.10.

240 Bourriau 2010, p. 276, fig. 65, 10.8.33, of early to mid-Nineteenth Dynasty date (p. 251), cream-slipped but no painted bands. The parallel has no painted decoration.

241 See Ruffieux 2009, pp. 124–25, figs. 3–4, where the decorative style and some parallels from Egypt and Nubia are discussed. The Dorginarti sherd shows no evidence of the figural decoration on the Kerma sherd. Ruffieux dates this group of pottery to the reigns of Hatshepsut and Tuthmosis III. Note the Eighteenth Dynasty example from Sai in Miellé 2014, p. 388, fig. 3.

242 The shape at least is very close to the marl forms in Aston 2014, p. 120, no. 125, and p. 154, nos. 458–60 (imitations of Canaanite forms). The parallels date between the late Eighteenth and Twentieth Dynasties.

243 Miellé 2014, p. 388, fig. 2. See also the assortment of decorated jugs and carinated vessels in Holthoer 1977, pl. 21 (JU ordinary) and pls. 30–32 (CV ordinary and broad-neck). However,

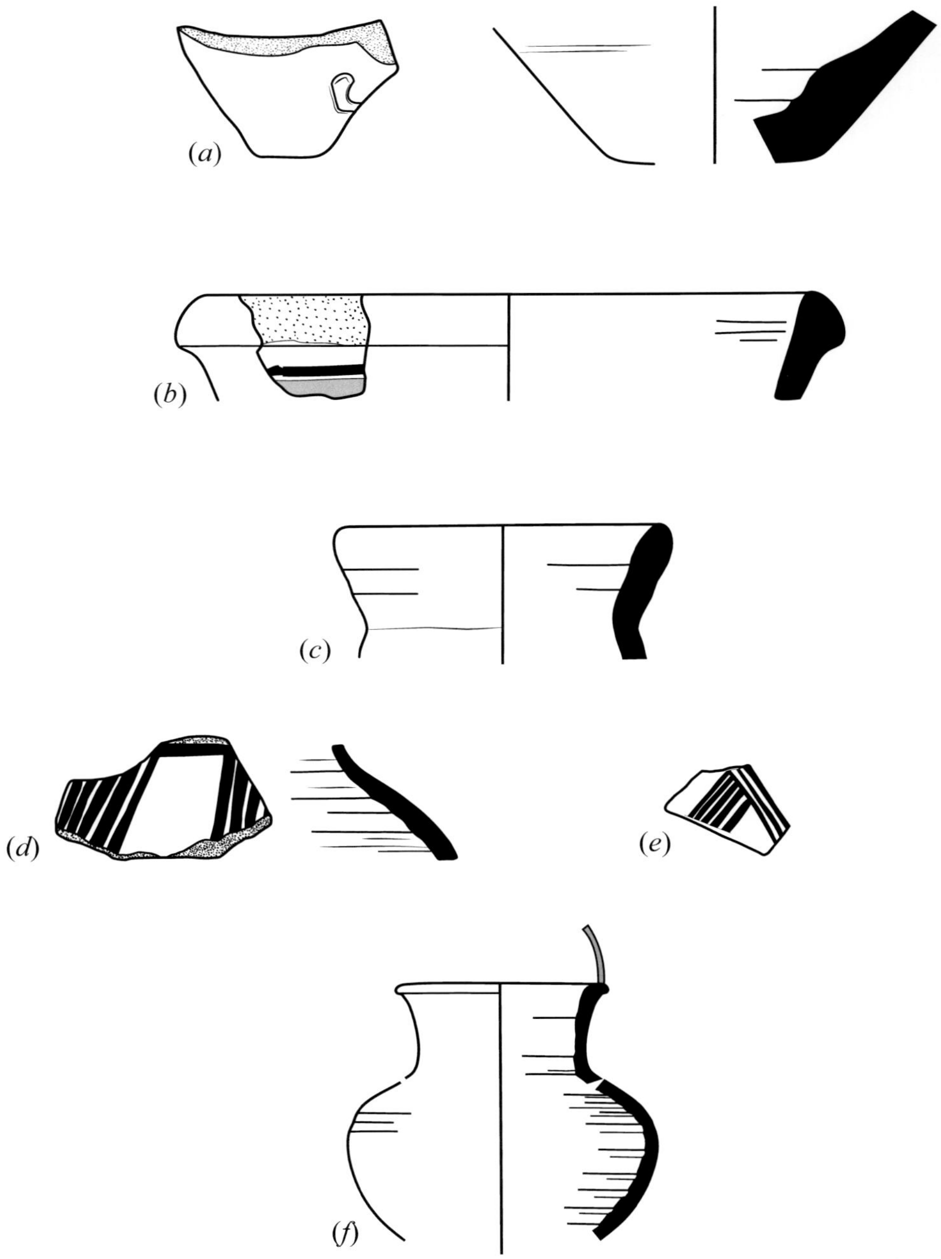

Figure 5.82. New Kingdom sherds (2:5)

(*a*) D 75; OIM E49873, prefiring incised potmark

(*b*) D 75; on drawing, speckled fill is white color, black fill is black color, gray fill is red color

(*c*) Under west relining wall of fortress at northwest corner of West Sector; similar example from D 20 (2), New Kingdom(?)

(*d*) D 200–D 202, surface and (1); Marl A4, black-painted decoration at shoulder

(*e*) East Sector, rubble outside north wall; marl jar sherd with brown-black diagonal bands

(*f*) One meter northeast of Level II stairs, 40 cm below surface; OIM E24345, New Kingdom(?), red paint atop rim

Various fragments of painted vessels with decorations or fabrics that are atypical of the Level III and Level IV assemblage are shown in figure 5.82e and on plate 74c–d.[244] This is not to say they are certainly New Kingdom in date, only that they do not share the typical decorative designs or marl fabrics of the vessels in the main assemblages at the fortress. The body sherd with two sets of red diagonal lines is composed of a reddish-pink marl with a white coating.

Finally, a small marl jar, with a red band of paint atop its rim, came from deposits east of the Level II platform (fig. 5.82f). The rim measures about 7 cm in diameter. The exterior reddish-yellow surface is smoothed and shows a slight luster, while the break and interior surface are pink. The vessel surface is lumpy, with sand, unmixed clay, and limestone chunks—which have erupted at the surface—used as temper. The form finds parallels in early Eighteenth Dynasty contexts, but there are also similarities to vessels from contexts attributed to the Third Intermediate Period.[245]

COMPARATIVE MATERIAL

When the pottery of Dorginarti was initially studied during the late 1980s, there were few relevant publications of well-stratified, contemporary materials from Upper Egypt or Nubia with which to date the Level III and Level IV corpus from the fortress. Except for the royal tombs in the Kushite region, the Sanam cemetery and temple, and some of the Medinet Habu pottery, none of this material was well dated by associated inscriptional material.[246] Similar pottery in Nubia and Upper Egypt was often found in reused New Kingdom tombs and was not clearly distinguished from the earlier tomb deposits.[247] On the other hand, the reuse of New Kingdom objects and pots in first-millennium BC tombs is another possibility that would confound our dating of tombs and the pottery and objects found within them.

The 1986 publication of the pottery from the Late Dynastic period found around the South Tombs at Amarna clinched the general dating for Dorginarti, since so many of the forms and fabrics are precise parallels, and although they are unstratified and lack independent dating evidence, the expertise of the author is unquestioned.[248] The publication of pottery from the fort of Abu ʿId in southern Egypt was also very timely.[249] The material from Amarna has been dated to around the time of the Twenty-Fifth Dynasty and that from Abu ʿId to around the time of Piye, although there is no independent evidence to prove either dating.

Other pottery publications from the late 1970s and 1980s with material comparable to that of Dorginarti include Ashmunein, Karnak, Elephantine, and several well-stratified Delta sites, all of which have been mentioned in previous footnotes.[250] Many of these works were also cited in the preliminary work on Dorginarti from the 1980s, along with many earlier nineteenth- and early twentieth-century publications of sites with relevant materials. These are not all referenced here. Comparative material was also found in the publication of the Twenty-Fifth Dynasty and later materials found at Kerma.[251] My access to the unpublished manuscripts of Helen Jacquet-Gordon from her work at Karnak North and the pioneering study on the pottery from post–New Kingdom phases in Egypt by David Aston were both indispensible for this earlier work.[252]

at Tanis, this group of jars, with variable rims and necks, is frequent in the Twenty-First and Twenty-Second Dynasties; see Laemmel 2011, p. 17, pl. 8, no. 54.

244 See again Miellé 2014 p. 388, fig. 3.

245 Holthoer 1977, pp. 137–38, pls. 30–32, type CV 1, and particularly note IR/O/c–d, which was "a rare find" in New Kingdom contexts. See also Jacquet-Gordon 2012, p. 108, fig. 48h (P.1030), Marl A2 of the Eighteenth Dynasty. However, see the marl example on p. 229, fig. 89r (P.469), which is dated between the Twenty-First and Twenty-Fifth Dynasties. Other parallels are from Medinet Habu and Tanis; see Hölscher 1954, p. 73, pl. 47, Q2 (apparently Twenty-Second to Twenty-Fifth Dynasties) and Laemmel 2011, p. 23, pl. 14, no. 81 (silt, Third Intermediate Period).

246 Dunham 1950 and 1955. See also Griffith 1923 and the foundation deposits of Taharqo at the Sanam temple in Griffith 1922.

247 Williams 1990; Vila 1980. For the Egyptian pottery evidence, see Aston 2009, pp. 317–48.

248 French 1986.

249 Aston 1996c.

250 For the Delta sites, see, for example, Holladay 1982 and Wilson 1982. See also Spencer 1993 and French 1986, both frequently cited in footnotes above. Many good drawings of the ceramics from the temple of Seti I at Qurna were published in Myśliwiec 1987, also cited frequently in the footnotes above. Lastly, an earlier publication of some of the pottery from Elephantine by David Aston (1990, pp. 236–42) was useful.

251 Initially published by Bonnet and Ahmed 1984. The final publication appeared in Mohamed Ahmed 1991. See also Bonnet 1999.

252 I would like to express my gratitude to David Aston for sharing his then-unpublished manuscript, which was published as *Egyptian Pottery of the Late New Kingdom and Third Intermediate Period (Twelfth–Seventh Centuries* BC*): Tentative Footsteps in a Forbidding Terrain* (1996a). Helen Jacquet-Gordon generously provided access to her unpublished manuscript on the Karnak North pottery, "From the Twenty-First Dynasty to the Ptolemaic Period," which also helped fine-tune the dating. References in the footnotes are from her 2012 published book, *Le trésor de Thoutmosis Ier: La céramique* (Karnak Nord 10).

The 1990s and 2000s saw a renewed interest in the pottery development of the first millennium BC in Egypt and Nubia. David Aston's work on the late New Kingdom through early Ptolemaic period ceramic assemblages from Elephantine, his work on the Third Intermediate Period pottery from Heracleopolis Magna, Helen Jacquet-Gordon's publication of the Karnak North pottery, and Elaine A. Sullivan's work on the pottery from the temple of Mut at Karnak were of great use.[253] While there is still limited independent evidence that precisely dates the pottery assemblages of the early first millennium BC, some refinement has occurred.[254] But ongoing work, such as that at Karnak in the area of the Osiris Wennefer Neb-djefau chapel and the temple of Ptah, will certainly help refine the ceramic chronology further.[255]

A number of tomb groups from Nubia contain objects with royal names associated with them and thus provide a *terminus post quem* dating. The chamber tomb (Unit 9) of a warrior from Tombos, where a scarab names King Shabaqo, and a number of the Sanam graves where scarabs with royal names were found in the tomb group, provide such anchors. Sanam yielded scarabs from earlier periods, but it also had tomb groups where objects with the names of Twenty-Fifth and Twenty-Sixth Dynasty kings were found.[256]

The Napatan pottery from the site of Qasr Ibrim shows numerous similarities, as mentioned above. A recent publication of some of the early first-millennium BC pottery from the site has been helpful, as was the access granted to me in the 1990s to study the ceramics excavated around the temple of Taharqo and from levels below the foundations of this temple.[257] The excavators at Kawa have recently published a selection of pottery from their recent excavations, and at least one jar form appears contemporary with examples from the fortress at Dorginarti, while the University of Cologne expedition to the fortress of Gala Abu Ahmed has uncovered both handmade and wheel-made pottery that is contemporary with that at Dorginarti.[258] The publication of the pottery from the 1938–39 excavations at Amara West, and the recent archaeological work at the site, provide evidence for first-millennium activity.[259] Survey and excavations in the Debba Bend have uncovered sites attributed to the proto-Kushite and Napatan periods, and some pottery comparisons mirror the handmade and wheel-made forms occurring at Dorginarti.[260] The 2006 publication of a number of tombs at Hillat el-Arab has also shed light on earlier first-millennium and Napatan burials in the Gebel Barkal area and provided comparative materials.[261]

Recent publications of stratified pottery from early first-millennium BC site contexts in Egypt, and the subsequent discussions of temporal changes in forms, fabric, and manufacturing technology, have indicated that there are vessels from perhaps two periods of occupation at the fortress of Dorginarti before its Saite reincarnation. Level IV vessels include the marl-slipped Nile D and related silt jars, perhaps from Lower Egypt, the MJ 1 and MJ 4, the marl two-handled flasks, the imported Phoenician flask, and the single sherd of a wide-mouthed marl decorated sherd (fig. 5.46c). The earlier silt forms include a few carinated silt bowls (fig. 5.11c–d) and the silt jars found below the Level IV residence stairs (fig. 5.32b). All the comparative dating is clearly shown in the discussion of the forms in this chapter, but the two pottery corpora were not divided here because contexts at the site mix many of the presumed Level IV types with Level III forms, and the chronological divisions are not always apparent. Whether any of the early categories of vessels, specifically the Nile D and related jars, were long-lived forms remains to be seen.

253 Aston 1996a, 1999, and 2010, pp. 28–86; Jacquet-Gordon 2012; Sullivan 2013.

254 For example, the proposed Sheshonq V scarab from levels at Elephantine; see Aston 1999, pp. 154–58 and n. 86.

255 The pottery, dating from the first-millennium BC levels and through the early sixth century, is discussed in Boulet and Defernez 2014.

256 See Smith 2008, p. 109 and fig. 11. For Sanam, see Griffith 1923 and Lohwasser 2012. It should be noted that both the chamber tomb at Tombos and some of the graves at Sanam have wheel-made beakers. This form is absent from the earlier Level III and Level IV assemblages, as well as the later Level II assemblage, at Dorginarti.

257 Rose 2019. I would like to thank Pamela Rose and Mark Horton for allowing me to study the pottery from James Knudstad's excavations around the Taharqo temple in 1984.

258 For the ceramics published from the recent excavations by the Sudan Archaeological Research Society and British Museum at Kawa, see Sjöström and Thomas 2011. The site of Gala Abu Ahmed has produced ceramics that parallel some at Dorginarti, particularly the handmade corpus; see Eger, Helmbold-Doyé, and Karberg 2010 and Jesse 2016.

259 Spencer 2002; Binder, Spencer, and Millet 2011; Binder 2011.

260 See the various chapters in Żurawski 2013.

261 Vincentelli 2006, mentioned in previous footnotes. The parallels to wheel-made bowls, marl jars, and marl pilgrim flasks are mostly from intrusive material in earlier graves.

6

The Level II Fortress and Its Pottery

The Central Sector was unevenly eroded and showed traces of the activity of *sebakhin* atop its surface. Below the remnants of the Christian building in Level I, and showing disturbance from the use of the platform for Meroitic burials, were the remains of Level II (pl. 75a–b; plans 9–10).[1] The walls consisted of the platform measuring 33 m per side, not including the corner projections, and rising to a maximum height of 3 m.[2] Atop this platform was a building or enclosure, which measured 2 m per side. The entire construction was built in one cohesive building program and set square with the general axis of the fortress beneath.[3] The wall faces all had traces of plaster and whitewashing, and the floors of the bays were also plastered and whitewashed.[4] The platform rested atop the north and south enclosure walls, which were leveled slightly for about 10 m in both directions in order to isolate the building (pl. 76a–b).

Limestone blocks were later placed within the bays to flatten the outer face of the platform wall, and the entire structure was then plastered and whitewashed again (pl. 77). This latter renovation was considered Level IIB.

THE PLATFORM

The builders of the platform used the high-standing walls of the Level III building as retaining walls for fill that had been deliberately thrown down into the area from atop the north fortification wall (pl. 78a). The loose layer of debris atop the Level III floors was the result of erosion and collapse of the residence. But atop it was a compacted layer of fill that in turn lay below loose fill consisting of stones, broken bricks, sherds, and dirt (plan 7).[5] The topmost fill also contained finer layers with fewer bricks and stones, indicating the use of the island's soil and sand deposits.

A photograph of the fill in D 23, north of the Level III residence, shows a different view of the debris than that from the north–south trench on plan 7, with the bottom layers tilting up from the enclosure wall toward the Level III wall (pl. 78b). The topmost layers, however, show the fill tipping downward from north to south over the bottom fill deposits.

The retaining walls of the rubble-filled Level II platform crossed the interior of the fortress on the east and west, and were constructed up against the rubble fill with no interior walling. The west wall for Level II was placed to the east of the earlier dividing wall between the West and Central Sectors, but the earlier residence walls were completely enclosed within the construction (pl. 79a).[6]

1 The Christian and Meroitic remains from Dorginarti will be discussed in a future publication. The Meroitic burials around the Level II structure, in the East Sector, and along the south wall were recorded in Pierce 1964. The descriptive loci used for the Level II remains are listed in table A in appendix III.

2 A good description of the Level II remains is found in Knudstad 1966, pp. 179–80. Details are provided in Knudstad 1964a, pp. 2–11.

3 The excavators initially designated the platform as IIA and the structure atop it as IIB, but this was eventually changed to simply Level II, with the later addition of limestone blocks within the bays and the plastering and whitewashing of the structure as a Level IIB activity. See Knudstad 1963–64, pp. 246–47. Although these projections are occasionally referred to as *towers* in the text, their floors were probably level with the walkway around the superstructure atop the platform. They served the functional purpose of both strengthening the walls at vulnerable points and serving as vantage points to guard the Level II structure.

4 Knudstad 1964a, p. 10.

5 Pierce 1964a, January 18; Knudstad 1963–64, p. 246.

6 Knudstad 1963–64, p. 218.

The platform was constructed in three horizontal layers, with a thick layer of mud separating them (fig. 6.1). The base was leveled using randomly laid stretchers, bricks set on edge, or tumbled brick and stones; this feature was also found elsewhere within the structure.[7] The lowest layer, in a north–south section taken along the west wall, measured 70 cm in depth and was composed of alternating headers and stretchers of bricks measuring 35–37 cm in length. The two layers above it exhibited alternating rectangular stretchers and square bricks (34 × 34 × 10 cm), with depths of 83 cm for the middle layer and 75 cm for the top layer. The rectangular bricks in these two top levels were a mix of sizes, mostly 35–37 cm long with some 40 × 20 cm bricks mixed in.

The façade of the platform was nearly vertical, with an incline of 1.15 degrees on the west.[8] The exterior of the wall on the north, east, and west sides incorporated the square 34 × 34 cm bricks, and perhaps on the south as well.[9]

The retaining walls were formed of a series of shallow bays alternating with piers and joined by square corner projections (pls. 79a–b, 80a–b). The preserved height of the walls does not prove that the corners were towers, although this is a probable interpretation. Knudstad suggested that the outer, buttressed face was carried above the terrace level to form a parapet, and he thus reconstructed a parapet-lined walkway around the platform (fig. 6.2).[10] Since the lower parts of wall slots were found in the upper terrace walls at both the southeast and southwest corners, he suggested that there had been archers' loopholes at regular intervals around the outer wall (see fig. 6.3; pl. 77).[11]

A stairway set into the middle of the east wall of Level II allowed access to the platform from the East Sector.[12] The three remaining steps led up a staircase that was 3 m wide, including the sidewalls, and a little more than 3 m long. The stairway was not bonded to the platform wall and rested on a 55 cm layer of debris. The sidewalls were not well preserved, with one standing just 13 cm above the step below it.

Wooden beams, 15–17 cm wide, served as the exterior treads of two preserved steps, which ranged from 32 to 45 cm in depth and rose 14 cm in height. The bricks were laid flat as headers underneath the treads, and the steps were formed of the wooden nose and bricks laid on edge. The top part of the staircase had no remaining steps, and the bottom step was also gone. The total rise of the construction was more than 88 cm as preserved, but the original floor surface next to the stairway was not preserved.

BRICK CONSTRUCTION OF THE PLATFORM AND BUILDING

The architectural drawings of the building and platform foundation detailed the bricklaying used in various areas, with rough sketches showing the following for both the platform and the building walls atop it:[13]

- *Southeast corner:* Building remains of three or four courses of alternating headers and stretchers atop the mud floor. Platform had two courses of rowlock bricks with thick mud layers between the layers.[14] Another wall or foundation area with three courses of headers and stretchers atop the mud floor, and then at least one course of tilted rowlocks in the foundation below (fig. 6.4).
- *North side:* Building remains of two rowlock courses with thick mud mortar levels between them laid atop some rowlock brick courses of the foundation.
- *Southwest corner:* Platform containing loose fill, on top of which were four courses of alternating headers and stretchers and then one course of tilted rowlocks below either a plastered floor or a layer of mortar. Lastly, the remains of the building consisted of one to four brick courses laid atop the platform (fig. 6.5).
- *West wall of the building:* Notes say that it was bonded at its face and shows a bottom level of tilted rowlocks mixed with flat headers(?), then three courses of alternating headers and stretchers. A course of slightly tilted rowlocks remained at the top and presumably below the platform surface or floor.

The preserved outer brickwork was noted as being bonded into the wall in only one place, and the

7 Knudstad 1964a, p. 10 reverse.

8 Knudstad 1964a, p. 10 reverse.

9 Knudstad 1963–64, p. 218.

10 Knudstad 1966, p. 180.

11 Knudstad 1964a, pp. 9–10. The south "tower" walls were at a higher elevation than many of the walls of Level II. The base of the arrow loophole is visible in the upper right quadrant of pl. 77, with the wall plaster below it eroded away by rainfall. Of course, these slots may well have been drain holes from the walkway around the structure.

12 Knudstad 1964a, p. 8.

13 Knudstad 1964a, pp. 4–7.

14 The term "rowlock" identifies bricks laid on their long narrow sides with their short ends visible.

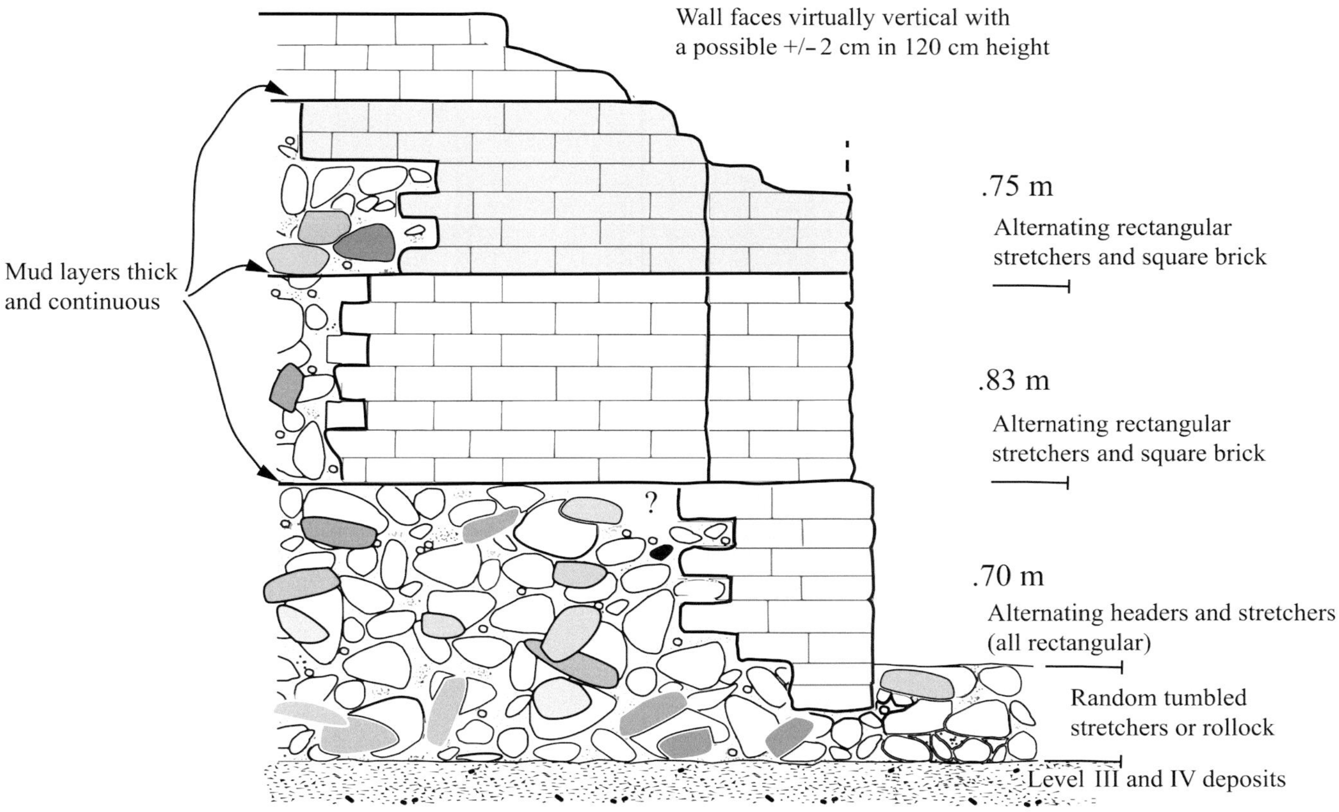

Figure 6.1. Profile of the west wall of the Level II platform, from the north (James Knudstad)

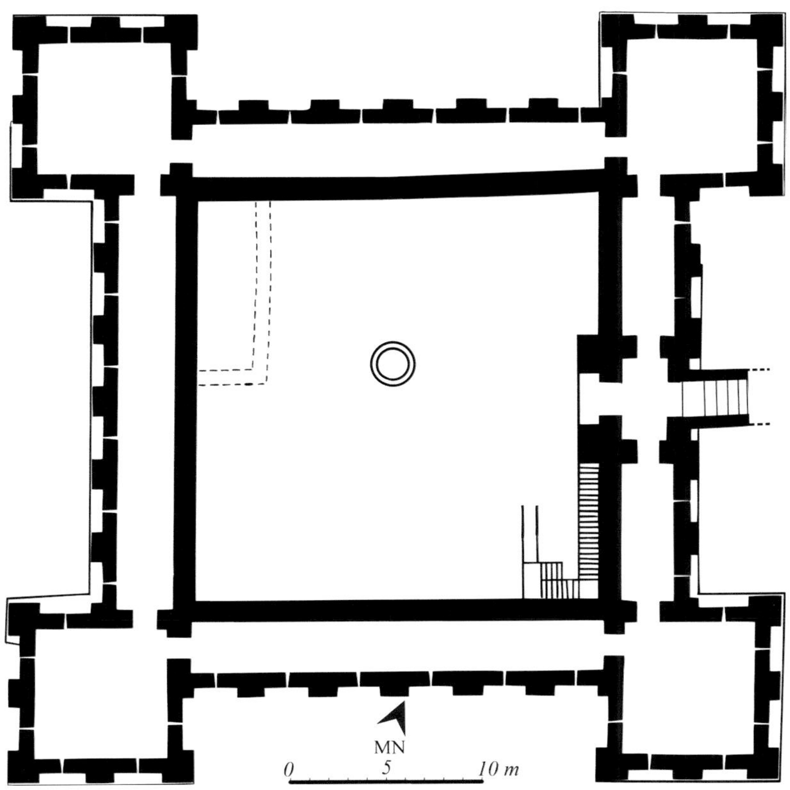

Figure 6.2. Level II reconstruction (James Knudstad)

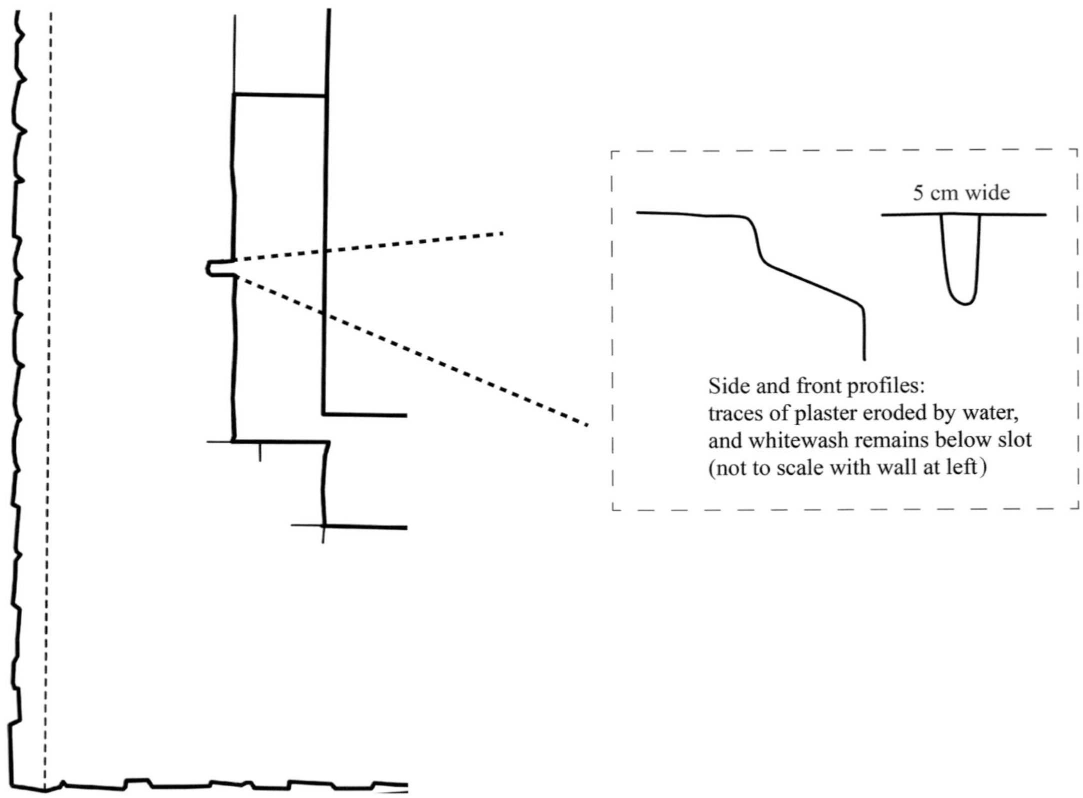

Figure 6.3. Wall slot at the southeast corner (James Knudstad)

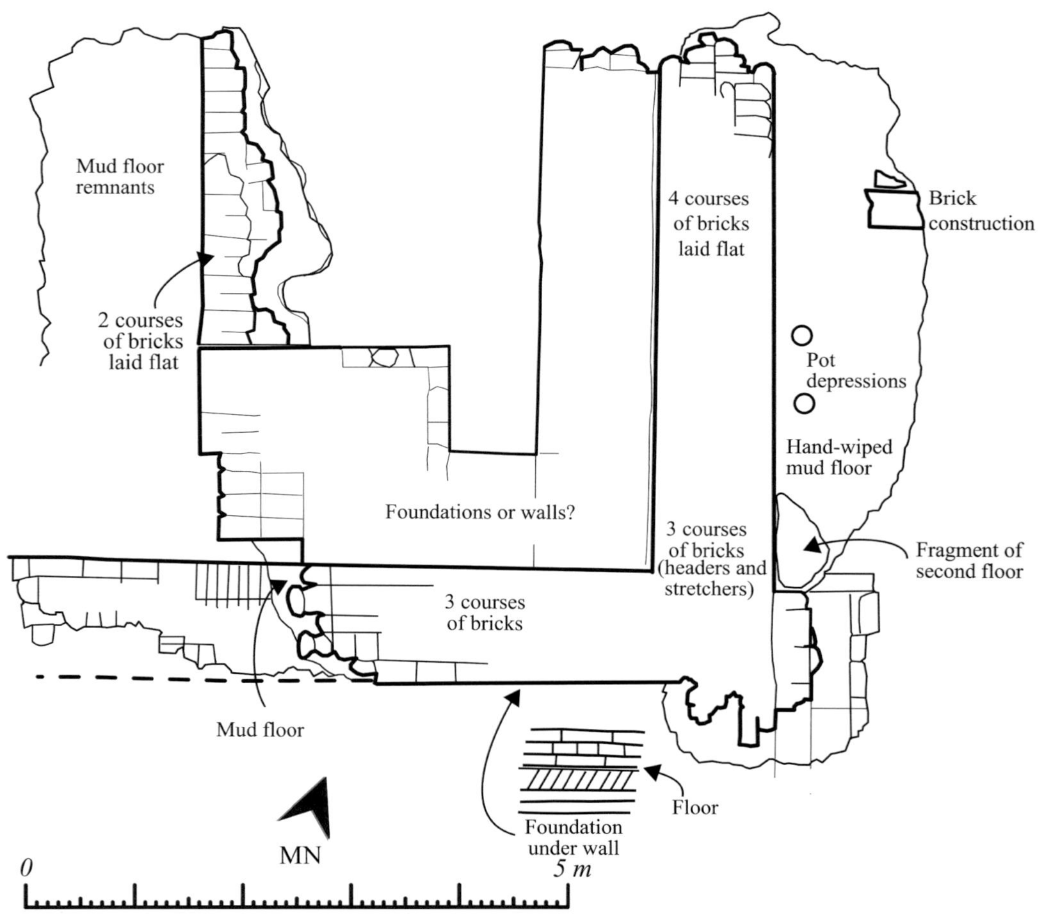

Figure 6.4. Southeast corner of the Level II building (James Knudstad)

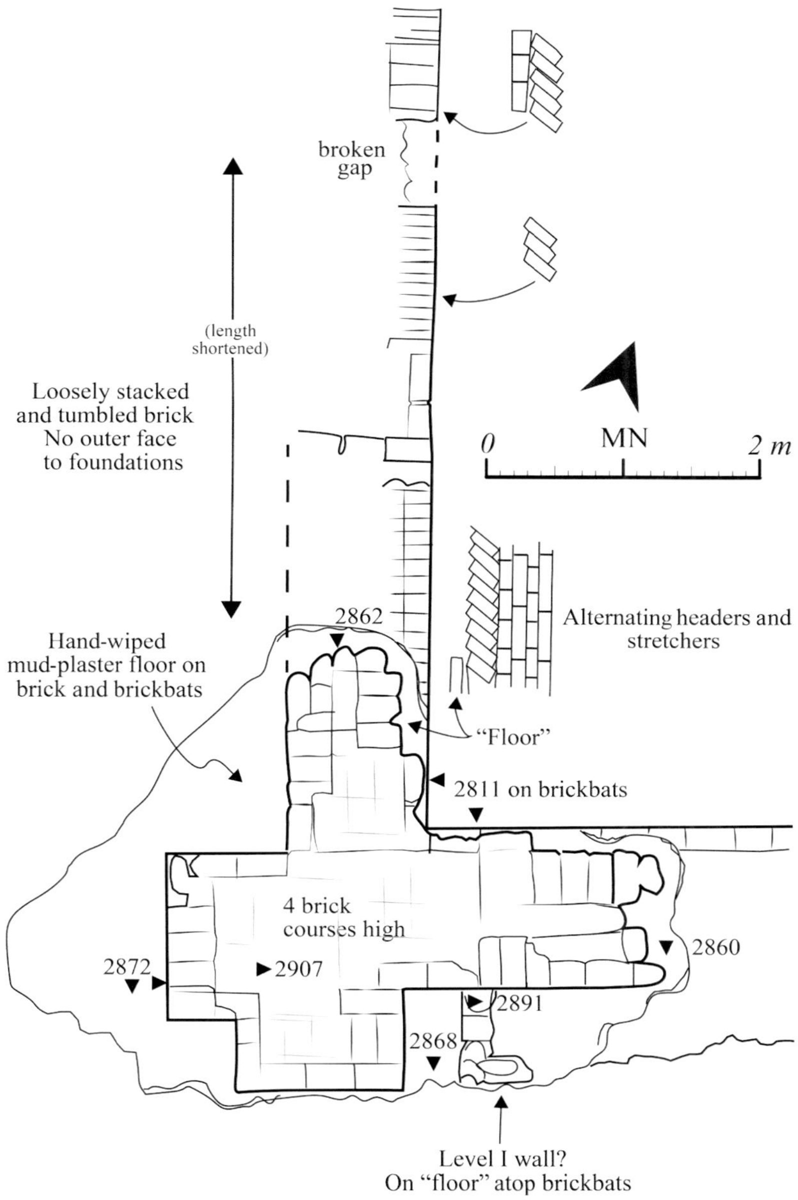

Figure 6.5. Southwest corner of the Level II building (James Knudstad)

style of the bricklaying was not uniform throughout the structure. But the builders had taken care to ensure the building was level, and the elevations of the floor levels within the outer bays were fairly consistent.

THE BUILDING

Atop the mud-plastered surface of the platform, and oriented with it, were the remains of a square building or enclosure. The walls were more than 1 m thick and were crossed at the corners. A doorway was apparent into the middle of the east wall. This entrance led across the walkway surrounding the structure to the stairway leading down to the East Sector.

The remains of the building atop the platform were very poorly preserved, since the Christian use of the area had destroyed or dismantled most remains (plans 9–10). At the northeast corner, however, was a remnant of the structure's east wall, extending south toward the entrance (see pl. 81a).[15] The preserved wall had four courses laid atop the brick surface of the platform. The photograph clearly shows the first course of rowlock bricks with headers atop it, and then another two courses laid in the same manner. The wall sloped down from north to south.

15 Knudstad 1964a, p. 4.

The only remains preserved in the interior of the structure were in the southeast and northwest corners (fig. 6.2; pl. 81b). These were either foundations or walls and were preserved two to four brick courses high on the southeast. They were laid either atop debris consisting mostly of fragmentary bricks or on the mud floor of the platform. The vestiges were largely indistinct so that no signs of joints were observed between the various thicknesses. Toward the north extension of the bricks along the southeast side, and just north of an entrance that was not preserved, the thick line of bricks ended. This mass of bricks was built using alternating headers and stretchers. Hand-wiped mud flooring was preserved on its west side and continued under the threshold area.

At the southeast corner outside the building, and near its east façade, the excavators found two pot depressions in the mud-plastered floor (fig. 6.4).[16] The hand-wiped floor surface went under the building atop the platform, and a second floor surface was preserved at the far southeast corner. A fragmentary brick construction was found atop the flooring just to the north of the pot emplacements, running at a slightly diagonal line to the main structure.

A floor paving of 34 × 34 × 9–10 cm bricks was preserved toward the northwest corner, forming an incomplete L-shape but not connected to the outer north or west walls. The area was, however, incorporated into the Christian construction that had preserved the flooring and reused the square bricks in its construction. (The Meroitic burials that were subsequently dug into the platform had also reused square bricks in their constructions.)

A circular pit was excavated through the plastered floor of the Level II platform in approximately the center. This was determined to be a well, which had been excavated down through the Level III and Level IV floors and was associated with the Level II fortress.[17] It was filled with small cataract rocks and fine, windblown yellow sand.[18] The fill also contained Meroitic and Christian sherds and therefore may also have been built, or was perhaps disturbed, during the Christian period. The well was described in the architectural drawings as a circular brick construction made of bricks 32–34 cm long.[19] It was lined with brick down to an elevation of 2659, or 1.02 m below the preserved height of its head. The diameter was more than 2 m, and the wall was 35–40 cm wide. It was associated with a hand-wiped mud pad atop a wind-laid sand layer, which may have been used for mortar during the construction of the well.

Pottery was found around the walkway surrounding the building, but there were rarely any architectural features except those that have already been noted. A marl bowl with multiple handles and a ring base (OIM E24344) was found in a bay atop the white-plastered floor to the north of the stairs (see fig. 6.12c later in this chapter). The sherds of a small bowl and a ring stand were found atop the floor and next to an L-shaped wall at the southwest corner and outside the confines of the upper structure (OIM E24369, E24338) (see fig. 5.56a in chapter 5 for the ring stand). A ring-based, carinated bowl with flared lip (OIM E24339) was one of the more complete vessels found on the walkway.[20] A faience fragment of a New Year's flask was found atop the west retaining wall (OIM E24336) (see fig. 7.3e in chapter 7, with pl. 90c), and a faience chalice fragment was uncovered atop the south wall debris (OIM E50806) (see fig. 7.3a in chapter 7). Some of these items date to the Level II phase at the fortress. Broken bits of Meroitic pottery and objects from the intrusive graves, which had been robbed, were also found strewn across the ruins, as were Christian sherds from the Level I reuse of the platform.

The pottery that is assigned to the Level II construction, which dates to the sixth century BC, is described after the next section. For the location of sites mentioned in the following discussion, see maps 1–3.

LEVEL II ARCHITECTURAL PARALLELS

The construction of platforms to support buildings has a long tradition in religious and palace architecture in Egypt, Nubia, and Israel, with the initial Egyptian stimulus attributed to a Levantine architectural tradition brought to Egypt during the time of the Hyksos.[21]

16 Knudstad 1964a, p. 5.

17 See Knudstad 1966, p. 179, where he does not associate the well with the Level II building. But his reconstruction of the platformed fortress included the well in the plan. It was sunk to a depth below the floor of the Level IV residence rooms, where it intruded. See Knudstad 1964a, p. 21, where it is questioned whether the well is part of the Level II construction.

18 Pierce 1964b, p. 6.

19 Knudstad 1964a, p. 21.

20 The feature is sketched and noted in Pierce 1964b, p. 5. This bowl is now thought to date from the Christian use of the sector in Level I.

21 Note the discussion and references to earlier publications in Bietak 2005. For the publication of the building, see Jánosi 1996. The New Kingdom complex at Kôm el-ʿAbd at Malkata, the precise purpose of which remains uncertain, is one parallel to the Level II structure that does not have a religious function and was not a typical royal palace; see Kemp 1977a.

Casemate platforms were also used for palaces, temples, and fortresses during the first millennium BC in Egypt and Nubia, when this style of building began to include platformed private residences, administrative buildings, and storage facilities.[22] The walls and architectural elements atop the platforms rarely survived, however, most often obscuring their precise purpose. Platform foundations for tower houses and palaces continued through the Persian era into the Ptolemaic and Roman periods in both Egypt[23] and Nubia,[24] but these—as well as the New Kingdom and Late Bronze Age palaces and official residences in Egypt and the Levant—are not within the scope of the present discussion.[25]

PLATFORMED BUILDINGS IN EGYPT

The square casemate platforms at Naukratis, Tell el-Balamun, and Tell Defenneh featured retaining walls built using concave courses of bricks, with foundation cells that were sometimes used for storage, as there is evidence for vaulting over the cells of the foundations at Tell Defenneh and also at the Levantine coastal site of Tell Jemmeh.[26] The platforms were sometimes placed within temple precincts, where they were typically built near the enclosure walls, while other examples from Egypt were built outside religious precincts.[27] Structural remains indicate that the platforms were accessed indirectly via a ramp situated at one corner of the structure, as is clearly demonstrated at Balamun. The sparse remains at Naukratis show that the walls atop the platform corresponded with those of the casemate walls in the foundation, with the outlines of a large central courtyard or hallway present in the building.[28]

The absence of superstructures and any associated archaeological deposits leaves the purpose of all these monuments uncertain.[29] However, the Palace of Apries at Memphis, which was composed of two platformed constructions accessed by ramps, and where the stone architectural debris associated with it included inscribed columns and the fragments of a sculptured gateway, was most certainly a palace.[30]

The foundation deposits of the platform at Tell el-Defenneh date to the reign of Psamtik I, and the foundation deposits in the temple near the platform at Tell el-Balamun also date to the reign of Psamtik I, thus pointing to a similar date for both. The similar plan of the casemate building at Naukratis suggests that it, too, was built in the Saite period.[31] Since the platforms were often reused, however, many of them are surrounded by archaeological materials of later date.[32]

THE SOUTHERN LEVANT

Close analogies to the casemate foundation buildings in Egypt were also common in the southern Levant during the eighth- and seventh-century Assyrian control of the region.[33] The courtyard or central hallway buildings that were built atop these foundations are typical of earlier architectural styles in the Near East,

22 The literature on this architectural type is reviewed in, for example, Leclère 2014; Mełecka-Drozd 2012; Smoláriková 2008b, pp. 45–123; and Spencer 1999. For the platformed foundations of Napatan structures at Kawa (building F1), the first-floor rooms of which were used for storage, and at Kerma, see Welsby 2014, pp. 54–56, fig. 1, and Mohamed Ahmed 1991, pp. 40–42, fig. 4. New Kingdom villas outside the town of Amara West also featured platformed residences; see Spencer, Stevens, and Binder 2014, p. 38.

23 For the development of the tower house in Egypt, see the range of articles in Marchi 2014; note particularly Lehmann 2014 and Marouard 2014.

24 Meroitic palaces on platforms are discussed in, for example, Baud 2011; Maillot 2014 and 2015.

25 But see, for example, Monnier 2010, p. 85 (Tell Abyad), pp. 86–88 (north Sinai fortins), and pp. 97–99, 102–3 (New Kingdom fortified residences at Beth Shan, Tel Mor, and Deir el-Balah).

26 Tell Jemmeh will be discussed below. For the defining characteristics of monumental platformed buildings in Egypt, see Leclère 2014, p. 21. For Tell Defenneh, see Petrie 1888, p. 53. See also Kemp 2006, pp. 353–59.

27 Leclère 2014, p. 21; Kemp 2006, p. 355. Note that Kemp (pp. 358–60) questions the attribution of these enclosures as strictly religious, since they encircle citadels and houses in addition to temples.

28 See, for example, Leclère 2014, pp. 12–14, and Petrie 1896, pp. 24–26, pls. XLII–XLIII.

29 Leclère 2014, pp. 21–22, prefers a storage purpose for the Delta structures, either for the cult of the temple or, since they are situated along trade routes into the Levant, for their economic role. Other opinions are presented in the works cited in the footnotes here, including their use as fortifications and official administrative centers.

30 The reconstruction of the building was undertaken by Barry Kemp and is published in Kemp 1977b.

31 For the Saite date of the enclosure wall and casemate platform at Naukratis, see, for example, Spencer 2011.

32 Perhaps this is true for the platformed storehouse or bastion at Ashmunein, which was dated to the Ptolemaic period despite the absence of dating material; see Bailey 1991, pp. 27–28, pl. 30.

33 This is not a comprehensive survey of the archaeological sites, but for important places near the "Brook of Egypt," see Na'aman 1979 and Thareani 2016, pp. 82–94, tables 2–3, with references to earlier studies. For these and other sites in the southern Levant, see Finkelstein 1992, pp. 160–62, 166; Faust and Weiss 2005; Master 2003; and Berlejung 2012, with sites discussed on pp. 40–43.

and they are also at home in the southern Levant in Iron Age II.[34] The courtyard design is sometimes referred to as a *lateral-access podium*—that is, a high casemate platform, typically 15–20 m square, built against an enclosure or city wall, and accessed indirectly by a ramp at one side.[35] The superstructures have usually been destroyed, but if the upper walls follow the lines of those in the foundations, as mentioned above, they can be reconstructed as a courtyard or hall surrounded by rooms on three or four sides. This central courtyard or hallway building is also surmised for the Delta examples, as previously mentioned.

Platformed buildings, representing official residences and fortified administrative buildings, are found in the southern coastal cities that were important to Phoenician, Philistine, and Egyptian sea trade in the ninth, eighth, and seventh centuries BC, and thus were also important to the Assyrians. Other courtyard buildings, some built on square platforms, include residences in provincial capitals and fortresses along the overland trade routes that stretched north from the Arabian Peninsula through Edom and Judah and over to the west along the Via Maris.[36]

Many of these administrative and military buildings resemble the platforms at Naukratis, Balamun, and Defenneh, where the influence of Levantine architectural tradition is apparent. For example, the site of Tell Jemmeh, at the juncture of the Negev trade routes and the coastal plain, was an Assyrian administrative center. W. M. Flinders Petrie excavated a brick platformed structure there that resembles the platforms at Tell Defenneh and Naukratis, sites that he had excavated previously, and thus he described the remains as those of a Twenty-Sixth Dynasty fortress.[37] More recent excavations at Tell Jemmeh uncovered part of a similar platform showing evidence of vaulted roofing and arched doorways in the foundation rooms, indicating the involvement of Assyrians and other foreign workers in its construction.[38] These remains are dated to the Iron IIB–C, or the eighth through seventh centuries BC. Once again, there are few traces of an upper building, although a central courtyard or hallway is surmised.

Although Ben-Shlomo and Van Beek note that the courtyard building at Tell Jemmeh resembles a typical palace or administrative building of Assyrian tradition, they admit that the building atop the platform may also have been influenced by earlier Levantine prototypes.[39] Locally made Assyrian-style pottery was also uncovered in the early seventh-century contexts associated with both structures.

The partially preserved, perhaps square, foundation of the fortress at Tell Qudadi, which is located along the coast near modern Tel Aviv and dates to the late eighth through early seventh centuries BC, preserves a row of cells arranged around the walls.[40] The foundation is composed of stone blocks. The courtyard building was probably built of mudbrick but is no longer preserved. The platform is 3 m high, with a ramp entrance in the center of the east side. The nearby fortress at Rishon LeZion, although built of mudbrick, shares the same traits.[41] There is no mention of the use of a pan-bed brick technique in its construction. Other fortifications farther south in the Levant show similar

34 Amiran and Dunayevsky 1958. For open-court buildings surrounded by a row of rooms, built on a square monumental platform with a stepped glacis, such as those at Tell Abu Salima, Ashdod, Rishon LeZion, and Tell Qudadi, see Fantalkin and Tal 2009, p. 200, n. 19. For similar monumental buildings in major cities, see the references in the following footnotes.

35 Sharon and Zarzecki-Peleg 2006, fig. 1. For the distinctions between the "lateral-access podiums" with open courtyards and the covered hall of the "central hall tetra-partite residences," see Lehmann and Killebrew 2010, with additional sites and parallels.

36 For example, there is at least one platformed building or citadel within a larger fortified enclosure at places such as Megiddo, Samaria, Hazor, Ramat Rahel, Arad, and Lachish (i.e., the "Palace-Fort"); see Sharon and Zarzecki-Peleg 2006, fig. 1, and Lehmann and Killebrew 2010, pp. 24–30, fig. 8. For references that discuss fortresses, way stations, and cultic sites along these routes, see Thareani 2016, pp. 87–95, tables 2–3, and Finkelstein 1992, pp. 159–64. See also Na'aman 2001. For a discussion of the Neo-Assyrian sites along the northern Sinai coast, see Finkelstein and Singer-Avitz 2001, p. 252.

37 Petrie 1928, pp. 7–8, pl. XI.

38 Ben-Shlomo and Van Beek 2014, pp. 470–526. A number of construction details are considered Assyrian or Median, indicating influence or involvement in the construction, such as rib vaulting over the doors and vaults built with keystone-shaped bricks, or voussoirs, among other details mentioned in the text (Ben-Shlomo and Van Beek 2014, pp. 498–526, 1057–64, fig. 34.2). For the use of "Median" rib vaulting, a technique confined mainly to sites on the Iranian plateau, and for vaulting built using voussoirs, see Van Beek 1987, pp. 84–85. See also Na'aman 2001, pp. 264–65.

39 That is, if the walls of the superstructure are reconstructed to match those in the casemate foundation. See Ben-Shlomo and Van Beek 2014, pp. 513–18, 1057–64, table 34.1.

40 Fantalkin and Tal 2009. The preserved length of the east wall was about 33 m. There is perhaps one small tower partially preserved at its northeast corner. This detail is uncertain and is not mentioned by the authors, but see p. 193, fig. 5 top.

41 For the plan, see Levy, Peilstöcker, and Ginzburg 2004, p. 93, fig. 3 (in Hebrew), with discussion in English in Wolff 1996, p. 744. The Iron Age IIC and Persian-period platform is built of mudbrick and measures 18 × 17 m. See also Fantalkin and Tal 2009, p. 200, n. 19, for a comparison of sites with courtyard buildings constructed atop square, or nearly square, platforms.

construction techniques and feature a square or rectangular raised podium.

Concave courses of brick are not used in the Level II structure at Dorginarti. Even though the Level II retaining walls at Dorginarti do not enclose a casemate construction, the walls of the Level III residence and the spaces between the residence and fortification walls functioned as interior chambers that could be filled with debris as support for the upper walls. The north and south sides of the platform sat atop the fortress enclosure walls, providing further support. The positioning of the building, within and atop the earlier fortified enclosure and the Level III walls, is a thoughtful reuse of the higher topography and remains found in the Central Sector. Despite the absence of a specially built cellular foundation, the Level II structure follows the architectural blueprint being discussed here.

None of the examples cited so far, however, exhibits the square corner projections or towers, buttressed retaining walls, central stairway access, or possible arrow slots of the structure at Dorginarti.[42]

FORTRESSES WITH CORNER BASTIONS OR TOWERS IN EGYPT, NUBIA, AND THE SOUTHERN LEVANT

Square platforms with corner projections are found in Upper Egypt and Nubia, such as the square fortresses at Elkab and near Kom Ombo in Egypt, which date to the Roman or Late Roman periods.[43] However, there is no indication of Roman or Late Roman archaeological materials around the Level II fortress at Dorginarti.

Two, or perhaps three or four, small fortresses in Nubia may indicate prototypes of, or later comparisons for, the Level II platformed structure. The ruinous fortress at Semna South and the small riverside fortress known as M IV at Mirgissa are close in design to the Level II building, and the mysterious burg at Semna West is similar in size though ruinous.[44] But the dating of these structures differs from that of Dorginarti or is unknown. The square fortifications at Semna South and Mirgissa's M IV are both attributed to earlier Middle Kingdom Egyptian involvement in the region.[45] M IV and the fortress at Semna South are both built on outcrops of bedrock, have square corner projections, and range in size between about 35 m^2 and 50–55 m^2, respectively. They show no evidence of buttresses on the exterior of the walls. The burg or tower at Semna West was built atop a casemate foundation north of the Taharqo temple, and probably at a later date.[46] It may thus represent a fortified platform construction such as that under consideration here.

The construction techniques used in these structures are all different, but their basic ground plans—or size, in the case of the burg at Semna West—resemble the Level II fortress at Dorginarti, despite the lack of buttresses. It would be terrific if they were contemporary with the Level II building, but in the case of M IV and Semna South the excavators' dating of the ceramic remains and objects, as well as the Middle Kingdom sealings found in a dump near the structure at Semna South, do not support a later date. In addition, the lack of published remains found around and under the ruinous burg at Semna West means it remains undated.

A number of New Kingdom fortifications in Nubia also display regular buttressing and small square corner projections, although—like nearby Buhen—they enclose vast areas of settlement and temples rather than being singular platformed buildings.[47]

42 Since many platformed structures in Egypt and the Levant do not preserve much of their upper foundations or superstructures, the parallels for the arrow loopholes at the fortress of Buhen, besides being of much earlier date, are perhaps the most relevant for those at Dorginarti. See the discussion below. However, similar actual or ceremonial elements must have been present in approximately contemporary fortified structures in Egypt, the Levant, and elsewhere in the Near East. The central entrance is a feature that has been considered a Persian-period trait; see Amiran and Dunayevsky 1958, pp. 30–31.

43 Grossmann 2003, p. 125, fig. 19. The fortress is only slightly larger than that at Dorginarti, but there are architectural differences with the Level II platform, which include the staircases in each corner tower. This feature is mirrored in the much larger fortress at Kom Ombo; see Mackensen and el-Bialy 2013.

44 Žabkar and Žabkar 1982, pp. 7–12; Dunham and Janssen 1960, pp. 7, 13, maps III–IV; Vercoutter 1970, pp. 201–4, fig. 10.

45 The main fortress at Mirgissa was reused during the Second Intermediate Period, New Kingdom, and Napatan period. A cemetery at the northeast angle outside the fortress is Napatan in date, and the New Kingdom sanctuary of Hathor, just inside the north gate, was perhaps also used during the Napatan period; see Williams 1990, pp. 36, 39, table 12. The north gate was renovated at a later stage using a brick size identical to that used in the small fort; see Vercoutter 1970, pp. 112–14.

46 As noted in Dunham and Janssen 1960, p. 7. The thick walls at the east edge of this brick mass are curious, with a wide opening or doorway. It is uncertain whether this construction was part of the foundation or superstructure, since no clear profile is shown (Dunham and Janssen 1960, maps III–IV). It cannot be dated because none of the pottery or objects from it, or from the walled rooms below it, are included in the publication. A description of the area is also found in Reisner 1929, p. 152, where Reisner mentions the foundations for a tower but notes that the area was partly overbuilt by the New Kingdom temple, a detail not mentioned in the final report or indicated on the plans.

47 For example, the New Kingdom town enclosure at Amara West, which is square (108 × 108 m), and the enclosure at Sai, which is vast and rectangular, show a similar style. For a

There are fortresses in the southern Levant that feature either multiple towers and bastions or four corner towers.[48] These Iron IIB–C fortified enclosures range from large examples with elaborate two-, four-, and six-chambered gateways to smaller fortresses.[49] It is thought that the Assyrians built the larger examples in the later eighth and seventh centuries BC in order to control the movement of people and goods along the trade routes, and a few examples display building techniques that are considered Assyrian. Tell Jemmeh, which is noted below, is one example.

The inner and outer fortified enclosures at ʿEin Hazeva, on the border between Edom and Judah, were both built of stone and are square in plan with corner towers and an offset-inset wall, with the inner fortification built on a raised platform foundation.[50] Successive mudbrick fortifications at Tell el-Kheleifeh, at the head of the Gulf of Aqaba, are of a similar plan, although the offset-inset wall does not have pronounced corner projections or buttresses.[51] The sequence of fortresses at the site is dated between the eighth and early sixth centuries BC. The later wall encloses an earlier, inner casemate fortress wall built around a platformed structure measuring approximately 12.5 m per side. The architectural plan of these two fortresses is similar, with their outer enclosures encompassing areas that include platformed structures. The inner fortification at ʿEin Hazeva especially resembles the plan of the Level II platform at Dorginarti, despite its less pronounced corner projections and wall buttressing, and it is also close in size (ca. 45 × 45 m).

Other types of fortresses built during the period of Assyrian domination in the Levant are represented by the 50 × 50 m buttressed enclosure with corner towers at Arad, Stratum XI,[52] and the nearby enclosure at Horvat ʿUza, which also features corner towers and buttresses or bastions on the exterior of its walls.[53] The Stratum XI fort at Arad was probably established in the ninth century BC and is thought to have been destroyed in the mid-eighth century BC, while the pottery from Horvat ʿUza is attributed to the seventh and early sixth centuries.[54] In addition, the Stratum VI fortress at Arad, the destruction of which most likely occurred in the early sixth century BC, was also built with a casemate-wall design having corner towers and additional towers or bastions along two of the preserved walls.[55] These fortresses were probably built by the Judaeans to guard the trade routes and their borders.[56] Other contemporary fortified structures in the southern Levant possess similar elements.[57] All these

---

complete plan of the enclosure at Amara West, with restored corner towers or bastions, see Spencer, Stevens, and Binder 2014, inside back cover. For Sai, see Azim 1975, pl. II. For the siting of the east enclosure wall farther to the west, see Budka 2014, fig. 2.

48 Many sites with fortification features are similar but not mentioned in this book, but for further examples see Aharoni 1976, pp. 1–17. The typologies include square and rectangular fortresses with corner and wall towers or bastions, such as those at Kadesh-Barnea, Horvat ʿUza, and Arad. See also Cohen 1979. The dating used in these publications has sometimes been revised by the fine-tuning of the ceramic chronology.

49 For the distinctions between the types of fortified buildings built during the Neo-Assyrian period, see Pratico 1993, pp. 29–31, and Finkelstein 2010, pp. 121–22.

50 The plan of the fortresses at ʿEin Hazeva is published in Cohen and Yisrael 1996, p. 79, pl. 1 top (in Hebrew); Cohen and Yisrael 1995, p. 18. The excavators dated the earliest fortress to the tenth century BC, and the last stratum to the early sixth century BC. A reinterpretation of the inner citadel's structural remains indicates that it was a platformed building, ca. 45 × 45 m, with only the foundation elements preserved; see Naʾaman 2001, pp. 267–68, for references to this view, as well as Ussishkin 2010, pp. 246–48, fig. 1. Ussishkin also suggests that the Assyrians constructed the fortress as a unified building project in the eighth century BC, with its destruction occurring perhaps in the early sixth century.

51 For the plan of this fortress, also of the Neo-Assyrian period, see Pratico 1993, pp. 32–34, pls. 3–5; Pratico 1985, plans 5–7. The main enclosure of the fortification featured only the slightest of corner and mid-wall projections; that is, it is an offset-inset wall. The earliest wheel-made pottery from the fortress dates to no earlier than the eighth century BC, and is no later than the early sixth century; see Pratico 1993, pp. 33, 35–50.

52 For the plans, see Herzog et al. 1984, p. 6, fig. 6, and Herzog 2002, p. 14, table 1. The Stratum XI fortress has a casemate wall that encloses a central space with buildings and has three preserved solid corner projections, additional buttresses or bastions along each side, and a gateway at its northeast corner. It measures approximately 55 × 50 m. For the revised dating of the pottery from Stratum XI to the mid-eighth century BC, see Singer-Avitz 2002, pp. 114–16, 162.

53 Beit-Arieh, Cresson, and Beck 2007, pp. 15–56, fig. 2.3. The 42 × 51 m fort has a solid wall built of stone and is dated to the end of the seventh and early sixth centuries BC (Stratum III). Note also the contemporary fort at nearby Horvat Radum, a square fortified enclosure with a small central tower (Beit-Arieh, Cresson, and Beck 2007, pp. 303–17).

54 Finkelstein 2010, pp. 121–22.

55 Herzog et al. 1984, p. 26, fig. 26, and for a discussion of the dating of the destruction of this stratum to ca. 587/586 BC, see again Singer-Avitz 2002, p. 182.

56 Other fortresses in the Beersheba Valley are discussed in Thareani 2002, pp. 70–74.

57 Not noted in the text above is the fortified enclosure at Kadesh Barnea, which has a rectangular casemate wall built atop a solid foundation, with corner towers and one buttress at the center of each side; see Ussishkin 1995, p. 120, fig. 2. It encloses buildings. Ussishkin reinterprets the fort (60 × 65 m in size) as a single construction project dating to the second half of the eighth and the seventh centuries BC, not two separate fortifications as the original excavators assumed. See also Finkelstein 2010. The wayside shrine and fortified storage facility

fortresses have corner towers and wall buttresses, but once again they represent fortified enclosure walls and not platformed structures.

There is, however, an example of a platformed fortification with two corner towers at Horvat Rosh Zayit, on the Plain of Acco.[58] It is 25 × 25 m in size, and the Stratum IIb fortress has two towers preserved on the west side, with possible ramp access to the south.[59] But it is earlier than Level II at Dorginarti and was possibly destroyed in the early ninth century BC.

SUMMARY OF THE LEVEL II PARALLELS

The platform foundation and corner towers of the fortified administrative building in Level II at Dorginarti are attested in older Egyptian royal and religious architectural traditions and in Middle Kingdom fortifications such as those at Semna South and Buhen. But the parallels above indicate that this building was most likely planned according to contemporary building schemes found in the Delta and farther afield in the southern Levant and Near East. The square ground plan mirrors examples at sites in the north of Egypt and in the Levant, while the fortified corner bastions or towers are additions perhaps inspired by earlier Egyptian fortifications in Nubia or by observation of similar fortifications in the Levant. Based on the parallels discussed above and the sparse architectural remains atop the Level II platform, the superstructure was probably a courtyard building with rooms along at least the north, west, and south sides.

The Level II structure at Dorginarti, therefore, was in all likelihood built by Egyptian officials and personnel who were familiar with building styles of the time in the north of Egypt and in the Levant, and who may have participated in military, diplomatic, or trade missions to neighboring regions. This architecture, as well as the nature of the Level II pottery discussed below, indicates that the cultural interchange between northeast Africa, the Mediterranean, and the Levant during the Twenty-Fifth and Twenty-Sixth Dynasties—and especially in the last third of the seventh century BC during Saite control of the Levantine coastal region after the Assyrians vacated the area—reached into Lower Nubia.[60] The exchange of architectural styles and aspects of material culture was a result not only of intensified trading and diplomatic ties, but also the fact that Egyptian civilians and soldiers traveled and worked in the Levant, and Judaeans and other foreigners participated in the military forces of Egypt during the seventh and sixth centuries BC.

INTRODUCTION TO THE LEVEL II POTTERY

There was no appreciable stratigraphy for Level II, an occupation mostly represented by a scattering of potsherds atop and around the contemporary Level II fortress, below and around the earlier dividing wall between the West and Central Sectors, and atop the north wall at the northwest corner in the West Sector, and with a high concentration found broken in the East Sector, along the North Gate passageway, and on the River Stairs—apparently dropped in haste as the occupants left, or perhaps destroyed after the occupants abandoned the fortress.

The discussion of the Level II pottery will begin with the imported jars from the Mediterranean region and move on to the Egyptian forms that can be assigned with certainty to the same period, some of which are influenced by Judaean forms. The production centers for the various East Greek amphorae from Dorginarti are not always certain, but an attempt at attribution is given.[61] However, these vessels were manufactured using clays that differed from the typical fabrics used for the majority of pottery at Dorginarti; they are not Egyptian imitations. Some Level II pottery may have been included in the discussion of the Level III and Level IV ceramics, since a number of unique sherds were found in upper layers in the West Sector that may well have been from the Level II occupation, but no clear parallels were found during a search for comparable sixth-century materials. None of the handmade sherds can be dated with certainty to this level, and therefore some handmade pottery

at Kuntillet ʿAjrud also had towers but is likewise not noted in the text above; see the plan in Meshel, Carmi, and Segal 1995, p. 206, fig. 2. The pottery from the site is now dated to the late eighth century; see Singer-Avitz 2006.

58 Gal, Alexandre, and Baruch 2000, pp. 12–22, 196–200, plans 3–5. This particular fortress is thought to date to the tenth and ninth centuries BC; the possibility of its establishment and destruction within the early to mid-ninth century, according to the low chronology for the period, is discussed in Finkelstein 2005.

59 Sharon and Zarzecki-Peleg 2006, p. 146, fig. 1, and p. 158.

60 For example, Spalinger 1977, p. 223; Spalinger 1978, pp. 51–52; Schipper 2011; Kahn 2008.

61 A discussion of this pottery is also found in Heidorn 2018b. Unfortunately, the fragmentary nature of the amphora sherds contained in the Oriental Institute collections does not allow any reconstruction of their complete morphology, which is an important indicator of date and origin. Munsell soil-color charts were used to describe the color of some of the pottery fabrics, or the color of coatings found on the vessels (Munsell Color 1988). These readings are given in the text or in the figures.

discussed in chapter 5 could belong to the later activity at the fortress.

EAST GREEK AMPHORAE

Chian and Clazomenian Amphorae

One substantial fragment of a Chian amphora, found in fragments near the North Gate, preserved its upper portion (OIM E24372) (fig. 6.6a; pl. 82a). There are morphological and decorative variations between Chian and "broad-banded" Clazomenian shapes, but the thin-painted red bands on this larger Dorginarti example, as well as the broad bands on a few of the Dorginarti sherds, may indicate the presence of both types of imports in Level II.[62] None of these sherds can be reconstructed to form a more complete shape to indicate otherwise.

Thin lines of red paint were applied directly to the tannish-brown surface. One line runs down along the exterior of the handles until it meets two thin bands around the shoulder. The fabric is hard, with smoothed surfaces that have a slight sheen. The temper includes some very small flakes of mica; small subangular dark, white, and translucent particles; small rounded sands; and dark-red particles that may be mineral or pulverized pottery.[63]

Similar Chian amphorae have been uncovered at sites throughout Egypt and the Levant, as well as at locations around the Black Sea and Mediterranean and in the Aegean, from the earliest slipped types of the later seventh and sixth centuries BC to the pronounced, bulbous-necked amphorae that were made from the end of the sixth century into the third quarter of the fifth century. The example from Dorginarti is comparable to a new range of Chian jars that appeared at the end of the second quarter of the sixth century—as the white-slipped series was vanishing—and became prevalent in the third quarter of the sixth century.[64] The type is referred to as a *Lambrino A1*.[65] Good comparisons to the form of our Chian jar, with its bead rim and long, slightly convex neck, are typically dated to circa 560–510 BC.[66] Both Chian and Clazomenian jars, which are mentioned below, are found at sites in Egypt,[67] including examples from Thebes.[68]

Several body and handle sherds that were uncovered in the Central Sector and also found scattered along the north and south walls in the west half of the fortress exhibit a fabric that may be Chian (figs. 6.6b–c, 6.7a). A very worn handle from atop the north wall may also belong to a Chian jar, but it is small and its surface is worn, exposing numerous black and reddish subangular particles and white bits in the temper. Portions of red-painted bands are noticeable on other, unslipped, body sherds and on two of the handles.[69] A short fragment of a wide strap handle was found in D 15 (1) exhibiting a thick black-painted band running down its exterior side. The tan to pale-brown fabric in this case is very micaceous. It may be that the reddish-brown paint has darkened.

The red-painted bands on the body sherds are much wider than the bands on the Chian amphora illustrated in figure 6.6a, while the preserved parts of the line running down the top of one handle are also quite wide (fig. 6.6b). The handle profile is flatter and is perhaps more akin to the profile of Clazomenian handles.[70]

Other East Greek Amphorae

A number of sherds from Dorginarti stem from imported jars with less certain origins and are here referred to more generally as *East Greek* (figs. 6.7b–e, 6.8a–c; pl. 82b–d, f).[71] They include single rim-and-neck fragments and one lower-body sherd from at least four different amphorae found in levels associated with the

62 See, for example, Dupont 1998, pp. 146, 151–52. However, the Chian jars could also have broader band decoration during the first half of the sixth century (Dupont 1998, p. 148). For the differentiation of the thin- or broad-banded decoration as Chian and Clazomenian, see also Dupont and Goyon 1992, p. 154.

63 The fabric resembles CHI-A-3 of Gassner 2011a.

64 Dupont 1998, p. 148.

65 Lambrino 1938, pp. 107–12, figs. 71–72.

66 Dupont 1998, pp. 148–50, fig. 23.2a; also Lawall et al. 2010, pp. 358–59, pl. 290, L-8, and p. 375, pl. 294, L-69, and dated ca. 530–500/480 BC.

67 A collection of all archaic transport amphorae in Egypt outside Naukratis may be found in Weber 2012. For the Chian amphorae from Elephantine and a note on parallels elsewhere, see Aston 1999, p. 212, and also p. 211, nos. 1911–13.

68 For Lambrino A1 sherds from sites around Thebes, see Marangout 2015, pp. 68, 71–73, 80–81. See also Defernez 2015, pp. 50–51, fig. 4.

69 The fabric of these other sherds typically contains lots of mica, large white chunks (limestone?), numerous dark subangular and rounded sands, medium-sized black to red-brown subangular inclusions, and white rounded sands. However, the handle shown in fig. 6.6b has very small amounts of mica present, or at least visible, and another handle stump—not illustrated—has the usual micaceous temper in addition to red particles.

70 The flattened handle profile is shown in Dupont 1998, p. 153, fig. 23.3. For references to Zeest's "broad-banded" amphorae, see p. 151, nn. 61–62.

71 Dupont 1998, p. 142.

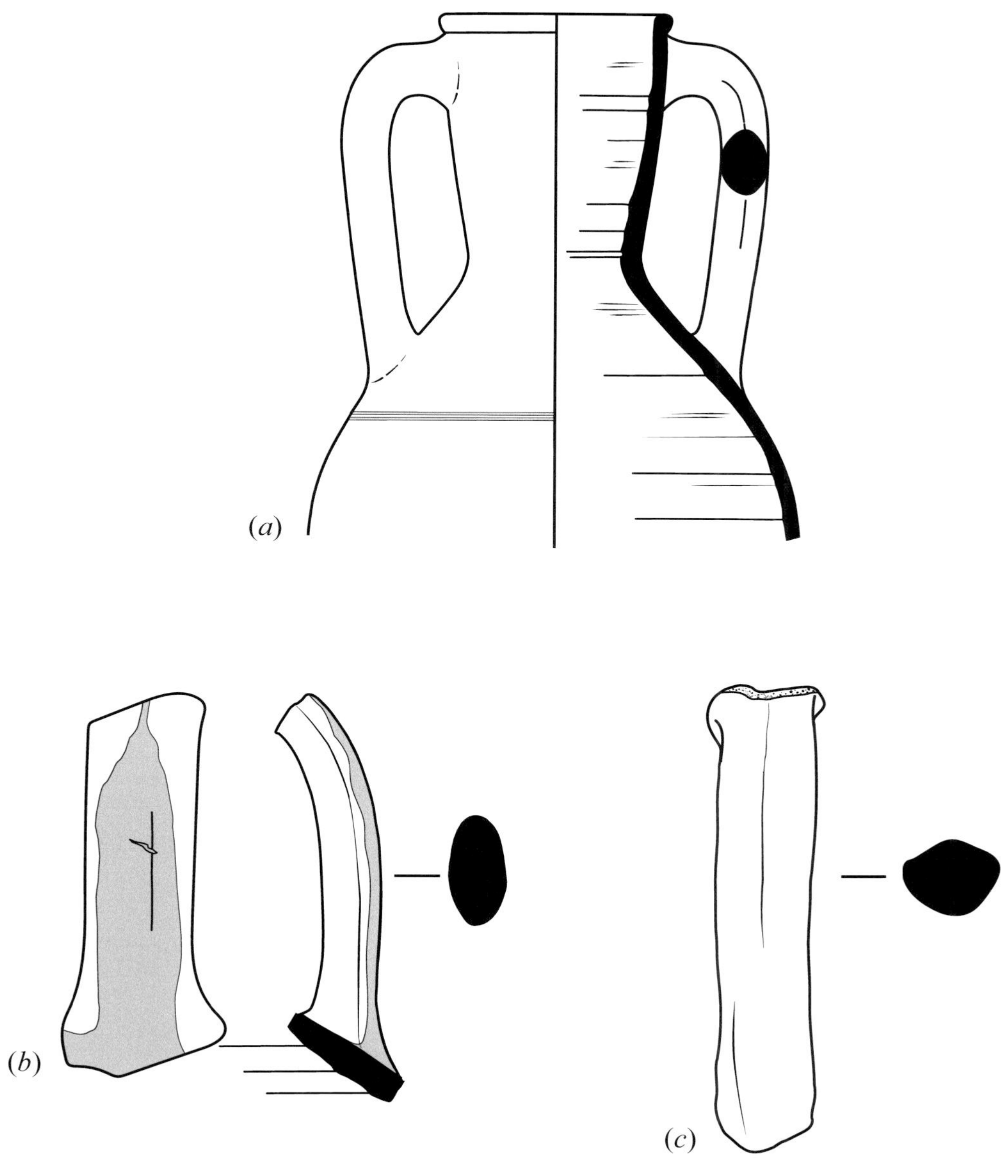

Figure 6.6. Chian and related amphorae (*a*, 1:5; *b–c*, 2:5)

(*a*) East sector, near east wall of Level II; OIM E24372, thin red-painted lines and one on exterior handle

(*b*) D 15 (1); light red (2.5YR 6/6) with thick gray core, 2.5YR 5/8 red paint

(*c*) River Stairs, first ten steps; Chian(?)

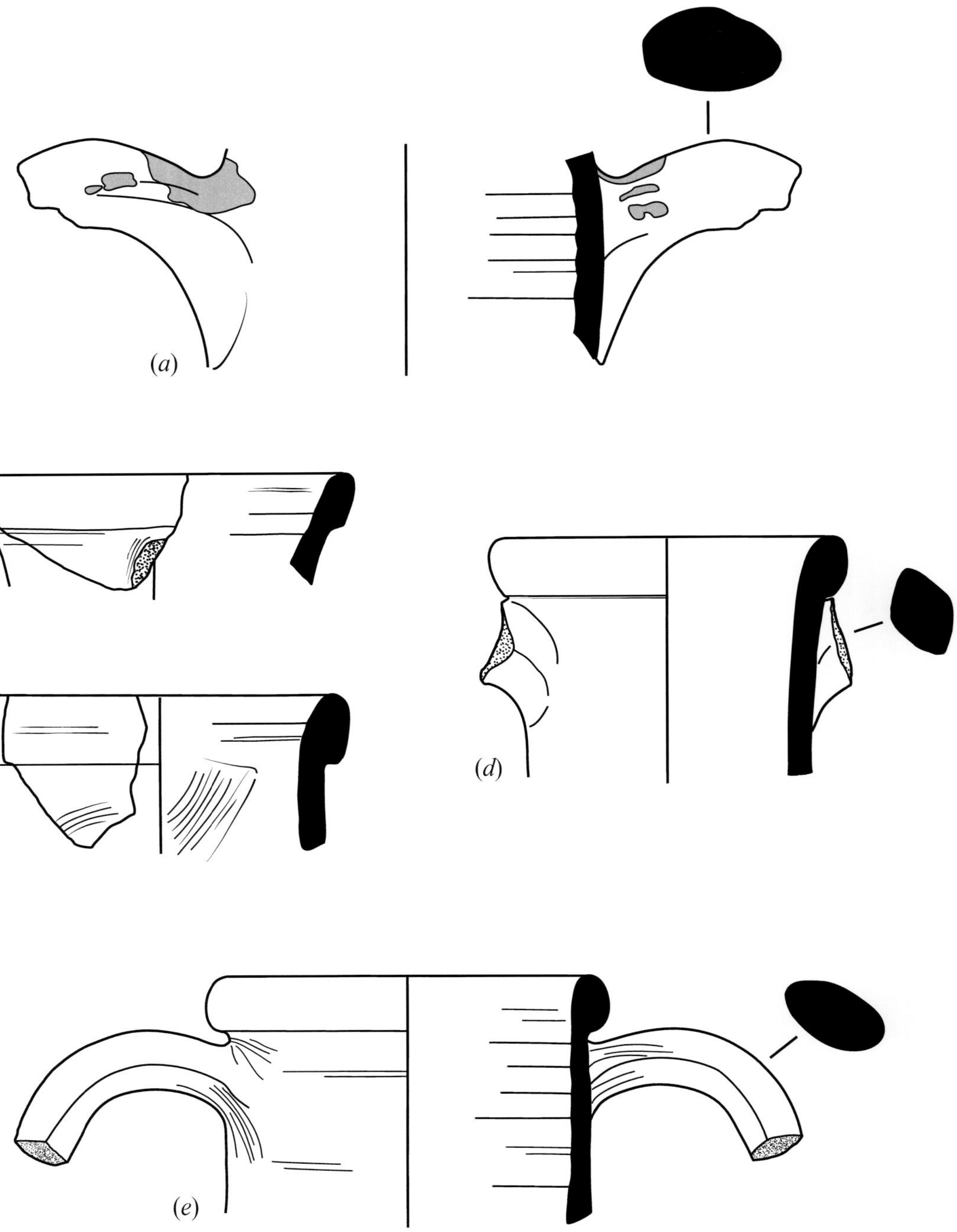

Figure 6.7. East Greek amphorae (2:5)

(*a*) D 24 (2); Chian(?), red paint splotches, stance and diameter uncertain

(*b*) Central Sector, west retaining wall for Level II, south corner; OIM E49507

(*c*) D 5 (1); OIM E49553 (see base in fig. 6.8c)

(*d*) River Stairs; OIM E49508, Samian, 12–13 cm

(*e*) East Sector, center, first 60 cm; OIM E49509

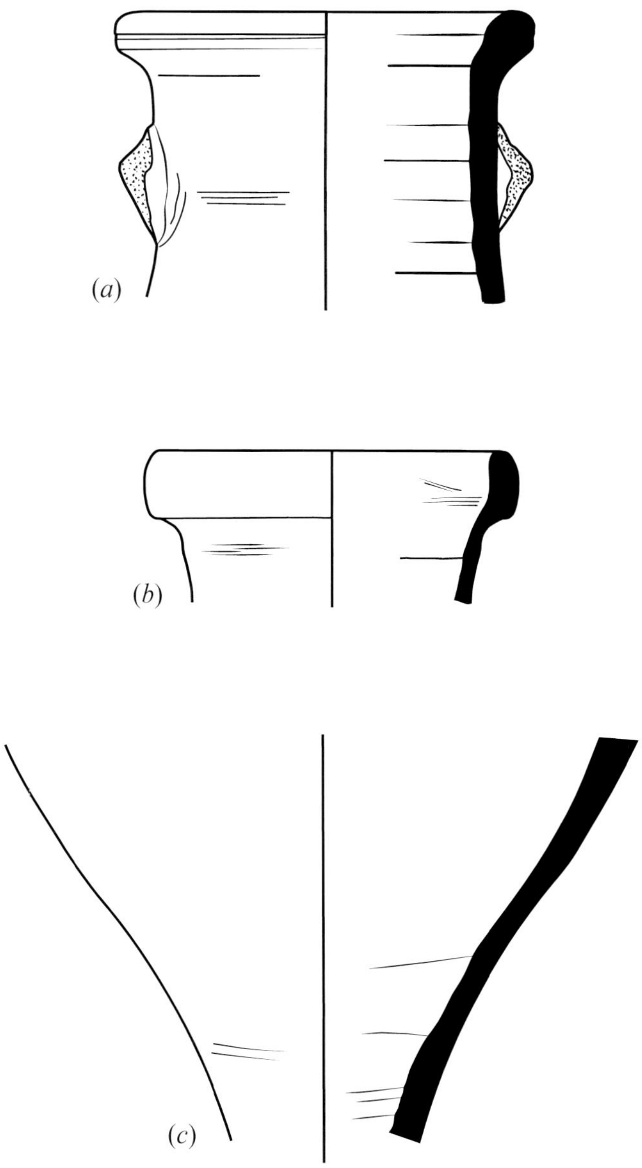

Figure 6.8. East Greek amphorae and base (2:5)

(*a*) D 15 (1); OIM E49506

(*b*) Outside north wall from stairs to second west buttress

(*c*) D 5 (1); uncertain diameter (plus handle fragment), OIM E49528, light reddish-brown fabric, see rim in fig. 6.7c

Level II occupation in the East and Central Sectors, around the River Stairs, and in various contexts in the West Sector. Despite the uncertainty of their points of origin, their forms are all quite typical of transport amphorae of the sixth century BC. Not every sherd has been drawn, but the rim variants are all represented in the following discussion.

The rims of the probable Samian vessels from Dorginarti are not as bulbous and everted as those on typical Samian jars, and their necks are also not as splayed. But there are parallels in the literature that resemble the straight necks, rims, and fabric of those shown in figure 6.7c–e.[72] Most of the sherds have yellowish-brown, reddish-yellow, or pale-brown surfaces, which sometimes display a slight sheen from being smoothed.[73] There is evidence for handles, or sometimes just the finishing marks indicating handles, on all these sherds. Another reddish-yellow body sherd, plus the sherd of a base, stems from the same locus as the rim in figure 6.7c, showing the profile below the widest point of the body and tapering down toward the base (fig. 6.8c). The lack of completely reconstructible vessels and bases hinders their attribution and finer dating.

An imported body sherd with a postfiring potmark was found in D 122 (1) and may belong to one of the types mentioned above (see fig. 7.16e in chapter 7). Its reddish-yellow to brown fabric has many dark, angular particles and is very micaceous. It bears an "X" incised on the exterior surface, while the interior surface is the only example in this collection with a coat of resin still present.

The rim in figure 6.7d (pl. 82f) is very worn and soft, exposing translucent gray-brown subangular particles, small black and white particles, and abundant mica. It also has a few larger biotite inclusions. Its exterior surface is reddish-yellow, and the color of the break is similar. The rim illustrated in figure 6.7e (pl. 82b) has a light-reddish-brown fabric, with abundant mica, dark subangular particles that may be biotite, small gray inclusions, lots of rounded sand, and small pink chunks of unmixed clay. The rim sherd in

---

72 Bourriau and French 2007, p. 130, fig. 1, no. 4 (Samian/Protothasian, dated ca. 500 BC). Other Samian jars are shown in Johnston 1993, pp. 364–66, fig. 9a, e–f. For examples from Israel, see the similar type SA 2 amphorae in Fantalkin 2001, pp. 90, 93, fig. 33, nos. 2–3; note also Waldbaum 2011, p. 311, types 505–6, and Lehmann 2002, pp. 235–36, fig. 5.95, no. 1 (late seventh to early sixth centuries BC).

73 The fired surfaces range in color from reddish yellow (5YR 6/6 to 7.5YR 6/6) to light brown (7.5YR 6/4) to reddish brown (2.5YR 5/6). A range of East Aegean fabrics from fifth century BC amphorae are discussed in Gassner 2011b.

figure 6.7b (pl. 82c) is also reddish-yellow, with lots of mica, sands and subangular particles, and shiny white particles in the temper.

The form and fabric of the amphora shown in figure 6.8a (pl. 82d) cannot yet be attributed to any known production region. The fabric is tannish to light brown and is filled with red and dark particles, shiny translucent sands, and white- to gray-colored inclusions that are probably limestone.

The amphora rim and upper neck shown in figure 6.8b is different again. The inner rim is concave, with a slight bump on the exterior of the neck, and its fabric is pale brown with possible traces of what is now a pinkish slip or, more likely, the original surface layer, since the sherd is very worn.[74] The form and fabric of this rim may be an earlier form of Samian jar.[75] The extreme concavity of the rim dates it earlier in the sequence, although a late seventh- or early sixth-century BC date is still within the time frame of the other imports of Level II, particularly given the possibility of reuse.

The excavations at Dorginarti also yielded a few gray-fabric body sherds, undoubtedly from the same vessel, which could potentially be those of a Lesbian amphora; however, no rim or handle sherds of similar fabric are found in the collection.[76]

Possible East Greek Import Sherds

Two handle sherds, toward the top and the shoulder of a single vessel, are made of a distinct clay fabric and were found in the "East Sector center first 60 cm" and the "East Sector, rubble over three southern silos," but they clearly belong to the same vessel (fig. 6.9a; pl. 82e). The top part of a handle and neck, and the bottom part of a handle and shoulder, allow a tentative reconstruction, with the stance and diameter uncertain. The sherds do not join. It is uncertain whether a second handle was present. The fabric is a light tan to pale brown, and the temper includes subangular inclusions, many small dark particles, and sand. The matte black paint has been sloppily applied in a band atop the handle and runs down onto the shoulder section. These sherds may have come from a table jar of East Greek tradition, although no parallel was found.

The attribution and date of the ring base with a flat foot shown in figure 6.9b is also uncertain. It was found in the debris over the three southern silos in the East Sector, where sherds of many dates were mixed with Level II types.[77] A handle with a prefired, incised V-pattern preserved at its top is also part of an imported jar (fig. 6.9c; see also fig. 7.16b in chapter 7).[78] Finally, a triple-ridged handle fragment is possibly also from a Level II import (fig. 6.9d).[79]

PHOENICIAN AMPHORAE

Phoenician amphorae originated from multiple production centers along the northern coastline of the Levant, with the form spreading and made in different localities by the seventh century BC.[80] The jars have low rims, carinated shoulders, and two large, skewed handles placed at the carination and on the body. There are a variety of shapes and fabrics. Phoenician amphora finds are distributed throughout Egypt, with the closest comparisons to the Dorginarti material being the sixth-century examples at Elephantine and Tell el-Maskhuta.[81]

74 The sherd's surface is worn, tending to crack, and spalling with pockmarks or burned-out limestone(?). The fabric is reddish yellow with grayish zones atop the rim and on the interior, and the exterior is a very pale brown. The large amount of temper includes unmixed clay lumps, some mica, subangular sands of all colors, white chunks, and white bits. The surfaces are gritty.

75 See Dupont 1998, pp. 164–71, on Samian and Milesian amphorae, which are difficult to distinguish in form and fabric. Note the discussion and illustrations on pp. 170–176, fig. 23.7a, c (Grace's Samian, but actually Milesian). Note the amphora belonging to the Samian group in Tzochev 2011, pp. 76, 79, fig. 4, no. 2, and pl. III, no. 4, which is dated to the early sixth century BC.

76 The fabric is gray throughout, with a lighter-gray to tan core, and contains multitudes of silver mica, shiny translucent particles, limestone bits, and some rounded sands.

77 The temper, or the clay itself, contains many rounded sands and mica particles, and the fabric is a pale brown with very eroded surfaces.

78 The fabric is light brown to yellowish brown, and the surface is very smooth. The temper includes many subrectangular gray particles, brown and yellow sand, and fine small red specks.

79 The worn surfaces show lots of dark angular particles, some white and gray limestone bits, mica, and perhaps unmixed clay pellets. Its exterior is gray, possibly from firing or having been burned.

80 Sites with sherds of Phoenician amphorae found farther west in the Mediterranean are not considered here. Numerous examples of various dates from sites along the coast of Syria and Lebanon are shown in Lehmann 1996, pp. 431–40, pls. 70–76. Note also Regev 2004, pp. 337–52. The Type 6 amphorae of Sagona 1982, pp. 80–82, fig. 2, nos. 1, 3, 4, are closest to a few of the Dorginarti examples. However, Sagona dates these "powdery, orange ware" amphorae only to the fifth and fourth centuries BC.

81 See the list of sites mentioned in Aston 1999, p. 232, and the comparisons to the Dorginarti sherds on pp. 232–39, especially nos. 2044–56 of fabrics P 12 and K6, dated between ca. 550 and 400 BC. For the distribution of fabric K6 amphorae,

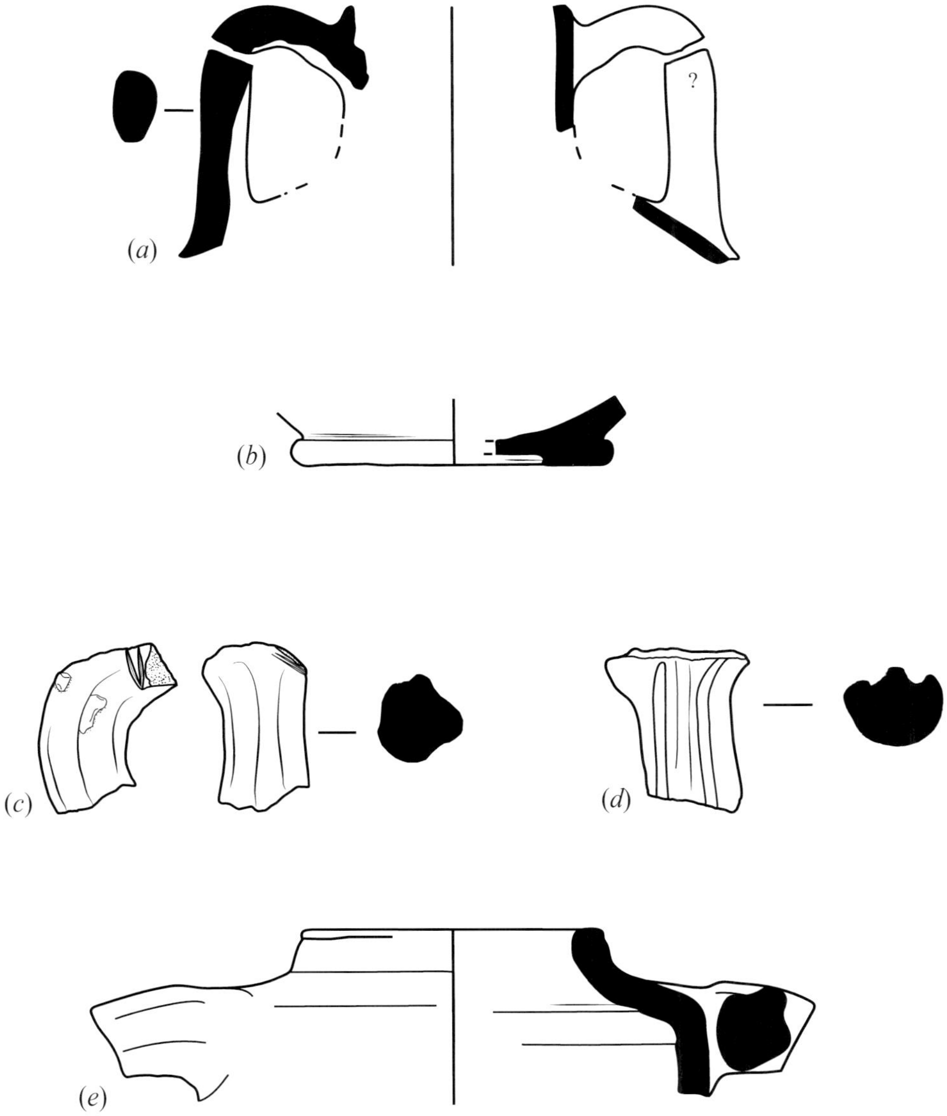

Figure 6.9. Handle fragments and Phoenician amphorae (2:5)

(*a*) East Sector, center, first 60 cm; OIM E49510, matte black paint exterior, two-handled(?), diameter uncertain

(*b*) East Sector, rubble over three southern silos; OIM E49510

(*c*) D 46; prefired potmark

(*d*) River Stairs; burned

(*e*) D 15 (1) and D 8, first 50 cm below Christian building; OIM E49543, tan fabric

Jar fragments were found scattered over the topmost levels in the easternmost half of the fortress, atop the River Stairs, and along the west side of the Level II construction. The shapes of the rims, shoulders, and handles indicate that a number of different types or periods of jars were in use at the fortress, although without evidence for more complete shapes it is difficult to make precise morphological comparisons. There are no complete forms, and only one base has been found among the materials in the Oriental Institute Museum (fig. 6.10b).

It should also be noted that two fragments of possible mud jar sealings were found atop the Central Sector Level II platform (see fig. 6.12a–b below and pl. 83d). But these are not spherical or dome-shaped covers, there is no evidence for rim impressions,[82] and they bear prefiring signs or letters rather than sealing imprints. Jar lids, which are ideally suited for these Phoenician amphorae, are part of the Egyptian repertoire from Level II and are discussed below.[83]

The jar fabrics are either orange or a pale-brown to tannish color, and their tempers appear to be very similar (pl. 84).[84] The temper includes many white limestone particles, dark particles, rounded to subangular sands, gray angular bits, and a few shiny bits of mineral or particles of quartz. Noticeable in some of the sherds are reddish particles that may be ocher.

A rim, shoulder, and handle sherd is paired with a base from the same context (fig. 6.10a–b) and may belong to an earlier phase than some of the others—that is, circa 580 BC, according to parallels that show a flattened rim and a longer shoulder with sharp carination (pls. 83a, c, 85e).[85] The rim has a slight channel on the interior. The compact fabric, with a smooth exterior surface, is fired to a reddish-yellow color. Close parallels to the rim and base of the jar range from the seventh century down to about 580 BC.[86]

The amphora rims and shoulders of the Phoenician amphorae discussed in the following paragraphs all have short, upright rims and narrow, sloping or slightly convex shoulders. There are color differences, ranging from reddish yellow and light brown to pink, although the temper used in each jar is similar. The closest parallels to the short shoulders and low rounded rims come from Elephantine, where they are dated circa 550–400 BC.[87]

The amphora in figure 6.10c has a bright yellowish-brown to orange fabric, and the rough exterior was fired to a creamy-yellow color, although it is encrusted with a deposit from becoming waterlogged (pl. 85d). The rim in figure 6.9e has a tannish color on its exterior (pl. 85c). Parallels to these fragments are, once again, found in contexts extending from circa 550 BC through the mid-fourth century.[88]

The rim and rounded shoulder fragment shown in figure 6.11b is again made from a reddish-yellow to orange fabric (pl. 85a). The comparisons are the same as those made for the previous jars, dating from the mid-sixth century into the fourth century BC.

The amphora rims and shoulders in figure 6.11e–f are different from the others, having bulbous rims projecting inward and resting flat atop the shoulder. The

see Aston 1999, p. 8. The closest parallels in Holladay 1982 are the later examples shown on pls. 14, 27. See also Paice 1986–87, pp. 97–99, figs. 1–2. The examples are dated between ca. 525 and 486 BC. The parallels to Dorginarti's Phoenician amphorae are closest to those at northern sites in Egypt and the Levant; no close parallels have been found, for example, from around ancient Thebes. For the examples from Tell el-Herr, with a discussion of the development of the form, findspots, and references, see Defernez 2001, pp. 367–86, with most examples from the site dated to the later Persian period.

82 Ben-Shlomo 2014, p. 853, fig. 19.9.

83 Lids are shown in French and Ghaly 1991, p. 116, nos. 73–74; Paice 1986–87, fig. 8:14 (Saite, marl and silt). Tell Qedua also has this form; see Oren 1984, p. 16, fig. 20, nos. 16, 20 (thick purple slips and burnished).

84 Fabrics were studied by visual examination with a magnification loop (10× and 20×). The descriptions of the fabrics of the Phoenician amphorae at Buto are informative; see Bourriau 2003, pp. 225–28. The fabric of the wider Sidonian-Tyrian region, Fabric Class IA, was identified by petrographic analysis and discussed in Bettles 2003, pp. 139–52, 206, fig. 6.4. For a note of caution about narrowing the production center only to Sarepta, see Lehmann 2006, p. 96, with a note on the dates of the amphorae shown in Bettles's fig. 6.4, some of which stretch back into the seventh century BC. See also the clay analyses of the Phoenician vessels discovered at Kommos, Crete in Gilboa, Waiman-Barak, and Jones 2015 (southern Lebanese origin, ca. 850–650 BC).

85 Patricia Paice gave me much-needed advice on the dating of the Phoenician transport jars when she visited Chicago in the 1980s. She noted that the rim was representative of types dated to ca. 580 BC or earlier and that the base was of sixth-century date. The jar compares well with examples excavated from the chambers of the "Qasr" at Defenneh and shown in Petrie 1888, pp. 64, 66, pl. XXXIII, no. 4. See also the long chronological span of such amphorae in Lehmann 1996, pp. 436–37, and see form no. 392 (ca. 700–580 BC).

86 Lehmann 1996, p. 436, form 392, and table 74. The shoulder of some examples is broader. They are also dated to Lehmann's Assemblages 3–4 (ca. 700–580 BC).

87 Aston 1999, pp. 232–33, nos. 2044, 2046, 2050–56, from Phase V.

88 Note the rim form in Lehmann 1996, pp. 437–39, form 394 of Assemblages 4–6 (ca. 650–440 BC). See also Lehmann 1996, form 395 of Assemblages 6–8 (ca. 540–330 BC). Note also Sagona 1982, pp. 80–82, fig. 2, nos. 1, 3–4 (Type 6, sixth to fourth centuries BC).

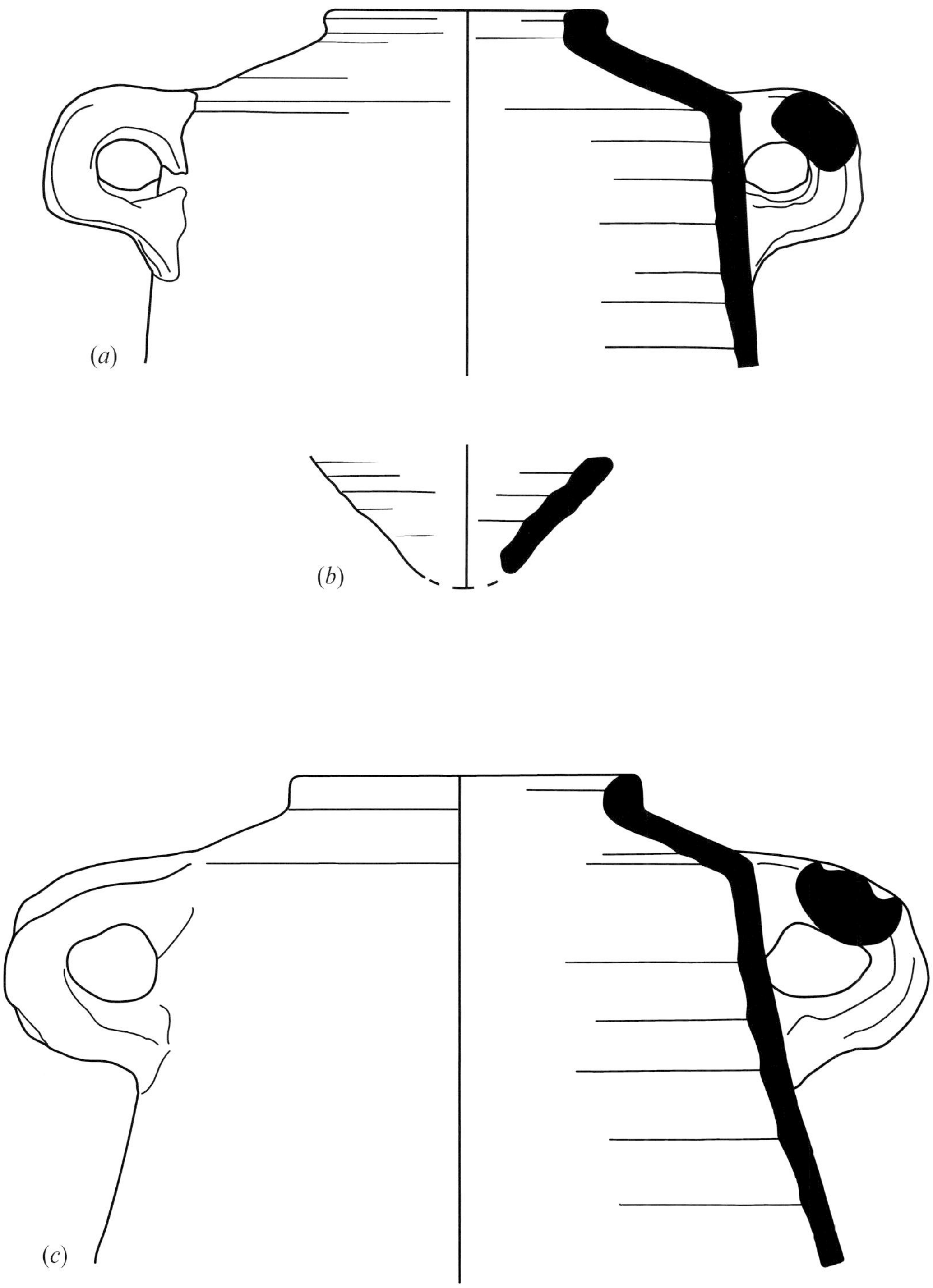

Figure 6.10. Phoenician amphorae (2:5)

(*a*) East Sector, north of stairs to Level II, first 40 cm; OIM E49541, 10 cm rim diameter
(*b*) East Sector, north of stairs to Level II, first 40 cm; OIM E49542
(*c*) River Stairs, first ten steps; OIM E49544

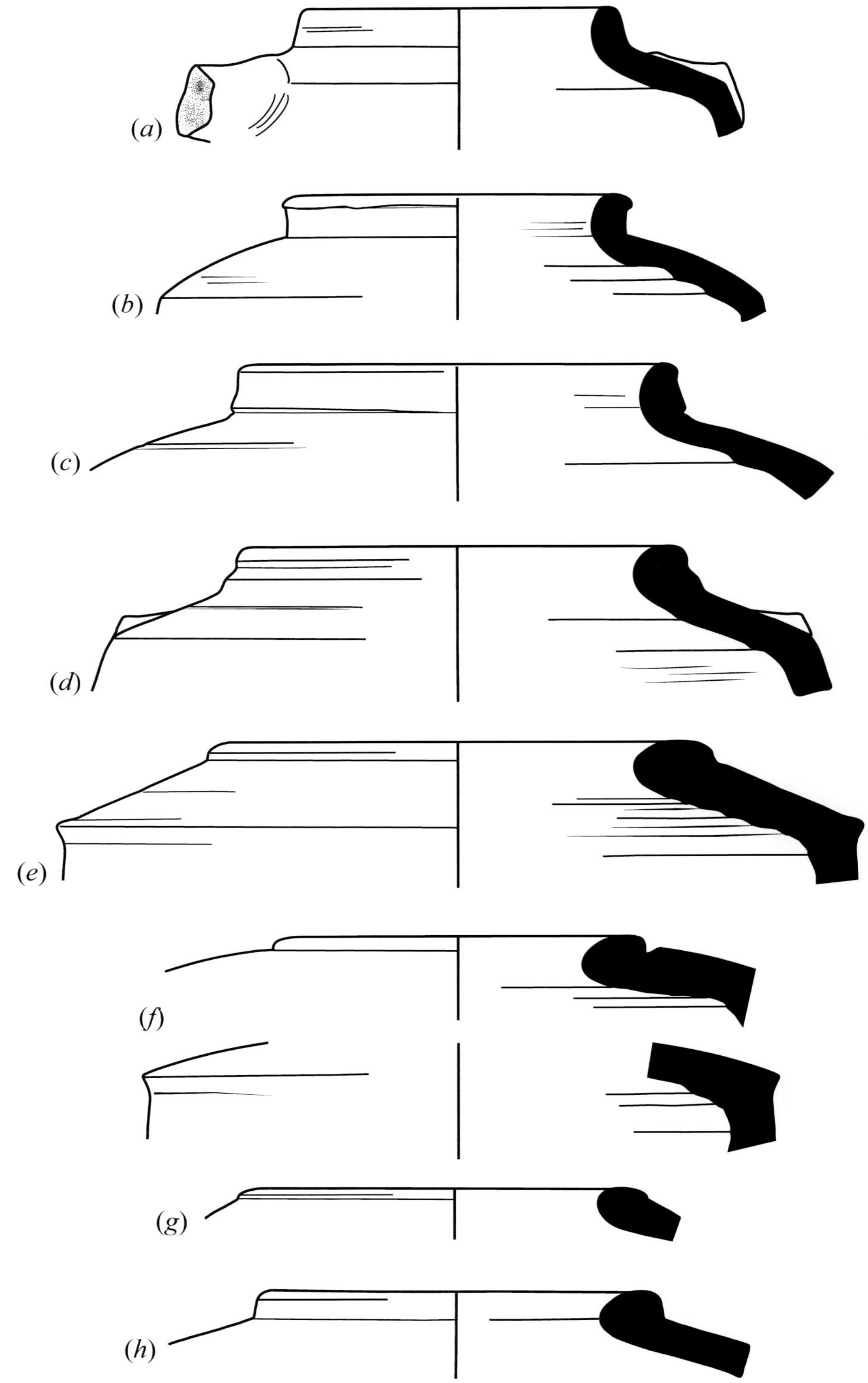

Figure 6.11. Phoenician amphorae (2:5)

(*a*) D 20 (2); OIM E49637

(*b*) D 15 (1); OIM E49536

(*c*) D 43; light-colored fabric

(*d*) D 15 (1); OIM E49540

(*e*) D 75; OIM E49538, tan to pink fabric, light-cream exterior surface

(*f*) D 75; OIM E49539 (two sherds), brown surface, stained black core

(*g*) Destruction of walls in D 4, D 47, and D 48

(*h*) River Stairs

carination at the shoulders is sharp, and the walls of the vessels are thick. The rim-and-shoulder fragment in figure 6.11e is a tannish to pink fabric with a light-cream-fired exterior surface (note the shoulder fragment shown on pl. 85b). Two sherds from perhaps the same jar, shown in figure 6.11f, have a light-brown to pinkish exterior surface and a black core (the shoulder fragment is shown on pl. 83b). They can be compared to materials dated from the late seventh through the sixth century BC.[89]

The rim fragments shown in figure 6.11g–h are too small to determine which amphora form they belonged to. Both had slightly tan fabrics, and their stances may be more horizontal.

The presence of East Greek and Phoenician amphora fragments at Dorginarti confirms at least the end date of the Level II occupation. Although many of the Phoenician store-jar forms and fabrics endured a long time, the East Greek pottery helps fine-tune the dating of the use and abandonment of Level II.

EGYPTIAN VESSELS

The pottery in Level II at Dorginarti also included Egyptian wheel-made forms comparable to bowls and jars from Egypt and Israel that are dated to the sixth century BC. Some of the vessels show Levantine influence, or at least there are parallels for the forms from this region, but they are made of silts or Upper Egyptian marls.

Four-Handled Bowl or Krater

The handled bowl shown in figure 6.12c is made of an Egyptian Marl A4 and follows the form of similar "kraters" in the Levant (pl. 86a). The sherds were found atop the plastered floor along the east side of the Level II platform. The present example has a light-colored fabric, varying from a pale brown on the top exterior to a darker pink at the bottom and on the interior.[90] It is very different in detail and finish from the smaller handled bowls of silt and marl from Levels III and IV. The vessel originally featured four handles, three of which are preserved. A string held the body together during its drying stage, with its impression visible below the shoulder, and a series of shallow vertical chattering marks are visible below the waist. Repair holes indicate that the vessel had broken in half at some point and been repaired.

Marl and silt two-handled bowls from the earlier levels at the site were discussed in chapter 5, but those forms were smaller and not so finely made. Analogies to the vessel under consideration here, although not exact, are found in the southern Levant, where this typical Judean form stems from eighth to seventh century BC contexts.[91] Similar types, however, although two-handled and with varying rims, were found in stratum E2 at Kabri, Israel, which is dated to either the late seventh or the early sixth century BC.[92] A similar shape in marl or silt has been found at sites in Upper Egypt, where such vessels have usually been excavated in contexts attributed to the late seventh or sixth century BC.[93] However, smaller examples from Karnak North were found in Ptolemaic levels and produced in Jacquet-Gordon's Marl A5 fabric.[94] But the form from Karnak is more like the smaller Level III and Level IV two-handled bowls discussed in chapter 5.

Cooking Pots

Numerous parallels were found at Dorginarti for deep bowls with wheel striations on both their interior and exterior surfaces and rims terminating in a slight knob (see fig. 6.13a–c). A horizontal line is found just below the rim on the exterior. The sherds were excavated in D 75, the surface clearance west of the Level II structure, as well as in a context along its west wall and in the East Sector rubble and the upper deposits. The Nile C fabric is reddish brown to yellowish red, with

89 Bikai 1987, p. 69, pl. XXIII, nos. 626–29, which are dated to the Amathus Horizon (from after 700 to after 600 BC). Tell Keisan yielded numerous examples of these "folded-rim jars"; see Briend and Humbert 1980, pp. 144–45, fig. 42, and, for example, pl. 125, no. 8. The form with a more horizontal shoulder is found through the sixth century BC. Briend and Humbert suggest that this amphora type may have been produced somewhere on the Plain of Acco, or perhaps even at Tell Keisan itself.

90 Unfortunately the vessel cannot now be located, and the photograph, used for study purposes in the 1980s, is a very old one that does not show the firing differences.

91 Singer-Avitz 1999, pp. 14–15, fig. 2, no. 5. Singer-Avitz dates Beersheba Stratum II to the end of the eighth century BC. See also Aharoni and Aharoni 1976, who note that the form continues until the end of the Iron Age, when its wheel burnishing is less pronounced, but disappears by the later seventh century BC.

92 Lehmann 2002, pp. 195, 205, fig. 5.78, no. 7. Stratum E2 is dated to the second half of the seventh century BC, with a possible destruction in either 604 or 585 BC (Lehmann 2002, p. 219).

93 For slightly similar two-handled bowls of marl, see Myśliwiec 1987, pp. 70–71, nos. 640–43 (handled examples). See also the silt example shown in Jacquet-Gordon 2012, p. 270, fig. 107s; the marl example from the site's later levels is shown on p. 295, fig. 121n (Ptolemaic). For the examples from Elephantine, see Aston 1999, pp. 228–31, nos. 2024, 2029–31 (ca. 550–400 BC).

94 Jacquet-Gordon 2012, p. 295, fig. 121n (P.568). For Karnak North's Marl A5 fabric classification, see pp. 6–7.

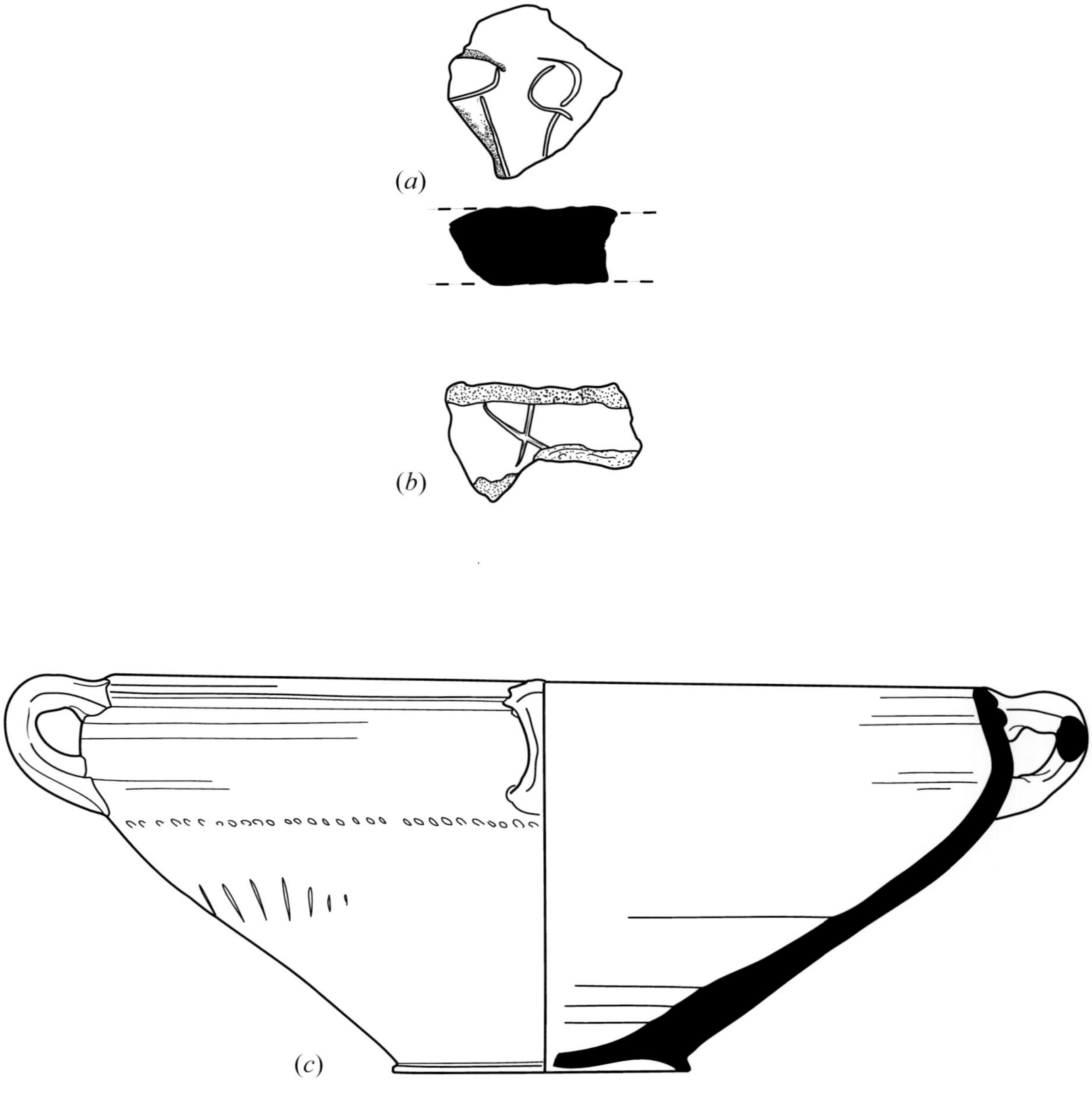

Figure 6.12. Mud sealings and Level II handled bowl (*a–b*, 2:5; *c*, 1:5)

(*a*) D 15 (1); mud fragment with incised letters

(*b*) D 15 (1); incised symbol on flat mud fragment

(*c*) Central Sector, Level II floor atop platform, north of Level II stairs, east side; OIM E24344, 32 cm diameter

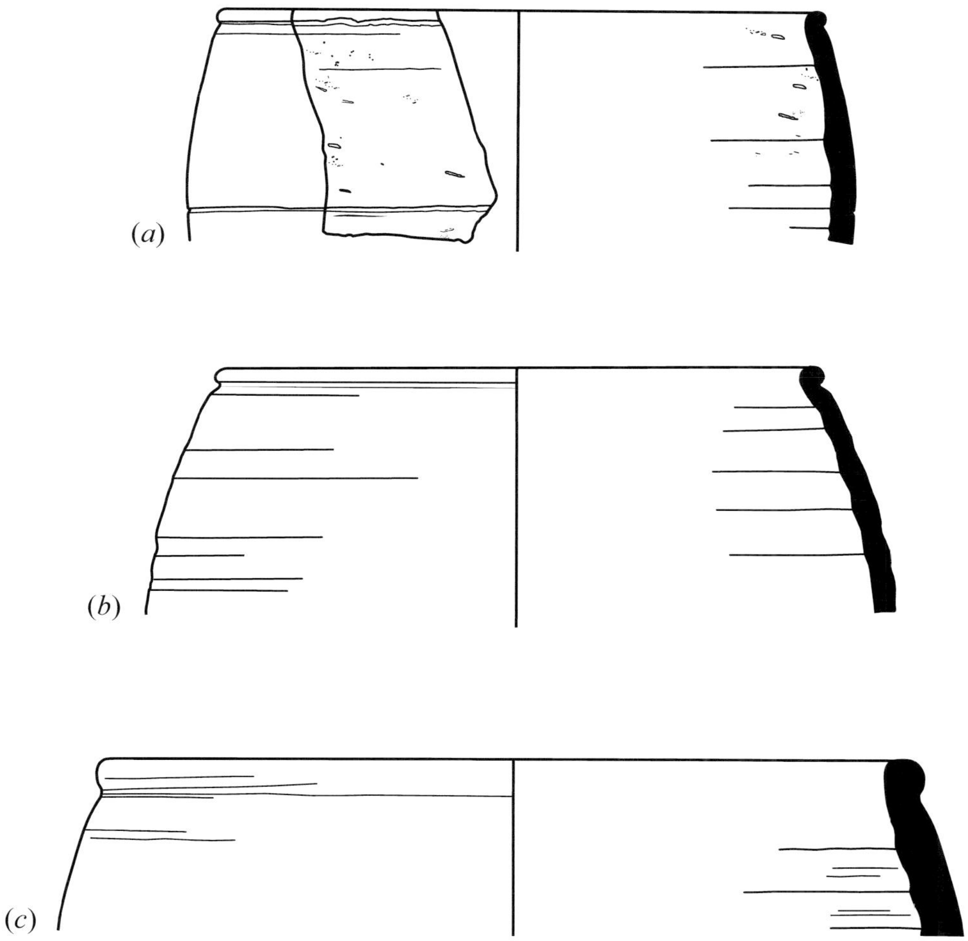

Figure 6.13. Cooking pots (2:5)

(*a*) D 75; pink slip exterior

(*b*) East Sector, rubble over three southern silos

(*c*) West Sector, rubble against retaining wall of Level II

chaff and brownish or reddish-brown sand added as temper and with small white particles evident in some sherds. Mica is present in all of them, and it was abundant in some sherds of this form. The sherds are sometimes blackened from use in cooking, but the sherd in figure 6.13a shows traces of a pink coating, perhaps an eroded red slip or a deposit, on a surface darkened by residue. None of the base sherds for these bowls has been identified in the collection.

Such vessels have been uncovered in Persian-period contexts in Egypt, where they are typically dated from the beginning of the sixth century BC to the fourth century.[95] The vessels are often red-slipped and become shallower over time, so the depth of the vessels from Dorginarti places them with vessels at the earlier end of their range.[96] The parallels to this form include a bowl found in a later level south of the Taharqo temple at Kawa, in Nubia, which has been described as being of "red ware."[97]

Silt Bowls

A number of other bowl forms are associated with Level II activity. A sharply carinated silt bowl form is shown in figure 6.14a–b. The shape compares well with seventh to sixth century BC Assyrian-style bowls found throughout the Levant, where they are made locally.[98] Levantine parallels show that this form continues into the Persian period.[99]

The bowl is shallow with an everted rim, and although no base sherds were present, parallels have either flat or ring bases. This type was a rare find in various East Sector surface debris and open contexts. The example in figure 6.14a is red-slipped on the interior and exterior, while another example (not illustrated) exhibits a cream coating or deposit on its upper exterior.

The dating of the silt bowl in figure 6.14c is less certain, although a few possible parallels of Persian-period date have been found in both Egypt and the southern Levant.[100] The Dorginarti bowl is made of a Nile B2 and has a red slip on its interior and exterior surfaces.

There are parallels in sixth-century BC contexts in Egypt for the silt bowl sherds illustrated in figure 6.14d–e. These parallels show that this form had a disk base.[101] The exterior surface of the two silt sherds is burned and has a light-reddish-brown surface color where it shows. They could be sherds of the same vessel, with one sherd having a thicker wall than the other, since they were found in contexts that were fairly close together. The clays are finely levigated and hard, and in the temper there are small rounded sands, many limestone bits, and small dark particles. Another very small sherd of possibly the same form (not illustrated) came from the rubble against the west wall of D 33, in the East Sector. It is pale brown on the exterior and brown in the break and has some limestone temper.

95 The best reference for this form's chronological development is Defernez 2001, vol. 1, pp. 64–76, and vol. 2, pl. I, 2a–b of Phase VII, and pls. VII–VIII, 22d–e of Phase VI, with references to parallels dated between the sixth and fourth centuries BC. All references to Persian-period pottery within Egypt, particularly vessels of northern influence, are listed in her compendium. The form is also found at Elephantine in Phase V, the late Saite and Persian periods of ca. 550–400 BC; see Aston 1999, pp. 223–25, nos. 1985–88.

96 Defernez 2001, pp. 64–65, fig. 1.

97 Macadam 1955, p. 224, and pl. XXXII, type N4 (2146). The vessel stems from the deposits in a tree pit south of the temple of Taharqo.

98 For parallels at Tell el-Maskhuta, see Holladay 1982, p. 111, fig. 17, nos. 1–3 (materials from the blocked-up Persian well, ca. 486 BC) and Paice 1986–87, pp. 100, 105, fig. 4, no. 10 (Saite contexts and also the 486 BC well deposit). Note also Oren 1984, pp. 14, 16, figs. 12 and 20, no. 9 (sixth century BC), citing parallels with the pottery assemblage from a foundation deposit of Amasis at Nebesheh in Petrie's publication. For a detailed discussion of these carinated bowls and other types of Assyrian-style pottery, along with the local copies within Israel and Palestine, see Ben-Shlomo and Van Beek 2014, pp. 732–48. These are dated to ca. 700 BC. For the origin of the shape and its spread throughout Israel, see Singer-Avitz 2007, pp. 183–85. For imitations of Assyrian ware in southern Palestine (perhaps a Transjordanian product), which are dated to the end of the seventh century and early sixth century BC, see Na'aman and Thareani-Sussely 2006.

99 For example, see Stern and Magen 1984, pp. 15–16, fig. 5, no. 1 (found throughout the Persian period, but especially from the mid-fifth through the end of the fourth centuries BC). Note also Fantalkin 2001, pp. 55, 58, fig. 23, no. 13 (red-slipped and burnished). Parallels date to the seventh century BC, and the site itself is now dated to the end of the seventh or beginning of the sixth century (pp. 128–36).

100 See Holladay 1982, p. 97, pl. 10, no. 2 (burnished, ca. 568 BC) and Fantalkin 2001, p. 55, fig. 23, no. 10 (dated by parallels to the eighth through seventh centuries, although the site itself dates to the late seventh or early sixth century BC).

101 See French 2007, pp. 106–7, fig. 14, no. 4. Examples from Karnak and Elephantine are found in Masson 2007, p. 610, pl. XXV, no. 8 (marl, Saite and Persian), while citations to the silt and marl parallels from Elephantine may be found in Aston 1999, pp. 210–12, nos. 1905–10 (Phase IV, sixth century), and pp. 242–43, no. 2110 (Phase V, ca. 550–400 BC). There is a parallel in the Twenty-Sixth to Thirtieth Dynasty levels at Karnak North in Jacquet-Gordon 2012, p. 253, fig. 98l (P.472 of Marl A5, variant 1).

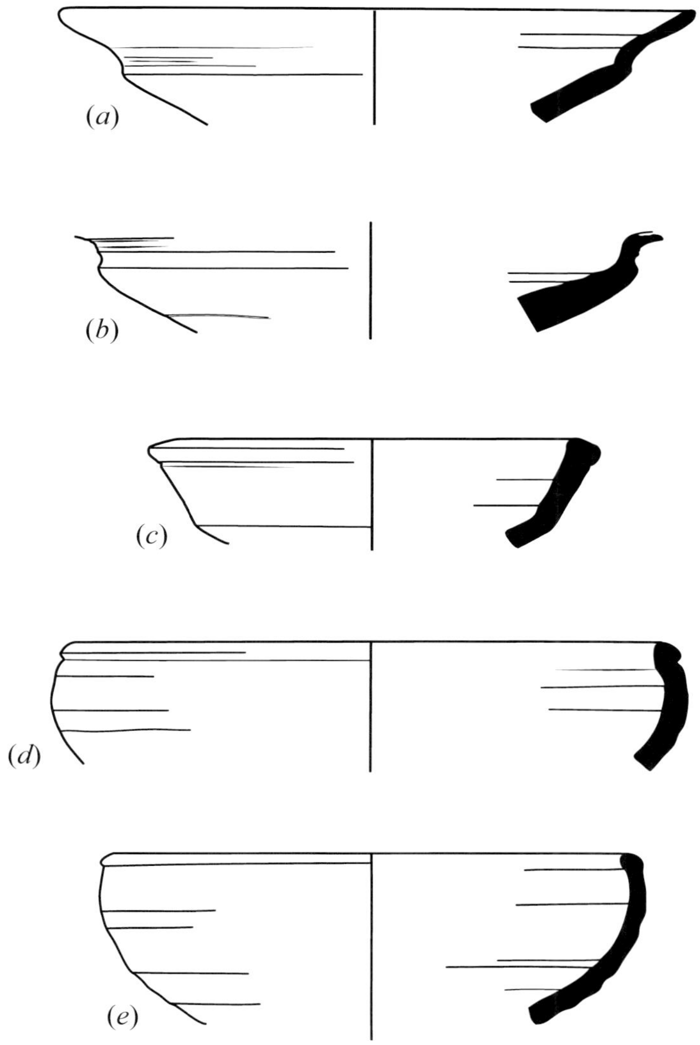

Figure 6.14. Cooking pots (2:5)

(*a*) East Sector center, first 60 cm; RS interior/exterior

(*b*) East Sector, rubble over southern silos; ca. 18 cm diameter at carination

(*c*) East Sector, rubble over three southern silos; RS interior/exterior

(d) East Sector center, first 60 cm

(e) East Sector, south of Level II stairway to Central Sector

Marl Mortarium(?)

A large marl bowl with a distinct folded rim may be a lid, but it may also belong to a mortarium, which was introduced into the Levant from Cyprus and East Greece beginning in the eighth and seventh centuries BC (fig. 6.15a).[102] The form in question lasted until at least the Hellenistic period. Both imported and locally made varieties are found at Naukratis, Tell Defenneh, Tell el-Balamun, Migdol, Tell el-Herr, and other Egyptian sites.[103] An example was possibly also uncovered at Karnak North, but no good parallels to the form or fabric have been found at Elephantine.[104]

Mortaria were used as bowls for the grinding of grain or other vegetal substances. The bowl from Dorginarti, however, does not have any extra additions to its temper to produce a gritty grinding surface. Moreover, it is not an import but is rather made from Upper Egyptian marl. Its wall is less thick than some examples in the literature, so it was probably intended for serving food rather than grinding it, although the interior surface is very worn.[105] It was excavated in D 80 (1), a layer below the north part of the fill west of the Central Sector's Level II platform (below D 75).

Lids

The lids found in Level II contexts at Dorginarti have slightly convex tops, a carination, and sometimes an incised line above the rim exterior (fig. 6.15b–c; pl. 86e). There are few sherds in the collection, all of them made of silt. They probably served as lids for the Phoenician amphorae. One example exhibits traces of a red slip on the exterior but is—like the other sherds of this form—discolored from use. The fabric is a light reddish brown to pinkish gray. Small-sized organic temper and sand have been added, although the mica is probably a natural clay inclusion. The diameters were about 16–18 cm, but one rim measured 21 cm in diameter. This form has many parallels from Egyptian sites and the Sinai dating to the Saite and early Persian periods.[106]

One of these lid sherds has a hole in the center of its top, and another has been repaired using drill holes to bind the fragments together with string or leather (OIM E24369) (fig. 6.15c).

Silt Cup

The sherds of a cup came out of the fill within D 66 (1) (fig. 6.15d; pl. 86b). There were later walls in the area above the Level III phase, and the fill contained other late sherds besides this cup. The form has parallels in sixth-century BC contexts in Egypt.[107] The fabric has been fired to a light brown, with a red slip on both surfaces and vertical burnishing strokes that covered the exterior from the rim to a point at the rounded base carination. The temper includes a moderate amount of small organics, small dark particles, and sand. Excess clay lays atop the interior surface toward the rim. A similar base came from an area at the southeast corner of the West Sector.

One-Handled Flasks or Decanters

The one-handled flasks from Dorginarti are comparable to types found in the north of Egypt and are sometimes referred to as *Judean wine decanters*, perhaps typically holding wine (fig. 6.16a–c, e; pl. 86d, f). They are also found in late Iron Age contexts in the Levant.[108] The decanters found at sites in Egypt may have been associated with the Judaean diaspora in Egypt.[109]

102 See, for example, the discussion and references in Villing 2006. A clay analysis was conducted on sherd samples from Naukratis, which are mostly Cypriote imports, with one identified as a local marl clay; see Spataro and Villing 2009, pp. 95–99. The fabric of one sample was identified as Marl D in Mommsen et al. 2006, p. 70, n. 29.

103 For example, Oren 1984, pp. 17–18, fig. 21, no. 10; Fantalkin 2001, pp. 79, 81–82, fig. 29, nos. 5–9 (Cypriote imports). See also Zukerman and Ben-Shlomo 2011, p. 88.

104 Jacquet-Gordon 2012, p. 264, fig. 105e, is perhaps an example. It is made from Marl A5 and comes from the Twenty-Sixth to Thirtieth Dynasty layers.

105 Fantalkin 2001, p. 82. Following other scholars, Fantalkin differentiates function according to the different wall thicknesses, with thin-walled examples having been used for serving food rather than for grinding vegetal matter.

106 See, for example, French and Ghaly 1991, p. 116, nos. 73–74, and Paice 1986–87, fig. 8, no. 14 (Saite, marl and silt). Tell Qedua also has yielded this form; see Oren 1984, p. 16, fig. 20, nos. 16, 20 (thick purple slips and burnished).

107 See Aston 1999, pp. 217–18, no. 1931, made from Nile C. These are in sixth-century BC contexts. Aston cites comparable forms from Qurna, Saqqara, Mendes, and Kafr Ammar on p. 218, nn. 60–63. Note also the examples in Oren 1984, p. 16, fig. 20, nos. 17, 21 (late seventh to sixth centuries BC).

108 See, for example, Fantalkin 2001, p. 73, fig. 27, nos. 1–8, some being red-slipped or of gray fabric and dated to the late seventh century; also Stager, Master, and Schloen 2011, p. 93, fig. 5.64 ("Judahite"), made of an inland fabric identified with parallels at Lachish and Meẓad Ḥashavyahu. The latter example also dates to the late seventh century BC. These two parallels are more similar to each other than to the Dorginarti examples. The dating of this form at Tell el-Maskhuta ranges between the late seventh century and 568 BC; see Holladay 2004, pp. 419–23.

109 Holladay 2004. The Egyptian parallels, and one Levantine parallel, are shown in Holladay's fig. 3. He suggests that the white-slipped examples from sites in southern Israel may have been intended to emulate an Egyptian marl fabric. For the

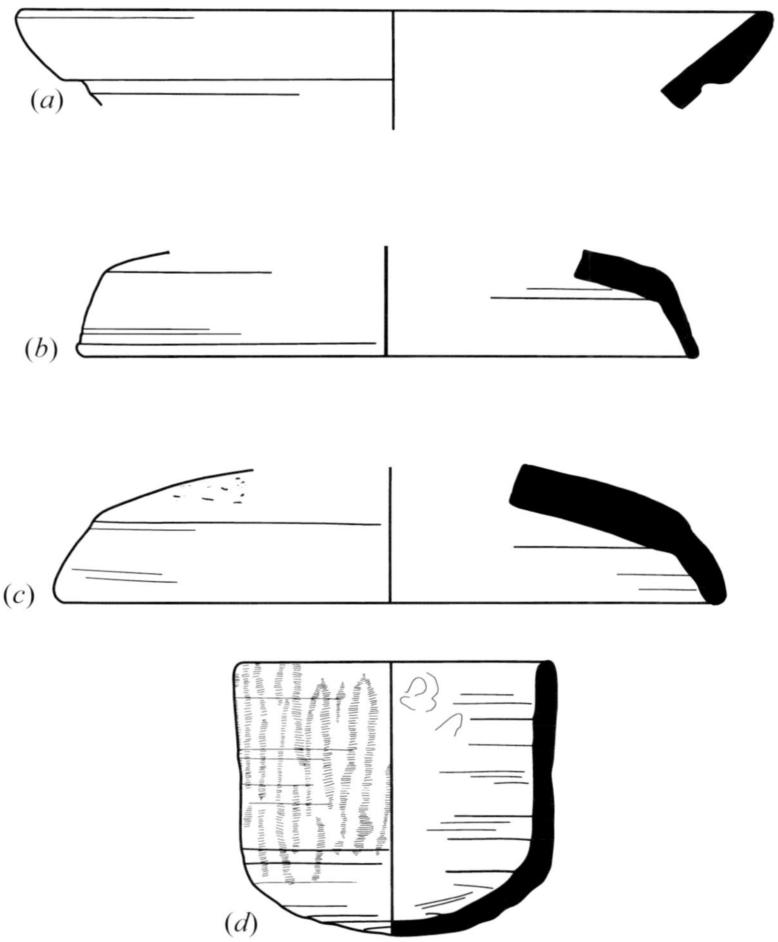

Figure 6.15. Marl bowl, lids, and cup (2:5)

(*a*) D 80 (1); Marl A4 mortarium(?), worn on upper interior

(*b*) East Sector center, first 60 cm; OIM E49565, silt

(*c*) Central Sector, against south wall of Level II; OIM E24369, silt

(d) D 66 (1); silt, RS interior/exterior, burnished exterior

The form is partly identified by a neck ridge, a handle that runs from the ridge to the shoulder of the vessel, and a ring or disk base, although no complete forms can be reconstructed from the sherds in Chicago. The Dorginarti examples are marl. There is no evidence of slip or burnishing on any of the sherds, a trait possessed by the examples from the Levant. The marl handle fragment that has been preserved—undoubtedly from one of these flasks—is oval, not ridged, doubled, or strap-like as on examples from Levantine and northern Egyptian sites (fig. 6.16e).[110] The center part of the base hangs below the rim line, and it is more of a disk base in one case (fig. 6.16b). The sherds all stem from areas identified with other Level II pottery deposits.

A rim and neck sherd, and a base sherd, were found in the first 60 cm of fill in the center of the East Sector (fig. 6.16a). The exterior surface of the rim and neck sherd is pale brown, while the marl base fragment is a bit lighter in color. Both are made from a Marl A4 fabric.

suggestion that the Upper Egyptian types were made by Judaean potters, see p. 416.

110 See Holladay 2004, pp. 409–15, for a listing of Egyptian and Levantine sites. Note the oval handle on one example from Lachish, which compares with the Dorginarti handle sherd.

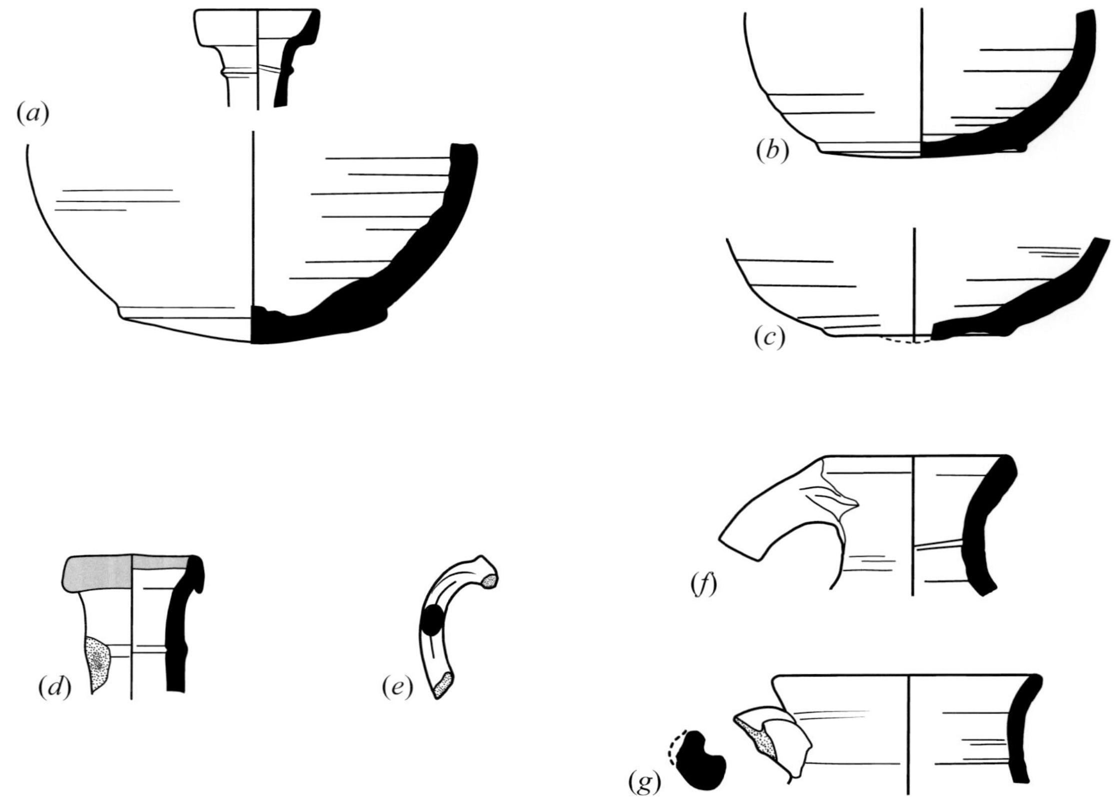

Figure 6.16. Marl decanters, flask, and marl jugs (2:5)

(*a*) East Sector center, first 60 cm; OIM E49511, marl, 4 cm rim diameter, marl base OIM E49512 from same locus

(*b*) D 351 (1), bin; marl

(*c*) East Sector, north of River Stairs, first 40 cm; marl

(*d*) East Sector center, first 60 cm; silt, RS rim, handle scar

(*e*) East Sector, rubble over southern silos; marl

(*f*) D 351 (1), bin; marl, two-handled?

(*g*) D 26 (1), first 25 cm; silt or mixed clay, one handle

A similar marl base was uncovered in the East Sector, north of the stairs to the Central Sector, within the top 40 cm of fill (fig. 6.16c). At least two other marl base sherds of this type were found in contexts associated with the Level II occupation, one atop the River Stairs and one in the fill of silo D 351 (1) (fig. 6.16b). The exterior surface of the latter example is tan in color with a greenish-gray tint. The fabric is light and porous, with many small voids, although with a typical marl temper of dark mineral particles, small sands, and limestone present. Although this particular fabric type remains uncertain, it may be Jacquet-Gordon's Marl A5, variant 1, fabrics as found at Karnak North.[111]

111 Jacquet-Gordon 2012, pp. 6–7.

These flasks have parallels from sites in northern Egypt, some of which have been noted above in previous footnotes.[112] Examples of marl sherds from this form were also found at Karnak North in the layers dated to the Twenty-Sixth through Thirtieth Dynasties.[113]

A silt flask rim (fig. 6.16d; pl. 86c), perhaps a copy of the marl decanters, is less well made and slightly larger. It has a red band around the rim interior and

112 But note the slightly similar rim in French 2007, pp. 118–19, fig. 22, no. 3, where the fabric is identified as an import of uncertain origin, and the vessel is covered by a thick red slip.

113 A possible disk base from one of these forms is shown in Jacquet-Gordon 2012, pp. 264–65, fig. 105k (marl), and the ridged neck and rim sherd, also marl, is shown on p. 265, fig. 105n.

exterior, and an exterior red slip, and was found in the first 60 cm of fill in the center of the East Sector. A handle scar lay atop the neck ridge.

The same locus also produced a one-handled rim sherd from a marl jug (fig. 6.16f; pl. 87c). Parallels for this example, and that in figure 6.16g, have not been found, and their dating remains uncertain. The first example is marl, while the latter is made of silt clay.

Silt and Marl Jars

A constricted jar, shown in figure 6.17a, has been reconstructed from a rim and body sherd. These two sherds are made of fine hard silt with red slip and have very fine burnishing on the exterior and the interior of the rim. Both have maximum diameters of 16–18 cm, but the stance of the rim may be tilted incorrectly in the drawing presented here. The temper is comprised of many small organics, sand, mica, and a few white bits that are probably from burned-out organic material. The surfaces of the sherds are quite eroded and flaking. This vessel may belong with the red-slipped, highly burnished vessels discussed in chapter 5, but for now it is placed with the Level II pottery because of its context; note, however, the mixed dating of the parallels mentioned here. The parallels include two sixth-century BC jars from Elephantine, a number of examples from seventh-century contexts in the Levant, but also a Twenty-First to Twenty-Fifth Dynasty vessel from Karnak North.[114]

Another unique form is shown in figure 6.17b. Sherds of this marl form were found in D 67 (1), D 66 (1), and perhaps D 7 (1), all in the area to the west of the topmost long wall in the West Sector, and probably all from the same vessel. The body is globular and the rim everted. A similar silt form was found at Tell el-Maskhuta among the remains from the debris in the Persian well, dated to about 486 BC.[115] However, this jar may also date to the Level III use of the West Sector.

Silt jars with tall necks, red slip, and burnish enjoyed widespread use throughout Egypt during the Persian period. The sherd shown in figure 6.17c is the only example of this type in the Oriental Institute collection, and it is not a precise parallel to most of its counterparts.[116] There is no burnish and no evidence of red slip on the surfaces. The fabric is tempered with organics, black particles, and perhaps additional sand and mica. The interior is a yellowish red, while the exterior is red to reddish brown, with a gray core present in the break. The ridges on the neck have been produced by incising multiple lines on the wheel, and therefore the profile has less defined bulges than most of the parallel examples. There are similar forms at the cemetery of Sanam in Nubia.[117]

The jars shown in figures 6.17d and 6.18a are unique at the site but are believed to belong to the Level II occupation of the Central Sector. The former was found along a retaining wall of the Central Sector, the latter in the fill in D 24 (2) in the Central Sector. Both were made from silt clay, although the second had limestone in its temper and may have been a silt-marl clay mix.

The jar in figure 6.18b came from D 24 (2), or the upper fill layers within a room of the Level III building, and was either marl or a silt/marl mix. No organics are observable in the temper, which includes many dark particles, some limestone, sand, and mica, the latter probably being a natural inclusion. The rim is similar in profile to the marl and silt jars shown in figure 6.18c, e. Figure 6.18c stems from the Central Sector "west retaining wall for Level II, south corner" and was made of a marl fabric (pl. 87b). The temper includes dark particles, limestone, small rounded sands, some translucent sparkling sands, and one or two gray chunks. Another marl sherd of the same form (same vessel?), which is not illustrated, came from D 15 (1)—that is, below the Christian building in the Central Sector.

The best parallels to all these jars come from Elephantine (where they are dated to the late Saite and Persian periods), Karnak North (where they are placed within the Twenty-Sixth to Thirtieth Dynasty assemblage), and the temple of Seti I at Qurna (the material of which is typically dated to the late seventh and early sixth centuries BC).[118]

114 Aston 1999, pp. 207, 209–11, 238–39, nos. 1890 (a black-slipped Nile B2 and burnished), 1901 (Nile C), 2075 (Nile C cream slip); Jacquet-Gordon 2012, p. 229, fig. 89r (P.469) (Marl A5, variant 1). The shape is similar to the Assyrian bowls in Phase 5 at Tell Jemmeh; see Ben-Shlomo and Van Beek 2014, p. 492, fig. 8.117d, with a number of other similar rim fragments. There is no slip and burnish on these examples.

115 Holladay 1982, pl. 25, no. 4 (with a red slip). However, note the dating of some of this material in Holladay 2004, pp. 421–23.

116 This form is also found in the Theban area; see Masson 2007, p. 608, pl. XXVI, no. 5 (red slip and burnish). Similar jars, although not precise in neck profile, are illustrated in French 2004, pp. 92, 94–95, pl. I, nos. 1–2. Note the parallels cited for all phases of this jar in Defernez 2001, pp. 116–29, and particularly pl. XIX.

117 Cited by Defernez 2001, p. 122. See also Griffith 1923, p. 95, pl. XVII, types Ia–b, IIa–b. These are silt, and rarely slipped or polished.

118 Aston 1999, pp. 231–32, no. 2041, pp. 242–43, no. 2111, pp. 244–45, no. 2130; and Jacquet-Gordon 2012, pp. 260–61, fig. 103e–f (late marl). The mortuary temple deposits of Seti I at Qurna are published in Myśliwiec 1987, pp. 63–64, nos. 431–35. All these examples were made of marl.

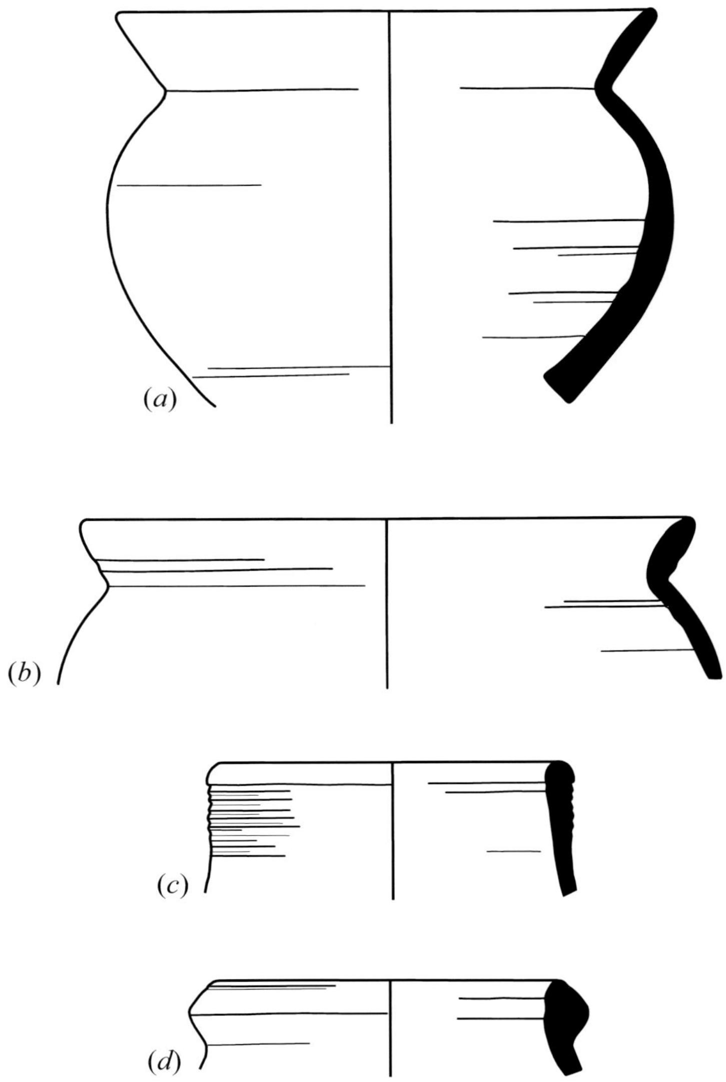

Figure 6.17. Silt and marl jars (2:5)

(*a*) East Sector center, first 60 cm; silt, RS and continous burnish on exterior and rim interior

(*b*) Sherds from D 66 (1), D 67 (1), and D 7 (1); same vessel(?), marl

(*c*) East Sector, south of stairs to Level II platform; silt

(*d*) Central Sector, retaining wall for Level II, south corner; silt

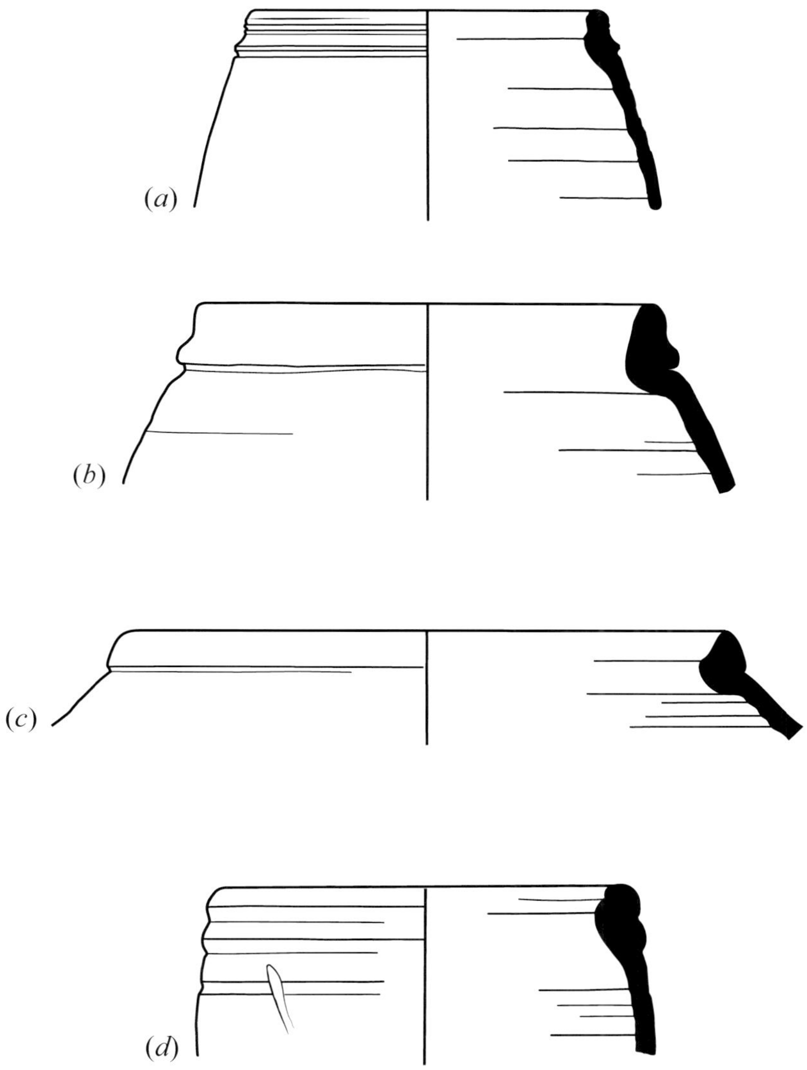

Figure 6.18. Marl and silt jars (2:5)

(*a*) D 24 (2), fill; silt or silt/marl mix

(*b*) Central Sector, south corner of Level II platform; OIM E49503, marl

(*c*) Outside north wall, between second and third buttresses from gate, below "ash" layer; silt

(*d*) D 23 (1), fill; silt, incised line accidental?

Another jar, which also finds comparison with a marl jar from Karnak North, was found in D 23 (1), the fill below the Level II platform surface (fig. 6.18d).[119] The Dorginarti example is made of orange-brown silt, with a medium amount of organics, and was blackened on both surfaces.

A single example of a channeled-rim jar, a form well known in the later eighth and earlier seventh centuries BC royal tombs at Kurru and Nuri, is shown in figure 6.19a.[120] It is made of sandy silt clay containing black particles and some chaff, but is not very micaceous.[121] The same jar form as that in figure 6.19a, but in marl, is typically considered to date to the late Twenty-Fifth Dynasty and Saite period, and parallels are found at various locations in Thebes.[122] The marl jar form is found in sixth-century deposits at Elephantine.[123] In Nubia, where it is either made of Upper Egyptian marl or is later imitated in local silt fabrics, the vessel has been uncovered at sites dated to the later Twenty-Fifth Dynasty, with the less well-made silt examples found in the royal cemetery at Nuri from the mid-seventh century BC and later.[124] The marl form, however, apparently continues on into later periods in Egypt.[125]

Silt jars with incised lines around their rims and upper bodies, such as those shown in figure 6.19b–e, were common in the later Third Intermediate Period and Saite era. The jar shown in figure 6.19c is like one from the tomb cache of Amenirdis at Medinet Habu and also other examples from elsewhere around Thebes.[126] While the jar from the magazines of Amenirdis's tomb chapel represents an earlier example, the other Theban parallels are dated closer to the sixth century BC. Similar rim sherds were found either atop the River Stairs or in the East Sector. They are perhaps from the same jar, although no joins have been found and their rim diameters are different. The red to reddish-brown fabric is chaffy, and there is a white deposit or slip on the exterior, which was probably an accretion left by water saturation. The core is dark gray. The small rim sherd shown in figure 6.19b is from the same or a similar jar. It has a red slip on its exterior, with a pale-brown exterior and pink interior. In addition, there is a lot of mica and limestone and many dark particles added as temper.

The other wide-mouthed jars with exterior channels find parallels in Egypt, particularly around the Theban region, but also at Tell el-Maskhuta (fig. 6.19d–e).[127] The examples from Dorginarti were found in D 3 (1), the intentional fill covering the walls of the Level III residence; in D 15 (1) atop the Level II platform, in fill over the three southern silos in the East Sector; and in D 13 (1) south of the West Gate entrance. The silt fabric is fired to a red color on all surfaces, with traces of a dark-reddish slip in the channels (pl. 87a). The fabric contains many organics, sands, and dark particles, but no additional mica. The rims of this type measure 20–29 cm in diameter.

---

119 Jacquet-Gordon 2012, pp. 260–61, fig. 103e (P.785), made from her late marl fabric.

120 Heidorn 2018a, pp. 329–30. It is also found in the early tombs at Meroe. Later silt copies, mostly poorly formed imitations of the original Upper Egyptian marl prototype, are found in the later royal tombs at Nuri, from the mid-sixth century and on through the reign of Harsiotef in the middle fourth century BC.

121 The fabric of copies of Upper Egyptian vessels was typically sandy and micaceous, with chaff and occasional limestone temper. See, for instance, Rose 2008, p. 205, and Sjöström and Thomas 2011, p. 64.

122 This jar is found in the Twenty-Sixth to Thirtieth Dynasty assemblage at Karnak North; see Jacquet-Gordon 2012, figs. 101c (P.478) and 103g (P.719). For other comparable jars from the Theban region dating between the late eighth and late seventh centuries BC, including marl jars from stratum 1 at the temple of Mut, see Sullivan 2013, pp. 195, 236, type 22-11. They are also found in deposits around the temple of Seti I at Qurna; see Myśliwiec 1987, pp. 63–64, nos. 434–35. The same form was also found in association with an intrusive burial in Theban Tomb 32, which has been attributed by Gábor Schreiber to the seventh century BC or perhaps the early sixth century based on pottery parallels (Schreiber 2008, p. 78, pl. LXXIX, no. 47). The jar is dated to the Twenty-Fifth and the beginning of the Twenty-Sixth Dynasties at Karnak; see Boulet and Defernez 2014, pp. 610–15, figs. 31-1(I), 31-2(C).

123 Aston 1999, p. 229, no. 2037, and p. 239, no. 2077. This form is found in Phase V, ca. 550–450 BC. The jar is not found in Phase III, and since marl jar forms that should accompany it are not found at Elephantine, there may be a late eighth- to seventh-century BC hiatus in the pottery series presented by Aston.

124 Heidorn 2018a, pp. 328–30. The recent publication of some early first-millennium BC pottery from Qasr Ibrim also presents these marl jars; see Rose 2019, p. 688, fig. 38.

125 For instance, note the late Saite and Ptolemaic marl examples from Karnak, with slight morphological differences between the earlier and later forms, in Masson 2011, pp. 273, 294–95, fig. 20 vs. fig. 25.

126 Hölscher 1954, p. 72, pl. 47, A3. See also Myśliwiec 1987, pp. 63–65, nos. 421–28, 455–57. A slightly similar form is also found in Jacquet-Gordon 2012, p. 275, fig. 110q. The jar is a Nile C with red slip and a white band around the neck. It is included in her Twenty-Sixth through Thirtieth Dynasty material and may represent a different type of jar.

127 Holladay 1982, p. 83, pl. 3, no. 3, silt with white slip on the exterior. It was found in levels attributed to ca. 609–605 BC. For Theban parallels dated to the Twenty-Sixth Dynasty and Ptolemaic period, see Masson 2011, pp. 271, 291, fig. 4, and Masson 2007, pl. XII, no. 2, dated to the Twenty-Sixth to Twenty-Seventh Dynasties. Similar types were also found at Karnak North and published by Jacquet-Gordon (2012, p. 272, fig. 108p [P.488], of Nile B2 or Nile C fabrics). They were found in layers ascribed to the Twenty-Sixth through Thirtieth Dynasties.

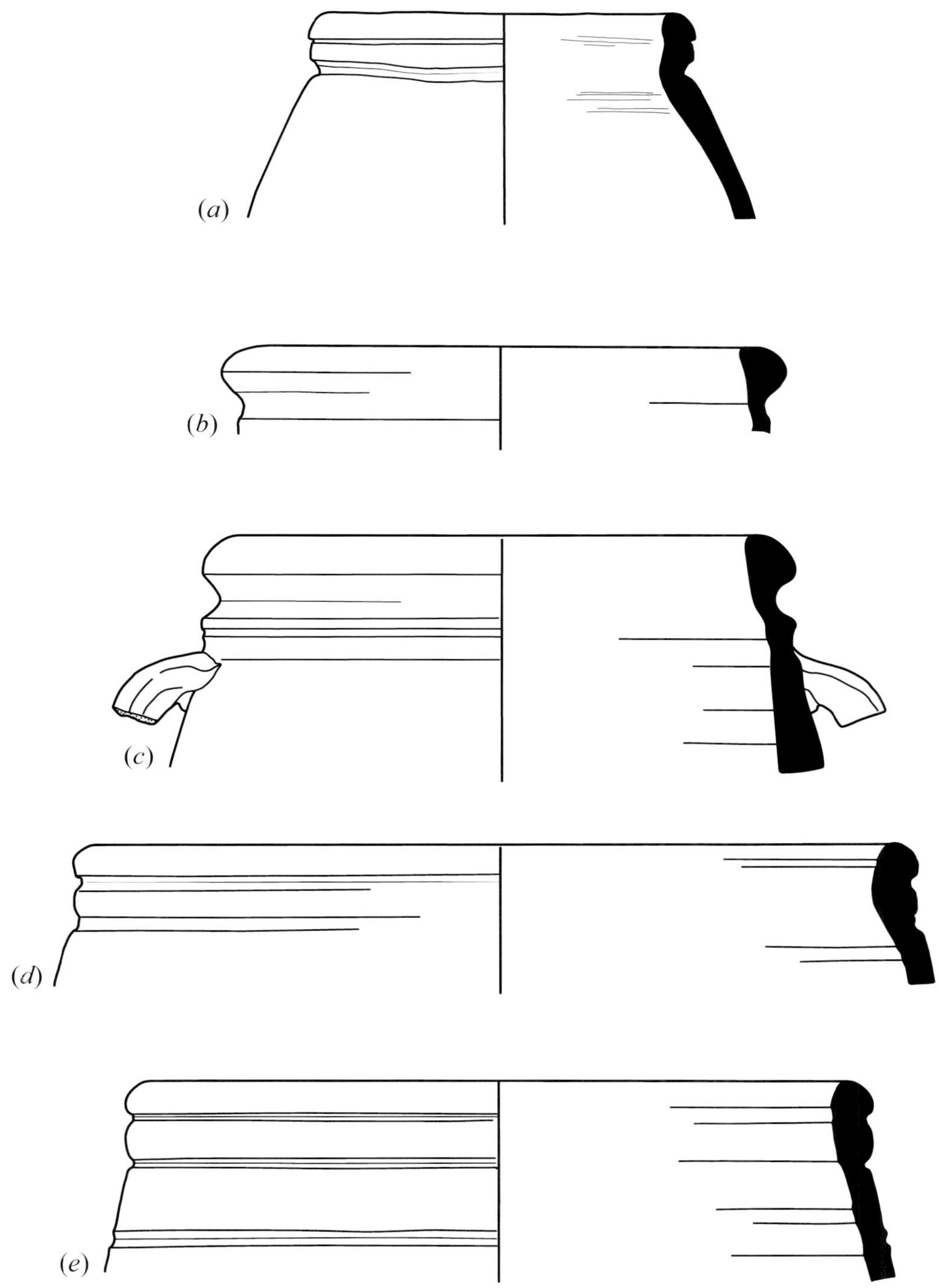

Figure 6.19. Silt jars (2:5)

(*a*) D 24 (2); sandy silt

(*b*) East Sector center, first 60 cm; RS traces on exterior

(*c*) River Stairs, above first ten steps; OIM E49505, white deposit or slip on exterior

(*d*) D 3 (1); OIM E49516

(*e*) D 15 (1); RS traces exterior

OASIS PRODUCTS

Larger flasks and kegs such as those illustrated in figure 6.20a–b (pl. 87e) were produced in the oases.[128] They were used to transport wine and were often reused for transporting water. A few sherds of oasis-ware vessels probably belong to the Level II use of the fortress (see below).[129] The shapes are not reconstructible, and the sherds may belong to either flasks or asymmetrical kegs. There are two major differences in fabrics. Some of the most distinctive sherds have a dark-gray exterior surface and a deep-pinkish-red interior (pls. 87e, 88b). Another fabric has tannish-brown surfaces (pl. 88a), but in one case the tanner fabric has a dark-gray exterior surface. One sherd of the latter fabric also has a red-slipped exterior. Almost all the darker-gray fabrics have large flakes of mica visible in their broken sections and on their surfaces.

These types are represented by rim, handle, or body sherds and were found in surface deposits of the East and Central Sectors, atop the River Stairs, in D 15 (1), D 24 (2), D 76 (1), D 87 (1) doorway, D 92 (1), and D 120. Body sherds of possible oasis ware were also found in D 200, D 201, and D 202 (1). D 9 (1), a fill deposit above the Level III residence and below the surface of the Level II platform, also yielded oasis-fabric sherds (see below). Many of these locations were deposits associated with other identifiable Level II sherds or were in fill deposits above Level III. However, one sherd of light-brown oasis fabric stemmed from the fill between the official residences of Levels III and IV—that is, from between D 310 (1) and D 314 (1). The later well dug into these two rooms also contained sherds of later date, including the Christian period. This is, however, the only indication that these oasis vessels may have been associated with the earlier activity at the fortress.

A large flask fragment that preserves the topmost part of the vessel was recovered from D 12 (1) atop the eroded wall and north of the West Gate. This fragment exhibits a dark-gray roughened exterior and a light-reddish-pink interior surface (fig. 6.20a; pl. 87e). Its fracture has a reddish-pink layer on both sides of a light-gray core. The spongy appearance of the fabric is due to the temper, which includes chaff, large subangular and rounded rock or mineral particles, rounded sands, and large limestone (or at least white) particles that do not appear to be shell but are perhaps rock particles. A few red bits are also noticeable, and the flakes of mica are most obvious on the dark exterior surface, where they may have been deposited by water or post-depositional debris. An inverted V-shaped potmark was incised next to one handle before the vessel was fired (see fig. 7.16a in chapter 7). Parallels to the Dorginarti flask shape have been found at Karnak North, the Dakhleh Oasis, and elsewhere.[130]

A neck-and-shoulder sherd from the East Sector is shown in figure 6.20b. The fabric of this particular example is reddish brown, and the surface is red with some dark-gray areas, while the core is gray and surrounded by light-red exterior and interior layers. The surface is coarse, and there are many white particles in the temper, along with some sand, pebbles, and probably grog or ocher particles.

The neck-and-handle fragment from D 9 (1), which corresponds to the fill of the Level II platform, represents a different type of bottle or flask, with only one handle preserved (fig. 6.20c; pl. 87d). The neck slopes inward at the top and widens out toward a shoulder that is no longer present. The thick core at the base of the neck is light gray surrounded by light-yellowish-tan interior and exterior layers, and the exterior is lighter gray. Many chunks and smaller particles of white limestone(?) are present in the temper, in addition to dark-gray particles, rock or mineral bits, and some brown sand. The fabric is full of elliptical voids that may have come from organic temper. The incomplete condition of the fragment makes it difficult both to draw and to find close parallels, but the form is reminiscent of a few of the flasks from Karnak North that were found in the Twenty-Sixth to Thirtieth Dynasty layers.[131]

Finally, the faience New Year's flask fragments found around silo D 32, south of the West Gate in D 13 (1), and along the west retaining wall for the Level II platform are also typically associated with the Saite period (see chapter 7, with fig. 7.3 and pl. 90).

128 Hope et al. 2000, with appendices by various authors.

129 I would like to thank Romain David (personal communication, April 20, 2015) for providing photographs of Karnak sherds made from these fabrics. The fine details separating the fabrics are nonetheless uncertain here, as is their origin.

130 Jacquet-Gordon 2012, fig. 97m (P.493) (end of the Third Intermediate Period; made of oasis fabric A29), and p. 286, fig. 118 (Twenty-Sixth to Thirtieth Dynasty). Jacquet-Gordon compares this form with shape A1 in Hope et al. 2000, p. 197, fig. 2a. Karnak's early examples have a dark-red or black slip, three concentric circles incised on the body, and a potmark. The later examples are made from oasis fabrics A1, B3, and B23.

131 Jacquet-Gordon 2012, p. 287, fig. 118k, p–r (all with red slips).

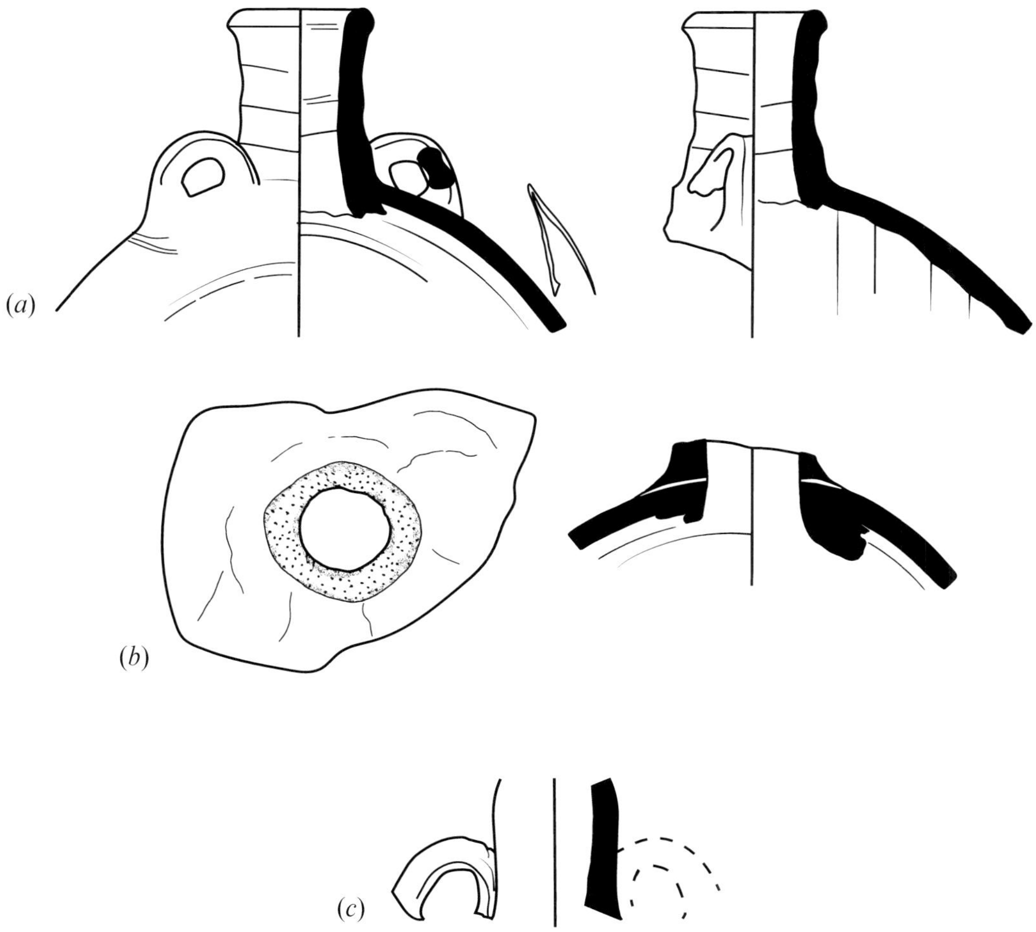

Figure 6.20. Oasis ware vessels (*a–b,* 1:5; *c,* 2:5)

(*a*) D 12 (1); OIM E36381; pre-firing potmark at shoulder, 7 cm diameter maximum

(*b*) East Sector, first 60 cm, and north of stairs to Level II, first 40 cm; burned in areas

(*c*) D 9 (1), south; stance and diameter uncertain, two-handled?

SUMMARY OF THE POTTERY AND CHRONOLOGY

The Level II sherds from Dorginarti include East Greek and Levantine amphorae, which date from the sixth century BC based on comparisons found in the Levant; in the eastern Mediterranean, including the Pontic region; and elsewhere in Egypt.[132] A number of items are slightly earlier in the sixth century, and these are noted in the discussion above. Many of these contexts were not sealed deposits, and they contained Meroitic and Christian materials in addition to sherds of earlier date. The evidence includes Chian and possible Clazomenian amphorae, as well as sherds of Samian amphorae. Other unidentified forms that have been described above will have to await a proper attribution.

Levantine jars from the northern coastal region are present, too, along with Egyptian vessels, some of which were influenced by Levantine styles. Parallels are found throughout Egypt, and especially at the

132 Various publications have been consulted for parallels to the foreign jars, and all of them are mentioned in the footnotes above. The dating comes largely from Dupont 1998. Note also the discussion of seventh-century East Greek imports into the Levant in Waldbaum and Magness 1997. There is a detailed compendium of Phoenician and East Greek amphorae from the Iron Age into the Persian period in Lehmann 1996, pp. 97–98, tables 94–95 (East Greek) and tables 70–76 (Levantine amphorae); and finally, for the site of Tell el-Herr and its Persian-period pottery, see Defernez 2001.

northern sites of Buto, Defenneh, Naukratis, and Tell Qedua, but also at Levantine sites, particularly those along the coast and in the south, such as Tell Jemmeh.

These jars entered Egypt as containers for food and beverages such as wine, olive oil, honey, fish sauces, and grains.[133] Wine was imported to various regions, and the oil from Samos and Miletus was prized, particularly in the western Mediterranean.[134] But these jars, as well as the oasis vessels from Dorginarti, were commonly reused for the transport of goods and water.[135] At least one of the East Greek amphora rims, and some of the Levantine amphora rims, appear slightly earlier in date, indicating the reuse of whole jars from various periods of production.

The date of the assemblage of imports should reflect the general date of the abandonment of the Level II fortress and not be too remote in time. The abandonment of the Level II fortified platform occurred perhaps no later than the end of the third quarter of the sixth century BC. The fortress was therefore probably established in the early sixth century BC, although a late seventh-century BC date cannot be ruled out. The archaeological and textual clues for contact between Egypt and Nubia during this period suggest a continued interest in the region and its trade circuit.[136] The Egyptian pottery from the Level II deposits does not contain, as had been thought, evidence for a continuation into the fifth century BC.[137]

133 The products carried in the jars are discussed in Dupont 1998. Note also the products listed under each class of amphora in Bourriau and French 2007, pp. 117–21.

134 Dupont 1998, pp. 143–45.

135 As often noted, Herodotus states that the Egyptians reused Greek and Phoenician wine amphorae for carrying water into the desert; see *The History*, Book 3.6–7.

136 For a brief discussion of this evidence, see chapter 8.

137 As, for example, in Heidorn 1991, 1992a, and 1992b.

7

Objects from the Fortress

The pottery and small objects found at Dorginarti provide glimpses of the cultural and social milieu of the garrison's inhabitants. The stone arrowheads, for example, are usually assumed to reflect a cultural tradition different from that of the Egyptian Nile Valley, as does the handmade pottery presented in chapter 5. The few finer objects found at the fortress are the type of possessions that were acquired by individuals from all backgrounds, Egyptian or Nubian, and that somehow got lost or broken and were left behind.

This chapter will describe and illustrate many of the artifacts found during the excavation of the site. The items registered in the field are also listed in table B in appendix III, the object register and division list at the end of the book. Some of the following objects stem from Level II contexts, although their dating is often uncertain. The mud jar sealings discussed in chapter 6 came from contexts that suggest they were part of the later activity at the fortress, but many of the other objects are less datable. The stone architectural details were discussed in chapter 4, part 2, and these are also listed in table B.

The survey will begin with the small objects, which have been grouped mostly according to the materials used in their fabrication. Martina Renzi and Thilo Rehren intended to analyze the technical equipment associated with the Levels III–IV metallurgical activities, but their laboratory analysis must await a future publication. Carol Meyer analyzed the glass, and Joanna Then-Obłuska studied the beads from all levels, including the Meroitic and Christian periods. Charcoal and wooden fragments are present in the materials brought back to the Oriental Institute Museum, but these have not yet been analyzed.

FAIENCE

The amulets among the Dorginarti finds are comparable to examples from Nubia, particularly those from the earlier tombs in the western and southern cemeteries of Meroe, Sanam, Missiminia, and Qustul. Very few precise parallels are found at those sites or from places in Egypt, but their styles are similar to objects dated to the Third Intermediate Period and Saite era.

FIGURINE, SEAL RING, AND HATHOR HEAD PENDANTS

Most of the faience objects from the site were found in broken and incomplete condition. They include the torso and upper legs of a figure with a broad-collar necklace and divine beard and with armlets and bracelets indicated on the upper arms and forearms (OIM E24332) (fig. 7.1a; pl. 89a; see also pl. 103e). The figure wears a pleated kilt. The area around the necklace preserves a dark-grayish color, perhaps from worn black paint, and the space between the arms and the torso preserves a bright-blue glaze that is not present elsewhere on the worn surface. This object was found atop the eroded south wall in D 9 (3) of the Level III residence. The pendant is included in the analysis by Then-Obłuska below.

A broken sealing ring, the bezel of which is shaped like a cartouche, bears a partial inscription that reads "[. . .] two lands, lord of heaven, and lord of the two lands" (OIM E24325) (fig. 7.1b; pl. 89g). It was uncovered in the surface debris in D 18 (1), a structure situated outside the fortress between bastion N and the southwest corner buttress. Similar rings are found in

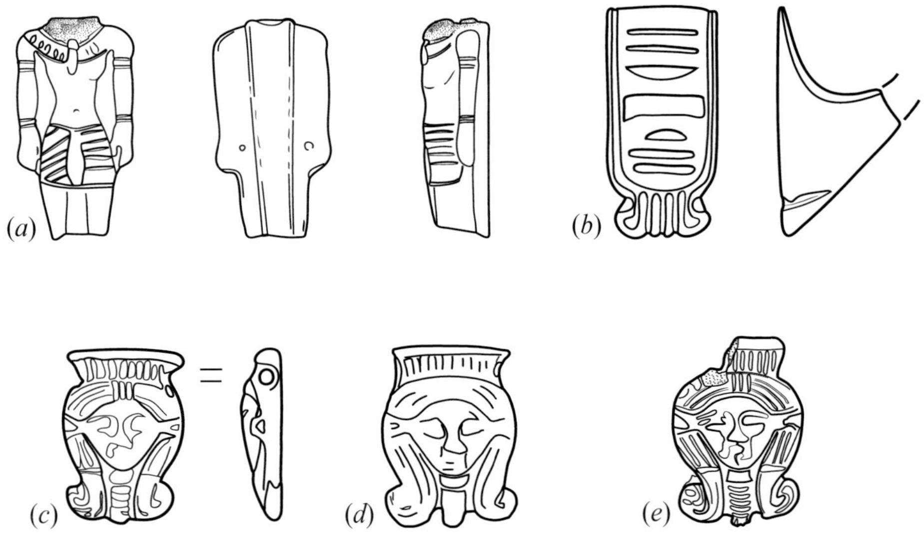

Figure 7.1. Faience objects (1:1)

(*a*) D 9 (3), on top of eroded south wall; OIM E24332

(*b*) D 18 (1); OIM E24325, faience cartouche ring fragment

(*c*) D 6 (1); OIM E24394

(*d*) D 6 (1); Khartoum 14270

(*e*) D 112 (1) fill; Khartoum 14271

the late New Kingdom and become common in the Third Intermediate Period, with a few comparisons dating to the Twenty-Sixth Dynasty.[1]

An example from Karnak is comparable, preserving the top part of the bezel, which reads "Amun-Re, ruler of the Ennead, lord [. . .]."[2] The Karnak ring is dated to either the Third Intermediate Period or the Twenty-Sixth Dynasty.

Two Hathor head pendants were found on the floor in D 6 (1) (OIM E24394, Khartoum 14270) (fig. 7.1c–d; pl. 89b; see also pl. 104b), and another example was found in the fill of D 112 (1) (Khartoum 14271) (fig. 7.1e). Hathor head pendants are found beginning in the New Kingdom and continue down into the Roman period, although they are stylistically distinct. They are quite common throughout the Third Intermediate Period in Egypt, and a number of examples have been found in Twenty-Fifth Dynasty and Napatan contexts in Nubia.[3] The technique of the Dorginarti examples is more fluid than the "Kushite bold" style found at sites and graves attributed to Napatan contexts, which are more angular.[4] Hathor head amulets are also used in the horse trappings from Kurru 201, dated to the reign of Shabaqo, where there are two styles of Hathor heads—a rounded rendering mingled with ones more angular in style.[5] The Dorginarti

1 Examples of this type of ring have been uncovered at Karnak (see the following footnote), Matmar, and Mit Rahineh. Note the comparable ring in Brunton 1948, pl. 44, no. 189 (Twenty-Second to Twenty-Fifth Dynasty). For the examples from Mit Rahineh, see Anthes 1965, pp. 138–40, fig. 21, nos. 359–64 (dated to the Twenty-Second Dynasty, possibly as early as the Twenty-First Dynasty).

2 See Masson 2007, pp. 617–18, pl. X, no. 10. Masson compares the Karnak ring to the chronological development indicated in Jaeger 1993, p. 117, no. 68, who would date this example to the Third Intermediate Period. Masson notes that the ring was made of the green color popular for faience in the Late Period; see Masson 2007, p. 612, nn. 134–35.

3 Note the Hathor column amulet from Qustul in Williams 1990, p. 74, fig. 25c, pl. 10e ("Kushite bold" style). There are examples with a softer, rounder rendering from Amara West, Tomb 211, in Binder 2011, p. 50, pl. 20. The niche grave with tumulus superstructure is post–New Kingdom, and the finds from the cemetery suggest it was in use from the late New Kingdom through the eighth century BC. See also the "Kushite bold" style of the example from Kawa in Macadam 1955, p. 184, pl. XCVIa (0865). This example came from the Eastern Palace (A 14/15), which was attributed to the Meroitic period (p. 115).

4 For example, see the Williams reference in the previous footnote. The style of his example matches the "Kushite bold" styles from Napatan graves at Missiminia and Meroe; see the references in Williams 1990, p. 27, nn. 36–37.

5 Dunham 1950, p. 112, pl. LXVII.C.

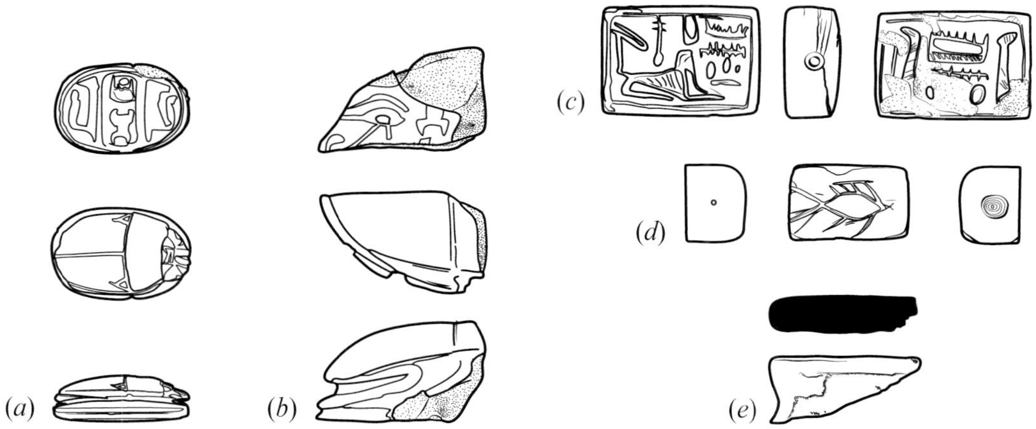

Figure 7.2. Faience objects (*a*, *b*, *d*, 1:1; *c*, *e*, 1:2)

(*a*) D 209 (1), fill; OIM E24396
(*b*) D 1 (1), surface; OIM E24310
(*c*) D 76, under "ash" layer; OIM E24386
(*d*) D 81, fill 10 cm above floor; OIM E24398
(*e*) West Sector, dump; OIM E24418

pendant in the Oriental Institute is also treated below under the beads and pendants section of Then-Obłuska.

SCARABS AND PLAQUE AMULETS

A scarab with the royal name Menkheperre inscribed in a cartouche on its base was found below the surface clearance and within the fill of D 209 (1), toward the northwest corner of the West Sector (OIM E24396) (fig. 7.2a; pl. 89f). The cartouche has a Maat feather and adjoined rearing cobra on each side.[6] Menkheperre was the prenomen of Tuthmosis III of the Eighteenth Dynasty but was also used by Piye from at least the time of his conquest of Egypt in 733/732 BC.[7] It is also the prenomen of Necho I (ca. 672–664 BC). The name is common on scarabs from the reign of Tuthmosis III but remained popular through the Third Intermediate Period and apparently on into the Saite period.[8]

Another fragmentary scarab preserves part of a protective *wedjat*-eye, representing the eye of Horus (OIM E24310) (fig. 7.2b). The bottom part of another sign is also apparent on one side. It comes from the surface debris inside the walls defining D 1 (1).

An inscribed faience plaque features a recumbent androsphinx with a false beard on its flat side, with an unidentified sign above the lifted tail on the left side and the recognizable *nefer*-hieroglyph over the sphinx's back (OIM E24386) (fig. 7.2c; pl. 89d). The object was found in D 76, in fill below the "ash" level. The glaze has largely been worn off, but the plaque is darkened from some type of deposit or burning. A crudely rendered *atef*-crown or possible sun disk is situated above the sphinx's head. No uraeus is apparent. On the other side, which is rounded in profile, are the hieroglyphs Amun-Re, with Maat feathers mirrored on each side. The motif of the king as a sphinx was found on scarabs starting in the late Middle Kingdom and continued on through the New Kingdom, although it becomes less common in the first millennium BC.[9] A

6 The rendering is not as fine as that on a scarab from Cemetery C at Amara West, which also has a joined feather and cobra on each side of the cartouche; see Binder 2011, p. 45, pl. 9, bottom left. G234 is a chamber tomb, and the finds are attributed to the late New Kingdom, with a possible extension into the Third Intermediate Period (p. 50).

7 For a convenient synopsis of some different views regarding the date of conquest, see Schneider 2010, pp. 377–78.

8 Jaeger 1982, pp. 238–53. The name Menkheperre is found on scarabs that typically have rearing cobras or feathers surrounding the name or cartouche. They are found in Twenty-Fifth Dynasty or Napatan tombs in Nubia at Qustul, Missiminia, and Sanam, and in tombs at Meroe's west and south cemeteries; see Lohwasser 2014. A similar scarab and inscribed design was found on an unpublished example from Tomb 102 at Amara West; see the photographs and discussion in Kormysheva and Idriss 2006, p. 253, no. 257, pl. 2 (Twenty-Fifth Dynasty or later). Although the design on the underside is different, the scarab profile and type is matched by two steatite scarabs from Missiminia's Napatan Tomb 186, shown in Vila 1980, pp. 62–63, fig. 51, nos. 2–3.

9 Sanhueza-Pino 2014.

very similar glazed steatite plaque comes from Hillat el-Arab, although it depicts a criosphinx.[10] Rectangular plaques are found in the New Kingdom in Egypt and became popular during the Nineteenth Dynasty. They continue on into the Third Intermediate Period[11] and are also found at Napatan sites in Nubia.[12]

An unfinished amulet with a Nile perch incised on one face came from debris below the "ash" layer in D 81 (OIM E24398) (fig. 7.2d; pl. 89c). Drill holes were started at each short end but never completed. The plaque is small, with one flat side and one convex side; its glaze has been completely worn off.

The last of the small amulets is the left corner of a *wedjat*-eye, which was found in an excavation dump in the West Sector (OIM E24418) (fig. 7.2e).[13]

FAIENCE VESSELS AND NEW YEAR'S FLASKS

Two faience vessel fragments display the remnants of a black-painted vegetal design. Figure 7.3a was found in the Central Sector rubble along the south wall of the structure. It is thick-walled and preserves a plant frond decoration (OIM E50806). Part of a goblet or chalice has an exterior painted with lotus leaves (OIM E24376) (fig. 7.3b; pl. 89e). It was found in the glacis rubble outside the fortress and west of the North Gate.[14]

Three shoulder fragments from New Year's flasks are shown in figure 7.3c–e and on plate 90. One fragment was found atop the plastered floor in the northeast corner of the west retaining wall for the Level II platform (OIM E24336), another is from the surface debris along the south fortification wall and east of D 32 (OIM E24350), and the third is from the eroded surface of the west enclosure wall south of the West Gate (OIM E24313). These flasks are typically considered Saite in date, although evidence indicates that they lasted at least through the Persian period.[15] The Level II contexts or the unsealed deposits of these fragments indicate that all three fragments probably belong to the Level II occupation phase. Parallels are found in both Egypt[16] and the Sudan.[17]

REUSED POTSHERDS, POTTERY, AND MUD OBJECTS

A number of potsherds were reused as tools such as scrapers or trowels, and broken bowls were reused as lids and shallow saucers. One marl jar, the rim of which had broken off, was smoothed down at the neck and reused. In some cases items such as beads, net weights, and stoppers were made from clay for a specific purpose. Some of these items are discussed below. The pottery crucibles and tuyères found at the site are also briefly considered below, while the inscribed signs on pottery sherds have their own section.

DISKS

Pottery disks were uncovered throughout the living spaces, as well as in many open contexts and fill deposits (fig. 7.4). They were found mostly in the structures along the north wall in the West Sector, but also within the Level IV residence and under a stairway into the

10 Vincentelli 2006, pp. 169–70, 176, fig. 3.4, A.19.909. Tomb 19 was originally a Nineteenth Dynasty tomb and remained in use throughout the Third Intermediate Period and into the Twenty-Fifth Dynasty.

11 Jaeger 1982, pp. 204–5.

12 See, for example, Williams 1990, p. 16, an oval plaque from tomb 43 with deeply recessed signs that depict a criosphinx. Williams refers to the Kushite bold style of the two plaques from this tomb and notes similar examples from Sanam (pp. 25–26, nn. 13–17). There are many plaque amulets from Sanam, although none of them is a parallel to the Dorginarti example; see Griffith 1923, pls. LII–LIV, which show the rectangular plaques bearing inscriptions on both faces.

13 According to the dating of the *wedjat*-eyes proposed by Claudia Müller-Winkler, the more rectangular-shaped, plain *wedjat*-eyes should date to the Twenty-Second through Twenty-Fifth Dynasties; see Müller-Winkler 1987, pp. 143–46, 153–56, 169.

14 See comparable fragments in Third Intermediate Period levels at the town of Ashmunein in Spencer 1993, p. 36, pl. 32, nos. 95–102. Spencer notes that faience chalices developed in the New Kingdom and continued through the Third Intermediate Period.

15 The Saite- and Persian-period examples from Egypt are referred to in Masson 2013–15, pp. 5–6. For a general discussion of Saite-period New Year's flasks, see Webb 1978, pp. 116–19, pl. 18, and Blanquet 1992.

16 For example, a faience New Year's flask from the Karnak priests' residences is published in Masson 2007, pp. 605, 614, pl. XXVII. They are found at many sites in Egypt, but note the Saite examples from the town of Ashmunein shown in Spencer 1993, p. 36, pl. 33, nos. 118–19. Saite flasks from graves at Tell el-Yehudiyeh are published in Petrie 1906, p. 19, pl. XXI (green glaze, Twenty-Sixth Dynasty). A faience flask fragment from the camp at Defenneh is shown in Petrie 1888, p. 75, pl. XL, no. 4 (sixth century BC). Faience New Year's flasks were also found at Mendes in Levels I and II; see Wilson 1982, p. 33, pl. 30, no. 3.

17 For example, from Napatan tomb 2-V-6/311 at Missiminia, which is shown in Vila 1980, p. 121, fig. 127, and p. 161, fig. 182. See also Kormysheva and Idriss 2006, pp. 254–56, no. 260 (Khartoum 23520). Faience New Year's flask fragments were also found at Gala Abu Ahmed; see Lohwasser 2004, pp. 151–56, tables 4f, 5, 6, and color pl. 12.

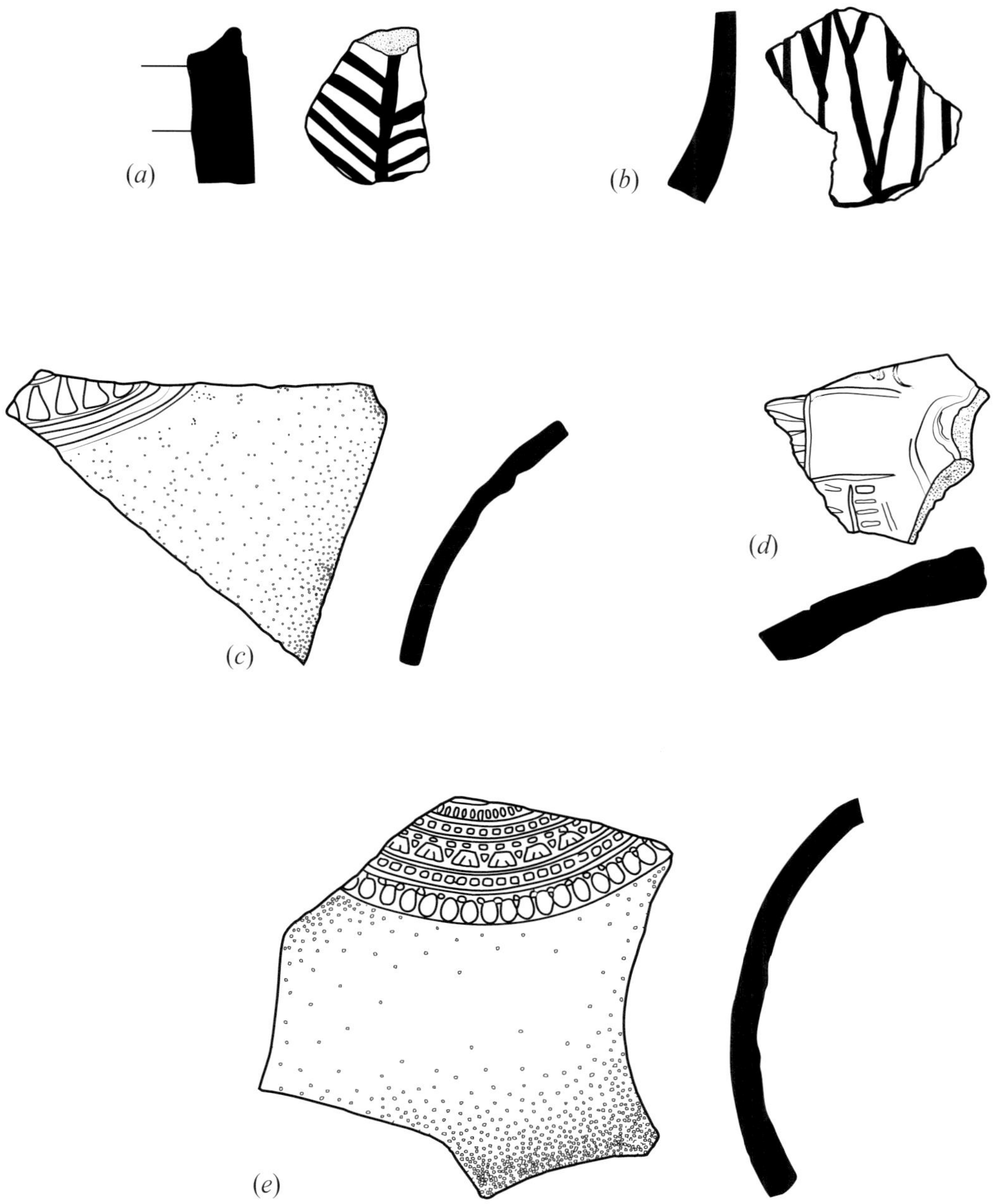

Figure 7.3. Faience vessels (*a*, *c–e*, 1:1; *b*, 1:2)

(*a*) Central Sector south wall, rubble; OIM E50806
(*b*) Glacis stones near west buttress of North Gate; OIM E24376
(*c*) D 32 (1), fill north of silo; OIM E24350, sherd laid flat for drawing
(*d*) Eroded surface of west wall of fort; OIM E24313
(*e*) Floor of bay of Level II structure, at west retaining wall; OIM E24336

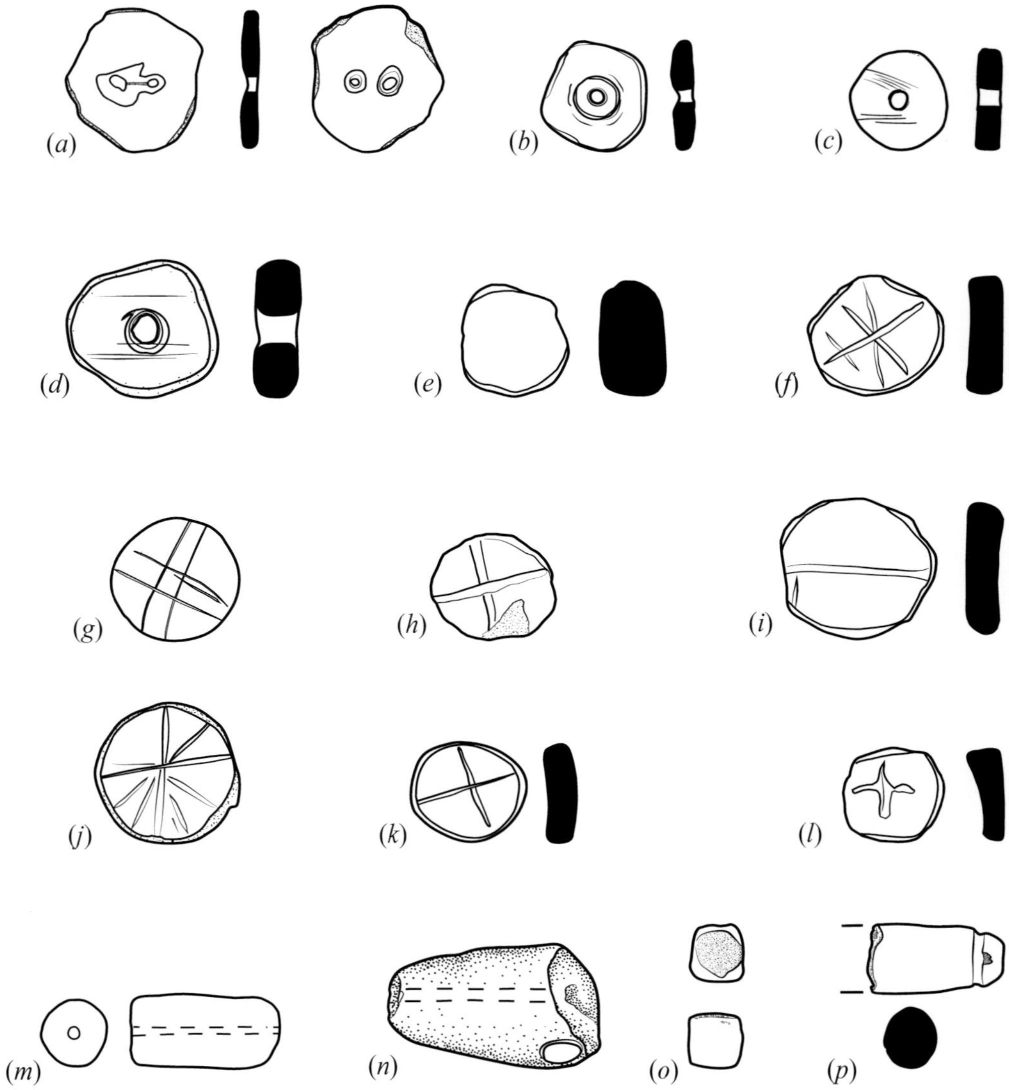

Figure 7.4. Pottery objects (1:2)

(*a*) East Sector rubble over three southern silos; marl
(*b*) D 12 (1); silt
(*c*) D 3 (1); OIM E49788, light wear lines, silt
(*d*) D 223 (1); OIM E49760, silt, blackened on one side, wheel striations
(*e*) D 218 (1); silt
(*f*) D 218 (1); marl
(*g*) D 12 (1); marl
(*h*) D 86 (1); marl
(*i*) D 12 (1); silt
(*j*) D 12 (1); silt
(*k*) D 221 (1); silt
(*l*) D 75; silt, RS, burnished interior/exterior
(*m*)D 3 (1); OIM E24351, silt
(*n*) East Sector, rubble over three southern silos; silt, broken, applied clay
(*o*) D 12 (1); silt, worn top
(*p*) D 112 and area; OIM E24410, silt, broken

Level III residence.[18] No examples were found in the East Sector. These disks were ground down from potsherds of both silt and marl, and typically measured 2–4 cm in diameter, averaging around 3 cm. They had unworked surfaces, but often one of the sides was divided into sections by incised, crossed lines. Some loci yielded multiple examples of disks, both decorated and plain, but they were often found alone. They have been referred to in archaeological literature as *counters*, *tokens*, *weights*, or *game pieces*, and they are found at sites throughout Nubia and Egypt and elsewhere in the Near East and Mediterranean region.[19] Although these prepared disks are not considered good dating evidence, the potential uses for them are interesting to contemplate. Whether they functioned as lots drawn for certain tasks, counters for a particular job done or contract established (buying or selling), voting indicators, meal tokens for particular types of food, or gaming pieces is uncertain.[20]

In some instances the reused sherd disks show a single or double hole drilled through the center, with the button-type example showing a clear wear pattern on one side from a string having been looped through the two holes (fig. 7.4a–d). The others sometimes have thin wear striations on at least one side.

The rubbed edges of the smaller disk tokens or counters are typically smoothed and rounded in profile, with some light scratching from the production process often found atop one of the surfaces or continuing over onto one of the surfaces from the edge. The incised lines are situated on either the exterior or the interior side of the potsherd, when this could be ascertained. The decorative lines are generally lightly incised into the surface. Although many of the disks are only roughly circular, their production nonetheless would have taken time and thus would have been considered a necessary or desired activity to perform.

MISCELLANEOUS CLAY OBJECTS

One pierced pottery bead and another possible example of a bead or pendant with an applied dot are shown in figure 7.4m–n. They were not made from reused sherds but rather were molded from clay.

A small cuboid produced from a reused sherd may have served as a weight or perhaps some other purpose (fig. 7.4o). One side had been ground or worn down to a flattened surface that was more or less circular when viewed from the top and exposed the black core of the sherd.

The only example of a pottery net weight is that shown in figure 7.4p. This type of object and its parallels are discussed below in the section on stone net weights.

A number of jar stoppers and two possible pulleys or lids are also made from clay (fig. 7.5a–d). Flasks were found in contexts associated with two of these clay objects (e.g., D 76, near D 90 where the flask was found), and these objects may therefore have functioned as lids for them. The sizes range from more than 3.5 cm to 5.0 cm in width.

MINIATURE BOWLS

Two handmade miniature bowls were found in D 18 (1), and another came from D 13 (1); all were cataloged under OIM E24399 (one is shown in fig. 7.5e). They are all of different shapes, but the rim of the first example is decorated with incised notches, and crossed lines and a double zigzag are found on its base exterior. The small handmade bowl, or perhaps a pottery fitting, came from D 19 (1) (fig. 7.5f). D 18 and D 19 were both extramural structures located at the southwest corner of the fortification between bastion N and the southwest corner. Miniature "toy" vessels are not usually datable, although the cultural influence is detected in the decoration of the incised example. Their presence, if they are toys, suggests that children and families were among the population of the fortress, at least at certain times. The child's footprint impressed into the mud floor of room D 20 of the Level III residence also proves their presence.

18 For example, there were at least seven of these disks, plain or with incised crossed lines, from D 69 (1); five disks with crossed lines were found in D 303 (1), the surface next to the Level III residence upon which the north stairs to the building were built; several were found in the Level IV residence, in D 312 (2); four disks came from the south part of D 208 (1); four came from D 80 (1), one of them bearing crossed lines; six came from D 200 (2), only one of them with no decoration; D 1 (3) had twelve disks, either plain or with incised crossed lines; and numerous examples were found in D 6 (1) and D 13 (1). The surface clearances of the fortress also produced many examples of these disks.

19 For pottery disks from other similarly dated contexts, see, for example, Wilson 1982, p. 35, pl. 34, no. 1, and Spencer 1993, p. 38, pl. 37, no. 166 (Third Intermediate Period and Late Dynastic), and from contexts of various dates at Tanis, see Brissaud and Cotelle 1987. They have been found at Gala Abu Ahmed and in building F1 of the very late eighth and early seventh centuries BC at Kawa (due to associated pottery). See Eigner and Jesse 2009, pp. 151–52, fig. 18, no. 2, and Welsby 2015, pp. 76–77, pl. 17.

20 D'Onofrio 2007, pp. 4–6. See also Manzo 2007.

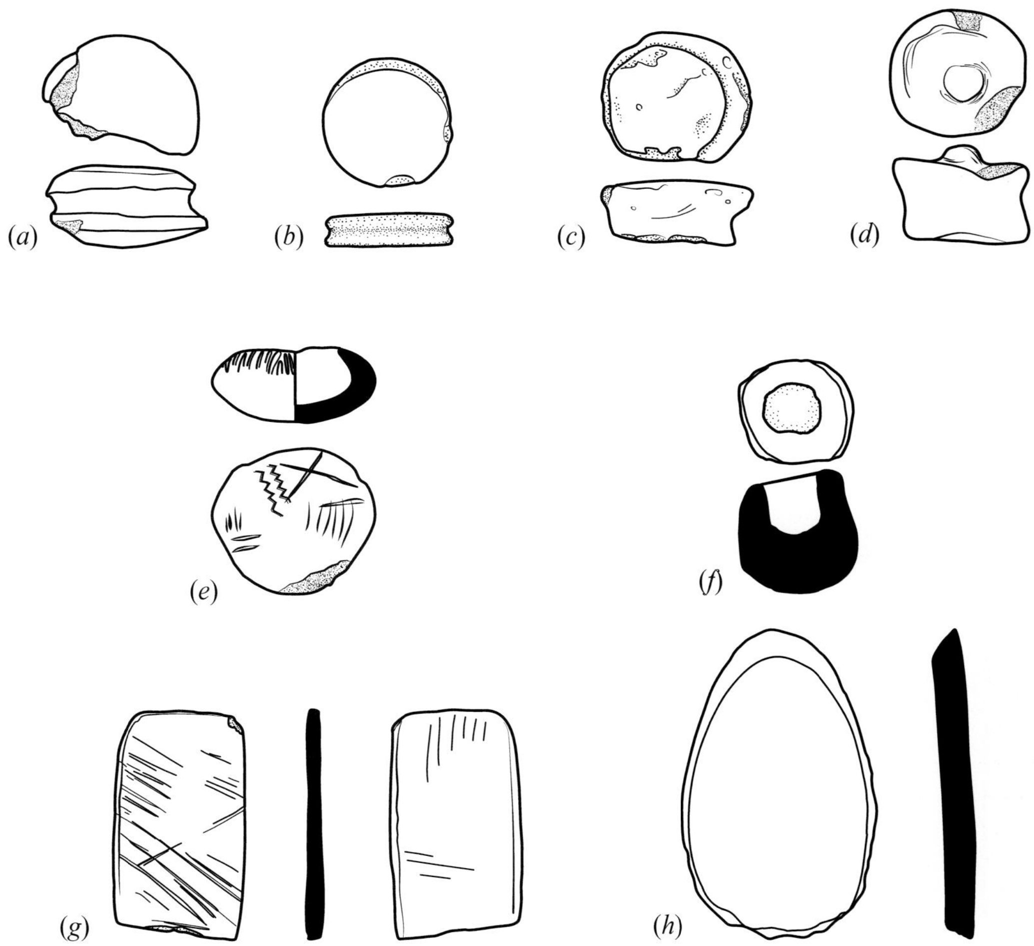

Figure 7.5. Pottery objects (1:2)

(*a*) D 218 (1); silt

(*b*) D 76, above "ash" layer; OIM E24381, silt

(*c*) D 221 (1), rubble; OIM E24413, silt

(*d*) D 12 (1); soft silt, worn

(*e*) D 18 (1) (one of three); OIM E24399, handmade silt

(*f*) D 19; handmade silt

(*g*) D 78 (1); reused marl body sherd

(*h*) D 4, when clearing silo; marl (top edge is worn, bottom is jagged)

TOOLS

Quite a few sherds of broken silt or marl vessels are among the objects reused as tools (figs. 7.5g–h, 7.6). They were used in very specific activities that are difficult to reconstruct from the contexts or associated materials.[21] They are described in detail below.

At least five sherds were ground down and retouched in the manner of those shown in figure 7.6a–b. They are body sherds of silt or marl, most often with light abrasion lines evident on the exterior surface below the preserved edge. Their reuse is at an angle placed either transversely or in line with the wheel striations of the original vessel. But the main indicator of their use was the worked edge of the sherd, where an incised line runs around the circumference, as though a string had been attached. The tops of the three best-preserved pieces from D 113 (1) and D 114 (1) are described as "thick sectioned body sherds, reworked on one edge to a smooth oval [top edge] surface, beneath which is a groove setting it off from the rest of the sherd" (OIM E24407, Khartoum 14280). The interior groove on the illustrated example in figure 7.6a is discontinuous, while that in figure 7.6b preserves no interior groove. The use to which these sherds were put is uncertain, and no parallels have been found.

Another reused sherd from a marl vessel has a worn edge at its "top" (fig. 7.6c). The profile of the worn edge is rounded, and the wear has created a concave surface along that edge. The interior of the sherd is slightly worn, while the exterior preserves the wheel corrugation and the excess clay pad of the original vessel.

The reused body sherd in figure 7.6d has a smoothed edge (on the left side as shown) with a rounded profile. The sherd exhibits evidence of a black-painted band on its exterior, with an incised line made after firing. The edge has been stained a darker gray-black color, although not from burning. The use of this sherd is also uncertain.

JAR SEALINGS

Two fragments of mud jar sealings were found in D 15 (1) and belong to the Level II phase at the site (see fig. 6.12a–b in chapter 6). Both are incised with symbols or letters on their exterior surfaces. They were uncovered in a locus atop the Level II platform, a context that also yielded sherds from imported Levantine amphorae dated to the sixth century BC.

21 The uses for similar or the same types of reused sherds from New Kingdom Qantir are discussed by Raedler 2007, pp. 45–50.

STONE AND MINERALS

Most of the stone objects in the Oriental Institute Museum collection, as well as some in the Sudan National Museum in Khartoum, are illustrated in the figures and plates in this section. It should be noted, however, that many of the grindstones and stone architectural fragments found at Dorginarti were left in the field after work at the site had finished, or were given to the Sudan National Museum in Khartoum and the Antiquities Service.[22] A few other worked pieces of stone—mostly sandstone fragments—have been omitted from the discussion because they are small and amorphous, all of them being smoothed-down fragments of stone with no distinct markings.

REUSED STELE FRAGMENT

A fragment of a stele was reused as a door socket in D 9, a room in the Level III official residence in the Central Sector (OIM E24424). The photograph on plate 91a provides two views of the object. The fragment measured 27 cm in height, while the inscribed surface was 19 cm in width. Since the object is presently housed in the Oriental Institute Museum collection, where a new photograph could be taken, it is the only one of the reused stone elements found at Dorginarti that has been included here.

The hieroglyphs that are still visible on the fragment are from a funerary text with a *ḥtp-di-nsw* offering formula. The text would have been hidden during the use of the stone as a door socket, which was sunk into the flooring and faced the jamb of the doorway.

The bottom side of the block, which was originally the side of the stele, shows chisel marks made during the quarrying of the light-gray sandstone, which created a herringbone pattern. A sandstone quarry was located across the river from Dorginarti on the west bank, and it is assumed that the reused blocks in the Level III residence were taken from New Kingdom monuments in and around Buhen.

22 Field register nos. D-128 and D-130 were left in the field, D-131 and D-133 were given to the Antiquities Service, and D-132 was given to the Oriental Institute Museum (Knudstad and Pierce 1964). For a description of the reused stone architectural fragments, see chapter 4, part 2. I did not locate the items in the Khartoum Museum in 1989, nor were they found by Bruce Williams on his visit in 2014.

Figure 7.6. Pottery objects (1:2)

(*a*) D 114 (1), rubble; OIM E24407 (one of three), silt, deep scratch lines on exterior, wheel lines on interior

(*b*) D 205 (1); Nile D sherd, white slip preserved, worn edges along top, chipped edge along bottom

(*c*) D 38 and D 39, surface clearance; OIM E49797, reused marl sherd, excess clay on exterior, upper edge worn and rounded

(*d*) East Sector, first 50 cm above north magazines; marl with painted band, incised after painting, deposit at left edge

TOOLS AND GRINDING EQUIPMENT

Among the stone objects found at the site are two whetstones made of soapstone from D 70 (1), and a possible third example from the fill layers in the Level II platform (OIM E24390A, E24354) (fig. 7.7a–b). The rock is also called *steatite* and is a soft stone useful for polishing or honing objects that may initially have been sharpened on a rougher stone. The two longer stones from D 70 range from 4.4 to 5.3 cm in length. The smaller, third plaque has a sunken dimple in the middle of one side, possibly the start of a drill hole for producing a bead or amulet. These objects display fine scratches running across all their surfaces, often lined up in the same direction, perhaps indicating their use for polishing or sharpening, or made during their production process.

The two whetstones from D 70 (1) were found next to a corroded copper blade(?) that had been broken into four fragments and is discussed below (OIM E24390B).

WEIGHT AND POTTERY-MANUFACTURING TOOL

A small sandstone fragment was probably a weight (fig. 7.7c).[23] One flat, reddish limestone rock from the fill of D 107 was prepared for use by incising one edge with at least six notches; it was broken in half (OIM E24405) (fig. 7.7d; pl. 92b). Parallels to this object are found at other Sudanese sites, and they are usually considered tools for producing handmade pottery.[24]

GRINDING AND POUNDING TOOLS

Querns, handstones, pounders, and other tools were found throughout the West Sector, concentrated mostly in rooms along the north wall (figs. 7.7e–g, 7.8a, c–e). A large whetstone or smoothing tool found in D 19 (1) was made of schist and had a hole drilled into its top (OIM E50785; fig. 7.8b).[25] One of its sides displayed thin scratch lines in the center. At least twenty-five saddle-shaped querns were found, along with a few other fragmentary examples that are neither shown in the photos nor listed in the field registration book.[26] All but one of the more complete stones were left behind due to their size. The photograph taken in the field shows that most of them were very worn, and some of them may have been used on both sides (pl. 91b).

A sandstone quern from D 104 (1) that was brought back to the Oriental Institute Museum was worn thinner at its midsection, both in thickness and width, was chipped along one side, and had been used on both sides (OIM E50793) (pl. 92c). This quern was approximately 42 cm long and had been worn down to a concave profile on both sides, ending up 2–3 cm thick in the center. Seven of the querns were found in contexts yielding typical Dorginarti pot stands and pilgrim flasks or jugs, suggesting a type of kit used for the preparation of food or other substance.

There were twenty-nine stone cobbles of sandstone, basalt, and quartzite, most of them showing marks from their use as pounders (fig. 7.8c).[27] Of those in the Oriental Institute Museum collection, an example from D 48 (2) (OIM E50777) measures 3.5 × 3.8 cm, an example from D 69 (1) (OIM E50799) measures 3.9 × 3.8 cm, and another (OIM E50776) measures as much as 6.6 cm in diameter. They are typically about 6 cm in diameter with varying measurements from one side to the other. Some display somewhat pitted or battered surfaces, either all over or along one continuous edge; in the latter case two of the sides preserve the original rock surface where it was held, often a more flattened surface. Most of them are in fairly pristine shape, in which case they are more spherical. The elongated example shown in figure 7.8e was 7.9 cm long and 5.9 cm wide (OIM E50780). The worn areas are at each end of its length, stretching from the ends over to one side of the stone.

One flattened grinder or pounder, with worn surfaces along its edges, was found in association with querns and other pounders in D 6 (1). With a red substance adhering to the surface of one end (OIM E50781; not illustrated), it is about 3 cm thick, 6.1 cm wide, and 6.9 cm long. Its two smooth surfaces were flat, and could be held between the fingers while the edges were used to pound or grind. The red substance

23 A similar weight is shown in Spencer 1993, p. 32, pl. 27, nos. 11–12, of uncertain date.

24 See the bone tool "rockers" in Addison 1949, pp. 151, 204, pl. LVIIB (bone, dating uncertain), and Crawford and Addison 1951, pp. 46–47, 65, pl. 41 (pottery, dating uncertain). See also the stone "peigne de potier" from a tomb at Amara East in Vila 1977, pp. 70–73, fig. 32, no. 5 (New Kingdom tomb 2-R-44, with probable later intrusions).

25 It measured 6.3 cm wide, with a preserved length of 6.7 cm.

26 Complete and fragmentary querns were found in the following loci, many of which had multiple examples: D 6 (1), D 10 (1), D 42 (1), D 53 (1), D 59 (1), D 60 (1), D 66 (1), D 69 (1), D 104, D 122 (1), D 208 (2), D 215 (1), D 216 (1), and D 221 (1). The handstones were found in D 67 (1), D 69 (1), D 216 (1), and D 19 (1).

27 There was one pounder from D 1 (1), and there were two from D 1 (3), five from D 6 (1), two from D 14 (1), one from D 32 (1), two from D 41 (1), one from D 48 (2) (round bin), seven from D 67 (1), three from D 69 (1), three from D 90 (1), one from D 125 (1), and one from D 216 (1).

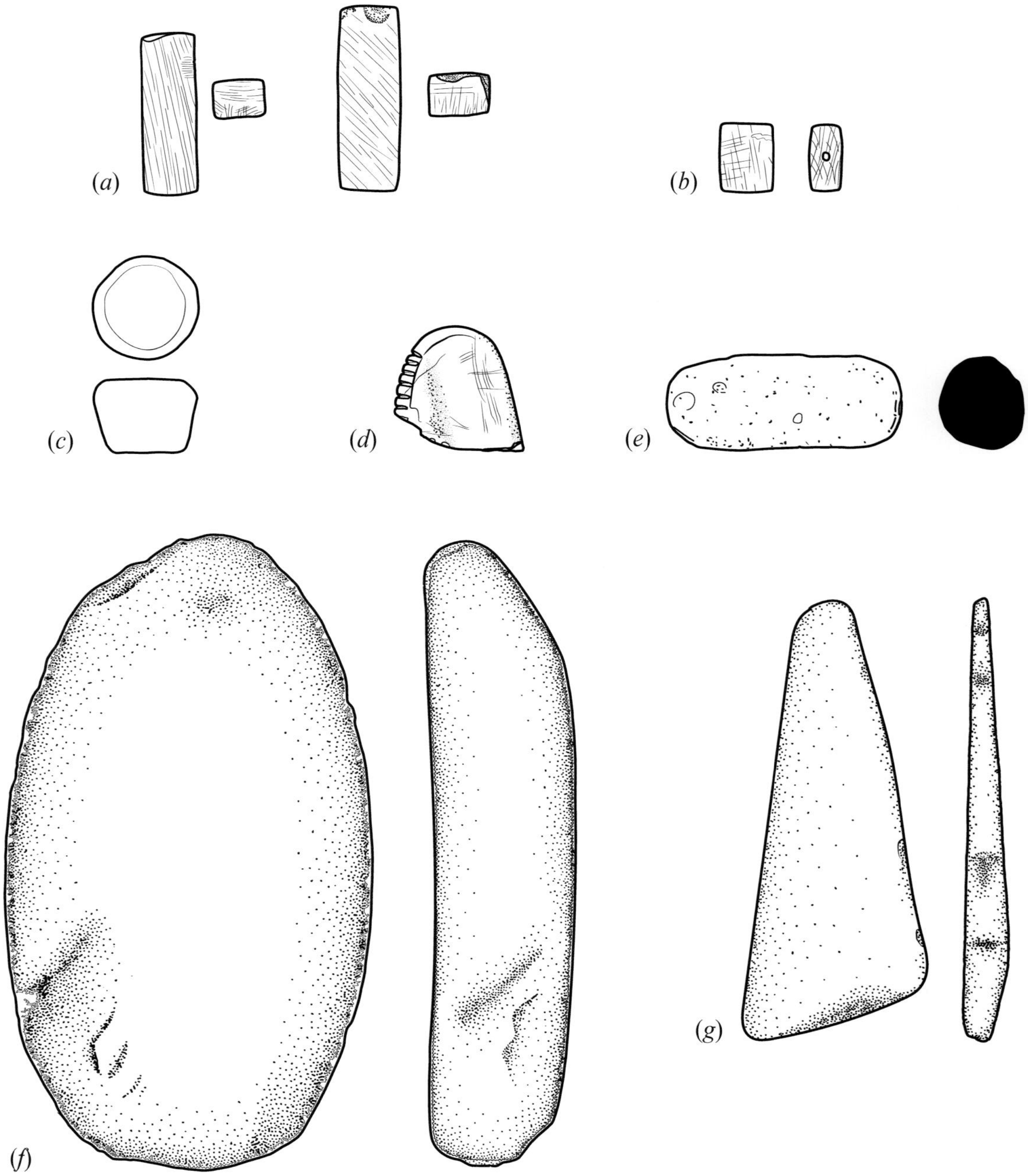

Figure 7.7. Stone objects (1:2)

(*a*) D 70 (1), 15 cm below "ash" layer at south doorway; OIM E24390A, gray soapstone

(*b*) D 8 (1), fill of Level II platform; OIM E24354, gray-green soapstone, drill hole incomplete

(*c*) D 76, fill below "ash" layer; OIM E24379, sandstone

(*d*) D 107 (1), fill; OIM E24405, ferruginous limestone, 0.4 cm thick

(*e*) D 35 and D 36; OIM E50783, sandstone

(*f*) D 69 (1); OIM E50773, gray sandstone (drawing by C. Abraczinskas)

(*g*) D 19 (1); OIM E50786, sandstone scraper with scraped outer edge (drawing by C. Abraczinskas)

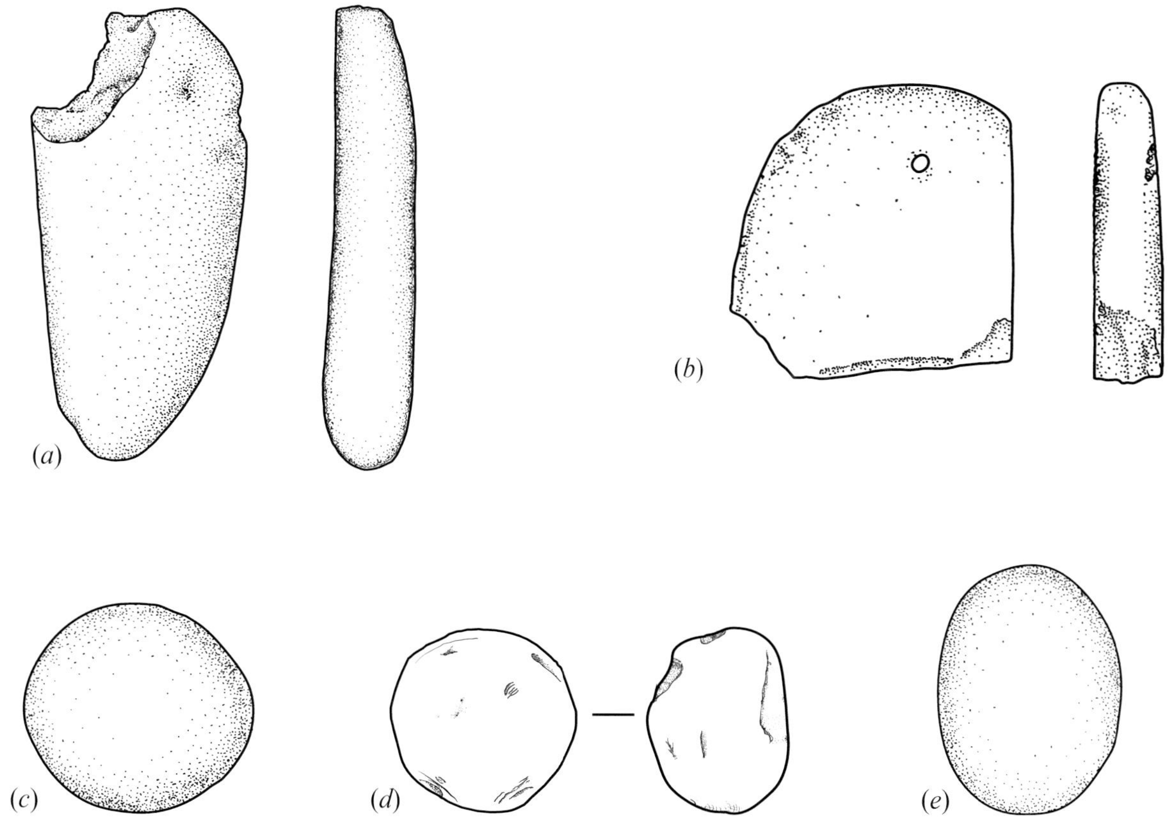

Figure 7.8. Stone objects (1:2)

(*a*) D 216 (1); OIM E50782, sandstone (drawing by C. Abraczinskas)
(*b*) D 19 (1); OIM E50785, schist, hole drilled at "top," flat edge along right side (drawing by C. Abraczinskas)
(*c*) D 67 (1); OIM E50778, basalt, lightly pounded
(*d*) D 75; OIM E50784, pounder? (friable and broken); white sandstone
(*e*) D 112 and area; OIM E50780, quartzite cobble, pounded at both ends

(ocher?) was found at the thinner edge, which displays most of the wear on the tool.

Four of the quern fragments, which were left in the field and were not available for study, are associated with one or more of these stone pounders. There was no evidence mentioned in the field notes, such as quartzite chips, that might indicate gold working. The only quern in the Oriental Institute Museum collection, mentioned above, has been abraded smoothly into concave profiles on both of its sides, with no evidence for impact marks left by pounding activities (OIM E50793).

Handstones were found along the north wall in D 6 (1), D 66 (1), D 69 (1), and D 216 (1) and in the surface clearance along the south wall—that is, in D 35 and D 36 (figs. 7.7f, 7.8a). The examples in the collection are made from sandstone. The example from D 69 (1), shown in figure 7.8a, is about 18 × 10 cm maximum.

The angular wear of the thin sandstone object in figure 7.7g (OIM E50786) may indicate its use as a scraper or polisher.

ARROWHEADS AND BLADES

Approximately twenty-seven bifacially flaked stone arrowheads with hollow or notched bases were found in various contexts at the fortress associated with Levels II–IV (grouped in fig. 7.9 and on pls. 93c, 94).[28]

28 Arrowheads were found in the East and Central Sectors, including around the Level II platform and the southern silos of the East Sector. Two were found in the fill outside the Level III and

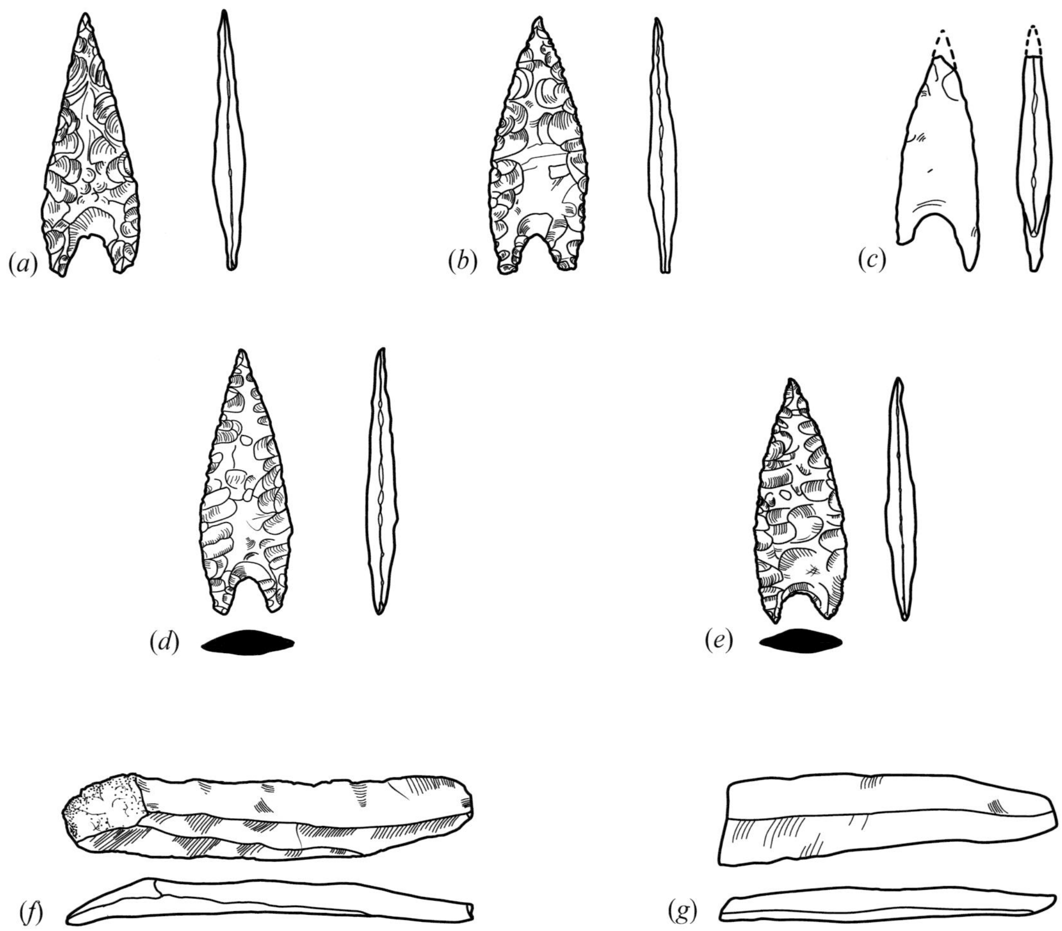

Figure 7.9. Stone objects (1:1)

(*a*) D 34 (1), in charcoal on floor; OIM E24349E, agate
(*b*) D 34 (1), in charcoal on floor; OIM E24349B, agate
(*c*) D 86 (1), rubble; OIM E24412, white silica/flint, worn
(*d*) Room east of D 34; OIM E24421, agate
(*e*) D 126, silo exterior; OIM E24419
(*f*) D 59C (1); OIM E24417, flint blade
(*g*) D 32 (1) and area; OIM E24363, flint blade

These arrowheads are usually 3–4 cm long and have a curvature or hollow at the base with downward-pointing wings at both sides.[29] They are made from agate or flint. The stone used to produce each arrowhead was carefully chosen, and their traditional significance as well as their beauty undoubtedly reflect their ceremonial and social importance to both maker and bowman.[30] The Dorginarti arrowheads are important cultural markers for at least one component of the fortress's population, and parallels have been found in regions to the south and southwest of the Second

Level IV official residences, ten cached arrowheads were found in the charcoal deposit resting on floor 1 of D 34 at the east end of the fortress, and one was found in D 33. The West Sector examples came from the fill of D 32, one on the floor and one near the door of D 73, and in the fill levels inside D 79, D 81, D 86 (1), D 210 (1), D 213 (1), the fill around D 126, and two in D 201 (1).

29 For the type, see Clark, Phillips, and Staley 1974, p. 358. The authors of this study note that type B, close to these examples from Dorginarti, went out of use before the Old Kingdom in Egypt but reappeared in the Saite and initial Persian periods. Interesting technical studies of stone arrowheads have been published in recent years, and one analysis that studies the use-wear evidence is found in Lemorini and Cocca 2013, p. 196. There has been no similar technical study of the Dorginarti arrowheads.

30 For the significance of the color of the stone chosen to produce the points, among many other topics, see Graves-Brown 2010, pp. 138–40.

Cataract, but also in Egypt. There was no certain evidence at the site for metal weaponry, but metal items would have been either taken by the garrison when the fort was abandoned or scavenged by others.

No debitage was found at the site—at least it is not mentioned in the records—and it is uncertain where the points were manufactured. Numerous knapping sites have been identified on the elevations to the east and west of the river in this region, but they are of much earlier age.[31] These work sites, or similar sites, could have been reused for producing blades and points during the early first millennium BC. Chalcedony and agate resources were readily available along the river.

There are precise or close parallels for the stone arrowheads from earlier periods in Egypt, for example from Tell el-Dab'a and Qantir.[32] During the Amarna period, arrows and archery gear are mentioned as items that came from Nubia.[33] But these stone arrowheads are not all considered Nubian products, since their shapes usually differ from their Nubian counterparts, having tangs, and are mostly made from flint.[34]

One stone leaf-shaped arrowhead (tanged?) was excavated at Medinet Habu in Egypt, while one (hollow-based?) arrowhead stems from late Twenty-Sixth or early Twenty-Seventh Dynasty contexts at Elephantine.[35]

The hollow-based arrowheads found at the fortress of Jebel Sahaba in Lower Nubia, as well as those found at Semna, Amara, and Kawa, are comparable to those from Dorginarti.[36] Both tanged and hollow-based points are found in the materials from New Kingdom tombs, in some cases clearly reused at a later period, at Sai and Soleb.[37] One burial at Sanam contained a tanged arrowhead, while recent excavations at the site have uncovered an example of a carnelian hollow-based arrowhead.[38] They are found in early and later levels at the town site of Meroe, and later types of stone arrowheads are found in tombs of the Meroitic period at Meroe.[39] Similar hollow-based arrowheads are found at Zankor in northern Kordofan and at the fortress of Gala Abu Ahmed in the Wadi Howar, and unpublished examples may also stem from sites in northern Darfur.[40] Lastly, similar arrowheads come from Jebel Moya, west of Sennar.[41]

Larger tanged and hollow-based arrowheads—the latter having longer wings and wider curvatures at

31 There is no evidence for the later reuse of these debitage sites in the published surveys. Note the mention of stone resources at the Second Cataract in Sandford and Arkell 1933, pp. 38–39, fig. 7.

32 See the listing in Graves-Brown 2015, pp. 40–42.

33 Cited in Graves-Brown 2015, p. 41. Graves-Brown cites the Nubian tribute of arrowheads, bows, and quivers mentioned in Zibelius-Chen 1988, p. 114.

34 The Egyptians produced flint arrowheads at least until the Nineteenth Dynasty; see Graves-Brown 2015, p. 41, who notes: "Nubian lithic material of this date, however, does not look like the Delta pieces [from Tell ed-Dab'a and Qantir] in either material or form." The Egyptian arrowheads are tanged and more often made of flint, not quartz or carnelian like most of the Nubian examples.

35 One example, the base of which cannot be seen in the preserved mounting, was found outside the enclosure wall at Medinet Habu; see Hölscher 1954, p. 6, pl. 3A (fifth from left). The Elephantine example is mentioned in Hikade 2001, p. 122. A silex arrowhead of a different shape, with "contracting stem," comes from Tell el-Ghaba Area I, L0001, a destruction level perhaps dated to the seventh century BC; see Fuscaldo 2005, pl. 33, no. 541 (F0511A).

36 A bifacially flaked, hollow-based arrowhead was found at the Lower Nubian site of Jebel Sahaba; see Säve-Söderbergh and Troy 1991, pp. 172–73, pl. 220 (Site 51, Twenty-Fifth Dynasty). The Semna example is described in Dunham and Janssen 1960, p. 25, 24-3-475b, from room 46, east of the temple of Taharqo. The agate arrowhead, Boston MFA 24.2069.2, is shown online at http://www.mfa.org/collections. It has a slightly recessed base. Stone arrows with recessed bases were also found in the latest levels at Amara West; see Fairman 1948, p. 10, and Shinnie and Bradley 1951, p. 11 (Napatan). The Kawa examples are illustrated in Welsby and Anderson 2004, p. 153, no. 133 (chalcedony, Kushite/Napatan from the surface of Q3).

37 The Soleb arrowheads are published in Schiff Giorgini 1971, p. 94, fig. 128. There are eleven examples, most of them like the hollow-based Dorginarti arrowheads, that came from the surface above T17 and the west chamber of T24, where a few vessels and a bead also belong to the intrusive burials. The tanged arrowhead from tomb 5 at Sai is published in Minault-Gout and Thill 2012, p. 398, pl. 175, T5P3.

38 Griffith 1923, p. 167, pl. XXXVI, no. 8 (above the body in tomb 1145); Vincentelli 2011, pp. 279–81, fig. 13.

39 Various types of white quartz and carnelian arrowheads were found in seventh-century(?) and perhaps early third-century BC contexts at Meroe; see Shinnie and Bradley 1980, pp. 191–92, 219, fig. 84 (especially example 1548 from the surface of the town site). The dates follow those in Bradley 1984, pp. 199–200. Other stone arrowheads of different types are found in Meroitic burials at Meroe, but only two examples in these later tombs are similar to those at Dorginarti; see Dunham 1963, p. 264, fig. 170, 4 (at the left and right of the photograph).

40 An arrowhead similar to the types found at Dorginarti was found at Zankor in North Kordofan at the west end of the Wadi el-Malik; see Edmonds 1940, pl. 1c. The Gala Abu Ahmed examples are published in, for instance, Eigner and Jesse 2009, p. 153, fig. 19. The Darfur examples are cited from a personal communication I received from Khidir Abdel-Karim Ahmed in January 1989, but no photographs or illustrations were made available.

41 Addison 1949, pp. 186–87, 194, pls. LXXXI–LXXXII. Of the four hundred flaked items preserved, it is noted that the majority were the hollow-based, leaf-shaped type of arrowhead, although they are smaller in size than the Dorginarti examples, being 1.5–2.5 cm in length. The dating of the site is uncertain, but faience amulets and other objects of Napatan type were found in some of the early tombs. For an overview of the current debate, see Sakamoto 2016.

the base than the arrowheads from Dorginarti—were also found in the earliest Kushite ancestral tombs at Kurru.[42] Lunate arrowheads were found associated with them.

The Nubians or southerners who served in the armies of Nubian, Egyptian, and foreign rulers used traditional stone arrowheads for generations. Herodotus, for example, notes that the "Ethiopians" fighting in Xerxes's army during the first part of the fifth century BC used stone arrowheads.[43]

There are three fragmentary flint blades from the site, two of which are shown in figure 7.9f–g (pl. 93a). They are undatable, but the tradition of making and using lithic tools continued into the Third Intermediate Period and later, as shown, for example, by the mostly Third Intermediate Period stone blades, flakes, and knives from Ashmunein.[44]

FISHING-NET WEIGHTS

Stone fishing-net weights or sinkers were common at the fortress (fig. 7.10a–d; pl. 93b).[45] These objects have been identified as cord ties, loom weights, and winding reels, but their resemblance to net weights from elsewhere makes their identification as net sinkers viable.[46] They are typically 2.8–6.8 cm long and are prepared for use by incising at least two indentations around the circumference of the stone that do not always line up from one side to another and may be discontinuous. Many of them show scratches on their surfaces from their use or manufacture.

The objects have been found in Egypt and Sudan, dating to a number of different periods but also more or less contemporary with the early fortification at Dorginarti.[47]

RECTANGULAR STONE WITH INSCRIPTION

A small sandstone block that was either a small stele or a name stone was found at the bottom of a test pit sunk nearly 1 m below the floor of D 5 in the Central Sector (OIM E24326) (pl. 92a).[48] This pit extended down below the floor level of room D 312 in the Level IV residence, and it is assumed the item represents a foundation deposit.[49] It may also have been used to stamp bricks or mud architectural details, as discussed below. The object's dimensions are 16.2 × 6.5 × 4.8 cm. The block was rounded at both short ends and thus resembles a cartouche in form. The back is rough and pitted, but the other surfaces have been smoothed.

The incised inscription names the fortress, at least its Level IV incarnation. The reading of *mnnw*, "fortress," is fairly straightforward.[50] This sign is possibly followed by *ḫw*, "to protect or safeguard."[51] Together with the following *nb* sign, the reading could be "the

42 There are hollow-based arrowheads from Kurru Tumulus 2 and Kurru 19, and arrowheads with tangs from Kurru Tumuli 1 and 4; see Dunham 1950, p. 14, fig. 1c, p. 16, fig. 2c, p. 18, fig. 3b, p. 75, fig. 24f, and pl. LXXI.C. The graves are dated between Generation A (ca. 860–840 BC) and Generation D (ca. 800–780 BC). Lunates were found associated with tanged arrowheads in Kurru Tumuli 1 and 4; see the photograph on pl. LXXI.C.

43 Herodotus, *The History*, 7.69.

44 Spencer 1993, p. 33, pls. 27–29.

45 These stone objects were found in the following loci: two in D 19 (1); one each in D 22 (3) and D 22 (4); inside south wall of West Sector, between hearths southwest of D 32, surface; D 26 (1), D 59, against west wall, north half, under "ash" layer; D 75, above "ash" layer; West Sector, from debris about 2 m north of breach in north wall (outside); and from the work dismantling the walls of the Level III building, from the area of the doorway between D 20 and D 9.

46 Bates 1917, p. 259, pl. 21, nos. 202–3. No. 203 is the one found by Petrie at Tell el-Retaba and mentioned in the following footnote, while no. 202 was found at Gammai in a context attributed to the late Predynastic or Early Dynastic period. Note that American Indian stone net sinkers, elongated stones of 11 cm in length, have incised lines at both ends and one down their lengths, and are used to troll the nets for catching salmon in the streams along the northwest coast of North America; see Stewart 2008, pp. 30–31, 86–87.

47 Note the stone object, called a *reel*, in Petrie 1906, pl. XXXVIC, no. 47, from Tell el-Retaba. The current excavations at the site of Tell el-Retaba uncovered limestone examples in late Twentieth Dynasty contexts; see Rzepka et al. 2014, p. 184, fig. 82. Another parallel, identified as a loom weight, comes from the Second Cataract site of Kor; see Smith 1966, pp. 241–42, fig. 21, no. 7. See also Vila 1978, p. 108, fig. 58 (2-V-28; occupation site of the Kerma–Middle Nubian period). The Kawa example is shown in Macadam 1955, p. 106, pl. LXIIb (0759), where it is called a *winding reel*. It was found in the northwest corner foundation deposit of the temple of Taharqo, but Macadam places the items in this disturbed deposit in the period after Anlamani but before Hariotef, when the temple was restored (p. 245). For New Kingdom examples, see Spencer 2014, p. 478, pl. 23.

48 I would like to thank the experts who kindly studied the inscription on the stone: Vivian Davies, Robert Ritner, Bruce B. Williams, and Karola Zibelius-Chen. The final interpretation was based on all their suggestions, and I apologize for any errors in my reading.

49 This suggestion is my own, but see Pierce 1964a, entry for January 30, and Hoerth 1964a, no. 2, p. 6.

50 For *mnnw* meaning "fortress," see Vogel 2004, pp. 21–22, and Morris 2005, pp. 809–14.

51 As proposed by Karola Zibelius-Chen (personal communication, February 23, 2017), who believes that the suggestion that the last cluster may be *tꜣ-sty* is improbable: "For the last sign to be n-ta-sty it must be noted that the sign does not at all resemble the sty-bow, although on the whole it would make sense to read 'the fortress: protecting the lord of eternity of Nubia.'"

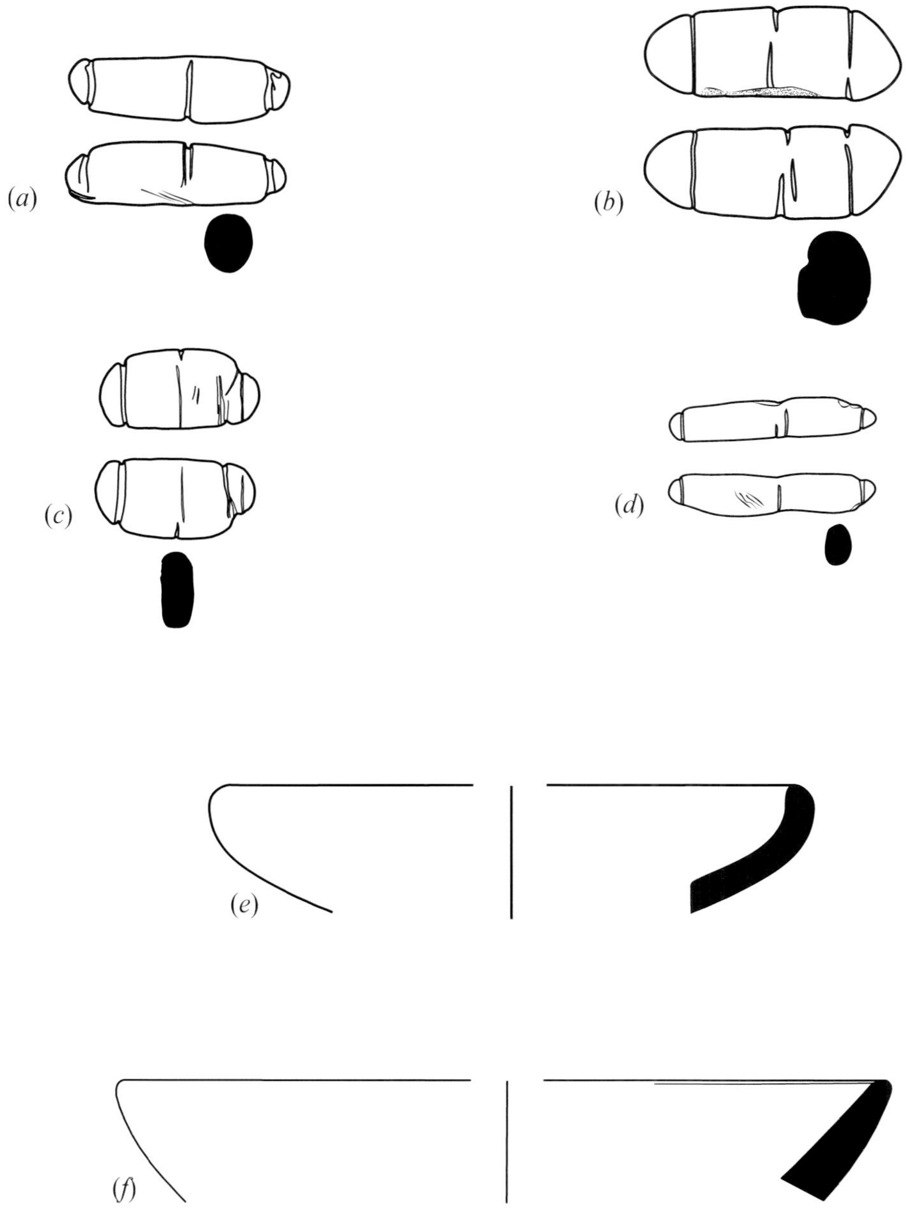

Figure 7.10. Stone objects (*a–d*, 1:2; *e–f*, 2:5)

(*a*) D 19 (1), below surface; OIM E24329, talc

(*b*) D 59, under "ash" layer; OIM E24380, soapstone

(*c*) Doorway between D 9 and D 20 (when dismantling walls); OIM E24411, gray soapstone

(*d*) D 75; OIM E24385

(*e*) D 8, fill at northwest corner of Level III residence; OIM E24402, travertine

(*f*) D 1 (1); travertine, diameter ca. 29 cm

fortress of the lord" or "the fortress for the protection of the lord." *Nb nḥḥ*, "the lord of eternity," could refer here to Amun, Osiris, or Re.[52]

The meaning of the last cluster of signs is more difficult, and a variety of readings have been suggested. The last of the horizontal lines below *nb nḥḥ* may belong to the top frame of a square *ḥwt* sign (O6) (less likely a *pr* sign), with the internal sign being Maat (C10) with a feather or an *ankh*-sign on her knee.[53] It may also represent a seated Asiatic man with a throw stick (A49), which is also used as a determinative for Nubian bowmen—that is, the *iwntyw-styw*—and generally for soldiers or the army.[54] With the initial reading mentioned above, the inscription would thus read "The Fortress for the Protection of the Lord of Eternity of the Estate-of-Maat."

The small dot at the bottom left of the inscription may be a hollow left by the manufacturing process, or part of a determinative.[55]

This stone object finds earlier New Kingdom parallels in large cartouche-shaped stamps, such as the stone example from Amara West and the cartouche-shaped impressions on mud roofing from both Sai and Amara West.[56] The oval stamps used in the mud roofing at Amara West measure approximately 10 × 4 cm. Those examples, however, bear the names of kings and not the names of places.

GALENA

Forty-seven small cubes were found below a layer of ash and on the floor in D 33 (1), at the east end of the fortress (OIM E24355; not illustrated). They were thought to be hematite or petrified wood, but they appear to be galena.[57] They are square to slightly rectangular, very light in weight, and average about 0.4 cm per side. The small blocks are more or less rectilinear in shape and have black, shiny surfaces. The fractures appear to be a natural aspect of the cubes, characteristic of galena.

OTHER STONE OBJECTS

A nodule of rock with a hollowed center, but no crystallization, was found resting on the floor in D 10 (1) (OIM E24371; not illustrated). It may have had some function in the house, or may just have been collected as a curiosity. It is a natural concretion common in sandstone formations.

Two travertine-vessel rims were also found (OIM E24402 and unnumbered) (fig. 7.10e–f). The example from D 1 (1) closely resembles stone plates or bowls from Heracleopolis Magna dated to the Third Intermediate Period through at least the seventh century BC.[58] But neither vessel comes from a very good context, their shapes are not distinctive, and they offer little dating evidence.

METAL

Twelve metal objects and metal residue are dated to either the Level III or Level IV use of the fortress. All these objects are discussed and illustrated here. In addition, Level I, an early to classic Christian occupation atop the Central Sector, exposed an iron key that had been stuck between the bricks of the structure (Khartoum 14821), while the Meroitic graves yielded an iron and bronze ring (OIM E24316, Khartoum unnumbered, both from burial 2) and an iron bracelet (OIM E24333, burial 4). These will be included in a future publication.

Only eight of the items represent finished, copper-based objects, although one registered item represented by two metal fragments of indeterminate shape was found in D 113 (1) (OIM E24403). Two spatulas were found in neighboring areas, D 78 and D 80. They ranged from 10.0 to 11.5 cm in length (OIM E24387, E24389) (fig. 7.11a–b; pl. 95a–b). Another folded-up piece of metal was found in D 5 (2)—that is, the fill within the Level II platform but within the walls of the Level III room (OIM E24357) (fig. 7.11c; pl. 95c). The original

52 I thank Zibelius-Chen again for her advice. See Leitz 2002, pp. 667–68.

53 For this reading, and the suggestions immediately following, I thank Robert Ritner. Zibelius-Chen suggests that if the square sign has two feathers within, then *nb nḥḥ n ḥwt-mꜣꜣty* may refer to Osiris, "lord of eternity of the house of the two Maats."

54 Robert Ritner (personal communication, September 1, 2014): "For the internal sign, I suggest Gardiner Sign-List (facing right) C 10 (Maat, but with a feather or ankh on her knee), or A 47–49. Note that the 'Syrian' with a throw stick (A 49) can also be used as a determinative for 'Nubian Bowmen.' Could it be an ideogram here?" Ritner also suggested a figure with *sꜣ* (V17) or *iry* (A48) upon the knee.

55 The block may have been clamped down at this spot to hold it in place while carving, although there is no corresponding mark on the back of the stone.

56 For the mud-impressed object at Sai, see Julia Budka at https://acrossborders.oeaw.ac.at/seti-i-at-sai-island/. For the stone from Amara West, see Spencer, Stevens, and Binder 2014, p. 23. The photograph of the stamp does not indicate the size, but the online discussion mentions a larger sealing dimension of 10 × 4 cm.

57 Used for eye paint or for medicinal applications, as pointed out by Forbes 1965, pp. 18–19. See also Nunn 1996, pp. 198–99, 201, from Ebers Papyrus 336–431.

58 Molinero Polo 1995, pp. 209–17, 266, pl. LXX, p. 261, pl. LXXI, and p. 282, pl. LXXVI.

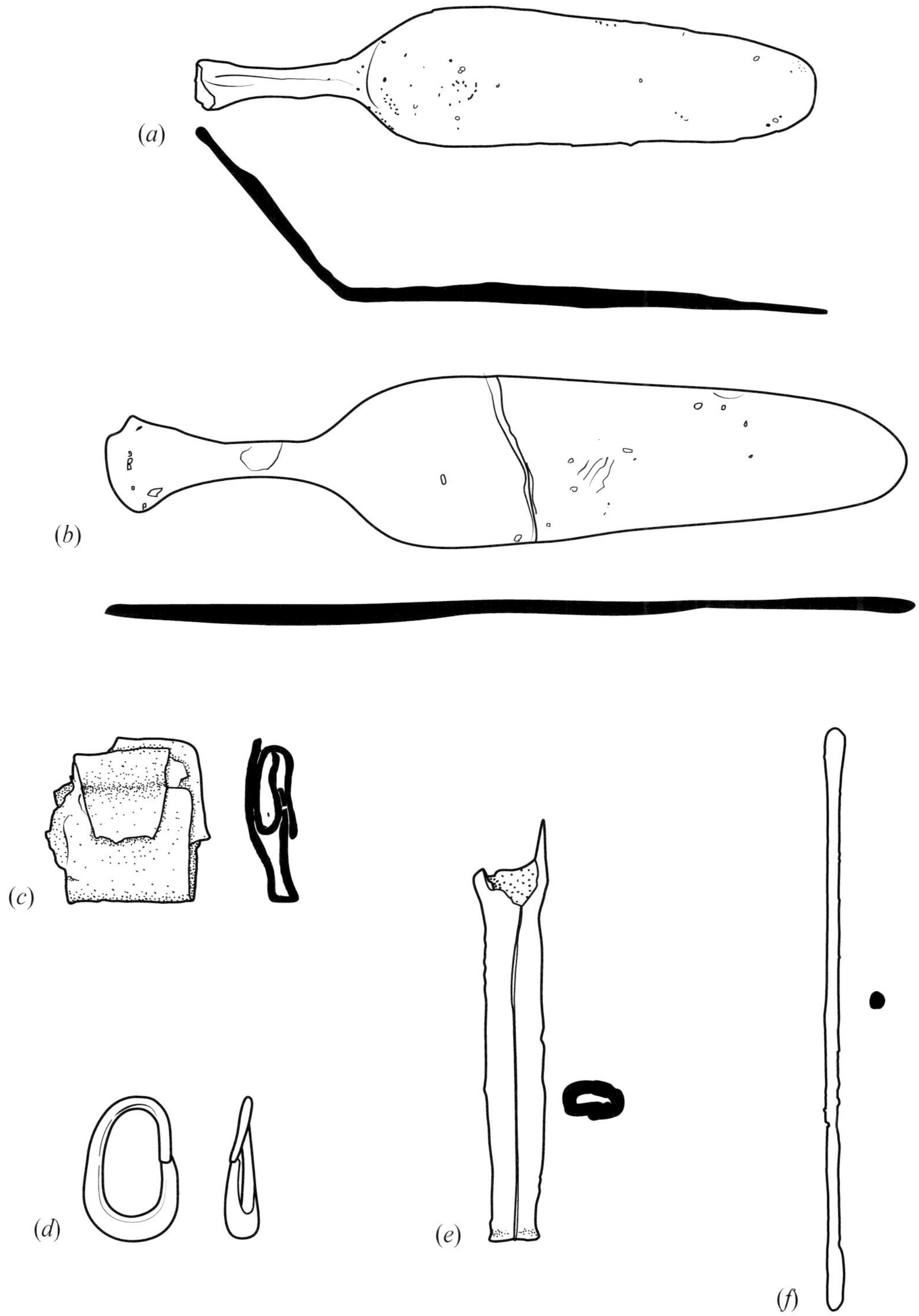

Figure 7.11. Metal objects (*a–c*, 1:2; *d–e*, 1:1; *f*, 2:5)

(*a*) D 78, fill below "ash" layer; OIM E24387, copper/bronze

(*b*) D 80, below "ash" layer, southwest corner of room; OIM E24389, copper/bronze

(*c*) D 5 (2); OIM E24357, copper/bronze

(*d*) D 13 (1), eroded wall of fort; OIM E24311, copper/bronze

(*e*) D 126, outside silo; OIM E24420, copper/bronze

(*f*) East of D 40, below surface and at wall; OIM E24415, copper/bronze

shape of the folded object cannot be distinguished, but it was either a spatula or some other thin-walled item, which was possibly folded for recycling. This type of spatula has parallels in both Egypt and Nubia.[59]

A tapered hoop or "boat" earring was found in the eroded fill deposit at D 13 (1), in the fill around the west wall of the fortification and south of the West Gate (OIM E24311) (fig. 7.11d; pl. 95d). It measures 2.2 cm in height. This type of earring has a long history, but the opening at the back is typical of earrings that date from at least the late New Kingdom through the first millennium BC.[60] A similar example comes from tomb 26 at Nebesheh. It was dated to around the Twenty-Sixth Dynasty.[61]

The function of the copper-based tube shown in figure 7.11e (pl. 95e) is uncertain, although it clearly resembles the perforated siphons or drinking tubes that are often found in burials and contexts from the New Kingdom and later (OIM E24420).[62] This example, however, has no remaining holes bored into it. It was found outside silo D 126 in the southwest corner of the West Sector.

A copper kohl applicator came from the fill within D 40, along the northern fort wall by the breach (OIM E24415) (fig. 7.11f; pl. 95f).

Another object, now represented by four fragments, was found in room D 70 next to the whetstones mentioned earlier (OIM E24390B; not illustrated). It is highly corroded, and therefore its original shape cannot be determined, but the field registration notes that it may be the remnant of a "blade." It is made of a copper-based metal and is 4.0 cm long and 1.3 cm wide.

BONE

A bone awl worked to a point at one end was found in the thick ash layers of D 123 (1), a passageway around the group of small bins at D 116 by the bedrock building in the West Sector (OIM E24426; not illustrated). The other fragment of bone is from a fish vertebra—of an unknown type of fish—that was found outside the north fortification wall between buttresses B and C (OIM E24397). It was resting in debris about 0.25 cm above the surface level of an oven with no associated structural remains. It is not illustrated.

The excavation reports mention a goat burial found near the surface of area D 1, before the walls of the room were found, but this was considered a fairly recent deposit and the bones were not kept.

POTMARKS

Numerous sherds and jars with pre- or postfiring potmarks were uncovered at Dorginarti and brought back to the Oriental Institute Museum. These objects are frequent in the collection, undoubtedly because of the interest in the signs incised on them; none of the marks is painted. There are numerous publications on the meaning of these marks, both on pottery vessels and on building materials, and many of the symbols found at Dorginarti are similar to incised and painted marks from a number of periods in Egypt.[63] The following summary is not a comprehensive study of all the sherds with potmarks, although many of them are illustrated here (figs. 7.12–7.16). The sherds are shown in the figures placed in line with the wheel striations.

Most of the marks, when possible to ascertain, were incised after firing. But the marks on the more complete jars discussed previously, and on handles, were usually applied before firing. Three of these marks, both pre- and postfiring examples, are dated to Level II and belong to an oasis-ware flask, an imported jar handle, and possibly a Samian-type amphora (fig. 7.16a–b, e). This set of potmarks is also discussed in chapter 6.

A bird is scratched into the shoulder area of a marl body sherd near the handle (fig. 7.12d); a "ship's sail" marking is incised on a number of sherds, including the silt and marl sherds in figure 7.13h–i; and three- to four-pronged symbols, either converging at the ends or spaced along a horizontal crossbar, were incised on a

59 Minault-Gout and Thill 2012, pp. 114–15, 332–33, pl. 129, T21Cc57. The original tomb dates to the Eighteenth Dynasty, with reuses in the Nineteenth and Twentieth Dynasties and later. Parallels to comparable items from sites in Egypt and Nubia are dated to either the New Kingdom or later and may be found on p. 333. See also Petrie 1888, pl. XXXVIII, no. 6 (Defenneh).

60 For example, the "leech earrings" of gold/copper alloy from a site in Jordan dated to the late Bronze Age and early Iron Age; see Ogden 1995, p. 73, no. 13, figs. 8.4 (second row, left), 8.25–8.26. These earrings were found along with other Egyptian-related objects at sites that were under Egyptian domination in the Twentieth Dynasty.

61 Petrie 1888, p. 21, pl. VIII, no. 18, from a "Cypriote" tomb.

62 Minault-Gout and Thill 2012, p. 395, pl. 175 top row. Most of these examples were from Sai tomb 20, with one from tomb 25. The parallels are most often New Kingdom in date. The authors refer to one from a Twentieth Dynasty tomb at Hillat el-Arab, published by Vincentelli 2006, pp. 38–39. Vincentelli also notes the two examples from Kurru 13, an ancestral tomb attributed to Generation C, published in Dunham 1950, pp. 51, 53, fig. 18d (19-3-268, from debris within the tomb enclosure). Examples are also found at Defenneh; see Petrie 1888, pl. XXXVIII, nos. 9, 10, 15–16, 19. However, the association of the Dorginarti object with a drinking siphon is very tentative.

63 Budka 2009, pp. 81–83; Gallorini 1998. See also the study by Ditze 2007.

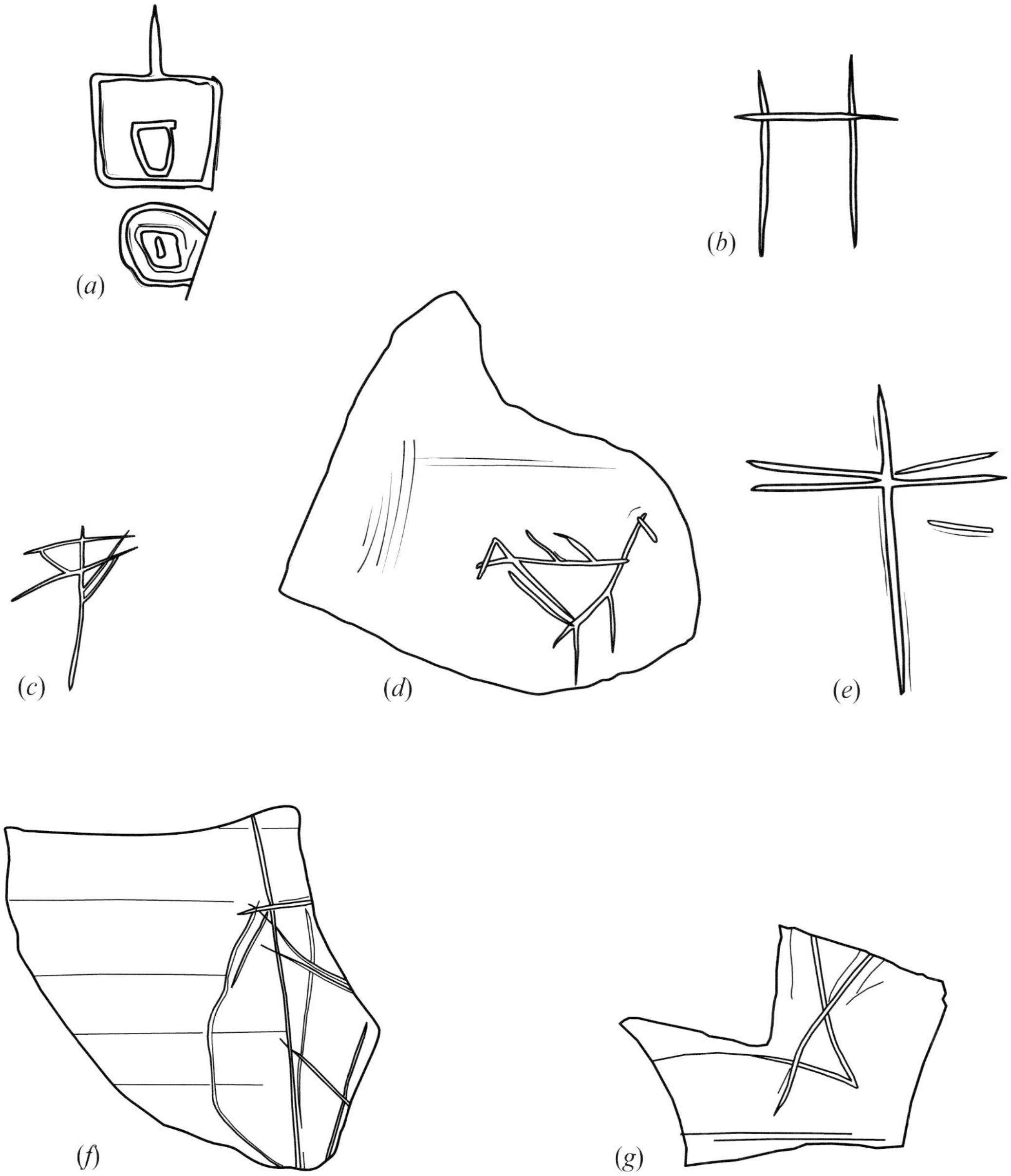

Figure 7.12. Potmarks (1:2)

(*a*) D 59; OIM E49257, marl jar (see fig. 5.36), prefiring

(*b*) D 59: OIM E49257, marl jar (see fig. 5.36), prefiring

(*c*) D 10 (1); silt jar (see fig. 5.31e), rim-and-neck sherd, prefiring

(*d*) D 208 (2); marl body sherd at shoulder, handle swipe marks(?), postfiring incised bird

(*e*) D 306 (2); potmark atop handle (see fig. 5.32b), prefiring(?)

(*f*) D 12 (1); marl body sherd, postfiring

(*g*) D 368 (1); marl body sherd, postfiring

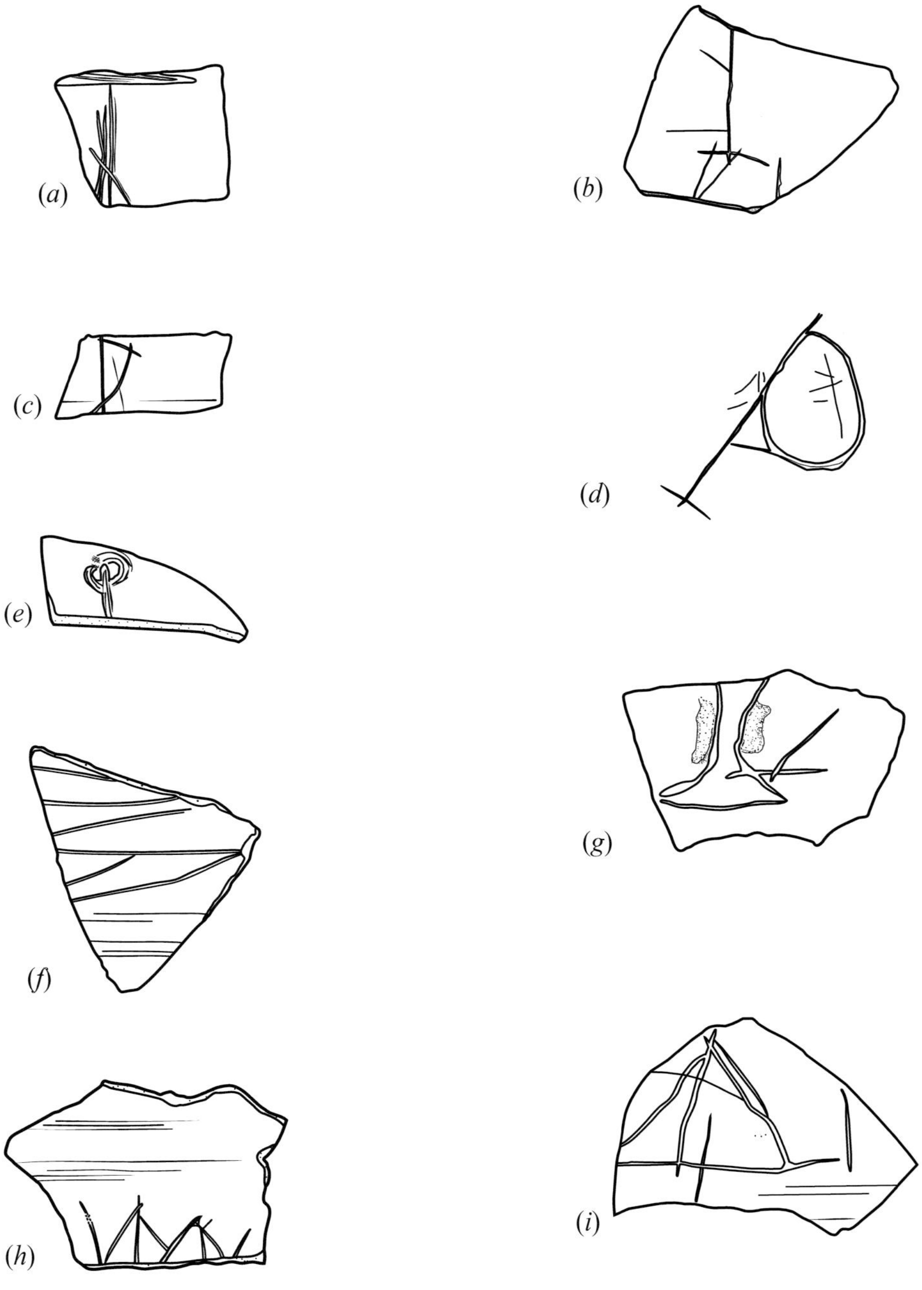

Figure 7.13. Potmarks (1:2)

(*a*) D 48 (2); marl body sherd, postfiring(?), incised line along top edge
(*b*) D 68; marl body, scratched postfiring
(*c*) D 75; marl body below shoulder, postfiring
(*d*) D 21 (1); marl body, scratches and sign postfiring
(*e*) D 13 (1); silt, prefiring
(*f*) D 38 and D 39, clearance; marl body sherd, postfiring
(*g*) D 221 (1); marl body sherd, prefiring
(*h*) D 10 (1); silt, shoulder sherd, pre- or postfiring, sails(?)
(*i*) D 75; marl shoulder, postfiring

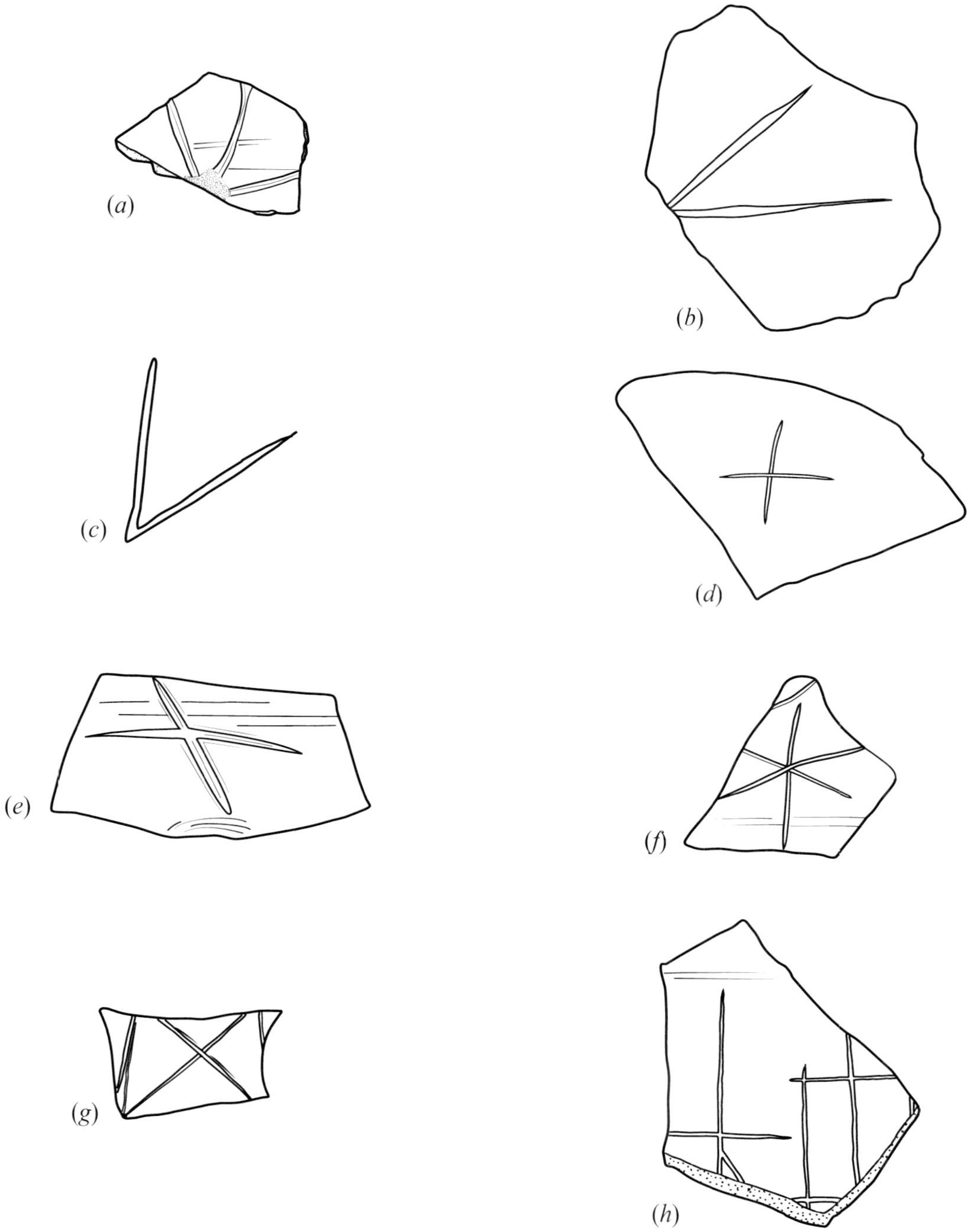

Figure 7.14. Potmarks (1:2)

(*a*) D 5 (1); marl body sherd, postfiring
(*b*) D 73 (1); silt body sherd, postfiring
(*c*) D 102; OIM E24377, marl jar (see fig. 5.37), potmark on shoulder
(*d*) D 118 (1); marl body sherd, below shoulder, postfiring(?)
(*e*) D 13 (1); marl, handle scar, prefiring
(*f*) D 35 and D 36, clearance; marl body sherd, postfiring
(*g*) D 73 (1); silt body sherd, cream slip, prefiring
(*h*) D 117 (1); marl body sherd, below shoulder, postfiring

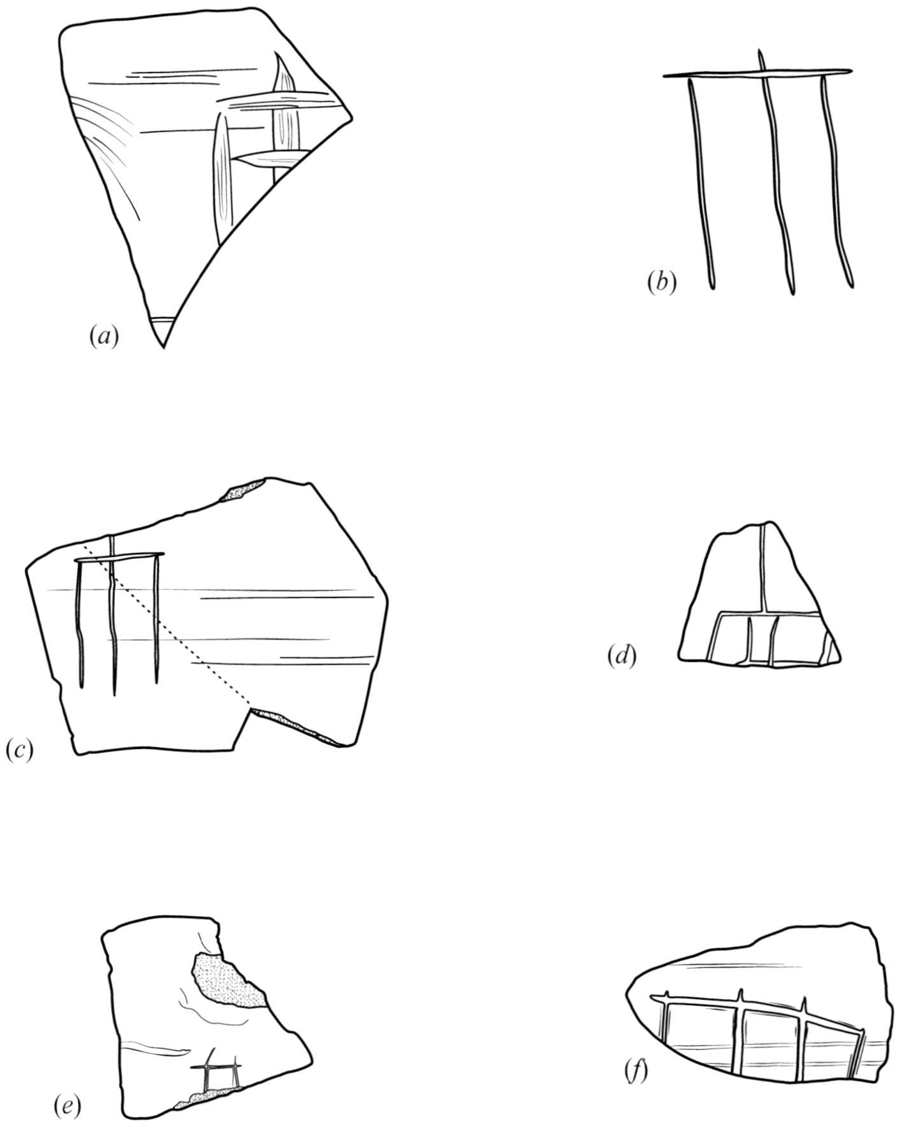

Figure 7.15. Potmarks (1:2)

(*a*) D 48 (2); Nile D, handle area
(*b*) D 221 (2), test pit below floor; marl body sherd, postfiring
(*c*) D 5, test pit through floor; OIM E24326, marl body sherd, prefiring
(*d*) D 35 and D 36, clearance; marl body sherd, pre- or postfiring
(*e*) D 59; marl, gouge to left intentional(?), handle scar, prefiring(?)
(*f*) D 68 (2); thick silt body sherd, prefiring(?)

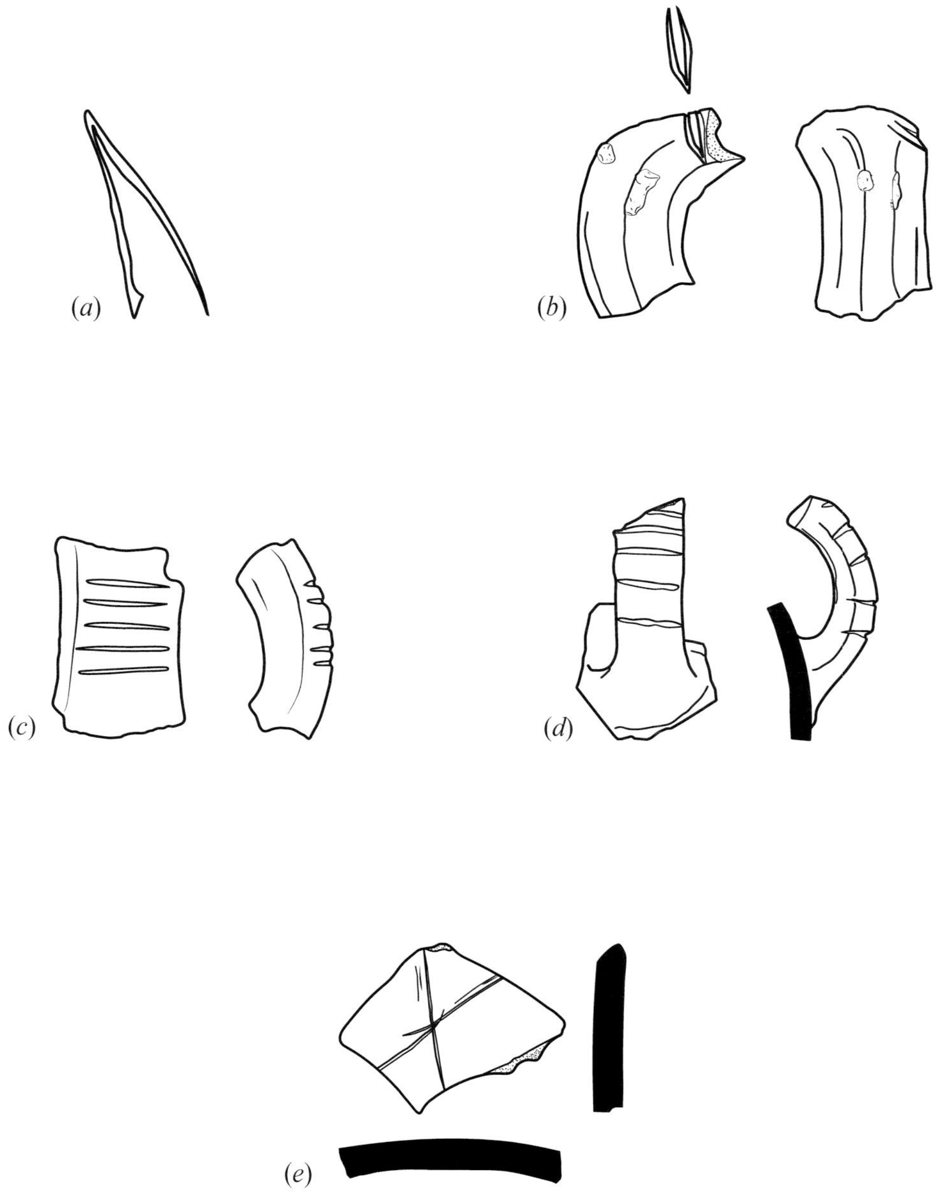

Figure 7.16. Potmarks (1:2)

(*a*) D 12 (1), Level II; OIM E36381, oasis-ware jar (see fig. 6.20a), shoulder, prefiring
(*b*) D 46, Level II; imported amphora (see fig. 6.9c), handle, prefiring(?)
(*c*) D 53, surface to "ash" layer; silt jar handle, incisions prefiring
(*d*) D 120; marl jar handle, cream exterior, incisions prefiring
(*e*) D 122 (1), Level II; resinated black interior, postfiring

number of other silt or marl sherds (figs. 7.14a, 7.15b–f). The prefiring(?) mark shown in figure 7.12c is identical to examples from Thebes and Elephantine, and has been described as a boat's sail and oar.[64] The postfiring marks include V shapes (figs. 7.14b–c), also found on the upper shoulder of an oasis-ware sherd from Level II (fig. 7.16a). Two to three crossed lines are found on both marl and silts (fig. 7.14d–f). The handles display prefiring, horizontal lines (fig. 7.16c–d). A red-slipped silt jar with ring base from a room in the Level IV residence has a prefiring(?) mark on its handle (fig. 7.12e).

The marl jar shown in figure 5.36 in chapter 5 displays multiple prefiring pot marks, including a group of signs that perhaps indicate a toponym below one handle (fig. 7.12a). The other prefiring marks on the jar are horizontal lines at the lower neck, with an upright V-shaped mark above these lines on the neck, and an H-symbol on the upper body between the handles (fig. 7.12b).

It should be noted that the marl jars with flaring necks (MJ 4) sometimes have a few haphazard notches incised perpendicularly atop the rim line, but these notches seem to have been designed to help tie on lids rather than to convey meaning. The marks on these jars were prefiring cuts and did not look particularly worn.

OSTRACA AND DIPINTO

The ostracon from D 1 (1), bearing an inscription in hieratic, comes from the fill below the surface clearance and within the walls of the building (pl. 96a). This ostracon, which records a delivery of grain to the fortress, is written in black and red ink on a Marl A4 body sherd (OIM E24308).[65] It was found in the fill within the walls of the room, but still near the surface clearance. In his preliminary reading of the inscription, George Hughes concluded that the paleography was that of the Nineteenth or Twentieth Dynasty. It reads:

1. arrived, the baker (*rtḥty*) to [. . .] 3

2. [. . .] 1 [. . .] 2 ¾ sacks [. . .]

3. [. . .] Deficit [. . .]

The shoulder region of a broken amphora from D 3 (1) has a few Demotic letters in black preserved on its shoulder (OIM E24309) (pl. 96b).[66] It was found within the top 50 cm of fill in the Level II platform. The amphora is made from Aswan clay and has a shape common in the first to second centuries AD (fig. 7.17).[67] The jar was from one of the Meroitic graves built into the platform. Aswan clay was used in the production of vessels during the Roman and Coptic periods in Egypt, and it was also used sparingly for pottery production earlier in the first millennium BC and in the Ptolemaic era.[68]

The smallest sherd with writing on it bears parts of two signs written in a Demotic script. It was found under the brick threshold of D 15 (1), a Christian building atop the Level II platform (OIM E24337) (pl. 96c).[69] This dipinto is found on the interior of a body sherd also made from Aswan clay and with a red slip.

METALLURGICAL EQUIPMENT

The remains from Dorginarti include about thirty pottery objects that indicate that metalworking activities took place during at least the Level III occupation of the fortress.[70] Rounded crucible fragments, some with

64 Note, for example, Budka 2009, p. 91, reg. 446n (Ptolemaic, although with similarities to an earlier painted sign on a bowl from Elephantine).

65 The ostracon is mentioned, but not described in detail, in Knudstad 1966, p. 186.

66 The ostracon came from the "uppermost fifty centimeters of the [surface] rubble"; see Pierce 1964b, p. 15. It was associated with pottery from one of the intrusive Meroitic burials dug into the Level II platform. The pink clay is coated on the exterior with a matte red slip, and the interior has traces of dark resin. I thank Brian Muhs for identifying the scripts on these sherds.

67 Adams 1986, pp. 533–36, fig. 300, form Z5 of Ware R30 (Aswan Greco-Roman Ordinary Red Ware). Note also the parallels in Rodziewicz 2005, especially p. 218, pl. 88, no. 1472.

68 That is, in the Third Intermediate Period and Ptolemaic era, with a possible extension back into the New Kingdom; see the discussion in Aston 1999, p. 7, nn. 70–71.

69 The Christian pottery from Level I has not yet been studied in detail. James Knudstad stated that the remains were late Christian in date; see Knudstad 1966, p. 179 (after William Y. Adams). The East Sector excavation also yielded the sherds of an amphora shoulder and handle fragment that resemble the form of an example from Thebes, shown in Mysliwiec 1987, pp. 163–64, no. 2016. See Adams 1986, p. 565, fig. 315, no. 3 (a silt amphora) for the shape.

70 Many of the fragments of pottery used for metallurgical activities were found scattered across the West Sector and in fill deposits of the Level II platform. Examples of loci with these metallurgical items, and the large stands, are D 1 (1); D 1 (2); in various deposits of fills within the Level II platform and atop the platform, such as D 24 (2); D 19 (1); D 21 (1); east of D 32; D 38 and D 39 (surface clearances); D 41; D 49 (level 2 of more than possibly four in northwest corner); D 59 (1); D 68 (2); D 73 (1); D 75 (1) (clearance of fill by Level II platform); D 79 (1); D 80 (1); D 81 (1); D 92 (1); D 93 (1); D 104 (1); D 111 (1); D 119 (1); D 208 (2); D 209 (1); D 215 (1); D 302 (1) (fill of the stairway platform of the Level III residence); D 303 (1) (structure north of the Level IV residence); and D 316 (1–2) (trench next to the Level IV

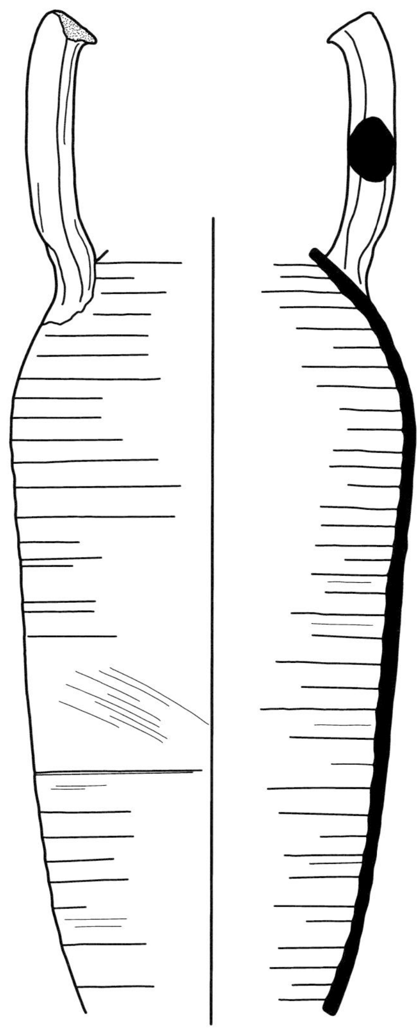

Figure 7.17. Aswan amphora with dipinto (1:5)

D 3 (1); OIM E24309, reconstructed, 2.5YR 5/6 RS exterior

thick deposits on their interiors, constitute the majority of the objects, but there are also slag and metal droplets, as well as the tuyères used in melting or smelting processes that occurred on-site (pl. 97). Large stands like those described in chapter 5 were found in the same loci with three of the crucible fragments mentioned here, perhaps functioning to support the pottery containers used in the remelting process or for the storage of materials used (see figs. 5.57–5.59). No certain evidence of pot bellows was identified, although such bellows may have been made of a perishable material.[71]

Cross sections prepared for reflected light microscopy and scanning electron microscopy with attached energy-dispersive spectrometry were initially used as the test samples. A subsequent analysis, using the noninvasive and nondestructive methods of pXRF analysis, was undertaken by Martina Renzi in 2015.[72] It is hoped the laboratory analysis will be published in a future article.

No installations for metalworking were uncovered at the fortress, and no molds were found in which to cast objects. Therefore, it is assumed that the metallurgical remains uncovered at Dorginarti were meant for the small-scale repair of equipment. No metal weaponry is evident, although the spatulas discussed above are sometimes identified as spear points in the literature. Numerous ovens throughout the fortress, as well as a couple of hearths, showed signs of multiple rebuildings, and some of these could have been used for metallurgical work such as remelting metals. Pit furnaces, if they were present, were probably located outside the fortress walls.

Evidence for metalworking has been found at a number of Saite-period sites in Egypt where garrisons were stationed, including Naukratis, Tell el-Maskhuta, and site T.21 or Migdol in the eastern Delta, as well as at sites in southern Israel, such as Meẓad Ḥashavyahu.[73]

Few metal artifacts were recovered, although they may have been carted off when the inhabitants left the fortress or were salvaged at a later time. The area with the biggest concentration of metallurgical items was in and around D 218 (1), which yielded halves of three different tuyères and the fragment of a large utility-vessel stand. No metal remains or crucibles came from the room, although a stone hearth was found against the room's west wall, and a crucible and large stand were found in the neighboring room, D 209.[74] Another tuyère fragment was uncovered in the clearance of disturbed deposits on the west side of the Level II platform, in D 75 (OIM E42948).

---

residence). Many fragments of the large stands were found atop the River Stairs, and some outside the fortress walls. Tuyères were found 10 cm below floor 1 in D 218 (1), and a crucible was found in the fill above the floor in the same locus, which may indicate Level IV or early Level III metallurgical activity.

71 This is suggested, for example, in the similar absence of ceramic pot bellows among the Iron Age metalworking remains at Megiddo; see Yahalom-Mack et al. 2017, pp. 62–64.

72 I would like to thank Drs. Thilo Rehren and Edgar Pusch for accepting the analysis of this material at the University College London facilities at UCL Qatar in 2012, as well as for allowing Martina Renzi to take this project forward in 2015–16.

73 See the references to Naukratis and Meẓad Ḥashavyahu, and the discussion of the materials at Migdol, in Oren 1984, p. 28. They are all, however, dated to the seventh century BC. For the brief discussion of furnaces at Meẓad Ḥashavyahu, see Fantalkin 2001, p. 29, fig. 12, and pp. 48, 127. For the materials from Tell el-Maskhuta, see Holladay 1982, p. 26, figs. 23a, 24a–b.

74 The crucible was found 10 cm below floor 1 in D 209 (2), and the pottery stand was found within the fill above floor 1, in D 209 (1).

Glass

Carol Meyer

Given the Level II date of the fortress at Dorginarti in the sixth century BC, a time when glass was still rare, it is not surprising that the excavations yielded very little of it. What there is, however, suggests the use of luxury vessels imported from the eastern Mediterranean for scented oils, perfumes, unguents, or cosmetics. None of the glass objects is assumed to stem from the earlier occupations at the fortress. The glass items appear to date from the intrusive Meroitic burials at the site, except perhaps for the marvered sherd, and their findspots suggest that this is a viable interpretation.

MARVERED SHERD

Locus D 3 (1) surface clearance revealed an opaque dark-blue sherd with yellow and white trailed decoration; pinprick bubbles; and thin, tough white weathering (OIM E50810) (pl. 98a). Judging from the curvature, the sherd came from a closed vessel such as a bottle or jar. Yellow and white threads were trailed over the blue vessel and dragged up and down into a herringbone design, or perhaps down only into an inverted festoon. Note the stripe of yellow on top of the dragged lines and the fact that the vessel was rolled on a slab, or "marvered," smooth.

This is a fragment of a core-formed vessel. The manufacturing process started with a solid core of mud, sand, or some other easily removed material packed around the end of a long rod. The body of the vessel was then formed by trailing hot glass around the core and subsequently rolling it on a slab to smooth it out. Dark blue was a common base color. Many kinds of core-formed vessels were decorated with fine threads of white, turquoise blue, yellow, and rarely other colors trailed on the hot vessel. The whole vessel was then marvered to smooth the threads into the surface. The threads were usually dragged up or down to make festoons or, more commonly, up and down to make a zigzag design. Other trailed threads above or below the dragged decoration were generally left as horizontal lines. When the vessel cooled enough, the rod was removed and the friable core was ground out. Handles and rims would have been applied at the end of the forming process, but are obviously lacking in the case of this sherd.

As to the date of this sherd, after a long gap in glass manufacturing following the New Kingdom, core-formed glass again became popular in the first five centuries BC. The Mediterranean Group I vessels of the late sixth to early fourth centuries BC are particularly relevant. These core-formed vessels occur in a limited number of shapes: slender alabastra, two-handled amphoriskoi with button bases, single-handled oinochoai with trefoil mouths, and round-bottomed two-handled aryballoi. All are small vessels used for perfumes, unguents, or cosmetics. Rhodes was one of the earliest (if not the earliest) production center for these vessels; certainly their Greek shapes and Aegean and Black Sea distribution argue for an Eastern Mediterranean origin.[75] Mediterranean Groups II and III were produced from the middle of the fourth century BC into the first few years AD, roughly the Ptolemaic period in Egypt.[76] The majority of these vessels were not completely marvered after the threads were dragged up or down; they tend to preserve slight furrows and noticeable bumps at the end of the row where the tip of the tool shoved the soft glass up or down.

From the Sudan, seven small fragments were excavated at Gala Abu Ahmed, an isolated desert fortress of the Napatan period (ca. 900–400 BC). They are core-formed dark-blue glass with trailed-on yellow and turquoise threads, some of which were dragged into zigzags.[77] Also found at Gala Abu Ahmed were a number of sherds of imported Greek vessels datable to the late sixth to early fourth centuries BC.[78]

A parallel in the Mediterranean Group I series is an amphoriskos in the Toledo Museum of Art, unexcavated, with similar white and yellow decoration on the shoulder.[79] Most of the many other amphoriskoi in the Toledo collection have turquoise-blue threads as well, but either the Dorginarti sherd had none or none are preserved. For parallels in the later Mediterranean Group II, see perhaps oinochoe no. 149 in the Toledo Museum of Art and hydriske no. 157, both unprovenanced, stylistically dated to the mid-fourth through early third centuries BC, and dark blue with opaque

75 Grose 1989, pp. 109–11.

76 Grose 1989, p. 131.

77 Fiedler and Jesse 2011, pp. 74–75. Friederike Jesse (personal communication, November 2, 2016) pointed out that "the glass fragments found at Gala Abu Ahmed might either come from Greek ateliers of Classical to Hellenistic times, or their Assyrian predecessors of the 8th to 6th century BC."

78 Fiedler and Jesse 2011, p. 77.

79 Grose 1989, pp. 146–47, no. 106.

white and yellow dragged-thread decoration.[80] There are some similar, excavated vessels from Dura Europos on the Euphrates, all dated stylistically to the second century BC, some of which may be ritual vessels.[81]

ARYBALLOS NECK

A sherd of translucent white glass, with many small bubbles and tough white weathering, was found in D 18 (1) (OIM E50808) (pl. 98d).

The sherd seems to be from an aryballos with a fairly wide neck and two handles. Aryballoi represented one of the few closed glass forms in the Hellenistic period, and they were a relatively popular kind of glass vessel. Certainly glass would have held oil without seepage better than most ceramic juglets. At first sight, the handle on this sherd looks as though it had been made in one piece with the body, which is very difficult to do, but it is actually smoothly applied. The weathering makes it seem even smoother.

BASE

A base fragment from D 8 is transparent, with blue-green tint, very few bubbles, and no weathering (OIM E24315A–C) (pl. 98b). It may be the base of a bowl with very thin walls. The drawing represents sherds A and C; sherd C is tiny, and B was not found.

Thin vessels with flat or nearly flat bases are not common in the repertoire of Hellenistic glass, but one parallel is a "clear dark blue glass" bowl from Meroe's North Cemetery, circa 116–99 BC (Beg. N. 21, 22-1-115b).[82] After the invention of glass blowing at the end of the first century BC and its rapid spread in the first century AD, however, thin vessels with flat or nearly flat bottoms were ubiquitous, and pale blue-green was a very common color.

ROD

This object came from the rubble west of silo D 32 (OIM E24365) (pl. 98c). Three Meroitic graves had been hollowed out atop the enclosure wall above this area, and had all been looted. One end of the rod is rounded, the other broken off. The rod is opaque black, even at the broken end, and has no bubbles; a few lighter specks give it some texture, and lengthwise striations may be seen at the fractures. The item is glassy but seems heavy for its size.

In general, rods of glass may provide raw material for manufacturing other items such as beads or thin canes, or used for thread in decoration, but black is unusual, and the ground end of this rod is intentional. It looks as though it had been deliberately prepared for use, perhaps as a dipstick to pull gummy unguents out of pots. Roman-period rods are not rare, but as a rule they are twisted. Most are bi- or multicolored, and most are thicker than the Dorginarti example. The use of the Roman-period rods is uncertain, but some were set into walls as architectural decorations, which does not seem to be the case here.[83]

Beads and Pendants

Joanna Then-Obłuska

This section treats the Napatan and Meroitic beads and pendants from Dorginarti (the Meroitic-period burials will be dealt with in a later publication).

In all, 461 beads and pendants that were found at Dorginarti are stored in the Oriental Institute Museum at the University of Chicago. They were found scattered throughout the fortress, and some of them were associated with the ten intrusive Meroitic burials atop the Level II platform, in the center of the East Sector, and atop the walls at the southwest corner of the enclosure.[84]

The beads and pendants are listed in table C in appendix III, where their archaeological context, shape, material, quantity, measurements, museum acquisition number, and additional data are noted.

Almost all the beads and pendants from the site were picked up as single items, although some Meroitic beads were found preserved on sections of their original string in burials at the site (pl. 99a). The beads and pendants from five of the Meroitic burials are treated separately, with many shown on plates 100–102.

80 Grose 1989, pp. 104, 162–63, no. 149, pp. 165–66, no. 157.

81 Clairmont 1963, p. 7, pl. 17, nos. 3, 6, pp. 147–48.

82 Dunham 1963, p. 84.

83 Grose 1989, p. 358.

84 The site contained ten graves or empty burials, three of them Meroitic, one possibly dated to this period, and the rest with dates that were assumed to be Meroitic. Some of these tombs are described and sketched in the field notes of Wenche Pierce in the Oriental Institute Museum archives. See Knudstad 1966, pp. 180–81.

Twenty-one beads were found on the neck of a child's skeleton in burial 3 (some shown on pl. 100b), and one near the feet of a child in burial 4 (pl. 101a); five beads were found in the pelvis region of one skeleton in burial 2 (pl. 100a), eight beads came from below the body in burial 6 (pl. 101b), and 190 beads came from below the body in burial 8 (pl. 102a).

The entire Dorginarti bead collection in the Oriental Institute Museum is presented here according to material and manufacturing technique, and also according to context and date.

MOLLUSK SHELL

Two Nile mollusk shells of *Bellamya* sp. (pl. 103g) and one of *Melanoides tuberculata* (pl. 102b) were found at Dorginarti. The shells lack traces of perforation. Thousands of perforated Nile mollusk shells, dominated by *Bellamya* sp. but including *Melanoides tuberculata*, have been recorded at Sedeinga necropolis.[85]

In addition, a perforated mollusk shell fragment of unidentified species was also found. It had been burned black (pl. 104l).

OSTRICH EGGSHELL

Two types of ostrich eggshell beads can be distinguished. One type is a regular disk cylinder with a double, parallel shape of perforation (pl. 104a) and measuring 6.2 mm in diameter. The other type, represented by three examples, is an irregularly cut short cylinder with a concave perforation (pl. 105a.1, 105e). The latter examples range in diameter from 5.8 to 7.0 mm.

Since ostrich eggshell beads rarely appear in Lower Nubian Meroitic contexts, those preserved from Dorginarti would seem to belong to adornments used during the life of the fortress in Levels III and IV and perhaps also Level II. Thousands of ostrich eggshell pieces, including unfinished and finished beads of various shapes, are already known from the surface layers inside the fortress of Gala Abu Ahmed and have been radiocarbon dated to between 1250 and 400 BC.[86] Napatan burials often contain strings of well-shaped disk beads.[87]

STONE

Stone beads and pendants were drilled from one end. They have a truncated conical shape of perforation. Both the sides and the ends—that is, the areas around the holes—are polished. The following materials and shapes can be observed among the stone beads. Standard bicones could be slightly faceted (pls. 105a.2, 106b), and they have been found in Napatan-period burials at Meroe (in Beg. W. 493).[88]

Large, spherical, translucent carnelian beads that are perfectly shaped are found in both Napatan (pl. 106g.1) and Meroitic contexts (pl. 100a.1).[89] A small, irregularly shaped, globular red agate bead (pl. 102a.8) is perfectly polished and is very common in Meroitic burials.[90]

Some faceting can be seen on the carnelian and diorite long cylinders (pl. 100a.2–3). The ends of the beads were simply cut off and left unpolished. These long cylinder beads have been found in Meroitic child burials at Sedeinga, Nag Gamus, and Sai, as well as in a grave at Berber.[91]

Stone teardrop pendants from Dorginarti exhibit the mark of a saw used in the drilling process. The saw mark that facilitated setting the drill in place is discernible adjacent to the larger opening of the truncated conical perforation. Three types of stone teardrop pendants can be distinguished: white chalcedony, red agate, and black quartz(?) pendants with globular bases (pl. 102a.1–2, 16); a white chalcedony long pendant with pointed base (pl. 105b); and a white chalcedony long pendant with rounded base (pl. 101b). In all cases, their thickness is almost comparable with their width. All above-mentioned pendant types are common finds

85 Then-Obłuska 2015a, pp. 29–30, fig. 5, d1/h and d1/i.

86 Lahitte 2013.

87 Griffith 1923, pl. XXVIII, nos. 65, 73, 74; Dunham 1963, p. 18 (23-M-710, from Beg. W. 503, Generations 3–4), p. 319 (23-3-568, from Beg. W. 774, Generations 2–6?). See also Vila 1980, p. 167, fig. 190, nos. 66–70, and Then-Obłuska 2014, cat. no. 127. Note also Lohwasser 2012, p. 110, fig. 47, ÄMP 2912.

88 Boston MFA 23-3-74a–e, published in Dunham 1963, p. 5, 354, fig. R, types Ia, IId, IIIh, and VIa, from Beg. W. 493 (ca. 743–664 BC).

89 For similar beads, see Boston MFA 24.1013, online at http://www.mfa.org/collections and published in Dunham 1950, p. 101, pl. LXX.A, f (19-3-1501 from Kurru 71, dated ca. 712–698 BC). See also Boston MFA 20.373, from Nuri, and published in Dunham 1955, p. 26, pl. CXV.A (18-2-268, 290a, from Nuri 59, dated ca. 664–653? BC). It is also online at http://www.mfa.org/collections.

90 Then-Obłuska 2016, fig. 2, nos. 1–4, and references therein.

91 Then-Obłuska 2015a, p. 33, fig. 4, T293 c1, and references therein.

from other Meroitic graves,[92] including the tomb of Amanishakheto.[93]

CLAY

While smaller beads and pendants made of clay have been recorded at Meroitic and Early Roman sites,[94] large elongated clay beads (pl. 103c) are very common at settlement sites in medieval Nubia, and their function is still being discussed.[95] Nevertheless, the dating of the Dorginarti clay object must remain an open question.

FAIENCE AND VITREOUS MATERIAL

A variety of beads and figurative pendants are made of faience, also called *glazed composition*.

An exceptionally tiny short cylinder, 2 mm in diameter, features a thick light-green glaze (pl. 99a.4). It was preserved together with other Meroitic beads on their original string fragment. Small, blue short cylinder beads were also found in Meroitic burial 8 (pl. 102a.10).

The most common beads at Dorginarti are disk cylinders. They range in diameter from 5.1 to 8.0 mm (pls. 103f.3, h; 99d–e; 104c, d.2, e, g, i, j; 106a, g.3; 105a.3, c, d, f). The biconical disk beads measure 4.2–9.7 mm in diameter and 2.0–3.9 mm in thickness (pls. 99b; 106c, g.2). A barrel bead is 6.6 mm in diameter (pl. 104n).

To another category belongs a blue long cylinder bead, from Meroitic burial 4 (pl. 101a), on which the glazing process did not cover the side of the bead resting on a surface and thus reveals the underlying core on that particular side.

A blue long cylinder bead was found on the eroded north wall of the fortress (pl. 99c). Long cylinder faience beads of diverse sizes have been observed elsewhere from both Napatan[96] and Meroitic burials.[97]

A triple-segment long tube with a white faience core is preserved together with a white disk cylinder, which was mentioned above (pl. 104d.1). Whether a glaze has worn off, or the cores were not glazed, remains uncertain. In contrast, smaller single- and multiple-segment tubes are characterized by a dark-blue glaze, especially discernible in any pitting or channels against the white cores (pl. 107a.1–4). They measure about 2 mm in diameter. The double-segment tube reveals a white core and preserves a thick blue glaze (pl. 104h). Another group of beads comprises larger single- and multiple-segment beads. They measure 2.1–2.5 mm in diameter (pl. 107a.5–7). They have coarse gray and blue matte bodies and are made of a vitreous material. Segmented faience beads can be observed from the Napatan cemetery at Missiminia.[98]

Among the figurative faience pendants are three Hathor heads. One of them was described by the excavators as blue and is one of two in the Sudan National Museum in Khartoum (Khartoum 14270, 14271). A single example from Dorginarti is in the collections at the Oriental Institute Museum (OIM E24394) and is one of two Hathor head pendants from D 6 (1) (fig. 7.1c; pl. 104b). Its faience core lacks any trace of glaze at the surface. It could be an unglazed object, or the glaze may simply have been lost. The face of the goddess Hathor is depicted in relief. She wears a two-tiered wig ending in two locks. At the hair parting may be three uraei. Her two ears are those of a cow, and she wears a crown. Under the chin, her long neck is marked by horizontal strokes, which could also be interpreted as a sistrum handle.[99] The back is flat, and a suspension hole runs horizontally through the headdress.

Similar in shape to this Dorginarti example is a gold plaque with a repoussé Hathor head inlaid with enamel and "backed on glazed composition," which is dated to the Meroitic period (ca. 300 BC or later).[100] Whether the Dorginarti example had once served as the backing for gold sheeting remains uncertain. In addition, a blue-glazed parallel can be observed in the Museum of Fine Arts in Boston.[101] Another similar specimen, although with a decorated back, has been found in a tomb at Hillat el-Arab.[102] What is more, the Hillat el-Arab goddess wears a crown formed by six uraei, each surmounted by a sun disk.

---

92 Woolley and Randall-MacIver 1910, pl. 40, 7759, G45; Dunham 1963, p. 356, fig. T, type Id, e, f; Baud, Sackho-Autissier, and Labbé-Toutée 2010, p. 183; Boston MFA 23.863, published in Dunham 1963, pp. 241, 356, fig. T (23-1-21, from Beg. W. 193, Generations 55–65[?]); Williams 1991, p. 236, fig. 49c, d, f; Then-Obłuska 2015a, p. 33, fig. 10, T196 d1; and Then-Obłuska 2016, fig. 2, nos. 7, 9, and references therein.

93 Wildung 1996, cat. nos. 365, 366.

94 Then-Obłuska 2015a, p. 35, fig. 5, and references therein.

95 Then-Obłuska 2013.

96 Vila 1980, p. 166, fig. 189, nos. 9–10, 26, 36–37, 43, 47.

97 From tombs at Qustul, OIM E22056 and E22159 (personal observation 2013–14).

98 Vila 1980, p. 166, fig. 189, nos. 6–8, 21–22, 51.

99 Vincentelli 2006, p. 161, from tomb ARA 13.

100 Andrews 1990, cat. no. 184f.

101 Boston MFA 72.2627, from the Robert Hay Collection, shown online at http://www.mfa.org/collections. The context is unknown, but the type is dated from the New Kingdom through the Hellenistic period (ca. 1150–30 BC).

102 Vincentelli 2006, pp. 161–62, 175, fig. 3.3, no. A.13:117, color pl. V.2.

A bearded god figure was found lying atop the eroded south wall of the Level III residence in D 9 (fig. 7.1a; pl. 103e). A whitish faience core preserves some traces of blue glaze in the deeper channels. The amulet is broken at two points—in the neck area, exactly at the height of the hole for suspension, and just below the kilt.[103] The kilt has details of corrugation. Traces of armlets and bracelets can be discerned at both sides of the body. What is more, around the neck area, on both sides of the protruding beard, a collar comprising tear-drop pendants has been incised. Similar collars can be observed on the amulet of a god in the Museum of Fine Arts.[104] On the other hand, the body position, beard, and adornment details (collar with teardrop pendants, bracelets, and armlets) are also discernible on a statue of Sebiumeker from Temple M 600 at Meroe City, dated to the first century BC.[105]

GLASS

Mandrel-Wound and Rod-Pierced Glass Beads and Pendants

Many monochrome glass beads were wound objects. The short and standard barrels dominate this category. Usually they are translucent and semitranslucent in diaphaneity. They are found in two general sizes. They measure from 7.0 to 9.1 mm in diameter and from 3.6 to 8.7 mm in length (pls. 103a–b, d, i; 104m; 106f). The smaller examples are 3.1–4.5 mm in diameter and 2.1–4.4 mm in length (pls. 104k, 105a.4–5). Although translucent and semitranslucent wound barrel beads are already present at Qustul in New Kingdom graves,[106] they are also recorded from Napatan sites.[107]

An exceptional opaque-yellow, wound-glass, long barrel bead measures 5.9 mm in diameter and 9.9 mm in length (pl. 106g.4).

A rod-formed and fluted-glass fragment produced a green melon bead (pl. 107c). The wound opaque-black body of a long barrel bead is decorated with a white spiral trail (pl. 106e). Similar beads can be observed from Meroe West Cemetery tomb Beg. W. 5.[108]

Many beads from Meroitic burials are made of faceted glass. They are both monochrome and polychrome. A long faceted hexagonal cylinder bead in cobalt blue (pl. 102a.21) has been recognized elsewhere at Meroitic sites also in other colors.[109]

Sixteen blue faceted-glass beads were found in burial 3, the tomb of a child (pl. 100b.2).[110] They measure from 7.6 to 8.4 mm in thickness and from 12.7 to 15.8 mm in length. These long, cornerless cuboid examples can be compared with Meroitic specimens from Karanog.[111] Smaller examples, made of blue and cobalt-blue glass, were found in burial 8 (pl. 102a.12, 19). Many similar, monochrome-glass cornerless cubes and cuboids have been found in Meroitic and post-Meroitic Nubia,[112] Roman Egypt, and the Levant.[113] Nevertheless, blue and cobalt-blue cornerless beads were recorded, among other beads, from a bead armlet of a 7.5-year-old individual in grave B 87 in Ballana that has been dated to early Phase IV (ca. third century AD).[114]

Some cornerless cuboids are made of blue glass with a central striped band in white and red (pl. 102a.18). They have not yet been paralleled from Nubia, at least to my knowledge, but they have been recorded from the Black Sea region, where they are probably dated to the second to third centuries AD.[115]

Five spherical, white-banded blue beads belong to another Meroitic type (pl. 100b.1).[116] They measure 7.4–9.1 mm in diameter and 6.7–7.8 mm in height. Similar specimens are known from Meroitic-period contexts at Karanog.[117] Other examples, which were found at Meroitic sites in Sedeinga, Sai, Qustul, and Ballana, and at early Roman Berenike, were additionally shaped into cornerless cubes.[118]

103 For spacers, figurative amulets with two suspension holes in the head and ankle area, see Griffith 1923, pl. XXII, nos. 2, 4.

104 Boston MFA 72.1928, from the Robert Hay Collection, shown online at http://www.mfa.org/collections. The context is not known, but the type is dated in the museum entry to the Third Intermediate Period or later. This example has an incised hieroglyphic text on the back.

105 Baud, Sackho-Autissier, and Labbé-Toutée 2010, fig. 61.

106 Personal observation of the Nubian Expedition assemblage at the Oriental Institute Museum (2013–14).

107 Vila 1980, p. 166, fig. 189, nos. 63–64.

108 Wildung 1996, cat. no. 394.

109 For example, Dunham 1963, p. 355, fig. S, type VIIId; Williams 1991, p. 234, fig. 47d, e.

110 Eight of the cuboid beads are kept in the Sudan National Museum in Khartoum (SNM 14272; personal observation, February 2015).

111 Woolley and Randall-MacIver 1910, pp. 28, 74, and pl. 40, 8010, G; 7837, G293; 7868, G527; 7766, G83.

112 Dunham 1963, p. 355, fig. S, type VIIIe, f; Williams 1991, p. 234, fig. 47c, f.

113 Then-Obłuska 2015b, fig. 5, no. 32; Arveiller-Dulong and Nenna 2011, cat. nos. 224:15, 230:4, 244–45, 246–47, and p. 176, n. 34; Brunton 1930, pl. XLV, no. 50, grave 521; see also Spaer 2001, p. 74, cat. nos. 48–49.

114 Williams 1991, pp. 133–34, and also OIM E22732 (personal observation, 2013–14).

115 Alekseeva 1978, pl. 27, no. 34, type 171.

116 Two of the beads are kept in the Sudan National Museum in Khartoum (SNM 14272; personal observation).

117 Woolley and Randall-MacIver 1910, pl. 40, 7765, G83.

118 Then-Obłuska 2015b, fig. 5, no. 25, and references therein.

Small glass teardrop pendants were found in burial 8, along with the stone examples mentioned above. While a blue pendant is flattened and has a lenticular-shaped base (pl. 102a.13), a white specimen is simply elongated (pl. 102a.3). They were most probably rod-pierced objects. The first type is known from Ballana,[119] Karanog,[120] and Sai.[121] The small, irregular white teardrop glass pendants have been recorded from Ballana grave B 112.[122]

Drawn and Segmented Glass and Metal-in-Glass Beads

Simple, monochrome single- and multiple-segment beads are the most common Meroitic and early Roman glass objects.[123] Molds probably used in segmenting long drawn glass tubes have been found in contexts at early Roman Alexandria.[124] Among monochrome single- and multiple-segment beads are translucent light-green ones (pl. 99a.6), green (pl. 102a.9), dark blue (pls. 99a.8, 102a.17), translucent purple (pl. 99a.9), opaque red (pl. 102a.7), and much larger translucent dark-blue beads (pl. 99a.12).

Among decorated drawn glass beads, there are red-over-colorless types (pl. 99a.5). The translucent colorless drawn tube was coated with a red layer and segmented into single or multiple beads. In some beads, there is a characteristic orange layer between the internal and external layers. The same types of bead were also found in Meroitic graves at Sai,[125] Sedeinga, Qustul, and Ballana[126] and were also noticed at the port site of Early Roman Berenike.[127]

Metal-in-glass beads constitute another decorated type among drawn and segmented beads. These are silver-in-glass (pls. 99a.11, 102a.4–5) and a few gold-in-glass (pl. 99a.10). The production of these tiny examples is attributed to a workshop in Alexandria,[128] with probable workshops at Elephantine and Meroe.[129] They are one of the characteristic finds in early Roman and Meroitic assemblages.[130]

A glass ear stud or earplug from burial 8 has a narrow shaft connecting two heads, one smaller than the other (pl. 102c). It was probably inserted into a hole in the earlobe or nose. Three kinds of ear studs have been distinguished, based on the shape of the larger head.[131] The larger head on the Dorginarti example is domical in shape and is similar to a glass specimen found in Meroitic Ballana tomb B 144-6.[132]

SUMMARY

Beads and pendants are associated with two main periods of Dorginarti occupation. While Meroitic bead types from the first to third centuries AD can be observed from intrusive burials or their vicinities, the remaining finds could not always be ascribed to a particular time period. Still, the majority of ostrich eggshell, disk cylinder, and segmented faience beads are here associated with Napatan or Third Intermediate Period contexts. What is more, many oblate and spindle-shaped beads made of monochrome wound glass are considered to be of Napatan origin. Although Hathor head amulets have parallels at other Napatan sites, the dating of the figurative pendant fragment remains uncertain.

Drawn glass and metal-in-glass beads are easily paralleled in Meroitic assemblages. Blue and white-banded globular beads have also been confirmed at other Meroitic sites. Cornerless cuboids appeared at Dorginarti in the form of monochrome beads and red/white-banded blue ones. The long faceted hexagonal cylinder, two types of small teardrop pendants, and the domical earplug complete the Meroitic glass adornment assemblage.

As for stone teardrop pendants, the trace of a saw in the form of a groove, which can be discerned next to the larger opening, would indicate the Meroitic period as a date of production. What is more, long cylinders with the truncated conical shape of perforation have also been found in other Meroitic graves. However, the biconical carnelian beads are Napatan in date.

119 OIM E22665 (personal observation, 2013–14).

120 Woolley and Randall-MacIver 1910, pl. 40, 7826A.

121 Then-Obłuska 2016, fig. 6, no. 12.

122 OIM E22529 (personal observation, 2013–14).

123 Then-Obłuska 2016, fig. 4, nos. 2–6.

124 Kucharczyk 2011, fig. 8, no. 1.

125 Then-Obłuska 2016, fig. 7, no. 2.

126 Personal observation (2013–14) of the Nubian Expedition assemblage at the Oriental Institute Museum of the following objects: OIM E20792A (ca. first century AD); OIM E22925B (second century AD); OIM E21508, a red-covered colorless long tube bead, not yet segmented, from a first-century AD grave at Qustul.

127 Zych 2011, fig. 12-69, cat. no. 72.

128 For example, see Spaer 2011, p. 131.

129 Gold-in-glass beads may have been produced at Elephantine in the early Roman period; see Rodziewicz 2005, pp. 34–35. Some doubts, however, have been expressed by Arveiller-Dulong and Nenna (2011, p. 175, n. 28). Similarly, a statement by Markowitz (2012) about the production of gold-in-glass beads at Meroe has not been supported with any details.

130 For example, Spaer 2011, pp. 130–39; Then-Obłuska 2015a, p. 38.

131 Williams 1991, pp. 110–11, table 23.

132 OIM E22677 (personal observation, 2013–14).

8
Conclusion

Egypt and Nubia were involved in a lucrative trade network throughout most of their history. The trade benefited whoever controlled a substantial portion of the primary or intermediary trade routes and could thereby directly exploit the resources so prized by rulers, their entourages, and other elites.[1] But control of Lower Nubia fluctuated between two powers, Egypt and Kush. When Egypt was decentralized and fragmented, or obsessed with guarding her northeastern frontier, the southerners controlled the trade moving along the river and land routes.

The history of Lower Nubia during the tenth and ninth centuries BC has not yet been comprehensively written due to incomplete and inconclusive textual and archaeological evidence that merely hints at circumstances. But partial histories of the period and its cultural milieu have been written using evidence that enhances that presented in this book.[2]

The earliest material evidence from Dorginarti provides important data for a reconstruction of post–New Kingdom activities in the region.[3] According to pottery parallels discussed in chapter 5, the range of the early assemblage may stretch back into the late Twenty-First Dynasty, but it can certainly be located within the early to middle Twenty-Second Dynasty (i.e., the mid-tenth to ninth centuries BC).[4] The later pottery is situated within the early eighth century BC. These periods coincided with the early Kushite or post–New Kingdom period in Lower and Upper Nubia, before the rise of the precursors of the Twenty-Fifth Dynasty in the late ninth or early eighth century BC.[5]

Some indications of the rising military and political strength of the Kushites, previous to their acknowledgment as the Twenty-Fifth Dynasty, are found in their later texts and in the slight archaeological evidence from Upper Egypt associated with Kashta.[6] A close alliance with Egyptians in Lower Nubia and

1 For the products flowing out of the south during the second millennium BC, see Zibelius-Chen 1988, pp. 69–135. For evidence during the Third Intermediate Period, see Zibelius-Chen 1989, pp. 339–40. See also Morkot 2016. African and Egyptian goods and exotic animals arrived in Assyria as gifts, booty, or tribute from Egypt; see, for example, Elat 1978. This begins during the reign of Ashur-bel-kala, in the late second millennium BC, and lasts until the Assyrians withdrew from Egypt and the Levant toward the end of the seventh century BC. For the Kushite horses given to the Assyrian rulers as gifts in the Twenty-Fifth Dynasty, see Heidorn 1997.

2 In addition to the works mentioned above, and many others not cited here, see Török 2009, pp. 285–309. Biometric methods are now used to study the remains of skeletons, such as those from the New Kingdom and Napatan burials at Tombos. These indicate that a multicultural group of people was buried at the site, with Nubian, southern, and Egyptian components recognizable in the population. See, for example, Schrader, Buzon, and Irish 2014, and Smith and Buzon 2014.

3 The pottery publications from Egypt provide the necessary evidence to prove the relative dating of the two earliest occupations of the fortress, as shown in chapter 5.

4 Jansen-Winkeln 2006. There is archaeological evidence for trade with Egypt at Dor and elsewhere in Israel at this time and slightly later (see chapter 5).

5 The term *early Kushite* is used by Pamela Rose to define the period from the end of the New Kingdom occupation in the eleventh century BC down to approximately the fifth century; see Rose 2019, p. 675. Florence Thill uses the term *post–New Kingdom* for the stretch of time from the end of the New Kingdom until the rise of the Twenty-Fifth Dynasty in the early eighth century BC; see Thill 2006–7, pp. 353–55.

6 A convenient compendium is found in Eide et al. 1994. While there is no inscriptional evidence for the presence of Alara in Egypt, Kashta is known by the stele he left at Elephantine, with a picture of himself as king of Upper and Lower Egypt (Cairo JE 41013). His presence in Egypt is also known from the Karnak priestly annals and a bronze aegis showing him being suckled by Mut (from an unknown site); for both items, see Ritner 2009, pp. 459–60.

Upper Egypt during the earliest periods of the first millennium, and the wealth gained from the continuing trade between south and north, and east and west, was undoubtedly the main driver for the formation of the Kushite state in later centuries.

It is likely that interaction with Theban priests, artists, and traveler-traders was ongoing in the post–New Kingdom period, and this gave impetus to the Kushites' adoption of Amun of Napata and Thebes as their source of power, a power permeating their influence in Nubia and the adjoining regions, and a power over their Egyptian compatriots.[7] Political accommodation and economic cooperation between Egypt and a number of stronger political leaders in Lower and Upper Nubia during the Twenty-Second Dynasty would make sense. No single group or state had the resources or influence to exclusively control the trade network in southern luxury goods, including gold, as well as agricultural and mineral resources and animal products. A similar political situation obtained during the Twenty-Second and Twenty-Third Dynasties in Egypt, when Lower and Upper Egypt were tenuously sharing Egypt's territory.[8] The rights over the southern trade goods could only favor the Egyptians due to the lack (or weakness) of a nascent Kushite federation and a strong Egyptian presence in Lower Nubia during the tenth and ninth centuries BC, when negotiations must have occurred frequently between the two states and their supporters. Trade in southern goods, however, occurred throughout this entire period, at least sporadically.

One of the Twenty-Second Dynasty kings may have established a base of power at the Second Cataract, building the initial Level IV fortification at Dorginarti, thereby encroaching again on Nubian territories. Whether their foray into the region was a coercive action or one welcomed by the southerners is unknown. But it is likely that the presence of Egyptians in Lower Nubia motivated the Nubians to exploit the highly fragmented political scene in Egypt during the early first millennium BC. The eventual success of the Kushite enterprise necessitated the exercise of authority and influence in areas beyond their native lands, using their own coercive control and the establishment of alliances in the regions along the Nile River and beyond.

An Egyptian fortress at the Second Cataract would eventually prove a hindrance to the functioning of the growing Kushite state, at least by the late ninth or early eighth century BC. In the scenario presented here, the struggle eventually led to the Egyptian abandonment of Lower Nubia, with the fortress at Dorginarti ceasing to exist during the Twenty-Fifth Dynasty. The repetition of cyclical power grabs in a region is typical for an important border and intermediary trade zone between two larger entities, as can be seen from the earlier history of Lower Nubia in the third and second millennia BC.

Lower Nubia and the Second Cataract region comprised a stretch of territory in which complete military control, or dynamic cooperation and compromise, existed between Egypt and Nubia during the early Kushite period. Clear evidence for such is lacking, so this is only one possible historical reconstruction.

TEXTUAL EVIDENCE

The troubling events described on a fragmentary stele of Queen Katimala at Semna, in a period before the rise of the Kushite dynasty at Napata, provide some insight into the loose control of the Second Cataract region and its consequences for the family of Katimala.[9] The text describes the criminal activities of a brigand who annually stirred up unrest in the region by robbing gold and silver from the Mountains of Gold and "sacrificing from the [cattle] herd of Amun." It is uncertain who Katimala was, whether an early member of the Twenty-Fifth Dynasty family or an unrelated group, but the text clearly indicates her tangible worship of the god Amun. The affiliation of the criminal, identified as Makaresh, and his conspirators is also not specified in the text. The linguistic identification of Katimala's name and that of Makaresh, and the dating of her stele and its historical context, have stimulated much discussion.[10]

7 Török 2009, pp. 285–309; Yellin 1995, pp. 243–48.

8 For the situation during the Twenty-Second and Twenty-Third Dynasties in Egypt, and the competing centers of control in both Upper and Lower Egypt, see Jansen-Winkeln 2006.

9 Comments on the dating of the text and its translation are found in Ritner 2009, pp. 456–59. Ritner notes that the artistic style of the tableau and the use of Late Egyptian preceded the archaizing classicism begun under Piye, as noted in unpublished lectures about the stele by Klaus Baer, who suggested an end date of around 850 BC. See also Caminos 1994. The text is retranslated and discussed by John Darnell, who dates the stele to the Twenty-First Dynasty (Darnell 2006, pp. 45–48). The same general date, stretching into the Twenty-Second Dynasty—as more hesitantly proposed by Darnell—is also offered in Collombert 2003–8, p. 209. For a response to Darnell, see Zibelius-Chen 2007.

10 See the relevant discussion in Eide et al. 1994, pp. 39–41, and Török 2009, pp. 294–98. Angelika Lohwasser has analyzed the iconographic and philological details and believes that the philology (Third Intermediate Period, Twenty-Second to Twenty-Fifth Dynasties) is later than the iconographic relief (perhaps Ramesside, but Twenty-First Dynasty at the latest) and that they were thus created in different periods; see Lohwasser 2018.

The stele is considered evidence for the continuing influence of Egyptian religion in Nubia and perhaps a close interaction between the priests of Amun at Thebes and the Kushites. Thebes is perhaps even "the city" mentioned in the text whose people condemn Makaresh for his sacrileges against Amun in Nubia.[11]

Events similar to those reflected in Katimala's stele may offer a reason for the establishment of the fortress at Dorginarti by either Nubians or Egyptians. Both Semna, which was situated at the strategic south end of the Second Cataract, and Dorginarti, located at the north end, were close to routes leading to the Mountains of Gold mentioned in the text, as well as to gold sources even closer to the Nile on the east.[12] Katimala and her family may have claimed control over the area, as intimated in her stele, but Egyptians and other tribal groups may also have been present in this area during the earlier first millennium BC, perhaps engaged in gold mining or trading expeditions. The existence of the fortress at Dorginarti, and a possible southern stronghold in the region of Semna, perhaps indicate that competing or allied groups were guarding the area's precious resources and the routes along the river and into the deserts.

Officials bearing the title *Viceroy of Kush and Overseer of the Southern Lands* are attested during the Twenty-First through Twenty-Third Dynasties, perhaps reflecting a situation akin to the use of the New Kingdom title, when these individuals had direct control over Lower Nubia and its gold mines.[13] If they did not directly control the Lower Nubian territories through which the trade moved at this period, their administrative duties would at least have entailed the collection of taxes on the products arriving into Egypt from the south. But they were most likely active participants in a continued engagement with Nubia and its products. The last known King's Son of Kush was Ankh-Osorkon, who held the office during the reign of Takeloth III.[14] Ankh-Osorkon may have died in about 757 BC, or around the time of Piye's accession. The titles associated with Egyptian control over northern Nubia disappeared with the arrival of the Kushites at Elephantine and Thebes, and most certainly by the reign of Piye, who conquered Egypt sometime in 734 or 733 BC.[15]

THE ESTABLISHMENT AND ABANDONMENT OF THE LEVEL III AND LEVEL IV FORTRESS

In both of its early first-millennium phases the fortress looks like an Egyptian establishment because of the prevalence of wheel-made Upper Egyptian marl and silt vessels, which suggests Egypt as the source for the fortress's provisions. If the Kushites established the fortress, or assumed control of it and maintained it intermittently for two centuries, then the pottery indicates that they had strong links with Upper Egypt in the early Kushite or post–New Kingdom period, and also during the Level III occupation of Dorginarti. This scenario also suggests that the Kushites had nominal or actual control over Lower Nubia, as well as the military prowess and resources to build and maintain large fortifications in the region, such as those at Qasr Ibrim and Dorginarti. The hypothesis is not excluded here, particularly considering a possible appropriation of the fortress during conflict.

But leaving aside the question of ownership, the events described on the stele of Katimala, discussed briefly above, show that the control of passage through the Second Cataract, and activities in the eastern desert's gold-bearing regions, was always precarious.

The evidence for strengthening the fortification walls with inner and outer relinings in Level III, the subsequent house renovations in the West Sector, and the building of a completely new official's residence in Level III could indicate a change of ownership at some point in time. But whether it was a change between Egyptian dynasts, or a change between Egyptians and Nubians, is unclear. The pottery forms do not provide us with a clear indication of the transformation and those responsible for it. The only burned stratum at the fortress was a limited fire in the southeast part of the Level III residence, after which its roof collapsed and the site was abandoned until the sixth century BC. The burning at the entrance to the Central Sector in the southeast corner of the West Sector, before it was

---

11 See, for instance, Zibelius-Chen 2007, pp. 383–86.

12 Expanding on the comments in Darnell 2006, p. 60 and n. 289, where the Mountains of Gold are identified as possible gold mines of the Kushite period—including the Meroitic phase—Darnell cites Klemm and Klemm 2013, pp. 15, 611. The distribution of possible Kushite-Meroitic gold-mining sites is shown on the map on p. 612, fig. 7.5, where they are identified by the reuse of apparent New Kingdom grinding stones. Some of the sites identified as earlier mines are located just west of Semna and Wadi Halfa, but the handmade pottery from these sites was not datable by the Klemms, and they do not mention any Egyptian imports. In another publication, they note that only handmade ceramics were found at these sites; see Klemm, Klemm, and Murr 2001, p. 656.

13 Kitchen 1986, pp. 566–67, 597–98, table *15; Zibelius-Chen 1989; and for a more recent survey of the office of King's Son of Kush, cited by Zibelius-Chen, see Lodomez 2005, pp. 82–84.

14 Lodomez 2005, p. 86; Zibelius-Chen 2007, p. 385, n. 27.

15 Kahn 2001, pp. 14–16.

reconstructed in a later Level III phase, also provides insufficient evidence to allow a historical reconstruction of the event.

Whatever the precise events surrounding the life cycle of Dorginarti, the fortress was home to a permanent garrison by its Level III phase, with houses that could accommodate families (a child's footprint was found in the floor plaster of the Level III residence). Although there is no evidence in the surviving architecture for barracks, and there is no indication for a naval installation in the nearby vicinity, the site served as a long-term base for guarding the river and its surrounding territory and was a center for the provisioning of administrative personnel, resident soldiers and their dependents, roving military patrols, and reinforcements. This is indicated by the large storage capacity for grain at the site.[16]

THE CONTEXT OF THE LEVEL III AND LEVEL IV FORTRESS

The lack of a substantial number of contemporary sites and large cemeteries in the region surrounding the fortress, which would suggest a thriving settlement associated with either its Level IV or Level III phase, is problematic.[17] There are archaeological remains in Lower Nubia that are dated to the Egyptian Third Intermediate Period[18] and early Kushite era, and also those identified with the Twenty-Fifth Dynasty, but thus far the published archaeological materials provide the barest hint of the history of the area. The sites are not those of large agglomerations, including those in the immediate environs of Dorginarti, consisting instead of burials,[19] building complexes of uncertain nature,[20] a quarry, a cooking site, and guard posts,[21] as well as the later temples of Taharqo at Semna West and Buhen.

The absence of a dedicated cemetery associated with the population during either of the early phases at Dorginarti is slightly perplexing.[22] Cemetery MF-NE at Mirgissa, cited by both Bruce Williams and André

16 The silos and their capacities are discussed in appendix II. The considerable grain storage facilities at the fortress may also indicate that Egyptian grain was used as one of the means of exchange for southern trade items and the area's resources. However, the contemporaneity of the complete corpus of silos is not easy to determine.

17 Bruce Williams compared the number of Twenty-Fifth Dynasty and Napatan tombs in Nubia to the number of burials from the Neolithic, A-Group, Middle Kingdom, Kerma, Pan Grave, and New Kingdom periods and concluded that "the extent to which Lower Nubia was abandoned in the late New Kingdom and Third Intermediate Period should be reassessed"; see Williams 1990, p. 44. The evidence for burials along the river, coupled with the lack of settlement remains, suggests a highly mobile population. For a list of sites that date to the general time period of the Napatan cemetery at Missiminia, see Vila 1980, pp. 176–78. The Napatan sites from the Batn el-Hagar are found in Edwards and Mills 2020, p. 396. The ceramics found during this survey have not been included in the present discussion. See also Osman and Edwards 2011, pp. 88–93.

18 The evidence includes a proposed Twenty-Second Dynasty stele from Wadi Halfa bearing only the images of Min and Khnum; see Porter and Moss 1962, p. 141. The stele is mentioned in Zibelius-Chen 1989, p. 341. There is no illustration of the stele, nor are there reasons for the determination of its date, but see Budge 1909, p. 217, no. 784 (BM 1045).

19 Williams 1990, pp. 31–41, and Nordström 2014, pp. 121–43, with the pottery on pls. 31–32, which show Egyptian vessels dating to the Third Intermediate Period, including one from burial 5-S18 at Mirgissa. Note also that "cog-wheel" beads, which are found in Third Intermediate Period and Late Period contexts in Egypt, are found at a grave (24-I-10, Faras West) and a work area (5-T-9) at Abusir, on Dorginarti's west bank; see Nordström 2014, pp. 129, 134. See also Geus 1975, p. 481, and, for examples, p. 488, fig. 10, p. 491, figs. 14–15, and p. 494, fig. 19.

20 For example, Site 6-G-9 at Gezira Dabarosa, which is mostly known for its Meroitic remains, appears to have earlier strata with remains dating to the pre-Twenty-Fifth or Twenty-Fifth Dynasty and Napatan period. See Adams 2004 and 2005, pp. 27, 90–91. The serrated faience disk bead from the site is mentioned in Adams 2005, p. 31 (no. 16522); see the previous footnote for the dating of this bead type. The wheel- and handmade pottery from the site is discussed in Lister 1967, pp. 61–64; it was dated to sometime before the early Meroitic period. See also Hewes 1964, pp. 176–79 (Napatan?). One radiocarbon date mentioned by Hewes (sample I-867) gave a calibrated date of around 800 BC; see Trautman and Willis 1966, p. 191. Radiocarbon dates for the early first millennium BC, however, are still unreliable since the Hallstatt Plateau flattens the readings between ca. 800 and 400 BC. For a good explanation of this radiocarbon "black hole," see Killick 2004, pp. 105–7, table 1, and fig. 1.

21 The guard posts at Abusir (5-T-11), and the quarry (5-O-6) and cooking area (5-O-13) around Abd el-Qadir, may be in part contemporary with the Level III or Level IV activity at Dorginarti; see Nordström 2014, pp. 121–23, 138–39. Although none of the pottery from the quarry or cooking sites is illustrated, the mention of a fragment of a faience sistrum pendant, with a Hathor head handle, from the cooking area is familiar at sites dating to the mid-first millennium BC and is not necessarily of New Kingdom date. The source of the stone used at Dorginarti was undoubtedly the sandstone quarry at site 5-O-6, as mentioned in previous chapters.

22 Although tentative in date, there are two cemeteries located northwest and south of the enclosure at Kor; see Smith 1966, pp. 223–24. Six of the brick graves to the northwest were excavated, with one of them containing a jar like the MJ 2 examples at Dorginarti; see Smith 1966, p. 238, fig. 19 bottom (N.W.1.1). There were about fifty graves in the south tumulus cemetery, but the dearth of distinctive pottery associated with these burials means they were not dated with certainty. Note the corridor "tomb" or water gate SJE 141 in Säve-Söderbergh and Troy 1991, pp. 323–24, and for the plan see p. 81, pls. 223–26. The pottery finds there were inconclusive, but perhaps New Kingdom.

Vila, contains child burials with faience beads and amulets dating to a period before, or during, the Twenty-Fifth Dynasty.[23] But it is assumed that this children's cemetery belonged to inhabitants of the site itself, as did the burials from this period at Gamai.[24] However, the remains from Mirgissa have not yet been completely published, and those from Semna are tantalizing but their significance is unclear.[25]

Although the region seems to have been less populated or active during the early first millennium BC than at other times, more remains may have existed that were never uncovered, lying below sand drifts; were incompletely excavated later structures that had earlier strata; or were simply misdated.[26]

As noted previously, Qasr Ibrim was active throughout much of the early first millennium.[27] Radiocarbon dates suggest that the early fortification wall at Ibrim may have been built within the time frame of the late New Kingdom, with the lower limit of these dates extending into the eighth century BC.[28] The later temple of Taharqo at Ibrim was surrounded by contemporary strata of late eighth- and early seventh-century BC pottery that was made of Upper Egyptian marl and local silt, and in shapes that were current in Thebes at the time. The excavation of the strata below Taharqo's temple and the set of fortifications built at the edge of the site in the early first millennium exposed jar sherds made from Upper Egyptian marl, with some examples parallel to types from Dorginarti.[29] Newer excavations at Qasr Ibrim indicate other structures that are dated by pottery to the Third Intermediate Period and perhaps the Late Period.[30]

The existence of military installations in Lower Nubia during the early first millennium BC is demonstrated by the evidence from Qasr Ibrim and the slight indications at the fortress at Gebel Sahaba, but there are also tantalizing hints of occupation at Buhen, Mirgissa, Semna West, and Gammai. The scattered sites around the Second Cataract, such as the quarry at Abd el-Qadir, the nearby cooking area, and some of the guard posts at Abu Sir, may also date to this period. Dorginarti was undoubtedly part of a wider network of social and political interaction, and contemporary evidence for either our Level III or Level IV materials at far-flung sites, including the Gala Abu Ahmed fortress in the Wadi Howar, around the northern Dongola reach, the Debba Bend, and Napata (el-Kurru), furnish a thread of proof.

THE ESTABLISHMENT AND ABANDONMENT OF THE LEVEL II FORTRESS

The Level II fortress at Dorginarti yielded Egyptian pottery along with Phoenician and East Greek amphora sherds dating to the sixth century BC.[31] The historical events surrounding its foundation are much more certain than those associated with the foundation of the earlier fortification, since there is textual evidence

23 Most of the burials were of young children and contained either no grave goods at all or only a few beads and amulets typical of the first millennium BC; see Geus 1975. See also Williams 1990, p. 36, and Vila 1980, p. 176.

24 Williams 1990, pp. 36–37, 48, nn. 55, 57–62.

25 There are unpublished Twenty-Fifth Dynasty remains from the site of Mirgissa, as noted in a personal communication from André Vila (December 22, 1988): "Mirgissa a livré des structures et du matériel datant de la 25e dynastie et plus tardif." The Hathor sanctuary at Mirgissa was also in use during this period, and a large vaulted tomb at Matuga Island (5-T-32) is attributed to this period as well; see Williams 1990, pp. 35–36, 39. The temple at Semna was renovated by Taharqo, but a number of tombs in cemetery S 500 were reused in earlier periods, as discussed in Williams 1990, pp. 36, 39, nn. 48–53. Other sherds in the fort at Semna West and in grave S716 belong to the early Kushite period, not the Napatan period or Twenty-Fifth Dynasty. These are discussed in chapter 5. A number of the sherds in the fortress that may belong to this earlier period also stem from loci below the temple of Taharqo, but they are mostly descriptive entries and not illustrated in Dunham and Janssen 1960.

26 The periodization used during the Aswan High Dam Campaign did not foresee the presence of early first-millennium BC remains in Lower Nubia, and thus the wheel-made Egyptian-type pottery was thought to be either (late) New Kingdom or Ptolemaic. Also, the handmade pottery from the early first millennium shows continuity with, although not a precise resemblance to, C-Group, Pan Grave, or Kerman pottery but was labeled according to the anticipated periods. Egyptian pottery dating to the early half of the first millennium BC was not yet well defined even by specialists working in Egypt, and the situation was compounded by the Aswan High Dam Campaign archaeologists, who were often experts on the material cultures in other countries and not familiar with Egyptian or Sudanese artifacts.

27 Rose 2008, p. 205; Rose 2019. Peter French mentions the marl pottery types found at Qasr Ibrim (French 1986, p. 166). The Taharqo temple contexts also included the channeled-rim marl jar that is associated with royal burials dating to the period around Taharqo's reign; see Heidorn 2018a. Earlier pottery resembling types from Dorginarti's Level III and Level IV are also found in strata below the Taharqo temple and around the gateway of the earliest fortifications (personal observation). See also Rose 2019.

28 Discussed above in chapter 3.

29 Discussed above in chapter 5.

30 Rose 2008, pp. 204–6; Rose 2019, pp. 675–96.

31 The Level II imports are discussed above in chapter 6. See also Heidorn 2018b.

demonstrating Saite involvement in Nubia.[32] The reestablishment of the stronghold was clearly undertaken as part of the Twenty-Sixth Dynasty campaigns and raids into the region, forays that are hinted at under Psamtik I[33] and Necho II[34] but are best attested by the campaign stelae of Psamtik II.[35] The evidence includes Greek, Carian, and Phoenician inscriptions left by Psamtik II's foreign troops at Abu Simbel,[36] and the Carian onomastical graffiti at Buhen and Gebel Sheikh Suleiman, which were left by mercenaries participating in the campaign.[37] The erasure of Kushite royal names on monuments in Egypt is attributed to Psamtik II,[38] and some scholars also believe that the damage and caching of royal statuary at Kerma, Sanam, and Gebel Barkal is associated with the arrival of the Saite army.[39] The best evidence for the southernmost extent of Psamtik's campaign, however, is the Carian graffiti and erasure of Taharqo's cartouches on the southern temple at Buhen.[40]

The site of Dorginarti was undoubtedly along one itinerary of Psamtik II's army, and a Greek graffito of one of the foreign mercenaries at Abu Simbel, located on the left leg of one of the Ramesses II colossi at the temple, indicates they went as far south as a place called *Kerkis*, where the river put a stop to their navigation.[41] The evidence for placing Kerkis in an area around Buhen and Dorginarti seems irrefutable, according to the graffiti at Buhen and the evidence from the Level II fortress at Dorginarti.

The fortress was undoubtedly still inhabited during the later part of the reign of Amasis II. A text from his forty-first year of rule (530/529 BC) mentions a military expedition or armed trade caravan setting forth southward from Elephantine carrying both military and ship personnel.[42] The latest ceramic evidence for the abandonment of the Level II fortification seems to cluster around a date in the third quarter of the sixth century BC, before the Persians conquered Egypt in 525 BC. However, if there was an actual campaign by Cambyses II into Nubia, as recounted in Herodotus (3.25), it may also have brought about the demise of the fortress sometime before 522 BC, the year Cambyses died.[43]

Transit through and around the Second Cataract area may then have declined when desert routes using

32 Theban scribes continued to date their texts by the Kushite king's regnal years until the eighth year of Tanwetamani (ca. 657 BC). By Psamtik I's ninth year of rule, however, the Saites were in control in Upper Egypt; see Kitchen 1986, pp. 403–6. For a brief analysis of this period, see also Török 2009, pp. 359–63.

33 Kitchen 1986, pp. 405–6; Sauneron and Yoyotte 1952, p. 201, nn. 3–4. Note also Burstein 1984. For a possible campaign against Kush under Psamtik I, with Judean soldiers present among the troops, see Kahn 2007.

34 A fragmentary inscription from Elephantine indicates that Necho stationed a naval fleet at the First Cataract or strengthened one already present there; see Junge 1987, pp. 66–67. Additional fragments of the text are published in Jansen-Winkeln 1989. The vessels of the fleet included transport ships or barges, freight and warships, fast boats or canoes, and other unidentified types. Since one of the last lines of the text mentions Nubian unrest, the ships may have been dispatched to patrol or pacify the border and frontier region to the south.

35 For the stelae of Psamtik II from Shellal, Karnak, and Tanis, see Der Manuelian 1994, pp. 333–71. Also note the discussion in Sauneron and Yoyotte 1952, pp. 159–87. The places visited by the troops of Psamtik II are also discussed in Zibelius 1972, pp. 61–62, and under her individual entries. The Nubian topographic names in the Sanam Historical Inscription, some of them still used in the early Saite period, are discussed in Pope 2014, pp. 59–145.

36 The Phoenician graffiti at Abu Simbel are discussed in Schmitz 2010. The author notes that the events may describe an expedition under Amasis, although the evidence strongly suggests it was the campaign of Psamtik II (Schmitz 2010, pp. 326–27). For the Greek names, see Eide et al. 1994, pp. 286–90. See also Sauneron and Yoyotte 1952, pp. 187–90.

37 The Carian graffiti on the South Temple at Buhen, and on the rock of Gebel Sheik Suleiman southwest of Buhen, are published in Masson 1978, pp. 50–54; Ray 1982, pp. 189–90; and note Vittmann 2001, p. 57. See also Sayce 1910, pp. 261–62, pl. XL. For the Doric graffito on the temple at Buhen, see Eide et al. 1994, pp. 289–90.

38 Yoyotte 1951.

39 Bonnet and Valbelle 2003, pp. 756–71; Bonnet 2011. For the statues at Gebel Barkal, see Haynes 2011, pp. 34–44. The broken statues at Dangeil, south of the Fifth Cataract, were more likely destroyed during the Meroitic period; see Anderson and Mohamed Ahmed 2009.

40 Eide et al. 1994, pp. 230–31, 286.

41 Eide et al. 1994, p. 288, citing Török, who notes that it seems to indicate the Second Cataract area. Other scholars, however, have placed it north of the Second Cataract or elsewhere, including a location south of Napata and even at the Fifth Cataract. For a discussion of some of these locales, see Sauneron and Yoyotte 1952, p. 169.

42 For Berlin Papyrus no. 13615, see Erichsen 1942. Other fragments of the papyrus found by Karl-Theodor Zauzich, and the most recent analysis of the text, are published in Zauzich 1992. The text lists various ethnicities and classes of men, including a large cohort from the Levant, who were accompanied by soldiers and oarsmen, and perhaps also foot soldiers.

43 For the tribute levied on the Ethiopians, or Nubians, by the Persians, and the issues surrounding an actual campaign by Cambyses into Nubia, see Eide et al. 1994, pp. 312–14. There is also an interesting discussion of Pliny the Elder's reference to the Lower Nubian place name *Forum Cambusis* in his *Natural History* (VI.35), captured during the course of Petronius's expedition to Ethiopia in 25 BC. The historical basis for the place name is uncertain, and its location is unknown.

camel caravans[44] and Red Sea shipping grew in importance for trade and travel.[45] In any case, the need for a military post on Dorginarti disappeared along with the changing political and economic circumstances.

As noted above, trade or diplomatic exchange between Egypt and Nubia continued after the Kushites had been pushed out of Egypt in the mid-seventh century BC. While the treasury at Sanam and its surrounding buildings were already active under the Twenty-Fifth Dynasty kings, as shown by the sealings of Shabaqo and Taharqo found in them, those of Senkamanisken and Anlamani were also found.[46] Recent excavations at Sanam have uncovered evidence for the acquisition and production of trade goods, including elephant tusks, beads of precious stones, rings, pendants of silver and gold, and the molds used to produce faience *wedjat*-eyes.[47] Imported Upper Egyptian marl jars and Phoenician amphorae dating to the seventh century BC were also found in these buildings.

A few East Greek and Levantine jars were found dispersed among the funerary goods in the royal tombs at el-Kurru and Nuri, extending in date from the reign of Senkamanisken in the later seventh century BC down to the reign of Amtalqa in the mid-sixth century BC. The presence of Upper Egyptian marl imports in the royal tombs, however, ceases after Senkamanisken.[48] The silt burial vessels may have been local copies, but the stone and other objects from these tombs indicate that social interaction and some exchange occurred.

There are also sporadic finds of Greek decorated vessels and sherds at other Nubian sites, including Kawa, Kerma, Meroe, and Gala Abu Ahmed, dating between the later sixth and the end of the fifth centuries BC.[49] However, the scarcity of Mediterranean and East Greek products in Nubia contrasts sharply with the abundant contemporary evidence for imports and foreigners in Egypt.[50]

44 The evidence for the introduction of camels into Nubia, perhaps indicating more extensive use of the western desert routes, is based on the evidence of a mandible and camel dung excavated at Qasr Ibrim; see Rowley-Conwy 1988. The dating of the mandible to 920–190 BC (95% confidence) and 810–390 BC (68% confidence) has a wide range, but the dating of the dung is more secure, i.e., 1040–770 BC (95% confidence) and 920–800 BC (68% confidence). For the current evidence of early first-millennium BC dating for camel domestication in other regions of the Near East, see Magee 2015. The author does not, however, mention the camel evidence from Qasr Ibrim.

45 Lemaire 1987; Bunnens 1985. See also Klotz 2015. The archaeological evidence for land routes leading to ports on the banks of the Red Sea is so far nonexistent for this period in the Sudan.

46 Vincentelli 2011.

47 Vincentelli 2011, p. 281.

48 East Greek and Levantine amphorae were found in the tomb debris of Chamber B and in the original stair filling of Nuri 21, perhaps the tomb of a queen of Anlamani (ca. 623–593 BC); see Dunham 1955, p. 66, fig. 42. Tomb Nuri 23, a queen of Senkamanisken, also contained what may be two hole-mouth jars of southern Levantine form; see Dunham 1955, p. 73, fig. 48 (ca. 643–623 BC). This vessel form begins in the eighth century BC but extends into the early sixth century. Sherds from Nuri 9, belonging to Amtalqa, appear to be the same Levantine form; see Dunham 1955, p. 123, fig. 91 (ca. 568–553 BC). Another possible East Greek amphora is from Nuri 25 (perhaps Generation 14, ca. 533–513 BC); see Dunham 1955, p. 160, fig. 120, 17-8-1822. However, I have not handled any of these sherds. For the cessation of Upper Egyptian marl vessels in the royal tombs, see Heidorn 2018a.

49 A compilation of the Greek pottery found in Nubia is in Fiedler and Jesse 2011, pp. 75–76.

50 See, for example, Vittmann 2003, particularly pp. 194–235. For the archaeological evidence, see, for example, Höckmann and Weber 2012, Villing and Schlotzhauer 2006, Leclère and Spencer 2014, and Gorre and Marangout 2015.

Appendixes

Appendix I

Stratigraphy and Pottery Types

The northwest corner of the West Sector had more than four levels of architectural remains, with the lowest stratum belonging to architectural fragments of Level IV and Level III residential units. The walls were sometimes founded at a very low level, and in these cases the Level IV walls were built up and reused in later renovation(s) to the area during Level III. For example, some of the earlier walls surrounding the rooms D 215–D 217 were still exposed in a later Level III phase, as discussed in chapter 4, part 1, although other walls took over their function to define the Level III building unit.

The most diagnostic sherds from this area, starting with the uppermost layers of D 4 and proceeding down to the early levels below the silo and over to D 1, are presented here. They are all drawn at the same scale relative to each other unless noted otherwise. Many of the sherds were not diagnostic and are not shown or mentioned.

UPPER LEVEL AT D 4 AND D 1

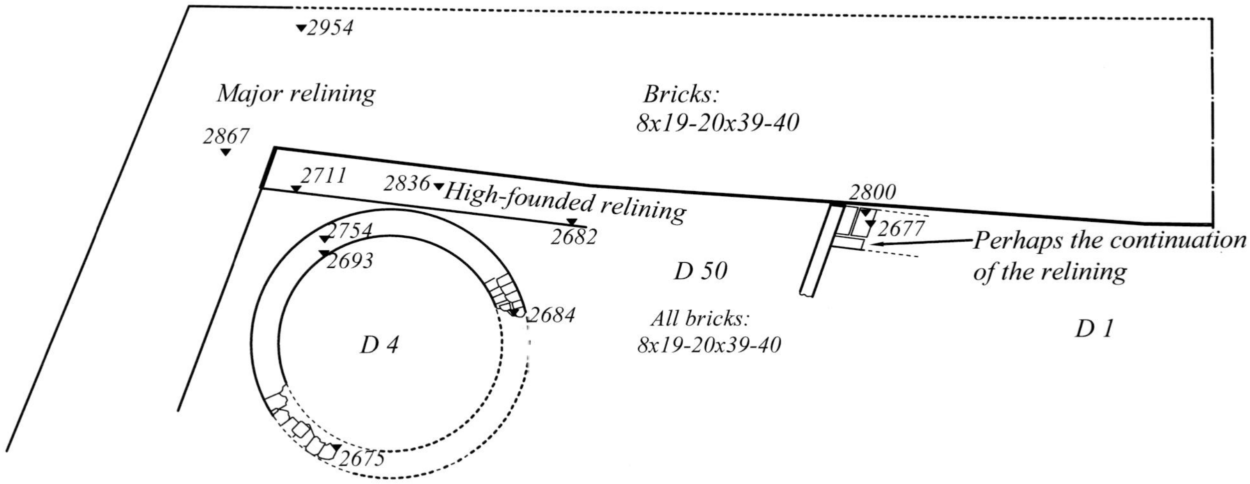

Figure A.I.1. Plan of upper level at D 4 and D 1

D 4 (1), join with sherd from D 47–D 50, below surface

- Marl jar body sherds with corrugated exteriors, and a small twisted loop handle
- Marl jar (type MJ 5), fig. A.I.2a
- Marl jar (type MJ 2), fig. A.I.2b
- Silt jar (type SJ 1) rim and neck, red slip traces exterior
- Silt bowl sherds (type SB 1), red rims and red slip interior/exterior
- Marl bowl ring base
- Nile D jar shoulder and neck fragments
- Handmade bowl, restricted mouth, burnishing traces and mat impressed

D 4, clearing fill

- Silt bowl sherds (types SB 1, SB 2)
- Low marl ring base, fig. A.I.2c
- Marl jar (type MJ 4), fig. A.I.2d
- Marl jar (type MJ 3), vertical stance, fig. A.I.2e
- Shallow silt bowl, fig. A.I.2f
- Marl bowl, rim folded to exterior, fig. A.I.2g
- Marl pilgrim flask sherds, some with black paint

D 4, sealed below

- Large silt bowl with string-impressed exterior, fig. A.I.2h
- Marl jar (type MJ 2), fig. A.I.2i
- Dorginarti pot stand, fig. A.I.2j
- Nile D or Nile B3 jar neck with cream slip exterior

D 4, removal

- Sherds of two marl pilgrim flasks with wide rims

Destruction of walls, D 4, D 47, and D 48 (context includes materials from the uppermost and intermediate levels in the area)

- Silt bowl sherds (types SB 1, SB 2, and shallow)
- Large silt bowl with string-impressed exterior
- Marl cup rim (as in fig. 5.16a–b)
- Dorginarti pot stand
- Dorginarti pot stand, reused after rim broken off, fig. A.I.2k
- Silt jar sherds (type SJ 3)
- Silt sherds and small loop handle
- Phoenician(?) amphora rim, fig. A.I.2l
- Marl jar (type MJ 1), fig. A.I.2m
- Marl jar loop handle, corrugated body, fig. A.I.2n
- Silt jar (type SJ 4), red slip exterior/red rim interior, fig. A.I.2o
- Nile D rim and neck, pink slip, fig. A.I.2p
- Handmade bowl sherds, coarse
- Handmade bowl with incised rim, fig. A.I.2q
- Silt bowl (type SB 2b) with traces of red rim or slip, fig. A.I.2r

D 48 (1), exterior floor around silo D 4

- Silt jar sherd (type SJ 1 or SJ 2)

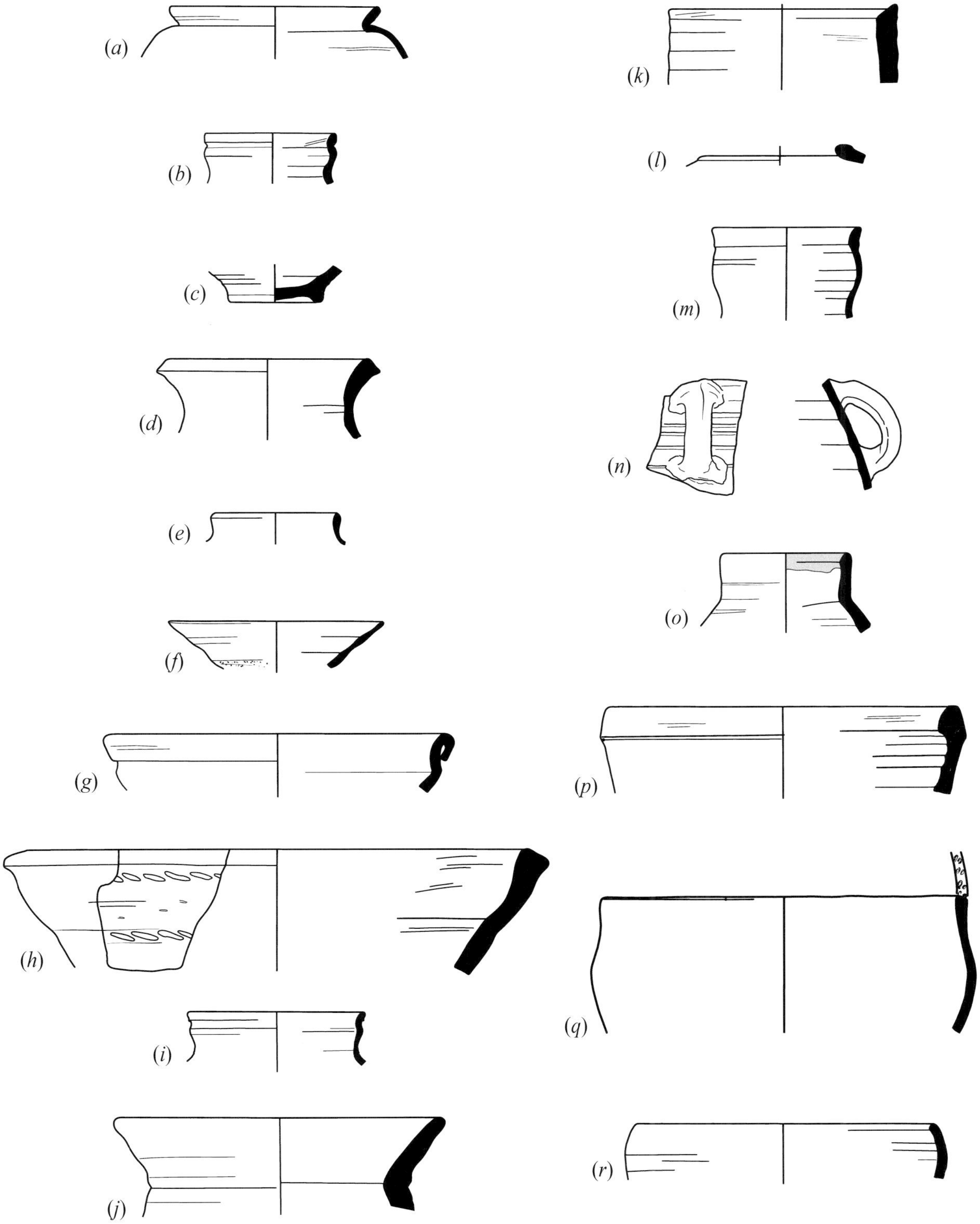

Figure A.I.2. Upper level diagnostic sherds

INTERMEDIATE LEVEL

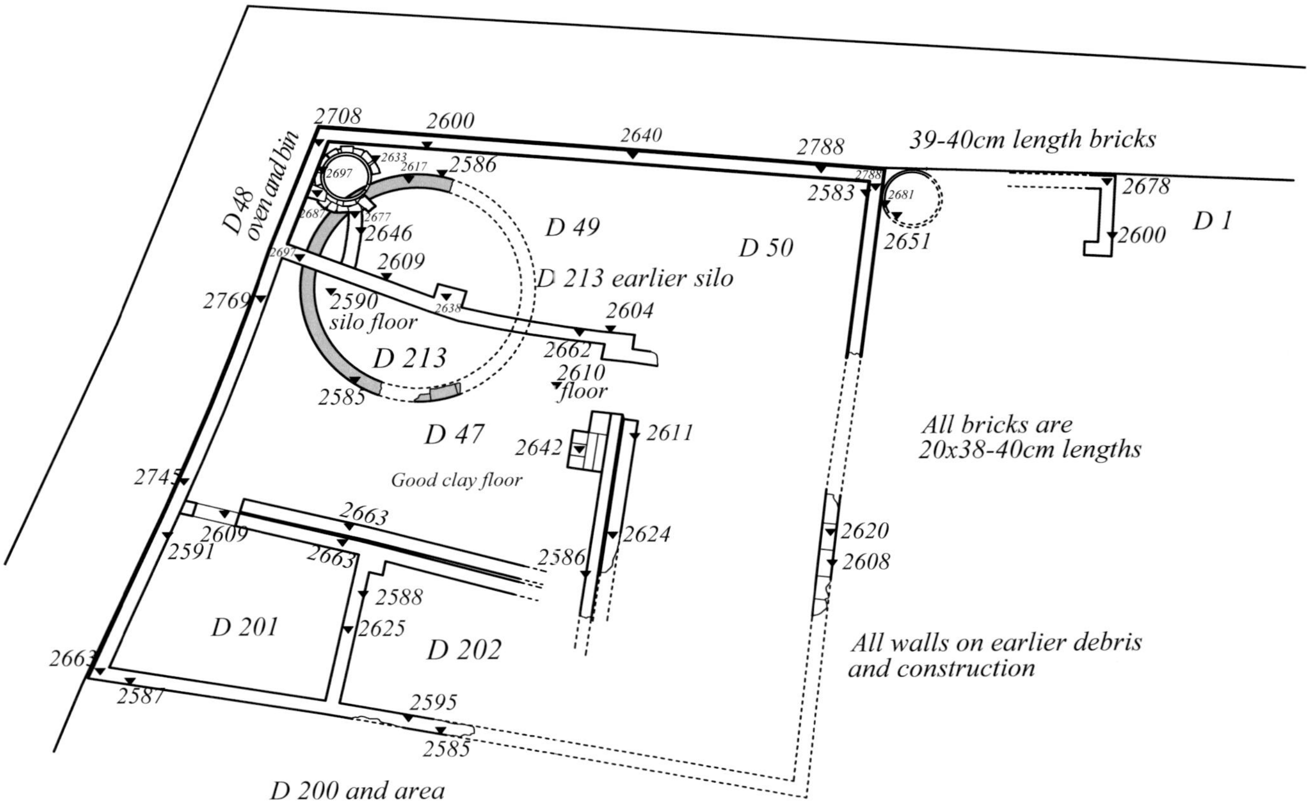

Figure A.I.3. Plan of intermediate level

D 47–D 50, below surface, testing for walls around D 4

- Marl jar rim sherds (MJ 2) with 10 cm diameters as in D 4 (1)
- Sherds of handled marl bowl, fig. A.I.4a
- Silt jar (form as type MJ 4), pink slip, fig. A.I.4b
- Marl wide-mouthed jar, thin walled, fig. A.I.4c
- Marl jar (type MJ 5), joins sherd from D 4 (1), fig. A.I.4d
- Marl jar with vertical neck (type MJ 3), fig. A.I.4e

D 47 (1)

- Dorginarti pot stand(?) base, fig. A.I.4f
- Marl jar (type MJ 1), fig. A.I.4g
- Marl bowl, fig. A.I.4h
- SJ 1 or SJ 2, red slip exterior, red rim interior, fig. A.I.4i
- Nile D or related silt jar variant, fig. A.I.4j

D 48 (2), fill in oven

- Handmade bowl, incised rim
- Dorginarti pot stand rim
- Marl loop handle and smooth marl body sherds
- Marl jar (type MJ 4), burned, fig. A.I.4k
- Marl(?) jar with slightly everted rim
- Large silt bowl with exterior string impressions
- Silt bowl sherds, including rounded base fragments
- Silt bowl sherds (type SB 2a), red rim exterior, red slip and streak-burnished interior, fig. A.I.4l
- Ledged silt jar, joins sherd from D 200–D 202, surface and (1), fig. A.I.4m
- Silt jar (type SJ 3), fig. A.I.4n
- Silt jar (types SJ 1, SJ 2), red slip exterior, fig. A.I.4o
- Marl jar (type MJ 3), fig. A.I.4p
- Wide-mouthed silt jar, fig. A.I.4q
- Marl bowl, fig. A.I.4r

D 48 (2), bin

- No diagnostic sherds

Under D 48, oven

- Silt bowl sherds (type SB 1a), some red rims and red slips present
- Large silt bowl, string-impressed exterior
- Handmade bowl sherds with blackened surfaces
- Dorginarti pot stand (as in D 4, sealed below)
- Marl jar base and loop-handle sherds
- Silt jar with thick cream slip on exterior
- Two different wide-mouthed silt jars
- Marl, silt, and Nile D jar sherds
- Marl handled(?) bowl, fig. A.I.4s
- Bread plate, fig. A.I.4t

Floor patches below oven

- Nile D jar, fig. A.I.4u
- Silt wide-mouthed jar, burned exterior rim, fig. A.I.4v

D 49 (east of D 48 oven and bin)

- Shallow silt bowl, fig. A.I.5a
- Crucible rim and base (OIM E42945, E42946)

D 200–D 202, surface and (1)

- Large silt bowl rim with string-impressed exterior
- Handmade black-topped bowl sherds, burnish interior/exterior, black interior, fig. A.I.5b
- Handmade coarse bowl, fig. A.I.5c
- Marl jar variant, as Nile D small jars (not MJ 1)
- Marl jar rim sherd (type MJ 2), fig. A.I.5d
- Marl jar neck and shoulder sherd, horizontal black bands and vertical lines in two groups, New Kingdom, fig. A.I.5e
- Ledged silt jar, joins one in D 48 (2) fill in oven, fig. A.I.5f
- Medium-sized silt ring base
- Silt jar (type SJ 4)
- Silt carinated bowl body sherd, fig. A.I.5g

D 200 (1) and D 201 (1)

- Silt jar sherds (types SJ 1, SJ 2)
- Silt jar sherds (type SJ 3), fig. A.I.5h
- Silt jar variant (type SJ 4), fig. A.I.5i
- Wide-mouthed silt jar sherds
- Pointed silt jar base and silt jar(?) ring base
- Silt bowl sherds (types SB 1a, SB 1b), red rimmed
- Small silt bowls (type SB 2), some red rimmed
- Marl handled bowl and bowl handle fragments
- Two large bowl sherds, string-impressed exteriors
- Handmade bowl sherds, black topped, burnishing
- Marl pilgrim flask sherd with black-painted rings
- Marl jar (type MJ 1), fig. A.I.5j
- Flat silt base, may be same vessel as in D 201 (1) and D 202 (1), though no join, fig. A.I.5q
- Slightly carinated marl jar shoulder and neck sherd
- Silt jar rounded base
- Silt ledged jar, fig. A.I.5k
- Dorginarti pot stand base rim
- Greenish marl jar sherds (type MJ 2)
- Nile D wide-mouthed jar sherds
- Two different wide-mouthed marl jars
- Large pithos with strap handle stump, fig. A.I.5l

D 200 (1) and D 202 (1)

- Handmade bowl with streak-burnished exterior, fig. A.I.5m
- Silt bowl (type SB 2a), fig. A.I.5n
- Wide-mouthed marl jar, fig. A.I.5o
- Marl flask sherd with black-painted bands
- Marl jar rim (type MJ 1)
- Silt jar sherds (type SJ 1)
- Carinated silt jar shoulder
- Silt jar with interior ledge
- Dorginarti pot stand base
- Flat silt base sherd as example in fig. A.I.5q

D 201 (1) and D 202 (1)

- Marl jar neck sherd
- Marl jar sherds (type MJ 2), one of greenish marl
- Wide-mouthed marl jar sherds
- Wide-mouthed Nile D, fig. A.I.5p
- Flat silt base with pierced hole in base, fig. A.I.5q
- Handmade globular bowl, streak-burnished exterior

D 201 (1)

- Dorginarti pot stand sherds
- Handmade bowl with groove below rim, fig. A.I.5r
- Silt jar(?) ring base, fig. A.I.5s

D 200 (2)

- Wide-mouthed silt jar sherds
- Marl handled bowl sherds, fig. A.I.5t

D 213 (1), intermediate level silo below D 47, D 50, and D 48

- Silt bowl sherds (type SB 1), some with red rim bands
- Marl handled bowl, slightly everted rim
- Wide-mouthed silt jar
- Marl jar loop handle

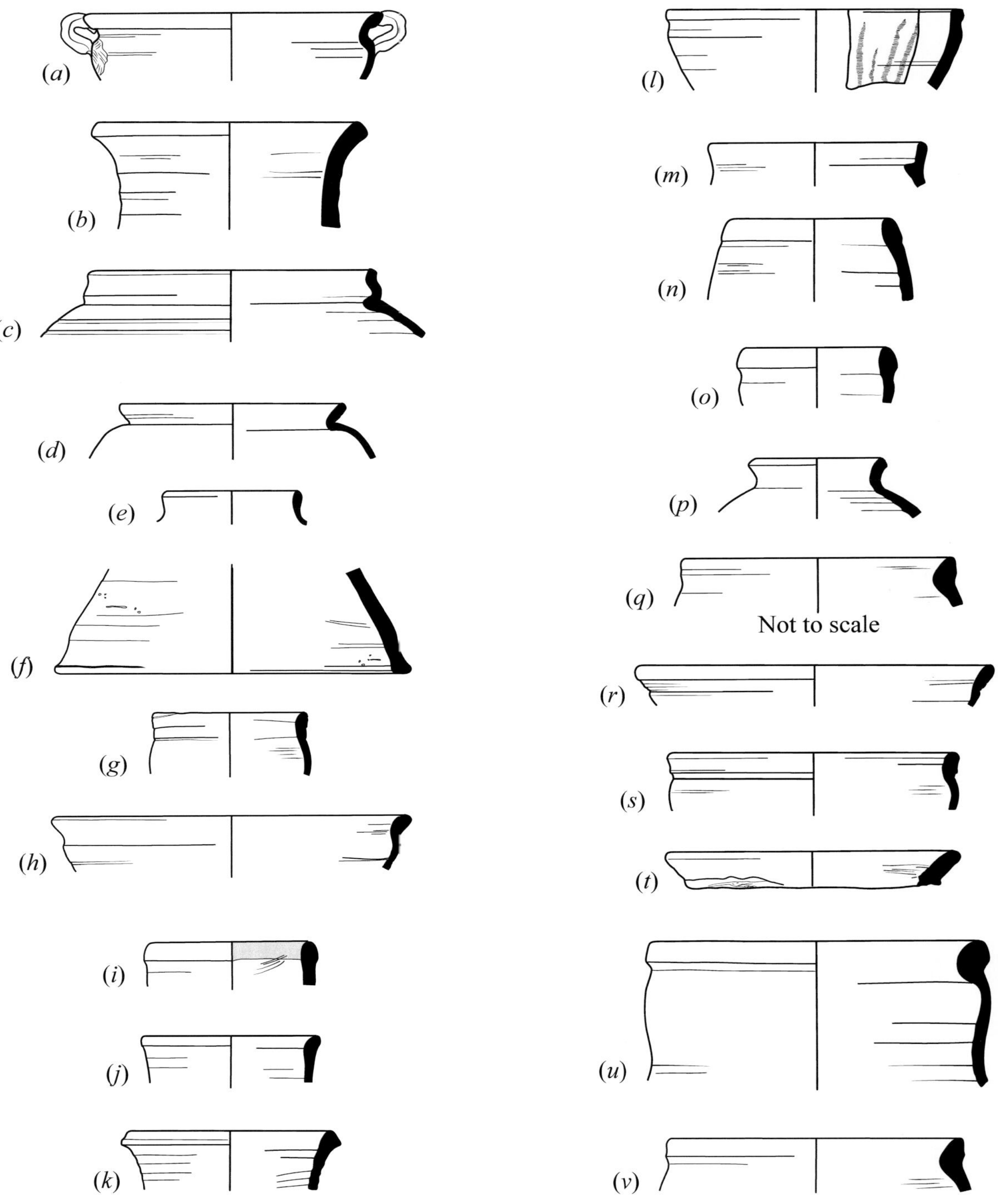

Figure A.I.4. Intermediate level diagnostic sherds

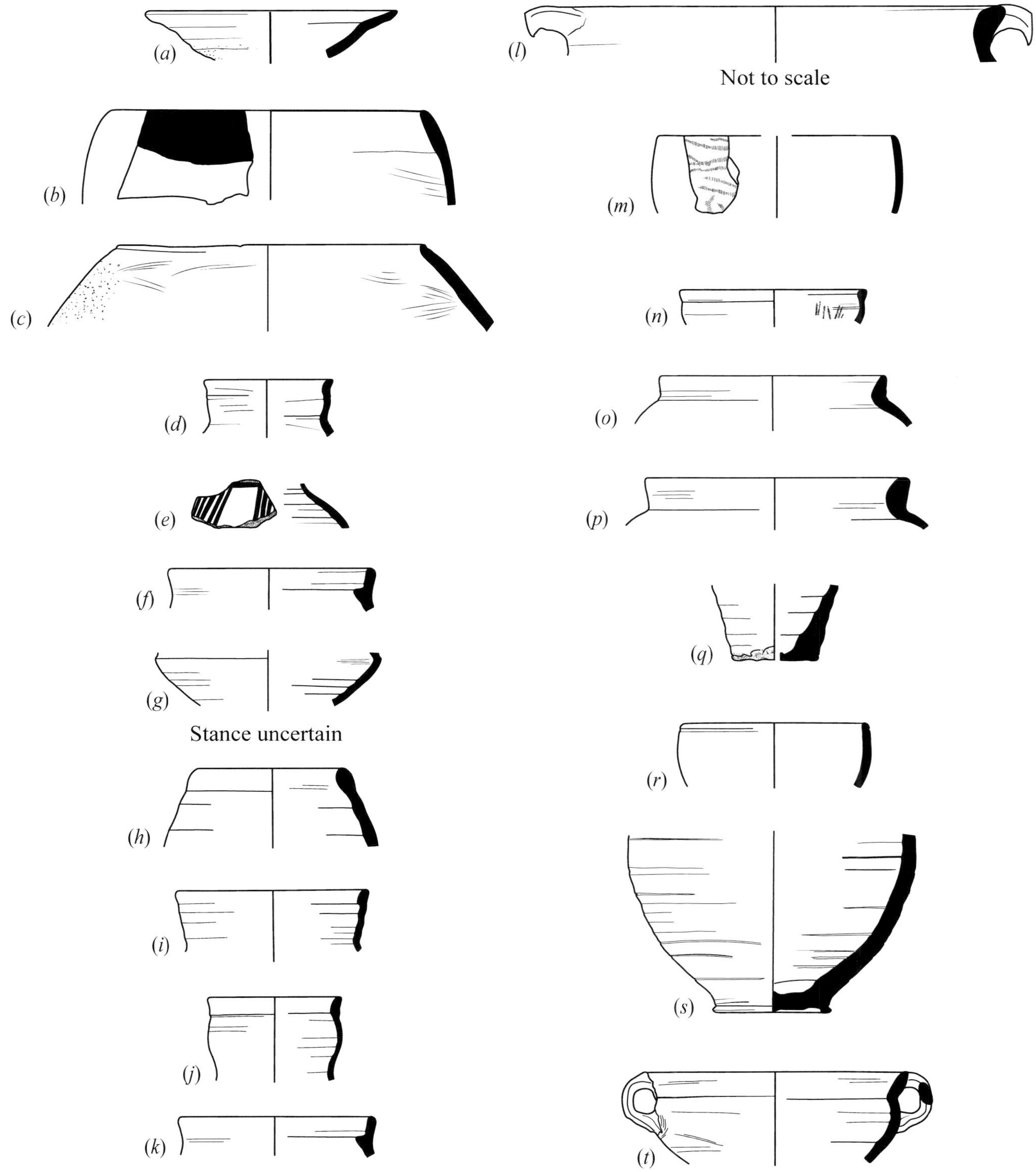

Figure A.I.5. Intermediate level diagnostic sherds

LOWER LEVEL

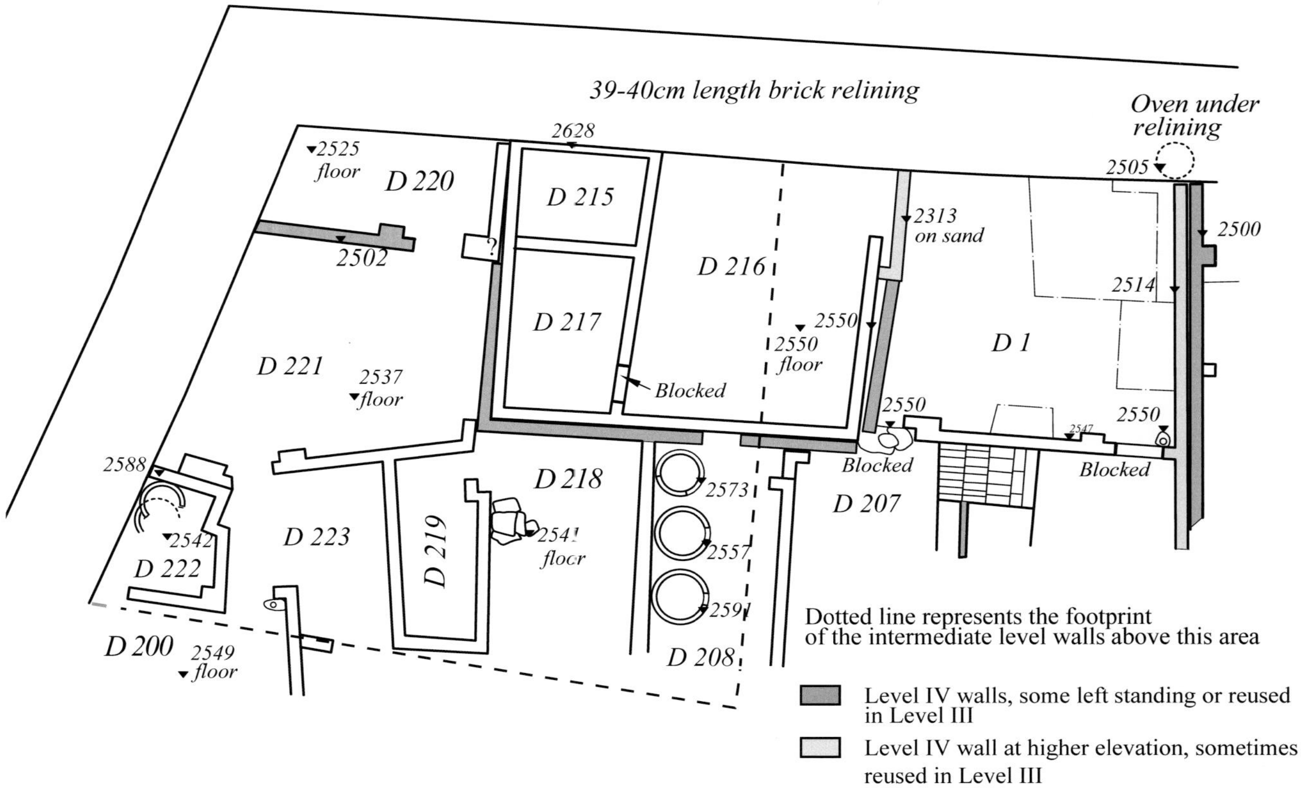

Figure A.I.6. Plan of lower level

D 208 (1)

- Silt bowl sherds (type SB 1a), some red rims or slips, and rounded, scraped base sherds
- Marl jar body sherds, smooth exteriors; loop-handle sherd
- Wide-mouthed marl jar
- Handmade bowl sherd
- Dorginarti pot stand sherds
- Large stand, fig. A.I.7a
- Silt jar variant, fig. A.I.7b

D 208 (2)

- Marl body sherd with potmark of a bird; see fig. 7.12d
- Nile D small jar, fig. A.I.7c
- Ledged silt jar
- Silt jar (type SJ 1 variant), traces of red slip exterior, fig. A.I.7d
- Silt jar (type SJ 4), traces of red slip, fig. A.I.7e
- Large stand with tilted rim, same as D 208 (1) sherd?
- Handmade bowl sherds, fig. A.I.7f–g
- Silt bowl sherds and globular bowl variant, fig. A.I.7h
- Large platter/bowl with string-impressed exterior
- Marl handled(?) bowl and ring-base sherds, fig. A.I.8a
- Marl jar (type MJ 4), fig. A.I.8b
- Marl jar body sherd with incised potmark (no illustration)

D 215 (1)

- Flat silt base sherd, hole in base; same as example shown in fig. A.I.5q
- Silt bowl sherds (types SB 1a, SB 1b, SB 2), red rims, streak burnishing
- Marl handled bowl with complete profile, fig. A.I.8c
- Marl jar sherds (types MJ 1, MJ 2), one shown in fig. A.I.8d
- Flat marl jar base sherds
- Silt jar (type MJ 2 variant); see fig. A.I.8m from D 221 (1)
- Marl wide-mouthed jar sherds
- Silt jar (type SJ 1) and silt base with pointed bottom
- Silt bowl variant
- Dorginarti pot stand rim
- Wide-mouthed silt jar
- Handmade black-topped sherds, burnished red exterior, fig. A.I.8e
- Plain handmade sherds, one burnished interior/exterior, most blackened
- Large stand with tilted rim, as example from D 208 (1) and D 208 (2); see fig. A.I.7a
- String-cut silt base, fig. A.I.8f

D 216 (1)

- Silt jug sherd, one handle preserved, fig. A.I.8g
- Handmade bowl, burned, fig. A.I.8h
- Silt bowl (type SB 2b) (as in fig. 5.6a–b), traces of red slip and interior vertical burnishing

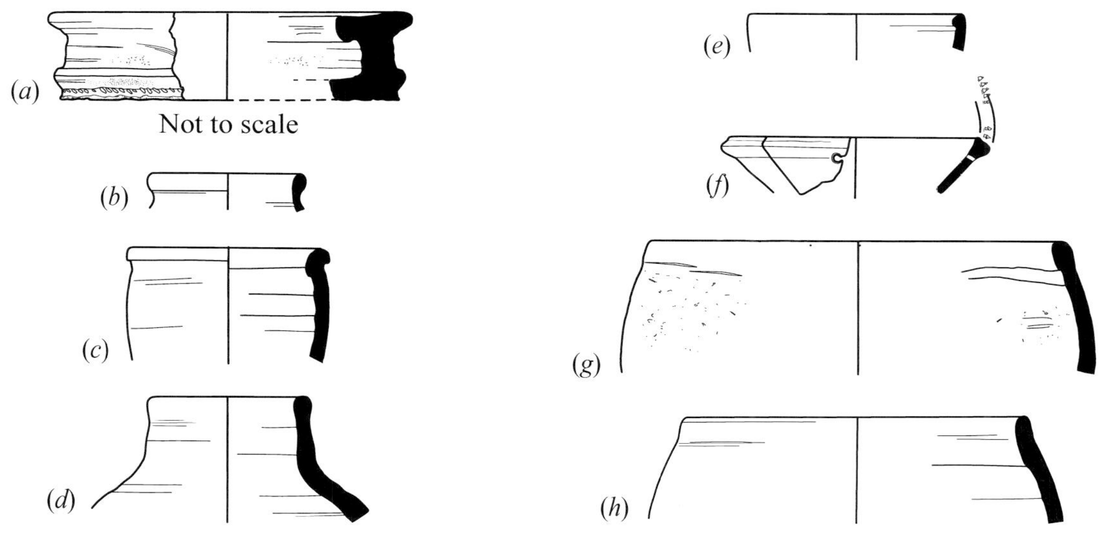

Figure A.I.7. Lower level diagnostic sherds

D 217 (1)

- Marl bowl, fig. A.I.8i
- Silt jar base, slightly pointed
- Handmade black-topped bowl with incurved rim
- Silt/marl mix(?) bowl with handle
- Marl jar sherds (type MJ 1)
- Nile D neck sherd with white slip exterior
- Silt ring base (jar or large bowl?)
- Silt bowl sherds (type SB 1a), some red rimmed
- Silt jar body sherds, streak burnished, string impressed
- Silt wide-mouthed jar
- Marl handled bowl sherds, handle fragments

D 218 (1)

- Ledged silt jar
- Large stand with median ridge and tilted rim
- Marl jar rim and neck (type MJ 1)
- Marl pilgrim flask, wide-mouthed rim sherd
- Handmade bowl with restricted mouth
- Handmade conical bowls with traces of streak burnishing
- Small marl jar neck with repair hole
- Silt bowl (types SB 2a, SB 2b), one with flattened rim, like example from D 221 (1)
- Large, wide-mouthed jar or vat rim with strap handle, red slip on handle

D 218 (2)

- Silt bowl (type SB 2b)

D 219 (1)

- Silt bowl (type SB 1a) with red rim
- Fragment of silt jug, fig. A.I.8j
- Rounded silt bowl bases with scraped finish

D 221 (1)

- Lots of silt bowl sherds (types SB 1a, SB 2b, SB 2a) and rounded base fragments
- Handmade bowl sherds, traces of burnishing
- Silt bowl (type SB 2b), flattened rim as in D 218 (1)
- Marl bowl sherds, as example from D 217 (1), no joins; see fig. A.I.8i
- Silt wide-mouthed jar sherds
- Handmade bowl sherds with incised rims
- Handmade rim and base sherds, burnished, blackened, fig. A.I.8k
- Marl body sherd with potmark; see fig. 7.13g
- Import body sherd with postfiring pot mark and resinated interior; see fig. 7.16e (Level II?)
- Large marl flask body sherds with black bands
- Marl jar rim and neck (type MJ 1)
- Marl wide-mouthed jar, small, fig. A.I.8l
- Silt jar sherd (type SJ 1 or SJ 2) and handle fragments
- Silt jar as example from D 215 (1), no joins, fig. A.I.8m
- Large silt jar/vat with one strap handle preserved

D 221 (2), test pit below floor

- Silt jar base with pointed tip
- Marl jar body sherd with potmark; see fig. 7.15b

D 220, D 222, and D 223

- Nothing found in the collection

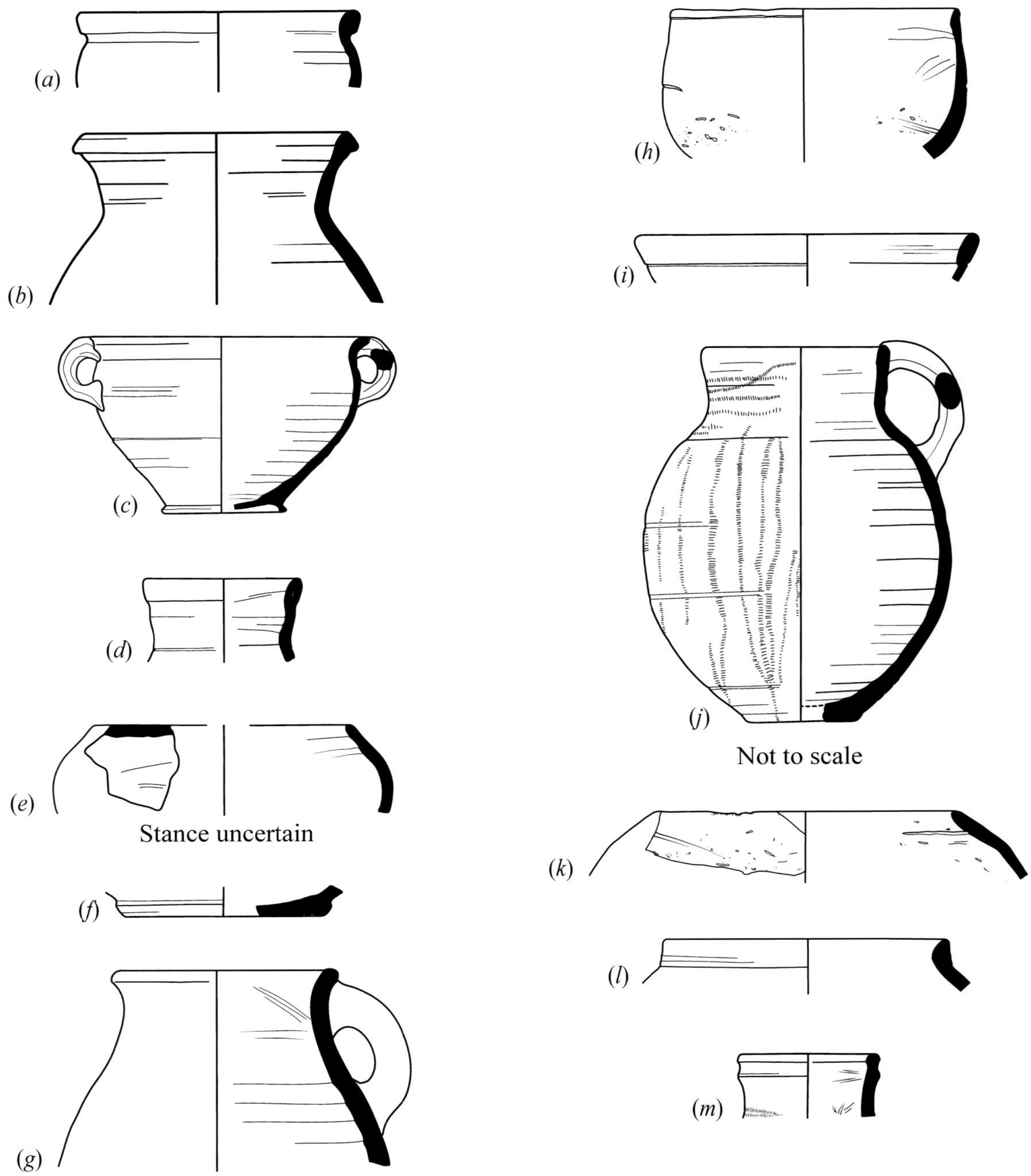

Figure A.I.8. Lower level diagnostic sherds

D 1 THROUGH UPPER AND LOWER LEVELS

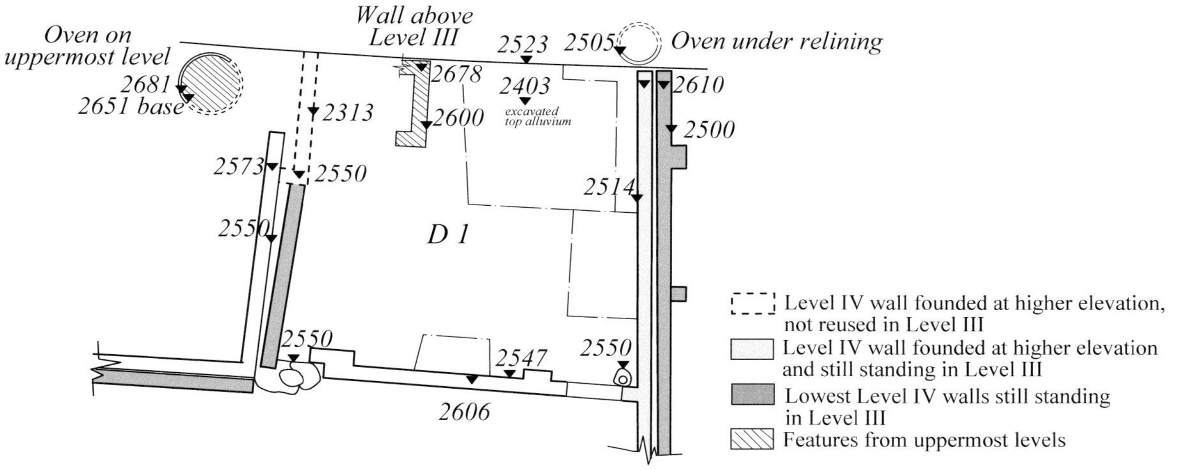

Figure A.I.9. Plan of D 1 upper and lower levels

D 1 (1), upper and intermediate levels

- Large strap-handled jar or vat, red slip and string impressed
- Large wide-mouthed silt jar rim
- Many silt bowl sherds (types SB 1a, SB 1b, SB 2)
- Shallow silt bowl with slightly everted rim; similar to fig. 5.7d
- Silt bowl (type SB 1), ribbed walls (OIM E24342)
- Marl flask with wide, flaring rim
- Bread plate
- Handmade jar with incised rim, black exterior, burnished, fig. A.I.10a
- Sherds of wide-mouthed marl jars
- Wide-mouthed Nile D and Nile B white-slipped rims and sherds
- Large stand with T-shaped base, median ridge, rim fragment in D 1 (2), fig. A.I.10b

D 1 (2), intermediate level

- Lots of SB 1 and SB 2 sherds
- Large silt ring base, finger-impressed interior
- Handmade cooking pot with mat-impressed exterior
- Silt jar rim and neck sherds (type SJ 1)
- Silt jar sherds (type SJ 3)
- Marl wide-mouthed jar rim
- Marl bowl sherds, ring base, rims folded over to interior
- Bread plate
- Marl wide-mouthed jar, black-painted decoration, fig. A.I.10c
- Tilted rim to large stand; see fig. 5.57d

D 1 (3), test pit that revealed oven under wall relining

- Silt jar sherds (types SJ 1, SJ 2), rim, body, and neck
- Pottery nozzle, misshapen
- Handmade sherds with traces of burnish
- Handmade bowl with incised rim

D 1 (5)

- Handmade bowl body sherds with black interior, red exterior, and finely burnished surfaces

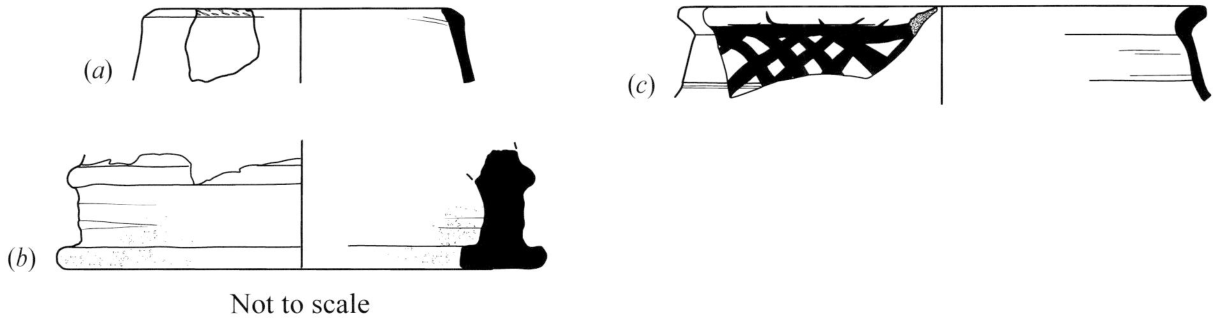

Figure A.I.10. Upper through lower levels at D 1

LEVEL III RESIDENCE

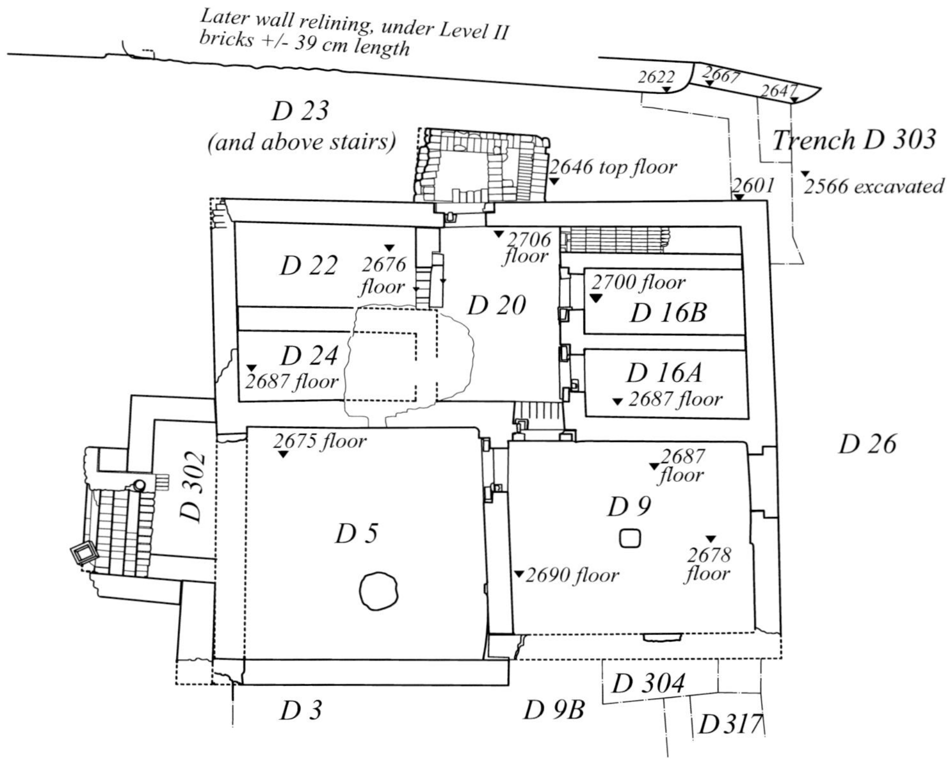

Figure A.I.11. Plan of Level III residence

STRATIGRAPHY

- D 5 (1) is 2 m of intentional fill. D 5 (2) is compact sand layer up to floor 1 (ca. 25 cm). D 5 (3) is ca. 5 cm ash lens above floor 1 in area by door into D 9. D 5 (4) is floor 1, uneven mud plaster of uncertain thickness. D 5 test pit went down below floor around column foundation, for 1 m, and through Level IV floor in D 312 and D 315.
- D 9 (1) is intentional fill. D 9 (2) is stratified sand layers mixed with brick rubble (ca. 40 cm). D 9 (3) is ca. 5 cm layer of ash, charcoal, and burned brick. D 9 (4) is ca. 1 cm maximum deposit atop floor 1. D 9 (5) is uneven mud-plastered floor of room. D 9 test pit excavated into D 312 and D 313 of Level IV.
- D 16A–B (1) is intentional fill within walls; more compacted fill of (1b) averaged 60–70 cm thick. D 16A–B (2) is charcoal, ash, burned brick of ca. 3–6 cm. D 16A–B (3) is thin, ca. 5 cm layer of sand over floor 1. D 16A–B (4) is uneven mud-plastered floor of both rooms.
- D 20 (1) is layer of sand filling intrusive pit (well from Level II) on west side. D 20 (2) is intentional rubble fill. D 20 (3) is stratified layers of sand and eroded brick and fallen stone lintels. D 20 (4) is 8 cm maximum layer of ash and charcoal. D 20 (5) is sand and brick debris. D 20 (6) is floor 1.
- D 22 (1) is layer of gray sand filling intrusive pit on southeast. D 22 (2) is intentional rubble fill (1 m max.). D 22 (3) is thin debris layer above floor. D 22 (4) is floor 1.
- D 23 is area both north of Level III residence and excavation around base of Level III north stairs. D 23 (1) is intentional fill sloping down from north fortress wall (uncertain depth). D 23 (2) is stratified sand and eroded brick sloping from north wall (ca. 28 cm). D 23 (3) is street surface, test pit in unit before excavation revealed floor at 2 m below top of rubble (43 cm above passageway floor of Level IV, elevation 2603).
- D 24 (1) is layer of sand filling intrusive pit on east (unclear depth); D 24 (2) is intentional fill; D 24 (3) is mud layer resting against south wall in area and sloping down to east; D 24 (4) is stratified layers of sand and brick of uncertain depth. D 24 (5) is floor 1 patch in southwest corner.

D 5 (1), mixed (Christian sherds not listed)

- Silt jar rim and body sherds (types SJ 1, SJ 3)
- Two silt jar loop handles
- Dorginarti pot stand, bottom rim
- Silt bowl rim/body sherds (types SB 1a, SB 1b, SB 2b)
- Shallow silt bowls, red rims
- Silt bowl variant, slightly carinated shoulder and red rim band; compare similar in fig. 5.11c–e
- Handmade black bowl
- Handmade bowl rims, undecorated
- Phoenician amphora body sherds, shoulder, handle stem; see fig. 6.10
- East Greek amphora rim and base sherds; see figs. 6.7c, 6.8c
- Marl jar sherds, smooth and corrugated exteriors
- Marl rims, bowls and unidentified jars

D 5 (2)

- Two shallow silt bowl rims

D 5, test pit through floor

- Marl jar (type MJ 1)
- Silt jar sherds (type SJ 3)
- Marl flask neck/handle area, wide rim; see fig. 5.52b
- Silt bowl sherds (types SB 1a, SB 1b)

D 5, removal of column foundation

- Marl jar rim (type MJ 1)
- Silt bowl sherds (type SB 1a)
- Dorginarti pot stand
- Silt jar sherds with handle fragment
- Silt jar rim sherds (type SJ 3)
- Silt jar(?) flat base

D 9 (1), mixed (Christian sherds not listed)

- Low silt ring base with concentric burnishing, fig. A.I.12a
- Marl jar rims, neck, loop handle (type MJ 1)
- Silt/marl mix jar variant (later?), fig. A.I.12b
- Silt loop handles, fig. A.I.12c–d
- Silt bowl sherds (types SB 1a, SB 1b, shallow), one variant (later?) shown in fig. A.I.12e
- Handmade bowl sherd with incised rim
- Marl jar with sharply everted rim (later?); see fig. 5.44a–b
- Marl bowl or lid (Level II?); see figs. 5.57b, 6.15a
- Oasis flask type (Level II); see fig. 6.20c
- Later(?) marl rim, fig. A.I.12f

D 9 (2), mixed

- Marl jar sherds (type MJ 4), fig. A.I.12g
- Samian(?) amphora body sherd

D 9 (3)

- No sherds found in the Oriental Institute Museum collection, but note faience pendant (OIM E24332) in fig. 7.1a

D 9 (4)

- Marl loop handle
- Silt jar sherds (types SJ 1, SJ 2)
- Marl jar body sherds and loop handles
- Silt bowl sherds, no rims
- Thin-walled ledged jar, fig. A.I.12h
- Handmade bowl, incised rim and mat-impressed exterior
- Handmade bowl sherds, red slip and burnished surfaces

D 16A–B (1)

- Dorginarti pot stand rim
- Marl bowl with rim folded to interior; as examples in fig. 5.16d–e
- Large bowl with string-impressed exterior
- Silt jar body sherd, toward base
- Marl jar shoulder, fig. A.I.12i

D 16A–B (3)

- Marl jar body sherd with black horizontal bands on exterior
- Thin-walled marl jar sherds with corrugated exterior

D 20 (2)

- Marl jar rim (type MJ 3, vertical stance), fig. A.I.13a
- Marl jar rim (type MJ 2) and other marl jar sherds
- Silt jar (types SJ 1, SJ 2)
- Variant marl jar, wide mouthed, fig. A.I.13b
- Shallow silt bowl
- Marl bowl rim and sherds, folded to exterior, fig. A.I.13c
- Large bowl or platter with string impressions, fig. A.I.13d
- Small silt bowl variant
- Shallow silt bowl sherds

D 22 (1)

- Wide-mouthed silt jar with red slip on exterior

D 22 (2)

- Wide-mouthed marl jar variant with everted rim, fig. A.I.13e

D 22 (3)

- Marl jar rim fragment (type MJ 1 or MJ 2)
- Marl jar (type MJ 3) vertical stance, fig. A.I.13f
- Silt jar (types SJ 1, SJ 2) with red slip
- Dorginarti pot stand

D 22 (3) *continued*

- Silt bowl sherds (types SB 1a, SB 1b, SB 2), some with red rims and streak burnish; fig. A.I.13g and see fig. 5.5c
- Handmade bowl sherd, black-burnished interior and red exterior

D 22 (4)

- Marl flask body sherd with black concentric bands
- Silt bowl sherd (type SB 1a), red rim
- Handmade bowl body sherds, as those from D 22 (3)

D 23 (1), mixed (from upper and lower fill)

- Wide-mouthed marl jar, fig. A.I.13h
- Marl jar rims (type MJ 2), fig. A.I.13i
- Flat silt base with mat-impressed bottom, fig. A.I.13j
- Silt bowl sherds (type SB 1b), red rim exterior, red slip interior, burnished, fig. A.I.13k
- Silt bowl variant, red slip exterior and burnish on interior, fig. A.I.13l
- Level II silt jar; see fig. 6.18e

D 24 (2), mixed (disturbed by later well)

- Handmade(?) cup or bowl rim fragment
- Silt bowl (type SB 1b), fine burnish strokes, fig. A.I.13m
- Silt jar with rim folded over to interior, bag shaped (date uncertain)
- Wide-mouthed ledged silt jar variant, red slip exterior, fig. A.I.13n
- Marl jar with everted rim (type MJ 5); see fig. 5.44a–b
- Silt/marl mix jar, Level II, fig. A.I.13o
- Silt copy of marl channeled-rim jar from Level II, fig. A.I.13p
- Silt strap loop handle, widely arched
- Silt bowl ring base, fig. A.I.13q

D 24 (4), above Level III floor

- Silt ledged jar variant, red slip exterior, fig. A.I.13r

D 302 (1), fill of Level III stairway

- Handmade sherds
- Silt bowl with everted rim; similar to fig. 5.7c–d
- Silt bowl sherds (SB 1a), including large bowl sherds with string-impressed exterior
- Bowl or base-rim sherd, fig. A.I.13s
- Silt jar body sherds, some with streak-burnished exterior
- Large stand with finger impressions on base, fig. A.I.13t
- Wide-mouthed silt jar (as MJ wide-mouthed type)
- Marl jar body sherds with smooth exterior surfaces

Removal of stairs north of Level III building

- Silt jar rim or pot stand base, shown as jar in fig. A.I.13u
- Marl jar sherds with variant rims; smaller sherd in fig. A.I.13v

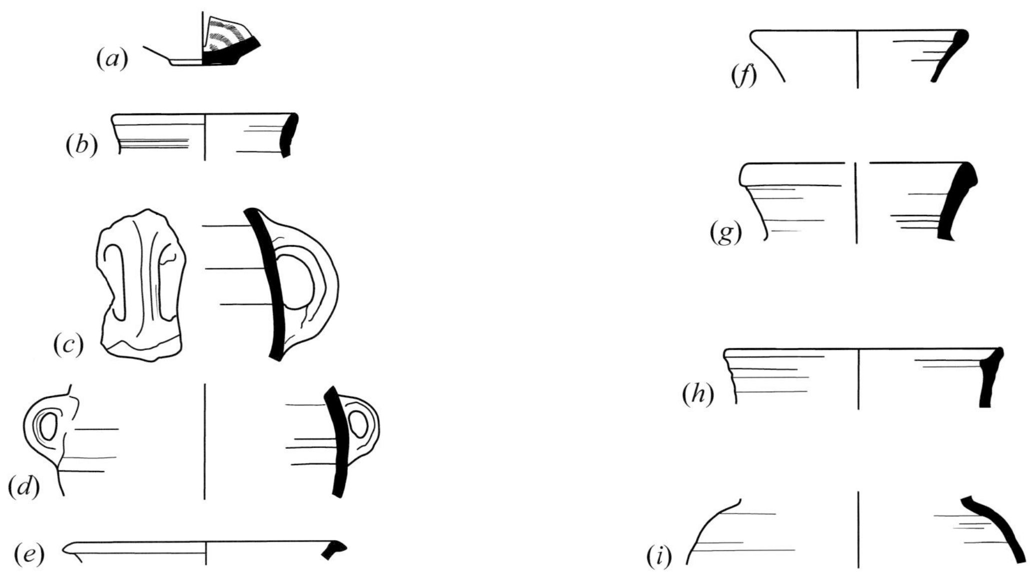

Figure A.I.12. Level III residence diagnostic sherds

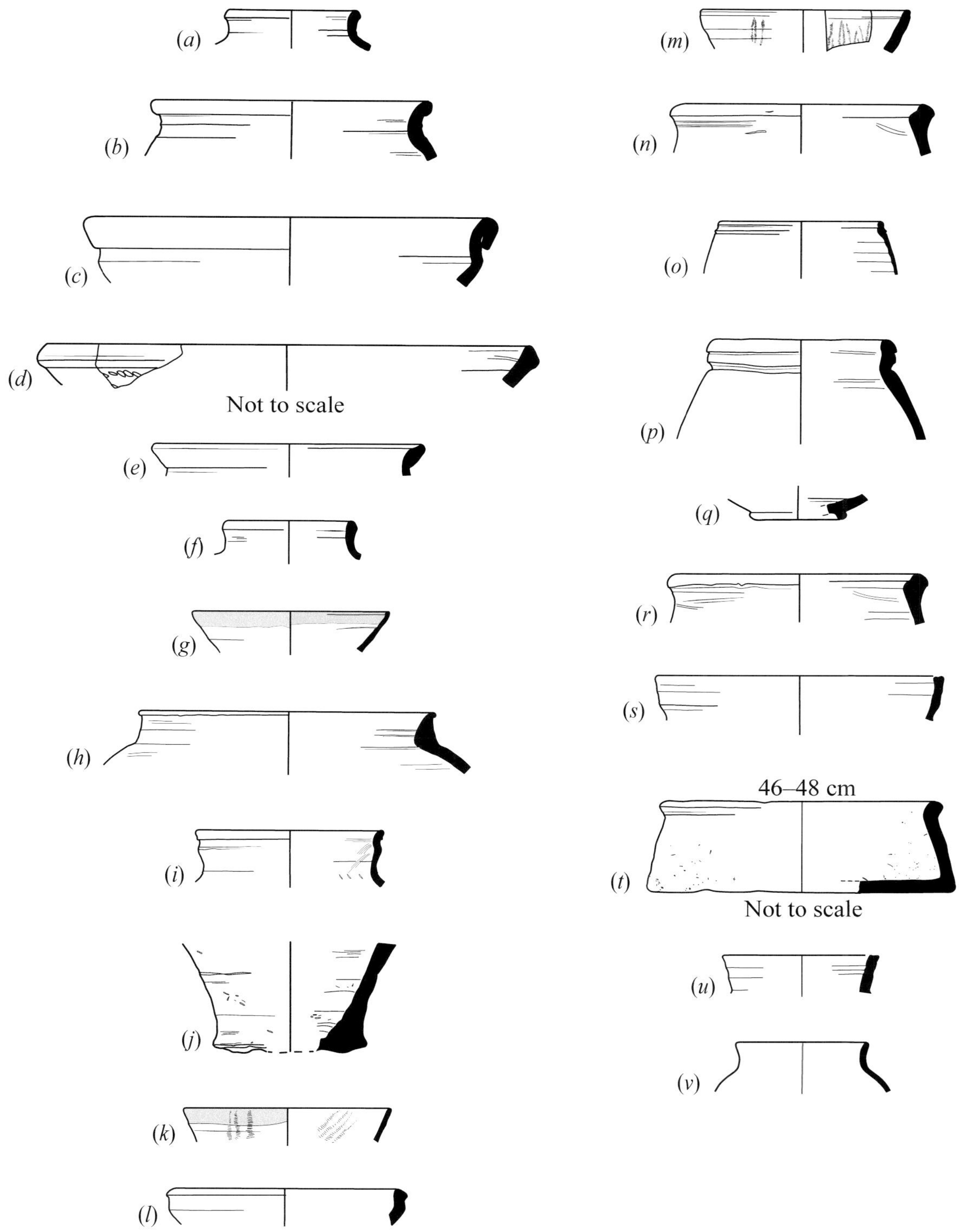

Figure A.I.13. Level III residence diagnostic sherds

LEVEL IV RESIDENCE

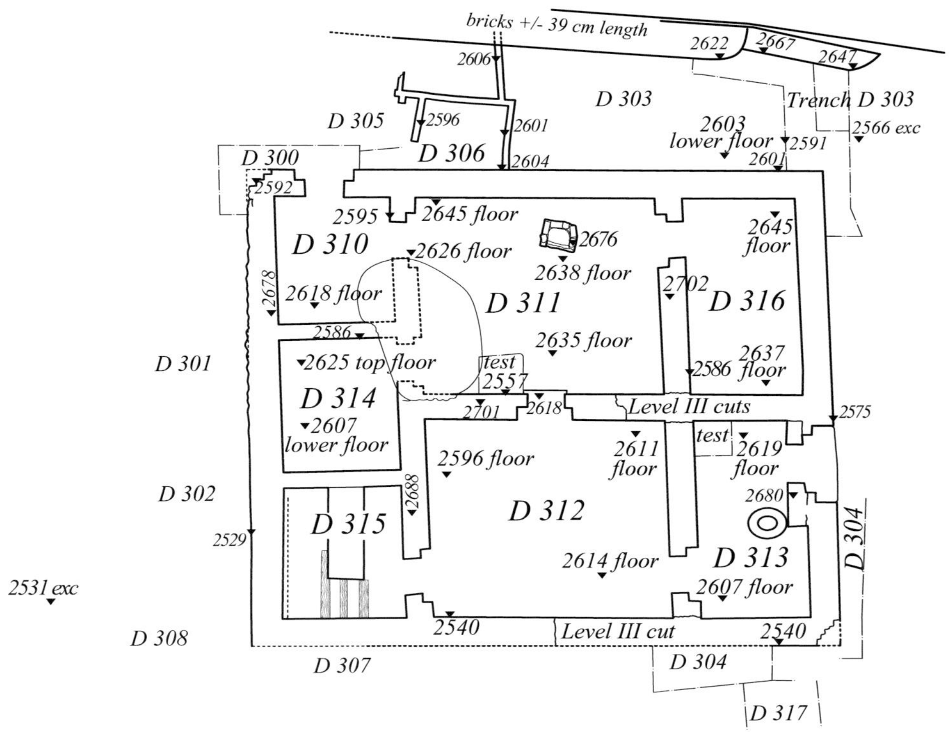

Figure A.I.14. Plan of Level IV residence

STRATIGRAPHY

- There is some evidence of multiple floors, for example in D 314 and D 316. Test pits below these floors exposed yellow sand laid for the foundation of the building, resting atop river sand and then alluvium. By the doorway from D 311 into D 312 a test pit was sunk to an elevation of 2387, or 2.48 m below the nearby floor level in D 311. See text for details.
- D 310 (1), fill within room below Level III remains. D 310 (2) was sand above floor. Below floor was rubble, then sand.
- D 311 (1), fill below Level III floor. D 311 (2) is ca. 15 cm debris and sand above floor. Below floor was rubble, then sand. D 311 test pit was sunk 2.48 m below floor and rubble, with layer of gray river sand and then brown alluvium below.
- D 312 (1) is ca. 60 cm brick and rubble debris; D 312 (2) is ca. 5 cm sand atop plastered floor. A series of floors totaling 15–18 cm in depth and above a brick-paved original floor. Below floor was ca. 30 cm of crumbly sandstone of different colors (red, pink, gray, purple, white to yellow, and orange), mixed in with, and sitting atop, yellow sand.
- D 313, original floor paved with irregular bricks, as in D 312. Test trench 62 cm below floor showed yellow sand and then gray river sand.
- D 314, upper floor of mud, and lower mudbrick floor, as in D 312 and D 313. Rubble below it and sitting atop yellow sand.
- D 316, foundations of Level III walls atop plastered floor in room, with Level IV floor ca. 50 cm below Level III floor. D 316 (1) is fill of bricks and debris above rough and pitted mud-plastered floor, which slopes down from walls. A number of floor replasterings were noted. D 316 (2) is ca. 5–10 cm of soft brown soil atop floor. Below floor was rubble.

D 301, under D 8

- Silt jar sherds (type SJ 3)
- Silt jar body sherds
- Marl jar body sherds

D 302 (1)

- Shallow silt bowl, fig. A.I.15a
- Silt bowl variant (type like SB 1, but thick rimmed), red slip, fig. A.I.15b
- Two different silt wide-mouthed jars, fig. A.I.15c–d
- Large silt bowl/platter with string impressions
- Large, poorly formed pot stand

D 303, by north stairs and trench D 303

- Silt bowl sherds (type SB 1), thin red slips, some red rims
- Silt sherds from large bowl, string impressed
- Shallow silt bowl rims, one as from D 302 (1), one with more everted rim, third shown in fig. A.I.15e
- Silt ring base, fig. A.I.15f
- Wide-mouthed marl jar variant, pink fabric, fig. A.I.15g
- Silt jar rim and sherds (type SJ 2), fig. A.I.15h
- Silt jar variant, fig. A.I.15i
- Silt jar rim and sherds (type SJ 3), red slip exterior
- Silt jar neck with creamy slip, fig. A.I.15j
- Marl jar rim and body sherds (type MJ 1)
- Handmade black-topped bowl, burnished, fig. A.I.15k
- Wide-mouthed silt jar, red slip, string impressions
- Marl body sherd (type MJ 4 or New Kingdom?) black bands, fig. A.I.15l

D 303 (1), floor under Level III staircase

- Handmade large bowls, some light burnish exterior
- Handmade nozzle or spout, fig. A.I.16a
- Silt cup, fig. A.I.16b
- Silt bowl sherds (types SB 1a, SB 1b), some with red rims and streak burnishing, fig. A.I.16c–d
- Shallow silt bowl sherds, red rim exterior or red rim interior/exterior
- Large silt bowl sherds, string impressed, fig. A.I.16e
- Silt jar body sherds, thin red slip, plus small silt loop handle
- Large-stand sherds, fig. A.I.16f
- Thin-walled marl jar body sherds
- Marl jar rim/body sherds (type MJ 1), fig. A.I.16g
- Marl bowl rim (variant), fig. A.I.16h

Trench D 304

- Handmade silt jar neck, sloping neck (not illustrated)
- Silt bowl sherds (types SB 1a, SB 1b), some with red slip
- Rounded and pointed silt jar base fragments
- Silt sherds with cream slip exterior
- Marl jar body sherds, plain or corrugated exterior surfaces
- Nile D jar sherd with pink slip exterior

D 307 (1)

- Small Nile D jar, see fig. 5.26a

D 308 (1) and (2)

- Shallow silt bowl, fig. A.I.16i

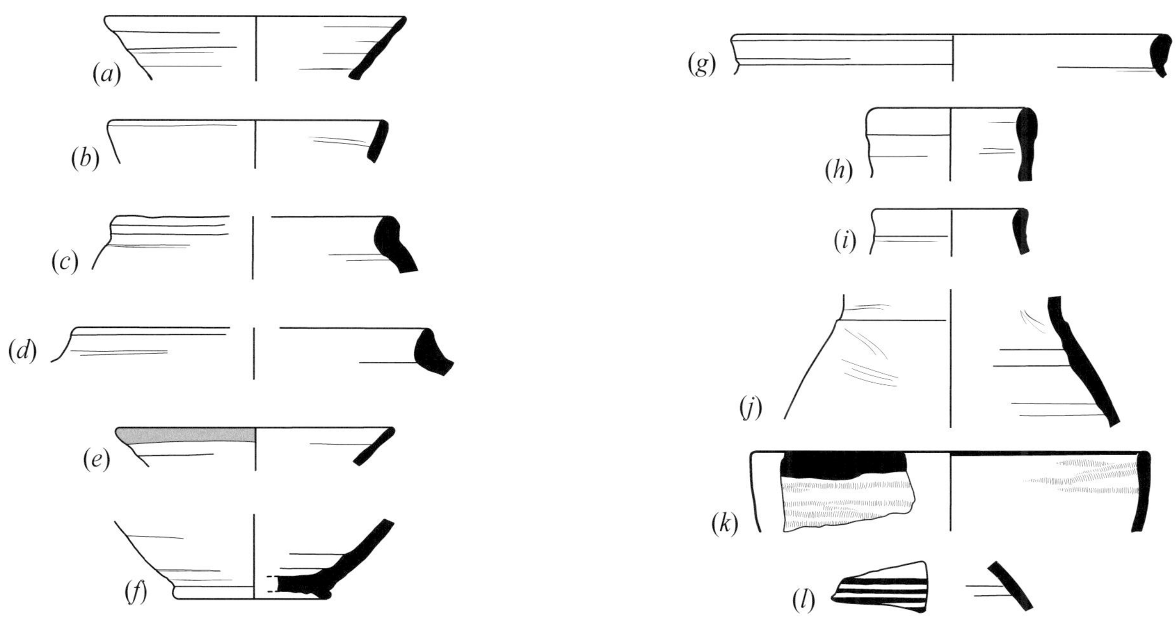

Figure A.I.15. Level IV residence diagnostic sherds

STRUCTURE NORTH OF RESIDENCE

D 305 (1)

- Shallower silt bowl variant, thickened interior rim, horizontal burnish strokes interior, fig. A.I.16j
- Silt bowl sherd (type SB 2a), vertical burnish interior, red slip traces, fig. A.I.16k
- Wide-mouthed silt jar, burned areas at rim, fig. A.I.16l
- Wide-mouthed marl jar, fig. A.I.16m

D 306 (2), test pit below wall foundation

- Large silt jar with handles and potmark, fig. A.I.16n

LEVEL IV RESIDENCE

D 310 and D 311, down to D 310/311 (1)

- Marl jar rim and sherds (type MJ 1)
- Silt bowl sherds (type SB 1a), red slip interior/ red rim exterior
- Silt bowl (type SB 2), red slip interior/exterior and streak-burnished interior

D 310 (1) and D 314 (1) (fill between Level III and Level IV residences)

- Handmade silt bowl sherds, some red slip interiors
- One rim of a silt bowl (type SB 1a)
- Small silt cups, one red rim and one red slip interior/ exterior
- Silt jar sherds, thin red slips on exterior
- Marl jar body sherds, one with thin incisions or corrugations on exterior, one neck, and one with string-impressed exterior

D 310 (1)

- Two silt bowl rims, one red rim, fig. A.I.16o
- Shallow silt bowl
- Marl bowl with rim folded over to interior
- Dorginarti pot stand rim
- Marl jar body sherd
- Silt burnished beaker (OIM E49523), fig. A.I.16p
- Silt jar (type SJ 3), red slip exterior
- Ledged rim silt jar

D 310 (2)

- Handmade bowl rim, roughly formed
- Silt bowl sherds (type SB 1a), red rims; see fig. 5.2d, g
- Sherds of silt bowls with red slips
- Silt jar rim (type SJ 1 or SJ 2), fig. A.I.16q
- Marl jar sherds, long sloping neck sherd

D 311 (1)

- Handmade base fragments, rounded
- Silt bowls sherds (types SB 1a, SB 1b)
- Ledged silt jar
- Small silt loop handle
- Rim from Dorginarti pot stand
- Marl jar neck and body sherds, and one base sherd with exterior ribbing
- Wide-mouthed marl jar

D 312 (1), D 313 (1), and test pit D 313 (2)

- Silt bowl sherds (types SB 1, SB2), red rim or red rim exterior and red slip interior
- Silt bowl with carinated shoulder, fig. A.I.16r
- Silt jar rim and body sherds (type SJ 3)
- Nile D and other related silt jar sherds, all creamy-pink slips; Nile D jar shown in fig. A.I.16s
- Wide-mouthed silt jar, fig. A.I.16t
- Marl body sherd with potmark (not illustrated)
- Wide-mouthed marl jar, thicker-walled variant
- Marl jar (type MJ 4), fig. A.I.16u
- Marl flask body sherds, black-painted concentric bands

D 312 (2)

- Silt bowl sherds (type SB 1a), one with red rim band
- Pointed silt jar base
- Wide-mouthed silt jar sherds, traces of red slip exterior
- Thin-walled marl jar body sherds, smooth exteriors

D 315 (1)

- Nothing found in the Oriental Institute Museum collection

D 316 (1)

- Silt bowl rims and sherds (type SB 1a), red rim bands, and rounded silt base with heavy scraping
- Large platter or bowl sherds, one with string-impressed exterior, red slip traces on interior

Trench D 316 (1) to D 316 (2)

- Silt bowl sherds (type SB 1a), large diameters, red rims
- Marl jar body sherds, part of potmark on one, some fired greenish

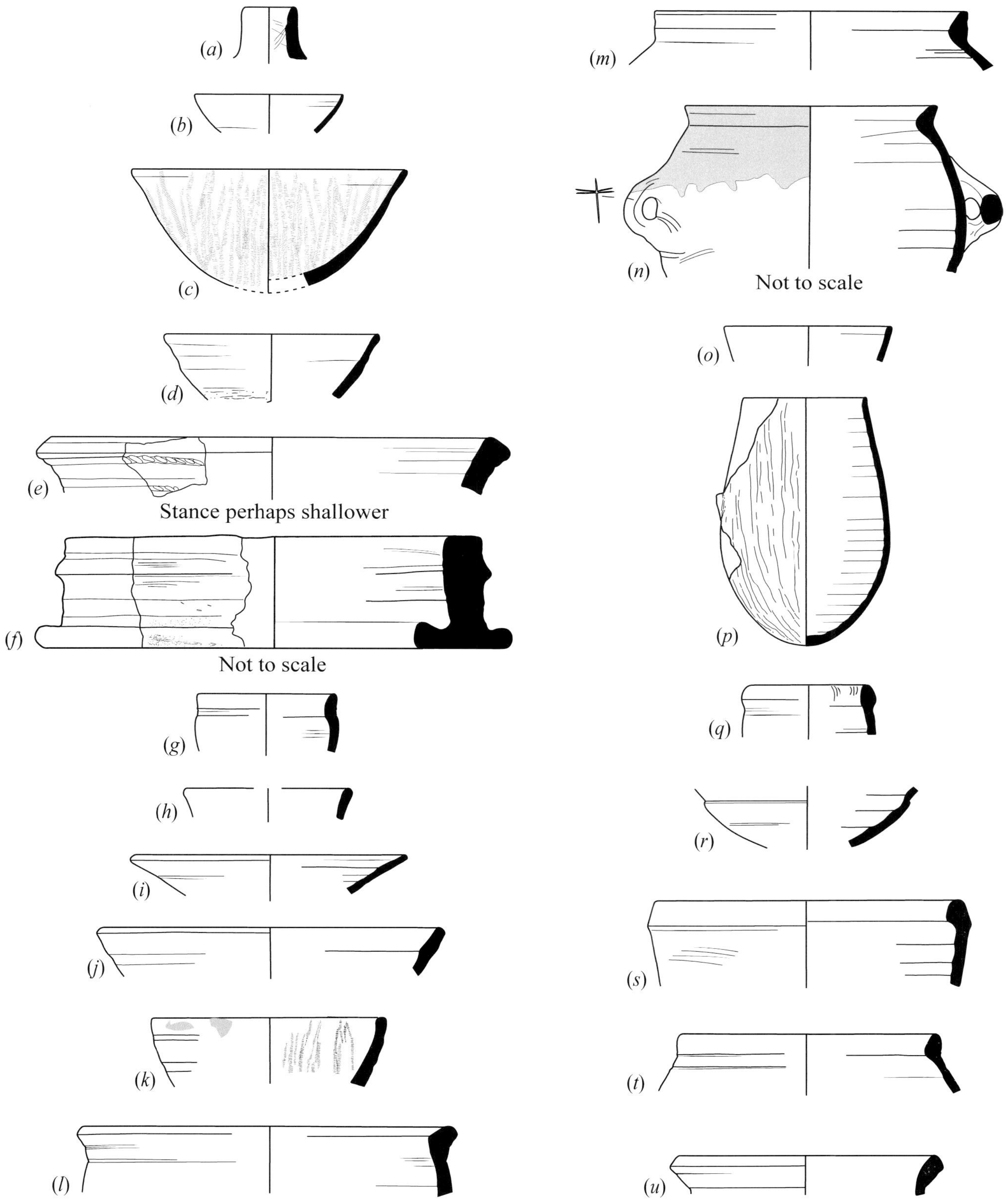

Figure A.I.16. Level IV residence diagnostic sherds

D 59, D 60, AND D 68

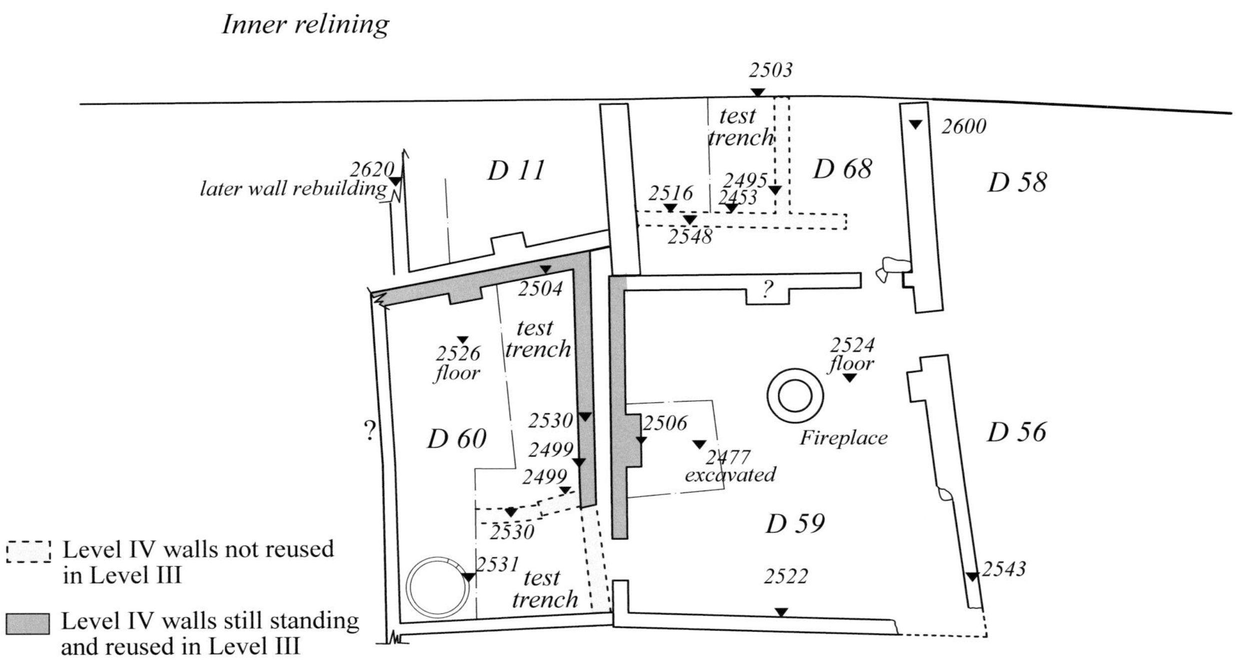

Figure A.I.17. Plan of D 59, D 60, and D 68

D 60

- Silt bowl sherds and bases (types SB 1, SB 2)
- Silt jar (type SJ 3), red slip exterior
- Small silt cup
- Marl bowl rim folded to interior
- Marl handled bowl rim
- Large complete silt jar profile found under west wall at north corner, red rim exterior and interior slip (OIM E24422), fig. A.I.18a
- Marl jar rim and body sherds (type MJ 1)
- Marl/silt mixed-fabric jar (type MJ 2)
- Marl jar rim (type MJ 4)
- Silt jar body sherds, globular or bag shaped
- Handmade bowl sherds, no decoration

D 60 (2), test trench below Level III floor and down to Level IV floor; level with D 68 (3)

- Silt jar sherds (types SJ 1, SJ 2), red slip exterior
- Silt bowls (types SB 1a, SB 2) red rims, traces red slip
- Shallow silt bowl
- Marl jar loop handle
- Bread plate rim
- Dorginarti pot stand rim with finger marks on exterior
- Wide-mouthed silt jar, traces red slip exterior
- Handmade body sherd with worn cross-hatched decoration

D 60 (3), test trench along east side of room (below Level IV floor)

- Silt jar (type SJ 3), red slip exterior
- Silt bowl sherds (types SB 1a, SB 1b, SB 2)
- Shallow silt bowl with red rim band
- Silt loop handles, one with traces of burnish and red slip

D 68 (1) and D 68 (1), third 50 cm clearance

- Lots of silt bowl sherds (types SB 1, SB 2), multiple joins with sherds from D 68 (2)
- Wide-mouthed marl jar sherds, with rounded shoulder
- Marl jar sherds (type MJ 2), rims and shoulder
- Handmade finger-impressed base section of cooking pot

D 68 (2) (stratum level with top of north–south earlier wall)

- Silt bowls (types SB 1a, SB 2), multiple joins with sherds from D 68 (1)
- Silt bowl variants, one with incurved rim, streak burnished
- Large bowl with channeled rim and associated ring base, streak-burnished interior, fig. A.I.18b–c
- Small marl ring base and rim of marl handled bowl
- Dorginarti pot stand sherds and pot stand variant
- Silt jar (type SJ 3), red splashes exterior and slipped interior
- Ledged jar variant, red slip exterior, fig. A.I.18d

D 68 (2) (stratum level with top of north–south earlier wall) *continued*

- Silt jar base with pointed tip
- Marl jar (type MJ 2), fig. A.I.18e
- Fragment of closed base for large stand(?); compare figs. 5.58c, 5.59c

D 68 (3), Level IV floor level associated with earlier cross walls

- Silt bowl sherds (types SB 1, SB 2)
- Silt bowl (type SB 2a) with channel under knobbed rim, streak-burnished interior, fig. A.I.18f
- Silt jar sherds (type SJ 3), red slip exterior
- Carinated silt bowl, traces of red slip on upper exterior, fig. A.I.18g
- Small silt jar with slightly everted rim

D 68 (4) below earlier floor, sandy and ashy fill

- Silt bowl sherds and bases (type SB 1a), four bowls?
- Large bowl/platter with string-impressed exterior
- Handmade black-topped bowl, fig. A.I.18h
- Marl jar sherds, pitted and spongy from water saturation

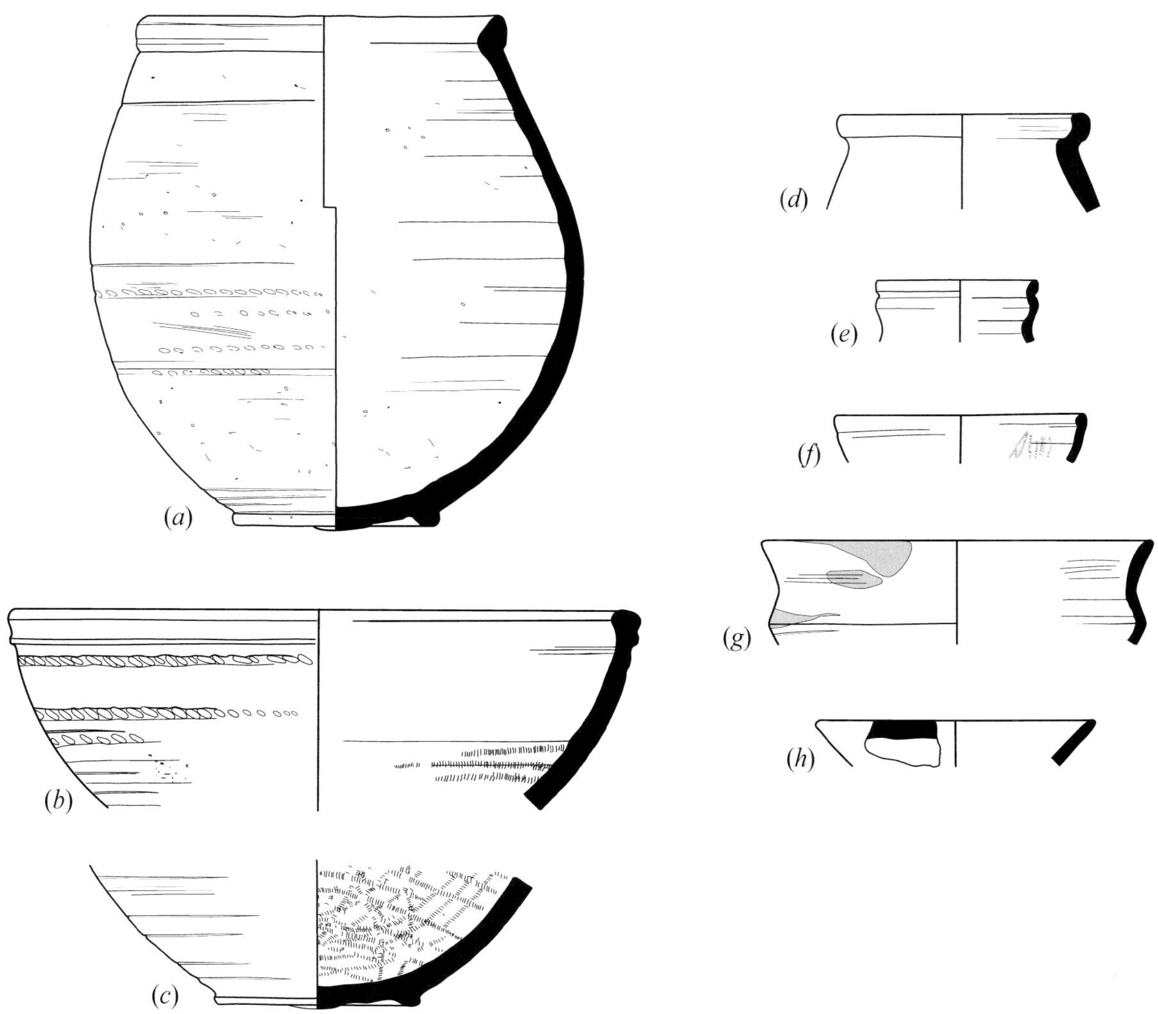

Figure A.I.18. Diagnostic sherds from D 60 and D 68

STRUCTURES BY THE NORTH WALL BREACH

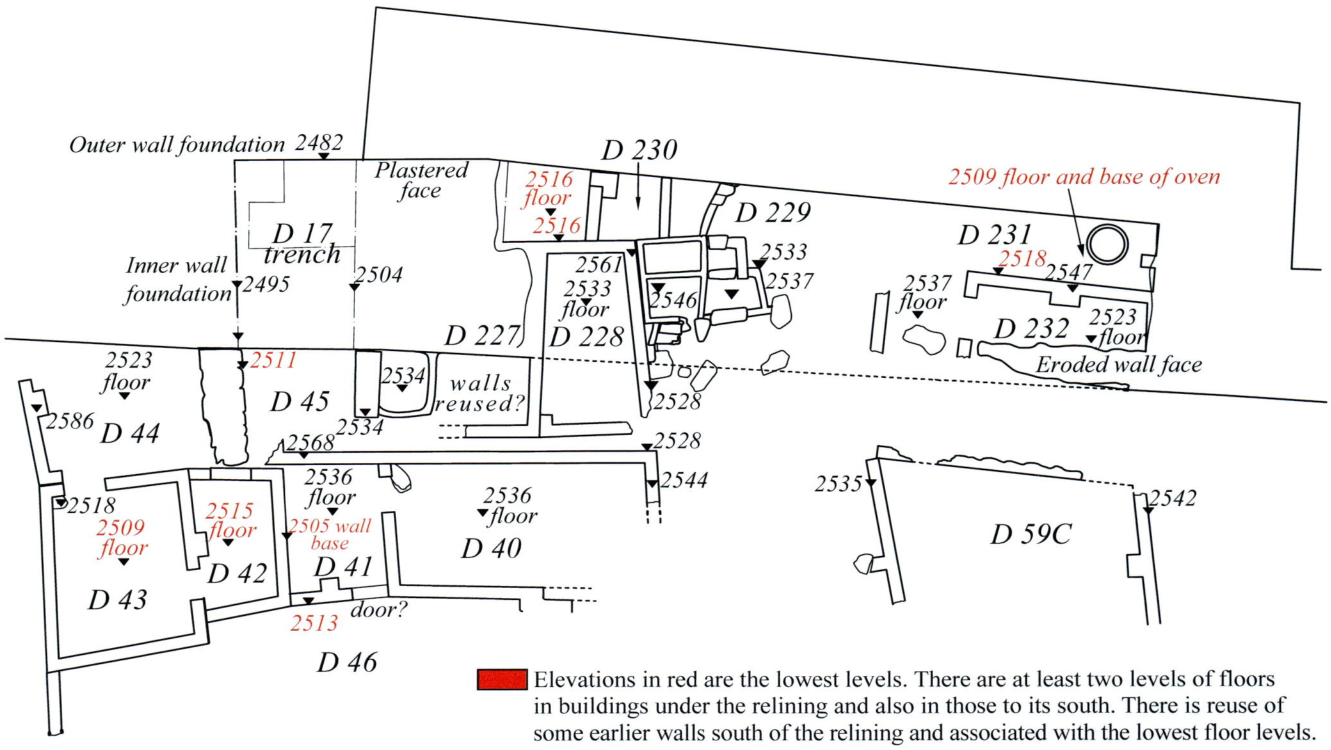

Figure A.I.19. Plan of structures by north wall breach

North wall at breach

- Marl jar variant, raised ridge at neck, fig. A.I.20a

D 17, test trench under wall

- Wide-mouthed silt jar, red slip exterior, red rim interior, fig. A.I.20b
- Silt bowl sherds (types SB 1a, SB 1b); variant shown in fig. A.I.20c with "stepped" rim exterior
- Silt body sherds of jars, both thin and thick walled, some traces of red slips

D 227 (1)

- Dorginarti pot stand rim
- Two different *dokka* rims and sherds
- Silt bowl sherds (types SB 1, SB 2)
- Shallow silt bowl
- Marl jar body sherds
- Silt jar body sherds, some with red slip exteriors
- Handmade finger-impressed base and rim sherds of blackened bowl, fig. A.I.20d (rim sherd)
- Large handmade bowl, red slip exterior and burnishing traces at base of exterior, fig. A.I.20e

D 228 (1)

- *Dokka* sherds, no joins
- Silt bowl sherds (types SB 1a, SB 1b), either red rim exterior and red slip interior or red rim bands, sometimes streak burnished
- Marl jar body sherds, mostly smoothed exteriors but two with faint ribbing
- Silt jar (type SJ 3), cream coating or deposit, fig. A.I.20f

D 229 (1)

- Silt bowl sherds (types SB 1a, SB 1b), red slips and some with red rim bands
- Marl bowl rim, folded over to interior
- Large bowl/platter, string-impressed exterior
- Silt jar conical base, pointed tip
- Carinated bowl form with red slip on upper exterior, fig. A.I.20g

D 230 (1)

- Silt bowl sherds (type SB 1a)
- Variant silt bowl with red rim band
- Marl jar sherds, thin walled, sloping neck and lumpy hand-formed base
- Silt jar rim and body sherds (type SJ 3), red slip
- Wide-mouthed silt jar sherds, rim shown in fig. A.I.20h
- Handmade black-topped bowl, mat impressed
- Handmade sherd with light burnishing

D 231

- Silt bowl sherds (types SB 1a, SB 1b), red rim or red slip interior, red rim exterior, some with burnishing
- Marl bowl rim sherds, folded to interior
- Large bowl/platter, no string impressions
- Marl jar body sherds
- Handmade bowl, light burnish on surfaces

D 232 (1)

- Silt bowl sherds (types SB 1a, SB 1b)
- Marl jar body sherds
- Handmade coarse body sherd

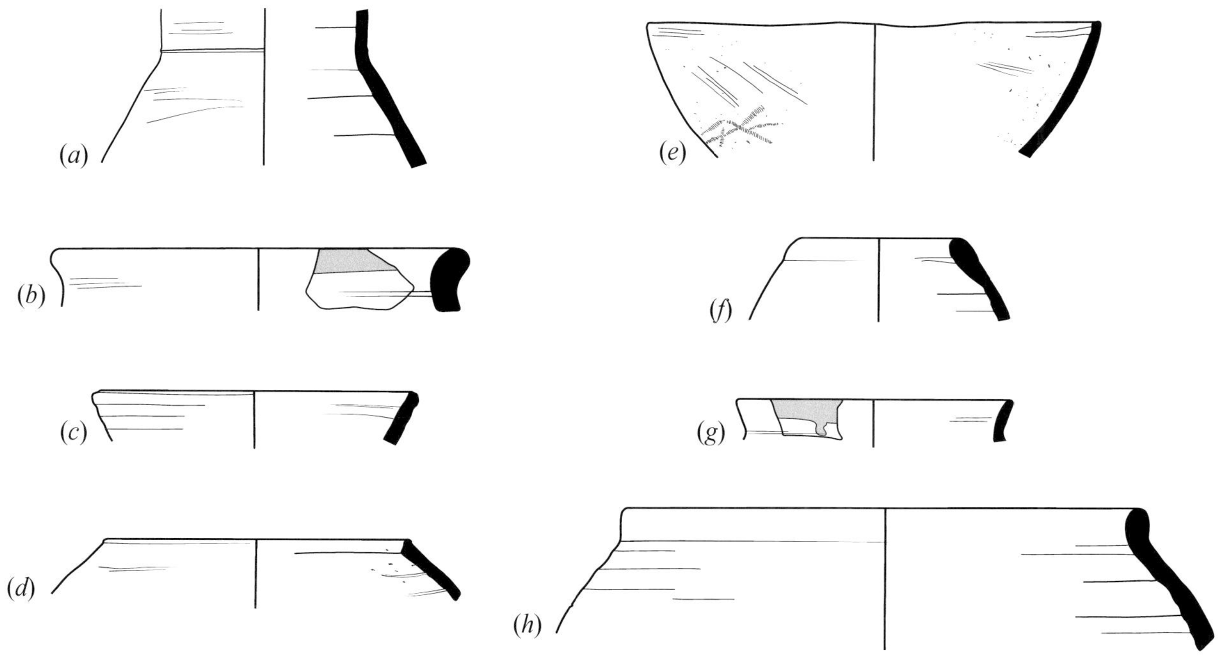

Figure A.I.20. Diagnostic sherds from structures under the fortification wall

DOORWAY D 87

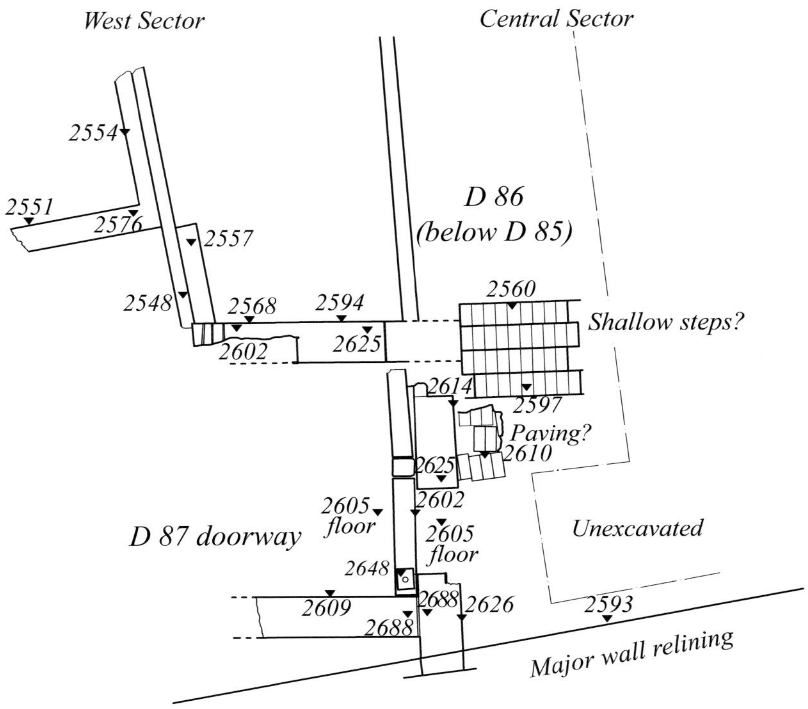

Figure A.I.21. Plan of doorway D 87

STRATIGRAPHY

- D 35 surface clearance above D 85 and D 86, along east side of west retaining wall for Level II platform. D 85 (1) down to "ash" layer (D 38 on north), and to D 87 on south.
- D 87 is large, stone-framed doorway at south end of dividing wall between West and Central Sectors. D 87 (1) is clearance of fill around doorway (uncertain depth). D 87 (2) is test pit below level of doorway. Appears to contain some of the fill of D 76–D 79, D 88–D 93, and room D 87 below later doorway addition. D 88 (1) is below doorway and above lower, incompletely excavated room D 87.

D 87 (1) doorway

- Mixed clay or Nile D jar sherd with thin pink slip
- Small marl handle, flattened oval
- Sherds of silt bowls (type SB 1)

D 87 (2)

- Silt jar with interior ledge, burned surfaces
- Silt bowl sherds (types SB 1a, SB 1b)
- Thin-walled handmade bowl, black-burnished surfaces
- Handmade cooking pot, blackened surfaces, possible potmark, fig. A.I.22a
- Wide-mouthed marl jar, fig. A.1.22b
- Large platter/bowl sherds

(*a*)

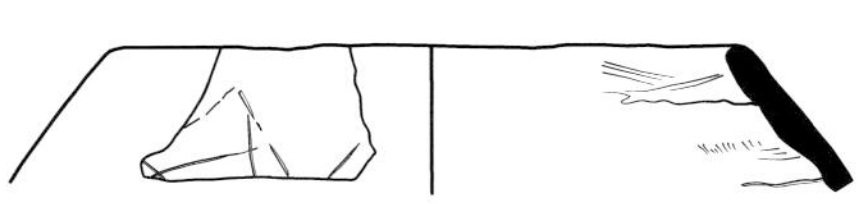

(*b*)

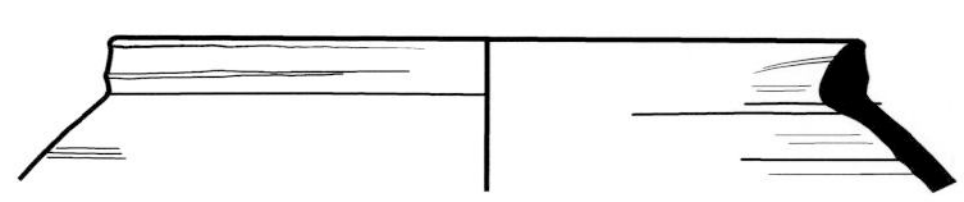

Figure A.I.22. Diagnostic sherds from the area around doorway D 87

STRUCTURES BELOW DOORWAY D 87

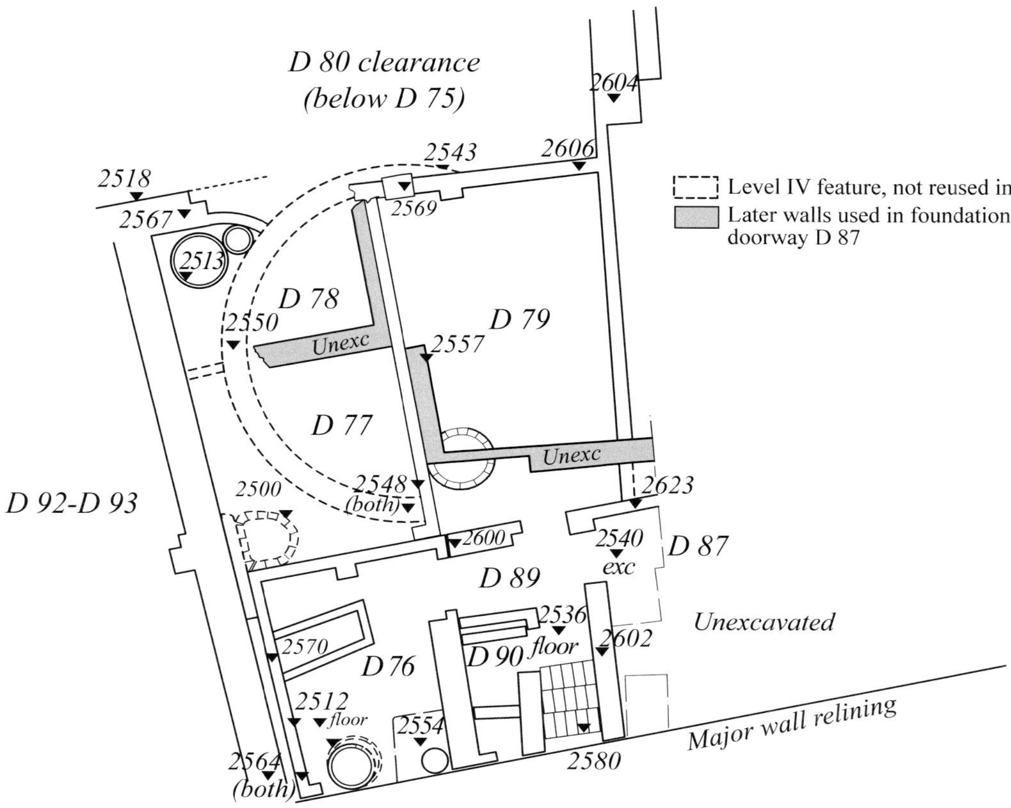

Figure A.I.23. Plan of structures below doorway D 87

D 76 (1)

- Handmade bowl with incised rim
- Large, low ring base for large jar
- Flat base of silt vessel
- Marl jar variant
- Wide-mouthed marl jar
- Marl jar rim and neck sherds (type MJ 1)
- Marl flask painted body sherds and wide rim
- Wide-mouthed silt jar, white slip exterior
- Handmade bowl with mat-impressed exterior, in northwest corner (OIM E49558), fig. A.I.24a
- Silt jar or vat with red slip, strap handle for same, in northwest corner
- Dorginarti pot stand rim
- Worn marl bowl ring base

D 78 (1) and D 78 (1) bin

- Silt bowl sherds (types SB 1a, SB 1b, SB 2), fig. A.I.24b
- Silt jar base with dimpled bottom, fig. A.I.24c
- Dorginarti pot stand rim sherds
- Two strap handles and shoulder fragment of large silt jar/vat
- Handmade bowl sherds, including bowl in fig. A.I.24d
- Silt jar rim (type SJ 3)
- Silt jar (type SJ 1), red slip and white slip traces(?), fig. A.I.24e
- Marl bowl sherds, rim folded to interior
- Marl jar rim sherd (type MJ 1)
- Small marl jar rim, variant, red slip atop rim, fig. A.I.24f

D 78 (2)

- Marl bowl rims, one slightly everted rim (see fig. 5.18c), another with rim folded to interior
- Rounded silt base with prominent wheel ridges

D 79 (1)

- Silt bowl sherds (types SB 1a, SB 1b), some with red rim bands
- Marl bowl sherds with rim folded to interior
- Marl flask rim with handle area, fig. A.I.24g
- Handmade body, rims, and base sherds; one with black interior and red exterior, burnish on both surfaces, fig. A.I.24h
- Silt jar sherds (types SJ 1, SJ 2 and variant), red slip
- Silt jar sherds (type SJ 3)
- Large silt loop handle fragment

D 79 (1) *continued*

- Small marl jar loop handle, twisted
- Marl jar sherds and wide-mouthed marl jar sherds
- Marl or mixed-clay wide-mouthed jar, fig. A.I.24i
- Large stand with tilted rim and median ridge, base missing

D 88 (1), fill below doorway D 87

- Silt bowl rims and sherds (types SB 1, SB 2) and rounded base
- Marl flask sherd with black concentric bands
- Handmade bowl with incised rim

D 89, fill below D 88 (1) containing "ash" layer? (above room D 87)

- Dorginarti pot stand sherds

D 90 (1)

- Silt bowl sherds of various types, some of type SB 1, one very finely burnished (OIM E50077); see fig. 5.3a
- Shallow bowl with incurved rim (variant), red slip, highly burnished; see fig. 5.8f
- Shallow silt bowl with knobbed rim
- Rounded silt jar base
- Dorginarti pot stand
- Handmade black incised bowl with burnishing (OIM E36383), fig. A.I.24j
- Handmade coarse bowl sherds (OIM E24414), fig. A.I.24k
- Marl jar rim (type MJ 2)
- Silt flask with long neck, red slip exterior ("room east of D 76"; either D 89 or D 90), fig. A.I.24l

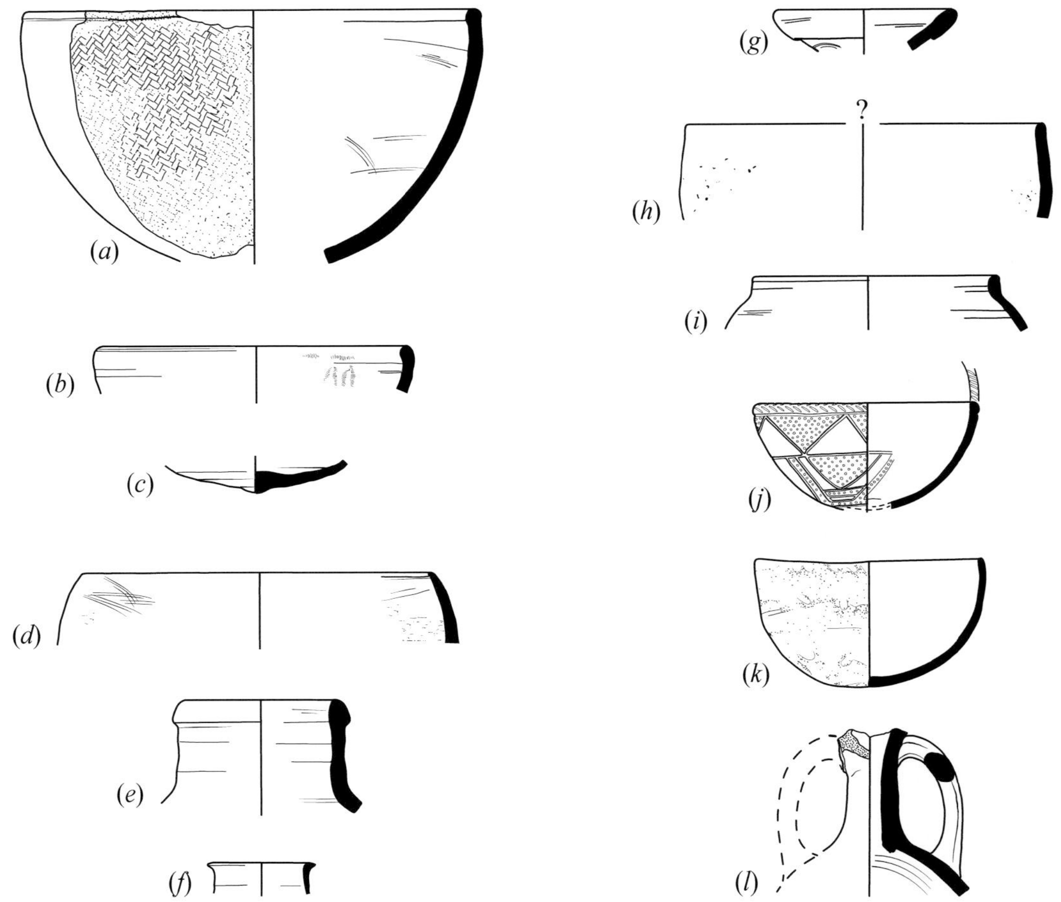

Figure A.I.24. Diagnostic sherds from the levels below doorway D 87

Appendix II

Silo Capacity and Population Estimates

The capacity of each silo and storeroom at Dorginarti depends on numerous variables, including its original height and the type of grain or products stored in it.[1] The contemporaneity of some of the storage facilities is uncertain, but a range of population estimates based on the possible grain rations at any one time, the number of men needed to guard and defend the walls, and the number of housing units and residents is suggested here for the Level III fortress. In addition, the daily caloric requirements for the individuals dependent on grain distribution from the fortress are discussed. This study is partly based on the publication of the silos at Gala Abu Ahmed and their capacities.[2]

The variables that go into this exercise—considering especially the uncertain phasing of the storage structures and whether they were all at full capacity, but also the amount of grain spoilage or loss to rodents, whether the silos and magazines were used for bulk storage or container storage, and the robustness of the population and how much grain made up their diets—are at best one set of estimates among a myriad of possibilities.[3] The averages used in the following calculations introduce further imprecision into the final figures, but they are the only practical solution to determining a feasible population estimate for both the earlier and later phases of the Level III fortress.

1 In this discussion, the term *silo* indicates a circular grain storage structure, while *magazine* and *storeroom* refer to the orthogonal structures for storage found in the East and West Sectors. See also the functional attributes of silos discussed in Ben-Shlomo and Van Beek 2014, pp. 558–59.

2 Flache 2012.

3 The factors involved in the study of grain storage at Mesopotamian sites of the third millennium, using computer simulation and statistical methods, are found in Paulette 2015, pp. 41–53.

THE SILOS

The height of the silo walls at Dorginarti was not preserved, but their exterior and interior diameters and wall widths are known. The height of each was thus determined to be that of the interior diameter, as found in the estimates for silo reconstruction by Dieter Arnold and used for the Gala Abu Ahmed silos.[4] It is also assumed that the silos at Dorginarti were dome shaped, like those at Gala Abu Ahmed. The calculations here do not include the varying circumferences of domed roofs, which were most likely not filled, or the compaction of the grain at lower levels within the silos.[5] Table A.II.1 below lists all the grain storage structures preserved at the fort, their general phasing, and their potential capacities. They are also illustrated in figures A.II.1–A.II.2. The sections following the table briefly discuss the variables involved in the calculations.

There are structures that cannot confidently be added into the sums for grain capacity during particular phases at the fortress. The two latest silos in the West Sector, D 4 and D 32, are not certainly associated with the latest East Sector silos or with each other;[6]

4 Flache 2012, p. 48. The heights of the Dorginarti silos may have been less in some cases; see the 2–3 m average reconstructed height of silos in Paulette 2015, p. 44.

5 The filling of modern silos entails leveling off, or finding the angle of repose, for the peak that occurs at the top of the grain when filling, to reduce the moisture migration tendencies of the grain toward the top peak and to allow for better airflow between the top of the silage and the domed roof. This prevents mildew. See McKenzie and Van Fossen 1995.

6 The brick sizes used in the construction of the three south silos in the East Sector are the larger 40 cm lengths, as are the bricks used for the silo wall of D 4. This size was also used in silo D 213, two levels below the uppermost silo of D 4, and the Central Sector silos D 38 and D 38 S.

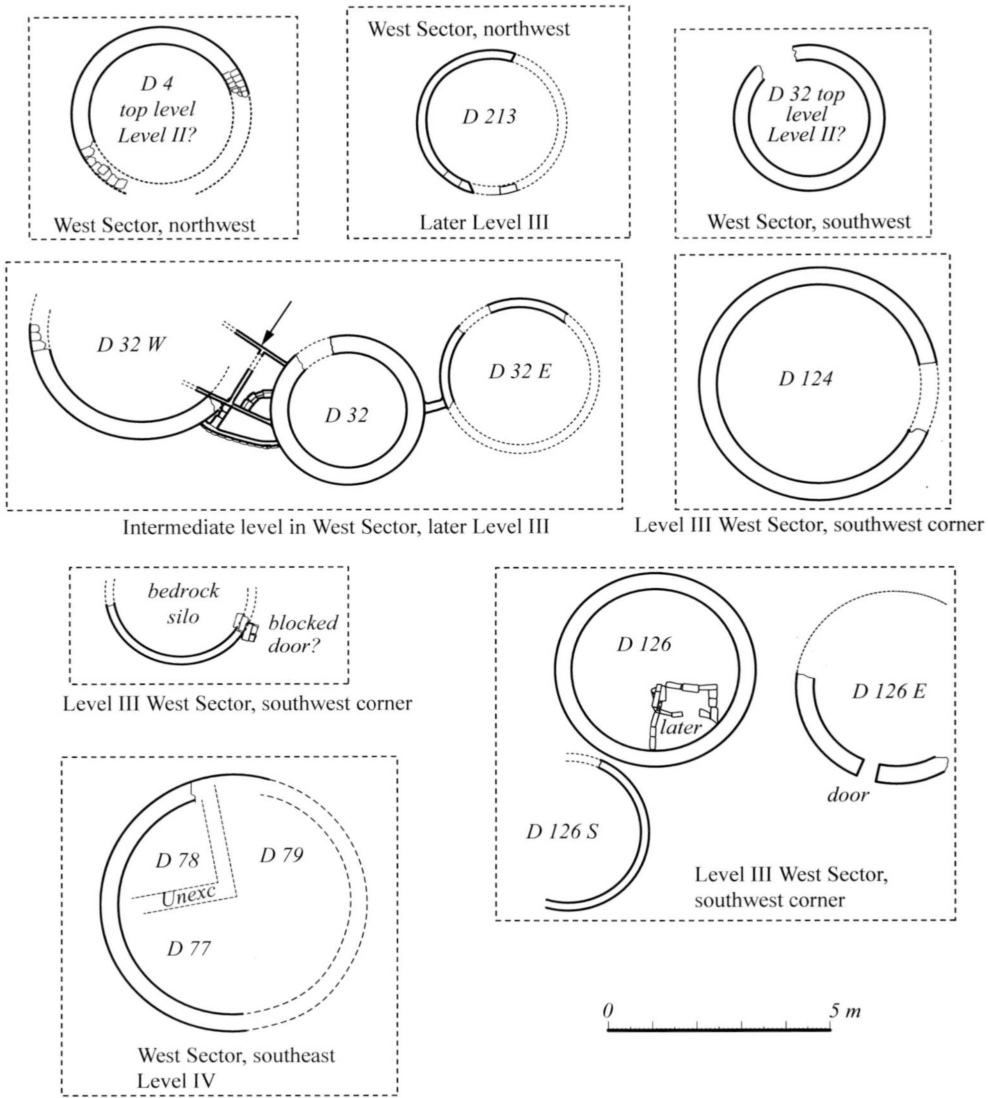

Figure A.II.1. Silos in the West Sector

silo D 77/D 78 was probably used only during Level IV; the capacity of the rectilinear storerooms or granaries in the East Sector, perhaps in use from Level IV on, is only briefly noted; and the Level III storage rooms or workrooms in the southwest corner of the West Sector are also only briefly mentioned.[7] Some of the latter units are workrooms, but container storage cannot be discounted, while those in the East Sector have no clear doorways and may have been used for bulk storage. All the examples just mentioned are not considered in the final calculations due to their imprecise phasing, although their capacities are noted in the tables below.

SILOS AND DENSITY OF GRAINS

The grain stored in the silos was almost certainly barley and wheat, along with other products provided by

7 The rooms with doorways, D 117–D 122, may have held special types of food, perhaps packed in jars or bags. They could have held 46.5 m^3 of grain, if an estimated height of 2.5 m is assumed, and about 34.9 m^3 if 25 percent is deducted for the interior space needed for movement and the stacking of containers. (If an additional 25 percent is applied for spoilage of grain, it would further reduce the capacity to 26.2 m^3.) The reduction used for container storage and interior walkways is that used by Paulette 2015, p. 45. The objects found in these rooms, plus the oven in D 122, suggest that at least some of the rooms along the south wall were workrooms. The structures in the East Sector were likely used for the bulk storage of grain and filled from the roofs. The total volume of these rooms, D 354–D 358, with a conjectural height of 2.5 m, is about 46 m^3. This includes the measurements reconstructed for D 354 and its possible capacity. Once again, deducting 25 percent for the spoilage of grain would result in a total capacity of 34.5 m^3 for the five rooms.

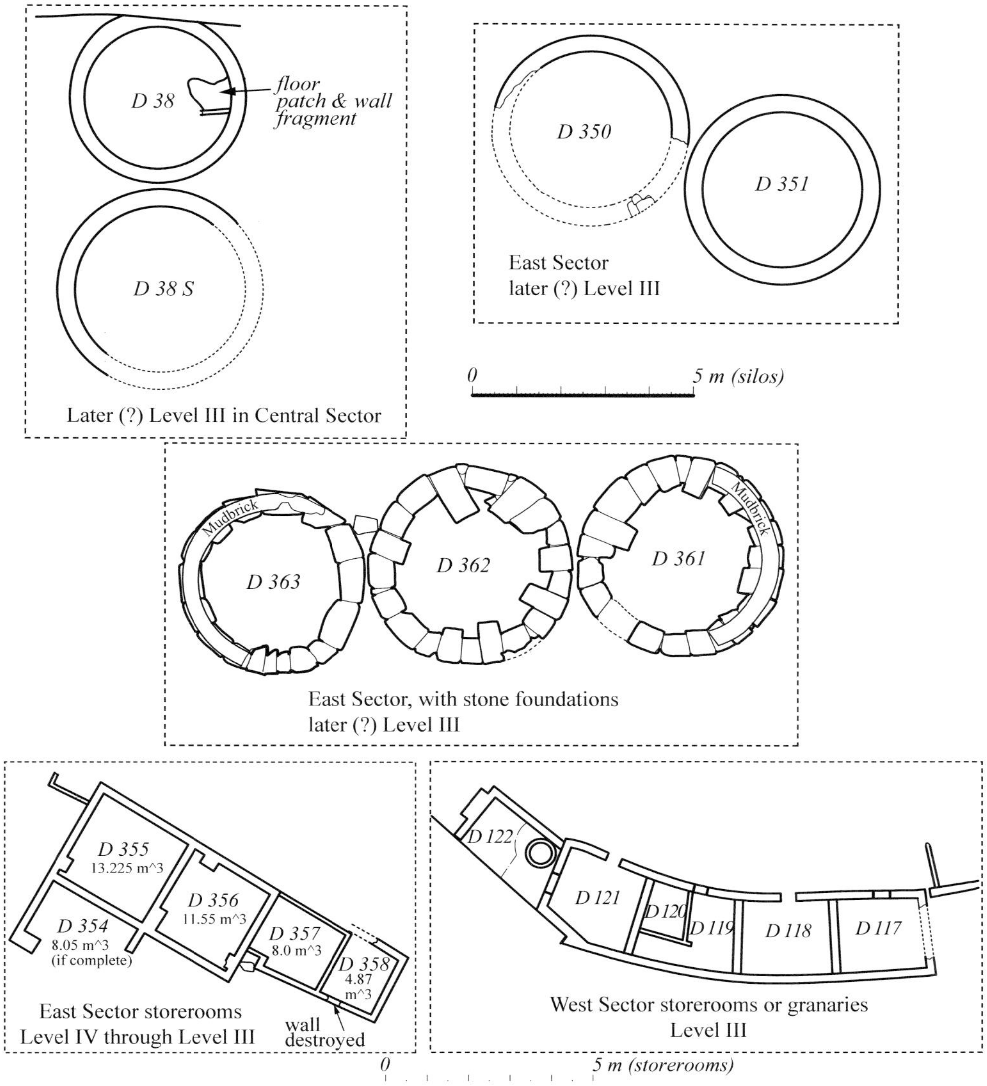

Figure A.II.2. Silos and storerooms in the West, Central, and East Sectors

the administration or acquired by individuals at their own initiative.[8] Rations likely covered some combination of wheat and barley, used to produce the vital staples of bread, porridge, and beer.[9]

The density of the grain in the silos at Gala Abu Ahmed was determined to be 800 kg/m^3, but according to the Food and Agricultural Organization of the United Nations, the density of (durum) wheat and barley is 770 kg/m^3 and 620 kg/m^3, respectively.[10] Using only the density of barley in the calculations shown in table A.II.1 results in approximately 19 percent lower capacities than if the silos were filled with wheat. The two densities were thus averaged: the capacities given in tables A.II.1–A.II.4 assume a grain density of 695 kg/m^3. The sums of the silo and magazine capacities are presented in table A.II.2 by location and the phase(s) in which they were probably used.

Silo D 4 and the topmost phase of D 32 are later and may have continued to be used in Level II at the

8 The types of grains and other foodstuffs found in excavations of early first-millennium BC sites in Nubia are discussed in chapter 1.

9 The southern inhabitants of the fortress may have preferred porridge and pancake bread, typically made later in the first millennium BC with sorghum and millet, but wheat and barley are also substituted for the more traditional grains in recipes from anthropological fieldwork; see Lyons and D'Andrea 2003, pp. 524–25. For more discussion on African versus Near Eastern dietary preferences, see the introductory paragraphs to chapter 5.

10 Emmer wheat is not listed, so the density of its close relative, durum wheat, is used; see Charrondiere, Haytowitz, and Stadlmayr 2012.

Table A.II.1. Silo capacities

Level(s)	Silo	Interior diameter (m)	Potential height (m)	Capacity (m^3)[a]	Capacity (kg)[b]	25% reduction for spoilage and loss[c]
IV	D 77/D 78	4.40	4.40	66.9	46,500	34,875
IV–III	East Sector magazines	–	–	45.7[d]	31,762	23,821
III	D 124	4.55	4.55	73.9	51,400	38,550
III	D 126[e]	3.70	3.70	39.8	27,700	20,775
III	D 126 E[f]	3.70	3.70	39.8	27,700	20,775
III	D 126 S	3.20	3.20	25.7	17,900	13,425
III	Bedrock silo[g]	2.90	2.90	19.1	13,300	9,975
III	D 117–D 122	–	–	46.5	32,318	24,238
Late III and II(?)	D 32	2.70	2.70	15.5	10,800	8,100
Late III	D 32 E	3.20	3.20	25.7	17,900	13,425
Late III	D 32 W	4.20	4.20	58.2	40,400	30,300
Late III	D 38[h]	3.20	3.20	25.7	17,900	13,425
Late III	D 38 S	3.90	3.90	46.6	32,400	24,300
Late III	D 213	3.00	3.00	21.2	14,700	11,025
Late III	D 350	3.70	3.70	39.8	27,700	20,775
Late III	D 351	3.70	3.70	39.8	27,700	20,775
Late III	D 361[i]	3.70	3.70	39.8	27,700	20,775
Late III	D 362	3.70	3.70	39.8	27,700	20,775
Late III	D 363	3.20	3.20	25.7	17,900	13,425
Latest III and II(?)	D 4	3.20	3.20	25.7	17,900	13,425

[a] The formula for calculating the volume of a cylinder is $V = \pi r^2 h$. The equation for orthogonal structures is $V = wlh$. All numbers are rounded to the nearest tenth.

[b] Assuming an average grain density of 695 kg/m^3.

[c] The spoilage for rodent and insect activity, as well as for mold, is considered here to be 25 percent; see Paulette 2015, pp. 49, 52. Paulette deducts 15–25 percent in his simulations, while a 30 percent loss to insects, pests, mold, and moisture was reported by United Nations World Food Programme 2016.

[d] See n. 7 above.

[e] Knudstad (1964a, p. 48) notes that the interior bin construction in D 126 is later. No special calculations were supposed for this silo, although the bin may indicate less storage capacity or container storage.

[f] D 126 E is the only silo at the fortress with clear evidence for a door.

[g] Silo D 32 W was built over the remains of the bedrock silo in a later phase of Level III (i.e., the intermediate level in the West Sector).

[h] The wall fragment shown on the plans may belong to an interior bin and appears to sit atop a patch of the floor associated with the silo. However, no elevations are given for the wall fragment, and an interior division of the silo is not considered here. See Knudstad 1964a, p. 40.

[i] The three south silos in the East Sector are measured according to the reconstructed brick walls sitting atop the stone foundations, so their actual diameters are smaller than those of the stone bases. No floors were found within these silos, as is the case for most of the silos found at the site.

fortress. The silos of the later Level III phase in the West Sector, D 32, D 32 E, and D 32 W, were built atop the earlier Level III silos along the south wall. That is, the bedrock silo associated with a level on par with silo D 124 was built over by D 32 W, while silos D 32 and D 32 E were built on the remains of silo D 126 S. The silos in the East and Central Sectors were built using the larger, 40 cm long bricks, which is indicative of a later renovation phase (Level III), while the bricks of the three silos at D 32 were not noted. Certainly other silos and storage facilities could have existed, but they did not leave any trace in the archaeological record.

RATION ESTIMATES

Table A.II.3 presents the numbers of rations that could be handed out according to the calculations given above and the ration amounts determined by other scholars. Once again, the earliest silo D 77/D 78 and the latest silos in D 32 and D 4 are not included, and the ration amounts used in table A.II.3 are based on an average grain density of 695 kg/m^3.

The two silos at Gala Abu Ahmed were assumed by Michael Flache to hold enough to feed about 140 people annually, based on their capacity and rations totaling 800 kg of grain per year.[11] Table A.II.3 also considers the rations used by Barry Kemp for personnel at the Middle Kingdom Nubian forts and the workmen's town of Kahun—allotments of 365 kg/year (maximum) and 219 kg/year (minimum).[12] Flache and Kemp also mention the 500–600 kg/year rations determined by Christian Tietze in his work on the Amarna silos, which is here averaged to 550 kg/year in table A.II.3.[13] Also used in table A.II.3 is the ration amount for soldiers of imperial Rome (310.25 kg/year), discussed by Jonathan P. Roth.[14]

POSSIBLE CALORIC REQUIREMENTS

According to Roth, the necessary caloric intake of a 30-year-old Roman soldier standing 170 cm tall and weighing 66 kg is assumed to be about 3,000 calories per day, the modern Recommended Daily Allowance for individuals as noted by the United States Army.[15] But, noting that the consumption of fewer calories would not lessen the soldier's ability to be effective in daily activities and fighting, he suggests 2,500 calories per day.[16] The Romans preferred wheat to barley, and the daily wheat ration of two sextarii is cited by various scholars as weighing anywhere between 0.7 and 1.4 kg.[17] According to Roth, the 0.85 kg/day grain ration would provide about 2,882 calories per day.[18] The textual sources Roth uses state that Roman rations also included meat, cheese, "vegetables" (usually legumes), oil, vinegar or sour wine, and salt.

Ration texts from the Middle Kingdom fort at Uronarti indicate that daily rations included both wheat and slightly lesser measures of barley, with a grain ration that amounted to about 2,136 calories per day for each person.[19] This figure is reached by assuming 354 calories per 100 g of wheat and 360 calories per 100 g of barley (see below). The Deir el-Medina texts mention wheat and barley in workers' rations, at an amount that would provide about 2,187 calories per day for four individuals in a household.[20] The modern daily caloric requirement for moderately active males aged 18–29 years and weighing 60–65 kg is 2,800–3,150 calories per day.[21] Males in the same age group and weight category, but engaged in more vigorous activities, need upward of 3,250 calories per day. The calorie requirements vary, of course, depending on height, robustness, and age. Once again, this caloric intake may be overly generous to apply to our population.

11 Flache 2012, p. 49; Eigner 2013, pp. 316–17. The other variables used for calculating population numbers at Gala Abu Ahmed also rely on the number of men needed to man the walls around Gala Abu Ahmed, which then results in a population estimate of about 225 soldiers.

12 Kemp 1986, pp. 132–34.

13 Tietze 1986, p. 72. Tietze, like Flache, assumes a grain density of 800 kg/m^3.

14 Roth 1999, pp. 23–24, 42–43. The ration of 1 liter per day weighs 0.85 kg (pp. 18–19).

15 The average height of a Roman soldier was regulated according to whether or not he served in an elite unit and is estimated at 162–71 cm for ordinary soldiers and 173–79 cm for elites; see Roth 1999, pp. 9–10.

16 Roth 1999, pp. 12–13, n. 41.

17 Roth 1999, pp. 23–24, table II.

18 Roth 1999, pp. 42–44, table III.

19 Miller 1991, pp. 257–58.

20 Miller 1991, p. 263. These numbers are based on the monthly rations of the lowest-paid workmen at Deir el-Medina.

21 See also the nutritional requirements according to age, weight, and level of physical activity in Food and Agriculture Organization of the United Nations 2004, pp. 35–42, tables 5.4–5.5. These initial numbers are calculated for only one age group—males weighing 60–70 kg with moderate activity levels (construction work, but less than daily agricultural work). They also consider caloric intake from a varied diet, not just the consumption of grain-based foods. Kemp cited army prison records from 1917 in Egypt, which showed that male military prisoners doing hard labor needed 3,200 calories per day (the type of food intake is not given); see Kemp 1991, p. 132.

Table A.II.2. Contemporary silos and capacities

Level(s):	IV	IV–III East Sector	III West Sector	III West Sector[a]	Late III West and Central Sectors	Late III East Sector	Latest III and II(?) West Sector
Silo or magazine	D 77/D 78	D 354–D 358	D 124, D 126, D 126 E, D 126 S, and bedrock	D 117–D 122	D 32, D 32 E, D 32 W, D 38, D 38 S, D 213	D 350, D 351, D 361, D 362, D 363	D 4, D 32
Capacity (m^3)	66.9	45.7	198.3	46.5	192.9	184.9	41.2
Capacity (kg)[b]	34,875	23,850	103,500	24,225	100,575	96,525	21,525

[a] If these rooms were used for container storage, it would reduce the stated capacities by up to an additional 25 percent. See n. 7 above.

[b] Assuming an average grain density of 695 kg/m^3 (less 25 percent).

Table A.II.3. Ration estimates

Level	Units 800 kg/yr (Flache/Eigner)	Units 550 kg/yr (Tietze)	Units 219 kg/yr (Kemp minimum)	Units 365 kg/yr (Kemp maximum)	Units 310.25 kg/yr (Roth)
III West Sector silos	12	188	473	284	333
III West Sector magazines (D 117–D 122)	30	44	111	66	78
Late III West and Central Sectors (plus D 213)	126	182	479	276	324
Late III East Sector	121	176	441	265	311
IV–III East Sector magazines (D 354–D 358)	30	43	109	65	77

PHYSIQUE

The height of a Roman soldier mentioned above, 170 cm, is the same as that of Seti I (170 cm) and close to the height of Ramesses II (170.7 cm), although the average height of all known New Kingdom pharaohs was 166.9 cm.[22] These are the measurements for an elite group whose diets were probably quite varied and nutritious compared to that of individuals from lower social classes. This height is comparable to taller male skeletons from the predynastic and Old Kingdom periods, as well as the skeletons of a higher-status group from Classic Kerma cemeteries in Nubia, where the mean height of males was 167 cm.[23] The skeletal material from Tombos also shows that Napatan individuals were taller than C-group or New Kingdom individuals, comparing with the Classic Kerma individuals

22 Raxter 2011, pp. 183–86, tables 31–32. The heights quoted here are mean heights only and do not consider the standard deviations quoted in the article.

23 The mean height of elite females was 155.7 cm; see Raxter 2011, p. 124, table 10, and pp. 130–31, table 15.

Table A.II.4. Level III ration estimates and caloric value

Level	Silo	Capacity (kg) less 25%	Personnel 800 kg/yr (Flache/Eigner)	Personnel 550 kg/yr (Tietze)	Personnel 365 kg/yr (Kemp maximum)	Personnel 210.25 kg/yr (Roth)
Early III	D 124	38,550	48	70	106	124
Early III	D 126	20,775	26	38	57	67
Early III	D 126 E	20,775	26	38	57	67
Early III	D 126 S	13,425	17	24	37	43
Early III	Bedrock	9,975	12	18	27	32
Totals	–	**103,500**	**129**	**188**	**284**	**333**

Level	Silo	Capacity (kg) less 25%	Personnel 800 kg/yr (Flache/Eigner)	Personnel 550 kg/yr (Tietze)	Personnel 365 kg/yr (Kemp maximum)	Personnel 210.25 kg/yr (Roth)
Late III	D 32	8,100	10	15	22	26
Late III	D 32 E	13,425	17	24	37	43
Late III	D 32 W	30,300	38	55	83	98
Late III	D 38	13,425	17	24	37	43
Late III	D 38 S	24,300	30	44	67	78
Late III	D 213	11,025	14	20	30	36
Late III	D 350(?)	20,775	26	38	57	67
Late III	D 351	20,775	26	38	57	67
Late III	D 361	20,775	26	38	57	67
Late III	D 362	20,775	26	38	57	67
Late III	D 363	13,425	17	24	37	43
Totals	–	**197,100**	**247**	**358**	**541**	**635**
Ration/day	–	–	2.19 kg	1.51 kg	1 kg	0.85 kg
Calories/day	–	–	7,402	5,104	3,380	2,873

just mentioned.[24] The evidence for Third Intermediate Period individuals in Egypt, living during the Twenty-Second to Twenty-Fifth Dynasties, shows that the height of adult males ranged between 162 and 169 cm, while females stood between about 153 and 157 cm.[25] The males at Dorginarti were probably no shorter or taller than the various studies indicate.

24 Gibbon and Buzon 2016, pp. 327–34, tables 1–4. See also Buzon 2014, pp. 4–6, table 6 (with 60 percent of samples being females). The size differences are thought to reflect the improved environmental conditions (less stress and better nutrition) and greater physical activity of Napatan individuals.

25 Eladany 2012, pp. 287–88. For more information on the determination of height, age, sex, and place of origin, see the catalog on pp. 390–471.

CALORIES PER RATION

There are approximately 339 calories in 100 g of wheat and 337 calories in 100 g of barley, for an average of 3,380 calories in 1 kg.[26] Gala Abu Ahmed's large ration of 800 kg/year, justified by the inclusion of other family members and privately owned animals, would provide about 7,400 calories per day. Tietze's average of 1.51 kg/day would provide 5,104 calories per day. Kemp's maximum ration of 1 kg/day would provide 3,380 daily calories, and his minimum would provide 2,028 calories. Roth's grain ration amount, as mentioned above, works out to 2,882 calories per day. Which types of foods and beverages were consumed besides bread or porridge and beer is uncertain, although fish, meat, fruits, and vegetables were undoubtedly eaten occasionally, either sourced/grown by the fortress residents or local inhabitants or supplied by the administration.

POPULATION ESTIMATES OF LEVEL III ACCORDING TO RATIONS

The total amount of grain at Dorginarti in the earlier phase of Level III could potentially have supplied anywhere from 129 to 333 ration units per year, depending on whose calculations are used, with an additional amount perhaps supplied by stores in D 117–D 122 and D 354–D 358 (these latter structures are not considered here). The silos from late Level III in the West, Central, and East Sectors could potentially supply between 247 and 635 rations per year. It must be considered whether the daily ration amounts varied during strenuous fighting or an expedition, or when an influx of soldiers was needed at the fortress, or whether the silos were meant to store grain for less than a year or multiple years. The ration amounts may have been intended to supply family members, in which case the size of the distribution might have been larger, as suggested by the 2.19 kg/day of Flache and Eigner, or the 1.51 kg/day of Tietze. The number of grain shipments arriving at the fortress may have varied depending on the administrative needs in the region.

If the large silos in the East Sector also belong to the earlier phases in Level III, the number of rations available would increase dramatically.

The question of where the grain shipments originated has to consider areas where grain could be grown, in which case either southern Egypt or the region to the south of the Second Cataract is likely. The question must also be asked whether the grain was meant for the military forces stationed at the fortress and nearby, or whether the shipments were part of a trade in grain that is not mentioned in any known texts.

In addition to looking at grain storage and the capacity for rations, scholars have used a number of methods to approximate the population of settlements. This discussion of population estimates for Dorginarti will now review a few area- or dwelling-based estimates.[27]

RESIDENTIAL UNITS, SPACE, AND POPULATION ESTIMATES

The Level III houses along the north fortification wall indicate that at least twelve to fourteen different residential units are represented. In addition, at least twenty-four more houses of similar size may have been present in the north half of the sector. Allowing room around the two larger buildings on the south side of the West Sector for open spaces or other administrative, work, and storage rooms still leaves space for at least ten to thirteen other houses of sizes similar to those along the north wall. A fortification would most likely accommodate the largest number of people possible, so these estimates may be too low, especially considering the lack of a known architectural configuration in the central part of the West Sector that most likely included barracks or smaller houses for single soldiers or staff.[28] The reconstruction shown in figure A.II.3 does not consider barracks, so an estimate of about fifty houses of single-story construction is illustrated.

Nine of the structures along the north wall in the West Sector were measured, with interior spaces that ranged from 11.8 m^2 (D 70) to 28.6 m^2 (house D 59), thus resulting in an average size of 21 m^2. The West Sector is approximately 2,394 m^2 or 0.24 hectare in size, but the space for roads, larger administrative buildings, courtyards, and silos should be discounted, and perhaps only the actual roofed areas should be considered, so the amount of available living space could be lowered to 70 percent of the total square footage (1,676 m^2). One hypothesis suggests that 625 people could be accommodated in a densely packed hectare, with a resulting Level III population of about 100 people.[29] But if the population was even denser, at 1,000 people per

26 United Nations World Food Programme 2016. The table does not include emmer wheat, so the caloric content of a close relative, durum wheat, is used.

27 See especially Kardulias 1992 and Zorn 1994.

28 See the comments about military settlements in Kardulias 1992, pp. 280, 284, and note also Zorn 1994, p. 41.

29 Kardulias 1992, p. 284. See also Zorn 1994, p. 36, table 2.

Figure A.II.3. Reconstruction of housing and roads in the West Sector

hectare, the population rises to an estimated 170. The higher estimate of residents per hectare becomes more viable within an enclosed military settlement.[30]

If the fifty houses projected for the West Sector each held between four and six people, as suggested by a few ethnographic analogies, the population could then have ranged between 200 and 300 residents.[31] Without knowing the architectural configuration in the central part of the West Sector, a true estimation of the Level III population remains elusive.

MANNING THE WALLS

It has been suggested that the Nubian fortresses would have needed one soldier stationed every 2 m around the fort enclosure in peacetime, and a soldier stationed every 1 m during combat, plus an additional 25 percent for administrative, supply, and reserve personnel.[32] These considerations would mean that the 340 m of walls around the fortress would have required 170–340 soldiers depending on the military situation, and thus 213–425 total military and civilian individuals. The larger number of people could not have resided at the fortress and were undoubtedly settled in camps outside the fortress or even on the western shore.[33] It is perhaps unlikely that an assault on Dorginarti would have needed such a large number of soldiers, since the river already protected the fortress from well-integrated land and naval attacks.

CONCLUSIONS

A daily grain ration at Dorginarti may have served the needs of a single soldier, or perhaps been shared with

30 The amount of enclosed floor space needed by an individual is cited in Kardulias 1992, pp. 278–80. The author discusses ethnographic examples and conditions at military establishments in detail (pp. 280–85). The personal space for each of the 170 individuals at the fortress, in Level III, is close to 10 m^2, which is perhaps too much according to Kardulias's discussion.

31 The average family size in a premodern, unwalled village ranges between 4 and 5.5 persons; see Zorn 1994, p. 33, but also p. 40, table 4.

32 Flache 2012, p. 49, n. 7, who cites Steiner 2008, pp. 81–83.

33 See, for example, Steiner 2008, pp. 81–82, where he suggests that as many as 450 soldiers may have been needed for a fortified enclosure of 300 m circumference. The spacing between the buttresses along the north and south walls is anywhere between roughly 8 and 18 m, with a typical distance of 10 m and an average of 12 m. Using this hypothesis, the 340 m wall circuit around Dorginarti would have required more than 500 men during a military encounter.

family members or used as fodder for livestock, pack, and riding animals. Additional provisions such as beer, fruit, vegetables, and meat are not considered here. The few house or communal-area ovens preserved along the north wall suggest that individuals did not always prepare their own bread.

If the storage facilities from both the earlier and later phases of Level III supported a permanent population of 170 resident soldiers and 30 civilians, who were each supplied with grain rations of 0.85–1.00 kg/day depending on their roles, then nearly 26 kg/day would have been provided to the 30 civilians, furnishing a daily intake of 2,873 calories per individual. Additionally, 170 kg of grain would have been supplied daily to the soldiers, providing 3,380 calories for each of them. The Level III silos alone provided almost 530 rations per year, considering the varying rations allotted to the civilians and soldiers mentioned above, while the late Level III silos would have provided the same population with 1,008 rations per year.

Thus, the Level III silos had to provide 71,358 kg of grain each year for this resident population, which leaves a surplus of 32,143 kg after one year. This is enough grain for an additional six months for the population estimate. The later Level III capacity needed to provide the same amount, but could provide two years' worth of rations for the 200 people mentioned above, with a surplus of 54,385 kg of grain. But the scenario would also include rations for nonresident soldiers and personnel who were active in the region.

Appendix III

Tables

Table A. Numbered and descriptive loci

NUMBERED LOCI

Locus	Sector	Stratum	Description
D 1	West Sector		Area/room along north fortress wall at west end of sector
		1	Surface to floor 2 (ca. 1.15 m)
		2	Floor 2 through "ash" layer to floor 3 (ca. 15 cm)
		3	Excavation of floor 3? Packed gray-brown soil below it to floor 4 (ca. 30 cm; stratum runs under inner lamination)
		4	Test pit in northeast corner; gray-white sand (ca. 1 m)
		5	Layer of burned alluvium
D 2	West Sector		Test pit in center of fortress (ca. 1.80 m deep)
D 3	Central Sector		Clearance below surface of Level II platform? Arbitrary unit south of residence
		1	Intentional rubble fill of Central Sector platform (ca. 3.0–3.5 m?)
D 4	West Sector		Surface clearance in northwest corner (uncertain depth)
		1	Tracing of silo walls?
		Clearing silo	Fill in silo to floor
		Removal of silo	Dismantling of silo walls
		Dismantling walls in area	Dismantling of walls in D 4, D 47, and D 48
		Northwest corner of D 4	Space behind silo and to the northwest of it
		Sealed under D 4	Deposit directly under D 4, before appearance of walls for D 47 and D 48 (same as "D 47 and D 48, under D 4")
D 5	Central Sector		Clearance below surface of Level II platform up to tops of Level III walls? (ca. 20–30 cm max.?)
		1	Intentional fill within room (ca. 2 m)
		2	Compact sand layer up to floor (ca. 25 cm)
		2 test pit	Test pit on south side, cutting through floor and into Level IV building (ca. 1 m depth) (not shown on plan)
		3	Ash lens above floor and in northeast corner by door to D 9 (ca. 5 cm)
		4	Uneven mud-plastered floor of uncertain thickness
D 6	West Sector		Room along north fortress wall
		1	Clearance of south half of room to floor (surface to top of walls is 66 cm and to floor 1 is ca. 1 m)
		2	Floor 1 down to, but not including, floor 2 (ca. 25 cm)
		3	Floor 2 down to, but not including, floor 3 (ca. 25 cm)
D 7	West Sector		Surface to "ash" layer clearance; above D 66, D 67, and D 69
		Surface to "ash"	Same as above
		1	Fill within space enclosed by walls

(continued)

Table A. Numbered and descriptive loci *(continued)*

Locus	Sector	Stratum	Description
D 8A–B	Central Sector		Clearance below surface of Level II platform, west of Level III official residence; main fortress wall to north and south, and Level II retaining wall to west; D 8A is the south half of this locus, and D 8B is the north half
		1	Intentional rubble fill (ca. 2.5+ cm?); above D 37; note locus "D 37, D 38, and D 8, mixed"
D 9	Central Sector		Clearance below surface of Level II platform? Main fortress wall to south; D 9A is the north half (north of south wall of Level III room), and D 9B is the south half
		1	Intentional rubble fill (1.30–1.45 m)
		2	Stratified sand layers mixed with brick rubble (40 cm)
		3	Layer of ash, charcoal, and burned brick (ca. 5 cm)
		4	Deposit atop floor (ca. 1 cm max.)
		5	Uneven mud-plastered floor of Level III room; column base in middle
D 10	West Sector		Surface and oven clearance to west of wall in intermediate level? Floor eroded away (ca. 15 cm?). Wall bricks measured 38 × 20 × 7–8 cm
		1	Surface to floor in a room or courtyard east of D 7 (ca. 55 cm from bottom of topmost oven to floor 1)
D 11	West Sector		East of wall and D 10 in intermediate level; perhaps clearance down to lower room walls (uncertain depth). Wall fragment in this upper level was constructed of bricks measuring 40 × 19 × 8 cm
		1	Fill within room
D 12	Outside walls		Surface clearance to west of eroded west fortification wall; between north buttress and West Gate (uncertain depth)
		1	Below surface clearance
D 13	Outside walls		Surface clearance to west of west fortification wall; between south buttress and West Gate (uncertain depth) (note also context described as "eroded surface of west wall of fortress, south of West Gate")
		1	Below surface clearance
D 14	West Sector		Room south of D 6; clearance of surface deposit to trace walls
		1	Fill within room to floor level (uncertain depth)
D 15	Central Sector		Area atop Level II platform
		1	Surface clearance of deposit in and upon Level I buildings (uncertain depth)
		2	All sherds built into, and immediately below, Level I buildings
D 16A–C	Central Sector		Clearance of Level II deposits until tops of walls of rooms below
		1	Intentional rubble fill; excavation of upper part of fill is of uncertain depth; lower compacted fill was ca. 60–70 cm thick
		2	Charcoal, ash, and burned-brick layer (ca. 3–6 cm)

Table A. Numbered and descriptive loci *(continued)*

Locus	Sector	Stratum	Description
D 16A–C	Central Sector	3	Thin layer of sand over floor (ca. 5 cm)
		4	Uneven mud-plastered floor
D 17	West Sector		Test trench through inner lamination of north fortress wall on west side of eroded section
		1	Beneath and south of lamination; several strata within D 17 (1) were not separately numbered, including surface debris (uncertain depth), a thin layer of ash running under fortress wall (4 cm), soil and ashy layers (40 cm, including 23 cm of grayish-black deposit at bottom), burned alluvium (6 cm), and alluvium of uncertain depth
D 18	Outside walls		Thin-walled structure built on cataract rock within large bay outside southwest corner of fortress
		1	Surface debris sloping away from fortress wall (1.12 m max.)
		2	Archaeological remains from atop and within cataract rocks of buildings
D 19	Outside walls		Thin-walled structure built on cataract rock within large bay outside southwest corner of fortress
		1	Surface debris sloping down from fortress wall (ca. 94 cm); mud-plastered floor atop cataract rocks
D 20	Central Sector		Hallway in Level III residence
		1	Layer of sand filling from intrusive well on west side of locus
		2	Intentional rubble fill
		3	Stratified layers of sand and eroded brick; fallen stone lintels
		4	Layer of ash and charcoal (8 cm max.); a thin lens between part of levels 3 and 5
		5	Sand and brick debris (uncertain depth)
		6	Uneven mud-plastered floor
D 21	Outside walls		Structure built in east corner of bay outside southwest corner of fortress
		1	Surface clearance of brick rubble (1.22 m max.); fireplace in northeast corner
D 22	Central Sector		Room in Level III residence
		1	Layer of gray sand filling intrusive pit on southeast
		2	Intentional rubble fill (ca. 1 m max.)
		3	Thin layer of stratified sand and eroded brick (ca. 58 cm)
		4	Uneven mud-plastered floor
D 23	Central Sector		Fill in area north of Level III residence; fortress wall on north; east and west boundaries defined by width of residence; excavation around base of Level III north stairs
		1	Intentional rubble fill sloping down from north fortress wall
		2	Stratified sand and eroded brick sloping down from fortress wall (ca. 28 cm)

(continued)

Table A. Numbered and descriptive loci *(continued)*

Locus	Sector	Stratum	Description
D 23	Central Sector	3	"Street" surface around Level IV residence (test pit in unit, before excavation, revealed floor level about 2 m below top of rubble)
D 24	Central Sector		Room in Level III residence
		1	Layer of gray sand filling intrusive pit on east
		2	Intentional rubble fill
		3	Mud pad resting against south wall and sloping down to east
		4	Stratified layers of sand and broken brick
		5	Rough, mud-plastered floor with mud pad in southwest corner
D 25	Outside walls		Bay on exterior of south fortress wall (east of D 18, D 19, and D 21); brick rubble on cataract rock; column base toward northwest corner perhaps in situ
D 26	Central Sector		Area east of Level III official residence; fortress wall to north and south, retaining wall on east
D 27			No information for this number; perhaps not used
D 28	West Sector		Circular kiln(?) built on plastered floor in northeast corner of bay on west side of Level II platform; ca. 50 cm below surface and ca. 65 cm deep; filled with potsherds and 25 cm of ash at bottom; Level I Christian structure
		Fill under D 28	Sherds from layer on which foundation of kiln was built (uncertain depth)
D 29	East Sector		East side of east retaining wall for Level II platform; north of stairs; clearance of deposit (uncertain depth)
D 30	West Sector		Surface(?) deposit north of kiln D 28 (uncertain depth)
D 31	West Sector		Surface(?) deposit south of kiln D 28 (uncertain depth)
D 32	West Sector		Silo along south fortress wall; immediately below surface; over area of earlier D 117 and D 118
		1	Clearance of silo; sand and broken bricks; ca. 50 cm from tops of silo walls is a plastered floor; tops of underlying walls below it (ca. 70–95 cm below surface); 40 cm below tops of these walls is another floor
		Dismantling walls in area	Removal of silo D 32 and various fragmentary walls on same level that lie above walls of D 118–D 121
			Walls to west of D 32 rest atop lower silo (= "west of D 32 [1]")
D 33	East Sector		Easternmost room built into convergence of fortress walls at east end
		1	Surface clearance in room (uncertain depth)
		2	Fill below surface clearance (82 cm)
		3	Grayish-black stain ("ash" layer in rest of fortress) (6 cm)
		4	Layer of rubble and series of thin deposits (47 cm)
		5	Floor 1 and, below, cataract stone foundation

Table A. Numbered and descriptive loci *(continued)*

Locus	Sector	Stratum	Description
D 34	East Sector		Room to west of D 33 at east end of fortress
		1	Surface clearance? (uncertain depth)
		2	Fill below surface clearance (uncertain depth; ca. 82 cm?)
		3	Dark stain ("ash" layer in rest of fortress) (8 cm)
		4	Layer of rubble and thick deposits (60 cm)
		5	Floor; atop cataract stones
D 35	West Sector		Clearance of surface to "ash" layer along inside of south fortress wall (uncertain depth and westward extent); westward from breach (atop D 110, D 111, etc.); part of clearance included upper layer in D 36, but sherd bags marked "D 35 and 36 mixed"
D 36	West Sector		"Ash" layer to tracing of walls of rooms D 100–D 103 (uncertain depth)
D 37	Central Sector		Area with floor level extending into silo D 38; below D 8; clearance of deposit?
D 38	Central Sector		Silo in northwest corner of sector; clearance of fill? (uncertain depth); cross walls in silo about 8 cm below D 37 floor patch; another floor about 16 cm below top floor patch; bricks used to build this silo, and silo to south, measured 40 × 19 × 7 cm
D 39	Central Sector		Space in northwest corner of sector; clearance of deposit in area of uncertain depth; mixed bags of remains from D 38 and D 39
D 40	West Sector		Room near north fortress wall at east end of West Sector
D 41	West Sector		Room near north fortress wall; west of D 40
		1	Room fill between walls and down to floor level
D 42–D 44	West Sector		Surface clearance of rooms along north fortress wall; D 42–D 44 have same floor level, ca. 25 cm lower than rooms to east and west; floors of D 40 and D 41 are on same (higher) level, but D 42 also has patch of floor on same level as D 41
		1	Deposits within rooms
		2	D 42 (2) is a floor at same level as that in D 43; deposit below floor?
D 45	West Sector		Room along north fortress wall at east end of West Sector
D 46	West Sector		Room near north fortress wall; toward central part of sector where walls were eroded away; "ash" layer below surface on same level as "ash" level in D 40 and D 41
		Pit	On west side of locus (uncertain depth); sherds from this pit joined with sherd from D 41
D 47–D 49	West Sector	Below surface	Testing for wall remains in area around D 4, before D 4 removed (clearance of uncertain depth)
		D 47 and D 48, under D 4	Same as "Sealed under D 4"
		D 47 (1)	Same level as D 48 (2)

(continued)

Table A. Numbered and descriptive loci *(continued)*

Locus	Sector	Stratum	Description
D 47–D 49	West Sector	D 48 (1)	Traces of floor at level of D 4 floor, but outside D 4; preserved against north fortress wall in west part of D 48; clearance of layer?
		D 48 (2) square bin	Deposit in bin
		D 48 (2) oven	Deposit in oven
		D 49	Part of D 48
		Under oven in northwest corner	Deposit below D 48 (2) oven
D 50	West Sector		Locus between D 1 and D 4 along north fortress wall; surface clearance
D 51–D 52			No information for these numbers; perhaps not used
D 53	West Sector		Room or courtyard near north fortress wall; surface to "ash" layer
		1	Room deposit to floor level; rectangular bin and hearth constructions
D 54	West Sector		Room near north fortress wall; clearance of surface to "ash" layer?
D 55	West Sector		Room near north fortress wall; clearance of surface to "ash" layer?
		1	Room fill to floor level
D 56	West Sector		Alleyway(?) between buildings along north fortress wall
D 57	West Sector		Stairway and foundation at north end of D 56
		1	Room deposit above floor
D 58	West Sector		North continuation of alleyway(?) D 56 and passage to west of D 57
		1	Clearance of deposit down to floor level
D 59	West Sector		House along north wall, hearth in main room; fill below surface before tracing of room walls
		1	Fill within room
		2	Fill above floor and floor(?)
		3	Below floor until base of wall discovered, and fill below this level
D 59A–F	West Sector		Clearance below surface deposit (D 75) and fill (D 81), but before tracing tops of walls of building(s); light architectural remains to the northeast (uncertain depth)
		1	Fill between walls
D 60	West Sector		Two-room building at midpoint of north fortress wall
		1	Fill between walls down to floor 1
		2	Deposit below floor 1 (ca. 20 cm); east–west cross wall found in this stratum and locus divided
		North (2)	Test trench north of east–west wall; floor 2 found ca. 40 cm down (elevation not on plan but in field notes); same level and phase as D 68 (3)

Table A. Numbered and descriptive loci *(continued)*

Locus	Sector	Stratum	Description
D 60	West Sector	North (3)	Deposit below floor 2 in northwest part of test trench (ca. 20 cm)
D 61			No information for this number; perhaps not used
D 62	West Sector		Room(?) near north fortress wall
		1	Fill within wall
D 63			No information for this number; perhaps not used
D 64	West Sector		Alleyway between buildings along north fortress wall
D 65	West Sector		Alleyway between buildings along north fortress wall; continuation of D 64
D 66	West Sector		Room or courtyard in building along north fortress wall; below D 7
		1	Fill within walls to floor level
D 67	West Sector		Room along north fortress wall
		1	Fill within room down to floor level
D 68	West Sector		Room along north fortress wall
		1	Deposit below surface clearance to discovery of tops of earlier walls; flat stone by door is higher than floor levels identified in room (not all information is on plan, but is in field notes)
		Test trench	In northwest corner of locus, north of east–west cross wall
		2	Fill from top of north–south cross wall down to earlier floor (floor elevation not known)
		3	Original floor associated with the two internal walls
		4	Gray sand deposit below floor 3, with some ash and sherds (uncertain elevations and depth)
D 69	West Sector		Small room along north fortress wall; below D 7 surface clearance
		1	Fill in room to floor; storage bin and small bench or ledge along wall founded on this floor (elevation not identified on plan or in field notes)
D 70	West Sector		Room in building near north fortress wall, but toward center; eroded walls (deposit of uncertain depth)
D 71	West Sector		Room(?) near north fortress wall (uncertain depth); toward center of sector and eroded walls
D 72	West Sector		Small room in building near north fortress wall; surface to "ash" clearance
		1	Clearance within walls and down to floor level (uncertain depth)
D 73	West Sector		Room in building along north fortress wall; surface clearance?
		1	Fill within walls down to floor level (uncertain depth)
D 74	West Sector		Room/courtyard in building near north fortress wall
		1	Fill in room down to floor level (uncertain depth); east–west wall of earlier construction runs below this locus

(continued)

Table A. Numbered and descriptive loci *(continued)*

Locus	Sector	Stratum	Description
D 75	West Sector		Surface to "ash" layer of brick debris in north–south trench west of Level II retaining wall (uncertain depth); corresponds to D 35 along south fortress wall and to D 85 to east of Central Sector retaining wall; above D 80–D 81 fill
Field record note on area of D 76–D 79 and D 88–D 93			Burned deposit covers tops of room walls and discolors them for ca. 25 cm down; signs of intense fire (and repeated moisture?) 0.40 cm above wall tops of D 76 and D 90; deposits of uncertain depth
D 76	West Sector		Courtyard or room at southeast corner of sector
		1	Wall tops ca. 40 cm below burned layer (see D 90); at level even with tops of walls, soil changes to yellow-brown color with no ash and few sherds; earlier floor level in room and earlier oven associated with this floor level; later oven atop earlier oven, with later column base reused for kitchen purpose?
		Northwest corner	Area by elongated bin or mastaba foundation in room D 76
D 77	West Sector		Area to south of intermediate level foundation wall for stairway D 87 construction, and consisting of south part of silo
D 78	West Sector		Area to north of intermediate level foundation wall for stairway D 87 construction, and consisting of north part of silo
		1	Fill within silo and fill outside and to west?
		Bin (1)	Deposit in one of the two ovens or bins west of silo D 77/D 78
D 79	West Sector		Room or building within sector; atop silo of lowest level
		1	Fill within walls; south part of room includes D 88 and D 91, south of intermediate foundation wall for stairway D 87 construction above area; "kiln" or oven at southwest corner
D 80–D 81	West Sector		Clearance below D 75, along west side of west retaining wall for Level II platform; D 80 on north (above D 59) and D 81 on south; layer may cover over some walls in southeast corner of sector?
		1	Deposit below surface clearance of area (D 75) and around stairway(?) construction along dividing wall between West and Central Sectors
D 82–D 84			No information for these numbers; perhaps not used
D 85	Central Sector		Surface clearance along east side of west retaining wall for Level II platform
		1	Clearance of deposit up to "ash" layer (uncertain depth)
D 86	Central Sector		Clearance of deposit below D 85 (uncertain depth)
		1	Removal of fill (uncertain depth)
Doorway D 87	West Sector		Large stone-framed doorway with door socket at south end of dividing wall between Central and West Sectors; patch of paved floor associated with sill on the east

Table A. Numbered and descriptive loci *(continued)*

Locus	Sector	Stratum	Description
Doorway D 87	West Sector	1	Clearance of deposit around doorway
		2	Test pit below level of doorway; may contain some deposits from lower-lying room D 87 (uncertain depth)?
Room D 87			Partially excavated area below doorway D 87
D 88	West Sector		South part of D 79 defined by higher-founded wall associated with doorway D 87
		1	Fill within walls of area and down to level associated with D 79 and associated rooms
D 89	West Sector		Hallway in southeast corner of sector; north of D 90
		1	Fill within walls and down to level associated with floor in D 90
D 90	West Sector		Staircase construction in southeast corner of sector; wall tops found 0.40 cm below burned levels
		1	Fill within walls and above floor
D 91			Above south part of D 79 and part of room; defined by later wall line associated with construction of doorway D 87
		1	Fill within walls and down to level associated with floor in D 90; below surface clearance D 35(?)
D 92	West Sector		Room in large building along south fortress wall; south of room D 93
		1	Clearance of fill within walls
D 93	West Sector		North room in large building along south fortress wall
		1	Clearance of fill within walls
D 94–D 99			No information for these numbers; perhaps not used
D 100	West Sector		Room(?) along south fortress wall
		1	Clearance of fill within space enclosed by walls of various phases (ca. 58 cm)
D 101	West Sector		Room(?) along south fortress wall; as with all constructions adjacent to south fortress wall, these walls seem to be part of a building covered by later wall relining
		1	Clearance of fill (between 8 and 14 cm); oven later or reused during later phase?
D 102	West Sector		Room(?) along south fortress wall
		1	Clearance of fill around walls
D 103	West Sector		Room or courtyard along south fortress wall; stairway on south may be later construction; wall fragment in center of area
		1	Clearance of fill
D 104	West Sector		Room near south fortress wall
		1	Clearance of fill between walls (uncertain depth)
D 105	West Sector		Area near south fortress wall, walls eroded away at north and west and near surface; east wall built atop small oven
		1	Clearance of fill around walls

(continued)

Table A. Numbered and descriptive loci *(continued)*

Locus	Sector	Stratum	Description
D 106	West Sector		Alleyway or hall between buildings along south fortress wall
		1	Clearance of fill (uncertain depth)
D 107	West Sector		Room next to stairway construction along wall
		1	Clearance of fill below remaining walls
D 108	West Sector		Stairway along south fortress wall belonging to room D 109
D 109	West Sector		Room(?) along south fortress wall; as with all constructions adjacent to south fortress wall, south wall seems to be part of a building covered by later wall relining
D 110	West Sector		Area or room along south fortress wall
		1	Clearance of fill to floor; walls to north, and line of stones on west, appear to belong to an earlier level
D 111	West Sector		Area along south fortress wall, west of D 110; silo or bin at southwest
		1	Clearance of fill around features; small silo or oven at southwest
D 112	West Sector	D 112 and area	Area atop eroded south fortress wall and immediately to its north; unclear what locus number used for surface clearance above this area
		1	Clearance of fill to north of south fortress wall (uncertain depth)
D 113 and D 114 surface	West Sector		Topmost stairway in southwest corner of sector
D 113	West Sector		Room below upper staircase construction
		1	Clearance of deposits below staircase and within walls of room
D 114	West Sector		Space at extreme southwest corner of sector
		1	Clearance of deposit below surface and west of room D 113 (1)? Above silo D 124 and west of staircase construction in D 113
D 115	West Sector		Clearance of area along south wall to east of D 113 and D 114 (uncertain depth); ovens southwest of D 32 upper level, along wall, with bin; ashy debris
		Oven	Fill of larger oven to the east
		Fill under ovens	Ashy fill layer above earlier oven, which then was labeled as D 115 (above D 119)
D 116	West Sector		Bin construction at west end of sector; four separate compartments
		D 116 and area	Clearance atop bins and surface area to north and west (uncertain depth)
		1	Upper half of ash and charcoal fill in bins (uncertain depth)
		2	Clearance of lower deposits in bins; ash and charcoal deposit of ca. 10 cm

Table A. Numbered and descriptive loci *(continued)*

Locus	Sector	Stratum	Description
D 117–D 120	West Sector	Fill in front of D 117–D 120	Deposits in front of rooms along south fortress wall by bedrock building (uncertain depth)
D 117	West Sector		Room along south fortress wall; covered over by later silo D 32
		1	Fill (ca. 25 cm); down to floor level
D 118	West Sector		Room along south fortress wall; covered over by later silo D 32
		1	Fill down to floor level
D 119	West Sector		Room along south fortress wall
		1	Clearance of room to lower floor level (65 cm)
D 120	West Sector		Small bin built into northwest corner of D 119; covered over by later silo west of D 32
		1	Clearance of bin (uncertain depth)
D 121	West Sector		Room along south fortress wall; covered over by later silo west of D 32
		1	Fill within walls (56 cm) down to bedrock floor
D 122	West Sector		Room along south fortress wall; filled with fine black ash mixed with charred wood; fine white ash divided the ca. 60 cm fill in half
		1	Test pit below east half of room by oven(?)
D 123	West Sector		Passageway to east and south of D 116
		1	Area filled with ash to depth of ca. 20 cm; fine white ash layer divided this layer in half; locus also known as "D 123, from thick ash fill" (uncertain total depth)
D 124	West Sector		Silo below part of D 113 (1) and D 114 (1)
		1	Fill within wall of silo (ca. 21 cm)
D 125			Gap between wall sections at southwest corner of fortress
		1	Fill in gap, not emptied to its base level
D 126	West Sector		Silo below preserved elevation below silo D 32 in southwest corner of West Sector; associated with two silos to east and southwest
D 127–D 199			No information for these numbers; perhaps not used
D 200–D 202	West Sector	Surface and (1)	Architectural remains at northwest corner of sector near West Gate; D 201–D 202 are rooms in an intermediate-level house; D 200 is clearance to south of this house and becomes designation for a room in lower levels
		D 200 (1)	Area along west fortress wall in northwest corner of sector; surface clearance(?) and down to floor 1 (ca. 10 cm) of lower-level room or courtyard
		D 200 (2)	Deposit below floor 1 in room associated with earliest walls in area (ca. 10 cm)
		D 201 (1)	Room in intermediate-level house; clearance of deposit within walls and down to appearance of wall in D 222 in lower level; fine black ash stemming from D 222

(continued)

Table A. Numbered and descriptive loci *(continued)*

Locus	Sector	Stratum	Description
D 200–D 202	West Sector	D 200–D 201 (1)	Mixed deposits from these two levels
		D 202 (1)	Room in intermediate-level house; clearance of deposit within walls and down to appearance of wall of D 219 in lower level
D 203			No information for this number; perhaps not used
D 204	West Sector		Courtyard or hallway near north fortress wall, leading into D 1 from D 65 passageway
D 205	West Sector		Large room with column base near north fortress wall
		1	Fill between walls to floor level (uncertain depth, no elevations given)
D 206	West Sector		Small space or staircase foundation near north fortification wall
		1	Clearance of deposit within walls (uncertain depth)
		2	Clearance of fill around lower wall? (ca. 10 cm according to field notes)
D 207	West Sector		Small room or entrance hall to stairway near north fortress wall
		1	Deposit within walls to floor (perhaps 20 cm?)
		2	Perhaps the fill below floor level (ca. 10 cm)? Field notes and plans unclear
D 208	West Sector		Alleyway or open courtyard with three ovens at different levels; base of south oven is above other two; middle oven is lowest and rests on same level as a grinding stone found to its east (deposit of uncertain depth); all reused and contemporary(?)
		1	Fill within walls down to foundation level of south oven
		1 bin area	Deposit around wall of south oven (not "bin")
		2 area of bins	Fill below D 208 (1) and to foundation level of lowest (middle) oven (34 cm)
		South of ovens in levels 1 and 2	Deposit in area to south of ovens and down through level 1 to base level of earliest oven (level 2)
D 209	West Sector		Large room in area near north fortress wall
		1	Clearance of fill within walls down to floor level (uncertain depth)
		2	Deposit below floor (another floor level may have been here according to field notes but is not on plan)
D 210–D 212	West Sector		Open area toward West Gate, 10–15 cm below surface; location of D 210 uncertain, but perhaps in area of wall fragments south of D 209; sherds labeled "south of D 210–D 212" found 10–50 cm below surface level
D 211	West Sector		Area within walls north of West Gate
		1	Clearance of deposit around walls (uncertain depth); walls near or on surface

Table A. Numbered and descriptive loci *(continued)*

Locus	Sector	Stratum	Description
D 212	West Sector		Area along west fortress wall, north of West Gate; wall fragments of various building phases
		1	Clearance of deposit around walls (uncertain depth)
D 213	West Sector		Silo in northwest corner of sector; resting on intermediate level below structure represented by D 47, and oven and bin of D 48; above locus D 214
D 214	West Sector		Area below D 213 and above rooms D 220 and D 221
		1	Deposit over east part of rooms D 220 and D 221, before earlier wall was found (8–9 cm); noted that fill is same as D 220 and D 221 (1)
D 215	West Sector		Small room in building along north fortress wall; no doorway and possibly part of earlier Level IV architectural remains
		1	Fill within walls, but no floor level found
D 216	West Sector		Large room along north fortress wall; reuse of Level IV walls in Level III
		1	Fill in area to floor (46 cm)
D 217	West Sector		Room in building along north fortress wall
		1	Fill within walls, probably down to floor level of neighboring room D 216
D 218	West Sector		Room or courtyard in building near north fortress wall
		1	Fill within walls and down to floor
D 219	West Sector		Small room in building associated with D 209 and D 218, near north fortress wall
		1	Fill in room to floor (uncertain depth)
D 220	West Sector		Room in northwest corner of sector along fortification walls
		1	Fill down to floor level (up to 52 cm)
D 221	West Sector		Large room in building associated with D 220
		1	Deposit within walls of room down to floor (ca. 40 cm)
		2	Test pit within room to find base of wall (51 cm below room floor level)
D 222	West Sector		Small courtyard along west fortress wall south of D 221; two superimposed ovens
		1	Deposit within walls, consisting of fine black ash (ca. 46 cm depth)
D 223	West Sector		Passageway and entry to east of D 222
		1	Clearance of deposit
D 224	West Sector		Room or courtyard in building near north fortress wall; tops of walls traced, since near surface
D 225	West Sector		Room to east of D 224; tracing tops of eroded walls, at or near surface
D 226	West Sector		Room to east of D 225; tracing tops of eroded walls

(continued)

Table A. Numbered and descriptive loci *(continued)*

Locus	Sector	Stratum	Description
D 227	West Sector		Area with mud bin along north fortress wall at breach; at least part of room covered over by later wall relining
		1	Clearance of area around walls and bin (uncertain depth)
D 228	West Sector		Room partially under north wall relining; its east wall founded at same elevation as north wall of room D 40
		1	Deposit within room down to floor
D 229	West Sector		Four garden plots(?) bordered by bricks and stones; below inner wall relining
		1	Clearance of deposit within walls (uncertain depth)
D 230	West Sector		Part of two(?) rooms below inner relining, and also main fortress wall(?)
		1	Clearance of deposit within walls and down to floor level upon which walls were founded (uncertain depth)
D 231	West Sector		Area below inner wall relining of north fortification, and also below main wall(?)
		1	Clearance of deposit down to floor (2 cm below main wall foundation level); oven at east end
D 232	West Sector		Part of room below inner relining of north fortress wall
		1	Deposit in area down to floor level (14 cm above floor in D 231)
D 233–D 299			No information for these numbers; perhaps not used
D 300	Central Sector		Trench excavated at northwest corner of Level IV residence; extended down 53 cm below preserved wall foundation
D 301	Central Sector		Trench along north side of west stairs into Level III residence and down to Level IV; below clearance D 8 outside Level III house
D 302	Central Sector		West stairway of Level III residence
		1	Fill inside west staircase for Level III residence; clearance down ca. 92 cm through fill to base of north staircase foundation
D 303	Central Sector		Fill below D 23, north of Level III residence; excavated down to Level IV residence layers, upon which north staircase of Level III residence was built
		1	Cleared down to street level on east and until tops of walls of D 306 were discovered on west; at east it rested 5–15 cm above yellow sand upon which residence was founded
		D 303, trench at northeast corner	Excavation at northeast corner of Level IV residence, which reached 37 cm below nearby floor level
D 304	Central Sector		Trench along south and east walls of Level IV residence; test pit sunk along south wall toward southeast corner of residence and went down 40 cm below Level IV wall foundation

Table A. Numbered and descriptive loci *(continued)*

Locus	Sector	Stratum	Description
D 305	Central Sector		Area with wall fragments below north stairway of Level III residence and contemporary with Level IV residence; below surface associated with Level III residence and below a white-plastered corrugated "floor"
		1	Deposit within and to west of walls; excavated down to wall base, which rested atop yellow sand foundation for Level IV residence (0.24 m fill)
D 306	Central Sector		Room east of D 305
		1	Deposit within room walls (0.28 m); a floor level found at same level as base of nearby Level IV wall (discussed in field notes but not on plan)
		2	Test pit below floor; deposit of gray sand with sherds of two complete jars in southeast corner (0.37 m depth)
D 307	Central Sector		Trench on south side of southwest corner of Level IV residence (not shown on plans)
		1	Fill within area (uncertain depth)
D 308	Central Sector		Area west of D 307, by southwest corner of Level IV residence
		1	Deposit in area (uncertain depth)
D 309	Central Sector		Test pits along south fortress wall, south of southwest corner residence; A is east pit and B is west; excavation went down ca. 1.5 m below D 8 clearance (noted in field notes and shown on Level IV residence plan); bottom levels were 28–71 cm below bases of interior relining
D 310	Central Sector		Room at northwest corner within Level IV residence; floor level lower than those in D 311
		1	Deposit within room walls down to floor (ca. 60 cm)
		2	Rubble below floor to yellow-sand foundation layer (ca. 33 cm)
D 310–D 311	Central Sector	Trench down to stratum 1	Trench across most of D 310 and into doorway with D 311 (field notes unclear)? Excavated 0.41 m below floor in D 311 and 0.33 below floor in D 310
D 311	Central Sector		Room in Level IV residence; according to field notes, doorway between this room and D 312 blocked with large, flat stones, but not on plan or photographs
		1	Clearance of fill down to packed-mud floor 1, ca. 15 cm silt and sand atop
		2	Exploratory pits below floor in various parts of room, reaching down 0.52–1.33 m below floor levels
		Test pit	Test pit at doorway into D 312, ending 2.48 m below floor and 1.70 m below wall foundation; floor-leveling layers of colorful sandstone chunks mixed with yellow sand; wall foundation level even with bottom of yellow sand layer; below yellow sand was 0.42 m of gray sand and then brown alluvium; noted that Level II(?) well went deeper than this excavated pit

(continued)

Table A. Numbered and descriptive loci *(continued)*

Locus	Sector	Stratum	Description
D 312	Central Sector		Room in Level IV residence
		1	Fill within walls down to paved floor, made of broken brick fragments; floor levels correspond to those in D 313 and D 315
		2	Excavation of fill below floor, resting atop 30 cm layer of colorful sandstone fragments placed in yellow sand
D 313	Central Sector		Room in Level IV residence
		1	Clearance of fill atop broken mudbrick-paved floor
		2	Test pit in room, below floor (see description of D 310–D 316); rubble fill below floor, yellow sand foundation, and then gray river sand
D 314	Central Sector		Room in Level IV residence
		1	Fill down to floor pavement of irregular, broken mudbricks; field notes mention two brick pavement levels below a top floor of brown compacted mud
		2	Levels below top floor and through the two brick-paved floors (field notes unclear)? Depth of ca. 0.15 m
D 315	Central Sector		Room in Level IV residence
		1	Fill above paved floor of irregular broken mudbricks
		2	Excavation of fill below floor, and through leveling layer of multicolored sandstone fragments placed in yellow sand
D 316	Central Sector		Room in Level IV residence
		1	Loose earth fill atop packed-mud floor with replasterings in certain areas of room (ca. 5–10 cm deep); 8 cm difference in floor levels from north to south
		2	Rubble fill below floor
		Trenches down to stratum 2	Two trenches in west part of room below floor levels, 0.64 m depth on south and 0.56 m depth in north trench; fill through yellow sand layer or below (field notes unclear)?
D 317	Central Sector		Test trench between D 304 area along south residence wall and south fortress wall (ca. 0.34 m); did not reach down to foundation level of south residence wall
D 318	Central Sector		Test trench at north end of Level II west platform wall (not on plan)
D 319–D 349			Not used
D 350	East Sector		Silo belonging to later level; built atop earlier walls
		1	Clearance of fill in silo, down to foundation levels of wall (45–48 cm)
		Below (1)	From bottom of wall foundation to floor (14–17 cm depth)
D 351	East Sector		Silo belonging to later level, east of D 350; built atop short wall fragment
		1	Clearance of fill within silo and down to floor (42 cm); wall of silo is 13 cm below floor level

Table A. Numbered and descriptive loci *(continued)*

Locus	Sector	Stratum	Description
D 352	East Sector		Area between D 350, D 351, and north fortress wall
		1	Deposit in area (uncertain depth)
D 353	East Sector		Area northeast of D 351, and between it and north fortress wall
		1	Deposit in area (uncertain depth)
D 354	East Sector		Part of room or open area south of square magazines or storerooms(?)
		1	Deposit in area between walls (uncertain depth)
D 355	East Sector		Storeroom along north fortress wall; south wall of room may be later?
		1	Clearance of fill (uncertain depth)
D 356	East Sector		Storeroom along north fortress wall and east of D 355; south wall may be later construction
		1	Clearance of fill (uncertain depth)
D 357	East Sector		Storeroom along north fortress wall and east of D 356
		1	Clearance of fill (uncertain depth)
D 358	East Sector		Storeroom along north fortress wall and east of D 357
		1	Clearance of fill (uncertain depth)
			Storeroom(?) along north fortress wall and east of D 358
D 359	East Sector		Clearance of fill in area with fragmentary walls ruined by later relining to east (uncertain depth)
D 360	East Sector		Area in center of sector; near surface was large area of burned limestone, perhaps a work area for whitewashing Level II platform after its renovation phase in Level IIB(?); field plans mention that final excavated levels went through series of "floors," perhaps in level 1 or below
		1	Clearance of deposit in area below Level II debris (uncertain depth)
D 361	East Sector		Easternmost silo along south fortress wall; foundation of stones, upper courses of bricks
D 362	East Sector		Middle silo along south fortress wall; see description of D 361
D 363	East Sector		Westernmost silo along south fortress wall; see description of D 361
D 364	East Sector	Pit	Area between south fortress wall, relining of east end, and silo D 361 (not on plan)
D 365	East Sector	Pit	Area along south fortress wall and between walls of silo D 363 and D 361 (not on plan)
D 366			No information for this number
D 367			No information for this number; perhaps not used
D 368	Central Sector		Area inside and to west of North Gate (below Level II platform); same locus as "inside North Gate"?
D 369	North Gate		Fill in stairway within west buttress of North Gate; same locus as "stairway within north wall on west"?

(continued)

Table A. Numbered and descriptive loci *(continued)*

DESCRIPTIVE LOCI

Locus	Sector	Description
East inner wall stairs	East Sector	Fill within stairway construction east of North Gate
Stairway within north wall	East Sector	Equivalent to D 369 (not on plan), western wall stairs at North Gate
Inside North Gate		Surface debris in area? May be same as D 368 on west side of North Gate inside East Sector
Center	East Sector	Surface debris above D 360
East end	East Sector	Surface debris above D 33 until walls reached
First 60 cm	East Sector	Deposit below surface clearance?
Rubble over three south circles	East Sector	Fill atop D 361–D 363
East of stairway leading up to Level II platform	East Sector	Surface debris
East retaining wall at Level II stairway, 1 m south of North Gate, 40 cm below surface	East Sector	Perhaps below "east of stairway leading up to Level II platform"?
On surface along north side of stairway leading up to Level II platform	East Sector	Surface debris
North of stairs leading to Level II platform	East Sector	Surface clearance
South of stairs leading to Level II platform	East Sector	Surface clearance
East retaining wall for Level II, on top of plastered floor of first bay north of staircase	East Sector	Below first 0.40 m of surface clearance to floor
First 50 cm above north magazines	East Sector	Surface debris
River Stairs	East Sector	Fill above stairs (uncertain depth); excavated at step intervals and bags labeled accordingly (e.g., "first ten steps")
Bottom of River Stairs	East Sector	Cleaning debris atop glacis stones to east and west
South wall, surface	Outside East Sector fortification	Surface debris atop cataract stones along south wall at east end
South of circles over main wall	East Sector	Clearance of debris atop south fortress wall at east end, south of silos
Outer face of south wall, in fill south of middle stone circle	Outside East Sector fortification	Outside south main wall and south of silo D 362
Glacis trench, south fortress wall	Outside East Sector fortification	Excavation of trench at southeast end of fortress

Table A. Numbered and descriptive loci *(continued)*

Locus	Sector	Description
Below flat brick of Christian level	Central Sector	Below threshold of Level I doorway
Debris of rebuilding of Level II, sealed between it and original platform	Central Sector	Between stone placed with bays along exterior buttresses of Level II platform (i.e., of platform)
Debris against rebuilding of Level II, south wall	Central Sector	See description immediately above
Rubble over east and central parts of south wall	Central Sector	Surface clearance along south fortress wall continued west from East Sector
South wall, rubble on west	Central Sector	Surface clearance atop wall, west of previous locus
Level II pocket against south wall, inside and parallel to retaining wall	Central Sector	Small brick construction (hearth?) against Level II building atop platform; clearance of sherd scatter
West side, black stratum	West Sector?	Uncertain location
Dismantling of stairway north of Level III building	Central Sector	Removal of staircase brick and fill; above D 305–D 306
Within Level II and 10 cm below walls	Central Sector	Clearance of top layer of intentional fill within retaining walls of Level II, before discovery of Level III residence walls
Two kilns along outside of north wall, between stairs and breach	Outside Central Sector	One oven along later relinings between bastions C and D; location of other oven uncertain
First buttress west of River Stairs	Outside fortification	Clearance around bastion D
Outside north wall, from River Stairs to second buttress	Outside fortification	Clearance between west bastion of North Gate and over west to bastion C
Deposit inside south wall	West Sector	Uncertain location
North, northeast, and northwest of D 32 (1)	West Sector	Clearance of upper debris around walls and down to level even with walls of intermediate-level silo along south fortress wall
West of D 32 (1)	West Sector	Clearance of debris
North–south trench at west end	West Sector	Trench excavated through center of West Sector, east of West Gate and running from D 208 to D 126 (ca. 0.45 m depth)
North wall at break	West Sector	Surface clearance? Same as "surface fill in breach of north wall" above D 228–D 232
Structure atop north wall, third buttress from west end	West Sector	Crenelated bastion(?), which is not third but second bastion if one counts later buttress added onto wall

(continued)

Table A. Numbered and descriptive loci *(continued)*

Locus	Sector	Description
Top of north fort wall, at west corner	West Sector	Clearance of debris around latest renovations atop northwest corner bastion
Debris ca. 2 m north of breach in north wall	Outside West Sector fortification	Clearance of surface debris along north wall face
Outside north wall and west of breach	Outside West Sector fortification	Clearance of surface debris along north wall face
Glacis trench, north wall	Outside West Sector fortification	Glacis trench between crenelated bastion and bastion A
Trench in bay between second and third buttresses from west end of north wall	Outside West Sector fortification	Surface clearance of east–west, 2 m wide trench in bay between crenelated bastion and bastion A
North wall, outside second and third buttresses from North Gate, below "ash" layer	Outside West Sector fortification	Clearance near bastions B and C
Outside north wall, east of second buttress, 30–60 cm above burned floor	Outside West Sector fortification	Clearance above darkened "ash" layer between bastions C and D (not a floor but a layer of saturated alluvial deposits)
Outside north wall, east of second buttress, 30 cm above burned floor, to another floor	Outside West Sector fortification	Fill below "ash" stratum between bastions C and D
Outside north corner of West Sector	Outside West Sector fortification	Surface clearance around fortification walls at northwest corner bastion
West wall, south of northwest outer corner	Outside West Sector fortification	Remains of extramural structures built between northwest corner bastion and West Gate; surface clearance
D 12 from bin	Outside West Sector fortification	Location of bin uncertain, but clearance of fill from one of the extramural structures
Eroded surface at west wall of fortress, south of West Gate	Outside West Sector fortification	Equivalent to D 13?
Outside south wall, west of breach	Outside West Sector fortification	Surface clearance along entire wall at bastions M and N; above D 18, D 19, D 21, and D 25
Outside south wall, from breach east to second buttress	Outside West Sector fortification	Clearance of debris around bastions G–L
Buttress on south wall of West Sector, tracing of glacis	Outside West Sector fortification	Clearance of surface debris at bastion N(?)
Outside south wall, between second and third buttresses from breach	Outside West Sector fortification	Clearance of surface debris at bastions J–K or bastions I–J

Table B. Object register and division list

Field no.	Division	OIM or Khartoum no.	Description and dimensions	Locus and associations	Comment
D-1	Chicago	E24308	Hieratic ostracon; two lines of script; black and red ink; max. dimensions 14.0 × 17.0 cm	D 1 (1), in surface debris	Grain delivery, Ramesside date (Knudstad 1966)
D-2	Chicago	E24309	Dipinto; on outer surface of partially mended Meroitic pot; traces of two lines of Demotic; max. dimensions 19.5 × 15 cm	Below surface east of D 5 [D 3 (1)—LAH]	
D-3	Chicago	E24310	Blue faience scarab; roughly one-third preserved; *wedjat*-eye discernible on underside; 2.1 × 1.4 cm	D 1 (1), in surface debris	
D-4	Chicago	E24311	Bronze earring; 2.2 × 1.4 cm	On eroded surface of west wall of fort, south of gateway [D 13 (1)—LAH]	
D-5	Chicago	E24312	Assorted beads	From eroded surface of west wall of fort, near gateway	
D-6	Chicago	E24313	Fragment of blue faience; max. dimensions 3.0 × 2.7 cm	From eroded surface of west wall of fort, south of gateway [D 13 (1)—LAH]	
D-7	Chicago	E24314	Partially mended goblet; impressed rows of half moons on body; unfired eggshell ware; D at rim 8.5 cm, H 9 cm	In surface debris of upper central portion of fort (Central Sector), south of D 5; looted	
D-8	Chicago	E24315	Fragments of pale-green glass	On surface over early Christian remains, upper central portion of Central Sector; D 15 (1)	Also a blue glass sherd with trailed decoration in one box
D-9	Khartoum	No number	Gilded bronze seal ring; D 2.1 cm	From burial 2, in pelvic region of skeleton C; found beside iron ring, field no. D-10	Intrusive burial
D-10	Chicago	E24316	Iron ring; very corroded and broken	From burial 2, in pelvic region of skeleton C; found beside bronze ring, field no. D-9	Intrusive burial, looted
D-11	½ Khartoum, ½ Chicago	E24317	String of 21 glass beads, 5 round (blue streaked with white) and 16 long faceted (blue-green)	Found on neck of skeleton of child, burial 3	Intrusive burial, looted

(continued)

Table B. Object register and division list *(continued)*

Field no.	Division	OIM or Khartoum no.	Description and dimensions	Locus and associations	Comment
D-12	Chicago	E24318	Meroitic bottle; shoulder decorated with two serpents painted in red and black; neck mended and incomplete; H 26 cm, body D 29 cm	Found at foot of burial 2	Intrusive burial, looted
D-13	Chicago	E24319	Goblet in eggshell ware; painted body covering pattern in red and black; mended, incomplete; rim D 9 cm, H 8.2 cm	Found at foot of burial 2; used as lid for field no. D-12	Intrusive burial, looted
D-14	Chicago	E24320	Bottle; shoulder painted with continuous frieze of stylized ivy leaves (white on black); pattern on body in white on black; probably an import from Egypt according to W. Y. Adams; H 21 cm	Found at foot of burial 2	Intrusive burial, looted
D-15	Chicago	E24321	Goblet in eggshell ware; red rim band and linear body covering pattern in black; rim D 8.5 cm, H 6.5 cm	Found at foot of burial 2; used as lid for field no. D-14; looted	
D-16	Chicago	E24322	Bottle; shoulder painted with three rows of large dots separated by thin black bands; top and bottom dots are red and middle ones are black; H 23.3 cm	Found at head of burial 2; looted	
D-17	Chicago	E24323	Goblet in eggshell ware; body decorated with pattern of alternating *ankh* signs and plants in red and black; rim D 8.7 cm, H 7.5 cm; broken and partially mended; incomplete	Found at head of burial 2; used as lid for field no. D-16; looted	
D-18	Chicago	E24324	Five beads, three tubular and two spherical	Found in pelvis of skeleton C of burial 2; looted tomb	
D-19	Chicago	E24325	Broken faience cartouche ring; handle on back also broken; the words ". . . two lands, lord of heaven" are legible; blue faience; L 3.0 cm	From surface debris outside south wall of West Sector of fort in D 18 (1)	
D-20	Chicago	E24326	Carved piece of sandstone; L 16.0 × W 6.0 × Th 4.0 cm [measurements are actually 16.2 × 7.0 × 4.7 cm—LAH]	From test pit through floor of D 5	

Table B. Object register and division list *(continued)*

Field no.	Division	OIM or Khartoum no.	Description and dimensions	Locus and associations	Comment
D-21	Chicago	E24327	Pilgrim bottle; spherical body, narrow neck, flaring lip, and single loop handle preserved; thick, heavy ware; painted decoration consisting of concentric body bands in red and black; H 17.5 cm	From debris outside south wall of West Sector of fort, just below surface; D 19 (1)	
D-22	Chicago	E24328	Small stone net sinker of green soapstone; max. D 5.5 × 1.7 × 1.2 cm; three incised lines	From debris outside south wall of West Sector of fort, just below surface and near pilgrim bottle (field no. D-21); D 19 (1)	
D-23	Chicago	E24329	Small stone net sinker of talc; max. D 5.8 × 1.7 × 1.7 cm	From debris outside south wall of West Sector of fort, just below surface near pilgrim bottle (field no. D-21) and field no. D-22; D 19 (1)	
D-24	Khartoum	Khartoum 14281	Iron key; incised "X" designs on side of handle(?); handle(?) perforated; L 19.7 cm	Built into wall of Christian building, D 15A of Level I, Central Sector	
D-25	Chicago	E24330	Broken net sinker of talc (see field nos. D-22 and D-23); max. dimensions 3.0 × 1.6 × 1.7 cm	Found on surface of densely packed rubble of D 22 (3)	
D-26	Chicago	E24331	Broken stone net weight of soapstone; max. dimensions 3.7 × 2 × 1.1 cm	D 22 (4)	
D-27	Chicago	E24332	Tiny faience figurine; head and legs below knees lost; bearded; H 3.1 cm	Lying atop broken wall on south side of D 9, D 9 (3)	
D-28	Chicago	E24333	Bent metal wire; 1–2 mm thick; formed into an oval of 2.5 × 4.0 cm; possibly an ornament [bracelet—LAH]	Found atop child's burial, burial 4	
D-29	Chicago	E24334	Tubular blue bead; L 0.5 cm, D 0.2 cm; noted that burial disturbed	Found near feet of child, burial 4	
D-30	Chicago	E24335	Fragments of two kinds of cloth; one coarse and thick, the other finer	Found in region of head and chest of child, burial 4	

(continued)

Table B. Object register and division list *(continued)*

Field no.	Division	OIM or Khartoum no.	Description and dimensions	Locus and associations	Comment
D-31	Chicago	E24336	Sherd of light blue faience(?); fine, thin bands of decoration survive; max. dimensions 7.0 × 6.5 cm [faience New Year's flask fragment—LAH]	Found on plastered floor in northeast corner of bay immediately south of late Christian kiln on west retaining wall for Level II of Central Sector	
D-32	Chicago	E24337	Small ostracon/dipinto with three Demotic letters preserved; max. dimensions 7.2 × 2.0 cm	Under brick threshold of doorway cut through west wall of Level II of Central Sector [D 15 (1), Christian building—LAH]	
D-33	Khartoum	Khartoum 14277	Flint arrowhead; L 3.8 cm	Found at depth of 0.5 cm, 1 m east of stairs on east side of Central Sector	
D-34	Chicago	E24338	Broken but reconstructible pottery ring stand; red fabric, straw-marked with black fracture; rim D 18.0 cm, H 11.0 cm	Found outside westernmost bay on south side of retaining wall for Level II of Central Sector, between edge of bay floor and light wall just to south	
D-35	Chicago	E24339	Broken footed bowl, straw-marked red fabric; rim D 18 cm, H 5.5 cm	From southernmost bay of retaining wall for Level II of Central Sector, above plastered floor	
D-36	Lost in field	E24340	Broken pottery jar stand; red-brown fabric, straw inclusions; H 10.5 cm	On surface of West Sector, near breach in north wall of fort	
D-37	Chicago	E24341	Eight white, drop-shaped beads; holes drilled through narrow end, vertical grooves worn down from holes	Found at bottom of burial 6	
D-38	Chicago	E24342	Sherd of silt bowl, straw-marked brown ware with slight ribbing; rim D ca. 18.0 cm, H 9.0 cm	D 1 (1)	
D-39	Chicago	E24343	Bowl sherd; brown fabric with straw; red rim painted; rim D ca. 20.0 cm, H 10.5 cm	D 7 (1)	

Table B. Object register and division list *(continued)*

Field no.	Division	OIM or Khartoum no.	Description and dimensions	Locus and associations	Comment
D-40	Chicago	E24344	Broken but reconstructible footed bowl; silt, horizontally burnished interior, grooved rim, upper exterior fired cream color, three of four handles preserved; rim D 35 cm, H 15.5 cm	Central Sector, east retaining wall for Level II; on top of white plastered floor of first large bay north of stairs on east side	Levantine influence
D-41	Chicago	E24345	Wide-mouthed necked sherd of pot; cream ware, lip painted; H as preserved 9 cm	1 m northeast of staircase of Level II Central Sector, 40 cm below surface	
D-42	Chicago	E24346	Six assorted beads	From surface of west wall of fort, south of gateway	
D-43	Chicago	E24347	Broken but reconstructible bowl, straw-marked brown ware; red rim; rim D 18.5 cm, H 9 cm	D 10 (1)	
D-44	Chicago	E24348	Assorted beads, 190 complete beads	From burial 8	
D-45	½ Khartoum, ½ Chicago	E24349	Ten bifacially flaked stone arrowheads; average length 4.0 cm; one arrowhead struck from agate	D 34, in charcoal resting on floor 1, East Sector at east end	
D-46	Chicago	E24350	Fragment of light-blue faience New Year's flask; remains of bands of incised design on exterior; max. D 5.2 cm, H 3.8 cm	West Sector, surface rubble inside main south wall of fort, east of D 32	
D-47	Chicago	E24351	Sun-dried clay cylinder, with hole drilled through its length; L 4.0 cm, D 1.7 cm	D 3 (1) Central Sector, intentional fill	
D-48	Chicago	E24352	One white teardrop bead; hole drilled though narrow end	East Sector, rubble on surface of south wall	
D-49	Chicago	E24353	Stone arrowhead; L 3.3 cm	In mixed sand and rubble fill in south section of D 32, above cross walls	
D-50	Chicago	E24354	One rectangular block of gray-green soapstone; small hole drilled into short ends; 1.8 × 1.4 × 0.7 cm	D 8 (1), intentional fill of Central Sector platform	

(continued)

Table B. Object register and division list *(continued)*

Field no.	Division	OIM or Khartoum no.	Description and dimensions	Locus and associations	Comment
D-51	Chicago	E24355	Forty-seven small rectangular cubes of hematite or petrified wood; average dimensions 0.4 × 0.4 × 0.4 cm; evidence of iron inclusions [probably galena—LAH]	D 33 (1), resting on floor 1 below "ash" layer	
D-52	Chicago	E24356	Net sinker of schist(?) with gold mica inclusions, L 4.1 cm	West Sector, inside south wall between hearths southwest of D 32, surface rubble	
D-53	Chicago	E24357	Folded copper or bronze item; L as folded 2.4 cm	D 5 (2), Central Sector	
D-54	Chicago	E24358	Two blue beads	D 9 (1), intentional fill of Central Sector Level II platform	
D-55	Chicago	E24359	One blue bead	Central Sector, immediately west of easternmost bay of south retaining wall for Level II	
D-56	Chicago	E24360	Two beads, one gray and white striped, one blue	Central Sector, from surface of stairway to Level II	
D-57	Chicago	E24361	Large blue glass bead; broken	Central Sector, surface of southeast corner of retaining wall for Level II platform	
D-58	Chicago	E24362	One light-blue bead	Central Sector, D 3 (1)	
D-59	Chicago	E24363	Flint blade; L 4.4 cm, W 1.1 cm	West Sector, surface rubble west of D 32	
D-60	Chicago	E24364	One large blue bead	West Sector, surface rubble west of D 32	
D-61	Chicago	E24365	Glass rod, broken at one end; L 5.6 cm, D 0.5 cm	West Sector, surface rubble west of D 32	
D-62	Chicago	E24366	Twenty-eight assorted beads: 17 on one piece of original string, 6 on another piece of original string, and 5 separate	West Sector, east end of space between D 32 and wall of fort, at depth of 35 cm, in rubble	
D-63	Chicago	E24367	Meroitic bottle; neckbands, shoulder band decoration and ivy pattern on body; H 25.0 cm; broken and mended	From burial 9 and rubble outside graves; West Sector	

Table B. Object register and division list *(continued)*

Field no.	Division	OIM or Khartoum no.	Description and dimensions	Locus and associations	Comment
D-64	Chicago	E24368	Broken but reconstructible Meroitic bottle; shoulder with half moons painted; H 20 cm; mended	From burial 9 and rubble outside; West Sector	
D-65	Chicago	E24369	Broken dish; straw-marked red ware; rim D 17.0 cm, H 3.5 cm	Central Sector, Level II, from pocket against south wall, inside and parallel to retaining wall	
D-66	Chicago	E24370	Small stone object of soapstone; broken; L 3.7 cm	D 26 (1) in Central Sector	
D-67	Chicago	E24371	Nodule of rock hollowed out to form a container; H 6.2 cm [natural formation—LAH]	West Sector, D 10 (1) on floor 1	
D-68	Chicago	E24372	Top of large amphora of hard tan fabric; handles and shoulders decorated with thin lines of red-painted bands; H as preserved 35.0 cm, rim D 14.5 cm	East Sector, east of retaining wall for Level II or Central Sector and 1.5 cm south of River Stairs; on surface of stained layer, 49 cm below surface	
D-69	Chicago	E24373	One turquoise faience bead	West Sector, just below surface of D 61	
D-70	Chicago	E24374	Buff-ware flask with faint traces of circles of black on body; encrusted and exfoliating; two handles and neck; half of one handle missing; H 17.5 cm, max. D 14.5 cm	West Sector, just below surface in southeast corner of D 61	
D-71	Chicago	E24375	Black-topped bowl; black interior and over rim, reddish on exterior; H 14.3 cm, rim D 23.5 cm	West Sector, southeast corner of D 44	
D-72	Chicago	E24376	Fragment of blue faience vessel with black line decoration, ca. 5.0 × 5.5 cm	Stone rubble outside fort wall, just west of first buttress west of River Stairs	
D-73	Chicago	E24377	Cream-ware jar, light irregular ribbing, warping, rim weathered, potter's mark; one or two pieces missing (including one handle); H ca. 53 cm, max. D ca. 25.5 cm	D 102, along south wall	

(continued)

Table B. Object register and division list *(continued)*

Field no.	Division	OIM or Khartoum no.	Description and dimensions	Locus and associations	Comment
D-74	Chicago	E24378	One blue faience bead	West Sector, D 42, northwest corner on bricks of wall	
D-75	Chicago	E24379	Stone disk of coarse sandstone; H 2.3 cm, D 2.0–2.8 cm [fishing net weight—LAH]	West Sector, from fill below "ash" layer of D 76	
D-76	Chicago	E24380	Stone object [fishing net weight—LAH]; L 6.8 cm, Th ca. 2.5 cm	West Sector, D 59, under "ash" layer, against west wall on north half	
D-77	Chicago	E24381	Pottery disk with groove around its thickness; D 3.1 cm, Th 0.7 cm	D 76, in fill below "ash" layer	
D-78	Chicago	E24382	Lumps of copper slag	D 75 area, north end	
D-79	Chicago	E24383	Turquoise-colored faience bead	Dump in West Sector	
D-80	Chicago	E24384	Blue faience bead	D 53, in fill below "ash" layer	
D-81	Chicago	E24385	Stone net sinker or toggle of talc; L 5.2 cm, Th 1.2 cm	D 75 area, above "ash" layer	
D-82	Chicago	E24386	Inscribed faience plaque with recumbent sphinx; 2.0 × 1.5 × 0.7 cm	D 76, just under "ash" layer	
D-83	Chicago	E24387	Corroded copper spearhead(?) or spatula, bent; L ca. 10.0 × W 2.1 × Th 3.0 cm	D 78, from redim under "ash"	
D-84	Chicago	E24388	Five white faience beads	D 66, from redim under "ash"	
D-85	Chicago	E24389	Corroded copper spearhead(?) or spatula, broken into two pieces; L 11.5 × max. W 2.6 × Th 0.2 cm	D 80, below "ash" layers, southwest corner of room	
D-86	Chicago	E24390A	Two small whetstones of gray soapstone; one L 4.4 × W 1.3 × Th 0.9 cm; other L 5.1 × W 1.7 × Th 1.1 cm	D 70, 15 cm below "ash" layer, in south doorway of room	
		E24390B	Also noted a corroded copper blade(?) broken into four pieces; L 4.0 × W 1.3 × Th 0.1 cm	D 70, 15 cm below "ash" layers in south doorway of room	
D-87	Chicago	E24391	Clay object; L 2.2 cm, D 1.1 cm	D 104 (1), from redim	

Table B. Object register and division list *(continued)*

Field no.	Division	OIM or Khartoum no.	Description and dimensions	Locus and associations	Comment
D-88	Chicago	E24392	Stone object of talc, split lengthwise, half missing; L 3.9 × W 2.2 × Th 0.6 cm [fishing net weight—LAH]	From redim ca. 2 m north of breach in north wall (outside fort wall)	
D-89	Chicago	E24393	Stone arrowhead, point missing; L 2.2 × W 1.5 cm; pale silicate with dark concretions	D 73 (1) on ash near door, 10 cm below surface	
D-90	Chicago (1) and Khartoum (1)	E24394, Khartoum 14270	Two faience Hathor head pendants; worn; L 2.2 × W 1.6 × Th 0.3 cm	D 6 (1), on floor; 80 cm south of fort wall, 1 m west of east room wall	
D-91	Chicago	E24395	Stone arrowhead; L 3.1 × W 1.4 × Th 0.3 cm	D 201 (1) from redim	
D-92	Chicago	E24396	Faience scarab, slightly chipped; L 1.8 × W 1.3 × H 0.8 cm	D 209 (1), from redim but below surface	
D-93	Khartoum	Khartoum 14274	Stone arrowhead; L 3.3 × W 1.7 × Th 0.4 cm	D 79, at jog in west wall ca. 25 cm below top of preserved wall	
D-94	Chicago	E24397	Wood object, chipped, burned and encrusted; D 3.8 cm, H 1.8 cm	Outside north wall between second and third buttresses west of North Gate, in debris against face of wall ca. 25 cm above a floor	Fish vertebra
D-95	Khartoum	Khartoum 14276	Stone arrowhead; L 3.2 × W 1.2 × Th 0.3 cm	From debris on surface of D 81, below “ash” layers; patches of a floor found at same level as D 38 farther north, and this object found ca. 10 cm below surface above this floor	
D-96	Chicago	E24398	Faience plaque with incised fish on one side; L 1.6 × W 1.0 × Th 0.6 cm	Same locus as field no. D-95	
D-97	Chicago	E24399	Three unbaked miniature clay bowls; max. dimensions D 4.5 cm, H 2.5 cm	Two from D 18 (1) and one from D 13 (1)	
D-98	Khartoum	Khartoum 14269	Stone arrowhead; L 2.6 × W 1.1 × Th 0.2 cm	D 210 (1), south half from redim	
D-99	Khartoum	Khartoum 14278	Stone arrowhead; L 3.2 × W 1.4 × Th 0.3 cm; point missing	D 213 (1) redim/fill	

(continued)

Table B. Object register and division list *(continued)*

Field no.	Division	OIM or Khartoum no.	Description and dimensions	Locus and associations	Comment
D-100	Chicago	E24400	Stone arrowhead; L 3.4 × W 1.4 × Th 0.5 cm	East Sector, outer face of south wall in fill between bricks and stone; south of middle silo on south	
D-101	Chicago	E24401	Stone arrowhead; L 2.8 × W 1.4 × Th 0.4 cm	D 8B, redim at outer northwest corner of central building's level 3	
D-102	Khartoum	Khartoum 14279	Stone arrowhead; L 3.3 × W 1.2 × Th 0.4 cm	Central Sector, D 301, trench under D 8; see picture drawn in field register: looks like east of north stairs into official residence?	
D-103	Chicago	E24402	Stone bowl rim fragment; L 3.8 × Th 0.7 cm; made of travertine/calcite	Central Sector, D 8B redim at outer northwest side	
D-104	Khartoum	Khartoum 14271	Blue faience Hathor head pendant; L 2.5 × W 1.6 × Th 0.6 cm; piece missing from top	D 112 (1) and area	
D-105	Chicago	E24403	Two copper fragments, incomplete and original shape uncertain; L 2.6 × Th 0.005 cm	D 113 (1) redim	
D-106	Chicago	E24404	Ca. 275 blue beads	West Sector, D 113 (1) on floor along west wall, north half	
D-107	Chicago	E24405	Worked oval object of ferruginous limestone; L 3.2 × W 3.1 × 0.4 cm	In rubble filling D 107, on a level just below remaining top of wall	
D-108	Khartoum	Khartoum 14275	Stone arrowhead; L 4.1 × W 1.7 × Th 0.3 cm	East Sector, on surface along north side of stairway leading into Central Sector Level II platform building	
D-109	Chicago	E24406	Seal-impressed rim sherd; L 6.5 × W 7.0 × Th 3.3 cm	West Sector, from rubble beneath surface, south of D 113, D 114	
D-110a, b, c	Chicago (2) and Khartoum (1)	E24407, Khartoum 14280	Three thick-sectioned body sherds, reworked on one edge to a smooth oval surface, beneath which is a groove setting it off from rest of sherd; dimensions of largest: L 10.5 × W 11.0 × Th 1.8 cm	West Sector, rubble of D 114 (1); a similarly reworked sherd was bagged from D 113 (1)	

Table B. Object register and division list *(continued)*

Field no.	Division	OIM or Khartoum no.	Description and dimensions	Locus and associations	Comment
D-111	Chicago (lost in field)	E24408	Base of red-burnished vessel reused as a "saucer"; D 10.0 × H 3.3 × Th 0.7 cm	West Sector, rubble of D 215 (1)	
D-112	Chicago	E24409	Brown silt bowl with faint trace of red paint on rim; three large chips missing from rim; warped; D 18.8 × H 9.0 × Th 0.5	West Sector, rubble of D 215 (1)	
D-113	Chicago	E24410	Incomplete baked-clay object; L 3.4 cm, D 1.8 cm	West Sector, from general clearing of surface to architecture level of D 112 and area	
D-114	Chicago	E24411	Stone net sinker of soapstone; L 4.0 × W 2.0 × Th 0.8 cm	Central Sector, tearing down walls of Level II building; from area of doorway between D 20 and D 9	
D-115	Chicago	E24412	Stone arrowhead, point and part of one tang missing; struck from white silica flint; L 2.9 × W 1.2 × Th 0.4 cm	West Sector, D 86 (1) rubble	
D-116	Khartoum	Khartoum 14267	Small red-burnished flask with two necks and sets of handles; H 12.8 × max. W 8.5 cm	West Sector, D 122 (1), found along east wall of this room; on floor beside oven and with a grinding stone on its other side along wall	
D-117	Chicago	E24413	Unbaked clay bottle-stopper(?), slightly chipped; H 1.4 × D 4.0 cm	West Sector, D 122 (1), rubble	
D-118	Chicago	E24414	Handmade brown-ware bowl, mended with one chip missing in base; H ca. 11.0 × D 19.5 × Th 0.5 cm	West Sector, D 90 (1)	
D-119	Chicago	E24415	Corroded copper kohl stick; L 17.9 × Th 0.4 cm at middle	West Sector, just east of locus D 40, below surface	
D-120	Chicago	E24416	Bone point, chipped; L 9.3 cm	West Sector, D 123 (1), from thick ash fill	
D-121	Chicago	E24417	Stone blade; L 5.5 × Th 0.5 cm	West Sector, D 59 (1)	
D-122	Chicago	E24418	Fragment of faience *wedjat*-eye(?)	Dump, West Sector	

(continued)

Table B. Object register and division list *(continued)*

Field no.	Division	OIM or Khartoum no.	Description and dimensions	Locus and associations	Comment
D-123	Chicago	E24419	Stone arrowhead; L 3.4 × W 1.2 × Th 0.4 cm	West Sector, found when outline of D 126 was being cleaned up	
D-124	Chicago	E24420	Length of corroded copper tubing; L 6.2 × D 0.5 cm	West Sector, found when outline of D 126 was being cleaned up	
D-125	Chicago	E24421	Stone arrowhead of agate; L 4.1 × W 1.4 × Th 0.3 cm	East Sector, on stone foundation at northeast corner of enclosed area east of D 34	
D-126	Chicago	E24422	Brown-ware jar with rope decoration on body; H 33.0 × D 29.0 × Th 1.0 cm	West Sector, D 60 (1) under west wall at north corner	
D-127	Chicago	E24423	One fragment of fine leather, folded; several fragments of leather straps, some with hair on one side; wood fragments [Quiver or bag?—LAH]	West Sector, D 125 (1)	
D-128	Left in field	No number	Lintel fragment, incomplete, light-gray sandstone, decorated with two symmetrically opposed scenes bearing figures in relief and hieroglyphic inscriptions; right scene incomplete; traces of yellow paint on relief and red, white, and blue bands on molding; for text, see locus D 5 notes; L 153.0 × H 53.0 × Th 18.0 cm	D 5 (2), in fragments in front (to west) of doorway to D 9; one fragment from center of room	Assumed to come from Buhen and reused; Nineteenth or Twentieth Dynasty
D-129	Chicago	E24424	Fragment of light-gray sandstone (stele or architectural fragment?) bearing beginning of hieroglyphic inscription; cut and reused as door socket; face dimensions 19.0 × 27.0 cm, Th 16.0 cm	D 9, reused as door socket in situ to east of south jamb of doorway from D 5 to D 9; inscribed on surface facing down	Assumed to come from Buhen and reused

Table B. Object register and division list *(continued)*

Field no.	Division	OIM or Khartoum no.	Description and dimensions	Locus and associations	Comment
D-130	Left in field	No number	Fragment of light-gray sandstone (stele or architectural fragment?) bearing incomplete hieroglyphic inscription; only a few signs preserved, perhaps prenomen of a Ramesside king; for text, see locus D 9 notes; 25.0 × 49.9 × 53.0 cm	D 9, reused as column base in situ on floor (stratum 5) in center of room; inscription found on side facing down	Assumed to come from Buhen and reused; has the cartouche of Ramesses IV
D-131	Given to Khartoum	No number	Lower portion of pink sandstone doorjamb bearing two columns of hieroglyphic inscription; incomplete jamb is left one of a pair (including field no. D-132) bearing identical but symmetrically opposed texts; for text, see locus D 9 notes; W 31.0 × H 54.0 × Th 18.0 cm	D 9, in situ as left (west) jamb for doorway from D 9 to D 20	Noted as Nineteenth or Twentieth Dynasty; assumed to come from Buhen and reused
D-132	Given to Khartoum	No number	Lower portion of pink sandstone doorjamb bearing two columns of hieroglyphic inscription; incomplete jamb is right one of a pair (including field no. D-131) bearing identical but symmetrically opposed texts; for text, see locus D 9 notes; W 32.0 × H 88.0 × Th 17.0 cm	D 9, in situ as right (east) jamb for doorway from D 9 to D 20	Noted as Nineteenth or Twentieth Dynasty; assumed to come from Buhen and reused
D-133	Given to Antiquities Service	No number	Fragment of light-gray sandstone (stele or architectural fragment?) bearing incomplete hieroglyphic inscription; for text (Ramesses IV?), see locus D 20 notes; cut and reused in doorsill; face dimensions 31.0 × 38.0 cm, Th 16.0 cm	D 20, as base in situ at north end of door sill for plinth under north jamb of doorway from D 20 to D 22; inscription laid face up	Noted as Ramesside; assumed to come from Buhen and reused

Table C. Beads

Locus	Find context	Field no.	OIM or Khartoum no.	Material	Shape	Color, diaphaneity	Technique	Quantity	Diameter/ length/hole diameter (mm)	Plate
MERIOTIC BURIALS										
Burial 2	Found in pelvis of skeleton C	D-18	E24324	Carnelian	Spherical	Red	Drilled from one end	2	6.2/5.6/1.0–1.3; 7.4/6.1/1.6–2.7	100a.1
				Carnelian	Long cylinder	Red	Drilled from one end	2	5.1/9.8/1.4; 5.4/9.8/2.4	100a.2
				Granite	Slightly faceted, hexagonal long cylinder	Black and white	Drilled from one end	1	5.2/11.3/1.7–2.5	100a.3
Burial 3	String of 21 glass beads found on neck of child	D-11	E24317, Khartoum (no number)	Glass	Spherical	Dark blue translucent and white opaque trail	Rod formed	3, additional 2 remained in Sudan	7.4–9.1/6.7–7.8/ 2.1, 2.9	100b.1
				Glass	Cornerless cube	Light blue, semitransparent	Rod formed and marvered	8, additional 8 remained in Sudan	7.6–8.4/12.7–15.8/ca. 3.6	100b.2
Burial 4	Found near feet of child	D-29	E24334	Faience	Long cylinder	Light-blue glaze		1	2.3/7.7/1.0	101a
Burial 6	Found in sand at bottom of burial	D-37	E24341	Chalcedony	Teardrop pendant with rounded base	White	Drilled from one end	8	Th 5.9, W 6.2, H 9.1–9.4, HD 0.8–1.5	101b
Burial 8	In fill and around skeleton	D-44	E24348	Agate	Teardrop pendant with globular base	Red	Drilled from one end	1		102a.1
				Quartz	Teardrop pendant with globular base	White	Drilled from one end	2		102a.2
				Glass	Teardrop	White	Rod pierced?	1		102a.3
				Silver-in-glass	Large conical		Drawn	1		102a.4
				Silver-in-glass	Long cylinder and small oblate		Drawn	2		102a.5

Gold-in-glass	Globular		Drawn	17	102a.6
Glass	Globular		Drawn	29	102a.7
Agate	Irregular globular	Red		1	102a.8
Glass	Globular	Green and bluish	Drawn	4	102a.9
Faience	Short cylinder	Blue		9	102a.10
Faience	Globular			1	102a.11
Glass	Cornerless cuboid	Blue, opaque	Faceted	27	102a.12
Glass	Teardrop with flattened base	Blue	Rod pierced?	1	102a.13
Glass	Barrel	Blue	Wound	1	102a.14
Glass?	Globular and spindle shaped	Appearing black		4	102a.15
Quartz?	Teardrop with globular base	Appearing black		1	102a.16
Glass	Oblate and globular	Dark blue	Drawn	29 and 1 broken	102a.17
Glass	Cornerless cuboid	White- and red-banded blue	Faceted	10	102a.18
Glass	Cornerless cuboid	Cobalt blue		45	102a.19
Glass	Globular	Cobalt blue		2	102a.20
Glass	Long hexagonal cylinder	Cobalt blue	Faceted	1	102a.21
Mollusk shell				1	102b
Glass	Ear stud with domical head	Blue		1	102c

(continued)

Table C. Beads *(continued)*

Locus	Find context	Field no.	OIM or Khartoum no.	Material	Shape	Color, diaphaneity	Technique	Quantity	Diameter/ length/hole diameter (mm)	Plate
FORTRESS										
D 3 (1)	Fill of Level II platform	—	E50815	Glass	Oblate	Light yellow	Wound	1	9.1/5.6–8.7/3.2	103a
		D-58	E24362	Glass		Green, semitranslucent	Wound?	1	7.2/3.6/2.8–3.0	103b
		D-47	E24351	Clay	Long barrel	Light brown		1	19.0/39.0/3.9	103c
D 6 (1)	On floor, 80 cm south of fortress wall	D-90	E24394, Khartoum 14270	Faience core	Hathor head pendant	Whitish core		1 (and 1 in Sudan)	W 15.3, Th. 4.7, H 23.2, HD 1.6	104b
D 8B (1)	Fill next to Level II platform on north	—	E50821	Glass	Oblate	Semitranslucent	Wound	1	7.6/6.0/2.4	103d
D 9 (3)	Level III residence, lying atop wall at south side of room	D-27	E24332	Faience	Amulet figurine, head and legs below knees lost	Light-blue glaze		1	W 15.5, Th 9.2, H 31.0, HD 1.1	103e
D 9 (1)	Fill of Level II platform	D-54	E24358	Faience	Standard tubular	Blue		1	3.4/3.4/1.2	103f.2
				Faience	Disk cylinder	Blue		1	6.1/2.3/1.9	103f.3
D 32 and area	From east side of D 32, surface	—	E50814	Freshwater mollusk shell			Not perforated	2	D 5.2, H 8.8	103g
				Faience	Disk cylinder	Blue		1	5.4/1.7/2.3	103h
				Glass	Short barrel	Translucent	Wound	1	7.0/5.3/2.0	103i
	Space between D 32 and wall of fortress, at depth of 35 cm	D-62	E24366	Faience	Short cylinder	Light-green glaze		1	1.9/1.1/0.9	99a.4
				Glass	Globular	Red/orange layer over colorless translucent layer	Drawn	2	2.8/2.2/0.8	99a.5
				Glass		Light green, translucent	Drawn	2	3.0/2.7/0.9	99a.6

				Glass	Triple segment	Light green, translucent	Drawn	2	3.3/7.7/1.2	99a.7
				Glass	Single segment	Dark blue, translucent	Drawn	1	2.4/2.0/1.1	99a.8
				Glass	Single segment	Dark purple, translucent	Drawn	1	2.3/1.2/1.1	99a.9
				Gold-in-glass			Drawn	8	2.0–3.3/1.0–2.5/1.1	99a.10
				Silver-in-glass	Globular		Drawn	1	3.3/1.0/1.1	99a.11
				Glass	Globular	Dark blue, translucent	Drawn	3	4.3–5.2/4.2/1.6	99a.12
	Surface rubble west of silo	D-60	E24364	Faience	Disk bicone	Blue glaze		1	8.9–9.7/3.9/2.6–3.5	99b
D 42	In bricks of wall	D-74	E24378	Faience	Long cylinder	Turquoise		1	3.0/13.9/1.2	99c
D 48 (2)	On floor	—	E50830	Faience	Disk cylinder	Blue		1	6.4/1.9/2.1	99d
D 53	In fill below "ash" layer	D-80	E24384	Faience	Short cylinder	Blue		1	7.2/1.4/2.0	99e
D 60 (2)		—	E50822	Ostrich eggshell	Disk cylinder			1	6.2/1.3/1.8	104a
D 61	Below surface in fill	D-69	E24373	Faience	Short cylinder	Blue glaze		1	6.7/3.2/2.0	104c
D 66	From fill under "ash" layer	D-84	E24388	Faience core	Quadruple segmented cylinder	White faience core		1	4.0/6.1/0.9	104d.1
				Faience core	Short cylinder	White faience core		1	6.2/2.1/1.8	104d.2
D 70 (1)		—	E50828	Faience	Disk cylinder	Dark blue		1	7.2/2.0/2.7	104e
D 73 (1)	Above floor along west wall of room	—	E50813	Quartzite	Standard bicone	Red	Drilled from one end	1	4.3/3.7/0.9–1.7	104f
				Faience	Short cylinder	Blue		2	7.7/2.8/2.8; 8.0/2.3/2.4	104g
D 74 (1)	About 30 cm below "ash" layer	—	E50827	Faience	Double segment	Blue glaze		1	3.7/3.7/1.7	104h

(continued)

Table C. Beads *(continued)*

Locus	Find context	Field no.	OIM or Khartoum no.	Material	Shape	Color, diaphaneity	Technique	Quantity	Diameter/ length/hole diameter (mm)	Plate
D 75	Surface clearance along west wall, Level II	—	E50831	Faience	Disk cylinder	Blue		1	6.3/1.1/1.7	104i
D 87 (1)	Fill in room or doorway?	—	E50823	Faience	Disk cylinder	Blue		1	6.3/1.5/2.0	104j
D 92 (1)	Clearance of fill in room	—	E50824	Glass	Oblate	Blue	Wound	1	3.1/2.1/1.7	104k
Outside north wall at North Gate	Ca. 25 cm above floor level	D-94	E24397	Shell fragment				1	3.8/1.8	104l
Fill of D 112 (1)	Meroitic burials 9 and 10 atop wall to south	—	E50817	Glass	Short barrel	Dark blue covered with whitish patina		1	7.2/5.9/2.4	104m
From thick wall between D 112 and D 113	Meroitic burials 9 and 10 atop wall to south	—	E50826	Faience	Standard barrel	Blue		1	6.6/6.3/2.2	104n
D 113 (1)	On floor along north half of west wall	D-106	E24404	Faience	Double-segment cylinder	Blue glaze, whitish core		44	1.9–2.1/2.1/0.6	107a.1
				Faience	Triple segment	Blue glaze, whitish core		23	1.9–2.1/3.2/0.6	107a.2
				Faience	Single segment	Blue glaze, whitish core		9	1.9–2.1/1.0/0.6	107a.3
				Faience	Quadruple segment	Blue glaze, whitish core		7	1.9–2.1/4.7/0.6	107a.4
				Vitreous material	Short cylinder	Blue-gray		74	2.1–2.5/1.4–1.7/0.7	107a.5
				Vitreous material	Double segment	Blue-gray		6	2.1–2.5/3.2/0.7	107a.6
				Vitreous material	Triple segment	Blue-gray		4	2.1–2.5/4.9/0.7	107a.7

D 123	From sifting of ash fill outside D 116	—	None	Glass		Blue, semitranslucent	Drawn	1½	4.7/3.7/0.9	107b
				Glass	Melon			1	6.3/7.6/1.9	107c
D 124	Found while tracing silo outline	—	E50825	Faience	Short cylinder	Blue glaze, white core		2	6.2/2.5/1.7; 7.1/2.0/2.7	106a
D 205 (1)	From fill at south side of room	—	E50820	Carnelian	Standard bicone	Red	Drilled from one end	1	5.1/5.0/0.9–1.9	106b
South retaining wall for Level II, easternmost bay		D-55	E24359	Faience	Disk cylinder	Whitish core, traces of blue glaze		1	5.1/1.3/1.6	106c
From surface of stairway to Level II platform		D-56	E24360	Glass?	Short barrel			1	4.2/2.0/1.6	106d
				Glass	Long barrel	Black body with white trail applied as spiral	Rod formed	1	6.5/12.3/2.6	106e
Surface at southeast corner of retaining wall for Level II		D-57	E24361	Glass	Standard barrel	Blue, semitranslucent	Wound	1	9.1/7.8/2.6	106f
From eroded surface of west fort wall near West Gate		D-5	E24312	Carnelian	Oblate	Red	Drilled from one end	1	7.7/6.8/2.2–2.9	106g.1
				Faience	Disk barrel	Light-blue glaze		2	7.6/3.0/2.4; 6.8/3.17/2.3	106g.2
				Faience	Disk cylinder	Light-blue glaze		1	7.2/2.3/2.1	106g.3
				Glass	Long barrel	Yellow, opaque		1	5.9/9.9/1.9	106g.4

(continued)

Table C. Beads *(continued)*

Locus	Find context	Field no.	OIM or Khartoum no.	Material	Shape	Color, diaphaneity	Technique	Quantity	Diameter/length/hole diameter (mm)	Plate
From eroded surface of wall, south of West Gate		D-42	E24346	Ostrich eggshell	Disk cylinder			2	5.8/1.8/1.4; 6.4/1.9/1.4	105a.1
				Quartzite	Slightly faceted, hexagonal bicone			1	5.0/4.3/0.8–1.6	105a.2
				Faience	Short cylinder			1	6.8/2.5/2.2	105a.3
				Glass		Green, translucent	Wound	1	4.5/4.4/1.3–1.6	105a.4
				Glass		Blue, translucent	Wound	1	3.6/2.8/1.4	105a.5
East Sector, rubble on surface of south wall		D-48	E24352	Chalcedony	Teardrop with pointed base	White		1	W 7.1, Th 6.4, H 13.1, HD 1.2–1.5	105b
Dump, West Sector		D-79	E24383	Faience	Disk cylinder	Turquoise		1	7.6/1.8/2.2	105c
Dump, West Sector		—	E50819	Faience core	Disk cylinder	White		1	6.3/2.1/2.1	105d
Dump, West Sector		—	E50818	Ostrich eggshell	Disk cylinder		Drilled from one end	1	7.0/1.7/1.7–2.7	105e
Dump, West Sector		—	E50829	Faience	Disk cylinder	Blue		1	6.7/1.4/1.6	105f

Maps

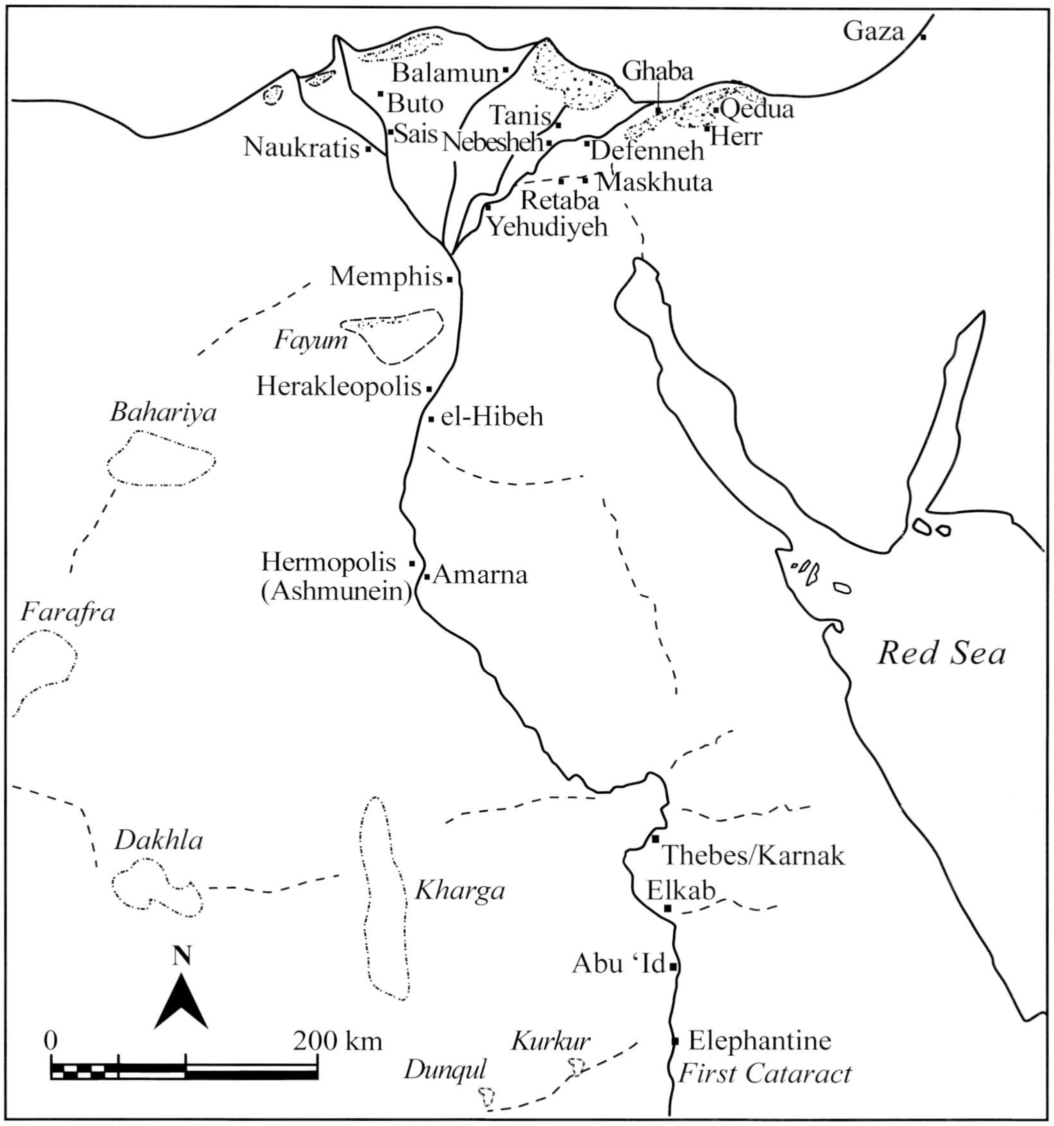
Gaza
Balamun
Ghaba
Buto
Tanis
Qedua
Sais
Nebesheh
Herr
Naukratis
Defenneh
Maskhuta
Retaba
Yehudiyeh
Memphis
Fayum
Herakleopolis
el-Hibeh
Bahariya
Hermopolis
(Ashmunein)
Amarna
Farafra
Red Sea
Dakhla
Thebes/Karnak
Kharga
Elkab
N
Abu ʻId
0
200 km
Kurkur
Elephantine
Dunqul
First Cataract

Map 1. Egypt

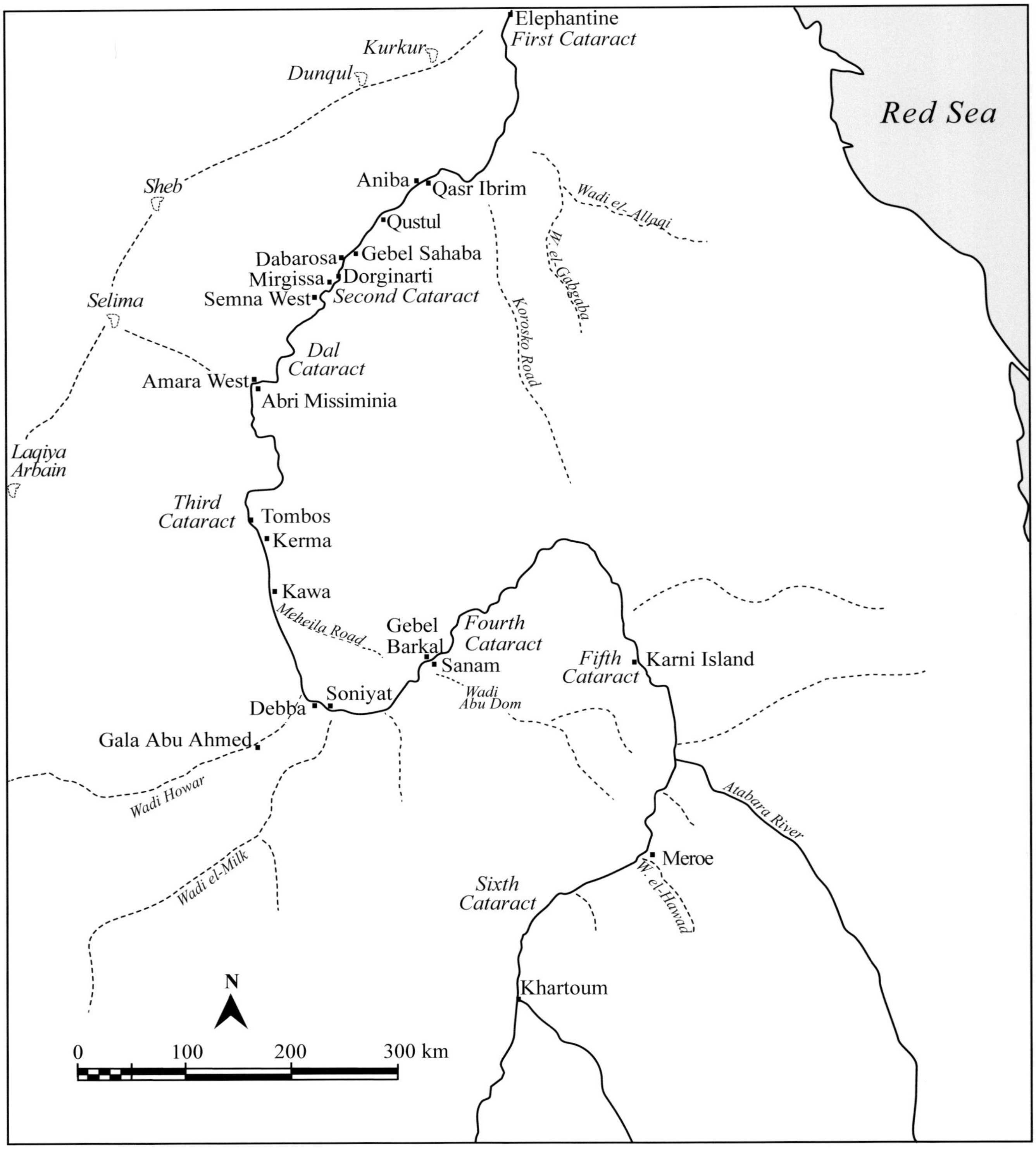
Elephantine
First Cataract
Red Sea
Kurkur
Dunqul
Sheb
Aniba
Qasr Ibrim
Wadi el-Allaqi
Qustul
Dabarosa
Gebel Sahaba
Mirgissa
Dorginarti
Semna West
Second Cataract
Selima
W. el-Gabgaba
Korosko Road
Dal
Cataract
Amara West
Abri Missiminia
Laqiya
Arbain
Third
Cataract
Tombos
Kerma
Kawa
Meheila Road
Gebel
Barkal
Fourth
Cataract
Sanam
Fifth
Cataract
Karni Island
Soniyat
Debba
Wadi
Abu Dom
Gala Abu Ahmed
Wadi Howar
Atabara River
Meroe
W. el-Hawad
Wadi el-Milk
Sixth
Cataract
Khartoum
N
0
100
200
300 km

Map 2. Nubia

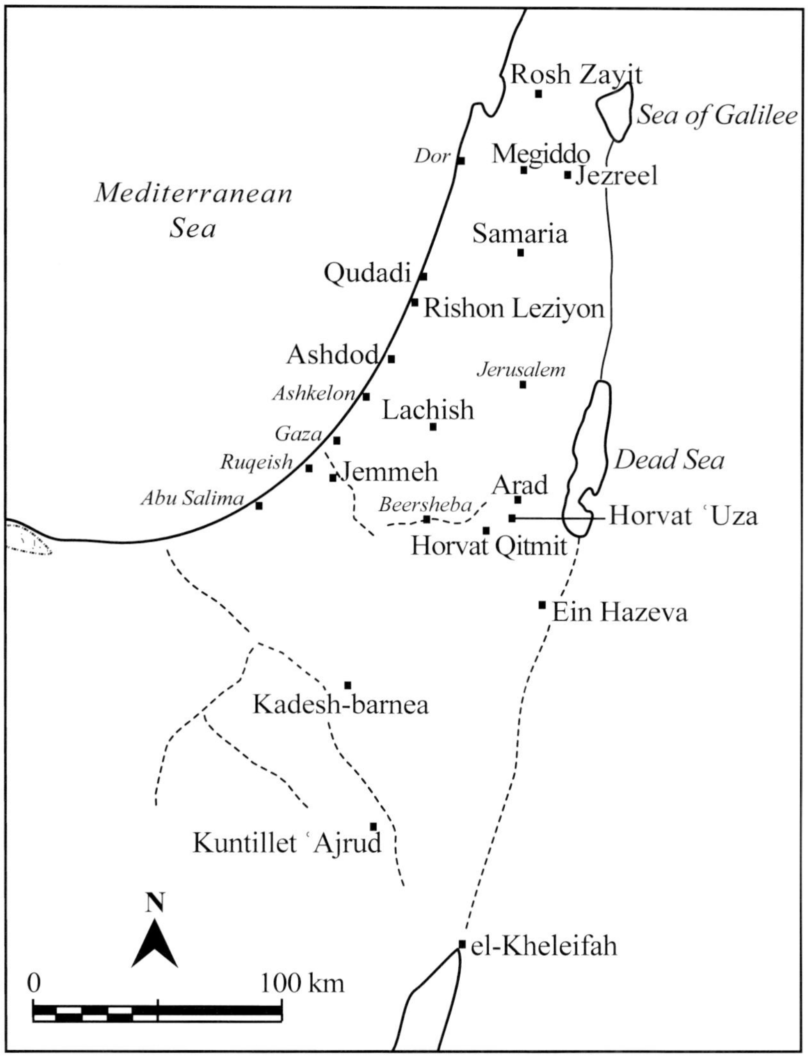

Map 3. The southern Levant

Plans

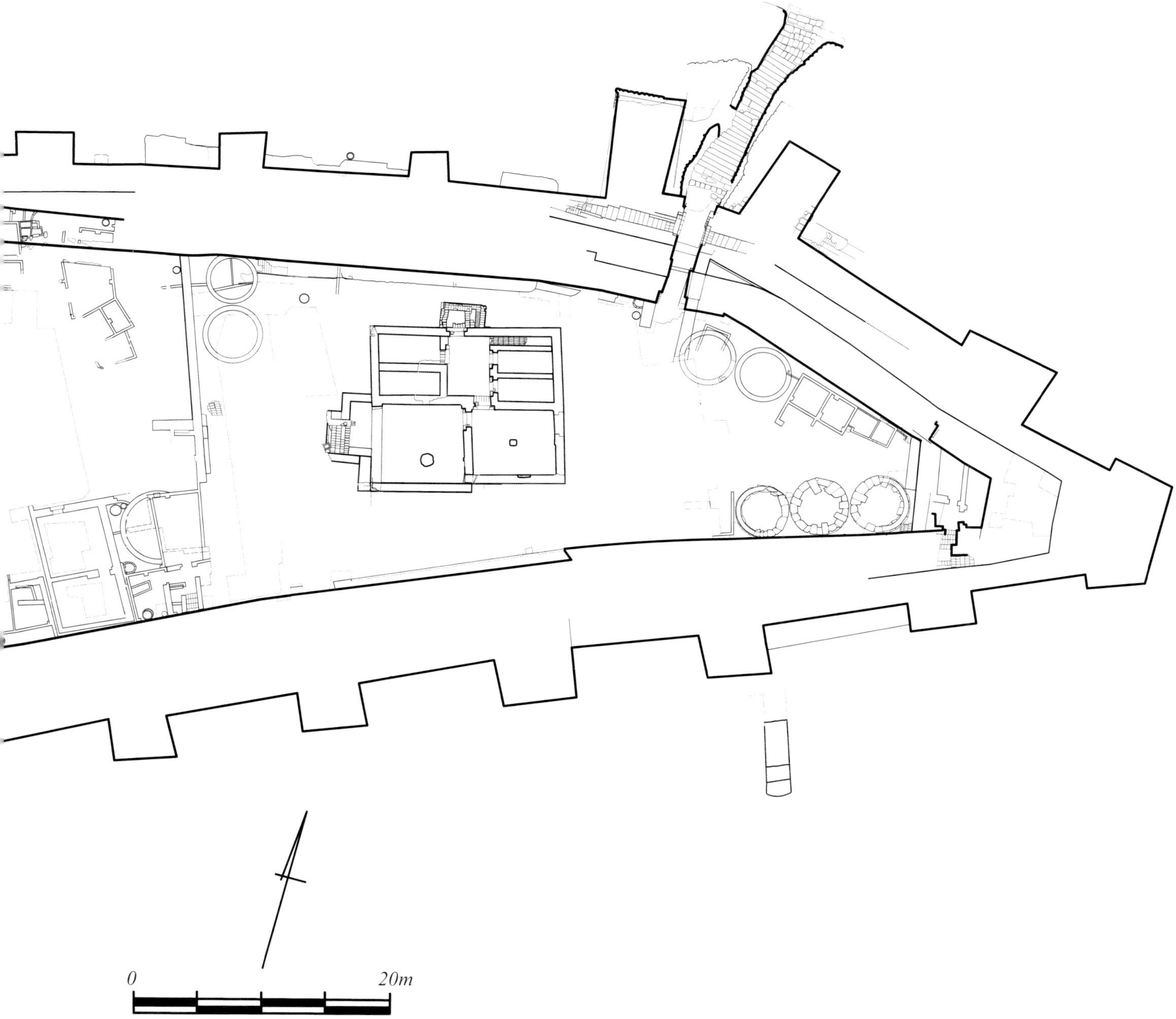

walls and fortress interior

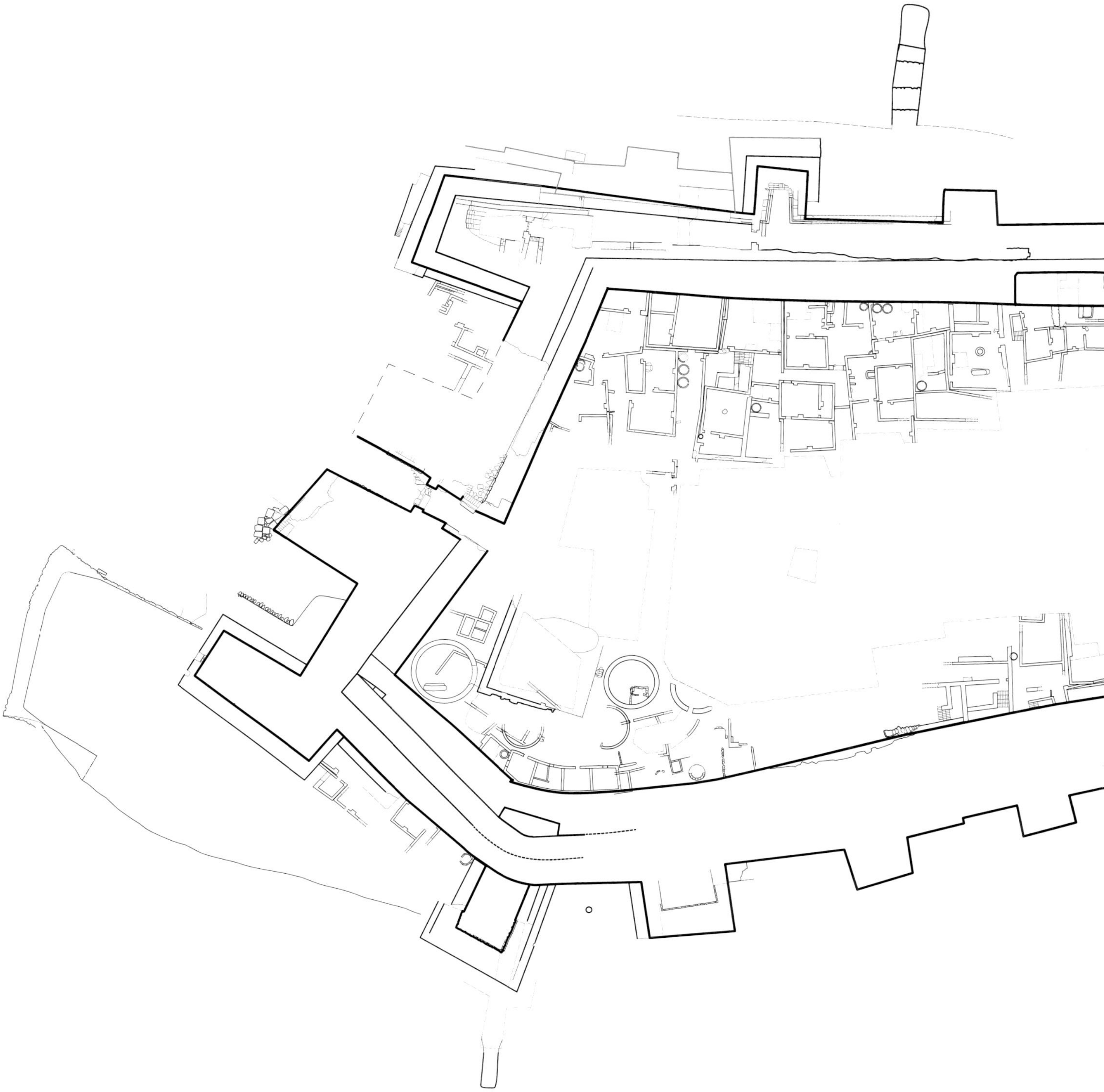

Plan 1. Enclosure

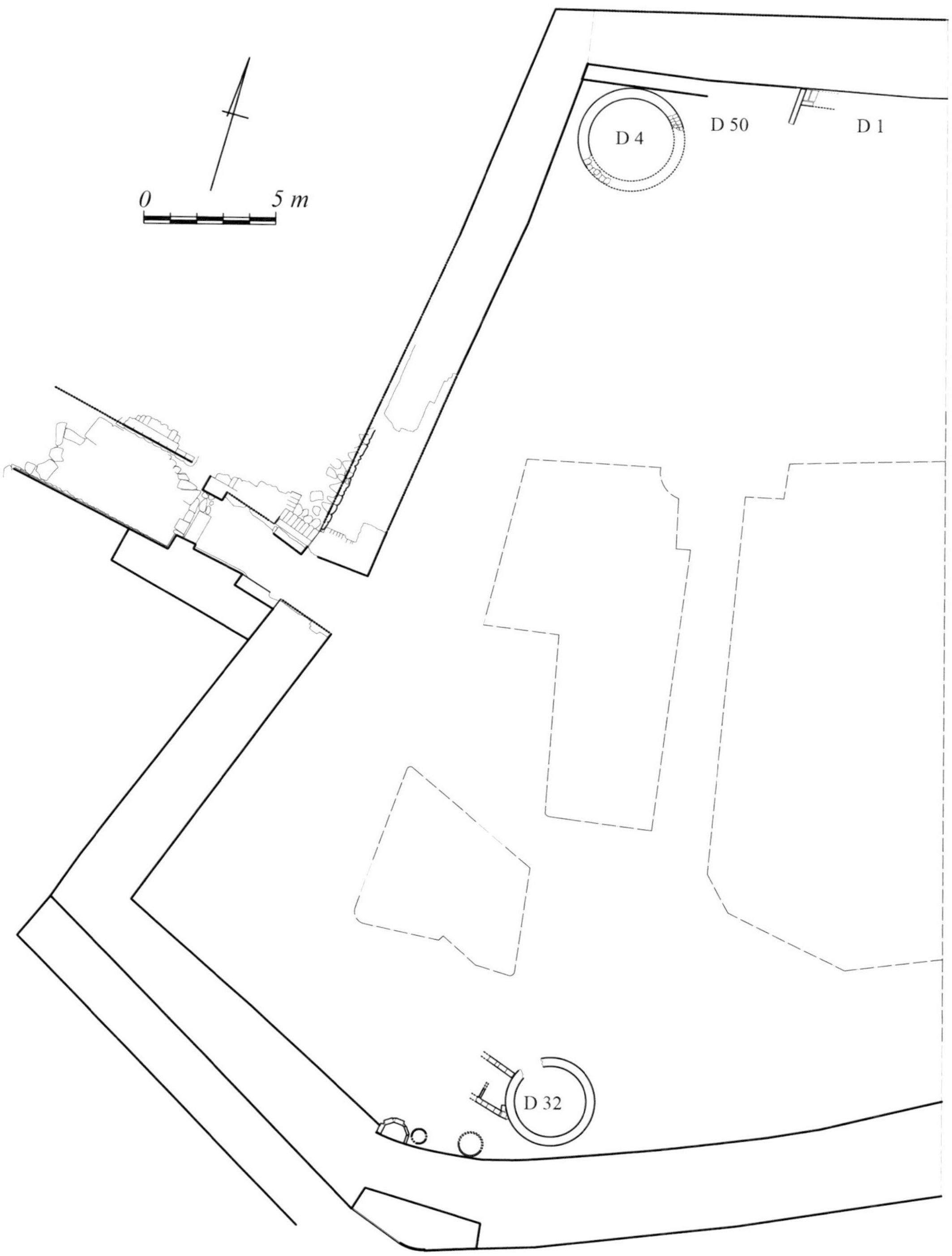

Plan 2. Top level in the West Sector

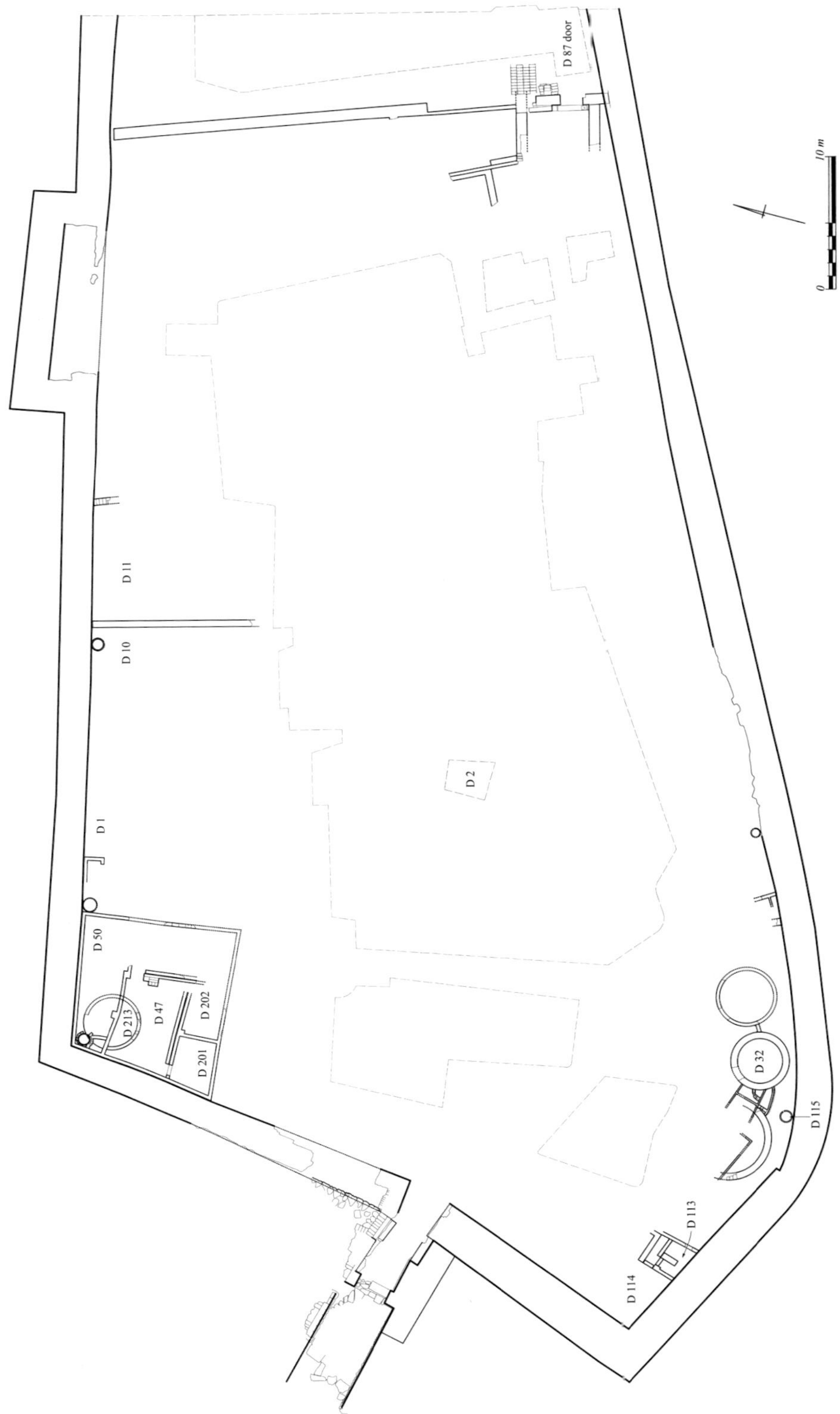

Plan 3. Intermediate level in the West Sector

vel IV walls in the West Sector

Plan 4. Level III and Le

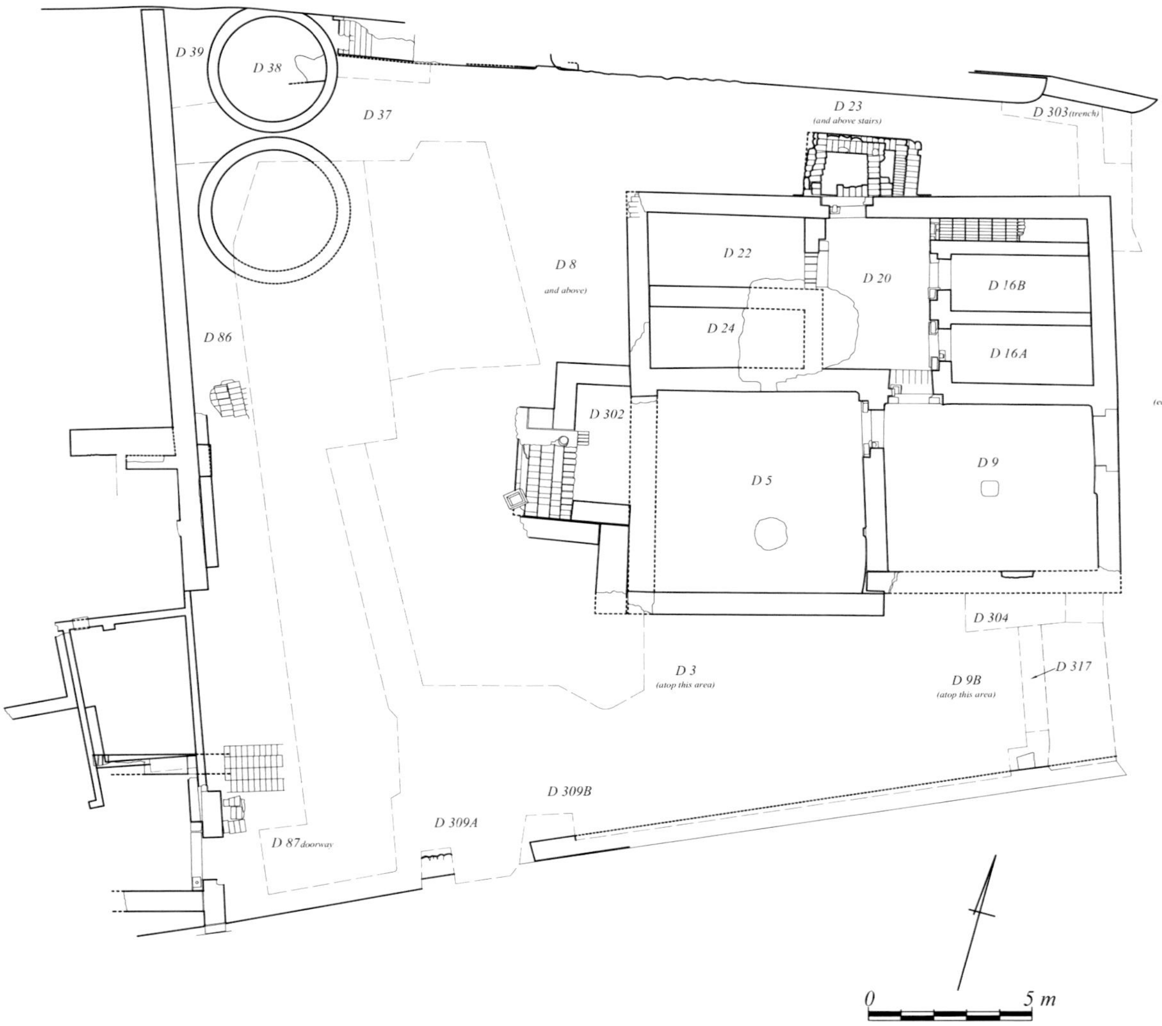

Plan 5. Level III residence in the Central Sector

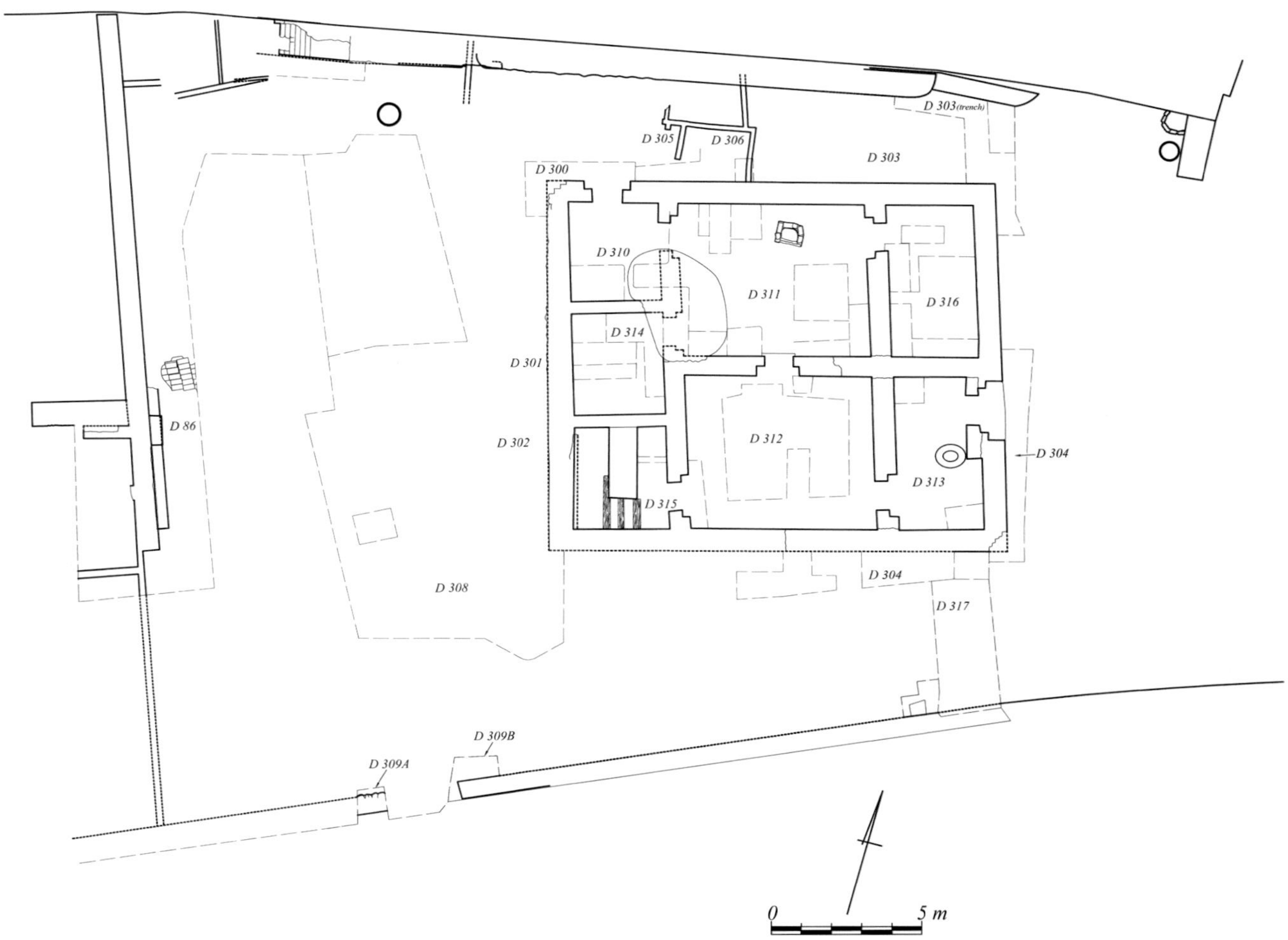

Plan 6. Level IV residence in the Central Sector

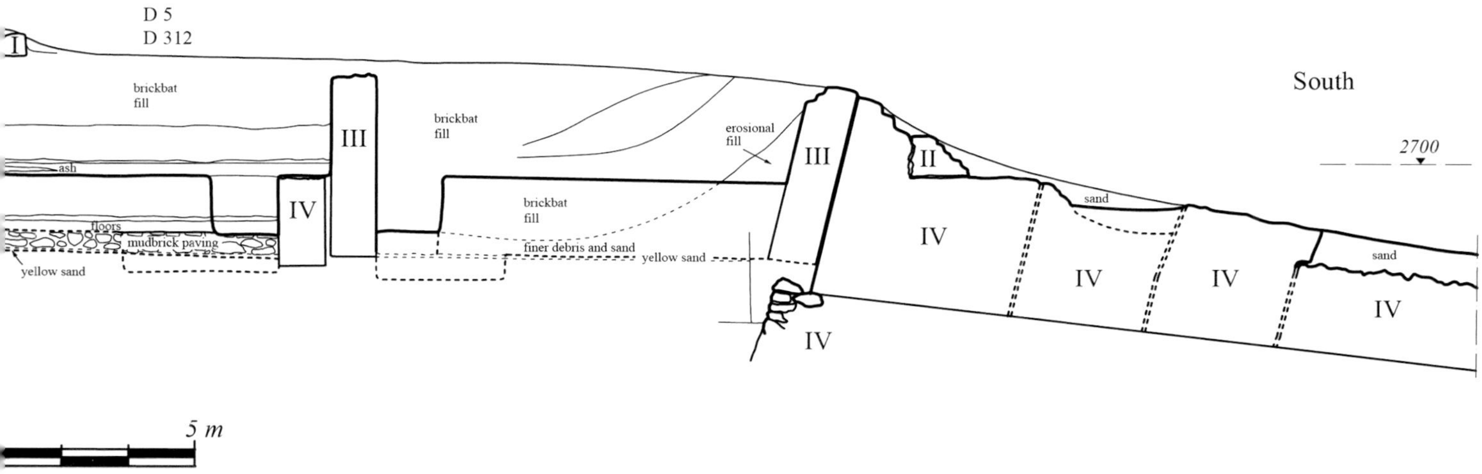

file through the Central Sector

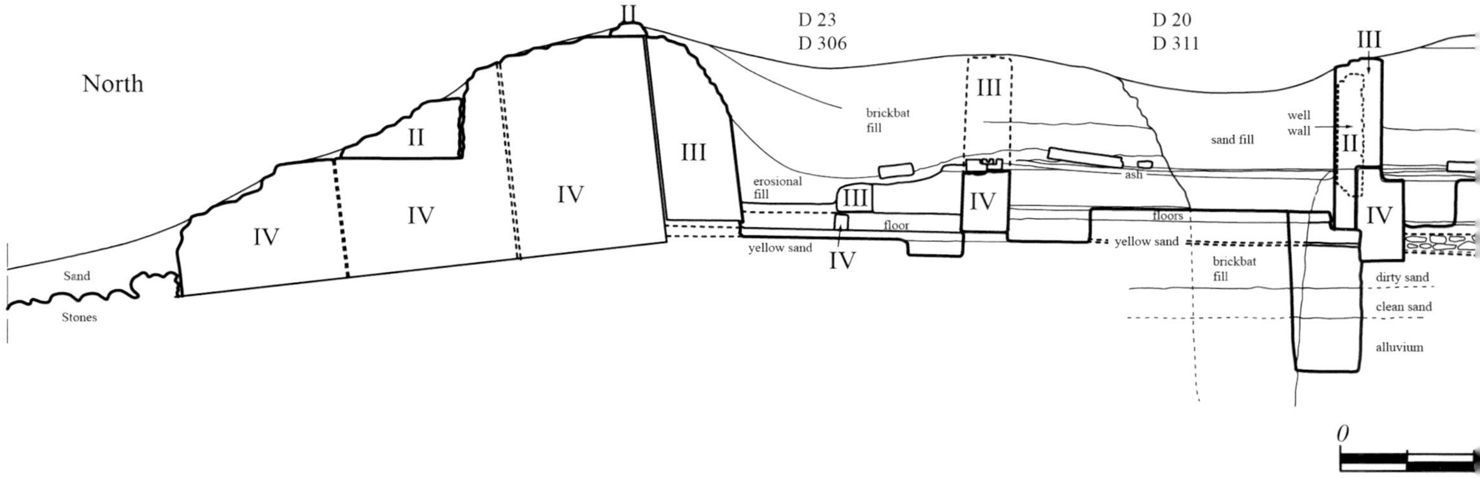

Plan 7. North–south pr

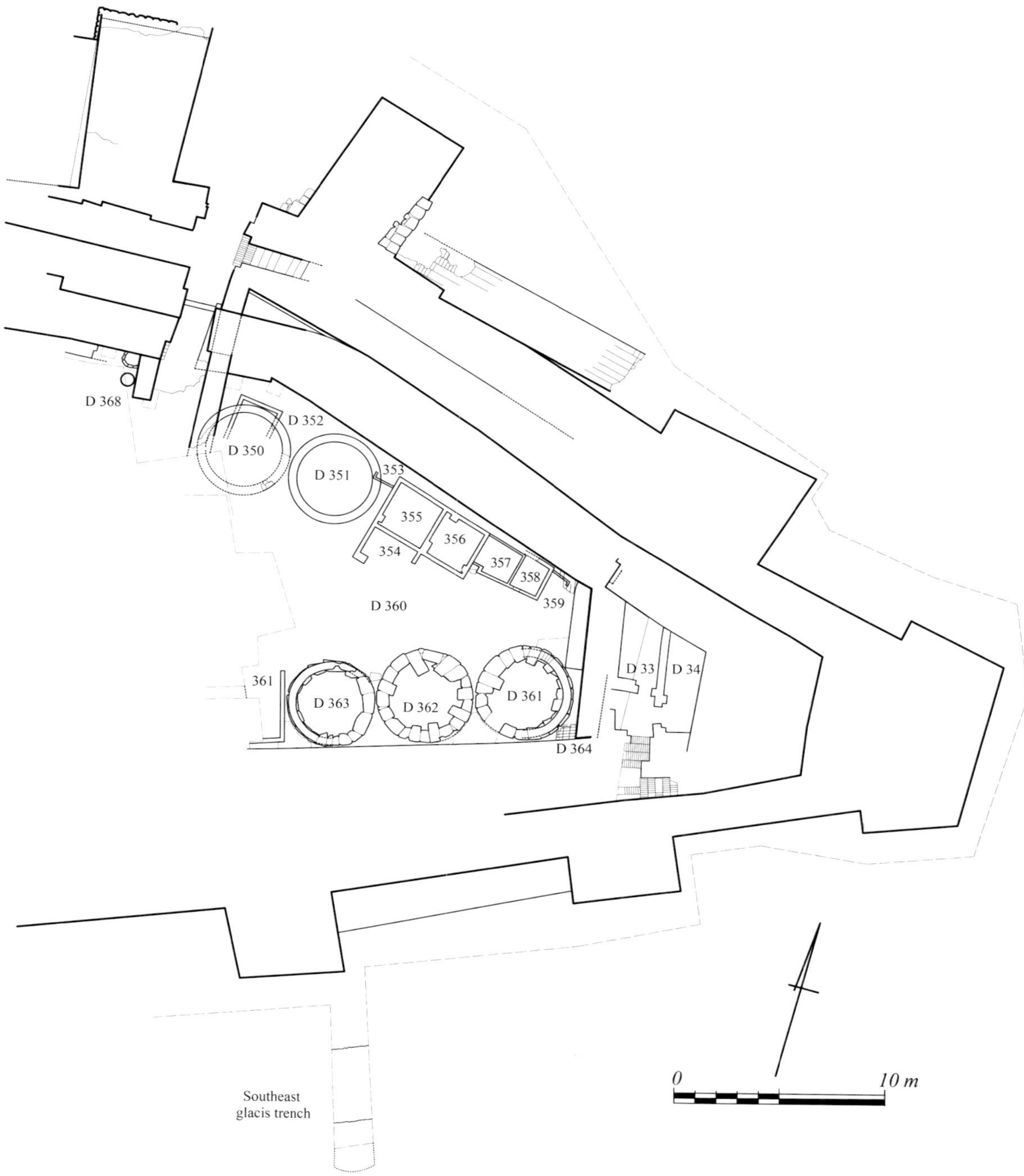

Plan 8. East Sector

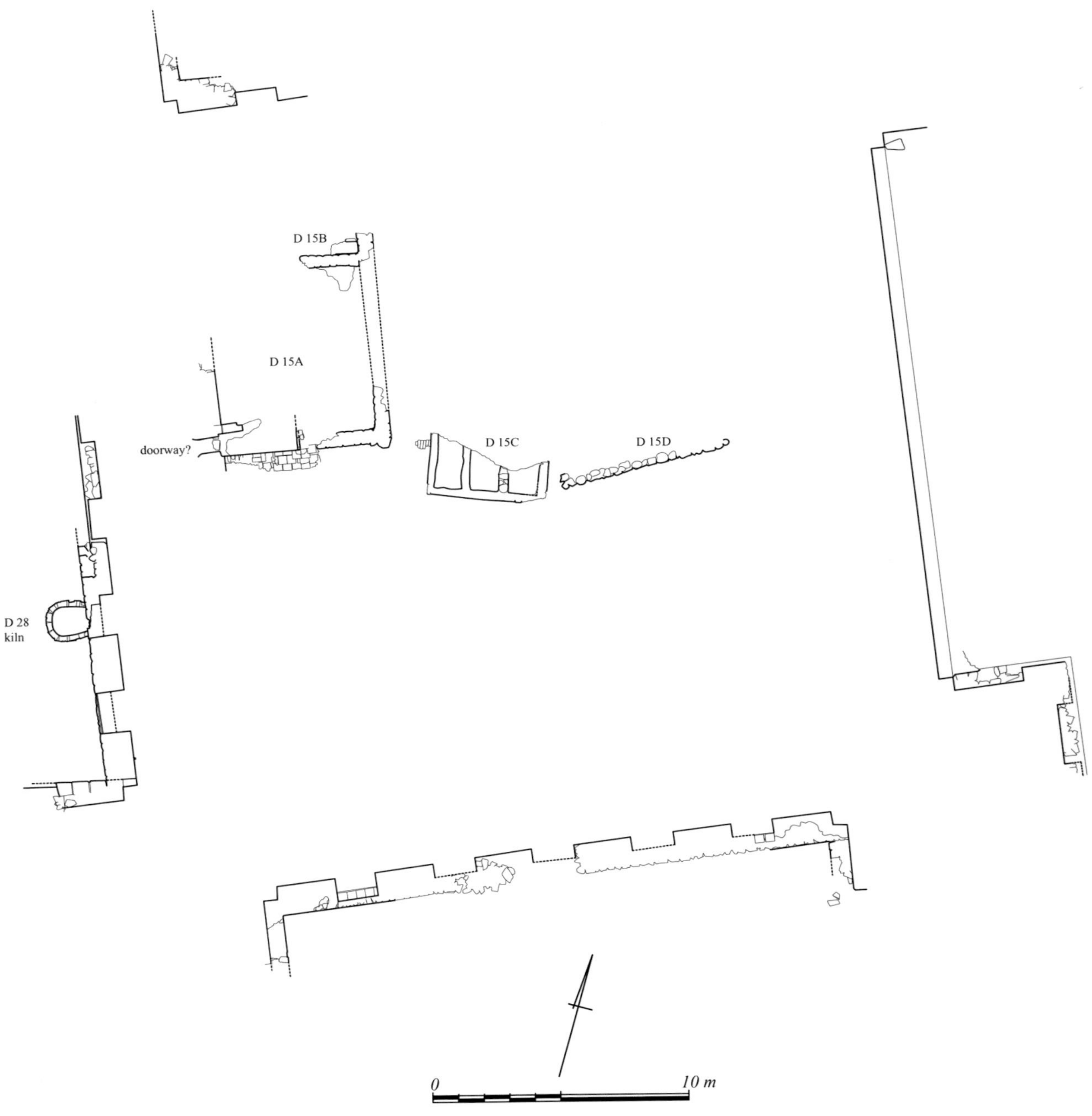

Plan 9. Christian remains atop the Level II platform

Plan 10. Level II remains in the Central Sector

Plates

(*a*) North wall at the breach, looking east toward the Central Sector

(*b*) James Knudstad and workmen excavating the River Stairs

Photos: (*a*) D-007, (*b*) D-05, Frame 12-12A

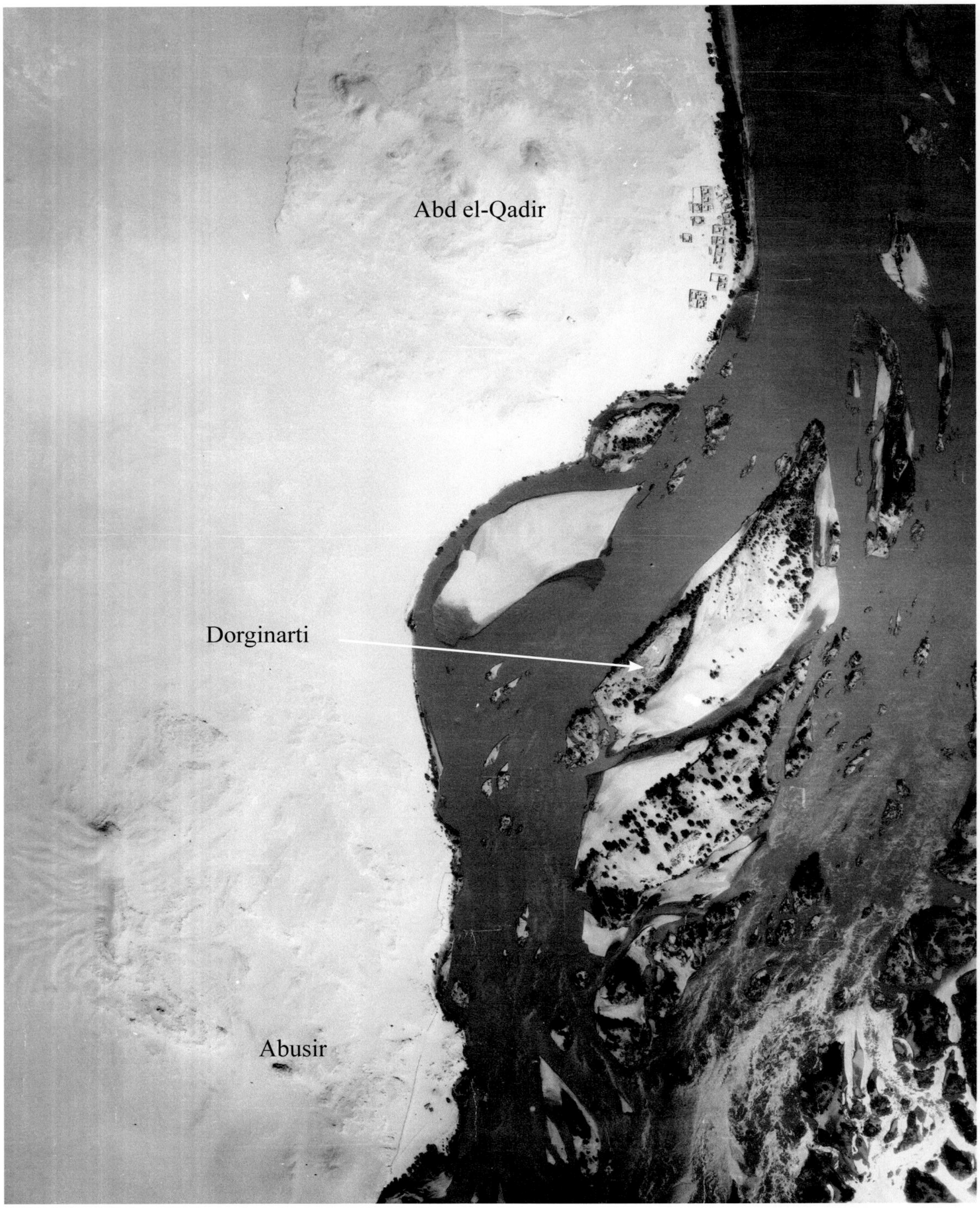

Aerial view of the north end of the Second Cataract (1964)

Photo: Oriental Institute 676-159

(*a*) Brickwork around the crenelated bastion, from the north

(*b*) North wall at the breach, from the east

Photos: (*a*) D-09, Frame 16A-17, (*b*) D-008

(*a*) West Gate, from the east

(*b*) South part of the Central Sector before excavation, from the north

Photos: (*a*) D-07, Frame 17A-18, (*b*) D-019

(*a*) Walls and relinings on the west side of the crenelated bastion, from the southeast

(*b*) Crenelation on the north wall, east side, before the removal of additions

Photos: (*a*) D-09, Frame 11A-12, (*b*) D-09, Frame 20A-21

(*a*) Crenelated bastion relinings, from the northwest

(*b*) Northwest corner bastion and north wall, from the northwest

Photos: (*a*) D-09, Frame 15A-16, (*b*) D-09, Frame 18A-19

(*a*) Crenelations, from the west

(*b*) East crenelations, from the southwest

Photos: (*a*) D-13, (*b*) D-10, Frame 13A-14

(*a*) Parapet relining on the east side, from the north

(*b*) Bottom half of the north glacis trench, from the south

Photos: (*a*) D-09, Frame 12A-13, (*b*) D-10, Frame 9A-10

Bastion west of the North Gate and River Stairs, from the northwest

Photo: D-05, Frame 05-05A

(*a*) Western wall stairs, from the east

(*b*) Eastern wall stairs, from the west

Photos: (*a*) D-10, Frame 17A-18, (*b*) D-10, Frame 19A-20

Later River Stairs above original paving

(*a*) North Gate with the later River Stairs in the background

(*b*) North Gate from the northwest, with the Level II fortress in the background

Photos: (*a*) D-11, Frame 02-02A, (*b*) D-11, Frame 07A-08

River Stairs, from the north

Photo: D-27

(*a*) River Stairs and surroundings, from the north

(*b*) East Sector and north wall before the excavation of the North Gate, from the west

Photos: (*a*) D-03, Frame 26, (*b*) D-02, Frame 35-35A

(*a*) Southern wall stairs from the north, with D 33 in the foreground

(*b*) Bastion H and the Level II platform atop the wall, from the south

Photos: (*a*) D-09, Frame 08A-09, (*b*) D-11, Frame 13-13A

(*a*) South wall and bastions K and L, from the west

(*b*) Bastion N, from the south

Photos: (*a*) D-23, (*b*) D-04, Frame 16

PLATE 16

West Gate from the west, showing piles of cataract stones and the eroded walls

Photo: D-16

(*a*) Oven in D 48, from the east

(*b*) Oven D 115 superimposed atop another oven, from the northeast

Photos: (*a*) D-06, Frame 07-07A, (*b*) D-06, Frame 11-11A

(*a*) Superimposed ovens in D 76, from the north

(*b*) Ovens in D 208, from the north

Photos: (*a*) D-11, Frame 09-09A, (*b*) D-08, Frame 04-04A

West Sector, from the northeast

Photo: D-24, Frame 07-07A

West Gate area with trench and test pit D 2, from the south

(*a*) Intermediate level wall in the West Sector, with an oven to its left, from the south

(*b*) House D 47 walls, with the uppermost walls in the foreground, from the east

Photos: (*a*) D-08, Frame 04-04A, (*b*) D-06, Frame 04-04A

(*a*) House D 221 and area, from the south

(*b*) D 221 and area, from the north

Photos: (*a*) D-07, Frame 20A-21, (*b*) D-73

(*a*) Wall of D 222 appearing under D 201, from the east

(*b*) House D 209 and the northwest corner buildings, from the south

Photos: (*a*) D-09, Frame 03-03A, (*b*) D-07, Frame 19A-20

Structures at the northwest corner of the West Sector, from the west

Photo: D-08, Frame 11-11A

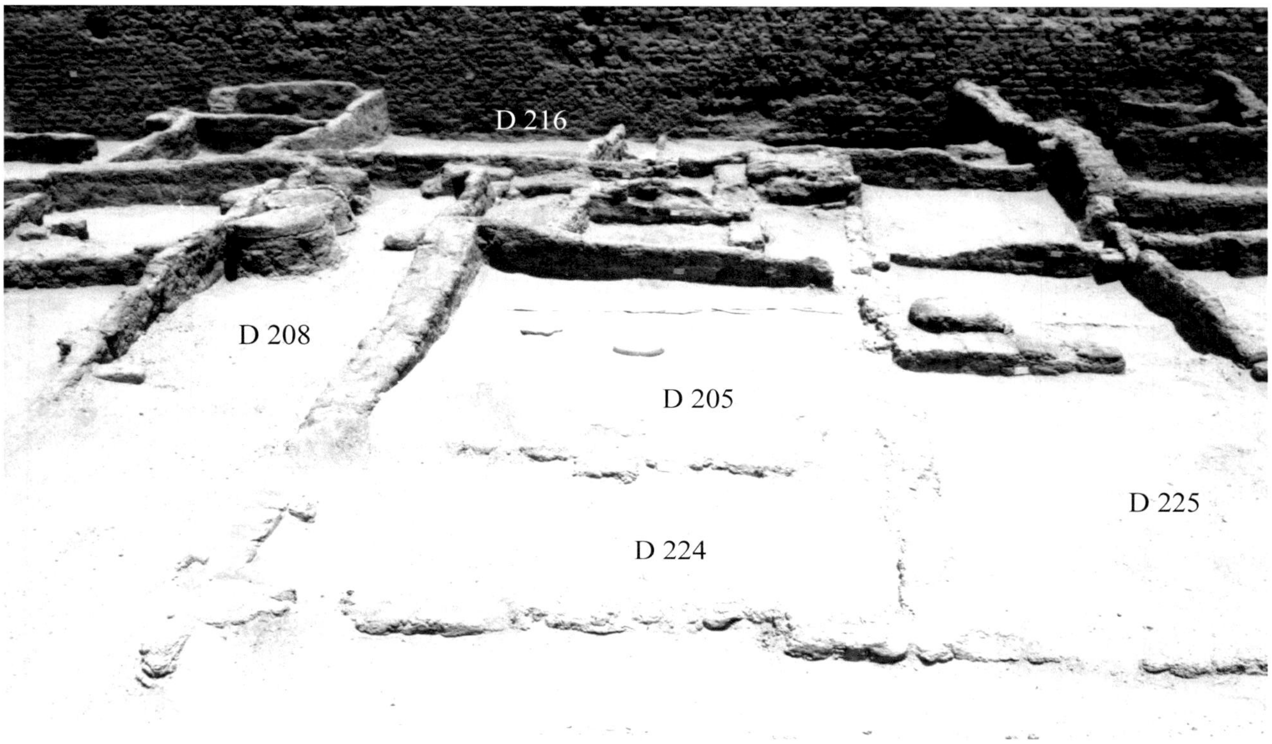

(*a*) D 224 and D 205, from the south

(*b*) Test pit in D 1, showing the base of the inner relining, from the south

Photos: (*a*) D-08, Frame 05-05A, (*b*) Pierce, IMG_0014

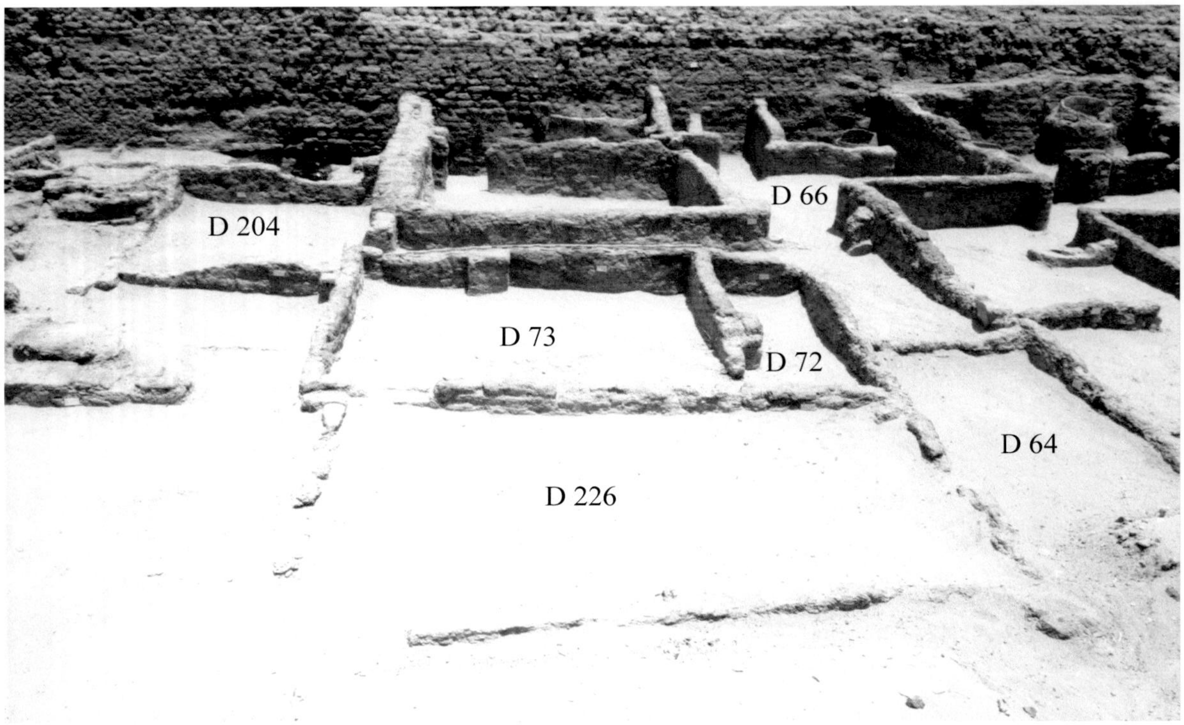

(*a*) D 226, D 72, and D 73, from the south

(*b*) D 6 and D 14, from the north

Photos: (*a*) D-07, Frame 06-06A, (*b*) Polaroid D-71

(*a*) Area from D 6 through D 10 along the wall, from the south

(*b*) D 67, from the south

Photos: (*a*) D-08, Frame 09-09A, (*b*) D-04, Frame 03

(*a*) D 53, D 66, D 54, and D 10, from the south

(*b*) D 10, from the east

Photos: (*a*) D-04, Frame 01, (*b*) D-04, Frame 05

(*a*) D 71, house D 59, and passageway D 56, from the south

(*b*) House D 59, from the north

Photos: (*a*) D-08, Frame 09-09A, (*b*) D-69

(*a*) House D 59 and D 68, from the east

(*b*) Stairway D 57 and D 40–D 44, from the south

Photos: (*a*) D-06, Frame 13-13A, (*b*) D-08, Frame 10-10A

(*a*) D 43, from the north, with staircase D 57 to the right

(*b*) Test trench D 17, from the north

Photos: (*a*) D-68, (*b*) Pierce, IMG_0023

(*a*) D 229 under the relining wall, from the east

(*b*) D 32 and area, with superimposed ovens and silos, from the northwest

Photos: (*a*) D-09, Frame 04A-05, (*b*) D-06, Frame 10-10A

(*a*) Southwest corner with D 113 along the wall and exposed bedrock, from the north

(*b*) Southwest corner remains, from the southeast

Photos: (*a*) D-016, (*b*) D-19

(*a*) Structures along the south wall, from the west

(*b*) D 117–D 121, from the east

Photos: (*a*) D-72, (*b*) D-08, Frame 13-13A

(*a*) South wall area, from the southwest

(*b*) View of D 101–D 103 at an early stage of excavation, from the west

Photos: (*a*) D-22, (*b*) D-12, Frame 25

(*a*) D 101, from the north in D 102

(*b*) Doorway D 87, from the west

Photos: (*a*) D-012, Frame 29, (*b*) D-07, Frame 13A-14

(*a*) Doorway D 87 and the earlier rooms at the southeast corner, from the north

(*b*) Dividing wall between the West and Central Sectors, from the north

Photos: (*a*) D-08, Frame 14, (*b*) D-05, Frame 14-14A

Level III residence, from the northwest

Photo: D-07

(*a*) Corridor north of the Level III residence, from the west

(*b*) West stairs of the Level III residence, from the west

Photos: (*a*) Pierce, IMG_0010, (*b*) D-11, Frame 07-07A

North stairs of the Level III residence, from the north

Photo: D-08

(*a*) Wall separating D 310 and D 314 below D 24, from the east

(*b*) Ramesside inscription on the column fragment in D 9

Photos: (*a*) D-04, Frame 13, (*b*) D-02, Frame 27-27A

(*a*) D 9 from the south, with the column foundation

(*b*) Doorways into D 5 and D 20, from the southeast

Photos: (*a*) D-03, Frame 30, (*b*) Pierce, IMG_0018

(*a*) Jambs of the door into D 20 from D 9, from the south

(*b*) Strata atop the floor in D 9, showing burned stratum 3, from the east

Photos: (*a*) D-02, Frame 24-24A, (*b*) D-26

(*a*) Jambs and fragments of the lintel and column in D 5, from the west

(*b*) South walls of D 9 and D 5, from the southeast

Photos: (*a*) D-01, Frame 26A-27, (*b*) D-03, Frame 2

North half of the Level III residence, from the west

Photo: D-09

(*a*) Stone sills of D 22 and the north entrance door, from the southeast

(*b*) Fallen lintel fragments in D 20, from the north

Photos: (*a*) D-03, Frame 90, (*b*) D-01, Frame 29A-30

(*a*) D 16A, D 16B, and D 16C, from the west

(*b*) D 16A, from the west

Photos: (*a*) D-04, Frame 10, (*b*) D-01, Frame 27A-28

(*a*) D 16C, from the west

(*b*) North entrance stairs with fallen architectural elements, from the north

Photos: (*a*) D-04, Frame 12, (*b*) D-06, Frame 17-17A

(*a*) Inscribed east jamb of the door into D 20, from the south

(*b*) Inscribed west jamb of the door into D 20, from the south

Photos: (*a*) D-02, Frame 26-26A, (*b*) D-02, Frame 25-25A

Graffiti on the sill between D 9 and D 20

Photo: D-13, Frame 14

Lintel from D 5

Photo: D-03, Frame 36

(*a*) Detail from the lintel in D 5

(*b*) Sandstone column fragment from D 5

Photos: (*a*) D-01, Frame 09A-10, (*b*) D-13, Frame 11

(*a*) Palm-leaf capital from the surface of the West Sector

(*b*) Lintel fragments near D 22

Photos: (*a*) D-13, Frame 10, (*b*) D-13, Frame 19

(*a*) Lintel fragments near D 22

(*b*) Ramesside inscription on the plinth at the doorway to D 22

Photos: (*a*) D-13, Frame 20, (*b*) D-01, Frame 12A-13

(*a*) Graffito of a boat on the sill between D 16A and D 20

(*b*) Graffiti on a stray stone from the Level III residence

Photos: (*a*) D-13, Frame 17, (*b*) D-13, Frame 18

Level IV walls appearing during the excavation of the Level III residence

Photo: D-12

(*a*) Doorway into D 311 from D 312, from the southeast

(*b*) East doorway into the Level IV residence, from the east

Photos: (*a*) D-07, Frame 06A-07, (*b*) D-08, Frame 16-16A

(*a*) D 313 from the west, with the blocked doorway and Level III walls in place

(*b*) Stairway in D 315, from the east

Photos: (*a*) D-04, Frame 08, (*b*) D-07, Frame 01A-02

(*a*) Brick feature in D 311, from the southwest

(*b*) Doorway into D 316, from the west

Photos: (*a*) D-07, Frame 10A-11, (*b*) D-07, Frame 12A-13

(*a*) East Sector with Meroitic burial 6 in the center, from the east

(*b*) East Sector with silos D 350 and D 351 in the foreground, from the northwest

Photos: (*a*) D-02, Frame 30-30A, (*b*) D-08, Frame 18-18A

East Sector, from the southwest

Photo: D-31

D 351–D 359, from the east

Photo: D-08, Frame 20-20A

D 33 and D 34, from the southwest

Photo: D-09, Frame 06A-07

D 33 and D 34, from the northwest

Photo: D-09, Frame 07A-08

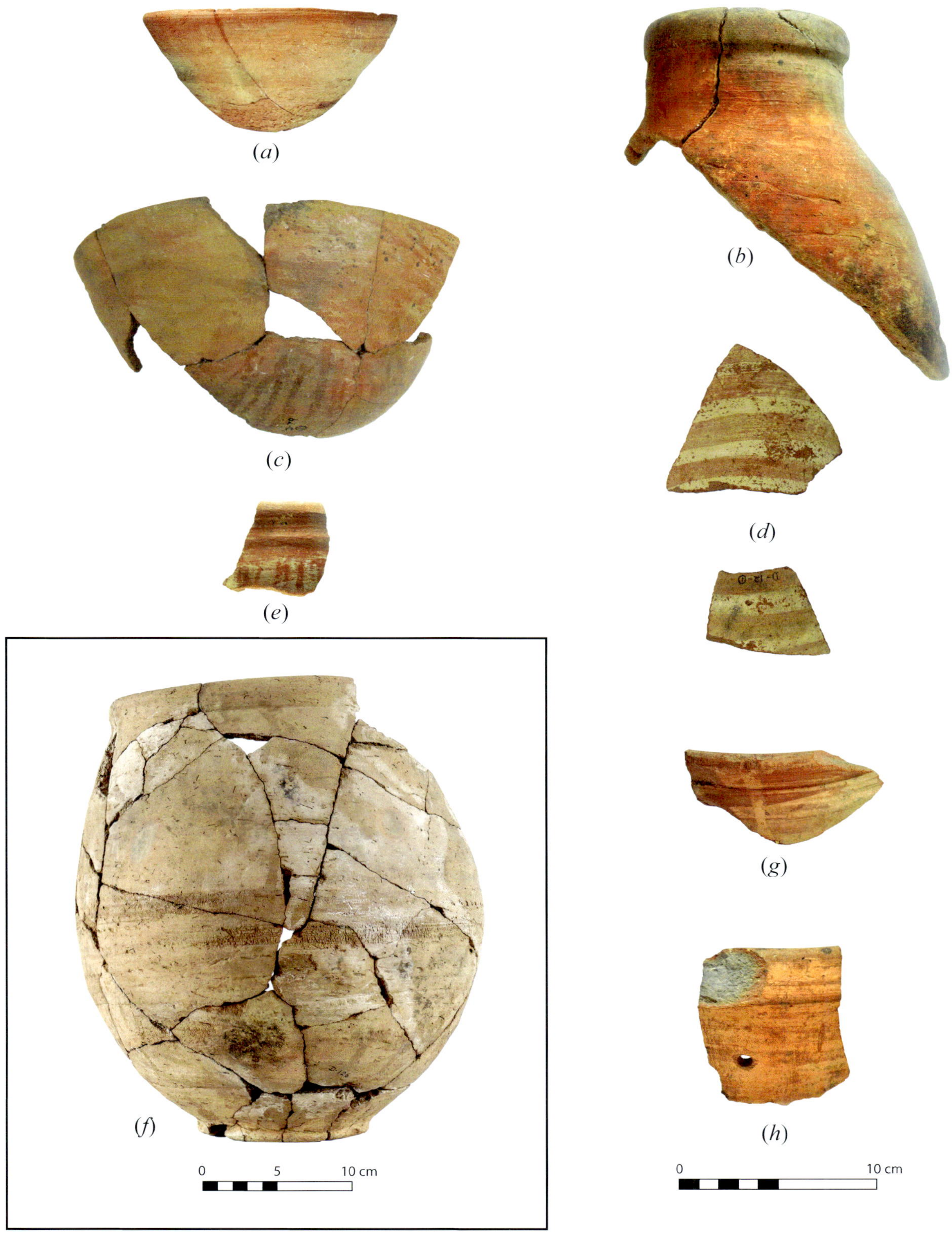

Silt pottery
(*a*) D 60 (1); OIM E50844, RR interior/exterior
(*b*) D 6 (1); OIM E49504, RS exterior
(*c*) D 79 (1); RS traces(?), streak-burnished
(*d*) D 12 (1); white bands on exterior, larger sherd warped
(*e*) East Sector, rubble over three south silos, RS exterior/RR interior
(*f*) D 60 (1); OIM E24422, RS traces on rim
(*g*) D 104 (1); OIM E49564
(*h*) D 1 (1); OIM E49562

(*a*)

(*b*)

(*c*)

(*d*)

(*e*)

(*f*)

Marl pottery
(*a*) D 215 (1); OIM E49517
(*b*) D 26 (1); OIM E49502
(*c*) D 59 (1); OIM E49527
(*d*) D 4 (1); OIM E49572
(*e*) D 208 (2)
(*f*) D 13 (1)

(*a*)

(*b*)

(*c*)

Silt jugs and flask
(*a*) D 8 clearance in Central Sector; OIM E50737
(*b*) Room east of D 76 (i.e., D 90); OIM E49515, RS exterior, black interior
(*c*) D 219 (1); OIM E49524, RS exterior, burnish strokes on exterior

Silt jugs and flask
(*a*) D 6 (1); black-painted decoration under handles on shoulder
(*b*) D 72 (1)
(*c*) D 61, below surface on southeast; OIM E24374
(*d*) D 102 (1)
(*e*) D 7 (1); OIM E49513

(*a*)

(*b*)

5 cm

Double flask and imported flask
(*a*) D 122 (1); Khartoum 14267, RS and burnished
(*b*) D 19 (1); OIM E24327

Highly burnished wheel-made vessels
(*a*) D 113 (1) + D 124 (1); OIM E50074, OIM E50078
(*b*) D 310 (1); OIM E49523, RS burnished exterior
(*c*) D 90 (1); OIM E50077 and OIM E50088, RS burnished interior/exterior

(*a*)

(*b*)

(*c*)

(*d*)

Handmade bowls
(*a*) D 90 (1); OIM E24414
(*b*) D 90 (1); OIM E36383, incised decoration exterior and rim, burnish traces interior/exterior
(*c*) D 10 (1); OIM E36385, incised decoration interior/exterior
(*d*) D 18 (1); OIM E36386, incised decoration exterior and cross-hatched atop rim, burnished rim on exterior and all of interior

Handmade jar and bowls
(*a*) D surface; OIM E36384, incised rim, burnished interior/exterior
(*b*) D 13 (1); OIM E49535, burnished interior/exterior, sooted, incised hatches on rim
(*c*) D 7 (1); burnished interior/exterior, incised cross-hatched rim
(*d*) D 92 (1); OIM E49532, burnished interior/exterior, incised Vs on rim
(*e*) D 214 (1); black-burnished interior/exterior, impressed wedges on rim
(*f*) D 200, D 201, D 202 surface and (1); OIM E49533, burnished interior/exterior
(*g*) D 215 (1); OIM E49568, burnished exterior

Handmade bowl
D 125 (1); OIM E36380

(*a*)

(*b*)

(*c*)

(*d*)

5 cm

New Kingdom sherds
(*a*) D 10 (1); OIM E50072
(*b*) East Sector, above three south silos along south wall
(*c*) East Sector, atop eroded surface of south wall; OIM E51977, red-painted concentric bands
(*d*) East Sector, rubble outside north wall; marl, reddish-brown bands on exterior

(*a*) Central Sector Level II, from the west (Christian oven or kiln in the center)

(*b*) Southeast corner of the Level II wall, from the east

Photos: (*a*) D-03, Frame 16, (*b*) D-01, Frame 17A–18

(*a*) Southeast corner of the Level II platform and the fortification wall, from the east

(*b*) South buttresses of the Level II platform, from the west

Photos: (*a*) D-02, Frame 32–32A, (*b*) D-07, Frame 14A–15

North side of the southeast corner buttress of the Level II platform, showing renovations (filled-in bays); arrow(?) slot visible to the right of the center of the photograph

Photo: D-03, Frame 42

(*a*) Superimposed walls, from the east

(*b*) D 23, section through the fill, from the west

Photos: (*a*) D-01, Frame 20A–21, (*b*) Dorginarti Polaroid 60

(*a*) West side of Level II, from the southwest

(*b*) East side of Level II, from the northeast

Photos: (*a*) D-05, Frame 13–13A, (*b*) D-05, Frame 20–20A

(*a*) East side of Level II with stairway, from the north

(*b*) West side of the Level II construction, from the north

Photos: (*a*) D-02, Frame 34–34A, (*b*) D-05, Frame 14–14A

(*a*) Northeast corner of the Level II building or enclosure, from the southwest

(*b*) Level II building or enclosure at the southeast corner, from the west

Photos: (*a*) D-01, Frame 21A–22, (*b*) D-01, Frame 19A–20

(*a*)

5 cm

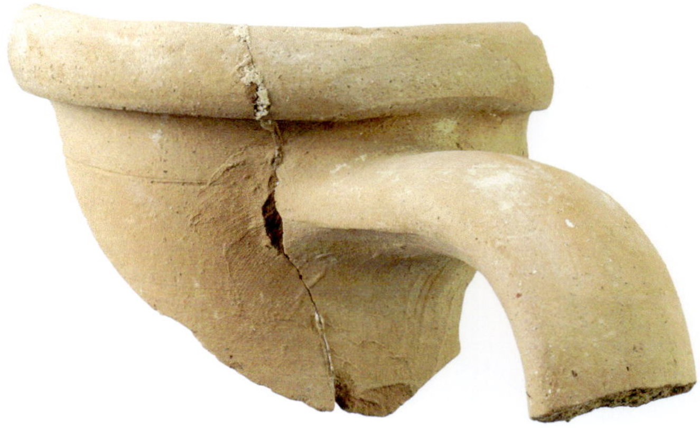

(*b*)

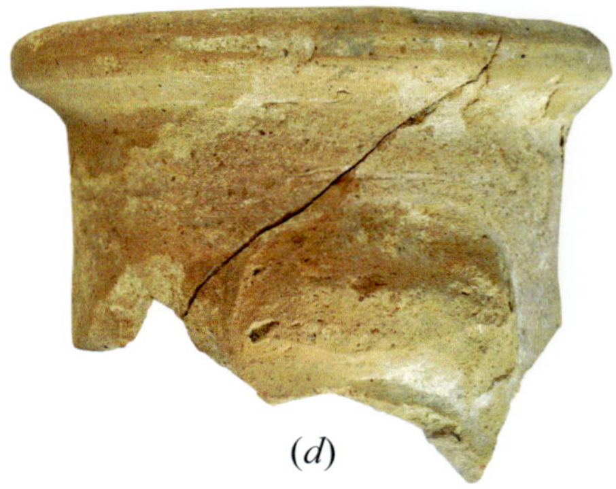

(*d*)

(*c*)

(*f*)

(*e*)

5 cm

Imported amphorae and black-slipped handle
(*a*) OIM E24372
(*b*) OIM E49509
(*c*) OIM E49507
(*d*) OIM E49506
(*e*) OIM E49510
(*f*) OIM E49508

(*a*)

(*b*)

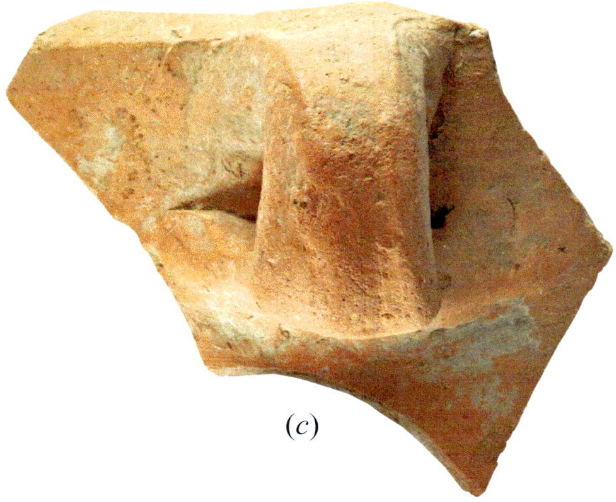

(*c*)

(*d*)

5 cm

Phoenician amphorae and mud sealing
(*a*) East Sector, north of stairs to Level II, first 40 cm; OIM E49541
(*b*) D 75; OIM E49539, top of shoulder
(*c*) East Sector, north of stairs to Level II, first 40 cm; OIM E49541
(*d*) D 15 (1); mud with incised symbols

(*a*)

(*b*)

Phoenician amphorae fabrics
(*a*) D 75; Phoenician amphora
(*b*) East Sector, over south enclosure wall (handle and shoulder fragment, not illustrated)

(*a*)

(*b*)

(*c*)

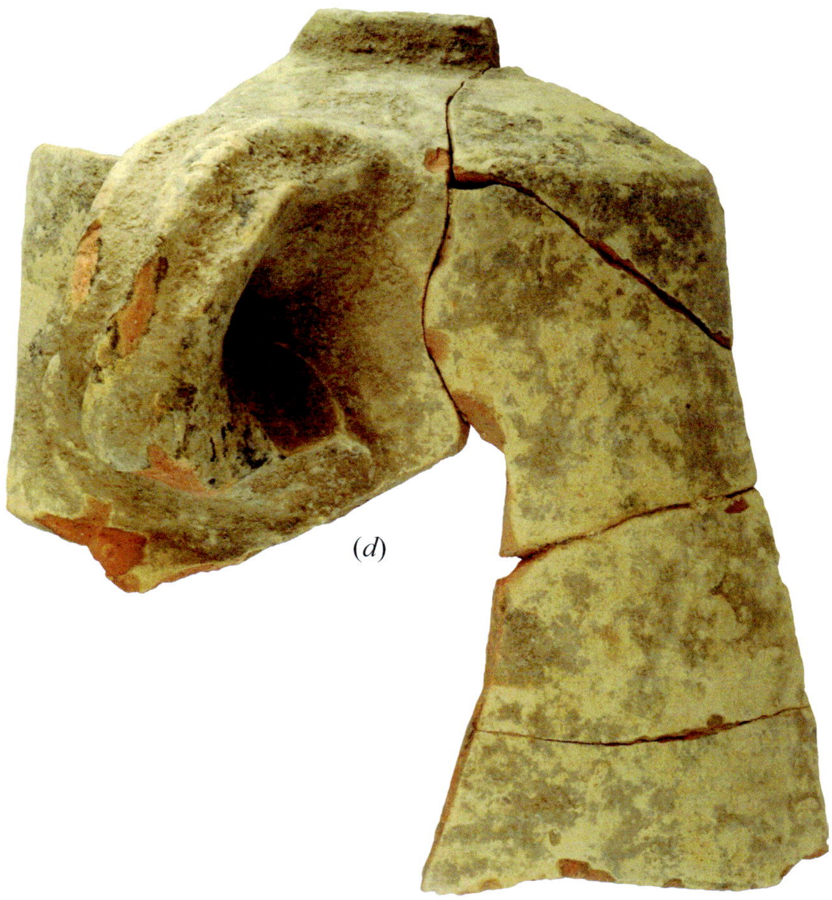

(*d*)

(*e*)

5 cm

Phoenician amphorae
(*a*) D 15 (1); OIM E49536
(*b*) D 75; OIM E49538
(*c*) D 15 (1) and D 8, first 50 cm below Christian building; OIM E49543
(*d*) River Stairs, first ten steps; OIM E49544
(*e*) East Sector, north of stairs to Level II, first 40 cm; OIM E49542, showing longer profile than fig. 6.10b

(*a*)

(*b*)

(*c*)

(*d*)

(*f*)

5 cm

(*e*)

Level II Egyptian vessels
(*a*) Central Sector, east retaining wall for Level II, atop plastered floor of bay north of stairs; OIM E24344, marl
(*b*) D 66 (1); RS interior/exterior, burnished exterior, silt
(*c*) East Sector, over three south silos; OIM E49567, silt, RS rim interior/exterior, RS exterior
(*d*) East Sector, first 60 cm in center; OIM E49511, marl
(*e*) Central Sector, south wall; OIM E49565, silt
(*f*) East Sector, first 60 cm in center; OIM E49512, marl

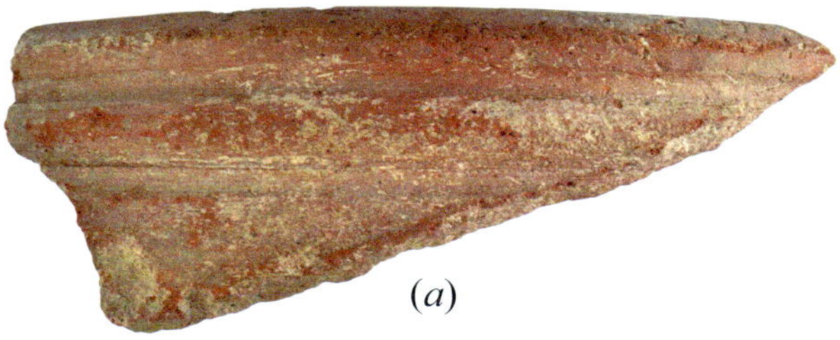

(*a*)

(*b*)

(*c*)

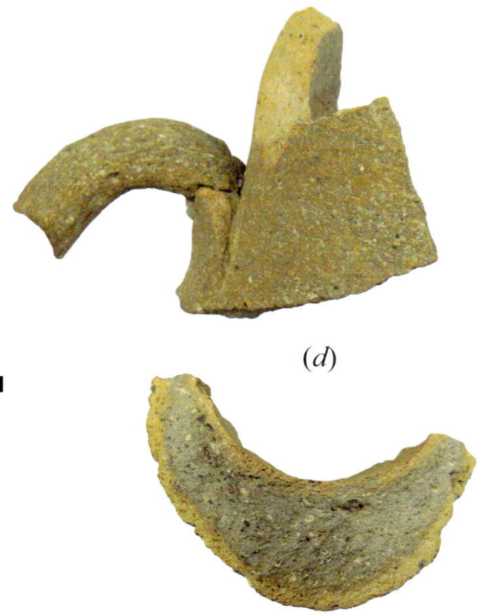

(*d*)

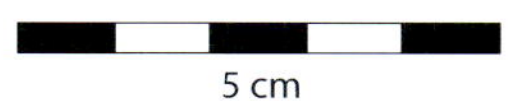

(*e*)

Level II vessels
(*a*) D 13 (1); OIM E49516
(*b*) Central Sector, west wall, surface; OIM E49503, marl
(*c*) D 351 (1); marl
(*d*) D 9 (1); small flask, bottom photo is break at bottom edge (not to scale)
(*e*) D 12 (1); OIM E36381, view of exterior and interior

Oasis-ware fabrics
(*a*) D 92 (1); exterior surface on left, interior on right
(*b*) Central Sector surface; exterior surface on left, interior on right

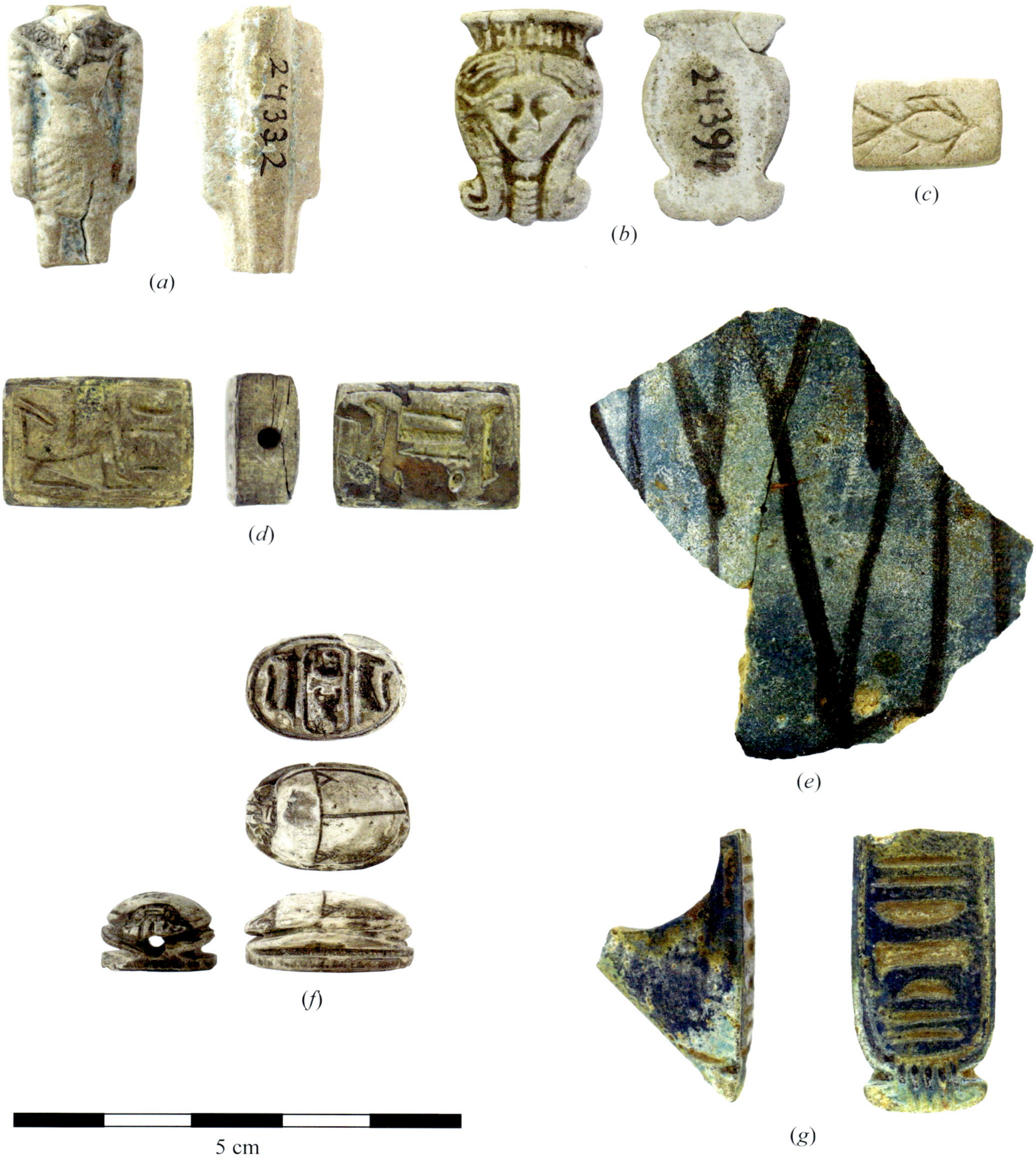

Faience objects
(*a*) D 9 (3), on top of eroded south wall; OIM E24332
(*b*) D 6 (1); OIM E24394, see also pl. 104b
(*c*) D 81, below "ash" layer above floor; OIM E24398
(*d*) D 76, below "ash" layer; OIM E24386
(*e*) From stone rubble outside north fortress wall, west of North Gate
(*f*) D 209 (1) fill; OIM E24396
(*g*) D 18 (1); OIM E24325

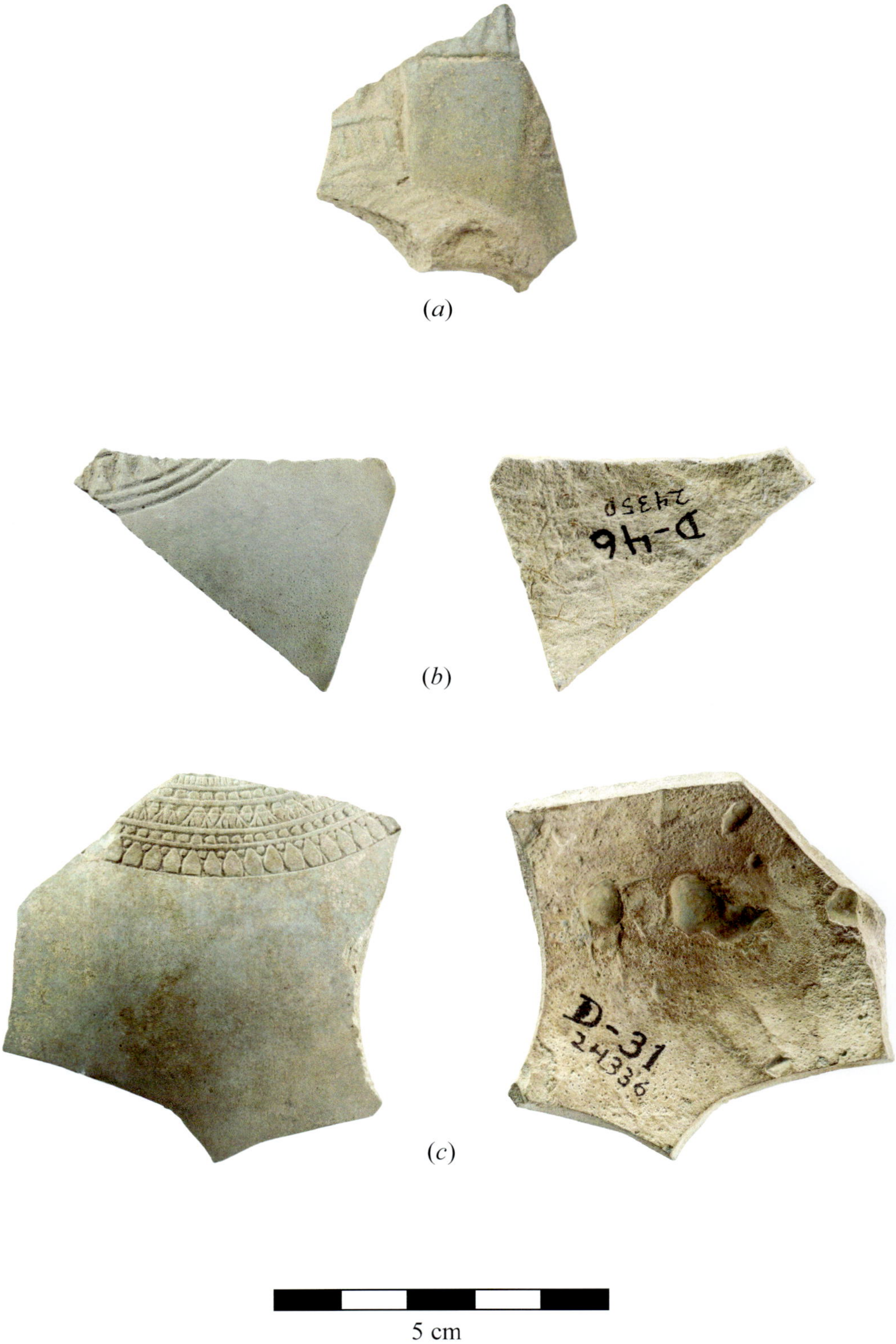

Faience New Year's flask fragments
(*a*) D 13 (1); OIM E24313
(*b*) D 32 (1), from fill north of silo; OIM E24350
(*c*) On plastered floor in bay of Level II platform at west retaining wall; OIM E24336

(*a*) D 9 and D 5 stone reused as a door socket; OIM E24424, light-gray sandstone

(*b*) Querns photographed in the field

Photo: (*b*) D-11, Frame 06-06A

(*a*)

5 cm

(*b*)

5 cm

(*c*)

5 cm

Inscribed stone block, pottery-working tool, and worn quern
(*a*) D 5 test pit; OIM E24326
(*b*) D 107 (1), fill; OIM E24405, ferruginous limestone
(*c*) D 104; OIM E50793

(*a*)

(*b*)

5 cm

(*c*)

5 cm

Stone blade, net weights, and arrowheads

(*a*) D 59C (1); OIM E24417, flint blade

(*b*) Top to bottom, left to right: D 22 (4), OIM E24331, gray soapstone; D 19 (1), OIM E24329, talc; D 19 (1), OIM E24328, green soapstone; D 32, southwest of silo, OIM E24356, schist(?) with mica; D 75, above "ash," OIM E24385, talc; D 26 (1), OIM E24370, gray soapstone; D 22 (3), OIM E24330, talc

(*c*) D 34, in charcoal on floor, OIM E24349A, agate; D 34, in charcoal on floor, OIM E24349B, agate; D 34, in charcoal on floor, OIM E24349D, agate; D 86 (1), OIM E24412, white silica/flint

Stone arrowheads
(*a*) East Sector, north of stairs to Level II; Khartoum 14275
(*b*) East Sector, 50 cm below surface by stairs to Level II; Khartoum 14277
(*c*) D 213 (1); Khartoum 14278
(*d*) D 79 (1), ca. 25 cm below top of wall; Khartoum 14274
(*e*) D 210 (1); Khartoum 14259
(*f*) Khartoum 14276
(*g*) D 201 (1); OIM E24395
(*h*) D 73 (1); OIM E24393, pale silicate with dark concretions/flint
(*i*) D 34, in charcoal on floor; OIM E24349E, agate
(*j*) Outside D 126; OIM E24419
(*k*) East of D 34; OIM E24421, agate

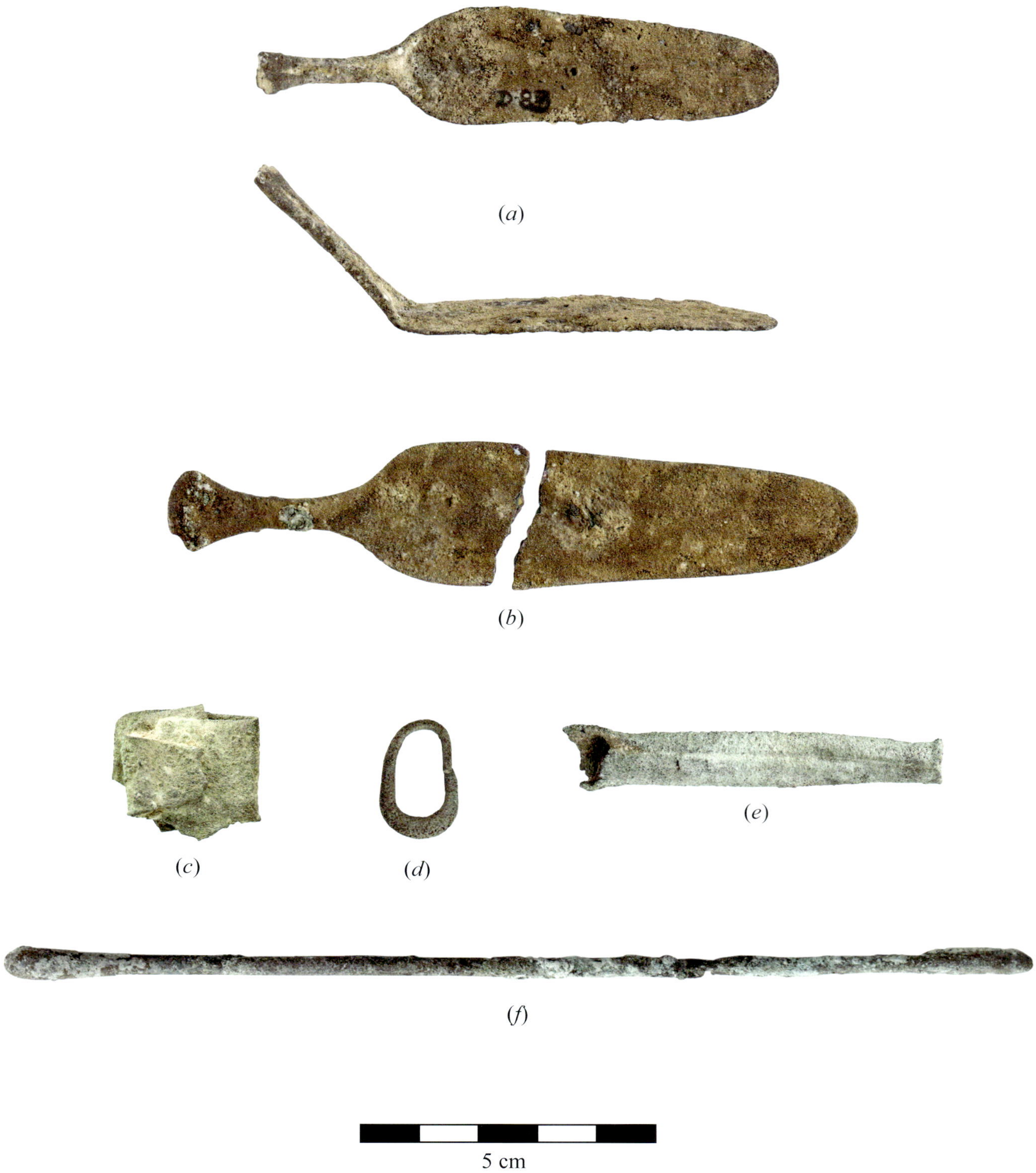

Metal objects
(*a*) Fill below "ash" layer in D 78; OIM E24387
(*b*) D 80, below "ash" layer, southwest corner of room; OIM E24389
(*c*) D 5 (2); OIM E24357
(*d*) D 13 (1), eroded wall of fortress; OIM E24311
(*e*) D 126, outside silo; OIM E24420
(*f*) East of D 40, below surface and at wall; OIM E24415

(*a*)

(*b*)

(*c*)

Ostraca and dipinto
(*a*) D 1 (1); OIM E24308, hieratic
(*b*) D 3 (1), below surface; OIM E24309, demotic
(*c*) Under threshold of Level I Christian building at D 15 (1); OIM E24365, demotic

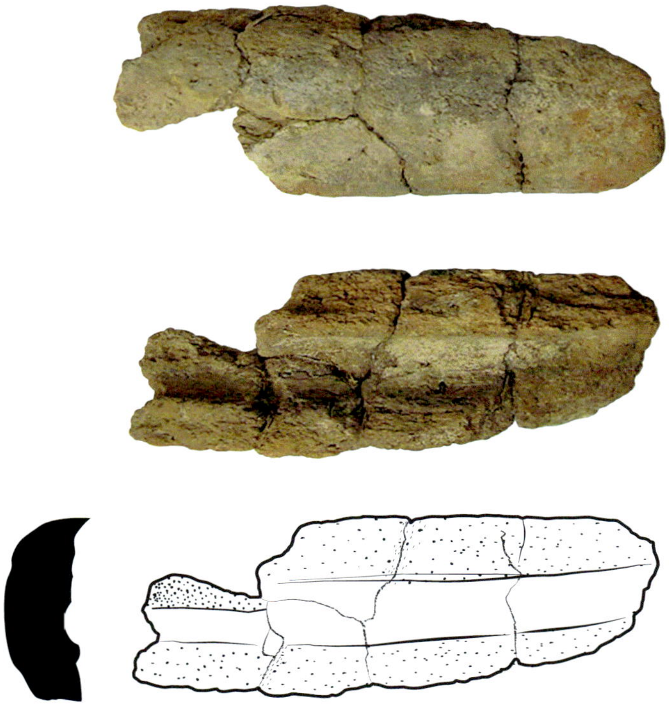

(*a*)

(*b*)

(*c*)

(*d*)

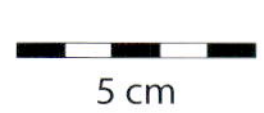

Tuyères
(*a*) D 218 (1); 22.2 × 7.9 cm maximum; channel shallower at broken end, 2.4 cm wall
(*b*) D 218 (1); 15.3 × 7.8 cm maximum
(*c*) D 218 (1); 17.4 × 7.9 cm maximum
(*d*) D 75; OIM E42948

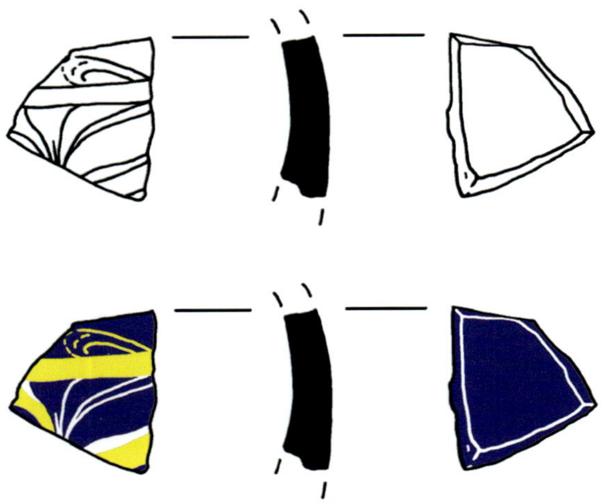

(*a*)

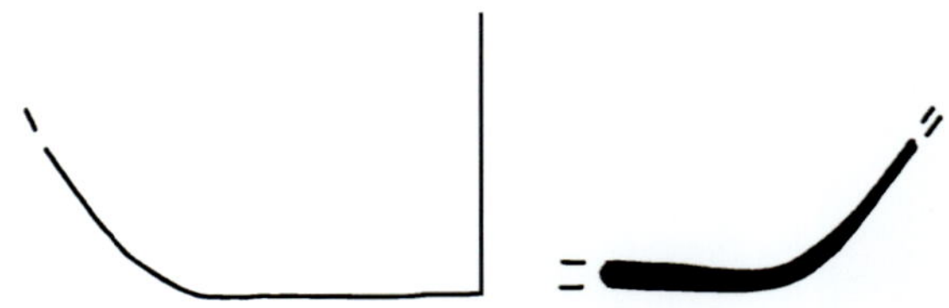

(*b*)

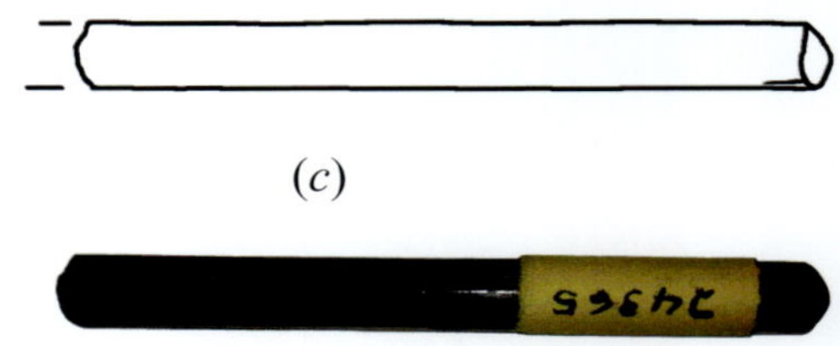

(*c*)

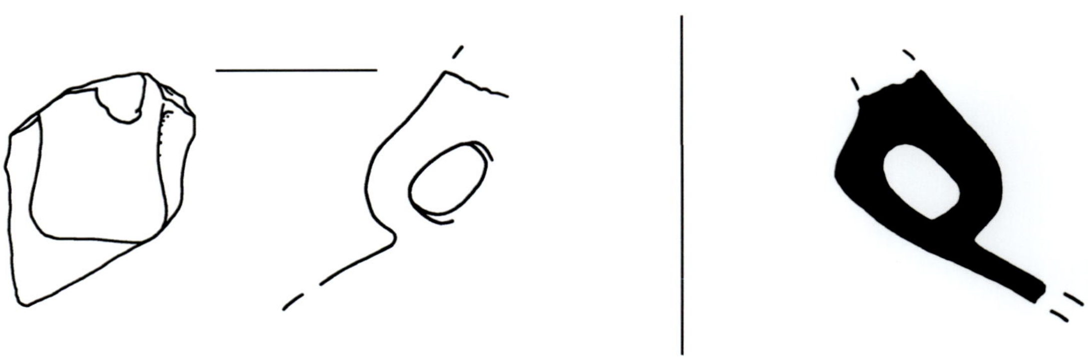

(*d*)

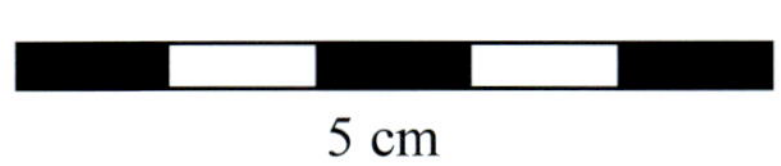

Glass objects
(*a*) D 3 (1); OIM E50810
(*b*) D 8; OIM E24315A–C
(*c*) D 32, rubble west of silo; OIM E24365
(*d*) D 18 (1); OIM E50808, white opaque glass

(*a*)

(*b*)

(*c*)

(*d*)

(*e*)

Beads
(*a*) D 32.4–12; original stringing
(*b*) D 32.13; modern stringing
(*c*) D 42, north wall of fortress; modern stringing
(*d*) D 48
(*e*) D 53; modern stringing

(*a*)

(*b*)

Beads from Meroitic burials 2 and 3
(*a*) D 18.1–3, burial 2; modern stringing
(*b*) D 11.1–2, burial 3; modern stringing

(*a*)

5mm

(*b*)

10mm

Beads from Meroitic burials 4 and 6
(*a*) D 29, burial 4; modern stringing
(*b*) D 37, burial 6; modern stringing

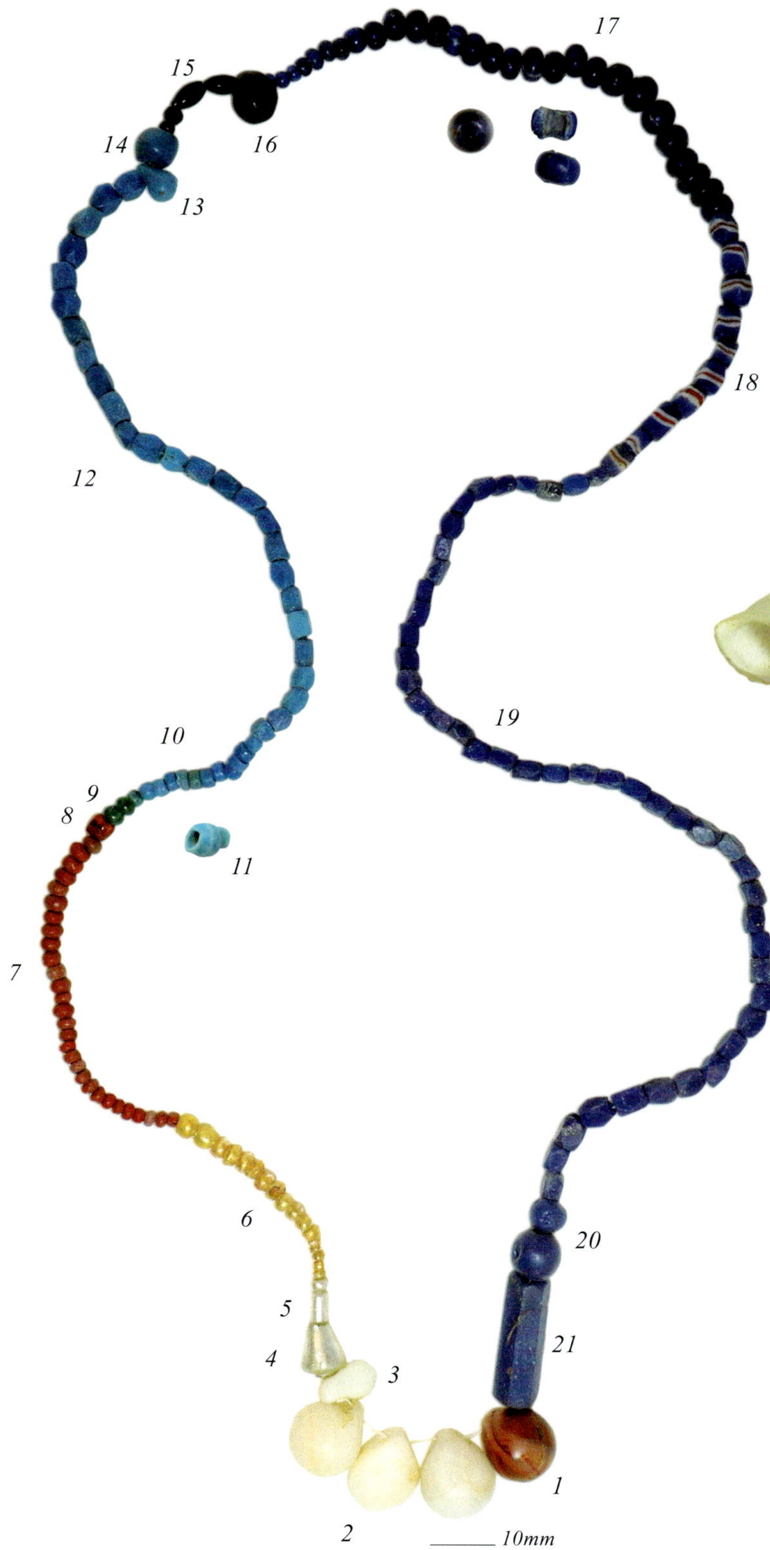

(*a*)

(*b*)

(*c*)

Beads, shell, and ear stud from Meroitic burial 8
(*a*) D 44.1–21; modern stringing
(*b*) D 44.22; mollusk shell
(*c*) D 44.23; glass

Beads, pendant, and shell

(*a*) D 3.1
(*b*) D 3.2; modern stringing
(*c*) D 3.3
(*d*) D 8
(*e*) D 9.1
(*f*) D 9.2–3; modern stringing
(*g*) D 32.1
(*h*) D 32.2
(*i*) D 32.3

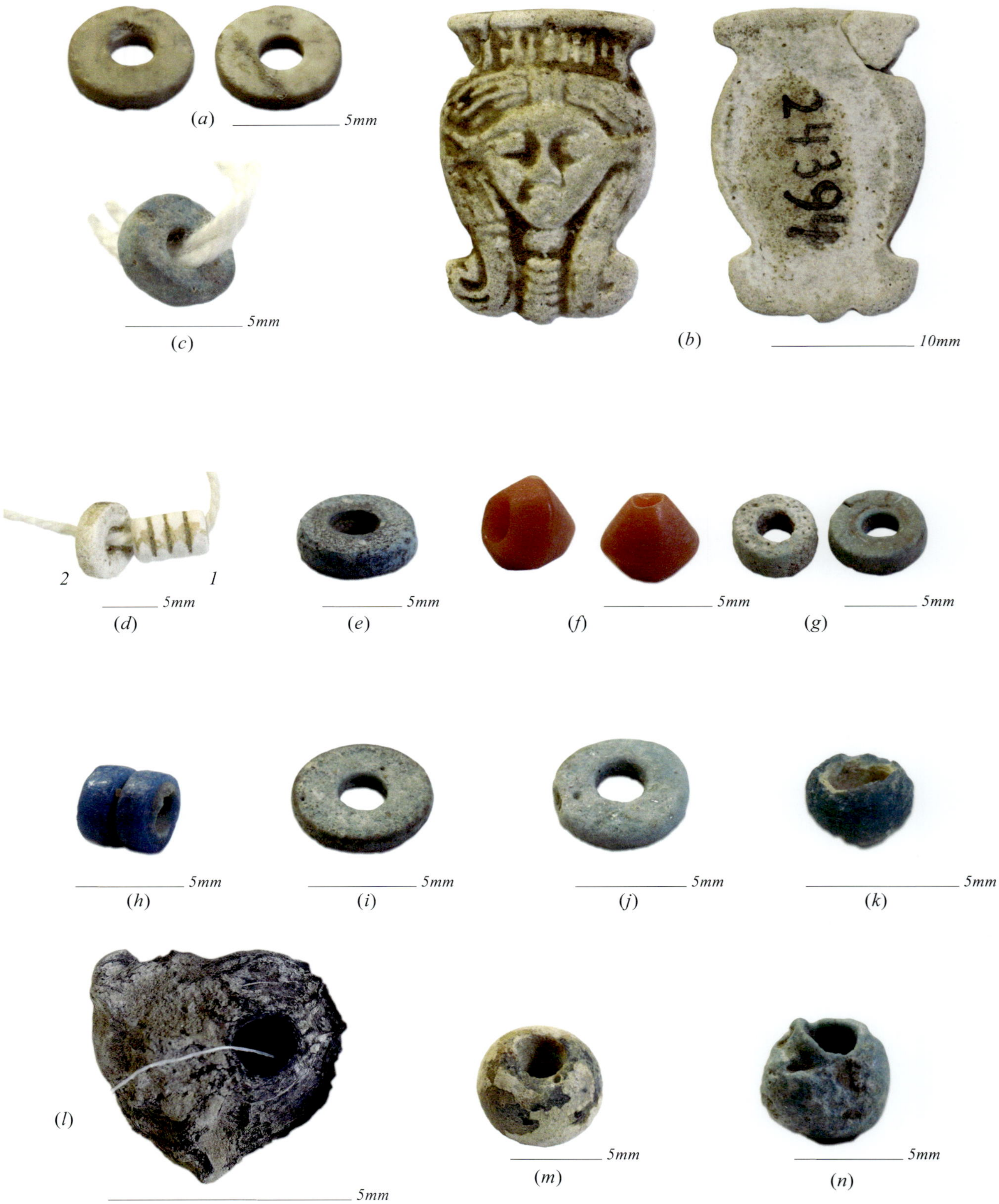

Beads, pendant, and shell

(*a*) D 60
(*b*) D 6 (1)
(*c*) D 61
(*d*) D 66.1–2; modern stringing
(*e*) D 70
(*f*) D 73.1
(g) D 73.2
(*h*) D 74
(*i*) D 75
(*j*) D 87
(*k*) D 92
(*l*) Outside North Gate; modern stringing
(*m*) D 112
(*n*) D 112 and D 113

Beads
(*a*) OC 2.1–5; modern stringing
(*b*) OC 3; modern stringing
(*c*) OC 4; modern stringing
(*d*) OC 5
(*e*) OC 6
(*f*) OC 7

(*a*)

(*b*)

(*c*)

(*d*)

(*e*)

(*f*)

(*g*)

Beads

(*a*) D 124
(*b*) D 205
(*c*) CS 1; modern stringing
(*d*) CS 2.1
(*e*) CS 2.2
(*f*) CS 3
(*g*) OC 1.1–4; modern stringing

(*a*)

(*b*)

(*c*)

Beads
(*a*) D 113.1–7; modern stringing
(*b*) D 123.1
(*c*) D 123.2